D0081001

Food Packaging

Principles and Practice

Second Edition

FOOD SCIENCE AND TECHNOLOGY

A Series of Monographs, Textbooks, and Reference Books

76. Food Chemistry: Third Edition, *edited by Owen R. Fennema*
77. Handbook of Food Analysis: Volumes 1 and 2, *edited by Leo M. L. Nollet*
78. Computerized Control Systems in the Food Industry, *edited by Gauri S. Mittal*
79. Techniques for Analyzing Food Aroma, *edited by Ray Marsili*
80. Food Proteins and Their Applications, *edited by Srinivasan Damodaran and Alain Paraf*
81. Food Emulsions: Third Edition, Revised and Expanded, *edited by Stig E. Friberg and Kåre Larsson*
82. Nonthermal Preservation of Foods, *Gustavo V. Barbosa-Cánovas, Usha R. Pothakamury, Enrique Palou, and Barry G. Swanson*
83. Milk and Dairy Product Technology, *Edgar Spreer*
84. Applied Dairy Microbiology, *edited by Elmer H. Marth and James L. Steele*
85. Lactic Acid Bacteria: Microbiology and Functional Aspects, Second Edition, Revised and Expanded, *edited by Seppo Salminen and Atte von Wright*
86. Handbook of Vegetable Science and Technology: Production, Composition, Storage, and Processing, *edited by D. K. Salunkhe and S. S. Kadam*
87. Polysaccharide Association Structures in Food, *edited by Reginald H. Walter*
88. Food Lipids: Chemistry, Nutrition, and Biotechnology, *edited by Casimir C. Akoh and David B. Min*
89. Spice Science and Technology, *Kenji Hirasa and Mitsuo Takemasa*
90. Dairy Technology: Principles of Milk Properties and Processes, *P. Walstra, T. J. Geurts, A. Noomen, A. Jellema, and M. A. J. S. van Boekel*
91. Coloring of Food, Drugs, and Cosmetics, *Gisbert Otterstätter*
92. *Listeria*, Listeriosis, and Food Safety: Second Edition, Revised and Expanded, *edited by Elliot T. Ryser and Elmer H. Marth*
93. Complex Carbohydrates in Foods, *edited by Susan Sungsoo Cho, Leon Prosky, and Mark Dreher*

94. Handbook of Food Preservation, *edited by M. Shafiur Rahman*
95. International Food Safety Handbook: Science, International Regulation, and Control, *edited by Kees van der Heijden, Maged Younes, Lawrence Fishbein, and Sanford Miller*
96. Fatty Acids in Foods and Their Health Implications: Second Edition, Revised and Expanded, *edited by Ching Kuang Chow*
97. Seafood Enzymes: Utilization and Influence on Postharvest Seafood Quality, *edited by Norman F. Haard and Benjamin K. Simpson*
98. Safe Handling of Foods, *edited by Jeffrey M. Farber and Ewen C. D. Todd*
99. Handbook of Cereal Science and Technology: Second Edition, Revised and Expanded, *edited by Karel Kulp and Joseph G. Ponte, Jr.*
100. Food Analysis by HPLC: Second Edition, Revised and Expanded, *edited by Leo M. L. Nollet*
101. Surimi and Surimi Seafood, *edited by Jae W. Park*
102. Drug Residues in Foods: Pharmacology, Food Safety, and Analysis, *Nickos A. Botsoglou and Dimitrios J. Fletouris*
103. Seafood and Freshwater Toxins: Pharmacology, Physiology, and Detection, *edited by Luis M. Botana*
104. Handbook of Nutrition and Diet, *Babasaheb B. Desai*
105. Nondestructive Food Evaluation: Techniques to Analyze Properties and Quality, *edited by Sundaram Gunasekaran*
106. Green Tea: Health Benefits and Applications, *Yukihiko Hara*
107. Food Processing Operations Modeling: Design and Analysis, *edited by Joseph Irudayaraj*
108. Wine Microbiology: Science and Technology, *Claudio Delfini and Joseph V. Formica*
109. Handbook of Microwave Technology for Food Applications, *edited by Ashim K. Datta and Ramaswamy C. Anantheswaran*
110. Applied Dairy Microbiology: Second Edition, Revised and Expanded, *edited by Elmer H. Marth and James L. Steele*
111. Transport Properties of Foods, *George D. Saravacos and Zacharias B. Maroulis*
112. Alternative Sweeteners: Third Edition, Revised and Expanded, *edited by Lyn O'Brien Nabors*
113. Handbook of Dietary Fiber, *edited by Susan Sungsoo Cho and Mark L. Dreher*
114. Control of Foodborne Microorganisms, *edited by Vijay K. Juneja and John N. Sofos*
115. Flavor, Fragrance, and Odor Analysis, *edited by Ray Marsili*
116. Food Additives: Second Edition, Revised and Expanded, *edited by A. Larry Branen, P. Michael Davidson, Seppo Salminen, and John H. Thorngate, III*
117. Food Lipids: Chemistry, Nutrition, and Biotechnology: Second Edition, Revised and Expanded, *edited by Casimir C. Akoh and David B. Min*
118. Food Protein Analysis: Quantitative Effects on Processing, *R. K. Owusu-Apenten*
119. Handbook of Food Toxicology, *S. S. Deshpande*
120. Food Plant Sanitation, *edited by Y. H. Hui, Bernard L. Bruinsma, J. Richard Gorham, Wai-Kit Nip, Phillip S. Tong, and Phil Ventresca*
121. Physical Chemistry of Foods, *Pieter Walstra*
122. Handbook of Food Enzymology, *edited by John R. Whitaker, Alphons G. J. Voragen, and Dominic W. S. Wong*
123. Postharvest Physiology and Pathology of Vegetables: Second Edition, Revised and Expanded, *edited by Jerry A. Bartz and Jeffrey K. Brecht*
124. Characterization of Cereals and Flours: Properties, Analysis, and Applications, *edited by Gönül Kaletunç and Kenneth J. Breslauer*
125. International Handbook of Foodborne Pathogens, *edited by Marianne D. Miliotis and Jeffrey W. Bier*
126. Food Process Design, *Zacharias B. Maroulis and George D. Saravacos*
127. Handbook of Dough Fermentations, *edited by Karel Kulp and Klaus Lorenz*

128. Extraction Optimization in Food Engineering, *edited by Constantina Tzia and George Liadakis*
129. Physical Properties of Food Preservation: Second Edition, Revised and Expanded, *Marcus Karel and Daryl B. Lund*
130. Handbook of Vegetable Preservation and Processing, *edited by Y. H. Hui, Sue Ghazala, Dee M. Graham, K. D. Murrell, and Wai-Kit Nip*
131. Handbook of Flavor Characterization: Sensory Analysis, Chemistry, and Physiology, *edited by Kathryn Deibler and Jeannine Delwiche*
132. Food Emulsions: Fourth Edition, Revised and Expanded, *edited by Stig E. Friberg, Kare Larsson, and Johan Sjoblom*
133. Handbook of Frozen Foods, *edited by Y. H. Hui, Paul Cornillon, Isabel Guerrero Legarret, Miang H. Lim, K. D. Murrell, and Wai-Kit Nip*
134. Handbook of Food and Beverage Fermentation Technology, *edited by Y. H. Hui, Lisbeth Meunier-Goddik, Ase Solvejg Hansen, Jytte Josephsen, Wai-Kit Nip, Peggy S. Stanfield, and Fidel Toldrá*
135. Genetic Variation in Taste Sensitivity, *edited by John Prescott and Beverly J. Tepper*
136. Industrialization of Indigenous Fermented Foods: Second Edition, Revised and Expanded, *edited by Keith H. Steinkraus*
137. Vitamin E: Food Chemistry, Composition, and Analysis, *Ronald Eitenmiller and Junsoo Lee*
138. Handbook of Food Analysis: Second Edition, Revised and Expanded, Volumes 1, 2, and 3, *edited by Leo M. L. Nollet*
139. Lactic Acid Bacteria: Microbiological and Functional Aspects: Third Edition, Revised and Expanded, *edited by Seppo Salminen, Atte von Wright, and Arthur Ouwehand*
140. Fat Crystal Networks, *Alejandro G. Marangoni*
141. Novel Food Processing Technologies, *edited by Gustavo V. Barbosa-Cánovas, M. Soledad Tapia, and M. Pilar Cano*
142. Surimi and Surimi Seafood: Second Edition, *edited by Jae W. Park*
143. Food Plant Design, *Antonio Lopez-Gomez; Gustavo V. Barbosa-Cánovas*
144. Engineering Properties of Foods: Third Edition, *edited by M. A. Rao, Syed S.H. Rizvi, and Ashim K. Datta*
145. Antimicrobials in Food: Third Edition, *edited by P. Michael Davidson, John N. Sofos, and A. L. Branen*
146. Encapsulated and Powdered Foods, *edited by Charles Onwulata*
147. Dairy Science and Technology: Second Edition, *Pieter Walstra, Jan T. M. Wouters and Tom J. Geurts*
148. Food Biotechnology, Second Edition, *edited by Kalidas Shetty, Gopinadhan Paliyath, Anthony Pometto and Robert E. Levin*
149. Handbook of Food Science, Technology, and Engineering - 4 Volume Set, *edited by Y. H. Hui*
150. Thermal Food Processing: New Technologies and Quality Issues, *edited by Da-Wen Sun*
151. Aflatoxin and Food Safety, *edited by Hamed K. Abbas*
152. Food Packaging: Principles and Practice, Second Edition, *Gordon L. Robertson*

Food Packaging
Principles and Practice
Second Edition

Gordon L. Robertson

A CRC title, part of the Taylor & Francis imprint, a member of the Taylor & Francis Group, the academic division of T&F Informa plc.

Published in 2006 by
CRC Press
Taylor & Francis Group
6000 Broken Sound Parkway NW, Suite 300
Boca Raton, FL 33487-2742

Printed in the United States of America on acid-free paper
10 9 8 7 6 5 4 3 2 1

International Standard Book Number-10: 0-8493-3775-5 (Hardcover)
International Standard Book Number-13: 978-0-8493-3775-8 (Hardcover)
Library of Congress Card Number 2005043706

Library of Congress Cataloging-in-Publication Data

Robertson, Gordon L., 1946-
Food packaging, principles and practice / Gordon L. Robertson.-- 2nd ed.
p. cm. -- (Food science and technology ; 152)
ISBN 0-8493-3775-5 (alk. paper)
1. Food--Packaging. I. Title. II. Food science and technology (Taylor & Francis) ; 152.

TP374.R63 2005
664'.09--dc22 2005043706

Taylor & Francis Group
is the Academic Division of T&F Informa plc.

Visit the Taylor & Francis Web site at
http://www.taylorandfrancis.com

and the CRC Press Web site at
http://www.crcpress.com

Author

Gordon L. Robertson is a consultant based in Brisbane, Australia. Previously, he was Vice President for Environmental Affairs in Asia for Tetra Pak for 11 years, and prior to that, Foundation Professor of Packaging Technology at Massey University. A member of several editorial boards, he is a Fellow of the Australian Institute of Packaging, and a Fellow and former president of the New Zealand Institute of Food Science and Technology. Dr. Robertson received B.Tech., M.Tech. and Ph.D. degrees in food technology from Massey University, Palmerston North, New Zealand.

Preface to the Second Edition

In the 12 years since the first edition of this book, there have been significant advances in many areas of food packaging. In particular, the areas of active and modified atmosphere packaging have seen considerable advancement and application since the first edition was published. In addition, the environment has placed new pressures and expectations on food packaging, and an understanding of this important area is required for all those involved in the development and selection of food packaging materials. Therefore, new chapters have been added on these topics, and other chapters updated as appropriate. Based on positive feedback from those using the first edition as a teaching text, additional historical aspects have been included to provide context and inspiration to today's students.

The book consists of five major parts. The first part (Chapters 2 to 9) deals with the manufacture, properties and forms of packaging materials in sufficient depth to enable those developing food packages to have a sound knowledge and appreciation of the available packaging materials. The emphasis in these chapters is on the properties of the packaging materials that can influence the quality of the food and its shelf life, rather than on the detail of the manufacturing process. Two new chapters have been added — one on edible and biobased food packaging materials, and one on printing processes, inks, adhesives and labeling. The chapter on optical and mechanical properties of thermoplastic polymers has been omitted as it was too specialized for a book on food packaging.

The second part (Chapter 10 and Chapter 11) reviews the various types of deteriorative reactions that foods undergo, and discusses the extrinsic factors which control the rates of these deteriorative reactions. This is followed by a chapter on factors controlling the shelf life of foods and shelf life testing — issues that are integral to food package development. The chapter from the first edition on food preservation and processing techniques has been omitted because there are now several excellent books that cover this area.

The third part (Chapters 12 to 15) contains chapters on aseptic packaging of foods, packaging of microwavable foods, active and intelligent packaging, and modified atmosphere packaging. The latter two areas have assumed increasing importance in the food industry over recent years and are worthy of individual chapters. The fourth part (Chapters 16 to 20) contains chapters on the packaging requirements of various food groups — flesh foods, horticultural products, dairy products, cereal, snack foods and confectionery and beverages. The final part (Chapters 21 and 22) consists of a chapter on the safety and legislative aspects of food packaging (again with particular emphasis on the relevant regulations and philosophy in the U.S. and the EU) and a new chapter on packaging and the environment, another area that has come to the fore since the publication of the first edition.

It would not have been possible to complete this book without assistance, encouragement and helpful advice from a number of people. I would especially like to thank the following: Dr. Raija Ahvenainen, Dr. Aaron L. Brody, Brigitte Cox, Brian P.F. Day, Dr. Hilton C. Deeth, Professor John D. Floros, Dr. Robert V. Holland, Dr. Robert Kawaratani, Professor Theodore P. Labuza, Dr. Roger D. MacBean, Dr. Thomas Otto, Dr. Otto G. Piringer, Michael L. Rooney and Dr. Robert J. Steele.

It is a real pleasure to acknowledge the tremendous assistance of my wife, Soozie, in rescuing most of the chapters from the first edition and converting them from WordStar into Word, as well as retyping and reformatting numerous tables and new material.

Finally, I would like to thank all those who provided feedback, constructive comments and suggestions for improvements on the first edition. I welcome further comments on this edition, which I will be happy to include in a third edition.

Gordon L. Robertson
kiaorala@bigpond.net.au

Preface to the First Edition

Food packaging is an area in which every practicing food scientist and technologist eventually becomes involved. The importance of packaging hardly needs stressing, because it is almost impossible to think of more than a handful of foods that are sold in an unpackaged state. Furthermore, the fact that, on average, around 25% of the ex-factory cost of consumer foods is for their packaging provides the incentive and the challenge for food packaging technologists to design and develop functional packages at minimum cost.

Food packaging is an area of study that draws on several disciplines including chemistry, microbiology, food science and engineering. It is my belief and hope that this book will provide a comprehensive background for those involved in the development of packages and packaging systems for foods. It is written for food scientists and technologists wishing to understand more fully those aspects of packaging technology that are relevant to the processing, preservation, distribution and marketing of a particular food; for packaging engineers wishing to know more about those aspects of food science and technology that influence the packaging process; and for students of food science and technology as well as packaging, who require an integrated approach to the subject and a textbook that could accompany a one or two semester course in food packaging.

The book consists of three major parts. The first part (Chapters 2 to 9) deals with the properties and forms of packaging materials in sufficient depth to enable those developing food packages to obtain a sound knowledge and appreciation of the available packaging materials. The emphasis in these chapters is on the properties of the packaging materials that could influence the quality of the food and its shelf life, rather than on the detail of the manufacturing process.

The second part (Chapters 10 to 12) reviews the various types of deteriorative reactions that foods undergo and discusses the extrinsic factors that control the rates of these deteriorative reactions. The various methods of food preservation are outlined, and the section concludes with a chapter dealing with factors controlling the shelf life of foods and shelf life testing —issues that lie at the very heart of food package development.

The third part (Chapters 13 to 19) begins with chapters on the aseptic packaging of foods and the packaging of microwavable foods, two areas that have assumed increasing importance in the food industry over recent years. Then follow chapters on the packaging requirements of various food groups including flesh foods, horticultural products, dairy products, cereal and snack foods as well as beverages. The book concludes with a chapter on the safety and legislative aspects of food packaging, with particular emphasis on the relevant regulations and philosophy in the U.S. and the European Community.

The temptation to suggest suitable packaging for different foods has been avoided for several reasons. Notably, such suggestions cannot be made without a full knowledge of the nature of the food, the characteristics of the distribution system and the required shelf life. Furthermore, suggesting a suitable package contradicts the very philosophy that underpins this book — that package development requires an intimate knowledge of the nature of the available packaging materials as well as the food, the distribution system and the required shelf life. A "cook book approach" to package development, where one finds the appropriate page in a book and then slavishly follows the author's suggestions, is a sure recipe for disaster, resulting in either under- or overpackaging. It is hoped that the approach adopted in this book will lead to the informed development of food packages that provide just the required amount of protection — no more and no less.

Full references are given at the end of each chapter so that particular aspects can be pursued further. In addition, an extensive index has been prepared. Where it has been considered appropriate to illustrate certain points with calculations, worked examples have been included.

This book is the culmination of over 20 years of experience teaching food packaging to fourth-year undergraduate food technology students. Through their feedback and reaction to the lectures, tutorials and laboratory classes, as well as their subsequent industrial experience in food package development, they have unwittingly contributed much to the nature, content and approach adopted in this book. In an attempt to make each chapter reasonably complete, some repetition of material across chapters has resulted; however, I believe that this will make the book easier to use.

It would not have been possible to complete this book without the assistance, encouragement and helpful advice from a number of people. I would especially like to thank the following: Marilyn Bakker, Dr. Aaron L. Brody, Professor Theron W. Downes, Professor Owen R. Fennema, Ron Goddard, Dr. Robert V. Holland, Dr. Leo L. Katan, Professor Theodore P. Labuza, Dr. Otto G. Piringer, Stanley Sacharow, Dr. Robert J. Steele, Margaret A. Turnbull and James D. Waters. A special mention must go to my supervising editor at Marcel Dekker, Inc., Henry Boehm. His encouragement and support over the last 5 years, along with his endless patience and good humor when deadlines were broken with almost monotonous regularity, have been very much appreciated.

I began formally writing this book in 1987, using WordStar 2.3 on an Osborne portable computer; I completed it using WordStar 6.0 on an IBM-compatible computer. The output was created on an HP LaserJet IIIP printer. It is a real pleasure to acknowledge the tremendous debt of gratitude I owe to my elder son Dean of whom it is true to say that without his assistance, this book would never have been completed. Dean acted as my computer and word processing expert and was a constant source of help, particularly with the final preparation of the camera-ready copy and diagrams. I was delighted that he was able to spend a university vacation assisting me complete this book.

Finally, I must acknowledge the wonderful support that I have received from the family dog Molybdenum (Mollie to her close friends). Her constant companionship during the writing of this book, and unquestioning agreement on important technical matters that I discussed with her, were much appreciated and a source of great comfort. I just hope that she appreciates and treasures the autographed copy of the book when she receives it.

Gordon L. Robertson

Table of Contents

1 Introduction to Food Packaging

CONTENTS

I. HISTORICAL

In today's society, packaging is pervasive and essential. It surrounds, enhances and protects the goods we buy, from processing and manufacturing, through handling and storage, to the final consumer. Without packaging, materials handling would be a messy, inefficient and costly exercise and modern consumer marketing would be virtually impossible. The packaging sector represents about 2% of Gross National Product (GNP) in developed countries and about half of all packaging is used to package food.

The historical development of packaging has been well documented elsewhere and will not be described in depth. However, an appreciation of the origin of packaging materials and knowledge of the early efforts in package development can be both instructive and inspirational and for this reason they are discussed briefly in this book. Suffice to say, the highly sophisticated packaging industries which characterize modern societies today are far removed from the simple packaging activities of earlier times.

Very few books can lay claim to be the first to expound or develop a particular area, and the present work is no exception. A number of books already exist with the words "food" and "packaging" in their titles[3,5–8,12,15,17,22,24–27] and this book complements the efforts of these earlier authors. The whole field of food science and technology has undergone tremendous development over the last 30 years and this has been reflected in a plethora of books, many of which address quite specific subject areas. In addition, there is the standard reference book on packaging[4] which is an essential resource for anyone working in the area of food packaging.

Food packaging lies at the very heart of the modern food industry and successful food packaging technologists must bring to their professional duties a wide-ranging background drawn from a multitude of disciplines. The interdisciplinary nature of food packaging is evident from the chapter headings of this book. Sufficient material has been included in the text for it to stand alone as a textbook for undergraduate and graduate students who are taking a two-semester course in food packaging. However, key references are included at the end of each chapter so that those who wish to pursue particular aspects in more depth will have some guidance to start them on their way.

II. DEFINITIONS

Despite the important role packaging plays, it is often regarded as a necessary evil or an unnecessary cost. Furthermore, in the view of many consumers, packaging is, at best, somewhat superfluous, and, at worst, a serious waste of resources and an environmental menace.[1,18] Such views arise because the functions which packaging has to perform are either unknown or not considered in full. By the time most consumers come into contact with a package its job, in many cases, is almost over, and it is perhaps understandable that the view that excessive packaging has been used gains some credence.

Packaging has been defined as a socioscientific discipline which operates in society to ensure delivery of goods to the ultimate consumer of those goods in the best condition intended for their use.[21] The Packaging Institute International defines packaging as the enclosure of products, items or packages in a wrapped pouch, bag, box, cup, tray, can, tube, bottle or other container form to perform one or more of the following functions: containment, protection, preservation, communication, utility and performance.[2] If the device or container performs one or more of these functions, it is considered a package.

Other definitions of packaging include a co-ordinated system of preparing goods for transport, distribution, storage, retailing and end-use, a means of ensuring safe delivery to the ultimate consumer in sound condition at optimum cost, and a techno-commercial function aimed at optimizing the costs of delivery while maximizing sales (and hence profits).[8]

It is important to distinguish between the words "package," "packaging" and "packing." The package is the physical entity that contains the product. Packaging was defined above and, in addition, is also a discipline. The verb "packing" can be defined as the enclosing of an individual item (or several items) in a package or container.

A distinction is usually made between the various "levels" of packaging. A primary package is one which is in direct contact with the contained product. It provides the initial, and usually the major, protective barrier. Examples of primary packages include metal cans, paperboard cartons, glass bottles and plastic pouches. It is frequently only the primary package which the consumer purchases at retail outlets. This book will confine itself to a consideration of the primary package.

A secondary package contains a number of primary packages, for example, a corrugated case or box. It is the physical distribution carrier and is sometimes designed so that it can be used in retail outlets for the display of primary packages. A tertiary package is made up of a number of secondary packages, the most common example being a stretch-wrapped pallet of corrugated cases. In interstate and international trade, a quaternary package is frequently used to facilitate the handling of tertiary packages. This is generally a metal container up to 40 m in length which can hold many pallets and is intermodal in nature, that is, it can be transferred to or from ships, trains, and flatbed trucks by giant cranes. Certain containers are also able to have their temperature, humidity and gas atmosphere controlled; this is necessary in particular situations such as the transportation of frozen foods, chilled meats and fresh fruits and vegetables.

Although the definitions above cover the basic role and form of packaging, it is necessary to discuss in more detail the functions of packaging and the environments where the package must perform those functions. The remainder of this chapter will address these issues.

III. FUNCTIONS OF PACKAGING

Packaging performs a series of disparate tasks: it protects contents from contamination and spoilage, makes it easier to transport and store goods and provides uniform measures of contents.[14]

By allowing brands to be created and standardized, it makes advertising meaningful and large-scale distribution possible. Special kinds of packages with dispensing caps, sprays and other convenience features make products more usable. Packages serve as symbols of their contents and a way of life and, just as they can very powerfully communicate the satisfaction a product offers, they are equally potent symbols of wastefulness once the product is gone.

A package must do three things: it must protect the contents, promote the product and inform the consumer. A fourth function, convenience, is closely related to promotion since convenient packages promote sales.[10]

The functions of a food package were defined by the Codex Alimentarius Commission in 1985 as follows:

"Food is packaged to preserve its quality and freshness, add appeal to consumers and to facilitate storage and distribution."

As succinct as this definition might be, it is inadequate for those responsible for designing and developing food packages. Four primary functions of packaging have been identified: containment, protection, convenience and communication. These four functions are interconnected and all must be assessed and considered simultaneously in the package development process.

A. Containment

This function of packaging is so obvious as to be overlooked by many, but, with the exception of large, discrete products, all products must be contained before they can be moved from one place to another. The "package", whether it is a bottle of cola or a bulk cement rail wagon, must contain the product to function successfully. Without containment, product loss and pollution would be widespread.

The containment function of packaging makes a huge contribution to protecting the environment from the myriad of products which are moved from one place to another on numerous occasions each day in any modern society. Faulty packaging (or under-packaging) could result in major pollution of the environment.

B. Protection

This is often regarded as the primary function of the package: to protect its contents from outside environmental effects, such as water, moisture vapor, gases, odors, micro-organisms, dust, shocks, vibrations and compressive forces, and to protect the environment from the product.

For the majority of food products, the protection afforded by the package is an essential part of the preservation process. For example, aseptically packaged milk and fruit juices in paperboard cartons only remain aseptic for as long as the package provides protection. Likewise, vacuum-packaged meat will not achieve its desired shelf life if the package permits oxygen to enter. In general, once the integrity of the package is breached the product is no longer preserved.

Packaging also protects or conserves much of the energy expended during the production and processing of the product. For example, to produce, transport, sell and store 1 kg of bread requires 15.8 MJ (megajoules) of energy. This energy is required in the form of transport fuel, heat, power and refrigeration in farming and milling the wheat, baking and retailing the bread and distributing both the raw materials and the finished product. To produce the low density polyethylene (LDPE) bag to package a 1 kg loaf of bread requires 1.4 MJ of energy. This means that each unit of energy in the packaging protects 11 units of energy in the product. While eliminating the packaging might save 1.4 MJ of energy, it would also lead to spoilage of the bread and a consequent waste of 15.8 MJ of energy.

C. Convenience

Modernization and industrialization have precipitated tremendous changes in life styles and the packaging industry has had to respond to those changes. Now an ever-increasing number of households are single-person, many couples either delay having children or opt not to at all and a greater percentage of women are in the workforce than ever before.

All these changes, as well as other factors such as the trend towards "grazing" (i.e., eating snack type meals frequently and on-the-run, rather than regular meals), the demand for a wide variety of food and drink at outdoor functions such as sports events and increased leisure time, have created a demand for greater convenience in household products; products designed around principles of convenience include foods which are pre-prepared and can be cooked or reheated in a very short time, preferably without removing them from their primary package, and sauces, dressings and condiments that can be applied simply through aerosol or pump-action packages which minimize mess. Thus packaging plays an important role in meeting the demands of consumers for convenience.

Two other aspects of convenience are important in package design. One of these can best be described as the apportionment function of packaging. In this context, the package functions by reducing the output from industrial production to a manageable, desirable "consumer" size. Thus, a vat of wine is "apportioned" into bottles, a churn of butter is "apportioned" by packing into 25 g minipats and a batch of ice cream is "apportioned" into 2 L plastic tubs.

Put simply, the large-scale production of products, which characterizes modern society, could not succeed without the apportionment function of packaging. The relative cheapness of consumer products is largely because of their production on an enormous scale and the resultant savings. But, as the scale of production has increased, so too has the need for effective methods of apportioning the product into consumer-sized dimensions.

An associated aspect is the shape (relative proportions) of the primary package with regard to consumer convenience (e.g., easy to hold, open and pour as appropriate) and efficiency in building into secondary and tertiary packages. In the movement of packaged goods in interstate and international trade, it is clearly inefficient to handle each primary package individually. Here, packaging plays another very important role in permitting primary packages to be unitized into secondary packages (e.g., placed inside a corrugated case) and secondary packages to be unitized into a tertiary package (e.g., a stretch-wrapped pallet). This unitizing activity can be carried a stage further to produce a quaternary package (e.g., a container which is loaded with several pallets). If the dimensions of the primary and secondary packages are optimal, then the maximum space available on the pallet can be used. As a consequence of this unitizing function, handling is optimized since only a minimal number of discrete packages or loads need to be handled.

D. Communication

There is an old saying that "a package must protect what it sells and sell what it protects". It may be old, but it is still true; a package functions as a "silent salesman".[16] The modern methods of consumer marketing would fail were it not for the messages communicated by the package.[9,11] The ability of consumers to instantly recognize products through distinctive branding and labeling enables supermarkets to function on a self-service basis. Without this communication function (i.e., if there were only plain packs and standard package sizes), the weekly shopping expedition to the supermarket would become a lengthy, frustrating nightmare as consumers attempted to make purchasing decisions without the numerous clues provided by the graphics and the distinctive shapes of the packaging.

Other communication functions of the package are equally important. Today the widespread use of modern scanning equipment at retail checkouts relies on all packages displaying a UPC that can be read accurately and rapidly. Nutritional information on the outside of food packages has become mandatory in many countries.

But it is not only in the supermarket that the communication function of packaging is important. Warehouses and distribution centers would (and sometimes do) become chaotic if secondary and tertiary packages lacked labels or carried incomplete details. When international trade is involved and different languages are spoken, the use of unambiguous, readily understood symbols on the package is imperative. UPCs are also frequently used in warehouses where hand-held barcode readers linked to a computer make stock-taking quick and efficient. Now the use of RFID tags attached to secondary and tertiary packages is beginning to revolutionize the supply chain. When their cost has decreased considerably, they will be applied to primary packages and revolutionize the way consumers shop.

IV. PACKAGE ENVIRONMENTS

The packaging has to perform its functions in three different environments.[21] Failure to consider all three environments during package development will result in poorly designed packages, increased costs, consumer complaints and even avoidance or rejection of the product by the customer.

A. Physical Environment

This is the environment in which physical damage can be caused to the product. It includes shocks from drops, falls and bumps, damage from vibrations arising from transportation modes including road, rail, sea and air and compression and crushing damage arising from stacking during transportation or storage in warehouses, retail outlets and the home environment.

B. Ambient Environment

This is the environment which surrounds the package. Damage to the product can be caused as a result of gases (particularly O_2), water and water vapor, light (particularly UV radiation) and temperature, as well as micro-organisms (bacteria, fungi, molds, yeasts and viruses) and macro-organisms (rodents, insects, mites and birds) which are ubiquitous in many warehouses and retail outlets. Contaminants in the ambient environment such as exhaust fumes from automobiles and dust and dirt can also find their way into the product unless the package acts as an effective barrier.

C. Human Environment

This is the environment in which the package interacts with people and designing packages for this environment requires knowledge of the variability of consumers' capabilities including vision, strength, weakness, dexterity, memory and cognitive behavior. It includes knowledge of the results of human activity, such as liability, litigation, legislation and regulation. Since one of the functions of the package is to communicate, it is important that the messages are clearly received by consumers. In addition, the package must contain information required by law such as nutritional content and net weight.

To maximize its convenience or utility functions, the package should be simple to hold, open and use. For a product which is not entirely consumed when the package is first opened, the package should be resealable and retain the quality of the product until completely used. Furthermore, the package should contain a portion size which is convenient for the intended consumers; a package which contains so much product that it deteriorates before being completely consumed clearly contains too large a portion.

V. THE FUNCTIONS/ENVIRONMENTS GRID

The functions of packaging and the environments where the package has to perform can be laid out in a two-way matrix or grid as shown in Figure 1.1. Anything that is done in packaging can

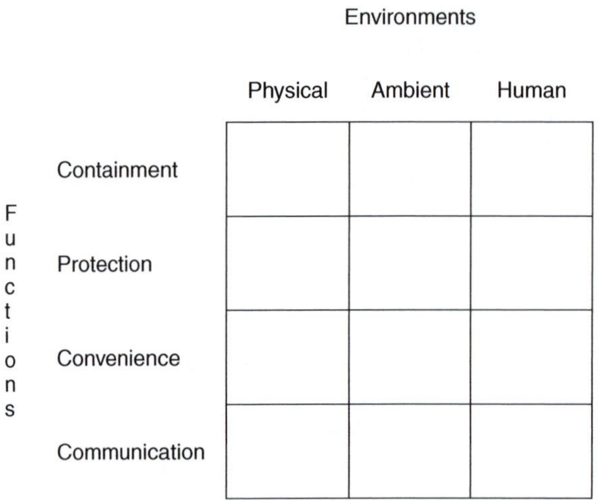

Environments

Functions	Physical	Ambient	Human
Containment			
Protection			
Convenience			
Communication			

FIGURE 1.1 The functions/environments grid for evaluating package performance. (Adapted from [21]. © John Wiley & Sons Limited. Reproduced with permission.)

be classified and located in one or more of the 12 function/environment cells. The grid provides a methodical yet simple way of evaluating the suitability of a particular package design before it is actually adopted and put into use. As well, the grid serves as a useful aid when evaluating existing packaging.

Separate grids can be laid out for distribution packaging analysis, corrugated packaging analysis, legal and regulatory impact or any mix of package-related concepts that are of interest.[21] In a further refinement of the grid, a third dimension has been suggested to represent the intensity of the interactions in each cell.[21]

Missing from the grid is an opportunity to evaluate the environmental impacts of the package. This aspect has now become such an important element in package design[20] that it should be considered fully in its own right and in addition to the evaluation carried out using the grid shown in Figure 1.1; it is the subject of the last chapter in this book.

VI. CONCLUSION

Knowledge of the functions of packaging and the environments where it has to perform will lead to the optimization of package design and the development of real, cost-effective packaging.

Despite the number of functions which a package must perform, this book focuses almost exclusively on the protective functions of the primary package and possible food and package interactions in relation to the ambient environment. Package performance in the physical environment is usually considered under the heading of packaging engineering.[13] The communication function of package performance in the human environment is the major concern of those with a primary interest in marketing and advertising, and several books specifically deal with this aspect.[9,16,23,28] For those focusing on the convenience-in-use aspects of packaging, books in the area of consumer ergonomics are the best source of information.[19]

REFERENCES

1. Abbott, D. L., *Packaging Perspectives*, Kendall/Hunt Publishing, Dubuque, IA, 1989.
2. Anonymous. *Glossary of Packaging Terms*, The Packaging Institute International, Stamford, CT, 1988.

3. Ahvenainen, R., Ed., *Novel Food Packaging Techniques*, CRC Press, Boca Raton, FL, 2003.
4. Brody, A. L. and Marsh, K. S., Eds., *The Wiley Encyclopedia of Packaging Technology*, 2nd ed., Wiley, New York, 1997.
5. Brody, A. L., Strupinsky, E. R., and Kline, L. R., *Active Packaging for Food Applications*, Technomic Publishing, Lancaster, PA, 2001.
6. Brown, W. E., *Plastics in Food Packaging: Properties, Design and Fabrication*, Marcel Dekker, New York, 1992.
7. Bureau, G. and Multon, J.-L., *Food Packaging Technology. 2 Volumes*, VCH Publishers, New York, 1996.
8. Coles, R., McDowell, D., and Kirwan, M. J., *Food Packaging Technology*, CRC Press, Boca Raton, FL, 2003.
9. Danton de Rouffignac, P., *Packaging in the Marketing Mix*, Butterworth-Heinemann, Oxford, England, 1990.
10. Doyle, M., *Packaging Strategy: Winning the Consumer*, Technomic Publishing, Lancaster, PA, 1996.
11. Eldred, N. R., *Package Printing*, Jelmar Publishing, Plainview, New York, 1993.
12. Han, J. H., Ed., *Innovations in Food Packaging*, Academic Press, San Diego, CA, 2005.
13. Hanlon, J. F., Kelsey, R. J., and Forcinio, H. E., *Handbook of Package Engineering*, 3rd ed., Technomic Publishing, Lancaster, PA, 1998.
14. Hine, T., *The Total Package: The Evolution and Secret Meanings of Boxes, Bottles, Cans and Tubes*, Little Brown, New York, 1995.
15. Jenkins, W. A. and Harrington, J. P., *Packaging Foods with Plastics*, Technomic Publishing, Lancaster, PA, 1991.
16. Judd, D., Aalders, B., and Melis, T., *The Silent Salesman*, Octogram Design, Singapore, 1989.
17. Kadoya, T., Ed., *Food Packaging*, Academic Press, San Diego, CA, 1990.
18. Kelsey, R. J., *Packaging in Today's Society*, 3rd ed., Technomic Publishing, Lancaster, PA, 1989.
19. Kramer, K., Kroemer, H., and Kroemer-Elbert, K., *Ergonomics: How to Design for Ease and Efficiency*, 2nd ed., Prentice Hall, Upper Saddle River, NJ, 2001.
20. Levy, G. M., Ed., *Packaging, Policy and the Environment*, Aspen Publishers, Gaithersburg, MD, 2000.
21. Lockhart, H. E., A paradigm for packaging, *Packaging Technol. Sci.*, 10, 237–252, 1997.
22. Mathlouthi, M., Ed., *Food Packaging and Preservation*, Blackie Academic & Professional, Glasgow, 1994.
23. Milton, H., *Package Design*, The Design Council, London, England, 1991.
24. Paine, F. A. and Paine, H. Y., *A Handbook of Food Packaging*, 2nd ed., Blackie, Glasgow, 1993.
25. Piringer, O.-G. and Baner, A. L., Eds., *Plastic Packaging Materials for Food, Barrier Function, Mass Transport, Quality Assurance and Legislation*, Wiley–VCH, New York, 2000.
26. Robertson, G. L., *Food Packaging: Principles and Practice*, Marcel Dekker, New York, 1993.
27. Rooney, M. L., Ed., *Active Food Packaging*, Chapman & Hall, London, England, 1995.
28. Stewart, B., *Packaging Design Strategy*, 2nd ed., Pira International, Surrey, England, 2004.

2 Structure and Related Properties of Plastic Polymers

CONTENTS

I. INTRODUCTION

The adjective *plastic* is derived from the Greek *plastikos*, meaning easily shaped or deformed; it was first introduced into the English language in the nineteenth century to describe the behavior of the recently discovered cellulose nitrate which behaved like clay when mixed with solvents. *The Oxford Dictionary* defines the noun "plastics" as a group of synthetic resins or other substances that can be molded into any form. From a technical viewpoint, *plastics* is a generic term for macromolecular organic compounds obtained from molecules with a lower molecular weight or by chemical alteration of natural macromolecular compounds. At some stage of their manufacture they can be formed to shape by flow, aided in many cases by heat and pressure. The term plastics can be used as a noun, singular or plural, and as an adjective.

The standard terms used for plastics are defined in ASTM D 883-00. Commonly, the word "plastic" is used to describe the easily deformable state of the material, and the word "plastics" to describe the vast range of materials based on macromolecular organic compounds. This chapter will describe plastics relevant to food packaging, with particular emphasis on their structure and related properties.

The utility of flexible sheet materials depends on properties of a special kind of molecular structure: long, flexible molecules interlocked into a strong and nonbrittle lattice. These structures are built up by the repeated joining of small basic building blocks called monomers, the resulting compound being called a polymer, derived from the Greek root "meros" meaning parts, and "poly" meaning many. Differences in the chemical constitution of the monomers, in the structure of the polymer chains and in the interrelationship of the chains determine the different properties of the various polymeric materials.

II. HISTORY

Although the chemical nature of polymers (and the fact that they consist of enormous molecules) was not understood until well into the mid-twentieth century, the materials themselves, and the industry based on them, existed long before that.

Since plastics include compounds obtained by chemical alteration of natural macromolecular compounds, then the earliest example of a plastics material would have to be hard rubber. In 1839 Charles Goodyear, an American inventor, found that rubber heated with sulfur retained its elasticity over a wider temperature range than the raw material and that it had greater resistance to solvents. The rubber–sulfur reaction was termed "vulcanization." The significance of the discovery of hard rubber lies in the fact that it was the first thermosetting (defined in Section III.A) plastics material to be prepared, and also the first plastics material which involved a distinct chemical modification of a natural material.

It is generally considered that the development of the plastics industry began in the 1860s. At the International Exhibition of 1862 in London, Alexander Parkes, an English chemist and metallurgist, displayed a new homemade material (which he later called "Parkesine") that he had made by treating cotton waste with a mixture of nitric and sulfuric acids. This was already a well-known process used for making the explosive called guncotton but Parkes found that by altering the proportions and then mixing the resulting material with castor oil and camphor, the consequent compound could be molded into decorative and useful articles. Unfortunately, the Parkesine Company, which he formed in 1866 to commercialize his process, went bankrupt in less than 2 years. In 1869 a collaborator of Parkes, Daniel Spill, formed the Xylonite Company to process materials similar to Parkesine. Once again, economic failure resulted and the company was wound up in 1874. Undaunted, Spill moved to a new site, established the Daniel Spill Company and continued production of Xylonite and Ivoride.

In the 1860s, Phelan and Collander, a U.S. firm manufacturing ivory billiard balls, offered a prize of $10,000 for a satisfactory substitute for ivory because the decimation of elephant herds had resulted in an enormous increase in the price of ivory. In an attempt to win this prize, the U.S. inventor John Wesley Hyatt solved the technical problems which beset Parkes by using camphor in place of castor oil. Although Hyatt did not win the prize, his product, which he patented with his brother in 1870 under the trademark "Celluloid," was used in the manufacture of objects ranging from dental plates to shirt collars. Despite its flammability and liability to deterioration under the action of light, Celluloid achieved notable commercial success.

Other plastics were gradually introduced over the next few decades. Among them were the first totally synthetic plastics, the family of phenol–formaldehyde resins developed by the Belgian–American chemist Leo Hendrik Baekeland and sold under the trademark "Bakelite." The first of his 119 patents on phenol–formaldehyde plastics was taken out in 1907. Other plastics introduced during this period include modified natural polymers such as rayon, made from cellulose products.

The first hypothesis of the existence of macromolecules was advanced by Kekule in 1877 when he proposed that many natural organic substances consist of very long chains of molecules from which they derive their special properties. The step from the idea of macromolecules to the reality of producing them took the genius of the German chemist Hermann Staudinger, who in 1924 proposed linear structures for polystyrene and natural rubber. In 1920 he had hypothesized that plastics were truly giant molecules or, as he called them, "macromolecules." His subsequent efforts to prove this claim initiated an outburst of scientific investigation that resulted in major breakthroughs in the chemistry of plastics and the introduction of large numbers of new products such as cellulose acetate and poly(vinyl chloride) in 1927 and urea–formaldehyde resins in 1929. Staudinger was awarded the Nobel Prize for chemistry in 1953 for his efforts in establishing the new polymer science.

Further developments occurred in the 1920s, but the years 1930 to 1940 were probably the most important decade in the history of plastics as today's major thermoplastics (polystyrene, poly(vinyl chloride) and the polyolefins) were developed. Further developments continue, with special purpose rather than general purpose materials having been discovered in recent years.

Although the possibility of discovering dramatically new polymers is remote, the plastics industry continues to grow despite a serious check to growth following the oil crisis of the 1970s. With modern society so dependent on plastics, and the food industry a major user of plastics packaging materials, continued research and development will provide new combinations of established plastics materials to perform specific functions in more efficient and cost-effective ways.

III. FACTORS INFLUENCING POLYMER STRUCTURES AND RELATED PROPERTIES

The properties of plastics are determined by the chemical and physical nature of the polymers used in their manufacture; the properties of polymers are determined by their molecular structure, molecular weight, degree of crystallinity and chemical composition. These factors in turn affect the density of the polymers and the temperatures at which they undergo physical transitions. The discussion which follows is necessarily brief, and readers are referred to standard textbooks on polymer chemistry[4,5,9,14] for more detailed treatments.

A. Molecular Structure

1. Classification of Polymers

Polymers are molecular materials with the unique characteristic that each molecule is either a long chain or a network of repeating units. This can best be understood by considering polyethylene (commonly referred to as polythene), which is one of the simplest polymers and most common food packaging film. A polyethylene molecule is built up by joining together many molecules of the monomer ethylene (C_2H_4), the chemical structures of which are indicated in Figure 2.1.

In terms of chemical composition, there are two broad types of polymers — homopolymers and heteropolymers. The former have the same repeating building-block unit throughout their molecules; the latter are polymers with two or more different building-block units regularly or irregularly distributed throughout their length. Heteropolymers are referred to as copolymers when two different monomers are polymerized together and terpolymers when three monomers are used.

Polymers such as the polyethylene used as an example in Figure 2.1 are called linear polymers and consist of a backbone of carbon atoms and a number of side groups which differ from polymer to polymer. An alternative way of describing a linear polymer is shown in Figure 2.2(a) in which it is pictured as a covalently bonded chain of monomer units. Linear copolymers may exhibit any of three combinational forms (Figure 2.2(b)–(d)). One is the form of a regular copolymer, in which two or more different repeating units occur alternately along the chain. It is actually equivalent to a homopolymer in terms of regularity. Another form is that of the random copolymer, in which, during polymerization, the units have taken up a statistically random placement along the length of the chain. The third type of structure is, in a sense, a combination of the previous two: it is the block copolymer, which is composed of alternating lengths of homopolymer along the molecular chains. Figure 2.2 shows two types of backbone units in the structure, although in practice there may be more than two.

Branched polyethylene is shown in Figure 2.3(a). If many such branches are formed, a network structure may result, as indicated by Figure 2.3(b), in which the long chains are connected together, often by relatively short cross-links.

Thus, plastic polymers can be divided into two broad categories: those polymers which extend in one dimension (i.e., they consist of linear chains), and those polymers which have links between the chains so that the material is really one giant molecule.

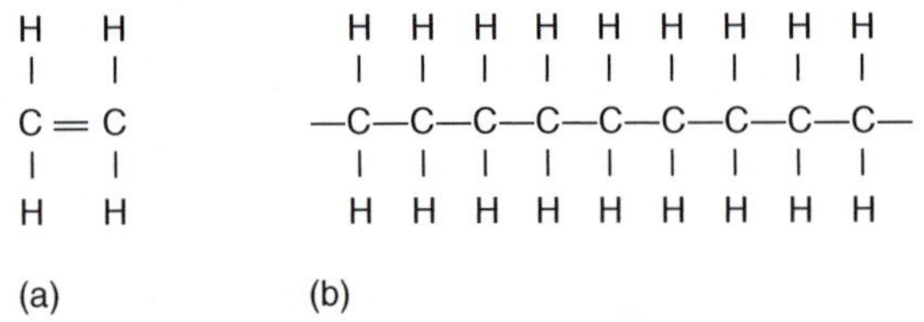

```
 H   H          H  H  H  H  H  H  H  H  H
 |   |          |  |  |  |  |  |  |  |  |
 C = C        —C—C—C—C—C—C—C—C—C—
 |   |          |  |  |  |  |  |  |  |  |
 H   H          H  H  H  H  H  H  H  H  H

  (a)             (b)
```

FIGURE 2.1 (a) The monomer ethylene and (b) the polymer polyethylene.

—M—M—M—M—M—M—M—M—M—
(a) Linear polymer

—M—L—M—L—M—L—M—L—M—
(b) Alternating polymer

—L—M—M—L—M—L—L—M—M—
(c) Random copolymer

—M—M—M—L—L—L—M—M—M—
(d) Block copolymer

FIGURE 2.2 Copolymers made with different structures. L and M are any monomers.

The first group is the linear polymers and they are thermoplastic; that is, they gradually soften with increasing temperature and finally melt because the molecular chains can move independently. They are characterized by extremely long molecules with saturated carbon–carbon backbones. Such polymers may be readily molded or extruded because of the absence of cross-links. If their temperature is raised, they become very flexible and can be molded into shape, even at temperatures below their melting point. Not surprisingly, their mechanical properties are rather temperature

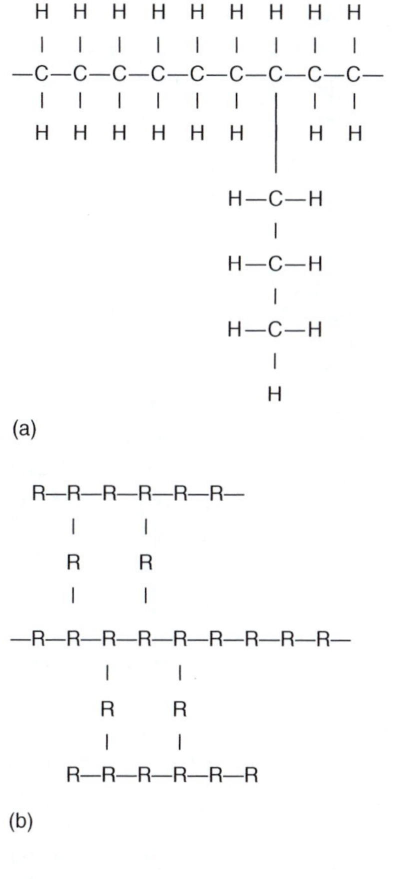

FIGURE 2.3 (a) Branched polyethylene and (b) cross-linked polymer.

sensitive. Thermoplastics are the most important class of plastics material available commercially and account for more than two thirds of all polymers used in the world today.

The second group is the cross-linked polymers and they are thermosetting. As the name suggests, these polymers become set into a given network when manufactured and cannot be subsequently remolded to a new shape. If the temperature is raised to the point where the cross-links are broken, then irreversible chemical processes also occur which destroy the useful properties of the plastic. This is called degradation. At normal temperatures the cross-links make the solid quite rigid. Thermosetting polymers do not melt on heating but finally blister (due to the release of gases) and char. Examples of thermosetting plastics are epoxy resins and unsaturated polyesters.

The above categorization of polymers is convenient because polymers in the two groups differ in their structure, their properties and the chemical processes used in their manufacture. The importance of thermosetting polymers in food packaging is minimal except for epoxy resins, which are used as enamels (lacquers) for metal cans.

2. Polymerization Processes

Thermoplastics can be made by joining together a sequence of monomers. Under suitable conditions of temperature and pressure, and in the presence of a catalyst called an initiator, the molecular chains grow by the addition of monomer molecules one by one to the ends of the chains. Branching can occur but cross-links are nearly absent.

The formation of thermoplastics by a process which involves the joining together of monomers to form polymers which have the same atoms as the monomers in their repeating units is called addition polymerization. A simple, low MW (molecular weight) molecule (which must possess a double bond) is induced to break the double bond and the resulting free valences are able to join up to other similar molecules. This reaction occurs in the form of a chain addition process with initiation, propagation and termination steps.

Under normal conditions with the usual catalysts, the spatial arrangements of the branches of the polymers are random; such polymers are called atactic. Some processes give products in which the branches are arranged in an orderly manner; these are called isotactic polymers. In the case of polyethylene this form of polymerization has the advantage of reducing the number of branches that are formed. Thus, the molecules in linear isotactic polyethylene can line up with one another very easily, yielding a tough, high density compound. Atactic polyethylene is less dense, more flexible than, and not nearly as tough as, the linear polymer, because the molecules are further apart.

Plastic polymers are also prepared by the process of condensation polymerization which involves two active sites joining together to form a chemical bond, a small molecule being removed in the process. In this case the starting monomers are not identical to those of which the chains are to be composed; the superfluous groups of atoms must be ejected when the monomer is added to the end of the chain. If there are enough groups of such superfluous atoms on each monomer molecule, some of them may be temporarily retained on the side of the chain as it grows. This promotes easy branching and leads rather rapidly to a highly cross-linked structure.

It follows that if monomers with few superfluous groups are used, cross-linking does not occur so that thermoplastic materials may also be made by the process of condensation. Polyesters and polyamides (see Sections III.F.4 and III.F.6) are examples of thermoplastic polymers made by the condensation process. However, it should be noted in passing that a thermoset polyester can be made if one (or both) of the reactant molecules has a double bond and a third compound is added; under these circumstances cross-linking occurs. Epoxy resins (basically condensation products of a dihydroxyphenol and a compound containing an epoxy group) are examples of thermosetting plastics formed by condensation.

To summarize, thermoplastic polymers can be formed either by addition or condensation polymerization, whereas thermosetting plastics are formed only by condensation polymerization.

The degree of cross-linking in a polymer may vary over a very wide range, thus blurring the boundary between thermosetting and thermoplastic materials.

B. Molecular Weight

The average number of repeating units in a single molecule of a polymer is known as the degree of polymerization or DP. At DP values of about 10 to 20, the substance formed is light oil (paraffin if formed from ethylene). As the DP increases, the substance becomes greasy, then waxy and finally at a DP of about 1000 it becomes a solid and is then a true polymer. The DP is almost unlimited and may increase to around 100,000 or so.

From knowledge of the atomic mass of the repeating unit and the DP, the MW of the polymer molecule can be calculated; it is equal to the DP multiplied by the MW of the monomer. Thus, polyethylene with a DP of 10,000 has a molecular weight of 280,000 because the molecular weight of the repeating unit [$-CH_2CH_2$] is 28. In the case of copolymers it is necessary to know not only the MW but also the relative proportions and arrangement of the two structural units in the chain.

In practically all industrial polymerization processes, the molecular chains that are produced vary considerably in dimensions. For this reason the MW of a polymer is actually a distribution of MWs. Hence a polymer system has to be characterized in terms of average molecular weight and broadness or molecular weight distribution (MWD).

Two average molecular weights are commonly employed: the number-average molecular weight (M_n) and the weight-average molecular weight (M_w). If the polymer molecules are averaged in terms of number fractions of various lengths, the M_n is obtained. Most thermodynamic properties are related to the number of particles present and are thus dependent on M_n. M_n values are independent of molecular size and are highly sensitive to small molecules present in the mixture. However, many bulk properties such as viscosity and toughness depend as much on the size of the molecules as on their number. In this case the appropriate function is the M_w, which is more dependent on the number of heavier molecules than M_n.

As the MWD becomes increasingly broad, the weight average assumes numerically larger values than the number average, so that the ratio M_w:M_n (known as the dispersity index (DI)) is taken as a measure of the spread of molecular chain lengths. For most commercial polymers the DI falls between two and eight.

C. Density

Density is a function of chemical composition, being dependent on the weight of individual molecules and the way they pack together. The hydrocarbon polymers do not possess heavy atoms and therefore the mass of the molecule per unit volume is rather low. Oxygen, chlorine, fluorine and bromine increase the density of polymers. For example, amorphous hydrocarbon polymers generally have densities of 0.86 to 1.05, while polymers containing chlorine have densities of 1.4 in the case of PVC and 1.7 for PVdC (poly(vinylidene chloride)).

D. Crystallinity

When a low molecular weight material such as a metal crystallizes from the molten state, nucleation occurs at various points, from each of which a crystal or grain grows. Likewise, when a molten crystallizable polymer is cooled, crystallization spreads out from individual nuclei. However, instead of individual grains, a considerably more complex structure develops from each nucleus.

The degree of crystallinity is an important factor affecting polymer properties. The great length of polymer chains means that a certain amount of entanglement normally occurs and this prevents complete crystallization on cooling as in the case of metals. This phenomenon is due to the difficulty of aligning every portion of each chain of the polymer. Thus, crystallinity in plastics consists of thousands of small "islands" of crystalline regions surrounded by amorphous material.

Areas where the chains are parallel and closely packed are largely crystalline, while the disordered areas are amorphous. These crystalline areas are known as crystallites. Unlike crystals of small molecules, crystallites are not composed of whole molecules or molecules of uniform size. High molecular weight, narrow molecular weight distribution, and linearity in the polymer backbone can yield high crystallinity. The crystallization rate is enhanced by the presence of impurities such as catalysts, fillers and pigments.

In crystalline polymers, the degree of crystallinity is normally limited and the crystallites are separated by amorphous supercooled regions. Thus, polyethylene, which crystallizes readily because of its rather simple molecular structure, is tough and flexible rather than brittle because of the presence of the amorphous fraction. Of course, the ultimate strength depends as much on the weaker, amorphous part of the solid as on the strength of the crystallites. PVC on the other hand does not crystallize so readily and is more of a supercooled liquid.

The presence of chain branching will tend to reduce the possibility of an ordered arrangement and so reduce the crystallinity. For example, the crystallinity of LDPE (low density polyethylene) usually varies between 55 and 70%, compared with 75 to 90% for HDPE (high density polyethylene).

Stretching a film orients the crystallites and realigns other molecules or segments of molecules, causing the total crystallinity of the film to increase. Such oriented films are generally tougher than either amorphous or unoriented crystalline materials.[6] Orientation (see Chapter 5, Section V) and crystallinity are related in that only polymers which are capable of crystallization can be oriented. During orientation, crystallites become aligned and the total amount of crystalline material increases. The crystallites disappear when the polymer is heated, and the temperature at which crystalline regions vanish is called the crystalline melting point. Noncrystalline, amorphous polymers have no melting point. They simply soften when heated, in much the same way as glass. To summarize, high shear during processing and rapid cooling inhibit crystallinity; annealing and orientation enhance it.

The transparency of unfilled plastics is a function of crystallinity, with noncrystalline polymers, such as PS (polystyrene) and PC (polycarbonate) having excellent transparency. Other polymers range from cloudy to opaque, depending on the degree of crystallinity, the size of spherulites or other forms of crystal aggregates and the ratio of the density between the crystalline phase and the amorphous phase. Thus, relatively thick films of polyethylene are translucent because of the presence of crystals. Long-chain, highly branched LDPE has low crystallinity and a broad melting point and LDPE blown films have very low haze and high gloss; HDPE chains are largely devoid of branching and so can be crystallized to a greater extent, resulting in translucent films.

E. Physical Transitions in Polymers

Simple molecules such as water can exist in any one of the three fundamental physical states; that is, solid, liquid and gas, according to the ambient conditions. The change from one state to another (transition) occurs at a specific, sharply defined temperature depending on the atmospheric pressure.

In polymers, changes of state are less well defined. There is no transition from the liquid to the gaseous state because the temperature required to completely separate the molecules from one another is far too high; decomposition occurs instead. To date, there is no universally accepted scheme for classifying transitions in polymers, partly due to the diversity of experimental methods used for the study of these phenomena and partly due to their complex nature.

Noncrystalline (amorphous) polymers are characterized by a glass transition at a temperature (called the "glass transition temperature" (T_g)); crystalline polymers are characterized by a melting transition at a temperature (called the "crystalline melting temperature" (T_m)). In addition, the latter polymers may show crystal–crystal transformations or transitions below T_m.

At a sufficiently high temperature a thermoplastic polymer is a liquid. In this state it consists of an amorphous mass of wriggling chain molecules. As it is cooled, the thermal agitation decreases and at the T_m the polymer may crystallize. In the crystalline state the molecules are all aligned and are packed together in a regular fashion. However, this state is not easily achieved because the molecular chains are normally entangled with each other. Therefore, many polymers will crystallize so slowly that supercooling is readily achieved. (Supercooling occurs when a material remains liquid below the temperature at which the crystalline solid melts). These supercooled polymers remain viscous until a still lower temperature is reached at which the material vitrifies; that is, becomes a glassy and relatively brittle solid. The temperature at which this occurs is the T_g. In the glassy state a polymer has physical properties similar to a crystalline solid but has the molecular disorder of a liquid.

T_g is the main transition temperature found in amorphous polymers. The underlying molecular process is that frozen backbone sequences begin to move at the T_g. Therefore, T_g is determined not only by the main chain architecture but also by its immediate surroundings. Chain ends and low molecular weight plasticizers lower the T_g of a polymer; a sufficiently large number of cross-links will increase T_g. Above T_g a few of the carbon atoms in each chain can still move with relative freedom, but below T_g nearly all the carbon atoms become fixed, and only side groups or very short chain sections can change position.

In linear amorphous polymers such as PS, two transition temperatures are observed: a rigid solid–rubber transition known as the T_g, and a very indefinite rubber–liquid transition referred to as the flow temperature, the term "melting point" being reserved strictly for crystalline polymers. The term "softening point" is associated with T_g for amorphous polymers and with T_m for highly crystalline polymers. In addition, there are many polymers that soften progressively between T_g and T_m and the value determined for the softening point depends considerably on the test method used.

The structure of crystalline polymers such as polyethylene and isotactic PP (polypropylene) is sufficiently regular so that they are capable of some degree of crystallization. The presence of crystalline regions in a polymer has large effects on such properties as density, stiffness and clarity. In a highly crystalline polymer, there is little difference in properties immediately above and below the glass transition. In fact, finding T_g can prove difficult with some highly crystalline polymers. This is largely because there is little scope for segmented motion as most segments are involved in a lattice formation in which they have low mobility.

For crystalline polymers an approximate relationship between T_g and T_m is:

$$T_g \approx 2/3T_m \quad \text{(for unsymmetrical chains)} \tag{2.1}$$

and

$$T_g \approx 1/2T_m \quad \text{(for symmetrical chains)} \tag{2.2}$$

when both temperatures are expressed in Kelvin. An important exception to this occurs with copolymers. Copolymerization also tends to broaden the temperature range over which T_g occurs, due to differences in chemical composition among the copolymer chains in the same sample. Generally, the crystalline copolymer is expected to have a lower melting temperature than the corresponding homopolymer. Ordered copolymers exhibit different transition behavior from random copolymers, generally showing a single, characteristic T_g and, if crystallizable, a single, sharp melting temperature.

The physical properties of a thermoplastic polymer depend on the values of T_m and T_g relative to room temperature. If both T_m and T_g lie below room temperature, the polymer is a liquid. If room temperature lies between T_m and T_g, the polymer is either a very viscous supercooled liquid or a crystalline solid. If both T_m and T_g are above room temperature an amorphous polymer is glassy and brittle in nature. Polymethylmethacrylate, known commercially as Perspex, Lucite or Plexiglass, is such a plastic.

TABLE 2.1
Glass Transition and Melting Temperatures of Some Polymers

Polymer	T_g (°C)	T_m (°C)	T_m/T_g[a]
Polyethylene			
High density	−125	137	2.77
Low density	−25	98	1.50
Polypropylene	−18	176	1.76
Poly(ethylene terephthalate)	69	267	1.57
Polystyrene (isotactic)	100	240	1.38
Poly(vinyl chloride)	87	212	1.34
Poly(vinylidene chloride)	−35	198	1.97
Polyacrylonitrile	104	317	1.56
Poly(hexamethylene adipamide) (nylon 6,6)	50	265	1.66
Poly(hexamethylene sebacamide) (nylon 6,10)	40	277	1.59
Poly(11-aminoundecanoic acid) (nylon 11)	194	–	–
Poly(tetrafluoroethylene)	126	327	1.50
Polycarbonate	150	220	1.16

[a] Ratio calculated from T_g and T_m temperatures in Kelvin.

Source: From Kaufman, H. S. and Falcetta J. J., Eds., *Introduction to Polymer Science and Technology: an SPE Textbook*, Wiley-Interscience Publications, New York, 1977. With permission.

The glass transition temperatures of the majority of the commercially important crystallizable polymers lie below 25°C as shown in Table 2.1, which also lists values for T_m and the ratio T_m:T_g. The T_g values are low for flexible, linear polymers, and relatively high for stiff chain polymers such as PET (poly(ethylene terephthalate)) and PC which require higher temperatures for the onset of molecular motions necessary for the glass transitions.

Bulky side groups decrease the mobility of the chain and thus raise T_g. For instance, substitution of alternate hydrogens in the polyethylene chain with methyl groups to give PP, or with phenyl groups to give PS, increases T_g from −120°C to −18°C and +100°C, respectively. Molecular symmetry tends to lower T_g. For instance, PVC has a T_g of 87°C, whereas for PVdC the T_g is −10°C.

The temperatures T_m and T_g are governed by the strength of the intermolecular forces (just as in inorganic materials) and by the degree of flexibility and length of the chains. Thus polar side-groups such as chloride and hydroxyl groups favor higher melting and glass transition points because they enhance the strength of the intermolecular bonds.

In some cases the T_g can be lowered by as much as 100°C by efficient use of plasticizers. For example, pure PVC has a T_g of 87°C and is quite brittle at room temperature. The addition of only 15% plasticizer lowers this to 60°C and, with addition of further plasticizer, PVC becomes tough and flexible at room temperature.

F. Chemical Structure

Polymer chains can and do align themselves in ordered structures and the thermodynamics of this ordered state determines such properties as the melting point, the glass transition temperature, other transition temperatures and the mechanical and electrical properties. However, it is the chemical nature of the polymer which determines its stability to temperature, light, water and solvents.

1. Polyolefins

Olefin means oil-forming and was originally the name given to ethylene. Today, olefin is a common term in the plastics industry and refers to the family of plastics based on ethylene and propylene. The term "alkene" is used for hydrocarbons containing a carbon–carbon double bond; for example, ethylene and propylene. Polyolefins form an important class of thermoplastics and include low, linear and high density polyethylenes and polypropylene. Industry commonly divides polyethylenes into two broad categories: HDPE $\geq$ 940 kg m^{-3}, and LDPE 915 to 939 kg m^{-3}.

a. Low Density Polyethylene (LDPE)

This is the largest volume single polymer used in food packaging in both the film and blow-molded form. It is a polymer of ethylene, a hydrocarbon gas available in large quantities as a by-product of petroleum refining and other processes. Polyethylene was first produced by Imperial Chemical Industries (ICI) Limited in 1933 during a research program devoted to the effects of extremely high pressures on chain reactions, and the basic patent relating to polymerization of ethylene was granted in 1937. It was produced on a pilot plant scale that same year with full commercial scale production commencing in 1939.[4] For the first few years of its production it was used in the electrical industry, particularly as an insulating material for underwater cables.

The polymerization of ethylene can occur over a wide range of temperatures and pressures but most commercial high pressure processes utilize pressures between 1000 and 3000 atmospheres and temperatures between 100 and 350°C (higher temperatures cause degradation of the polyethylene).

The simplest structure for polyethylene is a completely unbranched structure of $-CH_2-$ units, as shown in Figure 2.1(b). However, the vigorous nature of the high pressure process leads to a great deal of chain branching, with both short and long chains being formed. From Figure 2.3(a), it can be seen that the branch contains a terminal methyl ($-CH_3$) group. A convenient way of characterizing branching is by the number of methyl groups per 1000 carbon atoms.

The branch chains prevent close packing of the main polymer chains resulting in the production of relatively LDPEs (typically 910 to 940 kg m^{-3}). Molecular weights also tend to be relatively low. The great length of the polymer chains results in a certain amount of entanglement which prevents complete crystallization on cooling. Areas where the chains are parallel and closely packed are largely crystalline (the crystalline areas are known as crystallites), while disordered areas are amorphous. When the polymer melt is cooled slowly, the crystallites may form spherulites.

The crystallinity of LDPE usually varies between 50 and 70%. The softening point is also affected by chain branching. The attractive forces between the chains are reduced because they are unable to approach each other closely, therefore, less energy (in the form of heat) is necessary to cause them to move relative to each other and flow. The softening point of LDPE is just below 100°C, thereby precluding the use of steam to sterilize it in certain food packaging applications.

LDPE is a tough, slightly translucent material which can be blow extruded into tubular film or extruded through a slit die and chill-roll cast, the latter process giving a clearer film. It has good tensile strength, burst strength, impact resistance and tear strength, retaining its strength down to −60°C. While it is an excellent barrier to water and water vapor, it is not a good barrier to gases.

It has excellent chemical resistance, particularly to acids, alkalis and inorganic solutions, but is sensitive to hydrocarbons, halogenated hydrocarbons, oils and greases. These latter compounds are absorbed by the LDPE which then swells. Environmental stress cracking (ESC) is defined as the failure of a plastic container under conditions of both stress and exposure to a product, where neither condition causes the failure alone.[13] It is a phenomenon which occurs when a material is stressed multiaxially while in contact with certain polar liquids or vapors, resulting in surface cracks or even complete failure of the material. Essential and vegetable oils are capable of causing

ESC but the effect can be greatly reduced by using high MW grades of LDPE. ESC is strongly related to the crystallinity of the polymer; the higher the crystallinity the lower the environmental stress crack resistance (ESCR). To reduce the crystallinity, comonomers such as butene, hexene or octene can be used at less than 2%. One of the great attributes of LDPE is its ability to be fusion welded to itself to give good, tough, liquid-tight seals. It cannot be sealed by high frequency methods.

LDPE finds wide use as a rigid packaging material. It can be easily blow-molded into bottles where its flexibility enables the contents to be squeezed out. It is also widely used in the form of snap-on caps, collapsible tubes and a variety of spouts and other dispensers. The surface of polyethylene containers can be treated with fluorine after blow molding to form a very thin, polar, cross-linked surface which decreases the permeability of the polyethylene to nonpolar penetrants. It also eliminates the need for treating the surface by corona-arc discharge or flame techniques to improve printability properties. The process has been cleared by the U.S. Food and Drug Administration (FDA) for use with food containers.

At the present time there are many hundreds of grades of LDPE available, most of which differ in their properties in one way or another. Such differences arise from the following variables[4]:

1. Variation in the degree of short chain branching in the polymer.
2. Variation in the degree of long chain branching.
3. Variation in the average molecular weight.
4. Variation in the molecular weight distribution (which may in part depend on the long chain branching).
5. The presence of small amounts of comonomer residues.
6. The presence of impurities or polymerization residues, some of which may be combined with the polymer.

Polyethylene is one of the most inert polymers and constitutes no hazard in normal handling.

b. *Linear Low Density Polyethylene (LLDPE)*

The first production of linear LDPE (LLDPE) was made in a solution process in 1960. Attempts in the 1970s to produce LDPE, either by low pressure gas phase polymerization or by liquid phase processes similar to those used for producing HDPE, led to the development of LLDPE which has a similar molecular structure to HDPE. However, it is virtually free of long chain branches but does contain numerous short side chains. These arise as a result of copolymerizing ethylene with a smaller amount of a higher alkene, such as propene, butene, hexene or octene. Such branching interferes with the ability of the polymer to crystallize, and therefore results in LLDPE having a similar density to LDPE. The linearity provides strength, while the branching provides toughness. It was not until 1977 that LLDPE became available in commercial quantities.

The term "linear" in LLDPE is used to imply the absence of long chain branches. Owing to the linearity of its molecules, LLDPE is more crystalline and therefore stiffer, but less transparent, than LDPE, resulting in an increase of 10 to 15°C in the melting point of LLDPE compared to LDPE. LLDPE has higher tensile strength, puncture resistance, tear properties and elongation than LDPE.[8] LLDPE materials are now available in a range of densities from around 900 kg m^{-3} for very low density polyethylene (VLDPE) to 935 kg m^{-3} for ethylene–octene copolymers.

LLDPE combines the main features of both LDPE and HDPE, a major feature being that its molecular weight distribution is narrower than that of LDPE. Generally, the advantages of LLDPE over LDPE are improved chemical resistance, improved performance at low and high temperatures, higher surface gloss, higher strength at a given density, better heat sealing properties and a greater resistance to ESC. In film form, LLDPE shows improved puncture resistance and tear strength. At a density of 920 kg m^{-3} the melting points of LDPE and LLDPE are 95 and 118°C, respectively.

LLDPE commonly has a density of approximately 920 kg m^{-3} when butene is used as the comonomer. The superior properties of LLDPE have led to its use in new applications for polyethylene as well as the replacement of LDPE and HDPE in some areas; LLDPE is also often blended with LDPE.

c. *High Density Polyethylene (HDPE)*

Prior to 1950, the only commercial polymer of ethylene was the highly branched polymer LDPE. The technique for making a linear polymer was discovered by Nobel laureate Karl Ziegler of Germany in the early 1950s. Ziegler prepared HDPE by polymerizing ethylene at low pressure and ambient temperatures using mixtures of triethylaluminum and titanium tetrachloride. Another Nobel laureate, Giulio Natta of Italy (he shared the Nobel Prize for chemistry with Ziegler in 1963), used these complex co-ordination catalysts to produce crystalline PP; these are now known as Ziegler–Natta catalysts.[9]

HDPE is a nonpolar, linear thermoplastic that possesses a much more linear structure than LDPE. It has up to 90% crystallinity, whereas LDPE exhibits crystallinities as low as 50%. Although some branch chains are formed, these are short and few in number. HDPE film is stiffer and harder than LDPE and densities range from 941 to 965 kg m^{-3}. Its softening point is about 121°C, and its low temperature resistance is about the same as LDPE. Tensile and bursting strengths are higher but impact and tear strengths are both lower than LDPE. Of interest is the fact that due to the linear nature of the HDPE molecules, they tend to align themselves in the direction of flow and, thus, the tear strength of the film is much lower in the machine direction than the transverse direction. This difference can be accentuated by orientation to give a built-in tear tape effect.

The chemical resistance of HDPE is also superior to that of LDPE and, in particular, it has better resistance to oils and greases. The film offers excellent moisture protection and significantly decreased gas permeability compared with LDPE film, but is much more opaque. Heat sealing is considerably more difficult compared to LDPE film.

HDPE film has a white, translucent appearance and therefore tends to compete more with paper than transparent films. To be competitive with paper on a price-per-unit-area basis, it must be thin and consequently much of the HDPE film used is only 10 to 12 μm thick.

HDPE is blow-molded into bottles for a variety of food packaging applications, although its uses in this area have tended to be taken up by PET bottles which generally have better barrier properties than HDPE.

The basic properties of various polyethylene films are shown in Table 2.2. The melting point of polyethylenes is primarily a function of their densities, the melting point increasing with density as does the softening temperature.

TABLE 2.2
Basic Properties of Various Polyethylene Films

Type of Polyethylene	Moisture Vapor Transmission	Gas Transmission		Tensile Strength (MPa)	Softening Point (°C)	CH_3 Groups per 1000 Cs
		O_2	CO_2			
Low density (920 kg m^{-3})	1.4	500	1350	9–15	120–180	20–33
Medium density (940 kg m^{-3})	0.6	225	500	21	120–180	5–7
High density (960 kg m^{-3})	0.3	125	350	28	135–180	<1.5

d. *Irradiated Polyethylene*

Irradiated polyethylene is produced by passing ordinary LDPE film continuously under an electron beam accelerator which produces high energy β rays. This converts it to an infusible film, which causes cross-linking between the chains and gives it exceptional strength from the point of view of stretch resistance and shrink tension. Other effects include the evolution of hydrogen and a reduction in crystallinity. The process slightly reduces gas and water vapor transmission rates but increases the heat sealing range to make a practical shrink film from polyethylene. The film has good clarity. It is sealed by welding the overlaps together on a hot plate and shrunk by passing through a hot air tunnel at 220°C.

e. *Polypropylene (PP)*

Early attempts to polymerize propylene using the high pressure process used to make LDPE gave only oily liquids or rubbery solids of no commercial value. Work by Natta in Italy using Ziegler-type catalysts led to the development in 1955 of a stereospecific catalyst that controlled the position of each monomer unit as it was added to the growing chain, thus giving a polymer of regular structure. Today, typical processes take place at about 100 atmospheres and 60°C.

PP is a linear polymer containing little or no unsaturation. Depending on the type of catalyst and polymerization conditions, the molecular structure of the resulting polymer consists of the three different types of stereo configurations: isotactic, syndiotactic and atactic as shown in Figure 2.4. Industrial processes are designed to minimize the production of atactic PP (where the methyl groups are randomly distributed on either side of the chain), which results when polymerization occurs in the absence of stereospecific catalysts. This noncrystalline material has a density of about 850 kg m^{-3} and is soft, tacky and soluble in many solvents. It is a lower value product and mainly finds use in hot-melt adhesives.

The most regular crystalline polymer produced by stereospecific catalysts is known as the isotactic form, the name stemming from the original idea that the methyl groups were always above or below the horizontal plane. Isotactic PP, the most common commercial form of polypropylene homopolymer, is never perfectly stereoregular with the degree of isotacticity varying from 88 to 97%. It is a highly crystalline material with good chemical and heat resistance but poor transparency. Two other forms are syndiotactic where the methyl groups alternate above and below

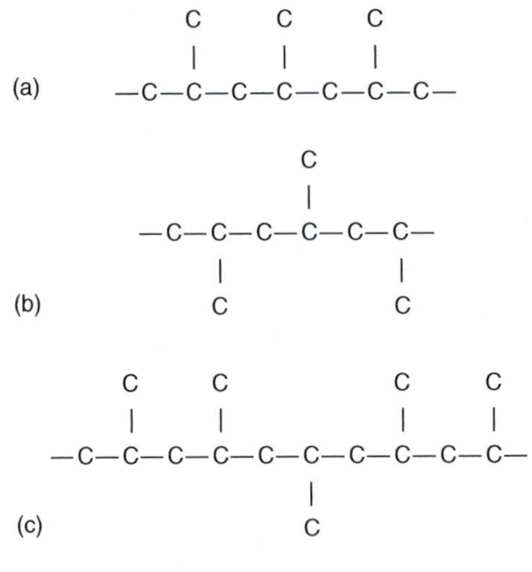

```
            C   C   C
            |   |   |
(a)     —C—C—C—C—C—C—C—

                 C
                 |
         —C—C—C—C—C—C—
             |       |
(b)          C       C

       C   C       C   C
       |   |       |   |
     —C—C—C—C—C—C—C—C—C—C—
                 |
(c)              C
```

FIGURE 2.4 Types of polypropylene: (a) isotactic, (b) syndiotactic and (c) atactic.

the horizontal plane and stereoblock where blocks of methyl groups are alternately above and below the horizontal plane. The regular helices of the isotactic form can pack closely together, whereas the atactic molecules have a more random arrangement.

While atactic PP is an amorphous, rubbery material of little value, isotactic PP is stiff, highly crystalline and has a high melting point. In commercial polymers, the greater the amount of isotactic material, the greater the crystallinity and, thus, the greater the softening point, tensile strength and hardness — all other structural features being equal.[4]

PP has a lower density (900 kg m^{-3}) and a higher softening point (140 to 150°C) than the polyethylenes, low water vapor transmission, medium gas permeability, good resistance to greases and chemicals, good abrasion resistance, high temperature stability, good gloss and high clarity with the two latter factors making it ideal for reverse printing.

PP can be blow-molded and injection-molded, the latter process being widely used to produce closures for HDPE, PET and glass bottles, as well as thin-walled pots and crates. The glass transition temperature of PP is placed between 10 and −20°C with the result that the polymer becomes brittle as subzero temperatures are approached. The T_m is in the range 160 to 178°C (sometimes quoted as 163°C), enabling foods inside PP containers to be sterilized by heat or reheated in microwave ovens. Copolymerization with 4 to 15% ethylene improves the strength and lowers the T_m and T_g slightly; such copolymers are often preferred to the homopolymer in injection molding and bottle blowing applications and also find use in shrink wrapping where the lower melting point is an advantage.

Although free from ESC problems, PP is more susceptible to oxidative degradation at elevated temperatures necessitating the inclusion of antioxidants in all commercial PP compounds. Whereas polyethylene cross-links on oxidation, PP degrades to form lower MW products. A similar effect is observed when PP is irradiated.

Nonoriented PP film is often referred to as cast PP film because it is generally made by the chill-roll cast process, although other methods can be used. PP film is a very versatile material being used as a thermoformable sheet in cast form for film and bags and as thin, strong biaxially oriented films for many applications. Cast and oriented PP are sufficiently different that they do not compete for the same end uses, the cost of cast PP being much lower than that of oriented PP. The cast form has polyethylene-type uses while the oriented form has regenerated cellulose film (RCF)-type uses and has largely replaced RCF in food packaging applications. Cast PP use in food packaging is limited owing to its brittleness at below-freezing temperatures and it is generally not recommended for use with heavy, sharp or dense products unless laminated to stronger, more puncture resistant materials. The relatively high temperature resistance of PP permits its use as the seal layer in retortable pouches, hot filled bottles and microwavable packaging.

In recent years there has been a large increase in the use of oriented polypropylene (OPP) for food packaging. Wide variations are possible in the extent of orientation in two directions, leading to a wide range of properties. However, biaxially oriented film (BOPP) has a high clarity because layering of the crystalline structures reduces the variations in refractive index across the thickness of the film and this in turn reduces the amount of light scattering. OPP can be produced by the blown tubular or high expansion bubble process or the tenter frame process.

BOPP film has a tensile strength in each direction roughly equal to four times that of cast PP film. Although tear initiation is difficult, tear resistance after initiation is low. Biaxial orientation also improves the moisture barrier properties of PP film and its low temperature impact strength. OPP film is not considered to be a gas barrier film but this deficiency can be overcome by coating with PVC–PVdC copolymer. OPP films often have a stiff feel and tend to audibly crinkle.

If heat sealing is required, PP is normally coated with a lower melting point polymer because shrinkage tends to occur when highly stretched film is heated. LDPE, PVC–PVdC copolymer and acrylic polymers are used as fusible coatings for OPP film. The LDPE is cheaper but the PVC–PVdC copolymer confers far better resistance to water vapor and O_2 permeability; acrylic polymers add no barrier properties to OPP film.

A relatively new addition to the family of OPP films is white opaque film, generally made by the tenter frame process and known as pearlized film because the diffusion of light gives the film the visual effect of pearlescence. Homopolymer resin is evenly mixed with a small amount of foreign particulate matter such as starch. In one product, when the thick filled sheet is oriented the PP pulls away from each particle creating an air-filled void or closed cell. After heat stabilization the OPP film is similar to a micropore foamed product. In the second product, the material produced is a filled film without voids, the opacity being a direct result of the amount of particulate material included in the film. The primary opacification is caused by light rays bouncing off the PP cell walls and the air within each cell. White opaque OPP films find application in snack food packaging, candy-bar overwraps, beverage bottle labels, soup wrappers and other applications that have traditionally used specialty paper-based packaging materials.[11]

f. Metallocenes

Metallocenes are a relatively old class of organometallic complexes, first discovered in 1951. They are based on a metal atom such as titanium, zirconium or hafnium. As early as 1957 Natta reported the (unsuccessful) polymerization of ethylene with a titanocene catalyst. The current interest in metallocenes originated with a discovery by Kaminsky at the University of Hamburg in the mid-1970s. While studying a homogenous polymerization system, water was accidentally introduced into the reactor leading to an extremely active ethylene polymerization system. Subsequent studies revealed that the high activity was due to the hydrolysis of the cocatalyst trimethyl aluminum. Because of the discovery of this new cocatalyst, metallocenes are sometimes called “Kaminsky” catalysts.

Metallocene-based catalyst technology is revolutionizing the immense polyolefin industry, particularly in the polyethylene and polypropylene markets. Metallocenes have been deemed to be the single most important development in catalyst technology since the discovery of Ziegler–Natta catalysts. Compared with conventional Ziegler–Natta technology, metallocenes offer some significant process advantages and produce polymers with very favorable properties. The development of metallocene polymers has brought forth the concept of single-site catalysis, of which metallocenes are just one example, albeit the first commercial success.

Since 1991, various polyolefins have been produced with the aid of single-site metallocene catalysts. By altering the metallocene structure, the types of polymers produced can be controlled. This characteristic of metallocenes is very important when producing polymers that can have different side branches such as isotactic and syndiotactic PP. These new polymers have features such as lower melting points, better optical characteristics, better heat stability, increased impact strength and toughness, better melt characteristics and improved clarity as films. These advantages are obtained through the control of polymer MW, MW distribution (elimination of both high and low MW fractions), comonomer distribution, comonomer content and tacticity.

Metallocenes are important in the production of polyethylene in that they allow the control of side branching due to the single activity site found at the metal center. Traditional Ziegler–Natta catalysts are hard to control because they have several active sites and polymers are produced by adding monomers to the end of the chain.

Using traditional catalysts, PP is produced as a mixture of the three forms consisting of 95% isotactic, some undesirable atactic and even less syndiotactic PP. With metallocene technology, the amount of each type of PP can be controlled through changes in the catalysts’ stereochemistry. New isotactic PP resins can be produced with improved stereoregularity and controlled comonomer content resulting in higher stiffness, clarity and melt strength. Comonomers employed to produce these enhanced PP resins include ethylene, butene and octene.

Metallocene-catalyzed resins have expanded beyond polyethylenes to polyolefin plastomers and elastomers, polypropylenes, polystyrenes, ethylene–styrene copolymers and cycloolefin copolymers.[13] After an almost 10 year gestation period, polymers made using metallocene catalysts are poised for significant growth. There is likely to be continuing penetration of metallocene-derived resins into the domain of commodity thermoplastics in a wide array of applications including food packaging films and stretch/shrink films. Recently, an intense search began for the next generation of single-site catalysts.

2. Copolymers of Ethylene

For copolymers of ethylene, the other comonomer can be an alkene such as propene, butene, hexene or octene, or a compound having a polar functional group such as vinyl acetate (VA), acrylic acid (AA), ethyl acetate (EA), methyl acrylate (MA) or vinyl alcohol (VOH). The polymer can be classified as either a copolymer or homopolymer if the molar percentage of the comonomer is less than 10%. LLDPE and LDPE–PP copolymer discussed earlier can rightly be considered as copolymers of ethylene. In addition there are four further copolymers of ethylene of particular interest in food packaging.

a. Ethylene–Vinyl Acetate (EVA)

EVA is a random copolymer whose properties depend on the VA content and molecular weight. EVA with a VA content of 3 to 12% is similar in flexibility to plasticized PVC and has good low temperature flexibility and toughness. The impact strength increases with VA content and MW. As the VA level increases, EVA becomes less crystalline and more elastic; as the crystallinity decreases, the permeability to gases, moisture, fats and oils increases and the clarity improves. EVA is totally amorphous (transparent) when the VA content reaches 50%. As the MW increases, the viscosity, toughness, heat seal strength, hot tack and flexibility all increase. The absence of leachable plasticizer provides a clear advantage over plasticized PVC in some food applications. The addition of antiblocking and slip additives reduces sparkle and clarity and increases haze.

EVA copolymers are not competitive with normal film because of their high surface tack and friction which make them difficult to handle on conventional processing machinery. However, they do have three advantages over LDPE: the heat sealing temperature is lower, the barrier properties are better and they have excellent stretch properties, the first 50% of extension at room temperature being elastic. Thus, they find use as a stretch film for food packaging (particularly fresh meat) and cling–wrap purposes and have replaced PVC for many stretch wrapping applications. Some EVA is used in coextrusion processes for the manufacture of laminated material. As a heat sealing layer, EVA is used in the extrusion coating of PET and BOPP films.

b. Ethylene–Vinyl Alcohol (EVOH)

EVOH copolymers were commercialized in Japan by the Kuraray Company in 1972 and in the U.S. and Europe in the early 1980s. EVOH copolymer is produced by transforming the VA group into VOH in a controlled hydrolysis of EVA copolymer; there is no VOH involved in the copolymerization. The vinyl alcohol base has exceptionally high gas barrier properties but is water soluble and difficult to process.

EVOH copolymers offer not only excellent processability but also superior barriers to contaminants such as gases, odors, fragrances and solvents. It is these characteristics that have allowed plastic containers containing EVOH barrier layers to replace many glass and metal containers for packaging food.

EVOH copolymers are highly crystalline in nature and their properties are very dependent on the relative concentration of the comonomers. Generally, as the ethylene content increases, the gas barrier properties decrease, the moisture barrier properties improve and the resins process more easily. For example, EVOH of 27 mol% ethylene offers a barrier to dry O_2 ten times as great as one of 42 mol % ethylene while providing a poorer barrier to moisture vapor.[3] When the VOH content ranges from 50 to 70%, the EVOH copolymers combine the processability and water resistance of LDPE with the gas and odor barrier properties of polyvinyl alcohol (PVOH).

The most outstanding characteristic of EVOH is its ability to provide a barrier to gases and odors. Its use in a packaging structure enhances flavor and quality retention by preventing O_2 from penetrating the package. In those applications where gas-fill packaging techniques such as MAP are used, EVOH effectively retains the CO_2 or N_2 used to blanket the product. EVOH also provides a very high resistance to oils and organic vapors but this resistance decreases as the polarity of the penetrant increases. For example, the resistance to linear and aromatic hydrocarbons is outstanding, yet for ethanol and methanol it is low, with EVOH absorbing up to 12% of ethanol.[8]

Due to the presence of hydroxyl groups in their molecular structures, EVOH resins are hydrophilic and will absorb moisture. As moisture is absorbed, the gas barrier properties are affected. However, through the use of multilayer technology to encapsulate the EVOH layer with high moisture barrier resins such as polyolefins, the moisture content of the barrier layer can be controlled.

EVOH resins have high mechanical strength, elasticity and surface hardness, very high gloss, low haze, excellent abrasion resistance, very high resistance to oils and organic solvents and provide an excellent barrier to odors. When used as the core of a multilayer material they provide excellent performance. EVOH copolymers are the most thermally stable of all the high barrier resins. This stability allows the regrinding and reuse of scrap generated during processing back into the package being produced. Many of the rigid packaging containers being produced today have regrind layers containing up to 15% EVOH.

Rigid and semirigid containers such as bottles, trays, bowls and tubes, flexible films and paperboard beverage cartons containing EVOH as the functional barrier layer are commercially available. Most multilayer structures have five or six layers although seven and nine layer structures are being produced for special applications, the EVOH always being surrounded by polymers such as polyolefins which provide a good barrier to water vapor. When using EVOH in multilayer structures, it is necessary to use an adhesive or tie layer to gain adequate bonding strength to the other polymers.[7]

For multilayer packages that will be retorted after filling, the structures outlined above may not provide a sufficient barrier to water vapor to prevent unacceptable quantities of O_2 permeating into the product. In these situations a desiccant is incorporated into the tie layer between the EVOH and the polyolefin to absorb any moisture which penetrates the polyolefin during retorting.[13]

c. *Ethylene–Acrylic Acid (EAA)*

By copolymerizing ethylene with acrylic acid, a copolymer containing carboxyl groups along the main and side chains of the molecule is obtained. EAA copolymers are flexible thermoplastics possessing chemical resistance and barrier properties similar to LDPE. However, EAA is superior to LDPE in strength, toughness, hot tack and adhesion. Adhesion strength increases and heat seal strength decreases with increasing AA content. Two major uses for EAA are for blister packaging and in adhesive lamination as an extrusion-coating tie layer between aluminum foil and other polymers. EAA films are used for the packaging of foods such as meat, cheese and snack foods, and in skin packaging, where they conform closely to the shape of the product.[8]

d. *Ionomers*

Ionomeric polymers with unique properties result when olefinic polymers are prepared in the presence of metallic salts of organic monomers. Ionomers were first discovered by researchers at DuPont in the 1960s, and are prepared by copolymerizing ethylene with a small amount (1 to 10% in the basic patent) of an unsaturated carboxylic acid such as acrylic acid or methacrylic acid using the high pressure process. Such copolymers are then neutralized to varying degrees with the derivative of a metal such as sodium, zinc, lithium methoxide or lithium acetate, causing the carboxylic acid to ionize. This leads to the formation of ionic cross-links which confer enhanced stiffness, transparency and toughness on the material at ambient temperatures (the puncture resistance of ionomer film is equal to LDPE film of twice the gauge) as well as higher melt strength. There are more than 50 commercial grades of ionomer resin with a wide range of properties.[8]

In comparison with LDPE, ionomers have excellent oil and grease resistance, excellent resistance to ESC, greater clarity, lower haze, greater abrasion resistance and higher moisture vapor permeability due to lower crystallinity. Ionomers are particularly useful in composite structures with films, paperboard or aluminum foil to provide an inner layer with good heat sealability. Ionomers have the ability, shared with certain other polymers, of bonding by heat sealing through particles of food that may be trapped between package layers during the filling process. Laminated or coextruded films with polyamides or polyesters are widely used for packaging meat and cheese where formability, toughness and visual appearance are important.[3] Their exceptional impact and puncture resistance (even at low temperatures) makes them ideal for skin packaging of sharp objects such as meat cuts containing bone. Another advantage is their high infrared absorption, which leads to fast heating during the shrink process. Their disadvantages include their poorer slip and block characteristics and relatively poor barrier to O_2.

3. Substituted Olefins

The simplest substituted olefins are those in which each ethylene group has a single substituent; these monomers are called vinyl compounds. Disubstituted monomers with two substituents on the one carbon are called vinylidene. Structures of some common monomers based on the ethylene chain are shown in Figure 2.5.

a. *Polystyrene (PS)*

If ethylene and benzene are reacted together with a suitable catalyst, ethylbenzene is formed and, by a process of catalytic dehydrogenation, styrene (commonly known as vinyl benzene) is

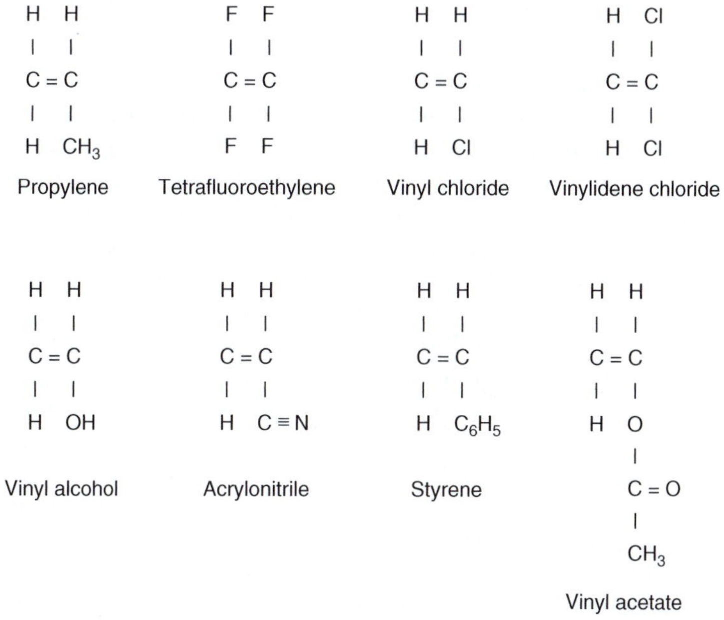

```
H   H          F   F          H   H          H   Cl
|   |          |   |          |   |          |   |
C = C          C = C          C = C          C = C
|   |          |   |          |   |          |   |
H   CH3        F   F          H   Cl         H   Cl

Propylene   Tetrafluoroethylene   Vinyl chloride   Vinylidene chloride

H   H          H   H          H   H          H   H
|   |          |   |          |   |          |   |
C = C          C = C          C = C          C = C
|   |          |   |          |   |          |   |
H   OH         H   C≡N        H   C6H5       H   O
                                                 |
Vinyl alcohol  Acrylonitrile  Styrene            C = O
                                                 |
                                                 CH3

                                             Vinyl acetate
```

FIGURE 2.5 Structure of monomers based on ethylene.

produced. Polystyrene (PS) is made by the addition polymerization of styrene. The polymer is normally atactic and is thus completely amorphous because the bulky nature of the benzene rings prevents a close approach of the chains. With the use of special catalysts and polymerization techniques, isotactic PS has been prepared but it reverts to the atactic form on melting.

PS was the first of the moldable clear rigid plastics to reach the commercial market in large volumes in the late 1940s.[3] It could be rapidly molded into finished shapes because of its ease of flow in the melt and its fast-setting nature from the melt. However, an increase in monomer content from 0 to 5% can cause a 30°C reduction in softening point.[4]

i. General Purpose Polystyrene (GPPS)

In this form, it is commonly referred to as crystal grade PS and is the unmodified homopolymer of styrene. At temperatures in the range at which food packages are stored it is glassy and noncrystalline given that its T_g is in the range 90 to 100°C because of the stiffening effect of the benzene ring. This results in a material which is stiff and brittle at room temperature with no melting temperature but excellent optical properties.

PS makes a distinctly metallic sound when dropped onto a hard surface. It has a high refractive index (1.592), which gives it a particularly high brilliance. Although acids and alkalis have no effect on it, it is soluble in higher alcohols, ketones, esters, aromatic and chlorinated hydrocarbons and some oils. Even if it is not soluble in the material, cracking and even chemical decomposition of PS may occur. The extent of the decomposition depends on the grade of PS, the time and temperature of exposure and the concentration of the reagent. In addition, many materials which do not attack PS individually do so in combination, so shelf life studies are essential to test for any synergistic effect before marketing a food product in PS. While a reasonably good barrier to gases, it is a poor barrier to water vapor.

Crystal grade PS can be made into film but it is brittle unless the film is biaxially oriented. The oriented film can be thermoformed into a variety of shapes although special techniques have to be used because orientation gives it a tendency to shrink on heating.

New applications of PS involve coextrusion with barrier resins such as EVOH and PVC–PVdC copolymer to produce thermoformed, wide-mouthed containers for shelf stable food products and multilayer blow-molded bottles.

ii. High Impact Polystyrene (HIPS)

To overcome the brittleness of PS, synthetic rubbers (typically 1,3-butadiene isomer $CH_2{=}CH{-}CH{=}CH_2$) can be added during polymerization at levels generally not exceeding 25% w/w for rigid plastics. The rubbers act by restricting propagation of microcracks formed during impact loading. At the start of the process, polybutadiene is dissolved in the styrene monomer. As the polymerization proceeds two phases are formed: a polybutadiene rich phase and a PS-rich phase with grafted polybutadiene. The grafting arises when some of the styrene-free radicals react with the polybutadiene. Controlling the sizes of the homopolymer blocks formed by each of the monomers determines the properties of the resulting copolymer. Blocks of PS in SB (styrene–butadiene) copolymers confer typical styrenic properties of stiffness, glossiness and ease of processing, while blocks of polybutadiene contribute great flexibility and extensibility. SB copolymer is sometimes referred to as K resin. Although copolymerization increases impact strength and flexibility, the transparency, tensile strength and thermal resistance is much reduced. The chemical properties of this toughened or high impact polystyrene (HIPS) are much the same as those for unmodified polystyrene.

HIPS is an excellent material for thermoforming. Because it is transparent, the use of radiant heat for thermoforming is inefficient and pigmented sheet is often used. It is injection molded into tubs which find wide use in food packaging, despite their opacity. Transparent unit packs are made from PS that has not been toughened.

iii. Expandable Polystyrene Sheet (EPS)

The properties of PS that make it useful for many applications as a solid polymer also make it very desirable as foam. PS foam has a high tensile strength, good water resistance, low moisture transmission, ease of fabrication and low cost. Closed cell foams have excellent thermal insulating capability, low weight and good cushioning characteristics. The combination of those properties provides a wide spectrum of products.

Extruded PS foam sheet is a closed cell, 0.13 to 6.4 mm thick sheet with densities ranging from 32 to 160 kg m^{-3}. It can be made by a variety of extrusion processes but is most commonly made using a tandem extrusion process. A large percentage of foam is thermoformed using matched metal molds. Most PS foam sheet is used for disposable packages such as meat and produce trays, egg cartons, disposable dinnerware and containers for take-away or carry-out meals.

b. Poly(Vinyl Alcohol) (PVOH)

When ethylene is reacted with acetic acid and pyrolyzed, vinyl acetate is formed. This can be polymerized to form poly(vinyl acetate) (PVA). Although plastic films could be made from PVA, they would be as brittle as PS and no outlet for them has arisen. Vinyl alcohol does not exist in a free state and all attempts to prepare it have led instead to the production of its tautomer, acetaldehyde. PVOH is made by alcoholysis of PVA, the PVOH having the same DP as the PVA.

PVOH is generally considered to be an atactic material but it can be made to crystallize. It is an unusual polymer in that it is soluble in water. It melts at 185 to 215°C depending on the method of alcoholysis. A water-soluble film offers the utility and convenience of packaging such materials as chemicals or dyes which need to be used at a controlled dosage in water without further handling. Suggestions have also been made for the packaging of convenience foods in PVOH, but an outer package of some sort is still required to maintain the PVOH in a hygienic form. This increases costs and is probably the major reason why the suggestions have not been taken up.

PVOH films are poor barriers to water vapor but excellent barriers to O_2 and greases. Wet film has little strength but the strength of dry film is high. Being water-soluble, it is difficult to process. If required, the film can be printed.

c. Poly(Vinyl Chloride) (PVC)

Ethylene dichloride is formed by an addition reaction of chlorine with ethylene, and the dichloride is then dehydrochlorinated or "cracked" to give vinyl chloride monomer (VCM). Addition polymerization of VCM produces PVC. From the structure of VCM it can be seen that addition of molecules to the growing chain can take place either head to head, head to tail or in a completely random manner. PVC polymerizing in either of the first two forms would be expected to be crystalline, while those containing the random arrangement would be amorphous. Generally, PVC polymerizes in the atactic form and thus is largely an amorphous polymer.

In 1974 there was widespread publicity resulting from the announcement that a rare and invariably fatal form of liver cancer had been diagnosed among workers in vinyl chloride polymerization plants in Britain and the U.S. Food and drug administrators around the world became particularly interested in the VCM content of PVC material used for food packaging and the likelihood of it migrating into the food. This topic is discussed further in Chapter 21 but suffice to say, the levels of VCM in PVC packaging material are currently extremely low.

Vinyl polymers and copolymers make up one of the most important and diversified groups of linear polymers. This is because PVC can be compounded to produce a wide spectrum of physical properties. This is reflected in the variety of uses to which it is put: from exterior guttering and water pipes to very thin, flexible surgical gloves. It is the second most widely used synthetic polymer after polyethylene and is commonly referred to simply as "vinyl."

A range of PVC films with widely varying properties can be obtained from the basic polymer. The two main variables are changes in formulation (principally plasticizer content) and orientation. The former can give films ranging from rigid, crisp films to limp, tacky and stretchable films. The degree of orientation can also be varied from completely uniaxial to balanced biaxial.

Unplasticized PVC tends to degrade and discolor at temperatures close to those used in its processing so suitable stabilizers have to be included in the formulation. The stabilizers used are generally the salts of tin, lead, cadmium, barium, calcium or zinc along with epoxides and organic phosphites, and these must be carefully selected for nontoxic applications. The use of lead and cadmium compounds as stabilizers is not generally allowed in food contact materials. Approval by the U.S. FDA of certain octyltin compounds for use in stabilizing PVC during blow molding of containers has expanded the use of this polymer for food packaging.

Extremely clear and glossy films can be produced having a high tensile strength and stiffness. The density is high at around 1400 kg m^{-3}. The water vapor permeability is higher than that of the polyolefins but the gas permeability is lower. Unplasticized PVC has excellent resistance to oils, fats and greases and is also resistant to acids and alkalis.

To a large extent the properties of plasticized PVC depend on the type of plasticizer used, as well as the quantity. For these reasons it is difficult to be very specific about the physical properties of PVC due to the wide range of plasticization possible. Rigid PVC has a $T_g = 82°C$, plasticizers decreasing T_g and the processing temperature.

The plasticizers used (about 80% of all plasticizers are used with PVC) are organic liquids of low volatility which facilitate internal movement of the molecular chains. Combinations of different compounds are used, the esters of phthalic acid being the most common. The amount can vary up to 50% of the total weight of the final material. Because plasticizers are not bound chemically, they tend to migrate to the surface where they are lost by abrasion, solution or slow evaporation, leaving a more brittle, stiffer material behind. Internal plasticization may be brought about by copolymerization of vinyl chloride with monomers such as vinyl acetate, ethylene or methylacrylate.[6]

Films with excellent gloss and transparency can be obtained provided that the correct stabilizer and plasticizer are used. Both plasticized and unplasticized films can be sealed by high frequency welding techniques. PVC is inert in its chemical behavior, self-extinguishing when exposed to a flame. Thin, plasticized PVC film is widely used for the stretch wrapping of trays containing fresh red meat and produce. The relatively high water vapor transmission rate (WVTR) of PVC prevents condensation on the inside of the film. Oriented films are used for shrink wrapping of produce and fresh meat but LLDPE films have increasingly replaced them in many applications in recent years.

Unplasticized PVC as a rigid sheet material is thermoformed to produce a wide range of inserts from chocolate boxes to biscuit trays. Unplasticized PVC bottles have better clarity, oil resistance and barrier properties than those made from HDPE. However, they are softened by certain solvents, notably ketones and chlorinated hydrocarbons. They have made extensive penetration into the market for a wide range of foods including fruit juices and edible oils, but in recent years they have been increasingly replaced by PET.

d. Poly(Vinylidene Chloride) (PVdC)

PVdC homopolymer has a melting temperature only a few degrees below the temperature at which it decomposes and yields a rather stiff film which is unsuitable for packaging purposes. Therefore, copolymers were synthesized in an effort to overcome these properties. Acrylates were found to be among the most useful comonomers, along with VC and AN.[3]

A soft, tough and relatively impermeable film results when PVdC is copolymerized with 5 to 50% (but typically 20%) of vinyl chloride. These copolymers were first marketed by Dow in 1940 under the trade name Saran® (supposedly an abbreviation of Sarah and Ann, the names of the inventor's wife and daughter, respectively). Although the films are copolymers of VdC and either

VC or AN, they are usually referred to simply as "PVdC copolymer," and such a convention will be followed in the remainder of this book. The specific properties of PVdC copolymer vary according to the degree of polymerization and the properties and relative proportions of the copolymers present.

The properties of PVdC copolymer film include a unique combination of low permeability to water vapor, gases, odors, greases and alcohols, and good ESCR to a wide variety of agents. It also has the ability to withstand hot filling and retorting, making it a useful component in multilayer barrier containers. Although highly transparent, it has a yellowish tinge. It is an important component of many laminates and is the best coating for RCF. PVdC copolymers can be sealed to themselves and to other materials.

By itself the copolymer is frequently used as a shrink film because orientation improves tensile strength, flexibility, clarity, transparency and impact strength. As well, gas and moisture permeabilities are lowered and tear initiation becomes difficult. The shrink film can be heat sealed using impulse sealing with Teflon-coated heating bars.

e. *Poly(Tetrafluoroethylene) (PTFE)*

The high thermal stability of the carbon–fluorine bond has led to considerable interest in fluorine-containing polymers as heat resistant plastics and rubbers. Chemically, the fluorine-containing polymers are set apart from other vinyl polymers because their monomers are the only ones which need not bear any hydrogen on the ethylenic carbons in order to be polymerizable. Physically, they are further set apart by their outstanding stability. PTFE was a chance discovery by Roy Plunkett while working for DuPont in 1938. He had a cylinder of TFE gas which, although apparently empty, had not yielded the theoretical amount of gas. When the cylinder was cut open, it was partly filled with a waxy white powder which was identified as a polymer. First marketed in 1945, today it accounts for about 80% of the fluorinated polymers produced and is commonly referred to as Teflon™.

The monomer TFE is made by reacting hydrofluoric acid with chloroform. Hydrogen chloride is displaced, forming $CHClF_2$, which can be pyrolyzed to TFE. PTFE is a linear polymer free from any significant amount of branching with a MW that can exceed 30 million atomic mass units, making it one of the largest molecules known. Because the C–C and C–F bond strengths are very high, PTFE has very high heat stability, even when heated above its crystalline melting point of 327°C.[4] Its chemical inertness, nonadhesive properties and excellent heat resistance and low coefficient of friction make PTFE ideal for certain specific applications.

In food packaging PTFE finds wide use as a nonstick separating surface between thermoplastic films and the jaws of heat sealers. It is common to use a band of PTFE (often reinforced with glass fibers) on continuous heat sealers. The main disadvantages of PTFE are its very high cost, unsuitability for the processing techniques conventionally applied to plastics, and poor scratch resistance and mechanical properties.

4. Polyesters

The polymers which have been discussed so far are all based on carbon-to-carbon links and are generally formed commercially by addition polymerization. In contrast, polyesters are based on carbon–oxygen–carbon links where one of the carbons is part of a carbonyl group and are formed by the process of condensation polymerization. In this process, two molecules are joined together through the elimination of a smaller molecule (typically H_2O) whose atoms derive from both the parent molecules. Simple polyesters are derived from condensation of a polyhydric alcohol and a polyfunctional acid and are sometimes described as alkyds (from *al*cohol and aci*d*). Each component needs a functionality (i.e., number of reactive groups such as –OH, –COOH, $-NH_2$ per molecule) of two to form a linear chain, while if one (or both) monomer has a functionality of at least three, cross-linkage can occur, resulting in a much more rigid 3D lattice structure.

Fiber-forming polyesters have been the subject of extensive investigations ever since the American chemist Wallace Carothers, who worked for DuPont, began his classical research that led to the development of nylons. However, while Carothers largely confined his research to aliphatic polyesters, John Whinfield and James Dickson working for ICI in England investigated aromatic materials. This led to the discovery in 1940, and subsequent successful exploitation, of poly(ethylene terephthalate) which DuPont was licensed by ICI to manufacture in the U.S.

a. *Poly(Ethylene Terephthalate) (PET)*

PET can be produced by reacting ethylene glycol (EG) with terephthalic acid (TPA), although in practice the dimethyl ester of TPA (dimethyl terephthalate or DMT) is used to give a more controllable reaction. Methanol is generated as a byproduct of the exchange reaction and can be recovered for further use; water is a byproduct if TPA is used. The overall scheme of the reaction is shown in Figure 2.6.

The product has a carboxyl group at one end and a hydroxyl group at the other, so it can condense with another molecule of alcohol and acid. The molecules grow to a molecular weight of up to 20,000.

EG is obtained from ethylene, mostly by the direct hydration of ethylene oxide. TPA is prepared by the catalytic partial oxidation of *p*-xylene, itself a by-product of the petroleum industry.

PET is a linear, transparent thermoplastic polymer with a T_m of 267°C and a T_g between 67 to 80°C. It has the capacity to crystallize under certain controlled conditions. PET is strong, stiff, ductile and tough while in the glassy state (i.e., at temperatures below T_g where side motion is restrained) and can be oriented by stretching during molding and extrusion, which increases its strength and stiffness still further.[3] PET bottles and films are largely amorphous with small crystallites and excellent transparency. However, crystallized PET containers have a higher degree of crystallinity, larger crystallites and are an opaque white.

PET films are most widely used in the biaxially oriented, heat stabilized form. There are virtually no applications for the material in its unoriented form because, if crystalline, it is extremely brittle and opaque, and if amorphous, it is clear but not tough. In a two-stage process, machine direction stretching induces 10 to 14% crystallinity and this is raised to 20 to 25% by transverse orientation. In order to stabilize the biaxially oriented film, it is annealed (or heat set) under restraint at 180 to 210°C, which increases the crystallinity to around 40% without appreciably

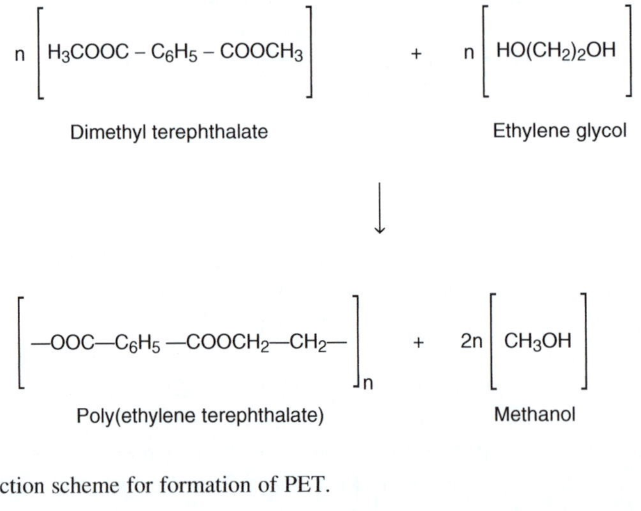

$$n\left[H_3COOC-C_6H_5-COOCH_3\right] + n\left[HO(CH_2)_2OH\right]$$

Dimethyl terephthalate — Ethylene glycol

$$\downarrow$$

$$\left[-OOC-C_6H_5-COOCH_2-CH_2-\right]_n + 2n\left[CH_3OH\right]$$

Poly(ethylene terephthalate) — Methanol

FIGURE 2.6 Reaction scheme for formation of PET.

affecting the orientation and reduces the tendency to shrink on heating.[4] Subsequent coatings are applied to obtain special barrier properties, slip characteristics or heat sealability.

PET film's outstanding properties as a food packaging material are its great tensile strength, excellent chemical resistance, light weight, elasticity and stability over a wide range of temperatures (−60 to 220°C). This latter property has led to the use of PET for "boil-in-the-bag" products which are frozen before use (the PET is usually laminated to or extrusion coated with LDPE and is typically the outside and primary support film of such laminations) and as oven bags where they are able to withstand high temperatures without decomposing.

To improve the barrier properties of PET, coatings of LDPE, PVdC copolymer or PVdCcoAN have been used. PET film extrusion-coated with LDPE is very easy to seal and very tough. It can be sealed through powders and some liquids and the integral seal will withstand sterilization by UV. Two-side, PVdC copolymer-coated grades provide a high barrier; a major special application is the single-slice cheese wrap.

Although many films can be metallized, PET is the most common. Metallization results in a considerable improvement in barrier properties. Reductions in WVTRs by a factor of 40 and O_2 permeabilities by over 300 are obtained.[10] Coextruded heat sealable films are frequently metallized and used as the inner ply of snack food packages. Rigid grades of metallized PET can be used in thermoformable applications.

PET is also used to make "ovenable" trays for frozen food and prepared meals where they are preferable to foil trays because of their ability to be used in a microwave without the necessity for an outer paperboard carton. These trays are thermoformed from cast PET film and crystallized, the crystallization heat setting the tray and preventing deformation during cooking and serving. In some instances such trays have been made with a two layer structure, with CPET for rigidity and APET for low temperature impact strength.[13] The properties of the various physical states of PET are summarized in Table 2.3.

In the 1970s the benefits of biaxial orientation of PET were extended from sheet film to bottle manufacture. Nathaniel Wyeth (brother of artist Andrew) and Ronald Roseveare were issued a series of patents (assigned to DuPont) in 1973 (filed in 1970) for a stretch blow-molded, biaxially oriented bottle prepared from PET. The first PET test market was held in New York State during 1975 to 76 and the first commercial production appeared on the market in 1977.[2] As a result, important new markets developed, particularly for carbonated beverages.

The bottles are stretch blow-molded, the stretching or biaxial orientation being necessary to get maximum tensile strength and gas barrier, which in turn enables bottle weights to be low enough to be economical. The presence of moisture during the extrusion of PET reverses the condensation reaction and produces some depolymerization. Therefore, before extrusion, PET must be dried to a moisture content of less than 0.005% to minimize hydrolytic breakdown and loss of properties.[8]

APET has nearly the same thermoforming characteristics as PS and has attracted some interest as a replacement for PVC. The properties of unoriented APET are similar to those of

TABLE 2.3
Properties of PET

Physical State	Property	Application
Amorphous	0–5% crystallinity, heat stable to 67°C, clear	Blister packs
Oriented amorphous	5–20% crystallinity, heat stable to 73°C, clear	Bottles
Crystalline	25–35% crystallinity, heat stable to 127°C, opaque	Food trays
Oriented crystalline	35–45% crystallinity, heat stable to 140–160°C, clear	Hot fill containers, films

Source: From Combellick, W.A., Barrier polymers, In *Encyclopedia of Polymer Science and Engineering*, 2nd ed., Vol. 2, Kroschwitz, J. J., Ed., Wiley, New York, p. 189, 1985. With permission.

semicrystalline oriented PET with the exceptions of strength and stiffness which are enhanced by orientation.[3]

b. Poly(Ethylene Naphthalate) (PEN)

In the late 1980s a high barrier polyester called PEN was developed. The precursor for PEN is 2,6-dimethylnaphthalene (DMN) which can be oxidized and esterified to give 2,6-dimethylnaphthalene dicarboxylate (NDC). This monomer undergoes polycondensation with EG to form PEN, a semicrystalline polymer similar to PET but with a double ring structure which makes it more rigid and improves its barrier and mechanical properties.

PEN provides approximately five times more barrier to CO_2, O_2, and water vapor than PET giving it obvious advantages as a flexible film, or as a rigid container for beer and carbonated beverages. It has a T_g of 120°C (43° higher than for PET) and is stronger and stiffer (but more expensive) than PET. The FDA gave final approval for food use of PEN homopolymers in April 1996 (the petition was filed in 1988), and for PET–PEN copolymers in March 2000.

Because the PEN homopolymer is priced at a significant premium to PET (PEN resin is about four times the cost of PET), attractive applications for PEN are those where its physical advantages outweigh its cost disadvantage. PEN extrusion blow-molded containers exhibit excellent properties in hot fill food (baby food, ketchup) and drink applications, reusable packaging and packaging of O_2 sensitive foods such as beer. PEN, like PET, has good clarity. Advantages of PEN over PET are its 35 to 55°C higher use temperature, 50% greater tensile strength, and fivefold better barrier performance. Additional markets for PEN encompass laminates, copolymers and blends. However, if PEN homopolymer is to penetrate large volume applications such as beer and single-serve carbonated beverage bottles, then the cost of making 2,6-DMN will need to be reduced. Like PET, PEN film can be vacuum metallized or coated with aluminum or silica oxides.

c. Copolyesters

The term copolyester is applied to those polyesters whose synthesis uses more than one glycol or more than one dibasic acid. The copolyester chain is less regular than the monopolyester chain and therefore has a reduced tendency to crystallize. As a result, some of the copolyesters are amorphous, some are crystalline and some can be made to be either crystalline or amorphous, depending on processing conditions.

PCTA copolyester is a polymer of cyclohexane dimethanol (CHDM), TPA and isophthalic acid. Primary use for this material is for extrusion into film and sheeting for packaging. For this type of application it is maintained as an amorphous polymer in its final form. Changing the ratios of the acids used in making the copolyesters can result in polyesters having uniquely different properties. One such copolyester, for example, is distinguished by its high crystalline melting point of 285°C and is useful as a material for ovenable trays that can withstand oven temperatures up to 250°C.

PETG copolyester (the G indicates the presence of a second glycol) is another example of the vast range of polymers attainable in the copolyester family. PETG is a clear amorphous polymer with a T_g of 81°C. It can be made by combining CHDM with TPA and EG. PETG can be readily molded or extrusion blow-molded and normally remains amorphous, clear, and virtually colorless even in very heavy thicknesses. It has high stiffness and hardness and good toughness, retaining an acceptable degree of toughness even at low temperatures. However, it is a poorer barrier than PET.

PET–PEN copolymers, which are made up primarily of PET with DMN as a comonomer at 10 to 25%, meet market hot fill requirements up to 95°C at a much more economical cost than PEN homopolymer which has a maximum hot fill temperature of 85°C.

PET and PEN can also be blended together. PEN homopolymers, copolymers and blends have not yet been used for any major, high volume, food packaging application but this situation is likely to change very soon.

5. Polycarbonates (PC)

PCs are polyesters of unstable carbonic acid and have carbonate (–OCOO–) linkages. They were originally produced by the reaction of phosgene (also known as carbonyl chloride [$COCl_2$]) with bisphenol A. While bisphenol A is still the most commonly used phenol, diphenyl carbonate has replaced phosgene. PCs have an outstanding combination of high temperature resistance, high impact strength and clarity, retaining their properties well with increasing temperature. Chemically they are resistant to dilute acids but strongly attacked by alkalis and bases such as amines. Permeability to both water vapor and gases is high and they must be coated if appreciable barrier properties are required. While PCs can be oriented, there is no decrease in permeability although tensile strength increases. PCs are not suitable as shrink films because the rate of shrinkage above their heat distortion point is extremely slow.

Because PCs are amorphous, they soften over a wide temperature range; the T_m of bisphenol A PCs is 220 to 250°C and the T_g 150°C. The melting point of PCs is decreased from 225 to 195°C when the methyl pendant groups are replaced by propyl groups. Thermoforming of PC film is readily carried out and deep draws with good mold detail are obtainable.

PCs are used to a very limited extent in food packaging as components of multilayer coextrusions and coinjection moldings to provide transparency and high strength to containers which undergo hot filling or hot processing after filling.[3] PC can be laminated or coextruded to PP, PE, PET, PVC and PVdC; coextrusions with EVOH or PAs are carried out with the help of adhesives.[8] Multilayered bottles typically contain a central EVOH layer to reduce O_2 ingress with the PC layers on either side providing strength. However, the relatively high cost of PC limits its use in these applications.

Another application for PCs is ovenable trays for frozen foods and prepared meals where low temperature impact strength adds durability and toughness; they are commonly coextruded with PET. The film has been used for boil-in-bag packs and, when coated with LDPE, skin packaging. Uses as retort pouches and microwave oven cookware are possible because of good stability at high temperatures. Vacuum metallizing gives good results because of PCs transparency, the finished product having a high gloss.

Although the film is biaxially oriented for electrical applications, it is not used in this form in food packaging. However, blow and injection molded PC is useful for reusable bottles (particularly baby bottles), which take advantage of PC's toughness and clarity. As well, PC has high resistance to staining by tea, coffee, fruit juices, tomato sauces, lipstick, ink, soaps and detergents.

6. Polyamide (PA)

PAs are condensation, generally linear, thermoplastics made from monomers with amine and carboxylic acid functional groups resulting in amide (–CONH–) linkages in the main polymer chain that provide mechanical strength and barrier properties. Their development was not fortuitous but the result of a long search for a family of polymers that would resemble silk, the most highly valued of all natural textile fibers.

The early development of the nylons (originally the DuPont brand name for the family of synthetic polyamides and said to be derived from a contraction of the names of the cities *N*ew *Y*ork and *Lon*don) is largely due to the work of Carothers and his colleagues at DuPont between 1928 and 1937. They first synthesized nylon 6,6 in 1935 after extensive research into condensation polymerization. Commercial production of this polymer for subsequent conversion into fibers was commenced by the DuPont in 1939.

In an attempt to circumvent the DuPont patents, German chemists investigated a wide range of synthetic fiber-forming polymers in the late 1930s. This work resulted in the successful introduction of nylon 6 (where a six carbon molecule contained an acid group at one end and an amine group at the other) and between them nylons 6 and 6,6 account for nearly all of the polyamides produced for fiber application.

Two different types of PA films are available based on their resin manufacture. One type is made by condensing mixtures of diamines and dibasic acids. These are identified by the number of carbon atoms in the diamine, followed by the number of carbon atoms in the diacid. The other type is formed by condensing single, hetero-functional amino acids, known as omega or ω-amino acids because the amino and carboxyl groups are at opposite ends of the polyamides. Identification is made by a single number signifying the total number of carbon atoms in the amino acid.

Although a large number of likely combinations have been tried on a laboratory scale, commercial production has been limited to those where the starting materials can be produced most cheaply. PAs did not become a commercial reality for packaging film applications until the late 1950s. Although considered a specialty film, many presently available food packages would not be possible without PAs.

- Nylon 6 refers to PA made from a polymer of ε-caprolactam. One material is used containing six carbon atoms.
- Nylon 11 refers to PA made from a polymer of ω-undecanolactam.
- Nylon 6,6 is formed by reacting hexamethylenediamine with adipic acid. Both materials each contain six carbon atoms.
- Nylon 6,10 is made by reacting hexamethylene diamine with sebacic acid ($HOOC-(CH_2)_8-COOH$). The diamine has six carbon atoms and is numerically first, followed by the acid which contains ten carbon atoms.

Most PA packaging films in the U.S. are produced from nylon 6 while European films are usually nylon 11, due to lower raw material costs. Films from nylon 6 have higher temperature, grease and oil resistance than nylon 11 films. The structures of the three most common PAs are depicted in Figure 2.7.

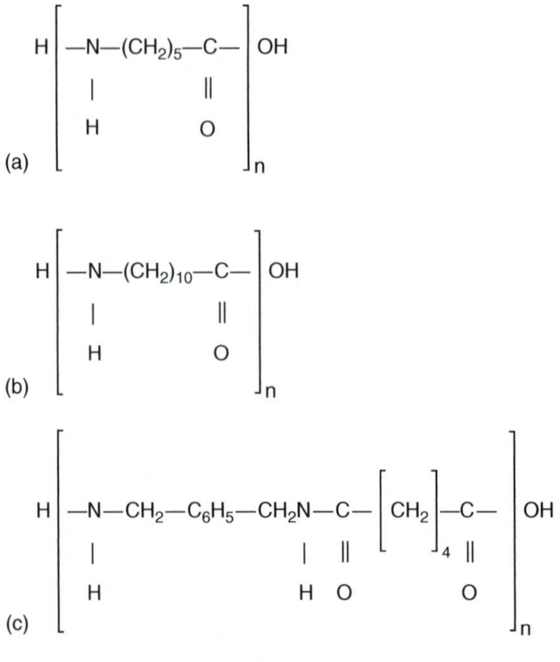

(a)
$$\mathrm{H}\left[-\underset{\mathrm{H}}{\underset{|}{\mathrm{N}}}-(\mathrm{CH_2})_5-\underset{\mathrm{O}}{\underset{\|}{\mathrm{C}}}-\right]_n\mathrm{OH}$$

(b)
$$\mathrm{H}\left[-\underset{\mathrm{H}}{\underset{|}{\mathrm{N}}}-(\mathrm{CH_2})_{10}-\underset{\mathrm{O}}{\underset{\|}{\mathrm{C}}}-\right]_n\mathrm{OH}$$

(c)
$$\mathrm{H}\left[-\underset{\mathrm{H}}{\underset{|}{\mathrm{N}}}-\mathrm{CH_2}-\mathrm{C_6H_5}-\mathrm{CH_2}\underset{\mathrm{H}}{\underset{|}{\mathrm{N}}}-\underset{\mathrm{O}}{\underset{\|}{\mathrm{C}}}-\left[\mathrm{CH_2}\right]_4-\underset{\mathrm{O}}{\underset{\|}{\mathrm{C}}}-\right]_n\mathrm{OH}$$

FIGURE 2.7 Three common polyamides: (a) nylon 6, (b) nylon 11 and (c) MXD6.

PAs produced with greater than six-carbon chains result in films with a lower melting point and an increased resistance to water vapor. Again, the key to successful commercialization is to find cheap sources of the monomers.

In addition to the homopolymers discussed above, copolymerization is used to introduce many additional varieties. Nylon 6,6 and 6,10 can be copolymerized to yield a film with a lower melting point than either of the homopolymers. Fillers, plasticizers, antioxidants and stabilizers can also be used with any or all of the many types of PA films. Over 100 different formulations are available in the production of PA film but most of those used in food packaging applications consist of nylons 6,6, 6 and 11.

The distance between the repeating polar amide (–CONH–) groups can considerably affect the properties of PAs. As the length of the aliphatic segment (i.e., the number of methylene groups between amide groups in the chain) increases, there is a reduction in melting point, tensile strength and water absorption, and an increase in elongation and impact strength. Thus, nylon 11 (where the distance is approximately twice that of nylon 6) has a lower interchain attraction, lower melting point, lower water absorption and is softer. Nylon 11 may in fact be considered to be intermediate in structure and properties between nylon 6 and LDPE.[4] The glass transition temperatures of the PAs appear to be below room temperature, while the T_ms are as follows: nylon 6 = 215°C, nylon 6,6 = 264°C, nylon 6,10 = 215°C and nylon 11 = 185°C.

As might be expected, copolymerization tends to inhibit crystallization by breaking up the regular polymer chain structure, resulting in lower melting points than the corresponding homopolymer. The properties of PAs are considerably affected by the amount of crystallization. Although each variety of PA film has its own characteristic properties, certain similarities exist. PA films are characterized by excellent thermal stability; that is, they are capable of withstanding steam at temperatures up to 140°C and dry heat to even higher temperatures. Low temperature flexibility is excellent and they are resistant to alkalis and dilute acids. Strong acids and oxidizing agents react with PAs.

In general, PAs are highly permeable to water vapor because the amide group is polar. The absorbed water has a plasticizing effect which causes a reduction in tensile strength and an increase in impact strength. Their permeability to O_2 and other gases is quite low when the films are dry. PVdC copolymer-coated PAs offer improved O_2, water vapor, grease and UV light barrier properties. Odor retention is excellent and the films are tasteless, odorless and nontoxic. Other important attributes of PAs are their excellent thermoformability, flex-crack resistance, abrasion resistance and mechanical strength up to 200°C.

For most applications PAs are combined with other materials such as LDPE, ionomer and EVA to add moisture barrier and heat sealability. Multilayer films containing a PA layer are used principally in the vacuum packing of processed meats and cheeses.

Biaxial orientation of PA films provides improved flex-crack resistance, mechanical strength and barrier properties. These films have applications in packaging foods such as processed and natural cheese, fresh and processed meats and frozen foods. They are used in pouches and bag-in-box structures. In some applications the PAs compete with biaxially oriented PET; although oriented PAs offer better gas barrier, softness and puncture resistance, oriented PET offers better rigidity and moisture barrier.

In the 1980s a new polymer, MXD6, was introduced. It is made from metaxylylene diamine and adipic acid, the 6 indicating the number of carbon atoms in the acid. It has a T_g of 75°C and a crystalline melting point of 243°C (between those of nylon 6 and PET). MXD6 has better gas barrier properties than nylon 6 and PET at all humidities and is better than EVOH at 100% RH, due to the existence of the benzene ring in the MXD6 polymer chain. For example, nylon 6 is five times as permeable as MXD6 at 0% RH and 12 times as permeable at 75% RH.[3] Biaxially oriented film produced from MXD6 is used in several packaging applications as it has significantly higher gas and moisture barrier properties and greater strength and stiffness than other PAs.

Together with its high clarity and good processability, the above properties make MXD6 film suitable as a base substrate for laminated film structures for use in lidding and pouches, especially when the film is exposed to retort conditions. Coextruded PET–MXD6–PET has been used as a container for beer and wine.

In the early 1990s amorphous polyamides (AmPAs) appeared in food packages as gas barriers intermediate to the higher barrier EVOH and PVdC copolymer and the lower barrier PET and PVC. Their manufacture involves introducing a ring into the linear nylon chain which prevents crystallization.[3] Consequently, they remain in the amorphous state and are therefore transparent. An unusual property is the fact that their gas barrier properties improve with increasing absorption of moisture in contrast to "normal" PAs.

7. Acrylonitriles

The ready availability of acrylic acid derivatives from propylene has led to their use in numerous industrial polymer products. Acrylonitrile (sometimes referred to as vinyl cyanide, especially by those who seek to raise fears about the safety of the polymer as a food contact material because the monomer is carcinogenic), b.p. 70°C, is produced mainly by the ammoxidation of propylene and contains a carbon–nitrogen triple bond.

a. *Polyacrylonitrile (PAN)*

PAN is produced in quantity as a fiber being commonly referred to simply as "acrylic" and often containing small quantities of other monomers, typically methyl acrylate, methyl methacrylate and vinyl acetate.

The pure nitrile polymer PAN is 49% nitrile and is an amorphous, transparent polymer. It has a relatively low T_g (87°C) and provides an outstanding barrier to gas permeation and exceptional resistance to a wide range of chemicals. However, it is of no commercial value in packaging, due largely to its inability to be melt processed because it degrades at 220°C. It is therefore copolymerized with other monomers that impart melt processability, thus making its desirable properties available in a packaging form.

AN can be copolymerized with styrene to yield styrene–acrylonitrile copolymer, and in combination with butadiene, the terpolymer acrylonitrile–butadiene–styrene (ABS) can be produced. However, these copolymers find limited use in food packaging apart from the thermoforming of some ABS sheet into tubs and trays.

b. *High Nitrile Resins*

i. *Acrylonitrile/Styrene (ANS)*

ANS copolymers are produced by combining AN and styrene in a 70:30 ratio. The backbone of ANS copolymers is characterized by a high degree of chain-to-chain attraction as a result of polarity, resulting in chain stiffness and immobility, high T_g and chemical inertness, most of these properties being attributed to the high nitrile content of the copolymer.

The realization of the excellent gas and moisture barriers of these high percentage AN copolymers, coupled with new and improved molding techniques, led to their development for carbonated beverage packaging in the 1970s but toxicological problems with AN monomer (trivial name vinyl cyanide) surfaced. The FDA withdrew their sanction for the use of high nitrile resins in beverage packaging in September 1977 because of concern about potential AN migration from the bottle into the beverage, but amended its position in 1984 to limit the residual AN content of the finished container to 0.1 parts per million. This limit can be met with modern processing methods. However, by this time PET had been successfully commercialized for carbonated beverages, gaining a commanding lead that proved impossible to overcome.[3]

ii. Rubber-Modified Acrylonitrile/Methyl Acrylate (ANMA)

ANMA copolymers have high barrier properties and are made by copolymerizing AN and methyl acrylate (MA) in a 75:25 ratio onto a nitrile rubber backbone. It was originally developed as a bottle blowing material for carbonated beverages, having good clarity, excellent gas barrier properties and a high resistance to creep. In addition, it has good impact strength and is insoluble in most organic solvents.

ANMA is also produced in film and sheet form and, when laminated with other materials such as LDPE, it is suitable for thermoforming into containers for products such as cheese and meat.

ANS and ANMA copolymers have similar barrier properties, their gas barrier properties being surpassed only by EVOH and PVdC copolymer. High nitrile resins show an affinity for water because of the polarity which the nitrile group imparts to the molecule, thus, their moisture barrier is lower than that of the nonpolar polyolefins. However, their polarity does mean that they are resistant to many solvents. ANMA is tougher than ANS because of the rubber content. Furthermore, ANMA can be blown into bottles using conventional bottle-blowing methods, whereas ANS requires stretching to improve toughness.

8. Regenerated Cellulose

In 1908 the Swiss chemist Brandenburger tried spraying a tablecloth with a viscose solution in order to give it a smooth surface. He found that not only was the surface smooth but it could be peeled off like skin. Three years later he designed a machine to produce a material he called "cellophane" from the first syllable of "cellulose" and the last syllable of the French word "diaphane" meaning transparent. Cellophane is now a generic term for RCF except in Britain and certain other Commonwealth countries where "Cellophane" is a registered trade name.

a. Manufacturing Process

Regenerated cellulose is manufactured from highly purified cellulose, usually derived from bleached sulfite wood pulp or cotton linters. The pulp, in sheet or roll form, is steeped in sodium hydroxide to form soda cellulose and then shredded. The alkali cellulose is then aged to begin the molecular depolymerization process, after which carbon disulfide is added to form sodium cellulose xanthate.

Viscose is formed by adding dilute sodium hydroxide to the xanthate, causing the cellulose to dissolve into solution. The viscose is then allowed to age or ripen, after which it is extruded through a slot-die into acid–salt coagulating and regenerating tanks to give a RCF. Subsequent tanks contain various solutions designed to complete the regeneration, wash out acid carried over from the coagulating bath, remove any elemental traces such as sulfur, carbon disulfide or hydrogen sulfide and bleach the now transparent but still slightly colored film. The film is then run through a bath containing glycerol or ethylene glycol which act as plasticizers and confer flexibility on the film. Finally, it is passed through a drying oven and wound up as plain, nonmoistureproof film.

RCF can be regarded as transparent paper. If uncoated it is highly permeable to steam and, when immersed in water, can absorb its own weight of water. Although impervious to gases when dry, in a wet state it is pervious to varying degrees. It is impervious to fats and oils and insoluble in organic solvents. It does not have the same tensile strength as plastic films and is nonresistant to strong acids and alkalis. As it is flammable it cannot be heat sealed. Therefore, it is not surprising that it has very few uses in its plain uncoated form.

For food packaging applications it is always used after various coatings have been applied to one or both sides. The type of coating largely determines the protective properties of the film. Where alternative thicknesses of the film are available, it is the thickness of the cellulose portion that is altered and only minor changes are made to the coating thickness.

b. Cellulose Film Types

There are four types of coatings which can be applied to regenerated cellulose to confer on it properties which make it a desirable film for food packaging applications. They are:

1. Nitrocellulose — these lacquers contain nitrocellulose, resins, plasticizers, waxes and agents to prevent surface blocking. This is the most common type of coating and provides a moisture barrier.
2. PVC — this is designed to provide excellent machinability and offers properties and costs between those of the nitrocellulose-coated and copolymer-coated cellophanes.
3. PVC–PVdC — this copolymer coating gives superior product protection, preventing volatile aromatic flavors getting out and O_2 from getting in to oxidize those flavors. It is commonly referred to as copolymer-coated cellophane.
4. LDPE — this coating is used for premium packaging such as fresh meats where no moisture loss but high O_2 permeability and heat sealability is required.

It should be noted that creasing of moisture-proof cellulose film can adversely affect its water vapor permeability because of damage to the coating. In recent years RCF has been replaced in many food packaging applications by plastic films (especially BOPP). However, its unique properties ensure that it is still used for the twist wrapping of sweets.

G. Additives in Plastics

Early in the development of the plastics industry it was realized that to obtain better products, additives needed to be added to the base polymer. Within the context of thermoplastics materials technology, the term additive is used to denote an auxiliary ingredient that enhances the properties of the parent polymer without appreciably altering its chemical structure.

In food packaging, all additives must, of course, have received clearance by the appropriate food regulatory authority. The problem of migration of additives from plastic packaging materials into foods is discussed in Chapter 21. In this section, the major additives that could be encountered in plastic food packaging materials will be briefly discussed, together with the reasons for their inclusion in the packaging materials.

1. Processing Additives

The degradation of polymers frequently involves oxidation reactions by a free radical mechanism, and at high temperatures, interaction of O_2 with C–H bonds leads to the formation of hydroperoxide groups. These decompose into very reactive ·OH radicals and lead to molecular scissions. Because it is practically impossible to eliminate O_2 from the system, additives are used to inhibit oxidation reactions.

This is accomplished by using primary stabilizers or antioxidants such as hindered phenols or aromatic amines which interrupt the chain reaction by combining with the free radicals; secondary stabilizers or peroxide decomposers such as organic thioesters, phosphites and metal thiocarbamates which react with hydroperoxides as they are formed; and chelating agents or metal deactivators such as organic phosphites and hydrazides which protect the polymer by immobilizing metal ions through co-ordination reactions.

In the case of PVC, heat stabilizers or acid absorbers which retard the decomposition of PVC into HCl and dark degraded polymers must also be used, and octyltin mercaptide, calcium–zinc compounds and methyltin have been given FDA clearance.

The tendency for polymers such as PVC and polyolefins to stick to metal parts during processing can be reduced by adding lubricants such as polyethylene waxes, fatty acid esters and amides, metallic stearates such as zinc and calcium stearate and paraffin.

2. Flexibilizers

Brittle polymers such as PVC must be plasticized to obtain flexible films and containers. The plasticizer also gives the material the limp and tacky qualities found in "cling" films. About 80% of all plasticizers are used in PVC. Typically phthalic esters such as dioctyl phthalate (DOP), also known as di-2-ethylhexylphthalate (DEHP), and dioctyl adipate (DOA), also known as di-2-ethylhexyladipate (DEHA), are used, as well as epoxidized oils and low MW polyesters. Safety concerns about the use of phthalates in food packaging have been raised and are discussed in Chapter 21. Internal plasticization may be brought about by copolymerization as in the case of PVC which can be copolymerized with vinyl acetate, ethylene or methylacrylate.

3. Antiaging Additives

Aging is the process of deterioration of materials resulting from the combined effects of atmospheric radiation, temperature, O_2, water, micro-organisms and other atmospheric agents (e.g., gases), indicating that a chemical modification in the structure of the material has occurred. Antioxidants have already been mentioned under processing aids, but they are also necessary in polymeric films such as PP, which degrade in the atmosphere. BHT has been cleared by the FDA and acts as a free radical scavenger. Organophosphites act as hydroperoxide decomposers. It is common for different antioxidants to be used together for synergistic effects.

UV stabilizers are used to prevent deterioration of polymeric films by photo-oxidation. They act by absorbing high energy UV radiation and releasing it as lower energy radiation.

4. Surface Property Modifiers

Static electricity is generated on a polymer surface by friction or by rubbing it against another surface. It can also be generated on fast moving film during converting operations or on filling lines. Antistatic agents are used to prevent the accumulation of electrical charges in polymeric films, an undesirable effect caused by the fact that polymers are nonconductors of electricity. Electrification of films results from a segregation of charges (electrons and ions) which occurs when two surfaces are parted after close initial contact. The addition of ethoxylated fatty amines, polyhydric alcohols and derivatives, and nonionic and quaternary ammonium compounds overcomes the problem; they migrate to the surface and form a conducting layer through the absorption of atmospheric moisture which permits the discharge of electrons.

In some food packaging applications, moisture tends to condense as small droplets on the internal surface and prevent a clear view of the package contents. The droplets are formed when the polymer surface tension is lower than the surface tension of water, which prevents the formation of a continuous layer of water.[8] The addition of nonionic ethoxylates or hydrophilic fatty acid esters such as glycerol and sorbitol stearate promotes the deposition of continuous films of moisture by increasing the critical surface tension of the polymer surface. These so-called antifogging agents can be applied on the surface of the material or compounded internally in the packaging material at levels ranging from 0.5 to 4%.

Many packaging films or sheets tend to stick together because they are nonconductors of electricity, a phenomenon known as blocking. Blocking (which may develop under pressure during storage or use) can be reduced by incorporating chemicals such as stearamide, calcium stearate, alkylamines and alkyl quaternary ammonium compounds in the polymer prior to processing. Other materials such as colloidal silica, clays, starches and silicones may be applied to the surface either during or after processing to reduce blocking. Antiblocking agents are used at levels ranging from 0.1 to 0.5%.

5. Optical Property Modifiers

The optical properties of a material from a technological aspect are normally described in terms of their ability to transmit light, exhibit color and reflect light from the surface (i.e., gloss). The majority of food packaging films are not pigmented, although some are colored by the addition of colorants that can be dyes (which are soluble in the plastic and tend to migrate, thus limiting their use) and pigments (which are insoluble in the plastic matrix). The principal pigments for use as colorants in packaging are carbon black, white titanium dioxide, red iron oxide, yellow cadmium sulfide, molybdate orange, ultramarine blue, blue ferric ammonium ferrocyanide, chrome green, and blue and green copper phthalocyanines.[12] The FDA has questioned the use of some of these colorants in packaging material in contact with food, their concern being with colorants that can migrate from the packaging into the food.

6. Foaming Agents

Foaming or blowing agents are used for the production of cellular products and are normally classified into physical and chemical types, according to whether the generation of gases to produce the cells takes place through a physical transition (i.e., evaporation or sublimation) or by a chemical process (i.e., decomposition reactions which result in evolution of gases). In food packaging applications, physical blowing agents are commonly used. Fluorocarbons were widely used but since the Montreal Protocol was signed in 1987, they have been eliminated in many countries to prevent further degradation of the ozone layer. Today, expanded and extruded PS foams are made using CO_2 or a light aliphatic hydrocarbon such as pentane as the blowing agent. Expanded PET, PP and PVC foams are produced using chemical blowing agents.

REFERENCES

1. Billmeyer, F. W., *Textbook of Polymer Science*, 3rd ed., Wiley, New York, 1984.
2. Brooks, D. W. and Giles, G. A., Eds., *PET Packaging Technology*, CRC Press, Boca Raton, FL, 2002.
3. Brown, W. E., *Plastics in Food Packaging: Properties, Design and Fabrication*, Marcel Dekker, New York, 1992.
4. Brydson, J. A., *Plastics Materials*, 7th ed., Butterworth-Heineman, Oxford, England, 1999.
5. Carraher, C. E., *Seymour/Carraher's Polymer Chemistry*, 6th ed., Marcel Dekker, New York, 2003.
6. Combellick, W. A., Barrier polymers, In *Encyclopedia of Polymer Science and Engineering*, 2nd ed., Vol. 2, Kroschwitz, J. J., Ed., Wiley, New York, pp. 176–191, 1985.
7. Foster, R. H., Ethylene–vinyl alcohol copolymers (EVOH), In *The Wiley Encyclopedia of Packaging Technology*, 2nd ed., Brody, A. L. and Marsh, K. S., Eds., Wiley, New York, pp. 355–360, 1997.
8. Hernandez, R. J., Food packaging materials, barrier properties and selection, In *Handbook of Food Engineering Practice*, Valentas, K. J., Rotstein, E. and Singh, R. P., Eds., CRC Press, Boca Raton, FL, 1997, chap. 8.
9. Kaufman, H. S. and Falcetta, J. J., Eds., *Introduction to Polymer Science and Technology: an SPE Textbook*, Wiley-Interscience Publications, New York, 1977.
10. Massey, L. K., *Film Properties of Plastics and Elastomers — A Guide to Non-Wovens in Packaging Applications*, 2nd ed., William Andrew Publishing/Plastics Design Library, Norwich, New York, 2004.
11. Mount, E. M. and Wagner, J. R., Oriented polypropylene, In *The Wiley Encyclopedia of Packaging Technology*, 2nd ed., Brody, A. L. and Marsh, K. S., Eds., Wiley, New York, pp. 415–422, 1997.
12. Rosato, D. V., Additives, plastics, In *The Wiley Encyclopedia of Packaging Technology*, 2nd ed., Brody, A. L. and Marsh, K. S., Eds., Wiley, New York, pp. 8–13, 1997.
13. Selke, S. E. M., *Understanding Plastics Packaging Technology*, Hanser, New York, 1997.
14. Van Krevelen, D. W., *Properties of Polymers*, 3rd ed., Elsevier, Oxford, England, 1990.

3 Edible and Biobased Food Packaging Materials

CONTENTS

I. INTRODUCTION

With the exception of paper-based products, food-packaging materials have traditionally been based on nonrenewable materials. This was not always so, and up until the beginning of the twentieth century, packaging materials, together with other industrial products such as inks, dyes, paints, medicines, chemicals, clothing and plastics were made from biologically-derived resources. During the twentieth century, petroleum-derived chemicals replaced biologically-derived resources for most of these industrial products.[28] Now, at the beginning of the twenty-first century, increasing attention is being given to sustainability (discussed further in Chapter 22) and the replacement of nonrenewable resources (particularly those derived from petroleum) with those from renewable sources, essentially plant-derived products and by-products from their fermentation.

At present, the market is based on biopolymers such as starch, cellulose, proteins and monomers produced from fermented organic materials. There is a large amount of research currently underway in both academia and industry to improve the performance of packaging materials made from renewable resources. Although commercial use of these materials for the packaging of food is still in its infancy,[13,15,16] this situation is likely to change over the next decade.[4,5,17]

Biobased packaging materials have been defined[16] as materials derived from primarily annually renewable sources, thus excluding paper-based materials since trees used for paper-making generally have a renewal time of 25 to 65 years depending on species and country.

Paper-based materials have been used for food packaging for decades. Edible films and coatings are included in this definition and are discussed below.

Although biobased materials may be compostable, this is not the main driving force behind development of biobased packaging materials. Rather, it is driven by the goal of replacing nonrenewable with renewable resources, thus leading to a more sustainable packaging industry. However, from a public point of view, the main drivers for the development of biodegradable packaging are the solid waste problem (particularly the perception of a lack of landfills), the litter problem, which the public feel would be solved if biodegradable packaging were used, and pollution of the marine environment by nonbiodegradable packaging.[3] The synthetic polymers currently used take 200 years to degrade if exposed to the atmosphere, and longer if placed in a landfill. Of course, glass and metal packaging never degrades, although metal packaging may oxidize and lose its mechanical integrity.

II. EDIBLE FILMS AND COATINGS

An edible film is defined as a thin layer of edible material applied on a food as a coating or placed (preformed) on or between food components.[21] Its function is to offer a selective barrier to retard migration of moisture, retard gas (O_2, CO_2) transport, retard oil and fat migration, retard solute transport, improve mechanical handling properties of foods, improve mechanical integrity or handling characteristics of the food, retain volatile flavor compounds and carry food additives such as antioxidants and antimicrobials.[18] Many of the functions are identical to those of synthetic packaging films. Although the most important functional characteristics of a particular application depend on the food product and its primary mode of deterioration, the resistance of an edible film or coating to the migration of water vapor is often a paramount characteristic.

There is no clear distinction between edible films and coatings, and the two terms are often used interchangeably. Usually, coatings are applied directly and formed on the surface of the food, while films are formed separately as thin sheets and then applied to or on the food.

The concept of using an edible film or coating to extend the shelf life of fresh food products and protect them from harmful environmental effects is not a novel one. In fact, the idea derives from the natural protective coating on some foods such as the skin of fruits and vegetables. Covering foods with lipid substances such as waxes and fats to retard desiccation is a very old practice. For example, hot-melt paraffin waxes became commercially available in the 1930s for coating citrus fruits to retard moisture loss, and carnauba wax oil-in-water emulsions were developed in the early 1950s for coating fresh fruits and vegetables.[18] Edible collagen casings for meat products such as sausages, and sugar or chocolate coatings for confectionery products, are both currently used commercially.

Edible coatings and films are not meant to, nor could they ever, replace nonedible, synthetic packaging materials for prolonged storage of foods. The utility of edible films lies in their capacity to act as an adjunct for improving overall food quality, extending shelf life and improving economic efficiency of packaging materials.[18]

The advantages of edible films over traditional synthetic polymeric packaging materials have been listed as follows[9]:

1. The films can be consumed with the packaged product, leaving no residual packaging to be disposed of.
2. Even if the films are not consumed, they could still contribute to the reduction of environmental pollution since they are likely to degrade more readily than synthetic polymeric materials, and are produced exclusively from renewable, edible ingredients.

3. The films can enhance the organoleptic properties of packaged foods provided that various components such as flavorings, colorings and sweeteners are incorporated into them.
4. The films can supplement the nutritional value of foods (this is particularly true for films made from proteins).
5. The films can be used for individual packaging of small portions of food, particularly products that are not currently individually packaged for practical reasons such as peas, beans, nuts and strawberries.
6. The films can be applied inside heterogeneous foods at the interfaces between different layers of components. They can be tailored to prevent deteriorative intercomponent moisture and solute migration in foods such as pizzas, pies and candies.
7. The films can function as carriers for antimicrobial and antioxidant agents. In a similar application, they can also be used at the surface of foods to control the diffusion rate of preservative substances from the surface to the interior of the food.
8. The films can be very conveniently used for microencapsulation of food flavoring and leavening agents to efficiently control their addition and release into the interior of foods.
9. The films could be used in multilayer food packaging materials together with nonedible films, in which case the edible films would be the internal layers in direct contact with food materials.

The majority of edible films and coatings contain at least one component that is a high MW polymer, particularly if a self-supporting film is desired. Long-chain polymeric structures are required to yield film matrices with appropriate cohesive strength when deposited from a suitable solvent. Increased structural cohesion generally results in reduced film flexibility, porosity and permeability to gases, vapors and solutes. As polymer chain length and polarity increase, cohesion is enhanced. A uniform distribution of polar groups along the polymer chain increases cohesion by increasing the likelihood of interchain hydrogen bonding and ionic interactions.[18]

A variety of polysaccharides, proteins and lipids derived from plants and animals have been utilized, either alone or in mixtures, to produce edible films; these are now briefly described.

A. Polysaccharide-Based Coatings

Polysaccharides and their derivatives that have been used as edible films or coatings include alginate, pectin, carrageenan, starch, starch hydrolysates and cellulose derivatives. Because of the hydrophilic nature of these polymers, they exhibit only limited moisture-barrier properties. However, certain polysaccharides applied in the form of high-moisture gelatinous coatings can retard moisture loss from coated foods by functioning as sacrificing agents, rather than moisture barriers.[18]

Amylose is the linear fraction of starch and forms coherent, relatively strong, free-standing films in contrast to amylopectin films, which are brittle and noncontinuous. Films made from hydroxypropylated derivatives of high-amylase starch are poor moisture barriers but good O_2 barriers at lower relative humidity (RH). Extruded edible film wraps based on hydroxypropylated high-amylose starch were commercially available in the late 1960s, and were used as protective coatings for frozen flesh foods, although no references to their effectiveness can be found in the literature.[10]

Alginates are extracted from brown seaweeds of the *Phaephyceae* family and are the salts of alginic acid, a linear copolymer of D-mannuronic and L-guluronic acid monomers. Films produced by evaporation of water from a thin layer of alginate solution are impervious to oils and greases but have high water vapor permeabilities. Calcium ions are used as gelling agents to bridge alginate chains together via ionic interactions.

The polysaccharide gum carrageenan is extracted from red seaweeds, the species known as Irish moss (*Chondrus crispus*) being the most well-known. Carrageenan is a complex mixture of at least five distinct polymers based on galactose.

Agar is a gum derived from a variety of red seaweeds of the *Rhodophyceae* class and, like carragenan, it is a galactose polymer. Agar forms strong gels characterized by melting points far above the initial gelation temperature. Agar coatings containing water-soluble antibiotics and the bacteriocin nisin have been used on flesh foods.

Dextrans are microbial gums formed from sucrose fermentation and composed solely of α-D-glucopyranosyl units with various types of glycosidic linkages. They have been applied as aqueous solutions or dispersions to preserve the flavor, color and freshness of food during refrigeration or frozen storage.

Cellulose ethers are polymer substances obtained by partial substitution of three hydroxyl groups at positions 2, 3 and 6 on the glucosyl units of cellulose. The most common cellulose ethers are methylcellulose, hydroxypropyl cellulose, hydroxypropyl methylcellulose and carboxymethyl-cellulose, and all have good film-forming properties. They have been used commercially to form edible films, which function as O_2, oil and moisture barriers on various foods.[10]

B. Lipid-Based Coatings

Lipid compounds have been utilized as protective coatings for many years, but since they are not polymers, they do not generally form coherent stand alone films. However, they can provide gloss and, because of their relatively low polarity, a moisture barrier. Waxes such as carnauba wax, beeswax and paraffin wax, and mineral and vegetable oils have been used commercially since the 1930s as protective coatings for fresh fruits and vegetables. Generally, wax coatings are substantially more resistant to moisture transfer than most other lipid or nonlipid edible coatings. Wax-, fat- and oil-based coatings can be difficult to apply due to their thickness and greasy surface, and may also confer a waxy or rancid taste.[10]

Mono-, di- and triglycerides are the mono-, di- and triesters of glycerin with fatty acids and have been used as coatings, as have acetylated gylcerides. Acetylated glycerol monostearate coatings are slightly more permeable to water vapor than PA and PS films and significantly more permeable than LDPE films; they are less permeable to O_2 than PS films. Certain application and sensory problems have been reported with acetylated monoglyceride edible coatings, including a tendency to crack and flake during refrigerated or frozen storage, to pick up foreign odors and to exhibit an acidic or bitter aftertaste. Unsaturated glycerides and acetylated glycerides may be susceptible to oxidation.[10]

C. Protein-Based Coatings

Films and coatings are made from animal and plant proteins including collagen, gelatin, wheat gluten, corn zein, soy protein, whey protein and casein. Owing to their inherent hydrophilicity, and the significant amounts of hydrophilic plasticizers such as glycerin and sorbitol incorporated into films to give them flexibility, protein films have limited resistance to water vapor. However, they do have good O_2 barrier properties at low RH.[10] When used as coatings on flesh foods, protein materials are susceptible to proteolytic enzymes present in these foods. In addition, given the increasing number of individuals who are allergic to specific protein fractions from milk, egg white, peanuts, wheat, soybeans, and so on, the use of protein films and coatings must be clearly identified on the label.

Collagen sausage casings are made from the regenerated corium layer of beef hides. Gelatin is derived from partial hydrolysis of collagen. Milk proteins used as edible films and coatings are made from casein as well as whey proteins. Cereal proteins used to form films include corn zein (made from the prolamin fraction of corn proteins) and wheat gluten (a mixture of the prolamin and glutelin fractions of wheat proteins). Edible films and coatings have been made from the globulin protein fractions of soybeans and peanuts.[10]

To improve the barrier and mechanical properties of protein films, physical, chemical and enzymic protein cross-linking treatments have been used.

D. Multicomponent Films

Multicomponent or composite films can consist of a lipid layer supported by a polysaccharide or protein layer, or lipid material dispersed in a polysaccharide or protein matrix.[19] To improve film flexibility and durability, plasticizers are generally added to edible films. These include sucrose, glycerol, sorbitol, propylene glycol, fatty acids, monoglycerides and (in the case of polysaccharide and protein films) water. Most multicomponent films studied to date involve a lipid as a moisture barrier and a highly polar polymer such as a polysaccharide or protein as the structural matrix.

Multicomponent films are formed using two basic techniques. The coating technique involves casting or laminating a lipid onto a dried, edible base film to form a bilayer or laminated film. The emulsion technique involves adding a lipid to a film-forming solution prior to film casting and creating an emulsified film.[29]

Considerable data are available on the water vapor and O_2 permeabilities of edible films.[8,9,14,20–22,29] However, comparison of edible film properties must be approached with caution as plasticizer content and test conditions (in particular RH and temperature) have a significant effect on film properties, and these are not always clearly indicated in the published literature.

Compared to LDPE, polysaccharide and protein films are poor moisture barriers. However, when they are combined with edible waxes or fatty acids, relatively good moisture-barrier properties are obtained (in the best cases approximately twice the WVTR of LDPE). Protein films appear to be better O_2 barriers than polysaccharide or lipid films, with permeabilities approaching that of EVOH at low to intermediate relative humidities.

III. BIOBASED PACKAGING MATERIALS

Biobased packaging materials have been divided into three types, reflecting their historical development; these are illustrated in Figure 3.1. First-generation materials consist of synthetic polymers such as LDPE with 5 to 20% starch fillers. Although these materials disintegrate or biofragment into smaller molecules when composted, they do not biodegrade.[1] Second-generation materials consist of a mixture of synthetic polymers and 40 to 75% gelatinized starch, and some of these materials are fully biodegradable. Third-generation materials consist of fully biobased and biodegradable materials.[12]

Before discussing biobased packaging materials in more detail, it is necessary to define biodegradability, a much overused and frequently misunderstood word. Biodegradable polymers constitute a loosely defined family of polymers that are designed to degrade through the action of living organisms. Biodegradability and compostability are different concepts. Although biodegradation may take place as a result of the disposal of a material in landfill, composting usually requires a pretreatment of municipal solid waste (MSW) to remove bulky noncompostable items and inorganic waste.[7]

A standard definition for biodegradable plastics can be found in ISO 472 and ASTM D 883 as "a degradable plastic in which the degradation process results in lower molecular weight fragments produced by the action of naturally-occurring microorganisms such as bacteria, fungi and algae."

Complete biodegradability of the product is measured through respirometric tests such as ASTM D 5338 (equivalent to ISO 14852). This method determines the degree and rate of aerobic biodegradation of plastic materials on exposure to a controlled composting environment under laboratory conditions. The test substances are exposed to an inoculum that is derived from compost

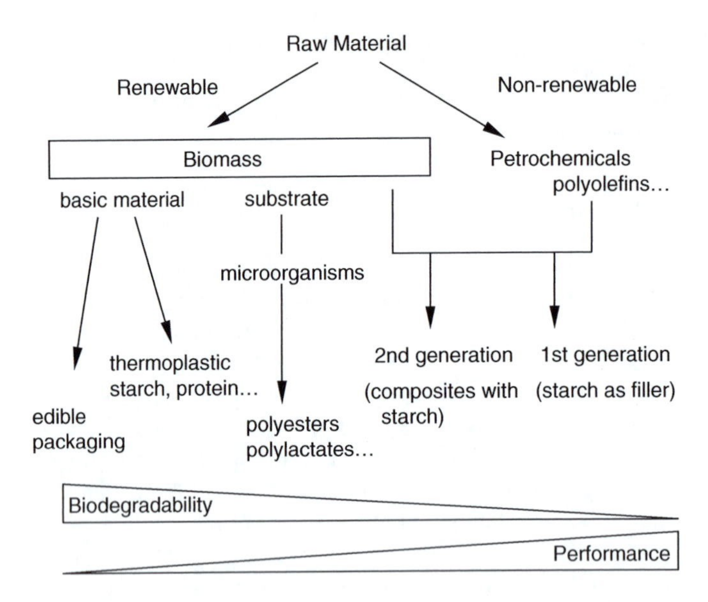

FIGURE 3.1 Different approaches to make biobased packaging using agricultural raw materials. (*Source*: From Gontard N. and Guilbert S., Bio-packaging: technology and properties of edible and/or biodegradable material of agriculture orgin, In *Food Packaging and Preservation*, Mathlouthi, M., Ed., Blackie Academic & Professional, London, 1994, chap. 9. With permission.)

from MSW. The aerobic composting takes place in an environment where temperature, aeration and humidity are closely monitored and controlled.

The challenge for the successful use of biodegradable polymer products in food packaging is achieving the desired shelf life, followed by efficient biodegradation after disposal. Obviously, premature biodegradation and insect infestation must be avoided. In addition, it is imperative that biodegradable plastics do not contaminate the recycling stream for nonbiodegradable plastics.

A. First Generation

The first generation of commercial, biodegradable polymers consists of 5 to 20% starch together with LDPE and pro-oxidative and auto-oxidative additives. They are mixed during the extrusion process, resulting in the starch granules being uniformly dispersed in the LDPE matrix without chemical interaction.[1] Biodegradation of the starch by microbial enzymes results in loss of mechanical properties and chemical degradation by O_2 of the LDPE. Higher levels of starch (up to 43%) can be incorporated if the surface of the starch granules are silylated to increase their hydrophobicity. Claims that these materials were biodegradable (it takes 3 to 5 years for them to degrade into dust) resulted in public controversy, and their behavior is now referred to as biofragmentation.[12]

B. Second Generation

The second generation consists of gelatinized starch (40 to 75%) and LDPE with the addition of hydrophilic copolymers such as EAA, PVOH and VA, which act as compatibility agents. Complete degradation of the starch takes 40 days, and degradation of the entire film a minimum of 2 to 3 years.

C. Third Generation

The third generation consists of completely biobased materials and can be classified into three main categories according to their method of production[16]:

1. Polymers directly extracted from biomass.
2. Polymers produced by classical chemical synthesis from biomass monomers.
3. Polymers produced directly by natural or genetically modified organisms.

The three categories are presented schematically in Figure 3.2. Although some materials from all three categories are already used for food packaging, others have considerable potential as packaging materials.

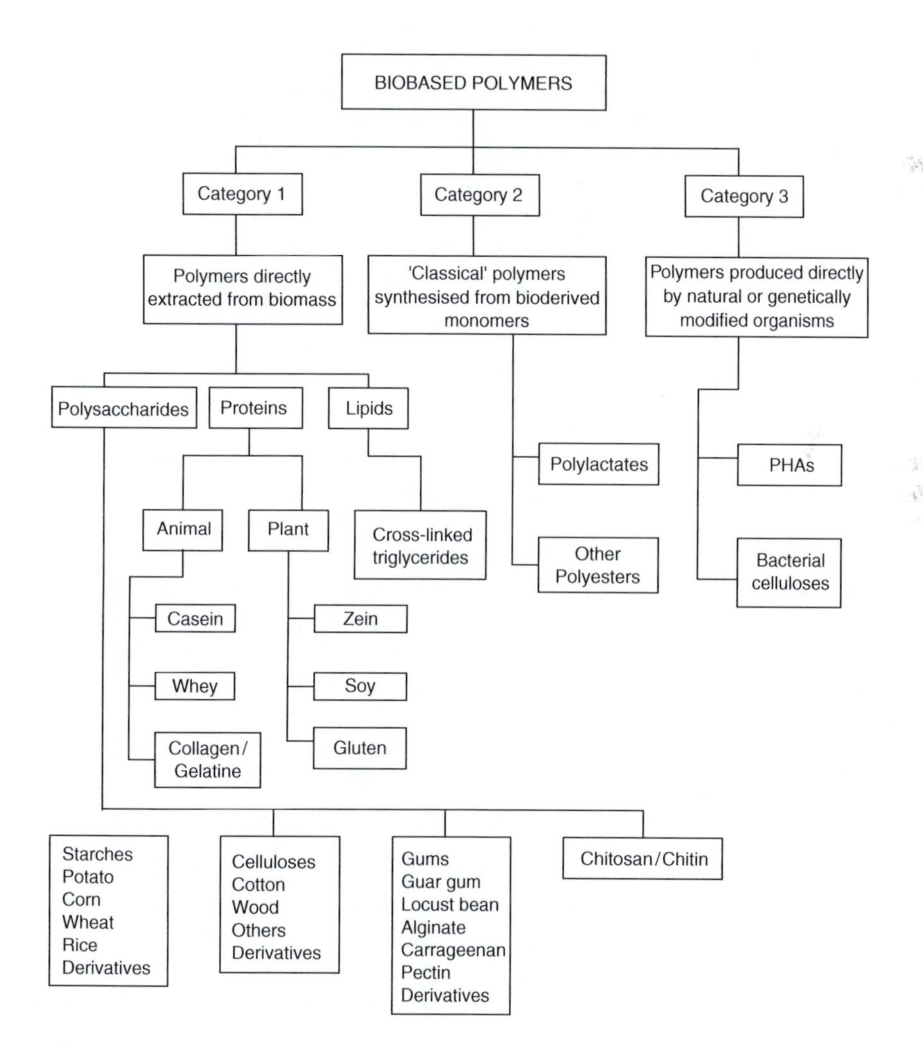

FIGURE 3.2 Schematic presentation of biobased polymers based on their origin and method of production. (*Source*: Reprinted from Haugaard V. K. and Mortensen G., Biobased food packaging, In *Environmentally Friendly Food Processing*, Mattson, B. and Sonneson, U., Eds., CRC Press, Boca Raton, FL, 2003, chap. 11. With permission.)

1. Category 1

Most of the commonly available Category 1 polymers are extracted from marine and agricultural products; examples include polysaccharides such as cellulose, starch and chitin, and proteins such as casein, whey, collagen and soy. The most widely used food-packaging material in this category is cellulose-based paper and board, which has been successfully used for decades.

2. Category 2

Of all the possible biopolyesters that have been produced from biobased materials, polylactic acid (PLA) has shown the highest commercial potential and is now produced on a comparatively large scale. PLA is a linear, aliphatic polyester synthesized from lactic acid monomers by polycondensation of the free acid, or by catalytic ring-opening polymerization of the lactide (dilactone of lactic acid).[3] The ester linkages in the polymer are sensitive to both enzymatic and chemical hydrolysis. Lactic acid can be produced cheaply by the fermentation of biomass such as corn or wheat, or waste products such as whey or molasses.

PLA can be made into films, coextruded into laminates, thermoformed and injection-stretch blow-molded into bottles. To date, the major application has been as food-service containers. Other current uses include thermoformed packaging for bakery products and bags for bread, fresh pasta and salads. PLA has important process advantages, with claims that 30 to 50% less fossil fuel is required and less CO_2 is emitted compared to petroleum-based plastics. In addition, PLA products are fully compostable in commercial composting facilities. Provided production costs can be reduced, it is expected to find packaging applications in areas such as candy twist wraps, coatings for paperboard beverage cartons, plastic film wraps for foods, blister packs, and plastic windows in boxes.

3. Category 3

Category 3 polymers consist mainly of the microbial polyesters poly(hydroxyalkanoates) (PHAs) of which poly(hydroxybutyrate) (PHB) is the most common. PHAs are a family consisting of renewable, biodegradable, biocompatible, optically active polyesters.[28] They are produced by many bacterial species in the form of intracellular particles, which function as an energy and carbon reserve. PHAs are linear aliphatic polyesters consisting of homo- or copolymers of β-hydroxyalkanoic acids that can be produced from the fermentation of sugars by *Alcaligenes eutrophus*, the sugar to polymer conversion yield being about 33%. By manipulating the growth medium, a random copolymer containing both 3-hydroxybutyrate (HB) and 3-hydroxyvalerate (HV) is obtained, which is biodegradable and thermoplastic, and can be formed by the same techniques as those used for synthetic polymers.

By changing the ratio of HV to HB, the resulting copolymer can be made to resemble either PP (low HV) or LDPE (high HV) with regard to flexibility, tensile strength and melting point.[6] PHB/V has good chemical and moisture resistance as well as good O_2- and aroma-barrier properties.

PHA has been sold in small quantities since being first developed by ICI in the 1970s. Despite its desirable properties, bacterial synthesis proved too costly and the focus is now on producing PHA plastics and derived chemicals directly in genetically-modified switchgrass, while providing biomass for alternative energy generation. It is hoped that direct production of PHAs in plants will be economically competitive with those of existing large-volume petrochemical polymers. For this to occur, the yield of PHAs will need to increase from 3 to 15% dry weight, and the proportions of the two building blocks must be able to be controlled.

PHAs are degraded on exposure to bacteria or fungi in soil, compost or marine sediment. Biodegradation starts when micro-organisms begin growing on the surface of the plastic and secrete enzymes that break down the polymer into hydroxyl acid monomeric units. In aerobic environments, the degradation products are CO_2 and CH_4. In composting trials, up to 85% of the PHA samples degraded within 7 weeks, and PHA-coated paper was rapidly degraded and incorporated into the compost.[23]

Bacterial cellulose produced by strains of *Acetobacter xylinum* and *A. pasterianis* is considered to have enormous potential within the food packaging industry, but, so far, it is largely unexploited.[28] The cellulose is identical in chemical and physical structure to the cellulose formed in plants, but has the advantage that it is not combined with lignin, hemicelluloses and pectin and so can be extracted without the need for harsh chemical treatment.

D. Properties of Biobased Packaging Materials

Data on the physical and mechanical properties of biobased materials have been published[26] but will be briefly discussed here.

1. Barrier Properties

The poor barrier properties (especially under conditions of high humidity) of the most widely used biobased materials (paper and RCF) are well known, and it is necessary for them to be coated with synthetic polymers in order to achieve the desired barrier properties necessary for the packaging of food.

WVTRs of starch-based films are four to six times greater than those of conventional films made from synthetic polymers. PLA has WVTRs four times higher than conventional films; PHA has WVTRs similar to those of LDPE. The gas-barrier properties of many biobased materials depend on the ambient humidity; however, two notable exceptions are PLA and PHA. PHB has better O_2-barrier properties than PET and PP, and adequate fat- and odor-barrier properties for applications with short shelf life products.[23] It is possible to improve the barrier properties of biobased materials using plasma deposition of SiO_x or applying nanocomposites from natural polymers and modified clays.[16]

The CO_2:O_2 permeability ratio of synthetic polymers is typically in the range of 4:1 to 6:1, depending on the nature of the polymer. The ratio for PLA has been reported as 7:1; for starch-PCL as ranging from 4:1 to 14:1[26] and for wheat gluten as 15:1.[2] The latter film has been successfully used in laboratory trials for the modified atmosphere packaging (MAP) of mushrooms.

2. Mechanical Properties

The mechanical properties of most biobased materials are similar to synthetic polymers, although some (such as PLA) are sensitive to heat and humidity. Orientation of PLA improves mechanical strength and heat stability, and varying crystallinity and MW results in films ranging from soft and elastic to stiff and high strength. PLA has a T_g of 60°C, a melt temperature of 155°C, a heat seal initiation temperature of 80°C and mechanical properties similar to those of PET. In general, PHB resembles isotactic PP with respect to mechanical behavior.[27]

E. Food Packaging Applications

Despite considerable research and development, the use of biobased packaging materials for the packaging of food remains limited.[25] The major scientific studies on the packaging of foods in (partly) biobased materials have been reviewed.[16] Although the number of published papers is few, they show that there are many potential applications.

A major use of PLA has been for one-way beverage cups for food-service applications; they have been used at recent Olympic Games, either as 100% PLA cups or paper cups coated with PLA, to promote an environmentally friendly image. Small niche markets for PLA trays and films have been found in Europe, often for organically grown food.

A recent review[16] suggested that in the short term, biobased materials will be used for short shelf life foods stored at chill temperatures due to the fact that the materials are biodegradable. Potential applications include fast-food packaging of salads, egg cartons, fresh or minimally processed fruits and vegetables, dairy products such as yoghurt and organically grown foods. The high CO_2:O_2 permeability ratio of certain biobased packaging materials suggests that they could find application in the packaging of respiring foods such as fruits and vegetables.

IV. ENVIRONMENTAL ASPECTS

As the preceding discussion has highlighted, development of biobased packaging materials is predicated on a widely held belief that such materials will have lower environmental impacts than existing petroleum-based materials. Without this advantage, there is less incentive for industry to adopt biobased packaging materials. A widely quoted article in *Scientific American*[11] questioned the utility of biodegradable and biobased plastics on the basis of energy and environmental analyses. Statements in the article such as "biodegradability has a hidden cost — the biological breakdown of plastics releases CO_2 and CH_4, heat-trapping greenhouse gases, that international efforts currently aim to reduce" and "total energy needed to produce one pound of PHA polymer from plants is equivalent to the consumption of 2.39 pounds of fossil resources as opposed to consuming 2.26 pounds of oil to manufacture polystyrene" suggested that biobased packaging materials offered no environmental advantages. The article made a sweeping indictment of all biodegradable and biobased plastics, vis-à-vis their energy and environmental benefits and concluded that "it is impossible to argue that plastic grown in corn and extracted with energy from fossil fuels would conserve fossil resources. What is gained by substituting the renewable resource for the finite one is lost in the additional requirement for energy."

Critics of the article have pointed out that it was based on the analysis of a single bioplastic technology, namely PHAs (a technology that is still not commercial and not optimized for scaled-up production), and only on analysis of the energy usage for production and not the entire life cycle. Since the article, detailed LCA studies on biobased products have been released, and comparative reviews of 20 published studies[24] give a completely different picture. Of the 20 LCAs reviewed, 7 dealt with starch polymers, 5 with PHA, 2 with PLA, 3 with other bio-based polymers and 3 with composites based on natural fibers.

In summary, the available LCA studies and environmental assessments strongly support the further development of biodegradable and biobased polymers. Careful monitoring of the various environmental impacts continues to be necessary both for decision makers in companies and those developing public policy. If combined with good-practice targets, then this may accelerate and focus the ongoing product and process innovation. For some materials, the environmental benefits achieved are already substantial. In many other cases, the potentials are very promising and need to be exploited.

In spite of some uncertainties and information gaps, the body of work analyzed overwhelmingly indicates that biodegradable and biobased polymers offer important environmental benefits today and for the future. Of all materials studied, starch polymers are considered to perform best in environmental terms, with some differences among the various types of starch polymers. Compared with starch polymers, the environmental benefits seem to be smaller for PLA (data were only available for energy and CO_2). For PHA, the achievable environmental advantage currently seems to be very small compared to conventional polymers (data were only available for energy use). For both PLA and PHA, the production method, the scale of production and the type of waste management treatment can decisively influence conclusions about the overall environmental balance.

REFERENCES

1. Arvanitoyannis, I. S., Totally and partially biodegradable polymer blends based on natural and synthetic macromolecules: preparation, physical properties, and potential as food packaging materials, *Rev. Macromol. Chem. Phys.*, C39(2), 205–271, 1999.
2. Barron, C., Varoquaux, P., Guilbert, S., Gontard, N., and Gouble, B., Modified atmosphere packaging of cultivated mushroom (*Agaricus bisporus* L.) with hydrophilic films, *J. Food Sci.*, 67, 251–255, 2002.
3. Bastioli, C., Biodegradable materials, In *The Wiley Encyclopedia of Packaging Technology*, 2nd ed., Brody, A. L. and Marsh, K. S., Eds., Wiley, New York, pp. 77–83, 1997.
4. Cagri, A., Ustunol, A., and Ryser, E. T., Review. Antimicrobial edible films and coatings, *J. Food Protect.*, 67, 833–848, 2004.
5. Cha, D. S. and Chinnan, M. S., Biopolymer-based antimicrobial packaging: a review, *Crit. Rev. Food Sci. Nutr.*, 44, 223–237, 2004.
6. Ching, C., Kaplan, D., and Thomas, E., *Biodegradable Polymers and Packaging*, Technomic Publishing, Lancaster, PA, 1993.
7. de Vlieger, J. J., Green plastics for food packaging, In *Novel Food Packaging Techniques*, Ahvenainen, R., Ed., CRC Press, Boca Raton, FL, 2003, chap. 24.
8. Debeaufort, F., Quezada-Gallo, J.-A., and Voilley, A., Edible films and coatings: tomorrow's packagings: a review, *Crit. Rev. Food Sci. Nutr.*, 38, 299–313, 1998.
9. Gennadios, A., *Protein-Based Films and Coatings*, CRC Press, Boca Raton, FL, 1997.
10. Gennadios, A., Hanna, M. A., and Kurth, L. B., Application of edible coatings on meats, poultry and seafoods: a review, *Lebensm-Wiss u-Technol.*, 30, 337–350, 1997.
11. Gerngrow, T. U. and Slater, S. C., How green are green plastics?, *Sci. Am.*, 283(2), 24–29, 2000.
12. Gontard, N. and Guilbert, S., Bio-packaging: technology and properties of edible and/or biodegradable material of agricultural origin, In *Food Packaging and Preservation*, Mathlouthi, M., Ed., Blackie Academic & Professional, London, 1994, chap. 9.
13. Guilbert, S. and Gontard, N., Edible and biodegradable food packaging, In *Foods and Packaging Materials — Chemical Interactions*, Ackermann, P., Jagerstad, M. and Ohlsson, T., Eds., Royal Society of Chemistry, Cambridge, England, pp. 159–172, 1995.
14. Guilbert, S., Redl, A., and Gontard, N., Mass transport within edible and biodegradable protein-based materials: application to the design of active biopackaging, In *Protein-Based Films and Coatings*, Gennadios, A., Ed., CRC Press, Boca Raton, FL, 1997, chap. 34.
15. Guilbert, S., Cuq, B., and Gontard, N., Recent innovations in edible and/or biodegradable packaging, *Food Addit. Contam.*, 14, 741–751, 2000.
16. Haugaard, V. K. and Mortensen, G., Biobased food packaging, In *Environmentally Friendly Food Processing*, Mattson, B. and Sonneson, U., Eds., CRC Press, Boca Raton, FL, 2003, chap. 11.
17. Haugaard, V. K., Udsen, A.-M., Mortensen, G., Høegh, L., Petersen, K., and Monahan, F., Potential food applications of biobased materials. An EU-concerted action project, *Starch/Starke*, 53, 189–200, 2001.
18. Kester, J. J. and Fennema, O. R., Edible films and coatings: a review, *Food Technol.*, 40(12), 47–59, 1986.
19. Krochta, J. M., Films, edible, In *The Wiley Encyclopedia of Packaging Technology*, 2nd ed., Brody, A. L. and Marsh, K. S., Eds., Wiley, New York, pp. 397–401, 1997.
20. Krochta, J. M., Baldwin, E. A., and Nisperos-Carriedo, M., *Edible Coatings and Films to Improve Food Quality*, Technomic Publishing, Lancaster, PA, 1994.
21. Krochta, J. M. and DeMulder-Johnson, C., Edible and biodegradable polymer films: challenges and opportunities, *Food Technol.*, 51(2), 61–72, 1997.
22. Morillon, V., Debeaufort, F., Blond, G., Capelle, M., and Voilley, A., Factors affecting the moisture permeability of lipid-based edible films: a review, *Crit. Rev. Food Sci. Nutr.*, 42, 67–89, 2002.
23. Nayak, P. L., Biodegradable polymers: opportunities and challenges, *J. Macromol. Sci.*, 39, 481–505, 1999.
24. Patel, M., Bastioli, C., Marini, L., and Würdinger, E., Life-cycle assessment of bio-based polymers and natural fiber composites, In *Biopolymers*, Vol. 10, Steinbüchel, A., Ed., Wiley-VCH, Toronto, 2003, chap. 14.

25. Petersen, K., Nielsen, P. V., Bertelsen, G., Lawther, M., Olsen, M. B., Nilsson, N. H., and Mortensen, G., Potential of biobased materials for food packaging, *Trends Food Sci. Technol.*, 10, 52–68, 1999.
26. Petersen, K., Nielsen, P. V., and Olsen, M. B., Physical and mechanical properties of biobased materials, *Starch/Starke*, 53, 356–361, 2001.
27. Södergård, A. and Stolz, M., Properties of lactic acid based polymers and their correlation with composition, *Prog. Polym. Sci.*, 27, 1123–1163, 2002.
28. Weber, C. J., Haugaard, V., Festersen, R., and Bertelsen, G., Production and applications of biobased packaging materials for the food industry, *Food Addit. Contam.*, 19, 172–177, 2002.
29. Wu, Y., Weller, C. L., Hamouz, P. H., Cuppett, S., and Schneff, M., Development and application of multicomponent edible coatings and films: a review, *Adv. Food Nutr. Res.*, 44, 348–394, 2002.

4 Permeability of Thermoplastic Polymers

CONTENTS

I. INTRODUCTION

In contrast to packaging materials made from glass or metal, packages made from thermoplastic polymers are permeable to varying degrees to small molecules such as gases, water vapor, organic vapors and other low MW compounds. As described in Chapter 2, a wide range of thermoplastic polymers is used in food packaging. The degree of protection or barrier which they provide to the transfer of low MW molecules ranges from low to high. Knowledge and understanding of the solution and transport behavior of low MW substances in polymeric materials has become increasingly important in recent years with the widespread use of polymer films and rigid plastics for food packaging. The selection or development of polymeric materials for food packaging applications with stringent design specifications relating to their solution and transport behavior requires knowledge and appreciation of the many factors which affect the transport of low MW compounds from either the internal or external environment through the polymeric package wall.

Unfortunately, there are many examples of foods packaged with an apparent lack of proper consideration of the effects of the end-use environment on properties, or of limitations imposed on performance due to unfavorable solution or transport characteristics. The plasticization of polymers by sorption of ambient vapors or liquids resulting in a decrease in mechanical properties, and the loss of components (e.g., CO_2, flavor compounds, etc.) from a beverage in a plastic bottle are just two of many examples that could be cited.

The protection of foods from gas and vapor exchange with the environment depends on the integrity of packages (including their seals and closures), and on the permeability of the packaging materials themselves. There are two processes by which gases and vapors may pass through polymeric materials:

1. A pore effect, in which the gases and vapors flow through microscopic pores, pinholes, and cracks in the materials.
2. A solubility–diffusion effect, in which the gases and vapors dissolve in the polymer at one surface, diffuse through the polymer by virtue of a concentration gradient, and evaporate at the other surface of the polymer. This "solution–diffusion" process (also known as "activated diffusion") is described as true permeability.

When sufficiently thin, most polymers exhibit both forms of permeability. Porosity falls very sharply as the thickness of a polymer is increased, reaching virtually zero with many of the thicker types of commercially available materials. True permeability, however, varies inversely as the thickness of the material, and hence cannot be effectively eliminated merely by increasing the material thickness.

The concept of permeability is normally associated with the quantitative evaluation of the barrier properties of a plastic. A plastic that is a good barrier has a low permeability. In addition to permeability, there are two other mass transport phenomena in package systems. Sorption (also called scalping) involves the take-up of molecules from the food product (e.g., flavor compounds) into (but not through) the package. Migration is the transfer of molecules originally contained in the packaging material (e.g., plasticizer, residual monomer, antioxidants) into the product and possibly to the external environment.

This chapter is concerned with aspects of the solution, diffusion and permeation of gases and vapors ("permeants") in effectively nonporous polymeric materials.

II. THEORY

The first recorded observation of the permeation of a gas through a membrane appears to be that of Scottish chemist Thomas Graham who, in 1826, described the permeation of CO_2 through a polymeric membrane. In 1831, J.K. Mitchell, an American physician and the inventor of the toy rubber balloon, discovered that his balloons collapsed at different rates when they were filled with different gases. In 1855, the German physiologist Adolf Fick, who was concerned with measuring the transport of O_2 in blood, proposed his law of mass diffusion by analogy with Fourier's law for heat conduction, Ohm's law for electrical conduction and Newton's law for momentum transfer.

In 1866, Graham postulated that the permeation process entailed solution of the permeant in the membrane, followed by diffusion of the dissolved species through the membrane as though through a liquid, a process Graham called "colloidal diffusion." His concept is similar to that which is still used today, known as the solution–diffusion model. The subject was placed on a quantitative basis in 1879 by the Polish physicist Szygmunt von Wróblewski, who showed that the solubility of gases in rubber obeyed Henry's law (named after the English physician and chemist who, in 1803, showed that, at a given temperature, the amount of gas dissolved in a solution is directly proportional to the pressure of the gas above the solution), and combined this with Fick's law to obtain the now familiar expression relating permeation rate and the area and thickness of the membrane.

Under steady-state conditions, a gas or vapor will diffuse through a polymer at a constant rate if a constant pressure difference is maintained across the polymer. The diffusive flux, J, of a permeant in a polymer can be defined as the amount passing through a plane (surface) of unit area normal to the direction of flow during unit time as follows:

$$J = Q/At \tag{4.1}$$

where Q is the total amount of permeant that has passed through area A during time t.

The relationship between the rate of permeation and the concentration gradient is one of direct proportionality and is embodied in Fick's first law

$$J = -D\frac{\delta c}{\delta x} \tag{4.2}$$

where J is the flux (or rate of transport) per unit area of permeant through the polymer, c is the concentration of the permeant, D is defined as the diffusion coefficient and $\delta c/\delta x$ is the concentration gradient of the permeant across a thickness δx. D reflects the speed at which the permeant diffuses through the polymer.[18]

Consider a polymeric material X mm thick, of area A, exposed to a permeant at pressure p_1 on one side and at a lower pressure p_2 on the other, as shown in Figure 4.1. The concentration of permeant in the first layer of the polymer is c_1 and in the last layer c_2.

If x and $(x + \delta x)$ represent two planes through the polymer at distances x and $(x + \delta x)$ from the high-pressure surface, and if the rate of permeation at x is J mL sec^{-1}, and at $(x + \delta x)$ is $J + (\delta J/\delta x)\delta x$, then the amount retained per unit volume of the polymer is $(\delta J/\delta x)$. This is equal to

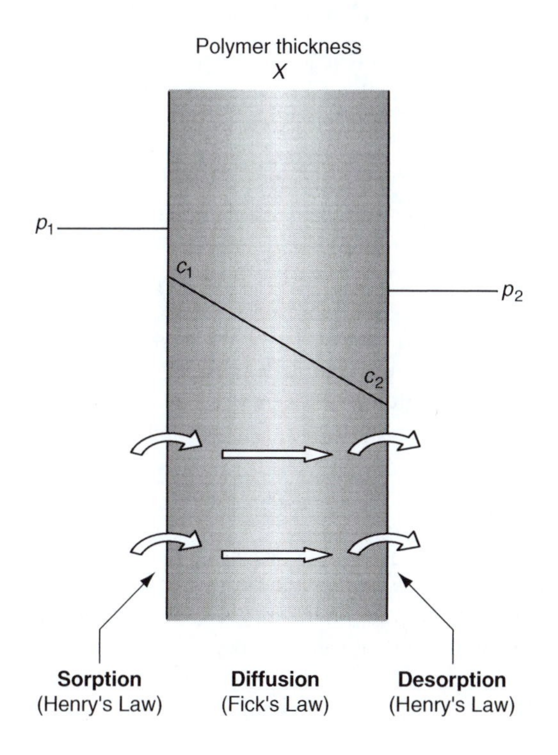

FIGURE 4.1 Permeability model for gas or vapor transfer through a polymer.

the rate of change of concentration with time:

$$\frac{\delta}{\delta x}(J) = -\frac{\delta c}{\delta t} \tag{4.3}$$

A negative sign is used because the concentration of permeant decreases across the material. Under steady-state conditions, $\delta c/\delta t = 0$ and $J =$ constant. When the concentration gradient is zero (i.e., $c_1 = c_2$), there will be no permeation.

If Equation 4.2 is substituted into Equation 4.3 then:

$$\frac{\delta}{\delta x}(J) = \frac{\delta}{\delta x}\left[-D\frac{\delta c}{\delta x}\right] = \frac{\delta c}{\delta t} \tag{4.4}$$

Rearranging the terms

$$\frac{\delta c}{\delta t} = \frac{\delta}{\delta x}\left[D\frac{\delta c}{\delta x}\right] \tag{4.5}$$

we obtain Equation 4.6

$$\frac{\delta c}{\delta t} = D\frac{\delta^2 c}{\delta x^2} \tag{4.6}$$

which is a simplified form of Fick's second law of diffusion and applies under circumstances where diffusion is limited to the x-direction and D is independent of concentration.

III. STEADY-STATE PERMEATION

When a steady state of diffusion has been reached, $J =$ constant and Equation 4.2 can be integrated across the total thickness of the polymer X, and between the two concentrations, assuming D to be constant and independent of c:

$$JX = -D(c_2 - c_1) \tag{4.7}$$

and

$$J = \frac{D(c_1 - c_2)}{X} \tag{4.8}$$

The above expression can be rewritten by substituting for J using Equation 4.1. This enables calculation of the quantity of permeant diffusing through a polymer of area A in time t:

$$Q = \frac{D(c_1 - c_2)At}{X} \tag{4.9}$$

When the permeant is a gas, it is more convenient to measure the vapor pressure p, which is at equilibrium with the polymer, rather than measure the actual concentration. At sufficiently low concentrations, Henry's law applies and c can be expressed as

$$c = Sp \tag{4.10}$$

where S is the solubility coefficient of the permeant in the polymer. S reflects the amount of permeant in the polymer.[18] By combining Equation 4.9 and Equation 4.10,

$$Q = \frac{DS(p_1 - p_2)At}{X} \tag{4.11}$$

The product DS is referred to as the *permeability coefficient* (or *constant*) or *permeation coefficient* or simply just the *permeability*[6] and is represented by the symbol P. Thus,

$$P = \frac{QX}{At(p_1 - p_2)} \tag{4.12}$$

or

$$\frac{Q}{t} = \frac{P}{X}A(\Delta p) \tag{4.13}$$

The term P/X is called the *permeance*.

There are four assumptions made in the above simple treatment of permeation. First, diffusion is at steady state. Second, the concentration–distance relationship through the polymer is linear. Third, diffusion takes place in one direction only (i.e., through the film with no net diffusion along or across it), and, fourth, both D and S are independent of concentration.

Like all simplifying assumptions, there are many instances when they are not valid, and, in such cases, the predictions made are not subsequently borne out in practice. Although steady state is usually attained in a few hours for small molecules such as O_2, larger molecules in barrier polymers (especially glassy polymers) can take a long time to reach steady state, this time possibly exceeding the anticipated shelf life.[7] Although D and S are independent of concentration for many gases such as O_2, N_2 and, to a certain extent, CO_2, this is not the case where considerable interaction between polymer and permeant takes place (e.g., water and hydrophilic films such as PAs, or many solvent vapors diffusing through polymer films).

The permeability coefficient defined above is independent of thickness, because the thickness is already accounted for in the calculation of P. However, the total amount of protection afforded by unit area of a barrier material (i.e., the permeance P/X) approaches zero only asymptotically. Consequently, as polymer thickness X is increased beyond a certain value, it becomes uneconomical to increase it further to obtain lower permeability.

IV. PERMEATION THROUGH PORES

The presence of pores in a plastic food package is generally considered undesirable, although it is not unusual to find pores in the area adjacent to the heat seal in plastic films. Pores, microholes and cracks not only allow mass transfer between the interior of a package and the ambient environment, but also permit microbe penetration into the package, with some bacteria penetrating holes as small as 0.4 μm in diameter.[12]

However, with the increasing commercialization of modified atmosphere packaging (MAP) technologies for fresh produce, porous plastic films have become widely used, particularly for those products having high physiological activity. Incorporating pores into films has extended the permeability range of plastic materials by several orders of magnitude, enabling them to be used to package a wide range of fruits and vegetables. The perforations range from laser-made micropores to mechanically perforated macropores up to 4 mm diameter.[10]

Permeant molecules move through pores by capillarity due to their thermal energy, with diffusion being dependent on both pore and permeant sizes. In the case of very small pores (diameter $d < 1 \times 10^{-7}$ m), collisions with the walls are more probable than collisions with other gas molecules, and gas flow through the pore follows Knudsen's law. For pores in the range

1×10^{-7} m $< d \ll 1 \times 10^{-5}$ m, the kinetics are governed by both diffusivity of gases and Knudsen's flow. In macropores, collisions with the pore walls can be disregarded and ordinary molecular diffusion predominates.[10] When there is a difference in absolute pressure across the porous membrane (e.g., a vacuum packaged food), the flow can be described by Poiseuille's law.[12]

V. PERMEABILITY COEFFICIENT UNITS

D has dimensions of length^2 time^{-1} and is usually expressed as cm^2 sec^{-1}. The dimensions used for P and S are much more varied and a source of much confusion. Consideration of Equation 4.12 shows the dimensions of P to be

$$P = \frac{\text{(quantity of permeant under stated conditions)(thickness)}}{\text{(area)(time)(pressure drop across polymer)}} \tag{4.14}$$

The quantity of permeant can be expressed in mass, mole or volume units. For gases, volume is preferred, expressed as the amount permeating under conditions of standard temperature and pressure (STP), which corresponds to the standard temperature of 273.15 K and standard pressure of 1.01325×10^5 Pa. The standard ambient temperature and pressure (SATP) are set at $T = 298.15$ K and $p = 1$ bar $= 10^5$ Pa $= 0.9678$ atm. Data in handbooks is still mostly expressed with $p = 1$ atm as standard pressure. For practical purposes, the difference between these two conventions is insignificant compared to the variability of the materials themselves.[25]

Over 30 different units for P appear in the scientific literature. Under the SI system, the units of P are mol m m^{-2} sec^{-1} Pa^{-1}. However, with two notable exceptions,[3,6] virtually no permeability data are expressed in these units. A logical suggestion made in 1975 to use cm^2 sec^{-1} [29] as the unit of P has gained little acceptance, and the recent use of atto seconds (1×10^{-18} sec)[12] appears unlikely to be widely adopted. An attempt to rationalize these units from a practical point of view was made in 1983[16] when the adoption of a single unit for P was recommended: 10^{-10} cm^3 (STP) cm cm^{-2} sec^{-1} (cm Hg)$^{-1}$. This unit has been adopted by ASTM[2] and given the name *barrer* after the New Zealander Richard Barrer (1910–1996), a pioneer in the study of diffusion through solids,[4] who first introduced the permeability constant P as a product of D and S.[9] In this book, the following metric units for the permeability coefficient will be used

$$\frac{10^{-11}\text{(mL at STP) cm}}{\text{cm}^2 \text{ sec (cm Hg)}}$$

This unit is more convenient than the barrer when small quantities of permeant are measured, which is usually the case with barrier packaging for foods.

It is important to note that the term milliliter (mL) is equivalent to and interchangeable with the term cubic centimeter (abbreviated cc or cm^3). A micron or micrometer (μm) is 10^{-6} m or 10^{-4} cm, and a mil when used to express thickness is one thousandth of an inch which is equivalent to 25.4 μm or 2.54×10^{-3} cm. The term gauge is also used to indicate the thickness of plastic films; its units are one hundred thousandths of an inch (i.e., 100 gauge is 1×10^{-3} in. or one "thou," which is one mil).

Factors for converting permeability coefficients from various units to the standard units used throughout this book are given in Table 4.1, and broadly representative permeability coefficients for a number of polymers to several gases and water vapor are given in Table 4.2. These values should be used with caution because insufficient details as to the precise nature of the film or the test conditions are often given. The most complete and detailed compilation based on data supplied by manufacturers can be found in Ref. [20].

TABLE 4.1
Factors for Converting Permeability Coefficients from Various Units to the Standard Units Used in This Book

Units	Multiplication Factor to Obtain P in $[mL(STP)\ cm\ cm^{-2}\ sec^{-1}\ (cm\ Hg)^{-1}]$
$[mol\ m\ m^{-2}\ sec^{-1}\ Pa^{-1}]$ (SI)	3.04×10^{5}
$[mL(STP)\ cm\ cm^{-2}\ sec^{-1}\ Pa^{-1}]$	1.33×10^{3}
$[mL(STP)\ mm\ cm^{-2}\ sec^{-1}\ (cm\ Hg)^{-1}]$ (barrer)	1.00×10^{-1}
$[mL(STP)\ cm\ cm^{-2}\ sec^{-1}\ atm^{-1}]$	1.33×10^{-2}
$[mL(STP)\ cm\ m^{-2}\ day^{-1}\ atm^{-1}]$	1.52×10^{-11}
$[mL(STP)\ cm\ m^{-2}\ day^{-1}\ bar^{-1}]$	1.54×10^{-11}
$[in.^{3}(STP)\ mil\ 100\ in.^{-2}\ day^{-1}\ atm^{-1}]$	9.82×10^{-12}
$[mL(STP)\ mm\ m^{-2}\ day^{-1}\ kPa^{-1}]$	2.33×10^{-12}
$[mL(STP)\ mil\ 100\ in.^{-2}\ day^{-1}\ atm^{-1}]$	5.99×10^{-13}
$[mL(STP)\ mm100\ in.^{-2}\ day^{-1}\ atm^{-1}]$	1.53×10^{-14}
$[mL(STP)\ mil\ m^{-2}\ day^{-1}\ atm^{-1}]$	3.87×10^{-14}

TABLE 4.2
Representative Permeability Coefficients of Various Polymers and Permeants at 25°C

Polymer	$P \times 10^{11}\ [mL(STP)\ cm\ cm^{-2}\ sec^{-1}\ (cm\ Hg)^{-1}]$				
	O_2	CO_2	N_2	SO_2	H_2O (90% RH)
Low density polyethylene	30–67	130–280	1.9–3.1	200	800
High density polyethylene	6–11	45	3.3	57	180
Polypropylene	9–15	92	4.4	7	680
Poly(vinyl chloride) film	0.05–1.2	0.3	0.0093	1.2	93
Polystyrene film (oriented)	15–27	105	7.8	220	12–18,000
Nylon 6 (0% RH)	0.12–0.18	0.4–0.8	0.95	22[a]	7,000
Nylon MXD6	0.01				
Poly(ethylene terephthlate)					
amorphous	0.55–0.75	3.0	0.04–0.06		
40% crystalline	0.30	1.6	0.07		
PETG film	1.0	4.6			
Polycarbonate film	15	64			
PVC/PVdC copolymer	0.003	0.3	0.009	–	93
EVOH copolymer					
27 mol% ethylene	0.0018	0.024			
44 mol% ethylene	0.0042	0.012			

[a] Nylon 11.

Difficulty and uncertainty are sometimes experienced in interpreting the permeability coefficients from tables and converting them into different units. The following examples are designed to demonstrate the use of the data in Tables 4.1 and 4.2.

Example 4.1. What is the permeability coefficient of polystyrene to O_2 at 25°C expressed in standard units and in barrer?

Taking the upper range value from Table 4.2:

$$P \times 10^{11} = 27[\text{mL(STP) cm cm}^{-2}\ \text{sec}^{-1}\ (\text{cm Hg})^{-1}]$$

and therefore

$$P = 27 \times 10^{-11}[\text{mL(STP) cm cm}^{-2}\ \text{sec}^{-1}\ (\text{cm Hg})^{-1}]$$

$$= 2.7 \times 10^{-10}[\text{mL(STP) cm cm}^{-2}\ \text{sec}^{-1}\ (\text{cm Hg})^{-1}]$$

To convert this to barrer, divide by 1.00×10^{-1}:

$$P = 2.7 \times 10^{-10}/1.00 \times 10^{-1} = 2.7 \times 10^{-9}\ \text{barrer}$$

Example 4.2 . What is the permeability coefficient of polypropylene to O_2 in standard units, given that it is 1.53×10^{12} [in.^{3}(STP) mil 100 in.^{2} day^{-1} atm^{-1}]?

From Table 4.1: Conversion factor is 9.82×10^{-12}

Therefore,

$$P \times 10^{11} = (1.53 \times 10^{12})(9.82 \times 10^{-12}) = 15[\text{mL(STP) cm cm}^{-2}\ \text{sec}^{-1}\ (\text{cm Hg}^{-1})]$$

and

$$P = 1.5 \times 10^{-10}[\text{mL(STP) cm cm}^{-2}\ \text{sec}^{-1}\ (\text{cm Hg}^{-1})]$$

Example 4.3 . What is the permeability coefficient of HDPE to SO_2 in [mL(STP) mil m^{-2} day^{-1} atm^{-1}]?

From Table 4.2:

$$P \times 10^{11} = 57[\text{mL(STP) cm cm}^{-2}\ \text{sec}^{-1}\ (\text{cm Hg}^{-1})]$$

and

$$P = 5.7 \times 10^{-10}\ [\text{mL(STP) cm cm}^{-2}\ \text{sec}^{-1}\ (\text{cm Hg}^{-1})]$$

The conversion factor from Table 4.1 $= 3.87 \times 10^{-14}$. This must be divided into the permeability coefficient obtained from Table 4.2:

$$P = \frac{5.7 \times 10^{-10}}{3.87 \times 10^{-14}} = 1.473 \times 10^{4}[\text{mL(STP) mil m}^{2}\ \text{day}^{-1}\ \text{atm}^{-1}]$$

Example 4.4. What is the permeability coefficient of PVC/PVdC copolymer to O_2 in standard units, given that it is 0.78 [mL(STP) mil m^2 day^{-1} atm^{-1}]?

From Table 4.1: Conversion factor is 3.87×10^{-14}

Therefore,

$$P = 0.78(3.87 \times 10^{-14}) = 3.01 \times 10^{-14} = 0.003 \times 10^{-11}\ [\text{mL(STP) cm cm}^{-2}\ \text{sec}^{-1}\ (\text{cm Hg}^{-1})]$$

or

$$P \times 10^{11} = 0.003\ [\text{mL(STP) cm cm}^{-2}\ \text{sec}^{-1}\ (\text{cm Hg}^{-1})]$$

which corresponds to the value in Table 4.2.

VI. POLYMER/PERMEANT RELATIONSHIPS

For noncondensable gases, the permeability ratio of two gases is relatively constant and independent of polymer type. Table 4.3 shows average values for a range of different polymers and is useful as a general guide. It can be seen that O_2 permeates about four times as fast as N_2, and CO_2 permeates about six times as fast as O_2 and 24 times as fast as N_2. Table 4.4 gives actual diffusion, solubility and permeability coefficients for LDPE and in this specific case CO_2 permeates about four times as fast as O_2 and almost 13 times as fast as N_2, illustrating that the values in Table 4.3 are averages of several different polymers.

The selectivity ratios are not constant for a given polymer but increase as the temperature is lowered. For example, Tolle[27] reported the CO_2:O_2 for LDPE ranged from 5.08 to 3.45 as temperature increased from 0 to 20°C. Others[8] reported the ratio for PVC films ranged from 6.0 to 4.0 as temperature increased from 5 to 40°C.

It may be thought strange that CO_2, the largest of the three gas molecules, has the highest permeability coefficient. In fact, it has a low diffusion coefficient as would be expected from its relative size, but its permeability coefficient is the highest because its solubility coefficient S is much greater than that for the other gases as illustrated in Table 4.4 for LDPE. When a value for one gas is known, then values for another can be estimated using the corresponding values from Table 4.3. These relative values are also valid to a lesser degree for the D and S values.[25]

TABLE 4.3
Relative Values of Permeability Parameters

Gas	P	D	S	E_P	E_D
N_2 (= 1)	1	1	1	1	1
CO	1.2	1.1	1.1	1	1
CH_4	3.4	0.7	4.9	(1)	(1)
O_2	3.8	1.7	2.2	0.86	0.90
He	15	60	0.25	0.62	0.45
H_2	22.5	30	0.75	0.70	0.65
CO_2	24	1	24	0.75	1.03
H_2O		5	–	0.75	0.75

Source: From Van Krevelen, D. W., *Properties of Polymers*, 3rd ed., Elsevier Scientific, Amsterdam, 1990, chap. 18. With permission.

TABLE 4.4
Diffusion, Solubility and Permeability Coefficients for Low Density Polyethylene Film at 25°C to CO_2, O_2 and N_2

	$D \times 10^6$ cm^2 sec^{-1}	$S \times 10^2$ mL(STP) mL^{-1} atm^{-1}	$P \times 10^{11}$ [mL(STP) cm cm^{-2} sec^{-1} (cm Hg)$^{-1}$]
CO_2	0.37	25.8	126
O_2	0.46	4.78	29
N_2	0.32	2.31	−9.7

Source: From Michaels, A. S. and Bixler, H. J., *J. Polym. Sci.*, 50, 393–412, 1961. With permission.

Factors that can affect the permeability coefficients of a polymer may be divided into those associated with the polymer itself, and those affecting the diffusion coefficient D and the solubility coefficient S.

VII. VARIABLES OF THE POLYMER

Not unexpectedly, the barrier properties of films depend on the specific molecular structures of the polymers involved. A structure that provides a good barrier to gases may provide a poor water vapor barrier. For example, highly polar polymers such as those containing hydroxyl groups (PAs and EVOH copolymers) are excellent gas barriers but poor water vapor barriers. Furthermore, their effectiveness as gas barriers is reduced when the polymer is plasticized by water. In contrast, nonpolar hydrocarbon polymers such as polyethylene have excellent water vapor barrier properties but poor gas barrier properties, the latter property improving as the density of the polyethylene increases. The magnitude of such effects is illustrated in Table 4.5.

Diffusion of a dissolved permeant in a polymer is viewed as a series of activated jumps from one vaguely defined cavity within the polymer matrix to another. Qualitatively, any agent that increases the number or size of cavities in a polymer or renders chain segments more mobile increases the rate of diffusion.

Therefore, to be a good all-round barrier material, the polymer must possess the following properties:

1. Some degree of polarity such as is found in nitrile, chloride, fluoride, acrylic or ester groups.
2. High chain stiffness.
3. Inertness to the permeant. Many polymers, particularly those containing polar groups, can absorb moisture from the atmosphere or from liquids in contact with the polymer. This has the effect of swelling or plasticizing the polymer and reducing the barrier properties as Table 4.5 shows.
4. Close chain-to-chain packing ability brought about by molecular symmetry or order, crystallinity or orientation. Linear polymers with a simple molecular structure lead to good chain packing and lower permeant permeability than polymers whose backbone

TABLE 4.5
Effects of Water Vapor on Oxygen Permeability Coefficients at 25°C

	$P \times 10^{11}$ [mL(STP) cm cm^{-2} sec^{-1} (cm Hg)$^{-1}$]	
Polymer	**0% RH**	**100% RH**
Poly(vinyl alcohol)	0.0006	1.5
Uncoated cellulose	0.0078	12
Nylon 6	0.06	0.3
Poly(vinyl acetate)	3.3	9
Acrylonitrile–styrene copolymer	0.06	0.06
Poly(ethylene terephthlate)	0.42	0.36
High density polyethylene	6.6	6.6
Low density polyethylene	28.8	28.8

Source: From Ashley, R.J., Permeability and plastics packaging, In *Polymer Permeability*, Comyn, J., Ed., Elsevier Applied Science, Essex, England, 1985 (reprinted 1994). chap. 7. With permission.

contains bulky side groups leading to poor packing ability. The higher the degree of crystallinity, the lower the permeability because the crystalline regions are relatively impermeable compared with the amorphous areas. Orientation of amorphous polymers decreases permeation by about 10 to 15%, while in crystalline polymers, reductions of over 50% can be observed.

5. Some bonding or attraction between chains. Cross-linking of polymer chains restricts their mobility and thus decreases their permeability, due mainly to the decrease in the diffusion coefficient. For example, in the case of polyethylene, one cross-link about every 30 monomer units leads to a halving of the diffusion coefficient. The effect of cross-linking is more pronounced for large molecular sized permeants.
6. High glass transition temperature (T_g). Below T_g the segments have little mobility and there is also a reduction in "free volume." Thus, not only are there fewer voids, but, in addition, a diffusing molecule will have a much more tortuous path through the polymer. Therefore, if a polymer has a high T_g, then it is likely that its temperature of use will be below its T_g and it will consequently have improved barrier properties.

A fairly systematic correlation has been established[28] between the activation energy for diffusion E_D, the size of penetrant molecules and the T_g of polymers for a wide range of gases and low molecular weight organic vapors. This correlation can be used to estimate values of the diffusion coefficient D. There is a greater dependence of both sorption and diffusion processes on the size and shape of the penetrant molecule in the glassy state than in the rubbery state.

The sorption of moisture by polymers lowers their T_g and causes them to soften at lower temperatures. In certain polymers, the water acts as an internal lubricant, decreasing the energy barrier for chain segment movements. For example, the T_g values of PAs and EVOH copolymers are greatly reduced by only about 1% of water. Strong hydrogen bonds in PAs act in a similar way to cross-links in raising T_g, while small amounts of water break these bonds and cause abnormally large decreases in T_g.

Table 4.6 shows the effect which different functional groups have on the O_2 permeability coefficient.

TABLE 4.6
Effects of Functional Groups on Oxygen Permeability Coefficients

Nature of R in $(-CH_2-CHR-)_n$	$P \times 10^{11}$ [mL(STP) cm cm^{-2} sec^{-1} (cm Hg)$^{-1}$]
$-OH$	0.0006
$-CN$	0.0024
$-Cl$	0.48
$-F$	0.9
$-COOCH_3$	1.02
$-CH_3$	9.0
$-C_6H_5$	25.2
$-H$	28.8

Source: From Ashley, R. J., Permeability and plastics packaging, In *Polymer Permeability*, Comyn, J., Ed., Elsevier Applied Science, Essex, England, 1985, (reprinted 1994) chap. 7.

VIII. FACTORS AFFECTING THE DIFFUSION AND SOLUBILITY COEFFICIENTS

A. Pressure

In the case of the permanent gases, the permeability coefficient is independent of the pressure of the diffusing gas. This is also true in many instances for other gases and vapors, provided there is no marked interaction between the polymer and the diffusing material. However, where there is strong interaction, the permeability coefficient is found to be pressure dependent and, in general, it increases as the pressure increases. This is due to an increase in the diffusion coefficient D promoted by the plasticizing effect of the sorbed permeant, and an increase in the solubility coefficient S.

B. Sorption

The term sorption is generally used to describe the initial penetration and dispersal of permeant molecules into the polymer matrix and includes both adsorption and absorption as well as cluster formation. Sorption behavior has been classified on the basis of the relative strengths of the interactions between the permeant molecules and the polymer, or between the permeant molecules themselves within the polymer.

The simplest type of sorption (known as Type I) arises when both polymer/permeant and permeant/permeant interactions are weak relative to polymer/polymer interactions; that is, when ideally dilute solution behavior occurs and Henry's law is obeyed. The solubility coefficient S is a constant independent of sorbed concentration at a given temperature and therefore the sorption isotherm shows a linear dependence of concentration vs. vapor pressure:

$$c = Sp \tag{4.15}$$

This type of behavior is observed when permanent gases are sorbed by rubbery polymers at low (<1 atm) pressure and arises from the very low solubility ($<0.2\%$) of the permanent gases in these polymers.

C. Temperature

The temperature dependence of the solubility coefficient over relatively small ranges of temperature can be represented by an Arrhenius-type relationship:

$$S = S_0 \exp(-\Delta H_s/RT) \tag{4.16}$$

where ΔH_s is the heat of solution. For the permanent gases, ΔH_s is small and positive and therefore S increases slightly with temperature. For easily condensable vapors, ΔH_s is negative due to the contribution of the heat of condensation, and thus S decreases with increasing temperature.

The temperature dependence of the diffusion coefficient can also be represented by an Arrhenius-type relationship:

$$D = D_0 \exp(-E_d/RT) \tag{4.17}$$

where E_d is the activation energy for the diffusion process. E_d is always positive and the diffusion coefficient increases with increasing temperature.

From the above two equations it follows that:

$$P = P_0 \exp(-E_p/RT) \tag{4.18}$$

$$= (D_0 S_0) \exp[-(E_d + \Delta H_s)/RT] \tag{4.19}$$

where E_p is the apparent activation energy for permeation. E_p, E_d and ΔH_s are expressed in kJ mol^{-1} and $R = 8.3145$ J mol^{-1} K^{-1} while T is the absolute temperature in Kelvin.

From the above, it follows that the permeability coefficient of a specific polymer–permeant system may increase or decrease with increases in temperature depending on the relative effect of temperature on the solubility and diffusion coefficients of the system. Generally, the solubility coefficient increases with increasing temperature for gases and decreases for vapors, and the diffusion coefficient increases with temperature for both gases and vapors. For this reason, permeability coefficients of different polymers determined at one temperature may not be in the same relative order at other temperatures.

A model has been developed[26] describing the water vapor permeances of three films (LDPE, PET and a laminate of the two) at 20, 30 and 40°C and 55 to 90% RH. Such models are essential for predicting moisture transfer and shelf life of dried foods under real life conditions.

Example 4.5. The permeability coefficient of a PET plastic bottle to CO_2 at 25°C is 1.6×10^{-11} [mL(STP) cm cm^{-2} sec^{-1} (cm Hg^{-1})]. Calculate the value of P at 45°C given that $E_p = 32$ kJ mol^{-1}.

Equation 4.18 needs to be written for each temperature:

$$P_{25} = P_0 \exp(-E_p/RT_{25})$$

$$P_{45} = P_0 \exp(-E_p/RT_{45})$$

On dividing the second equation by the first:

$$P_{45} = P_{25} \exp(-E_p/R[1/T_{45} - 1/T_{25}]) = 1.6 \times 10^{11} \exp(-32{,}000/8.314[0.003145 - 0.003356])$$

$$= 1.6 \times 10^{11} \exp(0.804) = 3.56 \times 10^{-11} \text{ [mL(STP) cm cm}^{-2} \text{ sec}^{-1} \text{ (cm Hg}^{-1})].$$

IX. TRANSMISSION RATE

The earlier treatment of steady-state diffusion assumed that both D and S are independent of concentration but in practice deviations do occur. Equation 4.12 does not hold for heterogeneous materials such as coated or laminated films, or when there is interaction such as occurs between hydrophilic materials (e.g., EVOH copolymers and some of the PAs) and water vapor. The property is then defined as the transmission rate (TR) of the material, where:

$$\text{TR} = \frac{Q}{At} \tag{4.20}$$

where Q is the amount of permeant passing through the polymer, A is the area and t is the time. Permeabilities of polymers to water and organic compounds are often presented in this way, and in the case of water, the term water vapor transmission rate (WVTR) is in common usage.

Because the transmission rate (TR) includes neither pressure of the permeant nor thickness of the polymer in its dimensions, it is necessary to know either the pressure or the concentration of permeant, and the thickness of the polymer, under the conditions of measurement. Because the TR is not a real constant, which is characteristic for a polymer, it should only be used as a means of comparing orders of magnitude. In this book the units of (g mm m^{-2} day^{-1}) will be used for transmission rate.

Data on WVTR's are commonly reported for 38°C and relative humidities on the high pressure side of 95% and on the low pressure side of 0%. Such data for a variety of polymers are given in Table 4.7.

TABLE 4.7
WVTRs at 38°C and 95% RH

Polymer	Transmission Rate (g mm m^{-2} day^{-1}) × 10^{-2}
Vinylidene chloride/vinyl chloride copolymer	4.1–19.7
Polypropylene	7.8–15.7
High density polyethylene	0.1–0.2
Poly(vinyl chloride)	19.7–31.5
Low density polyethylene	31.5–59
Poly(ethylene terephthalate)	31.5–59
Polystyrene	280–393
Ethylene–vinyl alcohol copolymer	546
Nylon 6	634–863

Source: From Ashley, R. J., Permeability and plastics packaging, In *Polymer Permeability*, Comyn, J., Ed., Elsevier Applied Science, Essex, England, 1985 (reprinted 1994, chap. 7); Karel, M., Lund, D. B., Protective packaging, *Physical Principles of Food Preservation*, 2nd ed., Marcel Dekker, New York, 2003, chap. 12. With permission.

Of considerable interest in many food packaging applications is the transmission rate of various organic compounds such as flavors, aromas, odors and solvents through polymers. This is particularly so where the package contents have to be protected against contamination from foreign odors, or where there is a requirement to ensure that volatile flavoring materials are not lost from the food. The significant off-flavors found in some food products may result from the packaging material itself, or may permeate through the packaging material from the outside environment. In other situations, foods may contain highly desirable but volatile flavor compounds whose loss from the packaged food would detract from its quality. In both situations, suitable tests must be undertaken to select materials that have the desired odor barrier properties.

Flavor and aroma losses from foods to package components occur by solution (sorption), with the rate being governed by the magnitude of the diffusion coefficient, followed by permeation if enough time elapses. However, the majority of such problems involve scalping, the industry term for solution of fugitive molecules into the package contact surfaces. In these situations permeation is of no consequence.[5] Scalping has been of particular concern in the packaging of beverages and fruit juices and is discussed in Chapter 20.

The permeation of organic vapors through polymer films is much more complicated than that of gases, due to the pressure-dependent solubility coefficient and the concentration-dependent diffusion coefficient. The values of D, calculated by different techniques and for films of different thicknesses, may differ by more than an order of magnitude. In addition, the rearrangement of polymer molecules in the presence of the permeant may proceed relatively slowly, so that the apparent diffusion coefficient is a function of both permeant concentration and time.

Although extensive studies have been made on the permeability, solubility and diffusivity of various organic vapors in polyethylene, much less data are available for other polymers. Because the solvent action of organic vapors varies from polymer to polymer, the permeability cannot be compared in a similar fashion as it can for permanent gases and water vapor. Table 4.8 gives some data for TRs through LDPE, and Table 4.9 presents some data on the TRs of seven multilayer packaging films at 0 and 75% RH to six organic compounds. These latter data show that moisture-sensitive films such as EVOH copolymer and PAs exhibit superior barrier properties even at elevated humidities, provided that they are surrounded by moisture barrier films in the laminate.

TABLE 4.8
Transmission Rate (g mm m^{-2} day^{-1}) of Various Organic Compounds through Low Density Polyethylene

Permeant	0°C	21.1°C	54.4°C	73.9°C
Acetic acid	0.14	1.22	25.9	119
Acetic anhydride	0.051	0.32	11.6	83.7
Amyl acetate	0.22	3.42	106	430
Benzaldehyde	0.15	2.67	81.0	417
n-Butyl alcohol	0.04	0.18	8.02	59
Carbon tetrachloride	20	240	3,080	17,700
Ethyl acetate	0.75	6.5	149	669
n-Heptane	19.1	106	1,040	3,200
Methyl ethyl ketone	1.45	4.95	128	550
Phenol	0.04	0.2	9.4	47.1
n-Propyl alcohol	0.03	0.2	8.8	66
Toluene	22.7	199	2,270	11,300
o-Xylene	14.2	101	1,420	6,530

Source: From Pauly, S., Permeability and diffusion data, In *Polymer Handbook*, 4th ed., Brandrup, J., Immergut, E. H. and Grulke, E. A., Eds., Wiley-Interscience, New York, 1999, Sec. VI-543.

TABLE 4.9
Transmission Rate (g m^{-2} day^{-1} 100 parts per $million^{-1}$) for Various Odors at 23°C; Top Figure is for 0% RH, Bottom Figure for 75% RH

Film	Ethyl Acetate	Toluene	Styrene	Limonene	β-Pinene	Ethyl Phenyl Acetate
A	<0.003	0.001	<0.003	0.0409	0.0120	0.0085
	0.0632	0.0199	0.0120	0.0009	0.0013	0.0053
B	0.30	0.027	0.0610	0.0012	0.0013	<0.0060
	0.0034	0.0050	0.0037	0.0037	0.0001	<0.002
C	<0.0004	0.002	0.0054	0.0014	<0.0004	<0.0080
	0.0041	0.0088	0.0060	<0.0003	0.0049	<0.002
D	<0.0004	0.001	<0.0003	0.0018	<0.0011	<0.0080
	0.0066	0.0008	0.0046	0.0076	0.0020	0.0076
E	<0.0004	0.0003	<0.0002	0.0400	<0.0036	<0.0070
	0.0092	0.0034	0.0338	0.0061	0.0031	0.0061
F	6.86	1.310	0.0018	0.0315	0.0088	0.234
	0.0040	0.0020	0.0096	0.0071	too fast	0.0071
G	0.52	0.470	0.0046	0.0400	0.0320	0.016
	0.0095	0.0007	0.0051	0.0060	0.1419	0.0060

A, HDPE-adhesive-PA-EVOH (3.175×10^{-3} cm); B, HDPE-adhesive-EVOH-EVA (3.175×10^{-3} cm); C, HDPE-adhesive-PA-EVOH-PA-adhesive-HDPE (3.556×10^{-3} cm); D, HDPE-adhesive-mineral filled PA-adhesive-HDPE (5.558×10^{-3} cm); E, oriented PP-adhesive-EVOH-adhesive-PP (2.54×10^{-3} cm); F, PP-adhesive-EVOH-adhesive-PP (2.54×10^{-3} cm); G, PVdC copolymer coated OPP (4.572×10^{-3} cm) (PA was nylon 6).

Source: From Hatzidimitriu, E., Gilbert, S. G., and Loukakis, G., *J. Food Sci.*, 52, 472–474, 1987. With permission.

X. MIGRATION

Migration is the release of substances initially present in the package into the food. Plastics contain numerous low MW substances that can migrate including monomers, oligomers and various additives such as plasticizers and antioxidants. These substances diffuse through the material until they reach the inside surface of the package where they are partially transferred to the headspace or dissolved in the food.[10] This may result in a loss of food quality due to flavor or color changes, or it may make the food toxic without perceptibly altering the organoleptic properties of the food. For this latter reason, regulations have been promulgated that establish maximum migration limits for substances found in plastics which are used to package food. Migration is considered further in Chapter 21.

XI. PERMEABILITY OF MULTILAYER MATERIALS

Many foods require more protection than a single material can provide to give the product its intended shelf life. Where increased barriers to gases or moisture vapor are necessary, it is more economical to incorporate a thin layer of barrier material than to simply increase the thickness of a monolayer. In most cases, increasing the monolayer thickness would be impractical as it would require too much material as this example[5] illustrates:

To equal the O_2 barrier of a 25 μm film of a high barrier material such as PVdC copolymer would require 62,500 μm of PP, or 4375 μm of PETG or 1250 μm of PET or 1250 μm of rigid PVC or 250 μm of nylon 6.

Multilayer materials can be considered as a number of membranes in series. Consider the case of three layers in series as shown schematically in Figure 4.2. The total thickness $X_T = X_1 + X_2 + X_3$. Assuming steady-state flux, the rate of permeation through each layer must be constant, that is,

$$Q_T = Q_1 = Q_2 = Q_3 \tag{4.21}$$

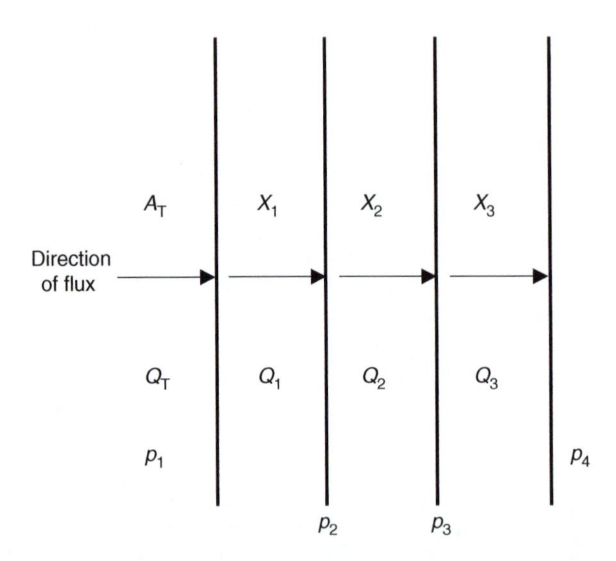

FIGURE 4.2 Schematic representation of permeation through three materials in series.

Likewise, the areas will also be constant so that

$$A_T = A_1 = A_2 = A_3 \tag{4.22}$$

Recall Equation 4.13:

$$\frac{Q}{t} = \frac{P}{X}A(\Delta p) \tag{4.13}$$

Then, on substituting

$$\frac{Q_T}{t} = \frac{P_1}{X_1}A_1(p_1 - p_2) = \frac{P_2}{X_2}A_2(p_2 - p_3) = \frac{P_3}{X_3}A_3(p_3 - p_4) \tag{4.23}$$

By rearranging Equation 4.23 and writing it for the case of permeation through the multilayer material

$$\frac{Q_T X_T}{tA_T P_T} = (p_1 - p_4) = \Delta p_i \tag{4.24}$$

Now, because

$$(p_1 - p_4) = (p_1 - p_2) + (p_2 - p_3) + (p_3 - p_4) \tag{4.25}$$

Therefore,

$$\frac{Q_T X_T}{tA_T P_T} = \frac{Q_T}{tA_T}\left[\frac{X_1}{P_1} + \frac{X_2}{P_2} + \frac{X_3}{P_3}\right] \tag{4.26}$$

and

$$\frac{X_T}{P_T} = \frac{X_1}{P_1} + \frac{X_2}{P_2} + \frac{X_3}{P_3} \tag{4.27}$$

or

$$P_T = \frac{X_T}{(X_1/P_1) + (X_2/P_2) + (X_3/P_3)} \tag{4.28}$$

Thus, if the individual thicknesses and permeability coefficients are known for each layer, and provided that the permeability coefficients are independent of pressure, then Equation 4.28 can be used to calculate the permeability coefficient for any multilayer material. If they are not independent of pressure, then differing permeability coefficients will be obtained for the multilayer material depending on the positioning of the layers.

The standard methods discussed below for gas permeability measurements through polymeric materials all specify dry gas. However, in practice, packaging films are almost always used in humid conditions, and for materials such as EVOH copolymers and PAs, the O_2 permeability is dependent on the humidity. In such cases, Equation 4.28 cannot be used directly because P_2 (the permeability coefficient of the center layer) will depend on the average partial pressure at the center.

Example 4.6. Calculate the total O_2 permeability at 30°C of a multilayer film with the following structure (units of P are mL(STP) cm cm^{-2} sec^{-1} (cm Hg)$^{-1}$).

	Polymer	P	Thickness (μm)
Layer 1	LDPE	55	50
Layer 2	Nylon 6	0.18	20
Layer 3	LDPE	55	50

First, the thicknesses must be converted into cm:

$$50\ \mu\text{m} = 50 \times 10^{-4}\ \text{cm} = 5 \times 10^{-3}\ \text{cm}$$

$$20\ \mu\text{m} = 2 \times 10^{-3}\ \text{cm}$$

Then, by substituting into Equation 4.27

$$\frac{0.005 + 0.002 + 0.005}{P} = \frac{0.005}{55} + \frac{0.002}{0.18} + \frac{0.005}{55}$$

and

$$\frac{0.012}{P} = 9.091 \times 10^{-5} + 1.111 \times 10^{-2} + 9.091 \times 10^{-5}$$

Solving for P:

$$P = 1.129 \times 10^{-2}/0.012 = 0.94 \times 10^{-11}\ [\text{mL(STP) cm cm}^{-2}\ \text{sec}^{-1}\ (\text{cm Hg}^{-1})]$$

Note that in this example, the nylon barrier layer is providing virtually all of the resistance to transmission. Given that the accuracy of P is seldom better than $\pm 5\%$, the contribution of the supporting layers in this example could have been ignored in the calculation without significantly affecting the final result.

An equation for predicting the average partial pressure at the center of a multilayer material containing a water sensitive center layer can be derived as follows. Consider again the case of three layers in series as shown schematically in Figure 4.2, but this time assume that the O_2 permeability of the center layer is humidity dependent and that the direction of water vapor flux is from the outside in.

Because the partial pressure of water vapor will not be constant across the multilayer, Equation 4.20 must be modified to include a term for the partial pressure difference and the thickness:

$$\text{WVTR} = \frac{Q}{AXt}\Delta p \qquad (4.29)$$

Now, because the area A and time t will be the same for all three layers, the equilibrium WVTR between the outside and center layers can be expressed as:

$$\frac{Q_1}{X_1}(p_1 - p_2) = \frac{Q_2}{X_2}(p_2 - p_3) \qquad (4.30)$$

Similarly, the equilibrium WVTR between the center and the inside layer will be

$$\frac{Q_2}{X_2}(p_2 - p_3) = \frac{Q_3}{X_3}(p_3 - p_4) \qquad (4.31)$$

The average partial pressure of the center layer (p_c) will be

$$p_c = \frac{p_2 + p_3}{2} \tag{4.32}$$

Simultaneous linear solution of Equation 4.30 and Equation 4.31 for p_2 and p_3 and substitution in Equation 4.32 yields

$$p_c = \frac{p_1\left[\frac{X_2}{Q_2} + 2\frac{X_3}{Q_3}\right] + p_4\left[\frac{X_2}{Q_2} + 2\frac{X_1}{Q_1}\right]}{2\left[\frac{X_3}{Q_3} + \frac{X_2}{Q_2} + \frac{X_1}{Q_1}\right]} = \text{average partial pressure of the center layer} \tag{4.33}$$

From knowledge of p_c, the permeability coefficient P_c of the center layer can be determined experimentally at this partial pressure, and Equation 4.28 can be used to calculate the overall permeability of the multilayer material.

XII. MEASUREMENT OF PERMEABILITY

A. Gas Permeability

There are many methods for measuring gas permeability; four major methods will be considered here. For a complete understanding of the principles behind permeability measurements, it is important that the meaning of two terms which are constantly used — total and partial pressure of gases in a mixture — is clearly appreciated.

In a constant volume, the total pressure exerted by the gases present is the sum of the partial pressures of each of the gases, a discovery made by English scientist John Dalton in 1801 and known as Dalton's law. The partial pressure of any one of the constituent gases is the pressure that would result if that particular gas occupied the same volume by itself. That is, each of the gases of a gas mixture behaves independently of the others.

The rate of permeation of a specific gas through a polymeric material is a function of the partial pressure differential of that gas across the material and not of the total pressure difference between the two sides.

1. Pressure Increase Method

The ASTM manometric method for measuring gas transmission rates and permeabilities of flat films is designated D 1434. Test gas (normally at 1 atm) is introduced on one side of the flat film or sheet, which is supported with a filter paper and sealed with an O-ring. The pressure in the receiving chamber is measured with an open-ended mercury manometer. Provided that the pressure on the high-pressure side remains much larger than that on the low-pressure side, the pressure difference remains essentially constant. Through equations relating the geometry of the cell with the rate of pressure rise in the manometer, the gas transmission rate can be calculated.

2. Volume Increase Method

In the ASTM standard volumetric method (also designated D 1434), the change in volume (at constant pressure), due to the permeation of gas through the film, is measured. Variable volume permeation cells are used for rapid measurement of relatively high steady-state permeation rates. Although the volume increase method is generally simpler to implement, it is less sensitive than the pressure increase method. Volumetric methods are used relatively infrequently compared with the use of the pressure increase or concentration increase methods. Recently, a simple whole bag method based on constant pressure/volume increase method has been described.[23]

3. Concentration Increase Method

In the ASTM D 3985 method (also known as the quasi-isostatic coulometric method because the total pressure on both sides is approximately equal), a partial pressure difference across the film with respect to the test gas is created without a difference in total pressure, thus obviating the need for rigid support of the film. A partial pressure difference is maintained by sweeping one side continuously with the test gas, and maintaining an inert gas on the other side into which the test gas diffuses. The concentration of the diffusing gas can be measured by chemical analysis, gas chromatography, thermal conductivity or special electrodes. A variation of this method (ASTM F 1307) can be used to determine the O_2 transmission rate through dry packages.

Instruments are in widespread commercial use for the measurement of O_2 and CO_2 transmission rates by the quasi-isostatic method, an advantage being that the permeability of not only flat film but also containers, bottles, pouches, tubes, and so on, can be determined, thus permitting the assessment of possible adverse effects from machine processing, printing and distribution. The use of these instruments is included in ASTM D 3985 and F 1307. The AOIR (ambient O_2 ingress method) is a modified version of the quasi-isostatic method, which can be used to test actual packages over a wide range of temperature and humidity conditions, including chill and freezing temperatures.[19]

4. Detector Film Method

A relatively new method for measuring permeabilities of films requires little equipment and is both rapid and accurate.[14] The basis of the method is a plastic detector film impregnated with a reagent that is sensitive to the gas being measured. The film has an absorption spectrum that changes as the gas or vapor is absorbed, and is thus suitable for spectrophotometric monitoring. The detector film is sealed between two pieces of test film in a simple cell so that the permeation rate of the penetrant gas or vapor can be readily measured.

The detector film can measure much less than the minimum detectable quantity of O_2 determined by most other methods, and therefore permits the use of either smaller film samples or more rapid permeability determinations.

The O_2 detector consists of a cast film of ethyl cellulose containing dimethylanthracene (DMA) and erythrosine. On absorbing blue light, the erythrosine can activate O_2 dissolved in the ethyl cellulose to form singlet O_2, a reactive form of O_2. Singlet O_2 (which has a lifetime of only a few microseconds in the film) diffuses to a neighboring DMA molecule and reacts with it. Thus, the disappearance of DMA (monitored in the UV) is a measure of the O_2 consumed. Because the ethyl cellulose detector is highly permeable to O_2, it is capable of measuring very low rates of O_2 permeation.

B. Water Vapor Permeability

The standard method to determine water vapor transmission rates (ASTM E 96) is to place a quantity of desiccant in an aluminum dish, which is covered with a sheet of the material being tested and sealed in position with wax. The dish is then placed in a closely controlled atmosphere (typically either 25 ± 0.5°C and $75 \pm 2\%$ RH for temperate conditions, or 38 ± 0.5°C and $90 \pm 2\%$ RH for tropical conditions), and the increase in weight noted as a function of time. If the points are plotted out, then they should fall more or less on a straight line because Δp is constant throughout the test. From Equation 4.20:

$$\mathrm{WVTR} = \frac{Q}{At} \tag{4.36}$$

$$= \frac{\text{slope}}{\text{area}} \tag{4.37}$$

$$= \frac{\text{g H}_2\text{O}}{\text{m}^2\ \text{day}} \tag{4.38}$$

To convert WVTR into permeance (P/X), it should be divided by the driving force Δp.

These methods have several disadvantages, including the length of time needed to make a determination (between 2 and 14 days) and the lower limit of the useful range (about 1 g m^{-2} day^{-1} for a typical packaging film). A further disadvantage is that, depending on the desiccant, Δp may not remain constant during the test period. In the case of anhydrous $CaCl_2$, the partial pressure of water vapor in the dish remains below 2% of the vapor pressure of water at the test temperature during the test, whereas in the case of silica gel, the partial pressure of water sorbed on it increases with coverage.

WVTR tests on flat sheets of film sealed across aluminum dishes do not always correlate closely with actual performance of the film when made up into complete packages. Therefore, it is often preferable to carry out WVTR tests where the desiccant is contained in a finished package that has been closed and sealed in the conventional manner. Details of the test procedures can be found in ASTM D 3079. A whole bag method for determining WVTR of hydrophobic films has been developed.[22]

In recent years, more rapid methods have been developed. Most of these depend on detecting small changes in the relative humidity of the atmosphere on the dry side of the film. Typically test cells consist of two sections separated by the material under test. The lower section contains water to give a saturated atmosphere, while the upper section contains a humidity sensor which is dried by purging with dry air. The movement of water vapor through the film raises the relative humidity of the air surrounding the humidity sensor, and the time for a given rise can be recorded; from this, the WVTR is calculated. Commercial instruments are available to determine WVTRs using either a pressure-modulated infrared detector or a mechanically modulated infrared detector.

A detector film to measure transmission rates of water vapor has also been developed.[15] It consists of transparent cellulose film that becomes bright blue when soaked in $CoCl_2$ solution and dried over $CaCl_2$, but rapidly turns pink on exposure to high humidities. A humidity cabinet is used to provide the partial pressure gradient across the test film, which is sealed in the same way and in a cell of similar design to that used to measure O_2 permeability (see Section XII.A.4). The change in absorbance of the detector film is measured at 690 nm, and from this, the quantity of water absorbed by the detector film, and hence the WVTR of the test film, can be calculated.

C. Permeability of Organic Compounds

Although the permeabilities of permanent gases and water vapor through many packaging materials are well known, there are limited data for the permeation of organic compounds. It was shown earlier in this chapter that, for gases, the permeability coefficient is independent of concentration. Thus, permeability measurements made at high concentrations can be reliably extrapolated to predict permeation rates where there are low concentration gradients across the barrier. However, in the case of many organic compound/plastic package combinations, the permeability coefficient is highly dependent on concentration. This effect occurs because the organic compound interacts with and swells the polymer, increasing the permeation rate.

Measuring the transport rates of organic compounds in plastics is more complicated than those of either water vapor or noncondensible gases, and elaborate equipment and sensitive analytical devices are required to obtain reliable results. The standard test method for measurement of the diffusivity, solubility and permeability of organic vapor barriers was published in 1997 (ASTM F 1769). A number of other methods have been described for organic vapor permeability measurements, although several of them are only suitable for use with saturated vapors. Various procedures developed for quantifying the rate of diffusion of organic penetrants through barrier membranes and the specific procedures employed have been described.[13,25] As with permeability studies of the permanent gases, procedures to study organic vapor permeability include the isostatic and quasi-isostatic methods.

One method[11] for the quantitative evaluation of the aroma barrier of packaging materials uses a permeation cell similar to that described for the concentration increase method (see Section XII.A.3). N_2 gas is bubbled through the liquid permeant and then passed with the permeant vapors through the cell. The concentrations of the permeating vapors and relative humidity are monitored by gas chromatography. The calculated permeation rates for seven films and six permeants were given in Table 4.9 for both 0 and 75% relative humidity.

Equipment is available commercially to measure the permeation of organic vapors through flat films from 5 to 50°C and finished packages at ambient temperatures. The system includes a sample holder with metal-to-metal seals to eliminate elastomeric gaskets, O-rings and PTFE parts, which would absorb and desorb organic vapors. The system also includes a dedicated gas chromatograph and computer. It is claimed that measurements of permeants in extremely low concentrations are possible.

Sorption experiments are usually carried out at equilibrium vapor pressure using a gravimetric technique in an apparatus that records continually the gain or loss of weight by a test specimen as a function of time. A recording electrobalance is commonly used for such studies. Sorption experiments when organic compounds are sorbed from liquids can be performed with thin, die-cut polymer sheets, which are stacked on a stainless steel wire support and spaced with glass beads to minimize occlusion of the specimen surface. The test specimens are placed inside a glass vial with a screw cap containing a double-sealing port. The fluid is sampled regularly to determine the extent of sorption by the polymer sheets.

A common, qualitative odor-penetration test involves packaging various odoriferous substances in pouches made from the test materials. The pouches are then placed in clean glass bottles and sealed by crimping with aluminum foil. After storage for a predetermined time, the bottles are sampled, either objectively using gas chromatography and maybe mass spectroscopy, or subjectively by sniffing using a sensory evaluation panel. From these results, it is possible to rank a range of packaging materials according to their odor barrier properties.

For the subjective method of evaluation, a much greater number of samples is required than for the objective (quantitative) methods. However, qualitative methods are much simpler than objective methods, which require that the offending odors must not only be known chemically, but must also be capable of identification at extremely low concentrations (typically parts per million or billion). To speed up qualitative tests, the glass bottles are sometimes held at elevated temperatures. Under these circumstances, it is important to be sure that the higher temperatures do not produce artifacts or give results that would not be applicable at ambient temperatures.

REFERENCES

1. Ashley, R. J., Permeability and plastics packaging, In *Polymer Permeability*, Comyn, J., Ed., Elsevier Applied Science, Essex, England, 1985, (reprinted 1994), chap. 7.
2. ASTM: American Society for Testing and Materials, Philadelphia, Pennsylvania. For the latest version of each standard visit www.techstreet.com.

3. Banks, N. H., Cleland, D. J., Cameron, A. C., Beaudry, R. M., and Kader, A. A., Proposal for a rationalized system of units for postharvest research in gas exchange, *HortScience*, 30, 1129–1131, 1995.
4. Barrer, R. M., *Diffusion In and Through Solids*, Cambridge University Press, London, 1968.
5. Brown, W. E., *Plastics in Food Packaging: Properties, Design and Fabrication*, Marcel Dekker, New York, 1992.
6. Combellick, W. A., Barrier polymers, *Encyclopedia of Polymer Science and Engineering*, 2nd ed., Vol. 2, Kroschwitz, J. I., Ed., Wiley, New York, pp. 176–192, 1985.
7. DeLassus, P., Barrier polymers, In *The Wiley Encyclopedia of Packaging Technology*, 2nd ed., Brody, A. L. and Marsh, K. S., Eds., Wiley, New York, pp. 71–77, 1997.
8. Doyon, G., Gagnon, J., Toupin, C., and Castaigne, F., Gas transmission properties of polyvinyl chloride (PVC) films studied under subambient and ambient conditions for modified atmosphere packaging applications, *Packag. Technol. Sci.*, 4, 157–165, 1991.
9. Franz, R., Permeation of volatile organic compounds across polymer films — Part 1: development of a sensitive test method suitable for high-barrier packaging films at very low permeant vapour pressure, *Packag. Technol. Sci.*, 6, 91–102, 1993.
10. Gavara, R. and Catalá, R., Mass transfer in food/plastic packaging systems, In *Engineering and Food for the 21st Century*, Welti-Chanes, J., Barbosa-Canovas, G. V. and Aguilera, J. M., Eds., CRC Press, Boca Raton, FL, 2002, chap. 33.
11. Hatzidimitriu, E., Gilbert, S. G., and Loukakis, G., Odor barrier properties of multi-layer packaging films at different relative humidities, *J. Food Sci.*, 52, 472–474, 1987.
12. Hernandez, R. J., Food packaging materials, barrier properties and selection, In *Handbook of Food Engineering Practice*, Valentas, K. J., Rotstein, E. and Singh, R. P., Eds., CRC Press, Boca Raton, FL, 1997, chap. 8.
13. Hernandez, R. J. and Giacin, J. R., Factors affecting permeation, sorption, and migration processes in package-product systems, In *Food Storage Stability*, Taub, I. A. and Singh, R. P., Eds., CRC Press, Boca Raton, FL, 1998, chap. 10.
14. Holland, R. V., Rooney, M. L., and Santangelo, R. A., Measuring oxygen permeability of polymer films by a new singlet oxygen technique, *Angew. Makromol. Chem.*, 88, 209–221, 1980.
15. Holland, R. V. and Santangelo, R. A., Spectrophotometric determination of water vapour permeation through polymer films, *J. Appl. Polym. Sci.*, 27, 1681–1689, 1982.
16. Huglin, M. B. and Zakaria, M. B., Comments on expressing the permeability of polymers to gases, *Angew. Makromol. Chem.*, 117, 1–13, 1983.
17. Karel, M. and Lund, D. B., Protective packaging, *Physical Principles of Food Preservation*, 2nd ed., Marcel Dekker, New York, 2003, chap. 12.
18. Krochta, J. M., Package permeability, In *Encyclopedia of Agricultural, Food and Biological Engineering*, Heldman, D. R., Ed., Marcel Dekker, New York, pp. 720–726, 2003.
19. Larsen, H., Oxygen transmission rates of packages at ambient, chill and freezing temperatures measured by the AOIR method, *Packag. Technol. Sci.*, 17, 187–192, 2004.
20. Massey, L. K., *Permeability Properties of Plastics and Elastomers: A Guide to Packaging and Barrier Materials*, 2nd ed., William Andrew Publishing/Plastics Design Library, Norwich, New York, 2003.
21. Michaels, A. S. and Bixler, H. J., Solubility of gases in polyethylene, *J. Polym. Sci.*, 50, 393–412, 1961.
22. Moyls, A. L., Whole-bag (water) method for determining water vapor transmission rate of polyethylene films, *Trans. Am. Soc. Agri. Eng.*, 41, 1447–1451, 1998.
23. Moyls, A. L., Whole bag method for determining oxygen transmission rate, *Trans. Am. Soc. Agri. Eng.*, 47, 159–164, 2004.
24. Pauly, S., Permeability and diffusion data, In *Polymer Handbook*, 4th ed., Brandrup, J., Immergut, E. H. and Grulke, E. A., Eds., Wiley-Interscience, New York, 1999, Sec. VI-543.
25. Piringer, O.-G., Permeation of gases, water vapor and volatile organic compounds, In *Plastics Packaging Materials for Food*, Piringer, O.-G. and Baner, A. L., Eds., Wiley-VCH Verlag, Weinheim, Germany, 2000, chap. 9.
26. Samaniego-Esguerra, C. M. and Robertson, G. L., Development of a mathematical model for the effect of temperature and relative humidity on the water vapour permeability of plastic films, *Packag. Technol. Sci.*, 4, 61–68, 1991.

27. Tolle, W. E., Variables affecting film permeability requirements for modified atmosphere storage of apples, *Technical Bulletin #1422*, USDA/ARS, Washington, DC, 1971.
28. Van Krevelen, D. W., *Properties of Polymers*, 3rd ed., Elsevier Scientific, Amsterdam, 1990, chap. 18.
29. Yasuda, H., Units of gas permeability constants, *J. Appl. Polym. Sci.*, 19, 2529–2536, 1975.

5 Processing and Converting of Thermoplastic Polymers

CONTENTS

I. EXTRUSION

A. MONOLAYER EXTRUSION

Extrusion is one of the most important plastics processing methods currently in use. Most plastic materials are processed in extruders, and commonly pass through two or more extruders on their way from the chemical reactor to the finished product. All thermoplastics are formed into sheet or film (commonly $<100\ \mu m$) by the process of screw extrusion. The first screw extruder designed specifically for thermoplastic materials appears to have been made by Paul Troesterin in Germany in 1935.

The heart of the extruder is the Archimedean screw, which revolves within a close-fitting, heated barrel. It is capable of pumping a material under a set of operating conditions at a specific rate, depending on the resistance at the delivery end against which the extruder is required to pump. The extruder resembles a mincer into which granules are fed, heated and compressed until they fuse into a melt, which is forced through a slit or circular die.

The standard single screw extruder (Figure 5.1) receives solid polymer in the form of powder, beads, flakes or granules through a hopper to a throat at one end of the barrel and delivers it to the compression zone of the screw. In this section, the diminishing depth of thread causes a volume compression and an increase in the shearing action of the material. It melts and is converted into a homogeneous mass by contact with the heated walls of the barrel and by the heat generated by friction. Generally, external heating is only required at the start of the run, and the frictional or exothermal heating is sufficient for steady-state operation. After the compression zone, the melt passes through the metering zone where the flow is stabilized, before being pumped through the die, which determines its final form. The output from the die is known as the extrudate.

The screw is the most important component of the extruder and different designs are used for extruding different polymers. Extruder screws are characterized by their length to diameter ratios (commonly abbreviated to L/D ratios) and their compression ratios — the ratio of the volume of one flight of the screw at the infeed end to the volume of one flight at the die end. L/D ratios commonly used for single screw extruders are between about 15:1 to 30:1, while compression ratios can vary from 2:1 to 4:1.

There are basically two processes by which the extruded thermoplastic can be converted into film: the flat film process and the tubular process. The first process is illustrated in Figure 5.1, while the latter process is illustrated in Figure 5.2.

In flat film (also known as cast film or slit die) extrusion, the molten polymer is extruded through a slit die into a quenching water bath or onto a chilled roller. In both cases, rapid cooling of the extruded film is most important. The ratio of the haul-off rate to the natural extrusion rate is referred to as the draw-down ratio. Draw-down ratios between 20:1 and 40:1 are typical.

In the tubular (or blown film) process, a thin tube is extruded (usually in a vertically-upward direction but sometimes in a horizontal or downward direction), and by blowing air through the die head, the tube is inflated into a thin bubble and cooled, after which it is flattened and wound up. The ratio of bubble diameter to die diameter is known as the blow-up ratio. Most LDPE blown films used in packaging are made using blow-up ratios of 2.0:1 to 2.5:1. Unless carefully controlled, blown film extrusion can produce defects such as variations in film thickness, surface defects, low tensile and impact strength, haze, blocking and wrinkling.

The properties of the film depend strongly on the polymer used and the processing conditions. The higher the density, the lower the flexibility and the greater the brittleness. The higher the MW,

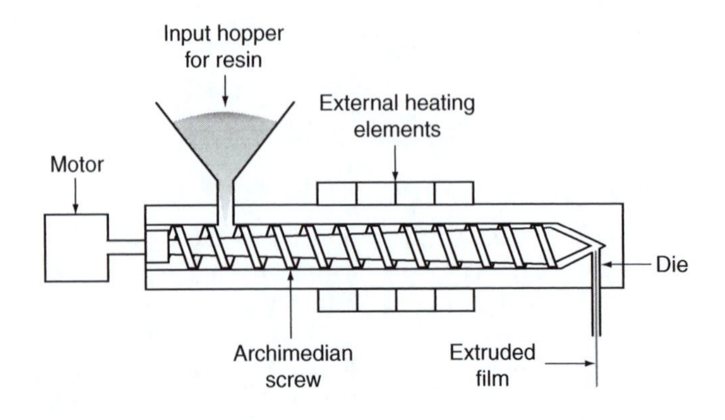

FIGURE 5.1 Single screw extruder.

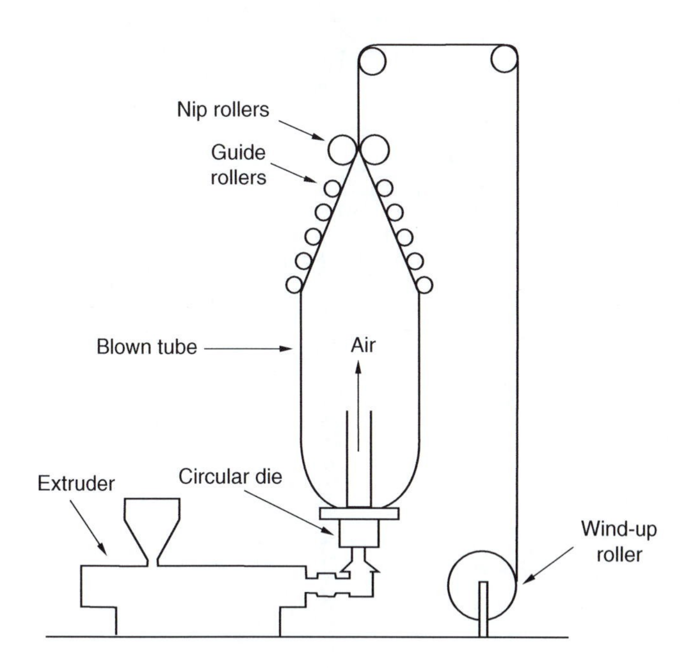

FIGURE 5.2 Blown tubular film extrusion.

the greater the tensile strength and resistance to film brittleness at low temperatures, but the lower the transparency.[4]

Advantages of tubular film are that the mechanical properties are generally better, and the process is easier and more flexible to operate. The cost for making wide tubular film is much lower than for wide cast film, due to the cost of precision grinding long chill rolls. Advantages of the flat film process include less thickness variation, very high outputs and superior optical properties. This latter advantage is a consequence of the quicker cooling, which can be achieved in the flat film process where cooling is by conduction, compared with the tubular film process, where cooling is by convection. Slower cooling permits the formation of more and larger crystals in the film, leading to haze that arises from scattering of light between the crystal interfaces.

B. Coextrusion

Coextrusion consists of coupling two or more extruders feeding different resins to a single die head to simultaneously extrude two or more different polymers, which fuse at the point of film formation into a single web. Such a process is known as coextrusion, and permits the production of a single web that has, for example, barrier properties not possessed by any one of the component polymers. A two-component slit die is capable of producing a two- or three-layer film from two materials, while a three-component die (such as the one shown in Figure 5.3) can produce a five-layer film from three materials. At present, seven- and nine-layer films and sheet are being produced for food packaging applications. Typical structures include:

HIPS-Tie-Barrier-Tie-HIPS (PP and HDPE may replace HIPS depending on the application)
PP-Tie-Barrier-Tie-Regrind-PP
PP-Regrind-Tie-Barrier-Tie-Regrind-PP (six-layer asymmetrical with one regrind layer)
LDPE-Tie-Regrind-Tie-Barrier-Tie-HIPS-GPPS (seven-layer symmetrical with two regrind layers)
PP-Tie-Barrier-Tie-Regrind-Tie-Barrier-Tie-PP (nine-layer with two barrier layers and one regrind layer)

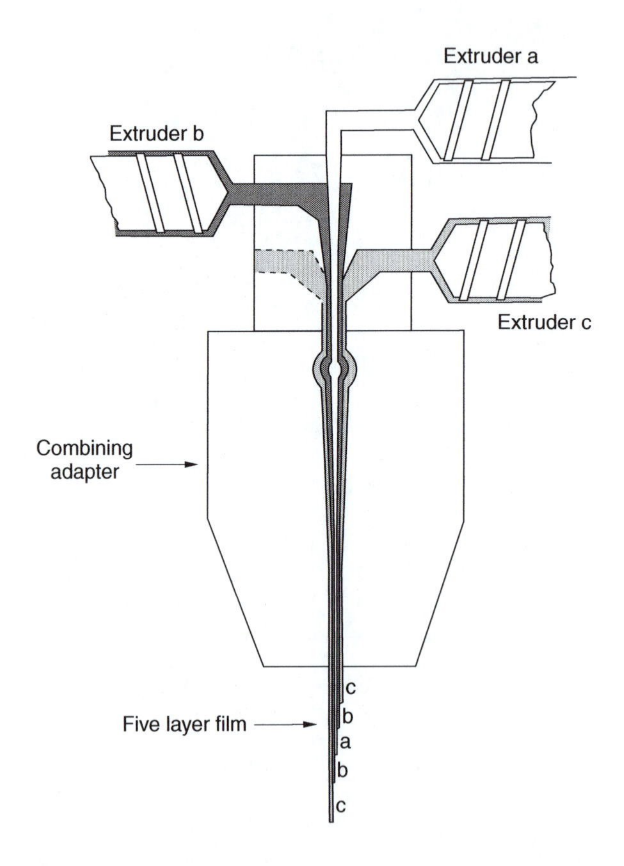

FIGURE 5.3 Three-component slit die.

The use of two barrier layers in the nine-layer structure decreases the potential loss of barrier due to pinholes or flaws in the barrier layer, the rationale being that there is a negligible probability of two adjacent holes in separate barrier layers.[3] The tie or adhesive layers used in the above structures vary depending on the nature of the layers that they are required to bind, and different tie layers may therefore be used in the same structure. The technology behind tie layers is complex and maintained as proprietary by coextrusion manufacturers. However, EVA copolymers (25 to 50% VA) are used extensively; the higher VA copolymers are more soluble and display greater initial bonding force (tack), better adhesion to polar surfaces and better low-temperature performance. Modified PE–PP copolymers (usually containing acrylates for increased adhesion) are also used.[3]

The main advantages of coextruded films over conventionally produced laminates (which they have displaced in many applications) are lower cost, less tendency towards delamination and a greater flexibility in obtaining a wide range of properties. The major disadvantage lies in the difficulty of utilizing the scrap produced during the extrusion process. Coextruded films can also be laminated to other films such as OPP and to paper and foil. Coextrusion has the ability to produce thinner layers than lamination, a distinct advantage when expensive barrier layers are used.

Rigid sheet makes use of resins such as HIPS, HDPE and PP for the bulk layers, while flexible films make more use of LDPE and LLDPE resins as well as PP. For barrier layers, both types use EVOH and PVdC copolymers. Films make greater use of PA as, apart from its gas barrier properties, it has excellent puncture, tear and abrasion resistance.[3]

II. CALENDERING

Calendering is a complimentary process to film and sheet extrusion, and involves the formation of continuous sheets of controlled thickness by squeezing a heated plastic material between two or more horizontal rollers. Although calenders were originally developed for processing rubber, they are now widely used for producing flexible PVC sheet and film.

Calenders consist of from two to five rollers, which can be steam heated or water cooled. The thickness of the final film or sheet depends on the gap between the last pair of rollers. The surface finish of the material is determined by the last roller and can be glossy, matt or embossed. On leaving the calender, the material is cooled by passing over cooling rollers and it is then reeled up.

In addition to the wide variety of PVC films and sheets that are processed by calendering, such materials as ethylene–propylene copolymers, EVA copolymers and rubber modified PS have been fabricated into sheet stock on calenders.

Calendering usually produces film with a better uniformity of gauge compared with that obtained by extrusion. Several factors contribute to this, including the precision engineering of the calender rollers. In extrusion, the gauge depends more on the blow-up ratio in the case of tubular film, or the draw-down ratio in the case of flat film.

III. COATING AND LAMINATING

Coating and laminating are two of the most widely used processes for transforming flexible films and sheets into products that have properties useful in food packaging. Coating is the process of applying one or more layers of a fluid or melt to the surface of a material, while laminating is the bonding of two or more webs.[3] A laminate is defined as any combination of distinctly different plastic film materials or plastic plus nonplastic materials (typically paper and aluminum foil), where each major web is generally thicker than 6 μm, regardless of the method of manufacture. There is no upper limit to the possible number of webs, but two is the obvious minimum and one of these must be thermoplastic.

There are two major techniques employed in the fabrication of laminates: adhesive lamination and extrusion coating. Basically, there are two types of coating processes: in one an excess coating is applied to the web and the surplus removed, while in the other a predetermined amount is applied to the web using rollers or other equipment.

Nitrocellulose and PVdC copolymer are the most common surface coatings used, but synthetic resins, acrylics and many other formulations are used for varnishing, barrier formation or heat sealing. The use of cold seal coatings, which are based on latex, has become increasingly common. They require only contact pressure for two coated surfaces to adhere, and permit packaging machine speeds to be considerably increased. In addition, damage to heat sensitive foods such as chocolate-coated confectionery is avoided.

A. Coating Processes

Extrusion coating (sometimes called *extrusion lamination*) was first practiced on a commercial scale in the production of LDPE-coated paperboard for milk cartons in the mid-1950s as a replacement for wax-coated stock.[11] Compared to wax, LDPE is superior with greater strength, seal integrity and resistance to cracking and flaking off. It also provides greater resistance to moisture, thus protecting the paperboard substrate from the damaging effects of water for much longer periods of time. Currently, almost all applications for wax-coated paperboard have been replaced by polyolefin-coated paper and board.[3]

In theory, there is no reason why any thermoplastic which is normally processed by extrusion techniques cannot be coated onto paper or other substrates. In practice, most extrusion coating technology that has been developed utilizes the lower density polyethylenes, although PP, PAs and

PET are also used. For example, PET-coated paperboard is used in dual ovenable trays, where a 38-μm PET coating is put on 500 to 625-μm paperboard.

Extrusion coating with polyethylene has several advantages over adhesive lamination of a prefabricated polyethylene film to paper. First, thin films of polyethylene are difficult to handle and maintain flat and handling them requires very low tensions, which are difficult to control at high speeds. Secondly, extrusion coating temperatures are sufficiently high so that good mechanical bonds are obtained by resin penetration into the porous paper substrate. The same adhesion level can be obtained only by the use of adhesives when free films are laminated to paper, thus making extrusion coating less expensive.

The development of adhesion between polyethylene and various substrates is an aspect of extrusion coating which has received a great deal of attention. Substrates, whether polymer films, paperboard, aluminum foil or RCF, require some type of surface pretreatment to obtain an adequate level of adhesion of the extruded polyethylene.

Two main approaches to adhesion enhancement have been developed. One is electrical treatment of the surface using corona discharge, where a high voltage (10 to 40 kV) at a low frequency (10 to 20 kHz) is applied between an electrode and an earthed roller, which carries the film. The air between the two surfaces ionizes and a continuous arc discharge (corona) is generated at the surface of the film. In continuous operation the discharge appears to be a random series of faint sparks in a blue-purple glow (UV radiation).

The corona treatment cleans, oxidizes and activates the surface by introducing polar groups into it. This results in the substrate surface being more compatible with a freshly oxidized polyethylene surface, thus promoting adhesion between the two. Corona-treated films should be used immediately for further applications such as extrusion coating or printing because of the diminishing effect of the improved properties with time.[9] A corona generator is usually mounted in-line and prior to the extrusion coating head; the degree of treatment is regulated by rheostat adjustment of the power fed to the discharge electrode. Ozone is a by-product of the corona discharge method, and provision must be made for its removal.

In flame treatment, the polymer surface is passed through a flame (1000 to 2800°C <1 sec) generated by the combustion of a hydrocarbon (typically natural gas). This produces an oxidized layer on the surface by a mechanism similar to that of corona discharge but it is more difficult to control; if the heat penetrates too far into the film, it degrades and becomes weak.

The second approach is based on the application of primer treatments in very thin coatings to the substrate. Optimum primer coating weight is usually of the order of a few milligrams per square meter. The primers seem to function by being of such a chemical nature that the oxidized, extruded polyethylene surface adheres strongly to them, and the primer, in turn, adheres strongly to the substrate. If the primer coat is too thick, then loss of adhesion results since the primers have low cohesive strength. Primers of polymers such as styrene-butadiene latexes are often applied to paper to prevent coatings from penetrating too deeply into the substrate.[3]

B. Laminating Processes

Methods that combine two or more webs by bonding them together are called laminating processes. Bonding is usually accomplished by thermal or chemical means with adhesives and curing systems. After the adhesive is adequately dried or cured, the coated web is combined with an uncoated web through the application of heat or pressure in a nip. A wide range of materials may be laminated to each other, and the process is continued if required until the laminate has the desired protective properties.

Thermal laminating is the joining of two webs with an adhesive that is first applied to and cooled and dried on one of the webs. The webs are heated before pressing them together in the nip of two rollers, which provide the force needed to establish the intimate contact required for the bond.[3] The adhesives most commonly used are polyolefins such as EVA, and the webs most commonly laminated this way include plastic films, and aluminum foil joined with heat seal coated film or paper.

Wet bond laminating uses solvent or aqueous-based adhesives, and can only be used when one or more of the webs are permeable to the water or other solvent used, thus allowing it to escape. Wet bonding is not generally successful with plastic films, even when laminating them to paper. Usually aqueous adhesives such as casein, sodium silicate, starch, PVA latex, rubber latex or dextrin are used.

Dry bonding is considerably more versatile in that any two materials can be laminated once an adhesive system has been developed. Either aqueous or solvent-based adhesives are used, and they are dried or cured if necessary by the application of heat, prior to laminating. Adequate drying of the solvent is particularly important where the solvents cannot be absorbed into the film as excess solvent is a major cause of delamination, and may permeate into the package and affect the food. However, the use of organic solvent-based adhesives has been largely phased out because of legislation limiting the release of VOCs (volatile organic compounds) into the atmosphere.

Solventless laminating consists of bonding together two webs by curing in the absence of solvents. It has now become the dominant laminating method in commercial use because of legislation limiting the release of VOCs. A reactive chemical system (either a single- or two-component system) is used to cure the adhesive. Because the adhesive layer is formed by curing (polymerization), it releases neither solvents nor water, although small amounts of CO_2 may be emitted. Single-component urethanes are the most widely used; polyester isocyanates are also used.

Extrusion laminating is a specialized use of extrusion coating, where a hot extruded film is trapped between two other webs and cooled. This process is used mainly for producing a triple laminate of such materials as paper, aluminum foil, RCF and PET with LDPE, where the latter material is extruded and acts as the bonding agent between the two substrates. As in the case of extrusion coating, this process is applicable to any thermoplastic material, but the technology has been highly developed mainly for polyethylene and associated copolymers, including ionomers.

IV. METALLIZATION

Metallization can be considered as a specialized type of coating. Films were first metallized in the 1930s and were used as decorative tinsel. Since the 1960s, metallized films have found a diverse range of applications in food packaging, initially in a purely decorative role, but, since 1975, as important barrier materials as well as microwave-heatable susceptors (see Chapter 13). These films have the appearance of thin metal foil, but are tougher than foil and usually have a more highly reflective surface. The end result of metallizing plastic films is a decorative, functional, durable and less expensive material than that made from solid metal. Aluminum usually gives the most highly reflective surface, and is the most common material used for metallization. Other materials that have been used for different effects include gold, silver, tin, copper, nickel and zinc.

Vacuum metallization, a physical vapor deposition (PVD) process, produces coatings by thermal evaporation of metals. The metal must be heated to a high enough temperature to effect vaporization, and for aluminum this involves temperatures in the range of 1500 to 1800°C.

Three basic heat sources are used in vacuum metallization.[2] In resistance heating, aluminum wire is fed onto a block of metal (usually tantalum), which is heated by holding it in a carbon crucible across which a high current is arced. Induction heating was used in the past, but has now been abandoned in favor of resistance heating. In electron beam heating, the heat generated is a result of an energy conversion process, care being taken to ensure that no secondary electrons impinge on the polymeric web.

Vaporization of the aluminum causes minute particles to be ejected from the surface in all directions. Because the distance traveled by evaporated aluminum molecules before encountering air molecules and being deflected is very short, a high vacuum (3 kPa) must be maintained in the metallization chamber. This high vacuum is also required to prevent oxidation of the metals being vaporized. If the vacuum is not sufficiently high, then the coatings are dull instead of being highly reflective, due to oxide contamination of the metal.

Metallization can be either a batch or continuous process. The batch process involves the unwinding and then rewinding of the film in a vacuum chamber as shown in Figure 5.4. The continuous process consists of passing the film through vacuum-sealed slits with unwinding and rewinding being performed outside the vacuum chamber.

Certain films have to be degassed prior to vacuum metallization because they contain moisture or other volatile constituents such as plasticizers, residual solvents or monomers. These may cause difficulties during the metallization process, because their continued out-gassing interferes with adhesion of the aluminum, giving dull and incompletely anchored coatings. Films which require degassing include RCFs (which contain water and glycol) and PAs. Given that degassing may also remove other desirable constituents of the film, it is often preferable to seal the surface of the film with a lacquer. If colored metallic effects are required, then the film can be self-colored or a colored lacquer applied after metallizing.

In order to increase the barrier properties of metallized film, more metal must be deposited on the film's surface; this can lead to adhesion problems and even flaking if the metallized layer becomes too thick. With thin films, the thickness of the coating that can be applied may have to be considerably less than the typical thickness of up to one micron, because the heat of condensation of thicker layers would cause melting of the film. These thinner coatings tend to be full of pinholes and may even be transparent to transmitted light to some degree.

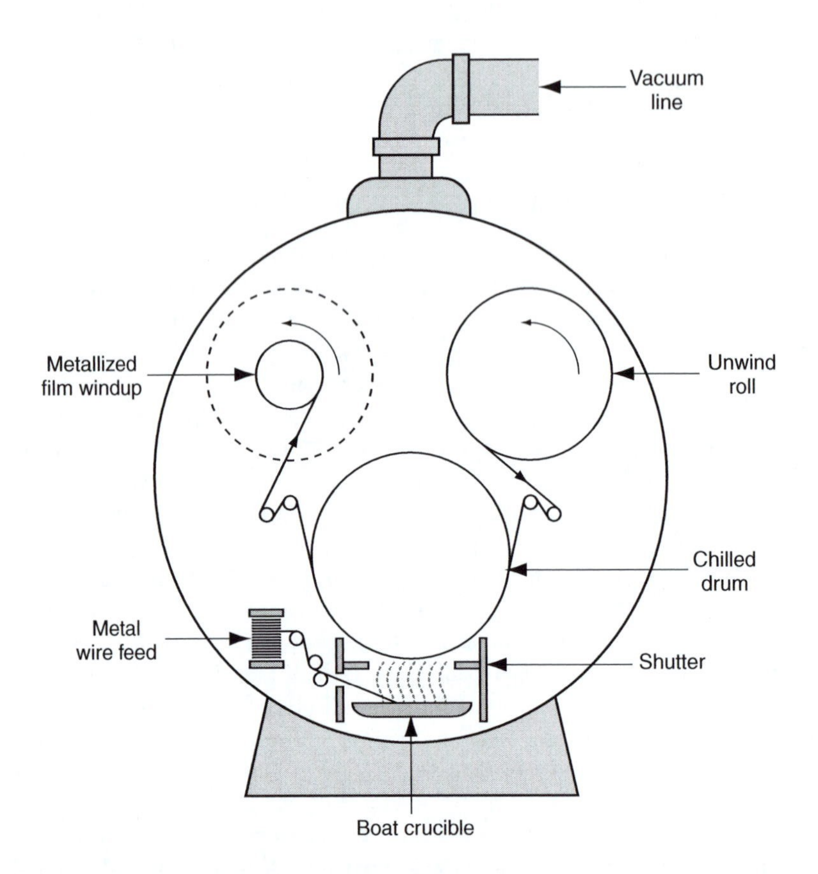

FIGURE 5.4 Schematic cross-section of a batch vacuum metallizer.

TABLE 5.1
Percentage Decrease in Water Vapor and Gas Barrier Properties of Aluminum Metallized Films

Material	Water Vapor (%)	Oxygen Barrier (%)
Polyester	98.5	99.0
PVdC copolymer-coated polyester	95.5	91.1
Cast polypropylene	93.4	98.7
Biaxially oriented polypropylene	75	98.7
Low density polyethylene	95.3	99.2

Thickness of aluminum layer 5×10^{-6} cm.

Many films have been successfully metallized, including PP, unplasticized PVC, PVdC copolymer, RCF, PS, PET and PAs. Paper can also be metallized if it is first given a coating of lacquer to ensure good adhesion. Metallized films are not often used by themselves; usually they are laminated with other materials to improve machinability. Although metallized films match the aesthetics of foil, they do not, and cannot, ever have the same functional barrier properties inherent in foil. Table 5.1 illustrates the magnitude of the improvement in barrier properties that can be expected from metallization of plastic films.

Thin glass-like SiO_x films ($x = 1.5$ to 1.8) have been produced by PVD of SiO_x since the mid-1980s, and by plasma-enhanced chemical vapor deposition (PECVD) of gaseous organosilane and O_2 since the early 1990s, on PET, PP and PA.[7] The thickness of the silicon oxide coating is 150 to 300 nm versus 20 nm for aluminum-metallized coatings.[5] They are sometimes referred to as quartz-like films (QLFs).

The PECVD process uses a mixture of helium, O_2 and the organosilicon compound hexamethyldisiloxane (HMDSO) with a plasma at low pressures. SiO_x films are transparent, retortable, microwavable and provide barriers comparable to those obtained by metallization as shown in Table 5.2. The main drawbacks of these films are the limited flex and crack resistance of the SiO_x layer and the relatively high production costs.[7] A SiO_x coating has been developed for the inside of PET bottles using microwave PECVD.

With PECVD, it is also possible to deposit hydrocarbon (HC) films (sometimes referred to as amorphous carbon) on different substrates using, for example, acetylene in a plasma. The coating is

TABLE 5.2
Barrier Properties of SiO_x and HC Plasma Films

Film/Coating	Oxygen Permeability $P \times 10^{11}$ [mL(STP) cm cm^{-2} sec^{-1} $(cm\ Hg)^{-1}$]	WVTR $\times 10^{-2}$ (g mm m^{-2} day^{-1})
PET	0.15–0.75	0.5–2
PP	7.6–15.2	0.2–0.4
PA	0.015–0.15	0.5–10
PET/SiO_x	0.0009–0.009	0.24–6
PP/SiO_x	0.003–0.06	0.2–2
PA/SiO_x	0.003	1
PET/HC plasma	0.0018–0.0036	
PP/HC plasma	0.15	

Source: From Lange, J. and Wyser, Y., *Packag. Technol. Sci.*, 16, 149–153, 2003. With permission.

applied on the interior of the bottle at a thickness of approximately 200 nm and has a high clarity but a slight amber tint. PET bottles with HC films have similar properties and barrier resistance to SiO_x but with a higher mechanical resistance. An HC coating on the inside of PET bottles reduces the O_2 permeability by a factor of up to 10.[7]

V. ORIENTATION

Orientation of polymer films is a means of improving their strength and durability in order to broaden their scope of application and make them serviceable in thinner gauges. Films may be oriented in either one direction (uniaxial orientation) or, more commonly, in two directions, usually at right angles to each other (biaxial orientation). Virtually all thermoplastics can be oriented to some extent, but amorphous films can be more readily oriented than crystalline films.

Orientation of thermoplastic film involves stretching the material in such a manner so as to line up the molecular chains in a predetermined direction. Once lined up, the ordered arrangement is frozen in the strained condition. Biaxially oriented (BO) films possess superior tensile and impact strengths, improved flexibility, clarity, stiffness and toughness, and increased shrinkability. Gas and water vapor permeability may also be reduced, generally by 10 to 50% depending on the type of polymer, and the degree and temperature of orientation.

Gas and water vapor permeabilities for amorphous polymers (e.g., PS and PET) appear to be nearly identical for both oriented and unoriented films. Crystalline polymers (e.g., PP and PVdC copolymer) show significant reductions in water vapor permeability when oriented. This difference is greatest at low degrees of crystallinity (10 to 15%) and gradually becomes less as the degree of crystallinity increases until, at 40 to 50% crystallinity, no differences are discernible. The gas permeabilities are largely dependent on the amorphous content, which outweighs any effect introduced by orientation.

Orientation generally has a detrimental effect on elongation, ease of tear propagation and the sealability of the film. The heat sealability range is narrowed and the film may vary in properties with age. Oriented film cannot be easily heat sealed because it shrinks and puckers at temperatures below the sealing temperature. A suitable solution is to apply a surface coating of some thermoplastic having a lower melting point. For example, OPP may be coated with a dispersion of PVdC copolymer, or a copolymer of PP with a small quantity of LDPE.

Among the more common commercially oriented films are PET, PA, PVdC copolymers, PP and LDPE, the latter commonly being irradiated before blowing into film. Because radiation cross-links the molecules (see Section VI), the film can be stretched without becoming fluid at the melting point of a nonirradiated film, resulting in greatly improved tensile strength and shrink tension compared to nonirradiated LDPE. HDPE is not oriented because its very rapid crystallization limits the extent to which it can be stretched. When the resin is blended 70:30 with LDPE, the rate of crystallization is slowed. The crystallinity imparts properties similar to those achieved through radiation cross-linking. The largest application of orientation techniques is in the manufacture of OPP, which results in a considerable improvement in its barrier properties.

In the case of crystalline polymers, the action of orientation induces additional crystallization, with the crystalline structure aligned in the direction of stretching. The induced crystallinity is general and does not occur in spherulite form; therefore, oriented films usually have a high degree of clarity, because of the relative absence of spherulites which cause light scattering.

For many applications, shrinkage is not desirable and a greater degree of heat stability is required. Films can be annealed by application of heat to partially relax the forces while maintaining the film in a highly stretched condition. It is then cooled to room temperature and the restraint on the film released. Such a film is referred to as heat set, and will not shrink if heated to below the annealing temperature. The procedure of annealing does result in some reduction in dimension in the stretched direction or directions.

A. Orientation Process

The most common method used to orientate a thermoplastic film is to stretch it after it has been heated to a temperature at which it is soft. This temperature is below the flow temperature at which the molecules would readily glide past one another when the material is stressed, but above the T_g. As a result of this stretching, the direction of the molecules changes towards that in which the material is stressed, and the molecules are extended like springs. The temperature is then dropped below the softening point of the material while the molecules are held in this configuration, so that the molecules are frozen in the strained position.

Films can be oriented using two processes: flat sheet and tubular. In the flat sheet or tentering process, thick (500 to 600 μm) cast film is fed to a system of differential draw rolls which are heated to bring the film to a suitable temperature below its melting point. The film is stretched in the machine direction and then fed to a tenter frame (Figure 5.5) where a series of clips (mounted side-by-side on endless chains) grasp both edges of the film and draw it transversely as it travels forward at an increasing speed. Draw ratios in both directions normally vary between 4:1 and 10:1. After tentering, the film is passed over a cooling roller and reeled up.

In the tubular or bubble process (Figure 5.6), molten polymer is extruded from an annular die and then quenched to form a tube. The tube is flattened by passing through nip rolls and reheated to a uniform temperature. The air pressure in the tube is increased to expand the film transversely, the draw ratio being varied by adjusting the volume of entrapped air. Pinch or collapsing rolls at the end of the bubble are run at a faster speed than rolls at the beginning of the bubble, thus causing drawing of the film in the machine direction. The film is then wound up.

The amount of orientation imparted to a film depends on the stretching temperature, the amount of stretching, the rate of stretching and the quench. Quenching is carried out either by extruding the web onto a chill roll or by passing it through a quench tank prior to orientation. Generally, orientation is increased by decreasing the stretching temperature, increasing the amount of stretch, increasing the rate of stretch and increasing the amount of quench. Films such as PVdC copolymer and PP, which have T_gs below room temperature, show an appreciable crystallization rate even at room temperature, and thus have to be quenched and oriented immediately after extruding.

The potential energy stored in the extended molecules is the *elastic memory*, characteristic of oriented, nonheat set thermoplastics. When such a film is reheated to its orientation temperature, it shrinks as the molecules tend to return to their original size and spatial arrangement. At elevated temperatures, but below the orientation temperature, some shrinkage will occur but to a lesser extent. BOPP is typically oriented 700 to 800%, while other films are oriented 200 to 1000% in either direction.

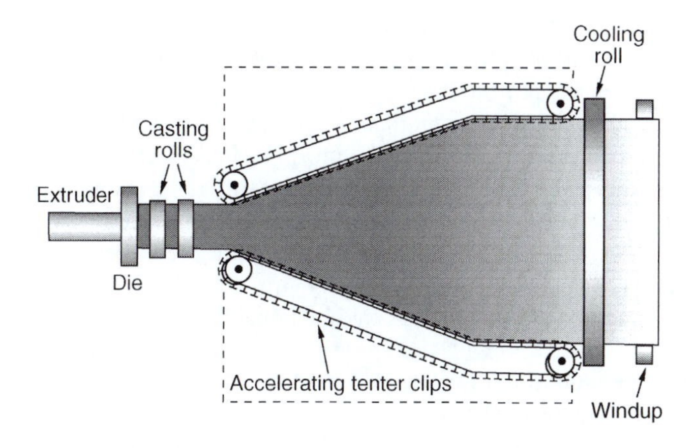

FIGURE 5.5 Flat sheet orientation process using a tenter.

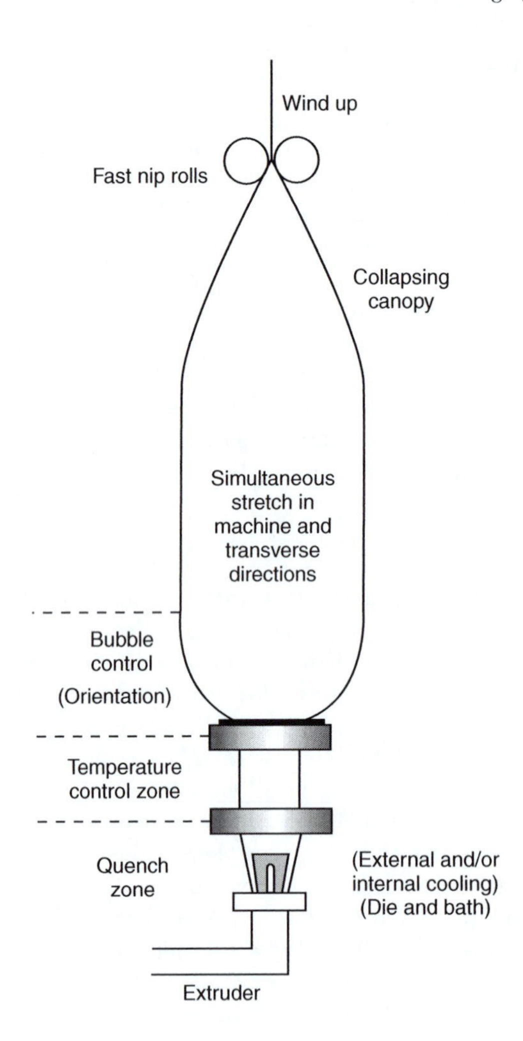

FIGURE 5.6 Orientation by the tubular or bubble process.

B. Shrink Films

Shrink films are composed of three basic categories: polyolefins, PVC and PVdC copolymer. With the exception of PVdC copolymer film which can be shrunk in hot water, most of the other shrink films require temperatures above 100°C to obtain a suitable degree of shrinkage, necessitating the development of hot air tunnels or heat guns.

Three properties of shrink films are important when selecting a film for a particular application. The first is the range of temperature over which a film will shrink. The lower the shrink temperature, the simpler and less expensive the shrink process. Films with a wide softening range are usually preferred as this makes temperature control of the heating equipment less critical.

The degree of shrinkage is also important, with some applications requiring a higher degree of shrinkage than others. The amount of shrink can vary from 15 to 80% depending on the polymer composition and manufacturing techniques. Of particular interest is the relationship between degree of shrinkage and temperature. Films with a steep shrink/temperature curve (e.g., PP, where a ±5°C variation in tunnel temperature could vary the degree of shrinkage by up to 20%) are more difficult to handle because of the closer temperature control necessary.

Shrink tension, the stress exerted by the film when it is restrained from shrinking at elevated temperatures, is the third important factor. Generally, the lower the temperature at which orientation was carried out, the greater the shrink tension. Tension of 300 to 1000 kPa is desirable in order to provide a tight package after shrinking. With shrink tension above 2000 kPa, care must be taken to prevent crushing and distortion of the package; this can be achieved by limiting shrink temperature and time.

Balanced orientation is especially important for printed films because uniform shrinkage is essential to avoid distortion of the print after shrinkage. Even a balanced biaxially oriented film may not shrink evenly in both directions if the product is of a very irregular shape. In these situations, it may be necessary to choose a print design that is not affected by such distortion.

Since the early 1990s, multilayer coextruded shrink films have become available, enabling films to be designed with greater control over desired characteristics.

C. Stretch Films

Although strictly unrelated, stretch films will be discussed here given that they compete in many end uses with oriented films. Stretch films were first introduced in the early 1970s, and have replaced shrink wrapping for large pallet loads and several other applications. In stretch wrapping, the film is stretched around the article and the loose end "wiped" against the underlying film, where the film-to-film adhesion or cling is sufficient to hold it in place. Other less common ways of attaching the end are heat sealing, adhesives, mechanical fasteners and tying. Orientation of the polymer chains occurs on stretching to give a stiffer film, improving the tightness of the wrap and the stability of the load. The residual tension in the film gives a tight contour wrap.

The main films used in stretch wrapping are LDPE, LLDPE, PVC, EVA copolymer and PP, with the choice depending on such factors as appearance (i.e., requirements for clarity, sparkle, etc.) and the protection required (gas and moisture barrier, or physical protection in preventing pallet loads from disintegrating).

LDPE is classed as a low stretch film with a stretchability of about 30%. Under ideal conditions, LLDPE can stretch up to 400%, although the practical limit is nearer to 200%. About 60% of the initial stretch is retained by this film, and the figure for LDPE is about 70%. The trend is towards lighter gauge films (20 μm is common) with some as thin as 13 μm.

An important property of stretch films is their cling, with blown LLDPE having much less cling than PVC or EVA copolymer. Additives such as glyceryl monooleate and polyisobutylene are added by film manufacturers to improve cling. Elevated humidity can sometimes enhance film cling because some cling additives function by attracting moisture from the atmosphere. Moreover, antioxidants, antistatic agents and antiblock agents (the latter preventing the layers of film on a roll becoming permanently bonded together) are also frequently added.

One problem is that different films retain different stress, since all films start to relax immediately after they have been stretched. Most of the relaxation (about 99%) occurs within 24 hours. The opposite of relaxation is stress retention, which is defined as the capacity of a film to maintain the applied tension during stretch wrapping.

VI. CROSS-LINKING

Cross-linking of crystalline thermoplastic polymers interferes with molecular packing, reduces the level of crystallization and consequently results in a polymer with a lower hardness and yield strength. Three main approaches are used for cross-linking polyethylene[4]:

1. Radiation cross-linking
2. Peroxide cross-linking
3. Vinyl silane cross-linking

TABLE 5.3
Properties of Standard and Irradiated Oriented Polyethylene Film

	Irradiated Polyethylene	Standard Polyethylene
Yield (m^2 kg^{-1})	42.7	42.7
Density	0.916	0.916
Tensile strength (kPa) 22°C	8000–16000	1500–3000
Tensile strength (kPa) 93°C	1500–3000	100–200
Elongation (%) at 22°C	100–200	50–600
Heat sealing range (°C)	150–300	110–150
Percent shrink at 98°C	20–80	0–60
Orientation release stress (kPa) at 96°C	100–500	0–10

Only the first will be considered further here. The effect of irradiation on polymeric materials has been widely investigated. Radiation cross-linking is normally carried out using an electron beam accelerator, which produces high energy β rays. The irradiation cross-links the C-C bonds such that the film can still be stretched but no longer becomes fluid at its original melting point. It therefore retains a relatively high orientation release stress which promotes tight shrink wraps.

One of the most successful applications has been the cross-linking of LDPE film. Irradiated LDPE film is made as follows: after extrusion and quenching, the thick walled tube is irradiated and then oriented using the bubble process. The cross-linking increases tensile strength, orientation release stress and heat sealing range to make a practical shrink film from LDPE. The film toughness is similar to that of PVdC copolymer film, while its flexibility at low temperatures is superior. A comparison of irradiated and standard oriented polyethylene film is given in Table 5.3.

VII. MICROPERFORATION

With the increasing commercialization of modified atmosphere packaging (MAP; see Chapter 15), a demand has arisen for so-called "breathable" films for high-respiring, fresh horticultural products. The permeabilities of normal polymeric films (even at very thin gauges) are too low to meet the requirements. Downgauging films to increase gas transmission rates is rarely the solution, since it makes machinability problematic. Therefore, many attempts have been made to develop more porous structures so as to increase their permeability properties.

Three generic approaches have been taken to create additional porosity. These are:

1. *Inorganic fillers*. Since the 1980s, commercial ceramic-filled polymer films have been available. The films typically contain about 5% of very fine ceramic powder, and the manufacturers claim that these films emit far-infrared radiation or absorb C_2H_4, but such claims have not been reported by other laboratories.[14] Gas permeabilities can be manipulated by adjusting the filler content, particle size of the filler and degree of stretching. The permeability of ceramic-filled LDPE to O_2, CO_2 and C_2H_4 is higher than plain LDPE, making these films more suitable for packaging produce with a high respiration rate.[8] Unfortunately, clarity decreases, thus limiting the commercial success of such films.
2. *Porous patch*. A porous patch containing a permeable membrane is placed over a die-cut hole in the film. The porosity of the patch can be changed to suit different respiration needs of various fresh produce items. This method is discussed further in Chapter 14 III.D.

3. *Perforations.* Punching large holes with hot needles or on-machine syringes in plastic films used to package fresh produce has been practiced for many years. However, this relatively unsophisticated approach results in holes that are too large to be useful in creating a modified atmosphere inside the package. Microperforation technology creates numerous, precise holes with diameters in the range of 40 to 200 μm but typically 50 to 60 μm. By altering the diameter and number per unit area of the microperforations, gas permeability through a package can be altered to meet well-defined product requirements.

Microperforation — a postextrusion converting process — was patented in England in 1989, and is now commercialized worldwide. It originally used spark technology, but, today, laser technology is used to make consistent, tiny holes in films at commercial speed. Although OPP is the most common polymer film used for microperforation, the technique can be applied to most polymer films using both monolayer and coextruded structures. With microperforation, the inherent permeability of the film is irrelevant and films are selected for other properties such as toughness.

VIII. INJECTION MOLDING

The injection molding process essentially consists of softening thermoplastic material in a heated cylinder, and injecting it under high pressure into a relatively cool mold where solidification takes place. It is a major technique for converting thermoplastics and is widely used for producing tubs and jars (such as those used for packaging yogurt and margarine), as well as various caps, spouts and dispensers. It is also used to manufacture PET preforms.

The injection molding process or cycle occurs in three consecutive steps:

1. Feeding and thermally plasticizing a stock of granules by means of an extrusion-type screw that rotates until the required amount of melt has been conveyed to the front of the barrel.
2. Injecting a metered amount of hot plastic into a cold cavity at high pressure.
3. Cooling and ejecting the molded component.

High production rates are possible, but because injection molds are expensive, short production runs are uneconomic. Most molds are made of hardened high-quality tool steel. Reverse tapers and sharp undercuts cannot be molded by this method because the finished articles would not be able to be removed from the mold. Injection molded products can be recognized by a small surface protrusion (know as the *gate*), which indicates the point of entry of molten plastic into the mold. It is clearly evident on the base of PET bottles.

Resins that are commonly injection molded include LDPE, HDPE, PP and PET, where the latter is a preform for stretch blown beverage bottles.

IX. BLOW MOLDING

Blow molding is a process to produce hollow objects. It was practiced with glass in ancient times, and the basic techniques used by the plastics industry have been derived from those developed by the glass industry. Currently, a wide range of blow molded bottles and containers are produced for use in food packaging.

In blow molding, a molten tube of thermoplastic (known as a *parison*) is surrounded by a cooled mold having the desired shape. A gas (usually air but occasionally N_2) is introduced into the tube, causing the molten mass to expand against the walls of the mold where it solidifies on cooling. The mold is then opened and the bottle or jar ejected.

Generally, the process for manufacturing plastic bottles and jars consists of three stages: melting the resin, forming the parison and blowing the parison to produce the final shape. The blowing step may take from a few seconds to more than a minute for large shapes; the rate limiting step is cooling of the molded shapes.

There are two techniques of plasticizing resin (i.e., making the material flow) and forming the parison:

1. Extrusion, which produces a continuous parison that has to be cut; this is the most common method used.
2. Injection molding, where the parison is formed in one mold and then transferred into another mold for blowing.

A. Extrusion Blow Molding

Extrusion blow molding (EBM) uses many arrangements for making and forming the parison.[6] In the simplest method, a mold is mounted under an annular die and the parison extruded between the open halves of the mold. When the parison reaches the proper length, the extruder is stopped and the mold closes around the parison. A blow pin mounted inside the die head allows air to enter and blow the parison into the final container shape. The shape of the bottle or jar is defined, but the distribution of material (and thus wall thickness) is less well controlled. The cycle restarts after the part has cooled and the mold opened, as shown in Figure 5.7.

Because the extrusion of the parison is a continuous process, numerous systems have been developed to use the full capacity of the extruder. One uses more than one mold, moving the filled mold away to cool, while another is moved into position to receive the next section of extruded tube. The molds can be reciprocating ones, or can be mounted on a rotary table. In several food packaging applications (e.g., fresh milk), the bottles are blow molded and filled on-line in a continuous operation.

EBM is widely used with the following resins: HDPE, PP, PVC and AN copolymers. The common grades of PET cannot be extrusion blown.

A related process is the production of coextruded bottles, where two or more extruders, each handling different plastic materials, produce a multilayer parison with the desired properties. For example, a high barrier, high cost material might be sandwiched between layers of a relatively low cost material to give a bottle with the desired barrier properties at an economical price. Coextruded structures with up to seven layers, of which one or more can be a barrier layer, are common today.

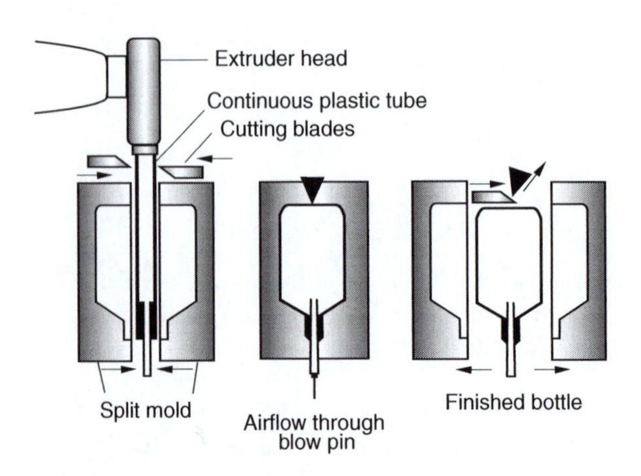

FIGURE 5.7 Extrusion blow molding of plastic bottles.

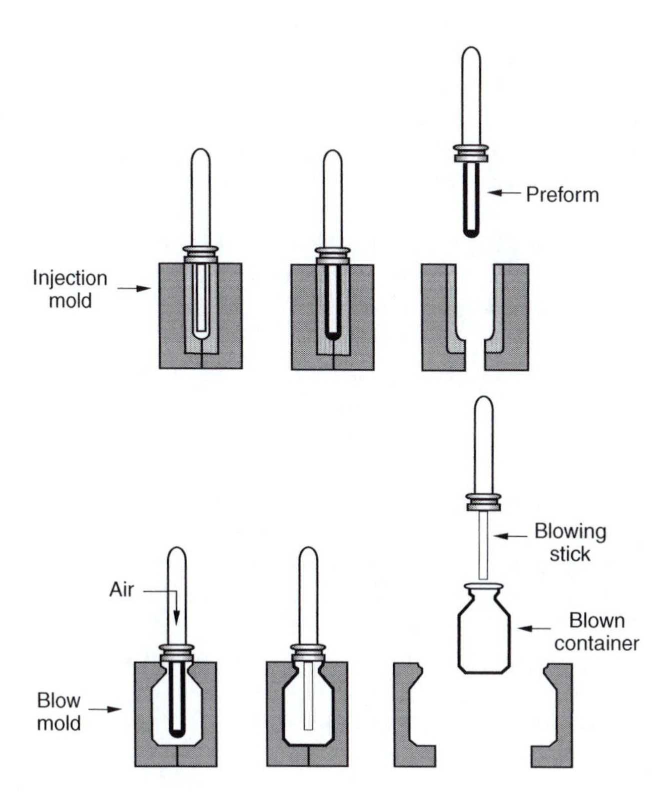

FIGURE 5.8 Injection blow molding of plastic bottles.

B. Injection Blow Molding

Injection blow molding (IBM) is a noncontinuous cyclic process shown in Figure 5.8, and most closely resembles the blowing of glass bottles. The parison is formed in one mold and then, while still molten, is transferred to a second mold where blowing with compressed air forms the final shape. After cooling, the mold is opened and the bottle ejected. Several molds must be available if the injection molding machine is to operate near full capacity.

The major advantage of IBM over EBM is that the process is virtually scrap-free, the finished parts usually requiring no further trimming, reaming or other finishing steps. In addition, the dimensions of the bottle (including the neck finish) show very little variation from bottle to bottle, and with some materials, improved strength and clarity are obtained due to the effect of a limited degree of biaxial orientation (see Section V).

The resins most commonly used for IBM are HDPE, PP, PS, PVC and PET. PET has replaced PVC in many countries, especially in Europe where PVC has a poor image among many consumers.

Coinjection blow molding has been developed using two or three injection units working with one mold to produce a preform, which is later blow-formed using compressed air inside a mold to make a bottle or jar.[10] The various component materials are metered into cavities in such an order that the barrier material flows through the main structural material to create a multilayer structure. This process is used to produce five layer retortable containers from three materials, typically PP as a structural layer and EVOH copolymer as a barrier layer with tie layers in between.

C. Stretch Blow Molding

Stretch blow molding (SBM) is a process where bottles with appreciable orientation in both longitudinal and transverse directions are produced; it is sometimes known as biaxial orientation blow molding.[6] Orientation in the transverse direction only is produced in normal EBM, while appreciable transverse, and some longitudinal orientation, is produced in IBM. True biaxial orientation produces bottles with improved properties including increased tensile and impact strength, improved surface gloss, reduced creep, improved gas and water vapor barrier properties and a reduction in haze in transparent bottles. As a result, lighter weight and lower cost bottles can be produced.

To produce a biaxially molded bottle, a preform or parison (produced either by injection molding or extrusion of a continuous tube which is then cut to the required length and closed at one end) is stretched longitudinally under heat and blown into a bottle shape with consequent transverse orientation.

The process of SBM is particularly important in the field of carbonated beverage packaging using PET. For best results, the resin molecules must be conditioned, stretched and oriented at just above the T_g where the resin can be moved without risk of crystallization. Although PET is the major stretch blown resin, PVC, PP and AN copolymer resins are also stretch blow molded.

Two different stretch blow techniques are available for SBM. In the one stage method, parison injection molding, temperature conditioning and blow molding take place in the same machine. This method is commonly used for widemouth jars and bottles with unusual cross-sectional shapes.[6] In the two stage method (also known as the "reheat and blow" method), the parisons are first injection molded in a completely separate stage and stored at ambient temperature until required. They are then reheated to between 90 and 110°C and blown to their final shape. No stretch occurs at the top and bottom of the bottle. Typical stretch ratios for a 2-L PET bottle are 2.3:1 in the axial (longitudinal) direction and 3.9:1 in the hoop (transverse) direction. Often, production of preforms and bottles are physically separated in different facilities, with the SBM often being conducted on-site by the filler or the packaging supplier in a hole-through-the-wall (HTW) operation.

For hot fill applications, heat resistant PET bottles are required, since a normal PET SBM bottle cannot be filled much above 65°C without causing bottle shrinkage.[6] Manufacture of a heat resistant bottle is called the "heat set process" and involves increasing the crystallinity to about 30% or more as opposed to 25% in a container produced by the conventional blow molding process.[13] Two methods are used. In the first, a single mold method, the preform is stretch blow molded into a hot mold where the mold halves are 125 to 145°C and the mold base is 90 to 95°C, resulting in the preform being heated to between 100 and 110°C, about 10°C hotter than in the conventional blow molding process. Heat setting takes place as the blown container encounters the surface of the hot mold while being constrained by high pressure against the mold.[13] The container must then be cooled down below its T_g before it can be removed from the mold.

In the second (dual mold) method, the preform is stretch blow molded into a first mold, followed by reheating in an oven to relieve stresses and shrivel the shape, followed by reblowing in a second mold to produce the final bottle shape.[6] To avoid unsightly bottle appearance as a result of vacuum formation inside the bottle after hot filling, hot fill bottles are molded with sidewall panels and base designs that move inwards as the product cools and ensure that the vacuum-induced forces do not distort the bottle. Generally, shrinkage of heat set containers after hot filling should be limited to $<10\%$.[13]

Production of multilayer coinjection preforms is the most economical way to achieve enhanced properties in a rigid PET container. Multilayer coinjection is the process by which one or more interior layers of material are totally encapsulated by outer virgin PET layers. It is the only technology able to provide any combination of clarity, gas barrier, gas scavenger and recycled PET (RPET) in a single process.[12] Multilayer parisons can be produced by either extrusion or injection molding techniques.

Coinjection SBM was a major breakthrough, enabling longer shelf lives for beverages to be achieved. It has been used to produce three-layer (PET–MXD6–PET) and five-layer (PET–RPET–PET–MXD6–PET) bottles, mainly in smaller sizes where the lack of a barrier layer would severely limit shelf life. Although EVOH copolymer could be used as a barrier layer, MXD6 is preferred as it has similar melt flow characteristics to PET. Although multilayer PET bottles have been commercialized, the focus has now turned to coating bottles with oxides of silicon or aluminum, or with hydrocarbons (see Section IV).

Aseptic blow molding is becoming increasingly popular for the packaging of beverages. It is usually based on the extrusion process where the bottle is blow molded in a commercially sterile environment, often with the product filler combined with the blow molder. In one system, the bottle is molded and sealed in the blow molding machine and then stored for minutes or days. At the filler, the outside of the bottle is resterilized and the top seal area cut off; after filling aseptically with sterilized product, the bottle is resealed. In an alternative approach, bottle blow molding, filling and sealing are all carried out in a commercially sterile environment.[6]

X. THERMOFORMING

In this relatively old and simple process, a sheet (generally 75 to 250 μm thick) of thermoplastic material is heated to its softening temperature, usually by means of an infrared radiant panel heater. Pigmentation of the sheet aids the heating process. By either pneumatic or mechanical means, the sheet is forced against the mold contours and, after cooling, is removed and trimmed. Typical thermoplastics used for thermoforming include HIPS, PVC, PP and PA.

There are two dominant means of thermoforming sheets for food packaging containers: the melt phase process and the solid phase pressure forming (SPPF) process.[1] The melt phase process is most applicable to monolayer structures, which have relatively high melt strength at thermoforming temperatures, for example, HIPS, PVC and PC. The SPPF process is primarily used to thermoform PP, a crystalline polymer that is difficult to thermoform uniformly in melt phase machines due to the sharp decrease in melt strength (viscosity) at its melting point.[3]

Thermally stable PET containers are commonly use for dual ovenable applications for chilled and frozen foods, as well as retortable, shelf stable applications where the food can be reheated in the package using either a microwave or conventional oven. These containers are known as crystallized PET (CPET) and are stable at temperatures up to 230°C, compared to amorphous PET (APET), which begins to soften at temperatures over 63°C.

The CPET process is based on conventional reheat thermoforming where an extruded PET sheet containing crystallization initiators is reheated to around 170°C when it softens. It is then thermoformed into a hot mold and held long enough for the optimum amount of crystallinity to develop, after which it is transferred into a second mold where it is cooled. CPET containers must be crystalline enough to be heat resistant but not so crystalline as to be too brittle at freezer temperatures.

The SPPF process involves forming at lower temperatures below the crystalline melting point; that is, 5 to 8% lower than melt phase forming depending on the material. For example, PP is melt phase thermoformed at 154 to 157°C and SPPF at 141 to 146°C where it is still virtually a solid with high viscosity, requiring the application of strong forces. In SPPF, the sheet is heated by infrared heaters and stretched into the mold cavity with a heated plug. Cold air at pressures of up to 700 kPa then force the hot sheet against the cooled inner wall of the mold, finishing the forming operation at a high speed. This process (developed mainly for PP) improves the strength of the containers as well as their clarity, and because lower temperatures are used, the containers are free from odor and taint, thus making them highly suitable for food packaging.

The ability to produce extruded PS foam sheet has provided additional packaging markets for thermoforming. The first of these was meat and produce trays, followed later by egg cartons. Other applications include fast-food carry-out cartons, institutional dinnerware and inserts for rigid boxes.

XI. FOAMED (CELLULAR) PLASTICS

Foamed or cellular plastics are defined as plastics whose apparent density is decreased substantially by the presence of numerous cells dispersed throughout their mass. The terms foamed plastic, cellular plastic, expanded plastic and plastic foam are used interchangeably. These materials have been used widely since the 1940s, largely due to their desirable properties, which include a high strength to weight ratio, and good insulating and cushioning properties. A wide range of plastic polymers can be foamed, including LDPE, HDPE, PP, EVA, PS and polyurethane. The market is dominated by the latter two polymers, with PS being used in preference to polyurethane for food contact applications.

Plastic foams are classified as either flexible or rigid, and may have an open or closed cell structure; in the former, the cells are interconnected, whereas in the latter, most cells are closed and separate. Cell formation is initiated by foaming agents. Physical foaming agents are compounds that change their physical state during cell growth and the most important are volatile liquids (typically aliphatic hydrocarbons and CO_2) with boiling points below 110°C at atmospheric pressure. Chemical foaming agents decompose under heat to at least one gaseous decomposition product, commonly N_2.

PS foams can be produced by two processes: injection molding and extrusion. With injection molded foam, machines similar to normal injection molding machines are used, except that steam is injected to heat the beads that contain a foaming agent. Extruded PS foam is produced by free expansion of hot PS, blowing agents and additives through the slit orifice of a high L/D ratio extruder to about 40 times the pre-extrusion volume. The amount and type of blowing agent control the density of the foam produced. The major food packaging uses for extruded PS foam sheet are egg cartons and meat and poultry trays.

XII. HEAT SEALING

The heat sealability of a packaging film is one of the most important properties when considering its use, and the integrity of the resultant seal is of paramount importance to the ultimate integrity of the package. A number of factors are involved in determining the quality of a heat seal. They can be conveniently summarized under three headings:

Machine factors: dwell or clamp time, temperature and pressure
Resin factors: density, MW and additives in the resin
Film factors: gauge, style or form (e.g., whether gusseted or not) and treatment for printing

All of these factors tend to interact in a complex way. For example, the amount of heat available may be limited by the capacity of the heating elements, by the rate of heat transfer of the sealing bar and its coating or by the type of product being packaged. Increasing the dwell time (i.e., the time during which heat is applied) will increase the heat available, but this may prove to be uneconomical since fewer units will be able to be handled per minute.

Heat sealable films are considered to be those films that can be bonded together by the normal application of heat, such as by conductance from a heavy heat-resistant metal bar containing a heating element. Nonheat-sealable films cannot be sealed this way, but they can often be made heat sealable by coating them with heat-sealable coatings. In this way, the two coated surfaces become bonded to each other by application of heat and pressure for the required dwell time.

A. Conductance Sealing

Conductance (also known as *resistance* or *bar*) sealers are the most common type of heat sealers in commercial use and typically consist of two metal jaws (often patterned or embossed to give

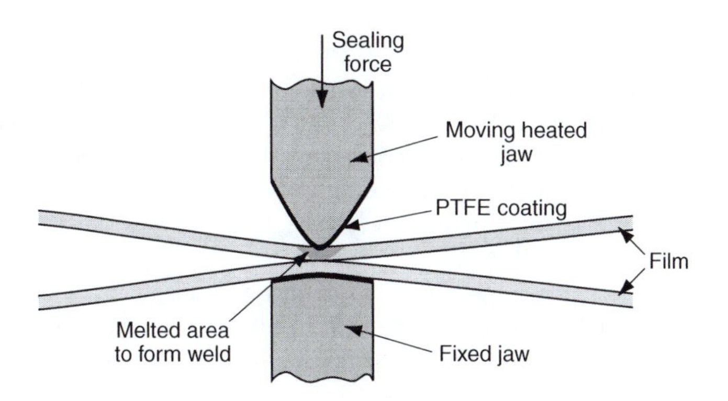

FIGURE 5.9 A simple heat sealer. (*Source*: From Proffit, P. J. G., Packaging of heat preserved foods in plastic containers, In *Processing and Packaging of Heat Preserved Foods*, Rees, J. A. G. and Bettison, J., Eds., Blackie, Glasgow, 1991, chap. 7. With permission.)

the seals extra strength), one of which is electrically heated, the temperature being controlled thermostatically. The second or backing jaw is often covered with a resilient material such as rubber to distribute pressure evenly and aid in smoothing out the film in the sealing area. Frequently, the unheated jaw is water cooled, although, in some situations, it may be heated to the same temperature as the first jaw to enable sealing through sheets of film in exactly the same way each time. A simple heat sealer is shown in Figure 5.9.

Conductance sealers are used for RCFs and other materials such as foil or paper with a heat seal coating. However, they are not suitable for unsupported materials such as LDPE film, which would simply melt and stick to the jaws. Serrated jaws can be used to ensure that the two webs are stretched into intimate contact with high local pressure; they also improve appearance. For all sealing jaws, a nonstick coating is desirable; PTFE is commonly used, either as a cloth-reinforced film or, in the case of serrated jaws, as a powder bound in a heat-resistant vehicle.

Dwell time should be able to be controlled to fractions of a second and be easily adjustable. Likewise, the pressure between the jaws should also be easily adjustable. Both of these factors need to be changed when different materials are heat sealed.

A variation on this type of sealer is the band sealer, where the films travel between two endless bands of metal which are pressed together by heated bars. The heat passes through the bands and seals the films. The bands are then pressed together by chilled bars to withdraw heat from the seal. Band sealers are widely used for sealing pouches and have the advantage of being continuous.

B. Impulse Sealing

In the impulse sealing method, the films (usually unsupported materials) are clamped between two cold metal bars and then fused by the effect of a short, powerful electrical impulse; cooling occurs under pressure. Much lighter jaws than for conductance sealers are used. A high current is sent for a short period through a nichrome resistance wire or ribbon covered with PTFE tape. The current heats the wire to the desired temperature, where the temperature is controlled with a transformer.

Dwell time of the heating impulse must also be controlled, as must the length of the cooling period. Thus, two timers are often found on impulse sealers, with the second timer controlling the cooling of the resistance wire to allow the film to harden under pressure and prevent deformation of the film in the sealed area. Sometimes, one of the jaws is water cooled to prevent excess heating and promote rapid cooling. If very heavy films are to be sealed, then both jaws may contain resistance wires. To prevent the film from sticking to the heated jaws, it is usual to cover them with a slip sheet such as glass cloth impregnated with PTFE. Generally, the seals produced by impulse sealing are of excellent quality.

C. Induction Sealing

Induction sealing is a noncontact method of heat sealing accomplished by exposing an aluminum foil coated with a thermoplastic adhesive substance to a magnetic field. The field induces eddy currents, which generate precise amounts of localized heat in the aluminum foil, melting the adhesive material without any physical contact between the coil and the package. Modern advances in solid-state technology have made induction heating a remarkably simple and cost-effective sealing method.

The basic components of an induction heating system are an AC power supply and an induction coil. The power supply sends alternating current through the coil, generating a magnetic field. When the package is placed in the coil, the magnetic field induces circulating (eddy) currents in the package.

Induction sealing is used to heat seal diaphragms or innerseals to bottles and jars. To be induction sealed, there has to be a thermoplastic heat seal layer laminated or coated onto aluminum foil (there is often a paper or plastic foam layer between the foil and the plastic). When the aluminum foil is heated by the induction coil, the heat is then transferred to the heat seal layer, which quickly reaches the melting point, and becomes an adhesive to bond the diaphragm to the bottle. An induction seal provides tamper evidence and a hermetic seal when used as an inner seal beneath a normal threaded closure. The torque of the cap holds the seal in place during the induction heating process.

Induction sealing is also widely used to heat seal both the longitudinal and transverse seals of laminated paperboard cartons containing aluminum foil. The induction system generates an eddy current in the aluminum layer inside the packaging material structure. The aluminum heats up and transfers heat to the LDPE layers that reach their melting temperature. After the sealing pulse is terminated, there is a cooling time that enables the LDPE to resolidify as a single layer, guaranteeing a strong and tight seal. The jaw pressure must be kept for sufficient time to allow suitable sealing and cooling, in order to ensure a tight closure of the package.

D. Ultrasonic Sealing

In this method of sealing (which is similar to high frequency induction heating), a generator feeds a 20 kHz signal into a transducer, which transforms electrical energy into mechanical vibrations of the same frequency. The mechanical energy is converted into heat at the interface, producing an almost instantaneous weld. Little overall heating is produced, thus enabling oriented films to be sealed without any change in dimensions.

A conical welding head acts as a focusing tool for the vibrations that pass through the film to an anvil. The pressure, along with temperature and dwell time (expressed in terms of the rate at which the film is fed between the welding head and anvil), are adjusted to obtain optimum seal quality.

E. Dielectric Sealing

In the dielectric sealing method, dielectric heating is used in which a high frequency current (typically 50 to 80 MHz) is passed through two or more layers of film. The electrodes (usually made of brass) are the top and bottom jaws, where the latter is the ground of the circuit. When the layers of film are in place between the closed jaws, a high frequency current is passed between the jaws, heating and liquefying the films. The pressure the jaws exert on the films helps to bring about thorough fusion and bonding.

This method of sealing is used principally with PVC and nylon 6/6 films as they are difficult to heat seal by direct means because they tend to degrade at temperatures close to their softening point. It is only applicable to materials that are polar and capable of forming a dipole moment.

F. Hot-Wire Sealing

Hot-wire sealing uses a thin wire or strip of metal with a radiussed edge, heated with a low voltage current. This method has found application for the manufacture of polyethylene bags and pouches in tubular form, with the cutting and sealing operations carried out in the one step. It is only applicable to thermoplastic films that can tolerate high temperatures for a short time and also have a low viscosity in the fused stage. When unsupported films are trim sealed by this method, they tend to form a strong bead in their seal areas due to surface tension and orientation. This method is also used to a limited degree with laminated constructions.

There is a tendency with this method of sealing to obtain what is known as "angel hair" — fine strands of polymer protruding from the seal area. This can be controlled by using proper temperatures and times. Generally, films thicker than 0.05 mm are difficult to seal through, especially if they have gusseted structures.

G. Testing of Heat Seals

The strength of heat seals is often determined by measuring the force required to pull apart the pieces of film that have been sealed together, either in a dynamic load test or a static load test. The latter test tends to be qualitative whereas the dynamic test (usually performed on some type of tensile testing equipment) is quantitative.

The ASTM F 88-00 standard test method for determination of the seal strength of flexible barrier materials describes the procedures to be followed, along with the equipment to be used and the methods to be employed, thus enabling useful comparisons to be made between the seal strength of different materials or different sealing methods. However, many other less formalized methods of testing heat seals are widely used in the food industry. For example, polarized light can be used with transparent materials, the birefringence patterns providing a visual guide to the consistency of the seal and highlighting any gaps or stresses.

REFERENCES

1. Anon. Thermoforming, In *The Wiley Encyclopedia of Packaging Technology*, 2nd ed., Brody, A. L. and Marsh, K. S., Eds., Wiley, New York, pp. 914–921, 1997.
2. Bakish, R., Metallizing, vacuum, In *The Wiley Encyclopedia of Packaging Technology*, 2nd ed., Brody, A. L. and Marsh, K. S., Eds., Wiley, New York, pp. 629–638, 1997.
3. Brown, W. E., *Plastics in Food Packaging: Properties, Design and Fabrication*, Marcel Dekker, New York, 1992.
4. Brydson, J. A., *Plastics Materials*, 7th ed., Butterworth-Heineman, Oxford, 1999.
5. Hill, R. J., Film, transparent glass on plastic food-packaging materials, In *The Wiley Encyclopedia of Packaging Technology*, 2nd ed., Brody, A. L. and Marsh, K. S., Eds., Wiley, New York, pp. 445–448, 1997.
6. Irwin, C., Blow molding, In *The Wiley Encyclopedia of Packaging Technology*, 2nd ed., Brody, A. L. and Marsh, K. S., Eds., Wiley, New York, pp. 83–93, 1997.
7. Lange, J. and Wyser, Y., Recent innovations in barrier technologies for plastic packaging — a review, *Packag. Technol. Sci.*, 16, 149–153, 2003.
8. Lee, D. S., Haggar, P. E. and Yam, K. L., Application of ceramic-filled polymeric films for packaging fresh produce, *Packag. Technol. Sci.*, 5, 27–30, 1992.
9. Ozdemir, M., Yurteri, C. U. and Sadikoglu, H., Physical polymer surface modification methods and applications in food packaging polymers, *Crit. Rev. Food Sci. Nut.*, 39, 457–477, 1999.
10. Proffit, P. J. G., Packaging of heat preserved foods in plastic containers, In *Processing and Packaging of Heat Preserved Foods*, Rees, J. A. G. and Bettison, J., Eds., Blackie, Glasgow, 1991, chap. 7.
11. Robertson, G. L., The paper beverage carton: past and future, *Food Technol.*, 56(7), 46–51, 2002.

12. Swenson, P., Injection and co-injection preform technologies, In *PET Packaging Technology*, Brooks, D. W. and Giles, G. A., Eds., CRC Press, Boca Raton, FL, 2002, chap. 6.
13. Tekkanat, B., Hot-fill, heat-set, pasteurization and retort technologies, In *PET Packaging Technology*, Brooks, D. W. and Giles, G. A., Eds., CRC Press, Boca Raton, FL, 2002, chap. 10.
14. Yam, K. L. and Lee, D. S., Design of modified atmosphere packaging for fresh produce, In *Active Food Packaging*, Rooney, M. L., Ed., Chapman and Hall, London, 1995, chap. 4.

6 Paper and Paper-Based Packaging Materials

CONTENTS

I. PULP

Pulp is the fibrous raw material for the production of paper, paperboard, corrugated board, and similar manufactured products. It is obtained from plant fiber and is therefore a renewable resource. Paper derives its name from the reedy plant papyrus, which the ancient Egyptians used to produce the world's first writing material by beating and pressing together thin layers of the plant stem. However, complete defibering, which is characteristic of true papermaking, was absent.[13] The first authentic papermaking — the formation of a cohesive sheet from the rebonding of separated fibers — has been attributed to Ts'ai-Lun of China in 105 AD, who used bamboo, mulberry bark and rags.[8]

Since then, many fibers have been used for the manufacture of paper including those from flax, bamboo and other grasses, various leaves, cottonseed hair and the woody fibers of trees. At present, about 97% of the world's paper and board is made from wood pulp, and about 85% of the wood pulp used is from spruces, firs and pines — coniferous trees that predominate in the forests of the north temperate zone. The influence of the raw material can be largely assigned to the length and wall thickness of the fibers rather than to their chemical composition.

There are three main constituents of the wood cell wall:

1. *Cellulose*. This is a long-chain linear polymer consisting of a large number of glucose molecules (weight-average degree of polymerization is 3500 for native wood cellulose *in situ*) and is the most abundant, naturally occurring organic compound. The fiber-forming properties of cellulose depend on the fact that it consists of long, relatively straight chains that tend to lie parallel to one another. Cellulose is moderately resistant to the action of chlorine and dilute sodium hydroxide under mild conditions, but is modified or dissolved under more severe conditions. It is relatively resistant to oxidation (e.g., with bleaching agents), and bleaching operations can thus be used to remove small amounts of impurities such as lignin without appreciable damage to the strength of the pulp.
2. *Hemicelluloses*. These are lower molecular weight (weight-average degree of polymerization 15) mixed-sugar polysaccharides consisting of one or more of the following molecules: xylose, mannose, arabinose, galactose and uronic acids, with the composition differing from species to species. The principal hemicelluloses are xylan in hardwoods and glucomannon in softwoods. Hemicelluloses are usually soluble in dilute alkalis. The quantity rather than the chemical nature of the hemicelluloses appears to determine the paper properties. Hemicelluloses are largely responsible for hydration and development of bonding during beating of chemical pulps.
3. *Lignin*. This is the natural binding constituent of the cells of wood and plant stalks. It is a highly branched, three-dimensional (3D), alkylaromatic, thermoplastic polymer of uncertain size, built largely from substituted phenylpropane or propylbenzene units. Hydroxyl or methoxyl groups are attached to the benzene carbon atoms. It has no fiber-forming properties, and is attacked by chlorine and sodium hydroxide with formation of soluble, dark brown derivatives. It softens at about 160°C.

A. Introduction to Pulping

The cell wall of softwoods, which are preferred for most pulp products, typically contain 40 to 44% cellulose, 25 to 29% hemicelluloses and 25 to 31% lignin by weight.[13] The average composition of soft- and hardwoods is shown in Figure 6.1. Compared with hardwoods, softwoods have fibers that are generally up to 2.5 times longer. As a result, hardwoods produce a finer and smoother, but less strong, sheet.

The purpose of pulping is to separate the fibers without damaging them so that they can then be reformed into a paper sheet in the papermaking process. The intercellular substances

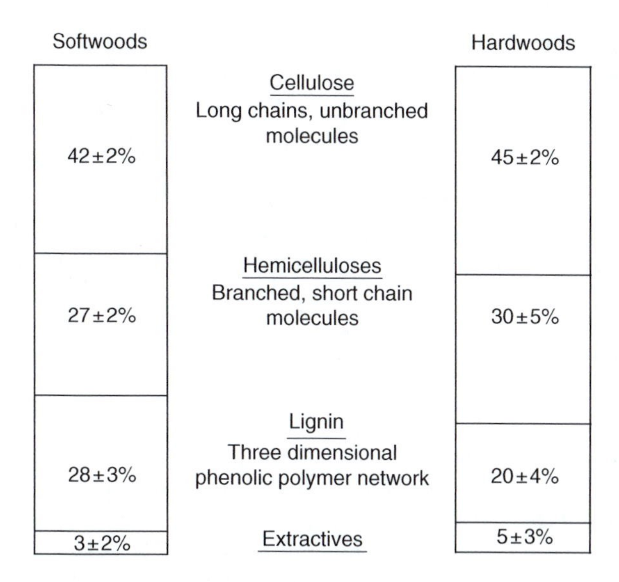

FIGURE 6.1 Average composition of softwoods and hardwoods. (*Source:* Adapted from Smook, G. A., *Handbook for Pulp and Paper Technologists*, 3rd ed., Angus Wilde Publications, Vancouver, BC, 2002. With permission.)

(primarily lignin) must be softened or dissolved to free individual fibers. Commercial pulping methods take advantage of the differences between the properties of cellulose and lignin in order to separate fibers. However, breaking and weakening of the fibers does occur at various stages during the pulping process.

Pulps that retain most of the wood lignin consist of stiff fibers that do not produce strong papers. They deteriorate in color and strength quite rapidly. These properties can be improved by removing most or all of the lignin by cooking the wood with solutions of various chemicals; the pulps thus produced are known as *chemical pulps*. In contrast, *mechanical pulps* are produced by pressing logs onto a grindstone when the heat generated by friction softens the lignin so that the fibers separate with very little damage. Mechanical pulps can also be formed by grinding wood chips between two rotating refiner plates. In addition, there are some processes that are categorized as *semi-chemical* and *chemi-mechanical*.

B. Mechanical Pulps

Groundwood pulp is produced by forcing wood against a rapidly revolving grindstone. Practically all the wood fiber (both cellulose and lignin) is utilized. This contrasts with several chemical processes where the lignin is dissolved to varying degrees. As a result, the yield of chemical pulp is about one half that of the mechanical process. The fibers vary in length and composition because they are effectively torn from the pulpwood.

Groundwood pulp contains a considerable proportion (70 to 80%) of fiber bundles, broken fibers and fines in addition to the individual fibers. The fibers are essentially wood with the original cell-wall lignin intact. Therefore, they are very stiff and bulky, and do not collapse like the chemical pulp fibers.[8]

Most groundwood pulp is used in the manufacture of newsprint and magazine paper because of its low cost and quick ink-absorbing properties (a consequence of the frayed and broken fibers). It is also used as board for folding and molded cartons, tissues and similar products. The paper has high bulk and excellent opacity, but relatively low mechanical strength.

Mechanical pulps can be bleached using oxidative (e.g., hydrogen peroxide and sodium hypochlorite) or reductive (e.g., sodium hydrosulfite) bleaching agents. The bleaching is conducted in a lignin-conserving manner called *brightening*, in which the chromophores are modified and little solubilization of the lignin occurs. Paper and paperboard containing mechanical pulps have poor brightness stability, even after bleaching, particularly in the presence of UV radiation.

In the 1950s the *refiner mechanical pulping* (RMP) process was developed, which produced a stronger pulp and utilized various supplies of wood chips, sawmill residues and sawdust. However, the energy requirements of RMP are higher, and the pulp does not have the opacity of groundwood fibers.[8]

Thermomechanical pulping (TMP) presteams chips to between 110 and 150°C so that they become malleable and do not fracture readily under the impact of the refiner bars. This material is highly flexible and gives good bonding and surface smoothing to the paper. The production of TMP pulps increased dramatically after its introduction in the early 1970s because these pulps could be substituted for conventional groundwood pulps in newsprint blends to give a stronger paper.[8]

Chemithermomechanical pulping (CTMP) increases the strength properties of TMP pulps even further by a mild pretreatment with sodium sulfite. CTMP pulps are obtained by a comparatively mild chemical treatment, followed by pressurized refining. In general, CTMP pulps have a greater long-fiber fraction and lower fines fraction than comparable TMP pulps.[8] CTMP is suitable for the middle layer of multi-ply boards where it adds bulk and rigidity (stiffness) at lower cost than kraft pulp.

C. Chemical Pulps

There are several chemical pulping methods, each of which are based, either directly or indirectly, on the use of sodium hydroxide. The objective is to degrade and dissolve away the lignin from the middle lamella to allow the fibers to separate with little, if any, mechanical action. The nature of the pulping chemicals influences the properties of the residual lignin and the residual carbohydrates. For production of chemical pulps, the bark is removed and the logs passed through a chipper. The chipped wood is charged into a digester with the cooking chemicals, and the digestion carried out under pressure at the required temperature.

1. Alkaline Processes

a. *Soda Process*

The first process for the manufacture of chemical wood pulp was invented by Hugh Burgess and Charles Watt in 1851, and was patented in 1854. The first successful soda mill commenced operation in 1866. The soda process consisted of boiling wood in 4 to 6% (by weight) sodium hydroxide liquor at a high temperature (170°C). A later patent in 1865 covered the incineration of the spent soda liquor to recover most of the alkali used in the process.[13] Less than 2% of the pulp currently produced uses this process, which is very similar to the sulfate process, except that only sodium hydroxide is used. Essentially, all former soda mills have converted to the sulfate process.

b. *Sulfate (Kraft) Process*

In 1879, German chemist Carl F. Dahl developed the sulfate method of pulping wood. The process was essentially a modification of the soda process, but instead of sodium hydroxide, sodium sulfate was the major chemical used as the cooking liquor. The new sulfate process produced a much stronger pulp, which is more commonly known as *kraft* pulp after the German and Swedish word for strength. Although Dahl commenced construction of a kraft mill in his native Germany, it was never completed due to lack of finance.[1] He moved to Sweden and the first kraft mill was built in Munksjo, Sweden in 1885. The location of the first kraft mill in the U.S. is disputed; one author[13] claims it was built in 1911 in Pensacola, Florida, while another[1] states that the Halifax Paper Company in Roanoke Rapids, North Carolina made the first kraft pulp.

Today, the sulfate process is the dominant chemical wood pulping process and uses solutions of sodium hydroxide and sodium sulfide (Na_2S) for cooking the chips. It takes its name from the fact that sodium sulfate (or bisulfate) was used as the make-up chemical in the recovery process, with sodium sulfate being reduced to sodium sulfide in the recovery furnace by reaction with carbon.[9] The sulfate process has the ability to pulp any wood species, particularly pines, which are more resinous than firs and spruces and not easily pulped by the acid sulfite process.

Pulp produced by this process is stronger than that produced from the same wood by the acid sulfite process, and the use of sulfate pulps in liner board has enabled the replacement of wooden cases by corrugated cartons. The sulfate and acid sulfite processes together account for over 90% of the chemical wood pulp currently produced in the world.

2. Sulfite Processes

The invention of the sulfite process is generally credited to the American chemist Benjamin Chew Tilghman, who, in 1857, found that wood could be softened and defibered with a solution of bisulfite and sulfurous acid. Swedish chemist C.D. Ekman treated wood with magnesium bisulfite in 1870, and constructed the first sulfite paper mill in 1872 in Bergvik, Sweden. A German modification of the American sulfite process was developed by the German chemist Mitscherlich, which involved cooking (using indirect steam heating) at lower temperatures and pressures and for longer times than previously. This was the dominant pulping process until 1937 when kraft pulping became the leading chemical pulping process.

Several pulping processes are based on the use of sulfur dioxide as the essential component of the pulping liquor. Sulfur dioxide dissolves in water to form sulfurous acid, and a part of the acid is neutralized by a base in preparation of the pulping liquor. The various sulfite processes differ in the kind of base used and in the amount of base added. These differences govern the resulting acidity or pH of the liquor. These processes depend on the ability of sulfite solutions to render lignin partially soluble. A recent modification involves the use of anthraquinone and methanol in an attempt to make the sulfite process more competitive with the kraft process.

D. Semichemical Pulps

Semichemical pulping combines chemical and mechanical methods in which wood chips are partially softened or digested with conventional chemicals, such as sodium hydroxide, sodium carbonate or sodium sulfate, after which the remainder of the pulping action is supplied mechanically, most often in disc refiners.

The object of this process is to produce as high a yield as possible to obtain the best possible strength and cleanliness. The hemicelluloses, mostly lost in conventional chemical digestion processes, are retained to a greater degree in semichemical pulping and result in an improvement in potential strength development. Although less flexible, semichemical pulps resemble chemical pulps more than mechanical pulps.

The *neutral-sulfite semichemical* (NSSC) process (applied mainly to hardwood chips) uses sodium sulfite and a small amount of sodium hydroxide or sodium carbonate to give a slightly alkaline liquor. The NSSC pulp is obtained in higher yield but with higher lignin content than in the other sulfite processes.

E. Digestion

The digestion process essentially consists of the treatment of wood in chip form in a pressurized vessel under controlled conditions of time, liquor concentration and pressure/temperature.

The main objectives of digestion are:

1. To produce a well-cooked pulp, free from the noncellulosic portions of the wood (i.e., lignin and to a certain extent hemicelluloses),
2. To achieve a maximum yield of raw material (i.e., pulp from wood) commensurate with pulp quality,
3. To ensure a constant supply of pulp of the correct quality.

Currently, most pulping processes are continuous, and to give an indication of the processing conditions encountered, the widely used Kamyr continuous digester is now briefly described. After steaming at low pressure, during which time turpentine and gases are vented to the condenser, the chips are brought to the digester pressure of 1000 kPa. They are picked up in a stream of pulping solution and their temperature is raised to 170°C over 1.5 h. After holding at this temperature for a further 1.5 h, the digestion process is essentially complete.[8]

After digestion, the liquor containing the soluble residue from the cook is washed out of the pulp, which is then screened to remove knots and fiber bundles that have not fully disintegrated. The pulp is then sent to the bleach plant or paper mill.

F. Bleaching

Pulps vary considerably in their color after pulping, depending on the wood species, method of processing and extraneous components. The whiteness of pulp is measured by its ability to reflect monochromatic light in comparison with a known standard (usually magnesium oxide). Brightness is an index of whiteness, measured as the reflectivity of a paper sample using light at 457 nm. Unbleached pulps exhibit a range of brightness values from 15 to 60. Cellulose and hemicellulose are inherently white and do not contribute to color; it is the chromophoric groups on the lignin that are largely responsible for the color of the pulp.

Basically, there are two types of bleaching operations: those that chemically modify the chromophoric groups by oxidation or reduction but remove very little lignin or other substances from the fibers, and those that complete the delignification process and remove some carbohydrate material.[8]

Chemical methods must be used to improve the color and appearance of the pulp; these are bleaching treatments and involve both the oxidation of colored bodies and the removal of residual encrusting materials (the principal one being lignin) remaining from the digestion and washing stages. Because bleaching reduces the strength of the pulp, it is necessary to reach a compromise between the brightness of the finished sheet and its tensile properties.

In 1986, the production process for bleached chemical pulp was identified as a major contributor of polychlorinated dioxins and dibenzofurans to the environment. Because these compounds are powerful toxins and carcinogens (see Chapter 21 for further discussion), much investigative activity was carried out in Europe and North America to identify point sources and suggest corrective measures.[13] Chlorine bleaching was identified as the major source of these compounds. In addition to dioxins and furans, a host of other chlorinated organic compounds (known collectively as *adsorbable organic halides* or AOX) are formed during chlorine bleaching. Strict regulations now limit the production of these chlorinated compounds, resulting in a move away from molecular chlorine bleaching to chlorine dioxide (so-called ECF or elemental chlorine free bleaching) and to oxygen and peroxide (so-called TCF or total chlorine free bleaching). These changes have been introduced to enable pulp and paper mills to meet tough new antipollution laws and regulations, and to conserve wood, chemicals, and energy.

1. Mechanical Pulps

The most effective bleaching agent for most groundwoods is hydrogen peroxide, and since the bleaching is performed in alkaline solutions, sodium peroxide is also used. The reaction requires 3 h at 40°C and is followed by neutralization and destruction of excess peroxide with SO_2.[8] These pulps may be improved in color to only a limited extent since they contain virtually all the lignin from the original wood. Peroxide bleaching allows brightness to be increased by nearly 20%.

2. Chemical Pulps

For chemical pulps, the reagents for full bleaching are mostly oxidative, and because the carbohydrates are also susceptible to oxidation, bleaching must be accomplished under the mildest conditions. Bleaching of chemical pulps is basically stepwise purification of colloidal cellulose, and bleaching can therefore be regarded as a continuation of the cooking process. The bleaching of pulp is achieved through the chemical reactions of bleaching agents with lignin and the coloring matter of the pulp. The bleaching is performed in a number of stages utilizing one or more of the following: chlorine dioxide, oxygen, ozone, and peroxide. Between these stages, the pulp is treated with an alkali to dissolve degradation products. Full details of these processes are given elsewhere.[6,14]

II. PAPER

Stock preparation is the interface between the pulp mill and the papermaking process in which pulp is treated mechanically and, in some instances, chemically by the use of additives, and is thus made ready for forming into a sheet or board on the paper machine. During the stock preparation steps, the pulps are most conveniently handled as aqueous slurries. However, in the papermaking process utilizing purchased pulps and waste paper, which are received as dry sheets, the first step is the separation of all the fibers from one another, and their dispersion in water with a minimum of mechanical work to avoid altering the fiber properties. This process is known as slushing or repulping, and is carried out in a machine such as the hydrapulper (see Figure 6.2), thus named because of the hydraulic forces that are developed. When the pulping and papermaking operations are adjacent to one another, pulps are usually delivered to the paper mill in slush form directly from the pulping operation.

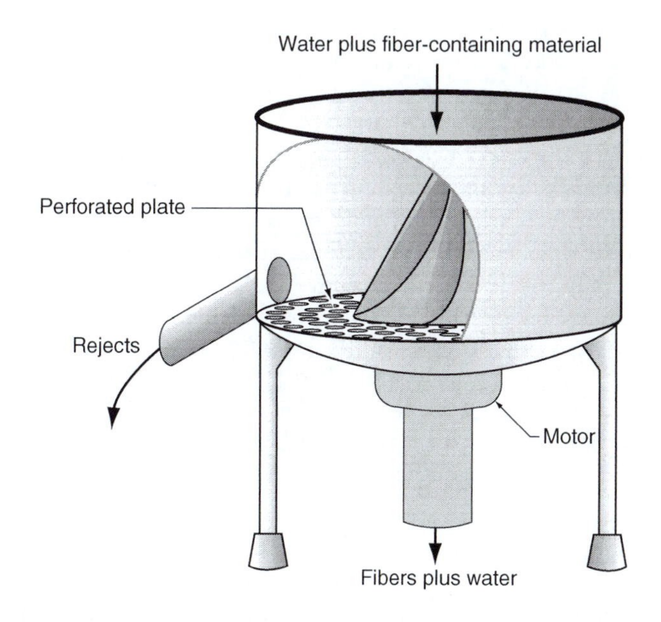

FIGURE 6.2 A batch hydrapulper.

A. Beating and Refining

Stock — as fibrous material is commonly called — is prepared through two main processes commonly referred to as *beating* and *refining*. Both operations are fundamentally the same; in many situations, the two terms are used synonymously. Beating and refining are used to improve the strength and other physical properties of the finished sheet, and to influence the behavior of the system during the sheet-forming and drying steps.[10]

The object of beating is to increase the surface area of the fibers by assisting them to imbibe water. As a result, additional bonding opportunities are provided for between cellulose molecules of neighboring fibers. The beating also makes the fibers more flexible, causing them to become relatively mobile and to deform plastically on the paper machine. The mixture of pulp (known as the *furnish*) is passed into the beater and brought to a consistency of 5 to 7%. The fibers are then beaten while suspended in the water so that they adopt many of the properties that will determine the character of the final product.

The quality and characteristics of the finished paper largely depend on the treatment in the beater. Because papermaking fibers are stiff and springy, the resulting paper would be flabby and weak if made into a sheet of paper without beating. There would be little adhesion between the fibers, and they could not be consolidated under the presses of the paper machine. A sheet formed from an unbeaten pulp has a low density, and is rather soft and weak, whereas if the same pulp is beaten, then the resultant paper is much more dense, hard, and strong. If taken to the extreme, beating produces very dense, translucent, glassine-type sheets.[10] Thus, beating can be controlled to produce paper types as widely different as blotting and greaseproof paper.

Since its invention in Holland around 1690, the principle of the batch-operated hollander beater has remained substantially the same. It consists of a cylindrical roll containing knives that revolves over a stationary bedplate, which also contains a set of knives. Circulating stock passes between the roll and the bedplate; the severity of beating is controlled by adjusting the load of one against the other. Circulation is continued until the pulp is considered ready to be made into the desired paper. In many papermills, beaters have been replaced by continuous refiners, including disc refiners (where rotary discs rotate against a working surface) and conical refiners. However, the batch beater is a convenient vessel for adding chemicals and mixing them with the pulp in order to give special properties to the final paper.

In papermaking, chemicals such as strength additives, adhesives, mineral fillers, and sizing agents may be added at the beater stage prior to sheet formation (i.e., internal addition), or to the resulting sheet after complete or partial drying, depending primarily on the desired effects. Strength additives are usually added internally if uniform strength throughout the sheet is desired, but they are applied to the surface if increased surface strength is needed. Fillers can improve brightness, opacity, softness, smoothness and ink receptivity, and are essentially insoluble in water under the conditions of use.[10] The main drawback is that the materials added may be lost through the wire of the paper machine in the large amount of water used. Therefore, if an additive cannot be retained efficiently from dilute pulp slurry, then it is better to apply it to the surface of the sheet.[7]

Sizing is the process of adding materials to the paper in order to render the sheet more resistant to penetration by liquids, particularly water. Rosin is the most widely used sizing agent, but starches, glues, caseins, synthetic resins, and cellulose derivatives are also used.[10] The sizing agents may be added directly to the stock as beater additives to produce internal or engine sizing. Alternatively, the dry sheet may be passed through a size solution to produce a surface size.

B. Papermaking

1. Fourdrinier Machine

The principle of operation of the modern paper machine differs little from that of the first Fourdrinier machine of 1804, named after its financiers Henry and Sealy Fourdrinier, two prosperous London stationers who purchased the patent interests of the Frenchman Didot Frères. He had obtained the patent rights from a clerk (one Louis Robert) at his mill who had built a paper machine in 1799. On behalf of the Fourdrinier brothers, British native John Gamble commissioned the engineering firm of Hall to build a prototype. One of Hall's apprentices, Bryan Donkin (who was also involved in the development of the tinplate can [see Chapter 7]), was assigned to the task. The Fourdrinier brothers spent the whole of their private fortune developing a practical paper machine, eventually becoming bankrupt and dying in poverty. Despite their misfortune, their name has been familiar to generations of papermakers for the development of a machine, the essential principles of which are still in use today.

Paper is made by depositing a very dilute suspension of fibers from a very low consistency aqueous suspension (greater than 99% water) onto a relatively fine woven screen, over 95% of the water being removed by drainage through the wire. The fibers interlace in a generally random manner as they are deposited on the wire and become part of the filter medium.

Although paper is the general term for a wide range of matted or felted webs of vegetable fiber that have been formed on a screen from a water suspension, it is usually subdivided into paper and paperboard. However, there is no rigid line of demarcation between the two, with structures $<300\ \mu m$ being considered paper regardless of the grammage or weight per unit area. ISO standards define paperboard as paper with a basis weight (grammage) generally above 224 g per square meter, but there are exceptions (see Section III).

The modern Fourdrinier paper machine essentially consists of an endless woven wire gauze or forming fabric stretched over rollers. The forming section of a Fourdrinier machine (illustrated in Figure 6.3) is made up of two essential parts: the headbox and the drainage table. The operation of both parts can influence the structure of the resulting paperboard web.[2] The concentration of the fiber suspension delivered to the moving screen is generally 0.4 to 1.2%, and increases as a result of free drainage through the screen. The relative speeds of the stock and wire affect the degree to which the fibers are aligned along the direction of travel. The concentration increases to between 3 and 4% further down the Fourdrinier table where a vacuum is applied in the suction boxes. For the production of multi-ply paperboard, a secondary headbox is often used. Fourdrinier machines are used to produce all grades of paper and paperboard throughout the industry.

2. Cylinder Machine

A second system was developed in 1809 by John Dickinson of England, and is known as the cylinder or vat machine process. A cylinder covered with a wire cloth is rotated partially submerged in a stock suspension. Because of a vacuum applied inside the cylinder, water drains inward through the wire cloth, and the paper web is formed on the outside. The web is picked up by a felt, which is pressed onto the top of the cylinder by a rubber roll. A series of vats provide individual plies of fiber, which are subsequently matted together. Cylinder machines are used to produce heavy multi-ply boards. They produce a sheet that is much stronger in the direction of flow than that produced on Fourdrinier machines.

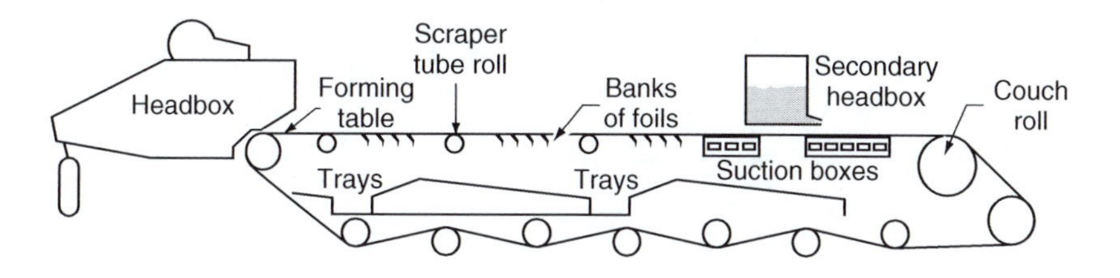

FIGURE 6.3 Fourdrinier machine. (*Source:* Redrawn from Anon., Paperboard, In *The Wiley Encyclopedia of Packaging Technology*, 2nd ed., Brody, A. L. and Marsh, K. S., Eds., Wiley, New York, pp. 717–723, 1997.[3] With permission.)

Cylinder–vat formation differs basically from Fourdrinier formation because the cylinder picks up individual fibers from the stock suspension, in contrast to the Fourdrinier wire on which fibers are deposited from an inflowing stock suspension where water is immediately removed by draining.

The advantage of the cylinder machine for the manufacture of boards is that a number of cylinder units can be arranged so that the fiber mat from each is deposited as a layer, and all the layers can be combined to make a multi-ply paperboard.

3. Twin-Wire Formers

The twin-wire formers method for forming paper and paperboard was developed in the U.K. in the 1950s. The paper web is formed between two converging forming screens by means of a flow box, and the water is drained from the slurry by pressure and later by vacuum. A typical twin-wire forming unit is the Inverform (shown in Figure 6.4), which was designed to provide a new method for the manufacture of single and multi-ply sheets at high speeds. Successive layers of fiber are laid down sequentially on the felt, with water being removed upwardly, overcoming the difficulty experienced in the conventional downward removal of water through several layers of board at high speed.

Twin-wire formers have replaced the Fourdrinier wet-ends on many machines, particularly for lightweight sheets, corrugated media and linerboard grades.

4. Presses and Dryers

After leaving the forming fabric of the papermaking machine, the sheet (which has a moisture content of 75 to 90%, depending on type) passes to the press and dryer sections for further water removal. Rotary presses (which may have solid or perforated rollers, often with internal suction) receive the sheets on continuous felts, which act as conveyers and porous receptors of water. On leaving the press, the moisture content is typically 60 to 70%, again depending on type.

The paper is then passed through a series of steam-heated rollers and dried to a final moisture content of between 4 and 10%. Other types of dryers are used for special products or situations. For example, the Yankee dryer is a steam-heated cylinder, which dries the sheet from one side only, and is used extensively for tissues and to produce *machine glazed* (MG) papers, the latter having a glazed or shiny surface from intimate contact with the polished dryer surface.

C. Converting

Almost all paper is converted by undergoing further treatment after manufacture, such as embossing, impregnating, saturating, laminating and the forming of special shapes and sizes such as

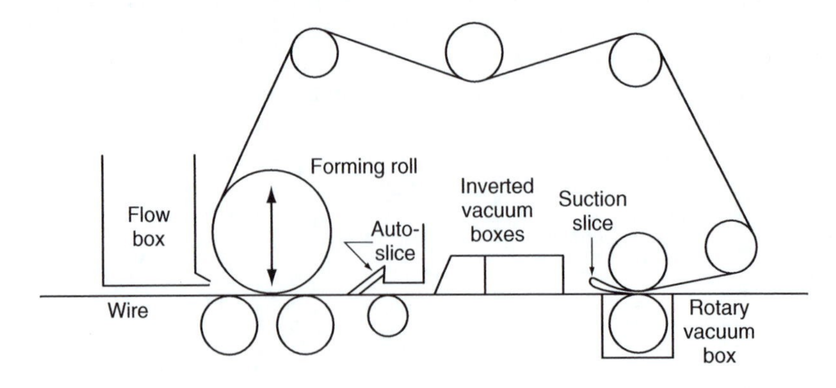

FIGURE 6.4 Inverform twin-wire forming unit. (*Source:* Redrawn from Anon., Paperboard, In *The Wiley Encyclopedia of Packaging Technology*, 2nd ed., Brody, A. L. and Marsh, K. S., Eds., Wiley, New York, pp. 717–723, 1997.[3] With permission.)

bags and boxes. Further surface treatment involving the application of adhesives, functional products and pigments are common, depending on the end use of the paper. The incorporation of strength additives, adhesives, fillers, and sizing agents throughout the whole paper has been briefly described in Sec. IIA. Surface treatment applies similar materials on the web.

Because of the widespread use of paper and paperboard in direct contact with foods, most mills use paper chemicals that have been cleared for use with food by regulatory authorities such as the U.S. Food and Drug Administration (FDA).

1. Calendering

In many applications, the surface of the sheet needs improvement so that any characters imposed on the sheet are legible. This is achieved by calendering, a process that reorients the surface fibers in the base sheet of paper (or the coating applied to the surface) through the use of pressure. This serves to smooth the surface, control surface texture and develop a glossy finish. Such papers are known as *machine finished* (MF).

Supercalendering, as well as smoothing the surface of the sheet by pressure, also alters the optical reflectance of the sheet by friction, giving a higher finish than that obtained on calenders. As the paper passes between steel or fiber rollers under great pressure, moisture is added. Only the bottom roller is powered; the others provide a certain amount of slippage, which irons the sheet. The slipping action of the sheet against the polished steel rollers results in a gloss while smoothing and leveling the surface contours.

2. Sizing

Surface treatments such as sizing and coating are extensively applied to improve the appearance of products. Paper may be coated either on equipment that is an integral part of the paper machine (i.e., on-machine coating), or on separate converting equipment. The most common method for the application of chemicals to the surface of a paper web is with a size press, where dry paper is passed through a flooded nip and a solution or dispersion of the functional chemical contacts both sides of the paper. Excess liquid is then squeezed out in the press and the paper is redried.[7]

Surface-sizing agents prevent excess water penetration and improve the strength of the paper. The sizing agent penetrates far enough into the paper to increase the fiber bonding and the dependent properties such as bursting, tensile, and folding strengths. An additional effect is an improvement in the scuffing resistance of the paper surface.

The most commonly used materials for surface sizing are starches, usually chemically modified (e.g., oxidized starches, cationic starches and hydroxy-ethylated derivatives). Other film-forming materials that can be used for surface sizing include carboxymethyl cellulose (CMC) and polyvinyl chloride (PVC), which provide oil and grease repellent coatings and improve paper strength. Other polymeric sizing agents such as polyurethanes are used as surface-sizing and strength-enhancing agents, but their cost is relatively high compared with other sizing agents.[7] Fluorochemical emulsion sizing agents can be applied to the surface of paper or paperboard to provide good oil and grease repellency. They find application for pet food bag papers, meat, fish and poultry wrap, cookie bags, and candy wrappers.[7]

3. Adhesives

The primary function of the adhesive in pigment coating is to bind the pigment particles together and to the raw stock. The type and proportion of the adhesive controls many of the characteristics of the finished paper, such as surface strength, gloss, brightness, opacity, smoothness, ink receptivity, and firmness of the surface. The strength must be sufficient to prevent the coating from being picked up by tacky printing inks.[10]

Animal glue was the first material used for bonding paper and as an adhesive in paper coating, but it has been replaced by casein and soy protein derivatives. Starches are used in many coated papers, and are the principal adhesives used in circumstances where resistance to moisture is not required. Various rubber latexes and other emulsions are also employed as adhesives. Acrylic-based emulsions are used mostly on paperboard, and their odor-free quality makes them ideally suited for use in food packaging.

4. Barrier Coatings

In many packaging applications, a barrier may be needed against water vapor or gases such as oxygen. A water barrier can be formed by changing the wettability of the paper surface with sizing agents. Coating the paper with a continuous film of a suitable material will confer gas or vapor barrier properties. Paraffin wax applied in a molten form was commonly used to produce a water vapor barrier, but polyethylene applied by extrusion gives a more durable and flexible coating (see Chapter 5, Section III).

5. Pigments

Pigments makes up 70 to 90% of the dry solids in paper coatings, and are generally designed to mask or change the appearance of the base stock, improve opacity, impart a smooth and receptive surface for printing or provide special properties for particular purposes.[10] The pigments that are used in paper coatings are similar to the materials that are used for paper, with kaolin clay being the largest-volume pigment for this application, followed by calcium carbonate. Plastic pigments based on PS are used in combination with mineral pigments to improve the gloss of coated papers.[7]

D. Physical Properties

Most properties of paper depend on direction. Paper has a definite grain caused by the greater orientation of fibers in the direction of travel of the paper machine, and the greater strength orientation that results partly from the greater fiber alignment and partly from the greater tension exerted on the paper in this direction during drying.[5] The grain direction is known as the *machine direction*, while the *cross direction* is the direction of the paper at right angles to the machine direction. The grain of paper must be taken into account in measuring all physical properties.

Papers vary in machine:cross direction strength ratios, with cylinder-machine papers having a higher ratio than Fourdrinier papers, the latter values varying from about 1.5 to 2.5.[10] Usually, there is less variation in paper properties in the machine direction than in the cross direction because variations occur slowly in the machine direction, whereas in the cross direction, they may occur quite suddenly for a variety of process-related reasons. In addition, the cross direction strength normally varies depending on how far the sample has been taken from the edge of the sheet. In general, papers should be used to take the greatest advantage of the grain of the paper.

E. Types of Paper

Paper is divided into two broad categories: (1) fine papers, generally made of bleached pulp, and typically used for writing paper, bond, ledger, book, and cover papers, and (2) coarse papers, generally made of unbleached kraft softwood pulps and used for packaging.

1. Kraft Paper

Kraft paper is typically coarse with exceptional strength, often made on a Fourdrinier machine and then either machine-glazed on a Yankee dryer or machine-finished on a calender. It is sometimes made with no calendering so that when it is converted into bags, the rough surface will prevent them from sliding over one another when stacked on pallets.

2. Bleached Paper

Bleached paper is manufactured from pulps that are relatively white, bright and soft, and receptive to the special chemicals necessary to develop many functional properties. It is generally more expensive and weaker than unbleached paper. Its aesthetic appeal is frequently augmented by clay coating on one or both sides.

3. Greaseproof Paper

Greaseproof paper is a translucent machine-finished paper that has been hydrated to give oil and grease resistance. Prolonged beating or mechanical refining is used to fibrillate and break the cellulose fibers, which absorb so much water that they become superficially gelatinized and sticky. This physical phenomenon is called hydration, and results in consolidation of the web in the paper machine with many of the interstitial spaces filled in.

The satisfactory performance of greaseproof papers depends on the extent to which the pores have been closed. Provided that there are few interconnecting pores between the fibers, the passage of liquids is difficult. However, they are not strictly "greaseproof" since oils and fats will penetrate them after a certain interval of time. Despite this, they are often used for packaging butter and similar fatty foods since they resist the penetration of fat for a reasonable period.

4. Glassine Paper

Glassine paper derives its name from its glassy, smooth surface, high density, and transparency. It is produced by further treating greaseproof paper in a supercalender where is it carefully dampened with water and run through a battery of steam-heated rollers. This results in such intimate interfiber hydrogen bonding that the refractive index of the glassine paper approaches the 1.02 value of amorphous cellulose, indicating that very few pores or other fiber/air interfaces exist for scattering light or allowing liquid penetration.[2] The transparency can vary widely depending on the degree of hydration of the pulp and the basis weight of the paper. The addition of titanium dioxide makes the paper opaque, and it is frequently plasticized to increase its toughness.

5. Vegetable Parchment

Vegetable parchment takes its name from its physical similarity to animal parchment (vellum), which is made from animal skins. The process for producing parchment paper was developed in the 1850s, and involves passing a web of high quality, unsized chemical pulp through a bath of concentrated sulfuric acid. The cellulosic fibers swell and partially dissolve, filling the interstices between the fibers and resulting in extensive hydrogen bonding. Thorough washing in water, followed by drying on conventional papermaking dryers, causes reprecipitation and consolidation of the network, resulting in a paper that is stronger wet than dry (it has excellent wet strength, even in boiling water), free of lint, odor and taste, and resistant to grease and oils.[2] Unless specially coated or of a heavy weight, it is not a good barrier for gases.

Because of its grease resistance and wet strength, it strips away easily from food material without defibering, thus finding use as an interleaver between slices of food such as meat or pastry. Labels and inserts in products with high oil or grease content are frequently made from parchment. It can be treated with mold inhibitors and used to wrap foods such as cheese.

Parchment paper with great shock-absorbing capability can be produced by wet creping, resulting in extensibility combined with natural tensile toughness. Special finishing processes provide qualities ranging from rough to smooth, brittle to soft, and sticky to releasable. It was first used for wrapping fatty foods such as butter, an application still used today.

Glazed imitation parchment (GIP) is made from strong sulfite pulp, which is heavily engine-sized and glazed to give the necessary degree of protection.

6. Waxed Paper

Waxed papers provide a barrier against penetration of liquids and vapors. Many base papers are suitable for waxing, including greaseproof and glassine papers. The major types are wet-waxed, dry-waxed, and wax-laminated. Wax-sized papers, in which the wax is added at the beater during the papermaking process, have the least amount of wax and therefore give the least amount of protection.

Wet-waxed papers have a continuous surface film on one or both sides, which is achieved by shock-chilling the waxed web immediately after application of the wax. This also imparts a high degree of gloss on the coated surface. Dry-waxed papers are produced using heated rollers and do not have a continuous film on the surfaces. Consequently, exposed fibers act as wicks and transport moisture into the paper. Wax-laminated papers are bonded with a continuous film of wax that acts as an adhesive. The primary purpose of the wax is to provide a moisture barrier and a heat sealable laminant. Frequently, special resins or plastic polymers are added to the wax to improve adhesion and low temperature performance, and to prevent cracking as a result of folding and bending of the paper. However, replacement of wax coatings by thermoplastics is a continuing trend.

III. PAPERBOARD PRODUCTS

Paper is generally termed board when its grammage exceeds $224\ g\ m^{-2}$. Various types of paperboard are manufactured, and a partial listing of paperboard grades is given in Table 6.1. Boards can be manufactured in a single Fourdrinier wire, a single cylinder former, on a series of formers of the same type or a combination of types.[13]

Multi-ply boards are produced by the consolidation of one or more web plies into a single sheet of paperboard, which is then subsequently used to manufacture rigid boxes, folding cartons, beverage cartons and similar products. One advantage of multi-ply forming is the ability to utilize inexpensive and bulky low-grade waste materials (mostly old newspapers and other postconsumer waste papers) in the inner plies of the board where low fiber strength and the presence of extraneous materials (e.g., inks, coatings, etc.) have little effect on board properties.[13] However, multi-ply boards containing postconsumer waste papers are not used for food contact purposes.

TABLE 6.1
Paperboard Grades

Linerboard: board having at least two plies, the top layer being of relatively better quality; usually made on a Fourdrinier with 100% virgin pulp furnish

Foodboard: board used for food packaging having a single- or multi-ply construction, usually made from 100% bleached virgin pulp furnish

Folding Boxboard (Cartonboard): multi-ply board used to make folding boxes; top ply (liner) is made from virgin pulp, and the other plies are made from secondary fiber

Chipboard: multi-ply board made from 100% low-grade secondary fiber

Base Board: board that will ultimately be coated or covered

A. Folding Cartons

Folding cartons are containers made from sheets of paperboard (typically with thicknesses between 300 and 1100 μm), which have been cut and scored for bending into desired shapes; they are delivered in a collapsed state for assembly at the packaging point.

The boards used for cartons have a ply structure and many different structures are possible, ranging from recycled fibers from a variety of sources (as in chipboards), through fibers where the outer ply is replaced with higher quality pulps to give white-lined chipboards, to duplex boards without any waste pulp and solid white boards made entirely from bleached chemical pulp. The three most widely used types of paperboard in the U.S. are[11]:

1. Coated solid bleached sulfate (SBS): this is 100% virgin, bleached, with a chemical furnish, and clay-coated for printability.
2. Coated solid unbleached sulfate (SUS): this is 100% virgin, unbleached, with a chemical furnish, and clay-coated for printability.
3. Coated recycled: multiple layers of recycled fibers from a variety of sources, and clay-coated for printability.

A number of steps are involved in converting paperboard into cartons. Where special barrier properties are required, coating and laminating are carried out. Wax lamination provides a moisture barrier, lining with glassine provides grease resistance, and laminating or extrusion coating with plastic materials confers special properties including heat sealing. The use of barrier materials in cartonboard is restricted by the inability of the normal types of carton closure to prevent the ingress of moisture directly.

Coating of the outerboard greatly enhances the external appearance and printing quality, and clay and other minerals are used for such purposes. The coating can be applied either during the board-making operation or subsequently. Foil-lined boards are also used for various types of cartons to (in certain applications) improve reheatability of the contents.

The conventional methods of carton manufacture involve printing of the board, followed by creasing and cutting to permit the subsequent folding to shape, the stripping of any waste material not required in the final construction, and the finishing operation of joining appropriate parts of the board, either by gluing, heat sealing or (occasionally) stitching. During creasing and folding, cartonboard is subjected to complex stresses, and the ability of a board to make a good carton depends on its rigidity, ease of ply delamination and the stretch properties of the printed liner. It is important that the surface layer on the top of the board is of an elastic nature and relatively high strength compared with the properties of the underlying layers since they will be in compression.

B. Beverage Cartons

The first records of paper being used to carry liquids on a commercial scale are found in reports, dated 1908, of a Dr. Winslow of Seattle. He remarked on paper milk containers, which were invented and sold in San Francisco and Los Angeles by a G.W. Maxwell as early as 1906. Paraffin wax was used to moistureproof the paper, but achieving a liquid-tight bond at the joins was more difficult. In 1915, John Van Wormer, owner of a toy factory in Toledo, Ohio was granted a U.S. patent for a "paper bottle" (actually a folded blank box) for milk that he called Pure-Pak. The unique feature was that this box would be delivered flat to be folded, glued, filled and sealed at the dairy. This offered significant savings in delivery and storage in comparison with preformed glass bottles, which were then the predominant package for milk, having being introduced in 1889. The challenge came with the design of the machinery to be sold or leased to the dairy to form, fill and seal the cartons. The detailed development history of the paper beverage carton has been described elsewhere.[12]

The carton normally consists of layers of bleached and (outside North America and Japan) unbleached paperboard coated internally and externally with LDPE, resulting in a carton that is impermeable to liquids and in which the internal and external surfaces may be heat sealed. There may also be a thin layer of aluminum foil, which acts as a gas and light barrier. The structure and functions of the various layers in an aseptic paperboard carton are described in Chapter 12, Section IIIA.

The modern gabletop carton retains the simple basic geometry of earlier years although flat- and plastic-topped versions are available. Added refinements such as plastic screw caps and reclosable spouts are also available. Incorporation of an aluminum foil layer allows a longer shelf life for chilled premium juice products; in some cases, the foil is replaced by a barrier polymer such as EVOH or SiO_x-coated PET. If the foil is replaced, then the carton must be sealed using ultrasonic sealing or heated jaws rather than induction sealing.

Liquid-tight, hermetically sealed brick-shaped cartons are widely used for the aseptic packaging of a range of liquid foods including milk, juices, soups and wines to give packs that will retain the product in a commercially sterile state for 6 to 9 months.

Recently, a blank-fed, retortable square-shaped paperboard carton for soups, ready meals, vegetables, and pet food has been released commercially. It is similar in structure to the aseptic carton (see Chapter 12), but with PP replacing LDPE, products packaged in it have a shelf life under ambient conditions of 18 months.

C. Molded Pulp Containers

The term "molded pulp" is used to describe 3D packaging and food-service articles that are manufactured from an aqueous slurry of cellulosic fibers, formed into discrete products on screened molds.[4] Typically, the raw materials consist of virgin mechanical and chemical wood pulp, and waste-paper pulps with or without the addition of the former materials.

The forming process is similar to the paper-making process, except that a mold fitted with a screen is used in place of the moving wire screen. Two molding processes are used. First, a pressure injection process in which air under pressure and at a temperature of approximately 480°C is used to form a pulp and water mixture in a mold. Second, a suction molding process is used where a partial vacuum is applied to one side of the mold screen after the pulp mixture has been placed in the mold. The pressure-molding process is most suitable for specialized shapes requiring a high degree of protection and a low tooling cost in relation to quantity, and the suction-molding process is most suitable for large quantities of moldings with a low unit cost.

Typical uses of pressure-molded containers include the packaging of bottled spirits where a pulp sleeve molded to the profile of the bottles enables them to be packed head to tail in a carton, thus saving a considerable amount of space. Well-known forms of molded pulp articles made by the suction-molding process include egg cartons, food trays and many other forms of tray-shaped articles for packing fruit and other commodities. Thin thermoplastic films such as PET can be laminated to one surface of a molded pulp tray, enabling it to function as a dual ovenability container (i.e., suitable for use in microwave and convection ovens) for such products as frozen dinners.

REFERENCES

1. Ainsworth, J. H., *Paper the Fifth Wonder*, 2nd ed., Thomas Printing & Publishing, Kaukauna, WI, 1959.
2. Anon., Paper, In *The Wiley Encyclopedia of Packaging Technology*, 2nd ed., Brody, A. L. and Marsh, K. S., Eds., Wiley, New York, pp. 714–717, 1997.
3. Anon., Paperboard, In *The Wiley Encyclopedia of Packaging Technology*, 2nd ed., Brody, A. L. and Marsh, K. S., Eds., Wiley, New York, pp. 717–723, 1997.

4. Anon., Pulp, molded, In *The Wiley Encyclopedia of Packaging Technology*, 2nd ed., Brody, A. L. and Marsh, K. S., Eds., Wiley, New York, pp. 791–794, 1997.
5. Brandon, C. E., Properties of paper, In *Pulp and Paper*, 3rd ed., Vol. 3, Casey, J. P., Ed., Wiley, New York, 1981, chap. 21.
6. Dence, C. W. and Reeve, D. W., Eds., *Pulp Bleaching: Principles and Practice*, Technical Association of the Pulp and Paper Industry, Atlanta, GA, 1996.
7. Dulany, M. A., Batten, G. L., Peck, M. C., and Farley, C. E., Papermaking additives, In *Kirk-Othmer Encyclopedia of Chemical Technology*, 4th ed., Vol. 18, Kroschwitz, J. and Howe-Grant, M., Eds., Wiley, New York, pp. 35–60, 1996.
8. Genco, J. M., Pulp, In *Kirk-Othmer Encyclopedia of Chemical Technology*, 4th ed., Vol. 20, Kroschwitz, J. and Howe-Grant, M., Eds., Wiley, New York, pp. 493–582, 1996.
9. Hanlon, J. F., Kelsey, R. J., and Forcinio, H. E., Paper and paperboard, *Handbook of Package Engineering*, 3rd ed., Technomic Publishing, Lancaster, PA, 1998, chap. 2.
10. Lyne, M. B., Paper, In *Kirk-Othmer Encyclopedia of Chemical Technology*, 4th ed., Vol. 18, Kroschwitz, J. and Howe-Grant, M., Eds., Wiley, New York, pp. 1–34, 1996.
11. Obolewicz, P., Cartons, folding, In *The Wiley Encyclopedia of Packaging Technology*, 2nd ed., Brody, A. L. and Marsh, K. S., Eds., Wiley, New York, pp. 181–187, 1997.
12. Robertson, G. L., The paper beverage carton: past and future, *Food Technol.*, 56(7), 46–51, 2002.
13. Smook, G. A., *Handbook for Pulp and Paper Technologists*, 3rd ed., Angus Wilde Publications, Vancouver, 2002.
14. Young, R. A. and Akhtar, M., Eds., *Environmentally Friendly Technologies for the Pulp and Paper Industry*, Wiley, New York, 1998.

7 Metal Packaging Materials

CONTENTS

I. INTRODUCTION

Four metals are commonly used for the packaging of foods: steel, aluminum, tin and chromium. Tin and steel, and chromium and steel, are used as composite materials in the form of tinplate and electrolytically chromium-coated steel (ECCS), the latter being somewhat unhelpfully referred to as tin-free steel (TFS). Aluminum is used in the form of purified alloys containing small and carefully controlled amounts of magnesium and manganese. Two other metals are used during the soldering or welding of three-piece tinplate and ECCS containers: lead and copper. However, because they are not used for the fabrication of containers in their own right, they are not discussed in detail in this chapter. The safety aspects of these different metals, together with can coatings, are discussed in Chapter 21.

The first commercial manufacture of tinplate occurred in England in 1699 and in France in 1720, where it was used for a variety of purposes including household utensils such as plates. Some time in about the middle of the eighteenth century, the Dutch navy began to use foods preserved by packing them in fat in tinned iron canisters.[13] After cooking and while still hot, the material to be preserved was placed into the canister, covered with hot fat and the lid immediately soldered on. Records show that from 1772 to 1777, the Dutch government supplied their navy (which had been sent out to Suriname [formerly Dutch Guiana in South America] to quell a revolt) with roast beef packed in this way.[4]

Before the end of the eighteenth century, the Dutch had also established a small industry to preserve salmon in a similar manner. Freshly caught salmon were cleaned, cooked in boiling brine, smoked over a wood fire for 2 days and then placed in a tinplated iron box. The spaces were filled up with hot salted butter or olive oil and a lid was soldered onto the box.[13] A famous London firm of snuff merchants supplied 13 tins of Dutch salmon to one of their clients in 1797.[4] Thus, a canning industry of sorts had been established in Holland independently of, and prior to, Appert's work.

The French confectioner Nicolas Appert discovered a method of "conserving all kinds of food substances in containers," and, in 1804, produced preserved meat for the French navy by packing it in glass champagne bottles, sealing them with a cork held in place with wire, and heating in boiling water for several hours. Appert received an *ex gratia* payment of 12,000 francs in 1810 from the Ministry of the Interior's Bureau of Arts and Manufactures on the condition that he publish details of his process; Appert obliged and 200 copies of his book were printed the same year. Appert deliberately avoided tinplate in his early work because of the poor quality of the French product, according to the fourth (1831) and fifth (1858) editions of his book. However, the quality of tinplate in England was good and it was freely available.

After almost two centuries of history, there is still controversy as to who introduced the tin can as a package. The latest account, based on extensive research of early nineteenth century archives,[3] has thrown additional light on those involved in the genesis of the canning industry and has revealed a new name: the French inventor Phillipe de Girard. It appears that he got Durand (a broker in London) to patent the process in 1810, the patent referring to the substitution of glass jars and bottles with tin cases. A successful trial with the Royal Navy was undertaken at Durand's request in 1811, and the patent was acquired by Bryan Donkin in 1812 for which Girard received £1000. Donkin had become interested in the tinning of iron as early as 1808 and, as mentioned in Chapter 6, was involved with John Gamble in developing the Fourdrinier papermaking machine.

Donkin applied to the British Admiralty for a test of his product and the first substantial orders were placed in 1814 with the London firm of Donkin, Hall and Gamble for meat preserved in tinplate canisters (John Hall was the founder of the famous Dartford Iron Works). By the 1820s, canned foods were a recognized article of commerce in Britain and France.

To complete the historical record, William Underwood left London and arrived in New Orleans in 1817. He traveled up to Boston where he started a business preserving food in glass jars by Appert's method. In 1819, Thomas Kennsett, also from England, started a similar business in preserved foods in New York in partnership with his father-in-law Ezra Daggett. The first offering of preserved provisions in tin cans in America is assumed to be the announcement by Daggett and Kensett in the New York *Evening Post* of July 18, 1822,[1] although it was not until 1825 that they took out a patent in which "vessels of tin" were mentioned.

The American Civil War provided the opportunity for canning to become a great industry, and by the end of the war in 1865, canners had increased their output sixfold. For many years, the cans were made slowly and laboriously by hand. Both ends were soldered to the can with a hole of about 25 mm in diameter left in the top. After the can was filled through this hole, a metal disc was soldered into place. The mechanical roll crimping (commonly known as double seaming) of the can ends onto a body with a soldered sideseam was patented in 1896 by Max Ams of New York,[13] making it possible to develop high-speed equipment for the making, filling and closing of these cans. In 1892, the first pineapple cannery was established in Hawaii. The first canned soup was produced in the U.S. in 1897.

Today, materials like tinplate and aluminum have become universally adopted for the manufacture of containers and closures for foods and beverages, largely due to several important qualities of these metals. These include their mechanical strength and resistance to working, low toxicity, superior barrier properties to gases, moisture and light, ability to withstand wide extremes of temperature and ideal surfaces for decoration and lacquering.

II. MANUFACTURE OF TINPLATE

The term tinplate refers to low carbon mild steel sheet, varying in thickness from around 0.15 to 0.5 mm with a coating of tin between 2.8 to 17 gsm (g m^{-2}) (0.4 to 2.5 μm thick) on each surface of the material. The combination of tin and steel produces a material that has good strength, combined with excellent fabrication qualities such as ductility (the capability to undergo extensive deformation without fracture) and drawability (these attributes arise from the grade of steel selected and the processing conditions employed in its manufacture) as well as good solderability, weldability, nontoxicity, lubricity, lacquerability and a corrosion-resistant surface of bright appearance (these latter properties are due to the unique properties of tin). Furthermore, the tin coating adheres sufficiently to the steel base so that it will withstand any degree of deformation that the steel is able to withstand without flaking.

Continuous demand for improved quality and more economic production has led to the development of highly sophisticated manufacturing techniques. These are outlined in this section, together with their effects on the metallurgical and mechanical characteristics. The chemical composition of the base steel has a very significant effect on the subsequent corrosion resistance and mechanical properties of the tinplate.

A. Manufacture of Pig Iron

The iron ores used are generally hematite (Fe_2O_3) with some magnetite (Fe_3O_4). Commercial extraction of iron from its ores is carried out in blast furnaces, where a mixture of iron ores, solid fuel (coke) and fluxes (limestone and dolomite) are heated to around 1800°C. This results in the reduction of most of the iron oxides to metallic iron (m.p. $\approx$ 1200°C). Today, modern blast furnaces are capable of producing molten iron of near constant composition at high rates.

B. Steelmaking

The pig iron from a blast furnace contains 3.5 to 5.0% carbon, 0.3 to 1.0% silicon and up to 2.5% manganese, 1% phosphorous and 0.08% sulfur, depending on the ore. These metalloids must be substantially reduced in the steelmaking stage, and this is commonly accomplished using a basic oxygen furnace. From the furnace, the steel is cast into ingots, which are subsequently rolled into slabs about 250 mm thick or, more commonly today, continuously cast into slab form.

The thick slabs are hot rolled down to about 2 mm, and, during this process, substantial layers (0.01 mm thick) of oxides or scale are formed as a consequence of the steel being heated to elevated temperatures for rolling. Next, the scale is removed by a process called "pickling," which uses a dilute aqueous solution of acid (typically 10 to 15% sulfuric acid) near its boiling point. After pickling, the strip is recoiled and coated with an oil to prevent rust formation and act as a lubricant in subsequent operations.

The final stage of thickness reduction (typically 90% from about 2 mm to 0.2 mm) is carried out by cold rolling. The effects of cold rolling are to increase the strength and hardness of the steel, but this is performed at the expense of ductility.

In the next stage (that of annealing), the steel is heated to temperatures of 600 to 700°C, causing recrystallization of the elongated ferrite grains into new fine grains. This results in a marked increase in ductility and a corresponding decrease in strength.

To reduce the possibility of severe fluting, paneling or creasing, and to impart the desired surface finish, the steel is given a final, very light cold rolling (generally a reduction of 0.5 to 2.0% in thickness) in a "Temper" mill. This imparts "springiness" to the steel but changes the temper or surface hardness only slightly.

At this stage, the uncoated steel sheet is referred to as *black plate*, thus called because some of the early production was covered with black iron oxide. It is the raw material for electrolytic tinplate (ETP) and ECCS. The four main grades of steel product for subsequent use in tinplate production are shown in Table 7.1.

C. Tinplating

The traditional method for tinplating involved dipping or passing the steel through a bath of molten pure tin, but, since the 1930s, the process of depositing tin by electroplating has been used. The introduction of the electroplating process enabled a different thickness of tin to be applied to the two surfaces of the steel. This "differential tinplate" is of economic benefit to the user because it enables the most cost-effective coating to be selected to withstand the different conditions of the interior and exterior of the container.

TABLE 7.1
Types of Steel Produced for Subsequent Use in Tinplate

	Composition % (maximum)							
Type	C	Mn	P	S	Si	Cu	Properties	Application
L	0.13	0.60	0.015	0.05	0.01	0.06	High purity, low in residual elements	Used where high internal corrosion resistance is required
MR	0.13	0.60	0.02	0.05	0.01	0.20	Similar to L but Cu and P maxima are raised. Is most widely used tinplate steel	Vegetable and meat packs where internal corrosion resistance is not too critical
N	0.13	0.60	0.015	0.05	0.01	0.06	Nitrogenized steel with up to 0.02% N to increase strength	Used where high strength and rigidity required (e.g., can ends)
D	0.12	0.60	0.02	0.05	0.02	0.20	Stabilized steel and therefore nonaging. Less C than other tinplate steels	Used for severe drawing operations (e.g., D&I cans)

There are a number of methods of electroplating, but the two principal methods are the acid stannous sulfate process (generally known as the Ferrostan process) and the halogen process.[10] Plating by either method is preceded by cleaning in a pickling and degreasing unit, followed by thorough washing to prepare the surface. After the plating stage, the coating is flow melted, passivated and then lightly oiled.

Flow melting consists of heating the strip to a temperature above the melting point of tin (typically 260 to 270°C), followed by rapid quenching in water. During this treatment, a small quantity of the tin–iron compound $FeSn_2$ is formed; the weight and structure depend on the time and temperature, as well as other factors such as the surface condition of the steel. The structure and weight of this alloy layer plays an important role in several forms of corrosion behavior.

Because the naturally formed oxide layer on the surface of the tin will readily grow in the atmosphere to form a yellow stain (especially when heated, e.g., during stoving after the application of organic enamels which are also known as lacquers or varnishes), the steel strip is given a passivation treatment to render its surface more stable and resistant to the atmosphere. Many types of treatments have been developed, but an electrolytic treatment in a sodium dichromate electrolyte is the most widespread. It results in the formation of a film (usually $<1\ \mu m$ thick), consisting of chromium and chromium oxides and tin oxides, the quantity and form of these basic constituents determining the varying properties of the film.

After passivation, the plate is given a light oiling (oil film weights are generally in the range of between 5 to 10 mg m^{-2}) to help preserve it from attack, and to assist the passage of sheets through container-forming machines without damaging the soft tin layer. It is obviously essential that the oil used is permissible for use in food packaging; cotton seed oil was used for many years but this has now been largely superseded by dioctyl sebacate (DOS) and acetyltributyl citrate (ATBC). These are applied by electrostatic precipitation or direct plate immersion. The level of application is carefully controlled because excessive oil film can cause dewetting of enamels and printing inks, which are often applied during subsequent container manufacture. Finally, the strips are sheared into sheets or coiled, and then packed for shipment to the can manufacturers.

The final structure of the completed coating is shown in Figure 7.1, and consists of a tin/iron alloy layer (principally $FeSn_2$) adjacent to the steel base, free tin, a film of mixed oxides formed by the passivation process and an oil film.

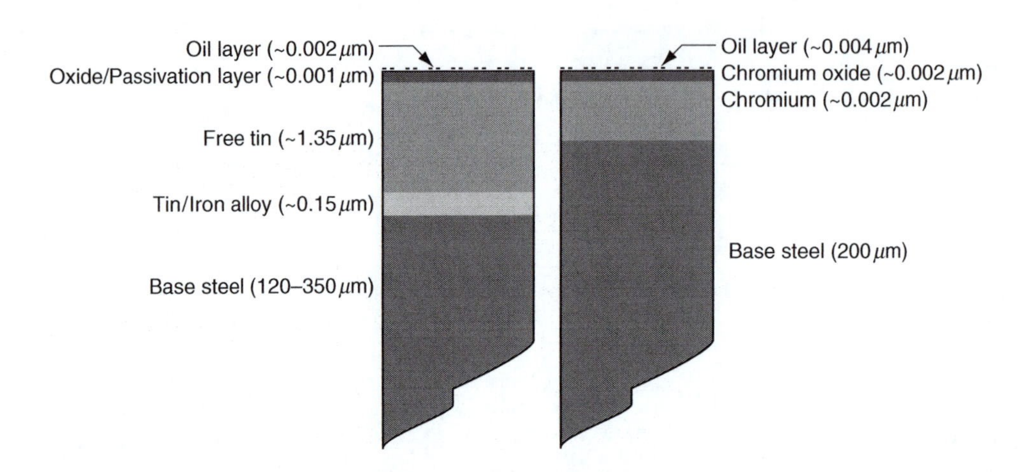

FIGURE 7.1 Schematic structure (not to scale) of tinplate and ECCS showing the main functional layers.

Tinplate sheets are described in terms of a base box, a hangover from earlier times when tinplate was sold in units of 112 sheets, each 356 × 508 mm (14 × 20 in.). Such a package was known as a base box, and the area it contained (20.2325 m^2 or 31,360 $in.^2$) survives today as the unit area for the selling of tinplate. In the original system, a 1-lb base box meant that 1 lb of tin was applied evenly to both sides of the plate, that is, each side received 0.5 lb (equivalent to 11.2 gsm) of tinplate. This was given the designation in the U.S. of No. 100.[15]

The standard grades of ETP available in most countries, together with their nominal tin coating masses, are given in Table 7.2. The designation of ETP with 11.2 gsm of tin on each surface of the sheet is shown as E.11.2/11.2, and this represents a thickness of tin of approximately 1.54 μm on each surface. Tinplate is now commonly graded using the metric unit SITA (Système International Tinplate Area), which is based on 100 m^2. The standard method for determining the mass of tin coating is the iodine titration procedure.

TABLE 7.2
Standard Grades and Nominal Masses of Electrolytic Tinplate

Code		Nominal Coating Mass Per Surface	
Euronorms (145–146)	**ASTM (624–626)**	**(gsm)**	**(lb/base box)**
	No. 10	1.1/1.1	0.05/0.05
E.2.8/2.8	No. 25	2.8/2.8	0.125/0.125
E.5.6/5.6	No. 50	5.6/5.6	0.25/0.25
E.8.4/8.4	No. 75	8.4/8.4	0.375/0.375
E.11.2/11.2	No. 100	11.2/11.2	0.50/0.50
D.2.8/0	—	2.8/0	0.125/0
D.5.6/2.8	No. 50/25	5.6/2.8	0.25/0.125
D.8.4/2.8	No. 75/25	8.4/2.8	0.375/0.125
D.11.2/2.8	No. 100/25	11.2/2.8	0.50/0.125
—	No. 135/25	15.1/2.8	0.675/0.125
D.8.4/5.6	No. 75/50	8.4/5.6	0.375/0.25
D.11.2/5.6	No. 100/50	11.2/5.6	0.50/0.25
D.15.1/5.6	—	15.1/5.6	0.675/0.25

III. MANUFACTURE OF ECCS

The production of electrolytically chromium/chromium oxide coated low carbon steel sheet (to give ECCS its full name) is very similar to electrotinning, the only essential differences being that, in the former case, flow melting and chemical passivation are not involved. The initial development work was carried out in Japan in the 1960s when tin was on occasions in short supply, and the price extremely variable.

The process involves cathodic deposition in a dilute chromium plating electrolyte (e.g., 50 g L^{-1} CrO_3 and 0.5 g L^{-1} H_2SO_4) at a temperature in the range 50 to 70°C. As shown in Figure 7.1, ECCS consists of a duplex coating of metallic chromium and chromium sesquioxide. The ideal range of coating weights for ECCS is between 0.07 and 0.15 gsm chromium metal and between 0.03 and 0.06 gsm of trivalent chromium present as oxide, giving a total coating weight of approximately 0.15 gsm. This is much thinner than the lowest grade of ETP, which has a tin thickness of 5.6 gsm.

The surface of ECCS is more acceptable for protective enamel coatings or printing inks and varnishes than tinplate, and the lack of a low melting point (232°C) tin layer means that higher stoving temperatures and thus shorter stoving times can be used for the enameling of ECCS. Unlike flow brightened tinplate, ECCS is a dull bluish color, which necessitates modification of decoration processes to allow for its poor reflection.

However, ECCS is less resistant to corrosion than tinplate as it has no sacrificial tin layer, and therefore must be enameled on both sides. In addition, ECCS containers cannot be soldered with traditional lead or tin solders, and therefore bonding of ECCS components must be by welding or the use of organic adhesives. If welded, then ECCS must be edge cleaned prior to welding to remove the chromium layer. This is a slow, costly and mechanically inefficient process. ECCS ends are commonly used with tinplate bodies.

IV. MANUFACTURE OF ALUMINUM

Aluminum is the Earth's most abundant metallic constituent, comprising 8.8% of the Earth's crust, with only the nonmetals O_2 and silicon being more abundant. Alumina or aluminum oxide (Al_2O_3) is the only oxide formed by aluminum and is found in nature as the minerals corundum (Al_2O_3), diaspore ($Al_2O_3{\cdot}H_2O$), gibbsite ($Al_2O_3{\cdot}3H_2O$), and most commonly as bauxite, an impure form of gibbsite.

Hans Christian Oersted, a Danish physicist and chemist, first isolated aluminum in 1825 using a chemical process involving potassium amalgam. Between 1827 and 1845, Friedrich Wöhler, a German chemist, improved Oersted's process by using metallic potassium. In 1854, a French chemist Henri-Étienne Sainte-Claire Deville obtained the metal by reducing aluminum chloride with sodium. Aided by the financial backing of Napoleon III, Deville established a large-scale experimental plant and displayed pure aluminum at the Paris Exposition of 1885, where it was considered a precious metal and used mainly for jewelry. In 1886, Charles Martin Hall in the U.S. and Paul Héroult in France independently but almost simultaneously discovered that alumina would dissolve in fused cryolite (Na_3AlF_6), and could then be decomposed electrolytically to a crude molten metal. A low cost technique (the Hall–Héroult process) is currently the only method still used for the commercial production of aluminum, although new methods are under study.

Owing to the chemical stability of its oxides, the energy requirements for smelting are extremely high. This has led to the production of aluminum in areas where cheap electrical power is available. Currently, a typical aluminum process can be described as follows. First, alumina is dissolved in cryolite in carbon-lined steel boxes called pots. Then, a carbon electrode or anode is lowered into the solution and an electric current of 50 to 150 MA is passed through the mixture to the carbon cathode lining of the pot. The current reduces the alumina into aluminum and O_2, the latter combining with the anode's carbon to form CO_2, while the aluminum (denser than cryolite) settles to the bottom of the pot.

TABLE 7.3
Chemical Composition of Some Commonly Used Aluminum Alloys

Alloy Type	Typical Usage	Si	Fe	Cu	Mn	Mg	Cr	Zn	Ti	Others
1050	Foils and flexible tubes	0.25	0.4	0.05	0.05	—	—	0.03	0.03	—
1100		1.0	0.20	0.05	—	—	0.10	—	—	—
3003		0.6	0.7	0.70	1.5	—	—	0.10	—	—
3004	Beverage can ends and D&I can bodies	0.30	0.7	0.25	1.5	1.3	—	0.25	—	0.15
5050		0.40	0.7	0.20	0.1	1.8	0.10	0.25	—	—
5182	Easy-open beverage can ends	0.20	0.35	0.15	0.5	5.0	0.10	0.25	0.10	0.15
8079		0.30	1.3	0.05	—	—	—	0.10	—	—

Most commercial uses of aluminum require special properties that the pure metal cannot provide. Therefore, alloying agents are added to impart strength, improve formability characteristics and influence corrosion characteristics. A wide range of aluminum alloys is commercially available for packaging applications, depending on the container design and fabrication method being used. The chemical composition and typical usage of some of the more commonly used aluminum alloys (the aluminum is at least 99% pure) are shown in Table 7.3. The alloys are identified by four-digit numbers where the value of the first digit indicates the alloy type and principal alloying ingredient. Commercially pure aluminum (Type 1100 and Type 1050) is used for the manufacture of foil and extruded containers since it is the least susceptible to work hardening. Type 5182 alloy contains 4 to 5% magnesium and 0.35% manganese, producing a very rigid material suitable for manufacturing beverage can ends.

The general effect of several alloying elements on the corrosion behavior of aluminum is as follows:

- Copper reduces the corrosion resistance of aluminum more than any other alloying element and leads to a higher rate of general corrosion.
- Manganese slightly increases corrosion resistance.
- Magnesium has a beneficial influence and aluminum–magnesium alloys have good corrosion resistance.
- Zinc has only a small influence on corrosion resistance in most environments, tending to reduce the resistance of alloys to acid media and increase their resistance to alkalis.
- Silicon slightly decreases corrosion resistance, depending on its form and location in the alloy microstructure.
- Chromium increases corrosion resistance in the usual amounts added to alloys.
- Iron reduces corrosion resistance and is probably the most common cause of pitting in aluminum alloys; a high iron content increases the bursting strength but reduces the corrosion resistance.
- Titanium has little influence on corrosion resistance of aluminum alloys.

Compared with tinplate and ECCS, aluminum is a lighter, weaker but more ductile material that cannot be soldered.

V. CONTAINER-MAKING PROCESSES

A. End Manufacture

The can end or lid is of complex design developed for optimum deformation behavior, the latter being dependent on plate thickness, the precise contour of the expansion rings and the countersink

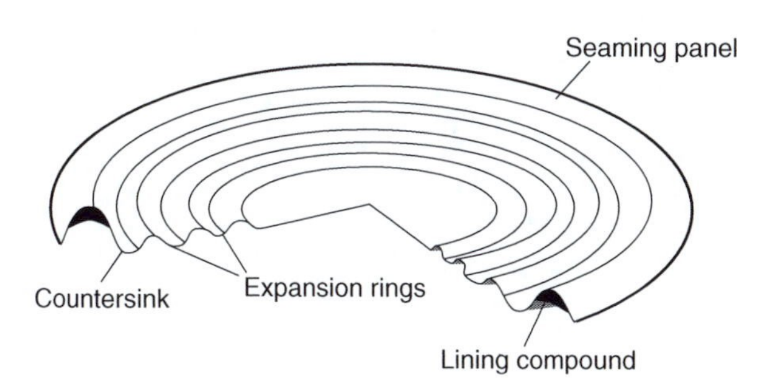

FIGURE 7.2 Profile of a typical can end.

depth. It is important that the ends are able to deform under internal and external pressure without becoming permanently distorted. In effect, they must act like diaphragms, expanding during thermal processing and returning to a concave profile when vacuum develops inside the can on cooling. The cross-section of a typical end design is shown in Figure 7.2.

The ends are stamped on power presses from tinplate sheet (generally of high temper grade), which has been previously enameled. After stamping, the ends fall through the press into the curler to form the outside curl and diameter.

A lining or sealing compound is then applied into the seaming panel; the sealant used is based on natural or synthetic rubber and is dispersed in water or solvent. Its constituents are subject to stringent food regulations. The purpose of the sealant is to assist the formation of a hermetic seal by providing a gasket between adjacent layers of metal.

Several types of easy-opening devices, such as the key-opening scored strip found in solid meat or shallow fish cans, have been available for many years. However, an increased demand for convenience features has seen the development of easy-open ends of two broad types: those that provide a pouring aperture for dispensing liquid products, and those that give a near full aperture opening for removing more solid products. The first easy-open end for canned beverages was developed in 1962 by Ermal Fraze of Dayton, Ohio, who had been caught at a picnic with beer cans but no opener. It was test marketed and emerged in 1965 as the familiar ring-pull tab. However, because it was detachable it resulted in litter and, in 1975, a patent was granted for a can end with an inseparable tear strip; this soon became the industry standard, and, in various formats, is known as the stay-on tab.

Most designs incorporate an easy-open end consisting of a scored portion in the end panel and a levering tab (formed separately), which is riveted onto a bubble-like structure fabricated during pressing. Most (but not all) of the entire aperture circumference is scored, leaving sufficient unscored portion to function as a hinge when the tab is pressed in. Close control of scoring conditions is vital to ensure adequate resistance to bursting without requiring an unduly high tearing load to open. Particular attention must be paid to metal exposure resulting from enamel fracture at the score.

Because of the greater ease of fabrication, integrated rivet ends have usually been made from aluminum. However, this presents problems when such ends are used with steel cans. For example, the corrosion of carbonated soft drinks in such cans is accelerated because of the bimetallic container. These problems have led to the development of steel easy-opening ends, but they are more difficult to fabricate.

B. Three-Piece Can Manufacture

1. Welded Sideseams

In developed countries, the majority of three-piece tinplate cans currently used for food have welded sideseams. Compared with soldered sideseams (see below), welding offers savings in

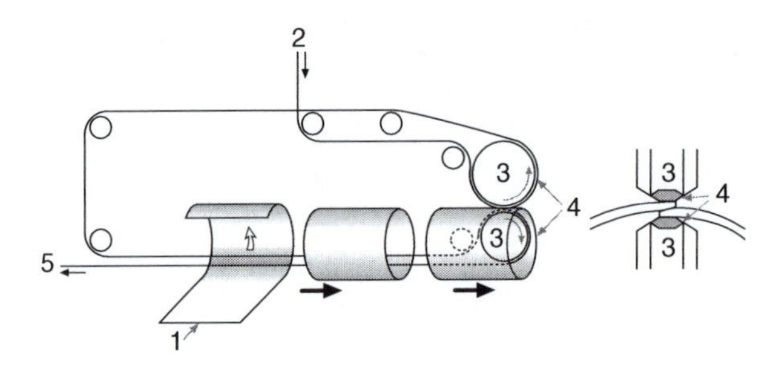

FIGURE 7.3 Stages in the formation of a three-piece welded can. 1, Blank rolled to cylindrical shape; 2, copper wire; 3, welding electrodes; 4, copper wire contacts; 5, used wire to recycling. (*Source*: From Turner, T. A., *Canmaking: The Technology of Metal Protection and Decoration*, Blackie Academic & Professional, London, 1998. With permission.)

material, since the overlap needed to produce a weld uses less metal than an interlocked soldered seam. In addition, the sideseam is stronger, it is easier to seam on the ends and a greater surface area is available for external decorating.

Prior to welding, sheets of steel are enameled and, if necessary, printed, with the area where the weld will be made left bare. The sheets are then slit into individual blanks. Each blank is rolled into a cylinder with the two longitudinal edges overlapping. The two edges are then welded together.

The wire welded operation currently used for the high-speed welding of tinplate and ECCS containers utilizes a sine wave alternating current (and, in the case of tinplate, a continuous copper wire electrode) to produce a weld with an extremely low metal overlap (0.4 to 0.8 mm).[15] The use of copper wire as an intermediate electrode is necessary to remove the small amount of tin picked up from the tinplate during the welding process, which would otherwise reduce welding efficiency.[7] High electrical resistance causes the interface temperature to rise rapidly to at least 900°C, resulting in solid phase bonding at all locations along the seam (see Figure 7.3). The tensile strength of a good weld is equal to that of the base plate.

To prevent traces of iron being picked up by some types of beverages and acidic foods, repair side striping (enameling) of the internal surface of the weld is required.

A system of chemical bonding of sideseams has been developed, mainly for dry or otherwise neutral products such as powders and oils. It utilizes a thermoplastic polyamide adhesive, which is applied to one edge of the preheated body blank before it is rolled into a cylinder, providing complete protection of the raw edges of the blank. A strong bonded lap seam is produced that is able to withstand the high in-can pressures generated by beers and carbonated soft drinks during can warming or pasteurization. This method can only be used with ECCS cans because the melting point of tin is close to the fusion temperature of the plastic. However, since the advent of high-speed welding operations, the use of chemically bonded sideseams has declined.

2. Soldered Sideseams

With the exception of some developing countries, very few food cans are currently produced with soldered sideseams, the concern of public health authorities being that lead from the tin/lead (2:98) solder would migrate into the food. Since the 1970s, most countries insisted that only pure tin solder be used on cans intended for baby foods, which adds significantly to the cost of such cans. The use of tin/lead solder ceased when the U.S. FDA issued a final rule in July 1995 prohibiting its use in food containers, and now tin/silver (96:4) solder is used.

The basic steps in the manufacture of three-piece cans with soldered sideseams are shown in Figure 7.4. The coil is first cut into rectangular sheets, which are then enameled and decorated as required, and cut into strips as wide as the body circumference (including the sideseam) on the first slitting machine. The slit strips are cut into body blanks of the required height, and fed into a

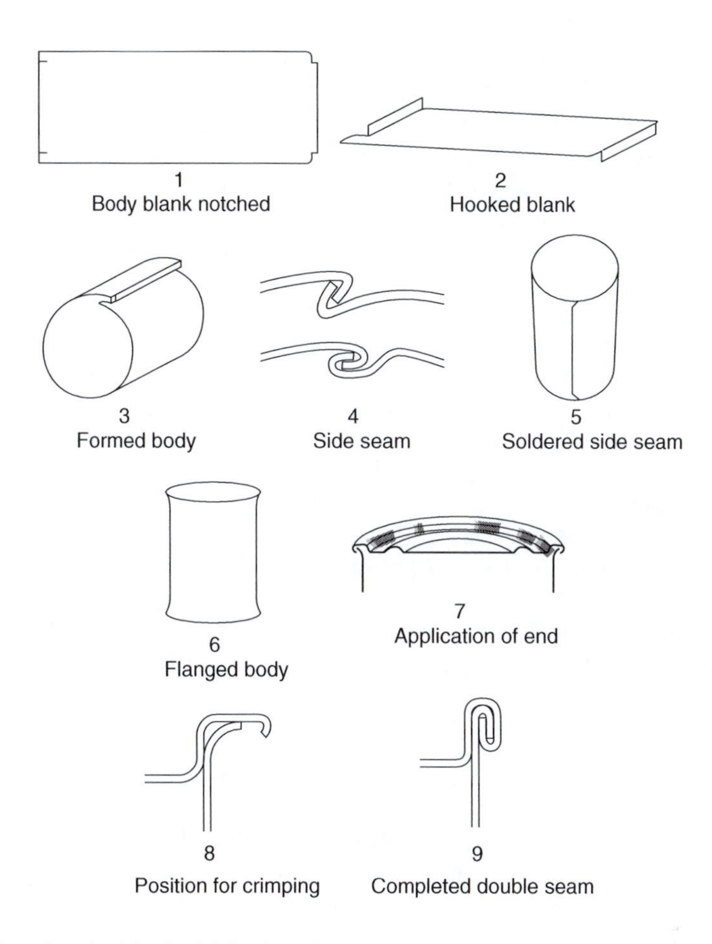

FIGURE 7.4 Steps involved in the fabrication of three-piece cans.

bodymaker where the corners are notched to avoid the extra thicknesses of metal where the sideseam is curled into the end pieces. The two short edges, which will form the sideseam, are bent to form hooks, and the hooked edge is coated with a thin film of flux before being bumped together to create the sideseam.

The seam area is then preheated by gas jets before passing through a bath of molten solder. Tinplate cans are easily soldered because the tin solder alloy readily fuses with the tin on the surface of the steel. The seam is reheated on leaving the soldering station, the excess solder being wiped off the outside by a rotating mop.

Enamel stripes are then sometimes applied to one or both sides of the seam ("side striping") in an attempt to repair damage made to the previously applied enamel by the heat of the solder. This is essential on beverage cans and those likely to contain highly corrosive products.

3. Double Seaming

After the sideseam has been formed, the bodies are transferred to a flanger for the final metal forming operation: necking and flanging for beverage cans, and beading and flanging for food cans. The can rim is flanged outwards to enable ends to be seamed on. The top of beverage cans is necked to reduce the overall diameter across the seamed end to below that of the can body wall, yielding savings in the cost of metal through the use of smaller diameter ends. This also allows more effective packing and stacking methods to be adopted, and prevents damage to the seams from rubbing against each other. Simultaneous creation of the neck and flange using a spin process is used. Double-, triple- and quadruple-necking is now relatively common, the latter reducing the end diameter from 68 to 54 mm for the common beverage can.

For food products where the can may be subjected to external pressure during retorting, or where they remain under high internal vacuum during storage, the cylinder wall may be beaded or ribbed for radial strength. There are many bead designs and arrangements, all of which are attempts to meet certain performance criteria. In essence, circumferential beading produces shorter can segments that are more resistant to paneling (implosion), but such beads reduce the axial load resistance by acting as failure rings.

The end is then mechanically joined to the cylinder by a double seaming operation. This is illustrated in Figure 7.5 and involves mechanically interlocking the two flanges or hooks of the body cylinder and end. It is carried out in two stages. In the first operation, the end curl is gradually rolled inwards radially so that its flange is well tucked up underneath the body hook, the final contour being governed by the shape of the seaming roll. In the second operation, the seam is tightened (closed up) by a shallower seaming roll. The final quality of the double seam is defined by its length, thickness and the extent of the overlap of the end hook with the body hook. Rigid standards are laid down for an acceptable degree of overlap and seam tightness. The main components of a double seam are shown in Figure 7.6. Finally, the cans are tested for leakage using air pressure in large wheel-type testers; leaking cans are automatically rejected.

C. Two-Piece Can Manufacture

A major innovation in canmaking was the introduction of the seamless or two-piece aluminum can in the 1950s and tinplate can in the 1970s. For many years, canmakers have manufactured in a single pressing, shallow drawn two-piece containers, such as the familiar oval fish can. However, the technology to produce deep drawn cans is a more recent innovation, although the basic concept dates back to the Kellver system for producing cartridge cases developed in Switzerland during the Second World War.

There are two main methods used commercially to create two-piece cans: the drawn and ironed (D&I) process, which can be adapted to produce a can for pressure packs (including carbonated beverages) and for food containers, and the drawn and redrawn (DRD) process, which is a

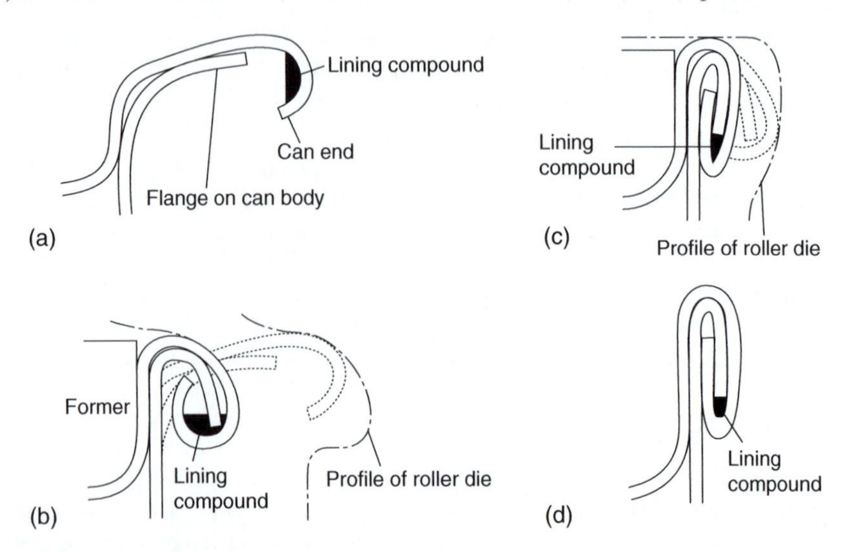

FIGURE 7.5 Double seaming of metal ends on to metal containers: (a) end and body are brought together; (b) first seaming operation; (c) second seaming operation; (d) section through final seam.

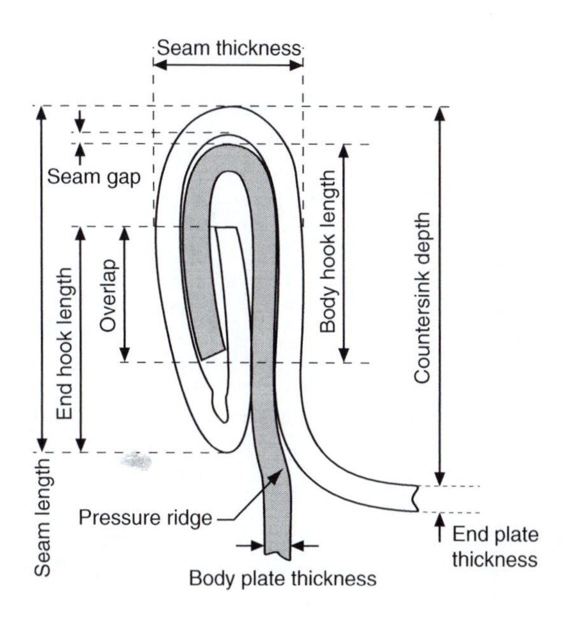

FIGURE 7.6 Main components of a double seam.

multistage operation and produces a can mainly suitable for food products. Both processes depend on the property of the metal to "flow" by rearrangement of the crystal structure under the influence of compound stresses, without rupturing the material.

The first aluminum D&I cans were introduced in the U.S. by a brewery in 1958, but it was not until 1971 that the first tinplate D&I can was launched. By comparison, the first three-piece soldered tinplate can for beer was introduced by the Krueger Brewing Company in the U.S. in 1935.

Two-piece cans have technical, economic and aesthetic advantages in comparison with soldered or welded three-piece cans. In terms of integrity, the two-piece can has no sideseam and only one double seam, which is more easily formed and controlled because of the absence of a sideseam lap juncture. The internal enamel does not have to protect a soldered sideseam or weld cut edge, and there are material savings in solder and (in the case of D&I cans) plate, the latter being up to 35% lighter than a standard three-piece can. Since 1970, through the conversion of three- to two-piece cans and subsequent lightweighting, the weight of a tinplate soft drink can has been reduced by 40% to 35 g, and that of the inherently lighter two-piece aluminum can by 24% to 18 g. Technology exists to continue this trend, especially with tinplate cans. Finally, the absence of a sideseam permits all-round decoration of the outside of the can, increasing the effective printing area and leading to a more aesthetically pleasing appearance.

1. Drawn and Ironed (D&I)

The D&I (also known as *drawn and wall ironed* [DWI]) tinplate or aluminum container is made from a circular disc stamped from a sheet or coil of uncoated plate, formed into a shallow cup with effectively the same side wall and base thicknesses as the starting material, as shown in Figure 7.7. The forming process involves a flat sheet being formed into a cup or cylinder by punch drawing it through a circular die,[12] the wall thickness of the cup being uniform throughout. The plate is covered with a thin film of water-soluble synthetic lubricant prior to forming.

The cup is transferred to an ironing press where it is held on a punch and passed through a series of ironing dies. As a consequence of the ironing process, the wall thickness is reduced (typically from 0.30 to 0.10 mm) and the body height is correspondingly increased. Concurrently, the integral bottom end is domed and profiled to provide added strength, with the end retaining essentially

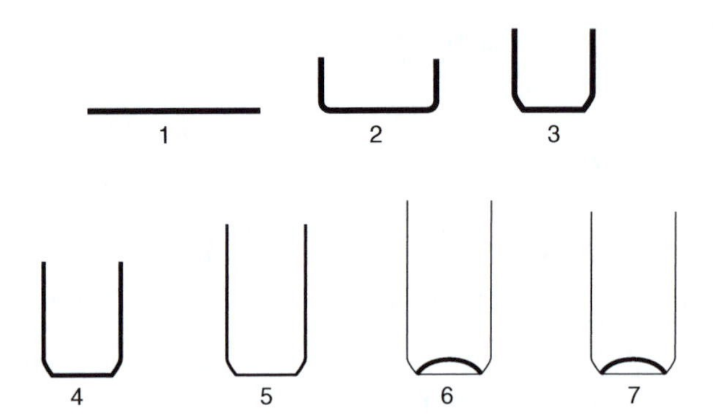

FIGURE 7.7 Sequential stages in the production of two-piece drawn and ironed (D&I) cans. 1, Disc cut from coil; 2, drawn into shallow cup; 3, redrawn into smaller diameter cup; 4, 5 and 6, wall thinning by ironing (diameter remains constant); 7, finished can trimmed to required height.

the original sheet thickness. Because the can wall may not iron to the same height all around the circumference due to slight variation in material properties, cans are "overdrawn" and then trimmed to the correct height.

The trimmed cans are chemically cleaned to remove drawing lubricants and to prepare the surface for receiving exterior and interior coatings. If the cans are to be used for beverages, they are then necked; D&I food cans are commonly beaded for added strength against body collapse under partial vacuum conditions. The cans are then flanged.

Tinplate is the best material for D&I cans as the tin coating is soft and ductile and imparts lubricity to the steel while remaining bonded to it throughout. Some aluminum is made into D&I cans for food packaging, but these are mainly shallow drawn containers. Most D&I aluminum cans are used for beverage packaging (i.e., beer and soft drinks). ECCS plate is not suitable for ironing as the chromium-based coating is too hard.

2. Drawn and Redrawn (DRD)

For many years, canmakers have manufactured shallow drawn containers. However, the novelty of the DRD process is the use of multistage drawing to produce a can with a higher height-to-diameter ratio. This process is essentially identical to the initial stages of the D&I technique, except that the final height and diameter of the container is produced by sequentially drawing cups to a smaller diameter — that is, causing metal to flow from the base to the wall of the container rather than ironing the container wall. As a consequence, the wall and base thickness, as well as the surface area, are identical to the original blank. This contrasts to the D&I can where the wall thickness is much less than the base thickness. A typical DRD process is illustrated in Figure 7.8.

In the D&I process, the internal diameter of the body remains constant throughout the ironing stages, while the internal diameter of the DRD can is progressively reduced as the height is

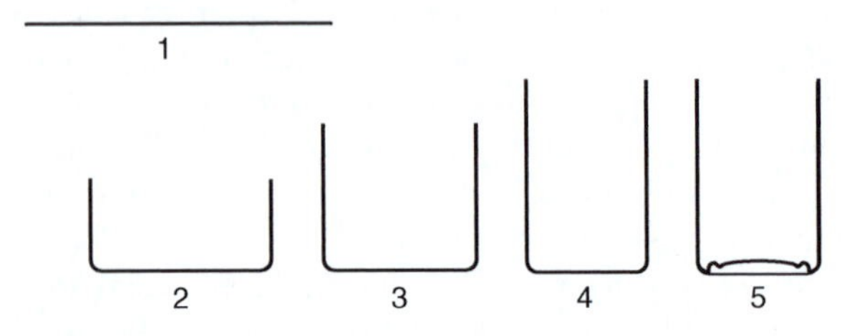

FIGURE 7.8 Sequential stages in the production of drawn and redrawn (DRD) cans: 1, body blank; 2, drawn cup; 3 and 4, diameter decreases as cup is redrawn; 5, finished trimmed can with profiled base.

increased during the various redrawing stages. Therefore, the DRD cans do not offer the same economies as D&I cans because, in the former, the metal cannot be selectively distributed as it can during wall ironing. Because the end is integral and is normally the thickest region, this governs the material gauge and the result is often excessive side wall thickness. Typically, 0.2 mm thick prelacquered tinplate and ECCS are used for the DRD process. DRD cans are currently used in the packaging of food rather than beverages because a greater wall thickness is required to withstand pressure reversals. The body is beaded, and ECCS is used more than tinplate as better enamel adhesion is achieved with the former.

D. Protective and Decorative Coatings

Container coatings provide a number of important basic functions[14]:

- They protect the metal from the contents
- They avoid contamination of the product by metal ions from the container
- They facilitate manufacture
- They provide a basis for decoration and product identification
- They form a barrier to external corrosion or abrasion

1. Protective Coatings

Internally enameled metal containers are used when the product and the plain container would interact to reduce the shelf life or the quality of the product to an unacceptable level. Thus, acidified beetroot, colored berry fruits, beer and soft drinks are packed in enameled containers; that is, containers in which organic coatings have been applied to the inside (and sometimes the outside) surfaces. The primary function of interior can coatings is to prevent interaction between the can and its contents, although some enamels have special properties that allow products such as meat loaf to be easily removed from the cans, while others are used merely to improve the appearance of the pack. Exterior can coatings may be used to provide protection against the environment (e.g., when the cans will be marketed in particularly humid or salt-laden climates), or as decoration to give product identity as well as protection. Generally, some external lacquering of tinplate and ECCS containers is necessary for products stored in hot humid atmospheres to prevent external corrosion, particularly at the sideseam region for three-piece cans.

There are several essential requirements needed in an interior coating: it has to act as an inert barrier, separating the container from its contents and not imparting any flavor to the contents; it must resist physical deformation during fabrication of the container and still provide the required chemical resistance; and the enamel must be flexible, spread evenly, completely cover the substrate and adhere to the metal surface. Adhesion failure may occur during, or as a result of, mechanical deformation during heat processing or undermining by corrosion.

For most containers, the enamel is applied to the metal in the flat before fabrication, typical film masses being in the range of between 3 and 9 gsm (4 to 12 μm thick). However, because of the considerable amount of metal deformation with substantial disruption of the surface that takes place in the D&I operation, such containers must be coated internally after fabrication. Control of the dry film weight is essential if the enamel is to function correctly. A dry film that is too thin may not cover the surface completely, while an overly thick film leads to brittleness and impaired protection, as well as being uneconomical. As the enamel film has imperfections and is damaged during container manufacture, it does not give complete protection to the can. Where it is essential to minimize product-container interactions, for example, for canned beer and soft drinks where metal pick-up can affect flavor and clarity, the cans are given a postfabrication repair lacquering.

Many types of internal enamel are available for food containers including oleoresinous, vinyl, vinyl organosol, acrylic, alkyd, polybutadiene, phenolic and epoxyphenolic (see Table 7.4).

TABLE 7.4
Steel Can Coatings

Canned Products	Coating Type	Comments
	Acrylic	
Applied as liner for cherries, pears, pie fillings, single-serve puddings, some soups and vegetables	For all exteriors and interiors where a clean, white look is desired	Expensive, but takes heat processing well and is suited to water-borne and high-solids coatings. Employed primarily on can exteriors because of flavor problems with some products. Makes an excellent white coat and assures color retention when pigmented
Acceptable for aseptic packs	As a high-solids sideseam stripe for roller and spray application to welded cans Clear wash coat for D&I cans Water-borne spray liner for three-piece beverage cans	
	Alkyd	
(Exterior use)	Quart oil can ends in clear gold or aluminum pigment Aerosol domes in white enamel Flat-sheet roll-coat decorating as a high-solids white basecoat and varnish	Low cost, used mostly as an exterior varnish over inks, because it would present flavor and color problems inside the can. Trend is towards supplanting conventional alkyds with polyesters, which are oil-free alkyds
	Epoxyamine	
Beer and soft drinks Dairy products Fish, ham and sauerkraut	D&I as beer and beverage basecoat DRD cans Can ends Overvarnish on aerosol cans and domes	Costly, but has excellent adhesion, color, flexibility, imparts no off-flavor, scorch resistant and abrasion resistant. Used in interior or as a varnish and sizecoat. Employed now in water-borne coating and, with polyamide, as a sideseam stripe in high-solids form for welded cans
	Epoxyphenolic	
Beer and soft drinks (as a basecoat) The coating for foods including fish, fruit, infant formula, juice, meat, olives, pie fillings, ravioli, soups, spaghetti and meat balls, tomato products and vegetables	All steels and cans	Big in volume for can interiors Used in Europe as a universal coating. Main attributes are retorting resistance, adhesion, flexibility and imparts no off-flavor. Especially suited for aggressive acid products. Has excellent properties as a basecoat under acrylic and vinyl enamels
	Epoxyphenolic with zinc oxide or metallic aluminum powder	
For sulfur-containing foods: fish, meats, soups and vegetables	Three-piece tinplate can ends Two-piece DRD cans	Used primarily to prevent tin sulfide staining; flexibility and clean appearance are the main attributes of these enamels

Continued

TABLE 7.4
Continued

Canned Products	Coating Type	Comments
	Oleoresinous	
Beer and soft drinks (as a basecoat) Fruit drinks A wide variety of fruits and vegetables, including acid fruits	All except DRD cans	A general purpose, gold-colored coating, least expensive of all. When additional protection is required, it can serve as the undercoat for another lacquer. Can be used in both high-solids and water-borne systems
	Phenolic	
Acid fruits, fish, meats, pet food, soups and vegetables	All steel cans where coating can be applied to flat plate	Low cost, exceptional acid resistance and good sulfur resistance. Film thickness restricted by its inflexibility. Tendency to impart off-flavor and odor to some foods
	Polybutadiene	
Citrus plus soups and vegetables (if zinc oxide is added to the coating) Beer and soft drinks	Restricted to three-piece can bodies because of its poor fabricability	Good adhesion, chemical resistance and ability to undergo retorting are its chief characteristics Resistance properties and cost comparable to the oleoresinous class.
	Sulfide stain-resistant oleoresinous — "C" enamel (with zinc oxide)	
Asparagus, beans, beef stew, beets, chicken broth, chocolate syrup, com, peas, potatoes, spinach, soups, tomato products and sulfur-containing foods	Soldered and welded tin plate Can ends	Low cost, flexible, often used as a topcoat over epoxyphenolic Not for use with acid fruits
	Vinyl	
Beer and soft drinks Fruit drinks Tomato juice	DRD, even in triple-draw cans, as coating is very formable Roller-applied topcoat for beer and beverage ends	Flexible and free of off-flavor; used for many years with beer Resistant to acid and alkaline products Not suitable for retorted meats and foods. Wide use as a clear exterior coating
	Vinyl organosol	
Beer and soft drinks Foods	DRD cans Topcoat on beer and beverage can ends Three-piece food can ends Thin tinplate which has to be beaded	Especially suited for thicker film application Good fabricability, superior corrosion resistance. May prove ideal for welded cans, which allow more severe beading. Still used in low-solids solvent-borne manner

Source: From Hernandez, R. J. and Giacin, J. R., Factors affecting permeation, sorption, and migration processes in package – product systems, In *Food Storage Stability*, Taub, I. A. and Singh, R. P., Eds., CRC Press, Boca Raton, FL, 1998, chap. 10. With permission.

The original can lacquers were based on oleoresinous products, which include all those coating materials that are made by fusing natural gums and rosins and blending them with drying oils such as linseed or tung (Chinese wood oil).[11] Although oleoresinous coatings are still used today (largely because of their low applied cost), their open micellar structure means that they are prone to corrosion/staining problems with sulfur bearing products unless they are pigmented with zinc oxide. In addition, they lack resistance to retorting processes and have poor color retention and taste characteristics.[14] For these reasons, a move has been made to synthetic phenolic resins dissolved in a blend of solvents.

Vinyl coatings are based on copolymers of vinyl chloride and vinyl acetate of low MW dissolved in strong ketonic and aromatic hydrocarbon solvents. The long carbon–carbon chains make them thermoplastic, and they can be blended with alkyd, epoxy and phenolic resins to enhance their performance.[11] The essential qualities of vinyl products are adhesion, high flexibility and a complete absence of taste.[14] Their flexibility allows them to be used for caps and closures as well as drawn cans. Their main disadvantage is their high sensitivity to heat and retorting processes, restricting their application to cans that are hot filled rather than retorted, and to beer and beverage products.

Epoxyphenolic coatings are made either by straight blending of a solid epoxy resin with a phenolic resin, or by the precondensation of a mixture of two resins in appropriate solvents. A three-dimensional structure (normally gold in appearance) is formed during curing and baking, which combines the good adhesion properties of the epoxy resin with the high chemical resistance properties of the phenolic resin. The balanced properties of epoxyphenolic coatings have made them almost universal in their application on food cans, with the exception of deep multistage DRD cans.[5]

Vinyl organosol coatings incorporate a dispersion of high MW PVC resins in hydrocarbon solvents with a plasticizer such as dioctyl phthalate to aid film formation. Typical epoxy materials include bisphenol A diglycidyl ether (BADGE), the safety aspects of which are discussed in Chapter 21. Organosols have all the desirable properties found in vinyl coatings, together with better process resistance. Soluble thermosetting resins (including epoxy, phenolic and polyesters) are added in order to enhance the film's chemical resistance, thermal stability and adhesion. These coatings are the most common on DRD cans in the U.S. and are typically white or buff colored due to the addition of titanium dioxide.[5]

Sulfur resistant enamels are used to prevent staining of tinplate surfaces by sulfur compounds released from foods such as meat, fish and vegetables with sulfur-containing amino acids that breakdown during heat processing and storage to release sulfides. These react with tin to form black tin sulfide, or accumulate in the headspace and give out an unpleasant odor. To overcome this problem, two approaches have been used. Enamels are pigmented with zinc oxide or zinc carbonate which reacts with the sulfur compounds to form white zinc sulfide (these are known as the sulfur-absorbing enamels), or the enamels are pigmented with aluminum powder or white pigment to obscure any tin sulfide that might form (these are known as sulfur-resisting enamels).

Special enamels with additives such as waxes to assist the release of the product from the can, or enamels pigmented with aluminum powder or other materials are also used. The latter were described above as sulfur-resisting enamels, but they are also used in premium quality packs (where sulfur staining is not a problem), simply to improve the appearance of the inside can surface.

Two methods are used for the application of protective coatings to metal containers: roller coating and spraying, the former being the most widespread. Roller coating is used if physical contact is possible; thus, it finds use in the coating of material in sheet and coil form, and the external coating of cylindrical can bodies. Spraying techniques are used if physical contact is impossible or difficult; thus, this method is used mainly to coat the inside surface of can bodies, including two-piece D&I and sometimes DRD cans.

Because the coating is generally applied wet (i.e., the resin is suspended in a carrier such as an organic or aqueous solvent for ease of application), it must be dried after application by solvent removal, oxidation or heat polymerization. This process (known as baking or curing) is usually carried out in a forced convection oven using hot air at up to 210°C for up to 15 min. Today's resin formulations require lower temperatures and shorter curing times, and this is achieved through the use of UV radiation to accelerate polymerization. Such UV-cured resins are virtually solvent-free. They contain photosensitive molecules that absorb the UV radiation and release free radicals, which polymerize or cross-link the liquid resin to form a solid coating. Because the substrate is not heated, it can be handled immediately for further operations, and substantial savings (up to 50%) in energy and space are possible.

Powder coating, where the resin is applied "dry" in the form of a fine powder under the direction of an electrostatic field, is used mainly where heavy coatings are required, such as in the protection of welded sideseams where the bare metal that exists in the weld area (most of the tin is removed during the welding process) must be covered. Curing is usually by infrared radiation or high frequency induction heating, since hot air could disturb the uncured coating. Powdered coatings contain little or no volatile effluents and require low energy consumption for application and cure.

Electrophoretic deposition, a process originally developed for protecting automobile bodies, has been used for coating both two-piece and three-piece cans.[12] A resin film is deposited electrically from an aqueous suspension, providing a far more even distribution than is obtained by spraying. Moreover, its throwing power enables it to coat regions inaccessible to spray. The process overcomes the current wasteful spray techniques with almost 99% coating utilization achieved.

Owing to environmental concerns and legislation concerning VOCs, the trend is for waterborne coatings to replace organic coatings, which can consist of up to 70% solvent. The aqueous coatings contain only a small amount of organic solvents, and can be applied using essentially the same coating and curing equipment.

2. Decorative Coatings

Although the primary purpose in decorating the external surface of a metal container is to improve its appearance and assist its marketability, it also significantly improves the container's external corrosion resistance. In many respects, decoration of the external surface is similar to the process used to protect the internal surface, the constituents generally being dispersed in volatile solvents, applied on roller coating machines (apart from the printed image) and baked in tunnel ovens.

Offset lithography (see Chapter 9 for a description) has been used for over a century for decorating sheet metal. Because metal is nonabsorbent, the coatings and prints cure by internal chemical reactions involving oxidation, polymerization or both. By adjusting the tack properties of the ink, it is possible to print one wet ink onto the previous wet ink ("wet-on-wet") without the second blanket picking off the first layer. Methods used to "set" the ink in between single color presses in tandem include high-temperature air blasts, flame treatment and, more recently, the application of UV radiation to the print. The latter approach has necessitated a new approach to ink formulation. The print process is usually carried out with sheet stock prior to slitting into can body blanks or scroll shearing into end stock. With the advent of two-piece cans, less elaborate designs are used on beer and beverage cans because of the necessity to use presses that can print completely fabricated cans.

VI. ALUMINUM FOILS AND CONTAINERS

A. Aluminum Foil

Aluminum foil is a thin-rolled sheet of alloyed aluminum varying in thickness from about 4 to 150 μm. It was first produced commercially in the U.S. in 1913 where it was used for wrapping Life Savers™, candy bars and chewing gum. In 1921, it was laminated on paperboard for folding cartons. Household foil was marketed in the late 1920s, and the first heat sealable foil was developed in 1938. Formed or semirigid containers appeared on the market in 1949.

Foil can be produced by two methods: either by passing heated aluminum sheet ingot between rollers in a mill under pressure and then rerolling on sheet and plate mills until the desired gauge is obtained, or continuously casting and cold rolling. This latter method is much less energy intensive and has become the preferred process.

Aluminum foil is available in a variety of alloys (see Table 7.3), with the alloys 1100, 1145 and 1235 most commonly used in flexible packaging and 3003 when heavier gauges are required for stiffness. In the softest temper, aluminum foil exhibits dead fold characteristics — that is, when wrapped around an object it will assume the profile of the object with no springback. Although this is frequently advantageous, soft temper foil also wrinkles very easily, which necessitates the use of great care during handling.

Aluminum foil is essentially impermeable to gases and water vapor when it is thicker than 25.4 μm, but it is permeable at lower thicknesses due to the presence of minute pinholes. For example, 8.9 μm foil has a WVTR of up to 0.3 mL m^{-2} day^{-1} at 38°C and 100% RH.

Aluminum foil can be converted into a wide range of shapes and products including semirigid containers with formed foil lids, caps and cap liners, composite cans and canisters, laminates containing plastic and sometimes paper or paperboard where it acts as a gas and light barrier, and foil lidding, the latter being sealed using inductive sealing (see Chapter 5). Processes involved may include converting, forming, laminating, coloring, printing and coating. It can also be embossed to provide textured surfaces.

B. Tube

The collapsible aluminum tube is a unique food package that allows the user to apply the product directly and in precise amounts when required. Typical applications include condiments such as mustards, mayonnaises and sauces, as well as dessert sauces, cheese spreads and pâté.

The aluminum tube is formed by the cold impact extrusion of an aluminum slug using a plunger. To relieve the hardness, the tube is annealed in an oven at 600°C, after which the inside is enameled with an epoxyphenolic or acrylic lacquer. Aluminum tubes are printed by a dry offset process (see Chapter 9) using either thermally or UV-cured inks. Aluminum tubes are closed by folding after application of a latex or heat sealable lacquer inside the fold area and heat applied, which ensures a hermetic seal. Currently, the aluminum tube is relatively rare with most food tubes being made of plastic.

Foil laminate tubes are sealed using a high frequency, which generates an eddy current in the aluminum, heating up the surrounding plastic layers and forming a hermetic seal. Although early plastic tubes contained aluminum foil as a barrier layer, it is now common to coextrude LDPE with EVOH to obtain a tube that provides an excellent barrier to air and moisture. Plastic tubes are also printed by a dry offset process.

C. Retort Pouch

Developed in the 1950s, the retort pouch is a flexible package, hermetically sealed on three or four sides and made from one or more layers of plastic or foil, each layer having a specific functionality. The choice of barrier layers, sealant layers and food contact layers depends on the processing conditions, product application and desired shelf life. Typical processing conditions involve temperatures of 121°C for up to 30 min (60 min for the large [3.5 kg] catering packs). One of the attractions of the retort pouch compared to the metal can is the thin profile of the package (12 to 33 mm for 200- to 1000-g pouches), enabling retorting times to be reduced by up to 60%, final quality to be improved, as well as rapid reheating prior to consumption. Other advantages include the ease of carrying, reheating and serving, as well as weight and space saving. Finally, disposal of the used pouch is much simpler than for the metal can as it can be easily flattened. For all of these reasons, retort pouches have found wide acceptance by military forces, the U.S. military term for this type of package being "Meal, Ready-To-Eat" (MRE). NASA began using retort pouch food for space missions in the 1970s and the U.S. Army began delivering large quantities of MREs to the troops in 1981.[8]

A typical three-layer pouch structure would consist of an outer layer of 12 μm PET for strength and toughness, a middle layer of 7 to 9 μm of aluminum foil as a moisture, light and gas barrier and an inner layer of 70 to 100 μm of CPP for heat sealability, strength and compatibility with all foods. An additional inner layer of 15 to 25 μm of PA is used when a longer shelf life is required. Traditionally a three-sided seal pouch was used for MREs and other commercial products, but a multilayer four-side-seal retort pouch has been recently developed. Stand-up pouch designs with a gusseted bottom have also been commercialized.

Transparent retort pouches can be produced by replacing the aluminum foil layer. Typical structures include PET-OPA-CPP, SiO_x PET-OPA-CPP, AlO_x PET-OPA-CPP, OPA-PVdC-CPP and OPA-EVOH-CPP. These materials allow the pouch to be reheated in a microwave oven.

Unlike the metal can, retort pouches are susceptible to rupture or seal separation during retorting if the internal pressure exceeds the external process pressure. This is most likely to occur at the start of the cooling cycle when the product is at its hottest. The use of superimposed air pressure to counterbalance the build up in internal pressure in the pouch and control pouch integrity is necessary. Retorts used in processing pouches can be batch or continuous, and agitating or static.[2] It is also necessary to use trays that support the pouches and ensure uniform thickness, and thus an adequate thermal process.

The shelf life of foods packaged in retort pouches is very dependent on storage temperature. If stored at 16°C, then they will last for about 130 months. If they are stored at 27°C, then they last for about 76 months; at 38°C, 22 months and at 50°C, only a month. Because of this, military MREs are stored in climate controlled warehouses where they can be kept for up to ten years before being used.

Pouches are reverse printed on the PET layer using any of the standard plastic printing techniques (see Chapter 9). However, the highest quality printing is generally obtained using rotogravure, with the high gloss of the PET adding a reflective sparkle to the surface of the package.

A recent development has been the incorporation of zippers into the pouch to make it easier to open and reseal. Conventional zippers are made of LDPE that will not withstand retort temperatures, forcing zipper manufacturers to design zippers made from PP resins and match them with compatible pouch film. Aesthetic considerations have also had to be addressed. Because food migrates to the edges of the pouch during retorting, a secondary barrier to shield the zipper and prevent food from crusting around it had to be designed.

Laser scoring (conducted on either the exterior or interior surface) can withstand the conditions in a retort and maintain pouch integrity. Currently, U.S. pet food suppliers are putting laser scores on retort pouches holding pet food. With laser-scored packages, the laser beam "burns off" a portion of the substrate, providing a line for easy, uniform tearing of the pouch. However, the laser should only remove enough substrata to allow easy opening while not providing a route for O_2 migration through the pouch which would decrease shelf life.

VII. CORROSION OF METAL PACKAGING MATERIALS

A. Fundamental Concepts

1. Introduction

Metals are important materials for the packaging of foods, combining properties of strength, toughness, ductility and impermeability. However, the chemical structure that gives them their valuable practical properties is also responsible for their main weakness — their susceptibility to corrosion. Corrosion is the term used to describe the chemical reaction between a metal and its environment to form compounds; it is a universal process affecting all metals to some extent. Because the reaction takes place at the metal surface, the rate of attack can be reduced and controlled by modifying the conditions at the surface.

Metals are chemically reactive and can be readily oxidized by O_2 and other agents to form largely useless corrosion products. This vulnerability to oxidation explains the fact that, with few exceptions (copper, silver and gold), metals do not occur naturally in the metallic state but are found combined with O_2 or sulfur in their ores. A considerable amount of energy is required to extract metals from their ores, and the reverse process (which releases energy) is strongly favored as the metal reverts back to its natural state. Generally, it can be said that the more difficult it has been to win the metal from its natural form, the greater will be its tendency to return to that form by corroding, but the rate of return will, of course, depend on the environment.

2. Electrochemical Corrosion

The reaction of metals in aqueous solutions or under moist conditions (known as wet corrosion) is electrochemical in nature, involving the transfer of electrical charges across the boundary formed by the metal surface and its environment. An electrolyte is a medium that conducts electricity by movement of ions, the cations (e.g., Fe^{2+}) and anions (e.g., Cl^-) moving in opposite directions. When an electrode reaction takes place at a metal surface, the electron flow in the metal corresponds to an ion flow in the electrolyte.

When a metal corrodes, atoms of the metal are lost from the surface as cations, leaving behind the requisite number of electrons in the body of the metal. This dissolution of the metal is called an *anodic reaction* and takes place at a surface termed an *anode*; an anodic reaction always involves the release of electrons or electrochemical oxidation. Thus, for the case of a metal, M,

$$\underset{\text{(reduced)}}{M} \rightarrow \underset{\text{(oxidized)}}{M^{n+}} + n \text{ electrons} \quad (7.1)$$

Simultaneously, reagents in the electrolyte solution react with the metal surface to remove electrons left behind by the departing metal ions. This removal of electrons is termed a *cathodic reaction* and takes place at a surface called a *cathode*. The cathodic reaction always involves consumption of electrons or electrochemical reduction.

Because practically all metals are covered with an oxide film, this must be removed before the metal can be exposed to an electrolyte. A metal covered with an oxide has different properties in solution from a bare metal, but in studying electrochemical corrosion, it is simplest to begin with the ideal case of a pure, bare metal electrode on which the only reaction occurring is metal dissolution.

3. Electrochemical Series

The removal of electrons from a metal to form an ion involves energy. When a metal with a high reaction energy is in an electrolyte, and is connected to one of lower reaction energy in the same electrolyte, electrons will flow from the high to the low energy level. If an infinite resistance voltmeter is placed across the system, then it will show the difference in potential (E) between the two metals; this is a measure of the relative tendencies of the metals to corrode in the particular environment involved.

Because only differences in potential are measurable and not absolute values, it is desirable to have some reaction equilibrium as a datum or reference zero of potential from which other potentials could be measured. This was chosen as hydrogen gas at a pressure of one atmosphere in equilibrium with hydrogen ions at pH $= 0$ and 25°C, and is taken as zero. Potentials measured

TABLE 7.5
The Electrochemical Series

Equilibrium Reaction	E_H (volts)	
Cathodic end		
$Au^{2+} + 2e \Leftrightarrow Au$	+1.50	Increasing potential to corrode ↓
$\frac{1}{2}O_2 + 2H^+ + 2e \Leftrightarrow H_2O$	+1.23	
$Cu^{2+} + 2e \Leftrightarrow Cu$	+0.34	
$2H^+ + 2e \Leftrightarrow H_2$	0.00 by definition	
$Pb^{2+} + 2e \Leftrightarrow Pb$	−0.13	
$Sn^{2+} + 2e \Leftrightarrow Sn$	−0.14	
$Fe^{2+} + 2e \Leftrightarrow Fe$	−0.44	
$Cr^{3+} + 3e \Leftrightarrow Cr$	−0.74	
$Al^{3+} + 3e \Leftrightarrow Al$	−1.66	
Anodic end		

relative to this reaction are indicated as E_H in volts. A list of equilibria and standard electrode potentials obtained from such measurements is usually called the *electrochemical series* or the *electromotive force series*, and an abridged version is shown in Table 7.5. The gold or noble end of the series is the cathodic end, and the aluminum or base end the anodic end. The further down the series a metal appears, the more readily it will give up its electrons — that is, it is more electropositive and has a greater potential to corrode.

The potentials shown in Table 7.5 refer to metals free from an oxide film. Aluminum covers itself with a highly protective oxide film, which can be difficult to remove, and commonly gives values much less negative than that indicated in the table, so that aluminum is frequently "nobler" than, for example, chromium.

Although oxide films on the metal tend to shift the potential in the positive (noble) direction, the presence of salts which form complex ions containing the metal in question renders the potential abnormally negative. This is particularly important in the case of tinplate, discussed further in Section VII.B.1.a.

In many ways, hydrogen behaves like a metal, and the chemical reduction of hydrogen ions can balance the corrosion reaction or ionization of any metal whose potential occurs below the hydrogen/hydrogen-ion equilibrium potential. Iron and tin will always tend to corrode in aqueous environments since the ionization or corrosion reaction can be balanced by hydrogen-ion reduction (i.e., evolution of hydrogen gas):

$$2Fe \rightarrow Fe^{2+} + 2e^- \quad (7.2)$$

$$2H^+ + 2e^- \rightarrow H_2 \quad (7.3)$$

If the concentration of hydrogen ions is increased (i.e., the aqueous environment has a lower pH), then the rate of the reaction tends to increase.

For those metals above the hydrogen/hydrogen-ion equilibrium (i.e., those metals which have a positive E_H), hydrogen-ion reduction will not give a balancing reaction for their corrosion, and some other reaction must be available if they are to corrode. The reduction of gaseous O_2 is such a reaction, except in the case of gold. If free O_2 were available, then the corrosion reaction would be balanced by O_2 absorption, and, in the case of tin, the balanced reactions would be

$$Sn \rightarrow Sn^{2+} + 2e^{-} \quad (7.4)$$

$$\tfrac{1}{2}O_2 + 2H^{+} + 2e^{-} \rightarrow H_2O \quad (7.5)$$

It follows that the reduction of any available free O_2 will assist hydrogen-ion reduction in balancing the corrosion of metals which are below the hydrogen/hydrogen-ion potential (i.e., have a negative E_H). This explains why, for example, in the canning of foods, positive attempts are made to remove O_2 from the can prior to seaming on the can end.

It is important to realize that the thermodynamic approach, as embodied in the electrochemical series, has severe limitations if used as a single basis for any theory of corrosion. This is because the effect of films that may form on metals under a variety of conditions is not taken into account by theories based only on the electrochemical series. For example, because chromium and aluminum are both more basic than iron, the electrochemical series predicts that they are both more liable to corrode. This is often interpreted as implying that they would both corrode more rapidly than iron. However, it is well known that these metals are far more corrosion resistant than iron in a wide variety of practical environments. In fact, chromium is added to steel in substantial quantities as an alloying constituent to produce corrosion-resistant stainless steels.

4. Factors Affecting the Rate of Corrosion

a. Polarization of the Electrodes

The potentials recorded in Table 7.5 represent equilibrium values. When a current flows, there is a change in the potential of an electrode; this is known as *polarization*. As the current begins to flow, the potential of the cathode becomes increasingly negative and the anode increasingly positive. Consequently, the potential difference between the anode and the cathode decreases until a steady state is reached when corrosion proceeds at a constant rate. Thus, the corrosion current and therefore the corrosion rate will be affected by anything that affects the polarization of the electrodes.

The potential at which the reaction takes place changes by an amount called the *overpotential*, η, which is defined as

$$\eta = E_{corr} - E_i \quad (7.6)$$

where E_i is the polarized potential. The anodic overpotential η_a drives the metal dissolution process, and the cathodic overpotential η_c drives the cathodic deposition process.[9]

The dominant polarization term controlling the corrosion rate of many metals in deaerated water is the hydrogen overpotential at cathodic areas of the metal. The hydrogen overpotential for iron at 16°C in 1N hydrochloric acid is 0.45 V, and for tin at 20°C in 1N hydrochloric acid, it is 0.75 V. The possible significance of this difference in hydrogen overpotential as it affects corrosion of tinplate is discussed in Section VII.B.1.a.

b. Supply of Oxygen

The rate at which O_2 is supplied largely governs the rate of corrosion, because corrosion by O_2 reduction requires the presence of O_2 for the cathodic reaction to proceed (see Equation 7.5). The rate of supply is proportional to the rate at which O_2 diffuses to the metal surface, and this depends on the concentration of dissolved O_2 in solution. This is further justification for the practice of attempting to remove all the O_2 from canned foods prior to seaming on the can end.

c. *Temperature*

The rate of corrosion generally increases with increase in temperature, as more reactant molecules or ions are activated and are able to cross over the energy barrier. Furthermore, increasing the temperature tends to increase the rate of diffusion of molecules or ions in a solution, although the solubility of O_2 in water decreases with increasing temperature.

5. Passivity

In Section VII.A.2, the dissolution process of a metal was described as an oxidation process of the general form $M \rightarrow M^{n+} + n$ electrons. If the metal can be oxidized to an oxide that is stable in the electrolyte, then the metal is rendered passive (i.e., passivated). Passivation usually requires strong oxidizing conditions.

Thus, corrosion-resistant metals and alloys can withstand an aggressive environment because of the presence of thin films of adherent oxides on their surfaces. The oxide layer will completely stop the anodic reaction which is the direct cause of corrosion, and if the film is insoluble in the electrolyte solution, then it will form an insulation barrier which will reduce the rate of the cathodic reaction.

For example, iron is readily attacked by dilute nitric acid, but is inert in concentrated nitric acid because a thin, protective film is formed. As a result, iron behaves in concentrated nitric acid like a much more noble metal than it actually is. Iron can also be passivated by chromate solutions, as can tinplate, the latter being a very important step in the manufacture of tinplate. Passivation of tinplate can be achieved using an aqueous solution of chromic acid, although an electrolytic treatment in a sodium dichromate electrolyte has gained widespread favor. The resultant film is composed of chromium and chromium oxides and tin oxide, its properties varying depending on the quantity and form of these basic components.

B. Corrosion of Tinplate

1. Corrosion of Plain Tinplate Cans

a. *Reversal of Polarity*

The tinplate surface consists of a large area of tin and tiny areas of exposed tin–iron alloy ($FeSn_2$) and steel as a result of pores and scratches in the tin coating. Although the now obsolete hot-dipped tinplate had a substantial tin–iron alloy layer, ETP has a much thinner layer, which is electropositive to the base and also to the tin, thus acting as a chemically inert barrier to attack on the steel base. The effect of this barrier is to prevent a significant increase in the steel cathode area. Thus, the density or degree of continuity of the alloy layer has a material effect on the rate of corrosion. The alloy-tin couple test gives a good indication of the continuity of the alloy layer in tinplate.

In the case of tinplate exposed to an aerated aqueous environment, tin is noble (i.e., cathodic) to iron according to the electrochemical series. Therefore, all the anodic corrosion is concentrated on the minute areas of steel and the iron dissolves (i.e., rusts). In extreme cases, perforation of the sheet may occur. This is the process that occurs on the external surface of tinplate containers.

However, inside a tinplate can, the tin may be either the anode or the cathode depending on the nature of the food. In a dilute, aerated acid medium, the iron is the anode and it dissolves, liberating H_2. In deaerated acidic food, iron is the anode initially, but later, reversal of polarity occurs and the tin becomes the anode, thus protecting the steel. Tin has been described in this situation as a *sacrificial anode*. This reversal occurs because certain constituents of foods can combine chemically with Sn^{2+}ions to form soluble tin complexes. Consequently, the activity of Sn^{2+} ions with which the tin is in equilibrium is greatly lowered, and the tin becomes less noble (i.e., more electropositive) than iron.

b. Rate of Tin Dissolution

Corrosion in deaerated acidic food cans comprises three stages as shown in Figure 7.9. During the first stage, the oil and tin oxide layers are removed from the can surface and the rate of tin dissolution is high. Oxygen and other depolarizers are reduced. This stage lasts from 4 to 15 days depending on the nature of the food. The mirror surface of the tin coating should change to one in which the shape of the individual tin crystals may be seen with the naked eye.

In the second stage, the corrosion rate is slow and almost constant. Continued dissolution of the tin causes enlargement of the existing pores and scratches, exposing the alloy layer and the steel. The exposed steel provides sites for cathodic reaction and the evolved H_2 is taken up by the depolarizers in the food. The area of exposed steel dictates the rate of H_2 evolution, and as the ratio between the areas of tin and steel decreases (i.e., as more steel is exposed), polarization decreases. This stage is slow and can last for over 2 years.

The third stage is characterized by a high rate of tin and iron dissolution. As large areas of steel become exposed, H_2 evolves at a faster rate and accumulates in the can causing swelling. Once the internal pressure in the can causes the ends of the can to bulge, the product is no longer saleable since consumers cannot distinguish between a swollen can caused by microbial spoilage or H_2 evolution. Moreover, the metal content of the food may have reached an unacceptable level by this stage. Therefore, this third stage is of little importance because, by this stage, the food will have reached the end of its acceptable shelf life.

c. Possible Tin–Iron Couple Situations

Four possible scenarios are possible in plain tinplate cans depending on the nature of the food and the presence of depolarizers.

i. Normal Detinning

Normal detinning is an essential process in plain cans of most foods. It has already been described above as the second stage and leads initially to etching and later detinning of the can. In this situation, the tin is anodic to iron and affords complete cathodic protection. Tin dissolution and H_2

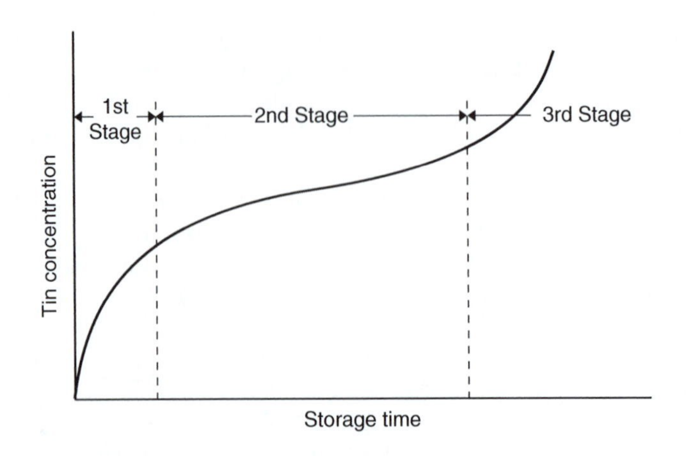

FIGURE 7.9 Schematic rate curve for tin dissolution in acidic foods. (*Source*: From Mannheim, C. and Passy, N., *CRC Crit. Rev. Food Sci. Nutr.*, 17, 371–407, 1982. With permission.)

evolution at the exposed steel are slow, with the area of exposed steel determining the couple current. The dissolved tin forms complexes with food components and the H_2 attaches itself to the depolarizers. Food products characteristic of this type of corrosion include low pH products such as citrus, pineapple, peach and apricot packed in plain tinplate cans.

ii. Rapid Detinning

Rapid detinning occurs when the tin is sufficiently anodic to protect the steel but the couple current is high. This leads to a rapid rate of tin dissolution and hydrogen evolution as shown in Figure 7.10a. Rapid detinning is caused by the use of tinplate with a tin coating mass that is too light, or by a product that is intrinsically too corrosive or contains corrosion accelerators (also known as *depolarizers*) such as dissolved O_2 or anthocyanins, which are chemically reduced. Food products characteristic of this type of corrosion include tomato and aggressive citrus products such as lemon juice, as well as berry fruits.

iii. Partial Detinning and Pitting

With partial detinning and pitting, the tin is anodic to iron, but protection is limited because local anodes on the latter are more anodic than the tin. Early failure occurs because of H_2 swelling or perforation as a result of the exposed steel continuing to corrode. This type of behavior is associated with products such as prunes or pear nectar, or with steel of inferior quality.

iv. Pitting Corrosion

Pitting corrosion is a reversal of the normal situation, with iron being anodic to tin. Thus, the tin does not corrode, but pitting corrosion of the base steel occurs at imperfections in the tin coating as shown in Figure 7.10b. In rare cases, pitting corrosion also occurs when the tin is in fact corroding but at too slow a rate to provide sufficient electrochemical protection to the exposed steel. Pitting corrosion used to be rare, appearing in highly corrosive products, such as pickles and carbonated beverages, formulated with phosphoric acid. However, it is becoming more common and for example is the single greatest cause of failure in canned pears and pear products.

2. Corrosion of Enameled Cans

As discussed earlier in this chapter, food cans with enamel coatings are used to protect against excessive dissolution of tin, sulfide staining, local etching and change in color of pigmented products such as berry fruits. However, the use of enamels will not guarantee the prevention of corrosion and, in some cases, may actually accelerate it. Therefore, careful consideration must be given before selecting an enamel system for a particular canned food.

The general pattern of corrosion in enameled cans is very different from that in plain cans, and is generally more complex. It depends not only on the quality of the base steel plate, the tin–iron alloy layer and the tin coating, but also on the passivation layers and the nature of the enamel coating. The only exposure of metal in an enameled can is at pores and scratches in the enamel coating and at cracks along the sideseam. Some of these discontinuities in the enamel coating may

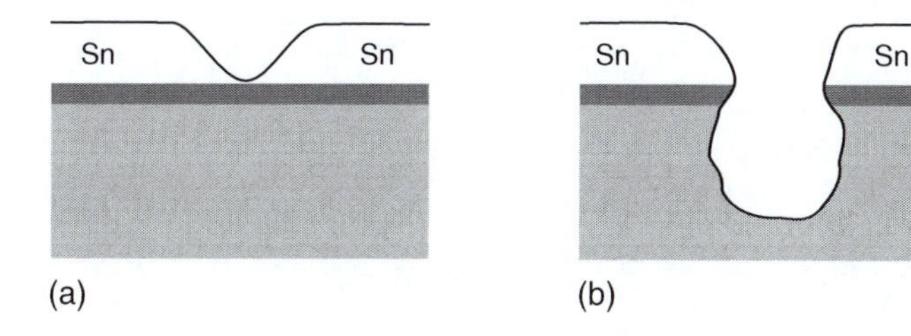

FIGURE 7.10 Schematic diagram illustrating two possible corrosion mechanisms of plain tinplate: (a) anodic tin; (b) cathodic tin.

coincide with pores in the tin coating, thus resulting in exposure of the steel. Even if defects in the enamel film expose only the tin coating, the availability of all the corrosion promoters in the can for attack on the limited areas of tin ensures that steel is soon exposed. Because these areas of exposed steel are virtually unprotected either by the electromechanical action of a tin coating or by dissolved tin, corrosion may proceed at a rapid rate, resulting in H_2 swelling or perforation of the can. Thus, it is possible to actually reduce the shelf life of a canned product by using an enameled can.

The effectiveness of an enamel coating is directly related to its ability to act as an impermeable barrier to gases, liquids and ions, thereby preventing corrosive action on the protected surface. The transport of ions through the enamel is governed by the electrochemical characteristics of the film. This contrasts to the transport of gases and liquids, which involves dissolution in, and diffusion through, the enamel film under a concentration gradient. Because ions are electrically charged, their transfer through the enamel coating (actually the flow of an electric current) is complex and depends not only on their electric charge, but also on the concentration of the electrolyte. If the transport rate of cations and anions through the coating differs, then the coating itself may become charged. Thus, the protection offered by the enamel coating depends on its resistance to ion transfer, which may take place even in the absence of pores, scratches or blisters.

The performance of enameled food cans is greatly affected by the thickness of the enamel coating. A thickness of 4 to 6 μm is sufficient for nonaggressive products such as apricots and beans, but aggressive products such as tomato paste require thicknesses of 8 to 12 μm, with the heavier coatings having much lower porosities.

A possible electrochemical corrosion mechanism in enameled cans is presented as Figure 7.11. In a plain can, the product would attack the tin layer, causing it to dissolve. The presence of an enamel coating protects the surface, and tin dissolution occurs only from both sides of a scratch under the enamel or through a pore, causing anodic undermining (Figure 7.11a). Because only a very small area is in contact with the product in the can, dissolution is slow. With time, the exposed area increases, and detachment of the enamel may be observed, resulting in the appearance of enlarged pores in the enamel coating. The tin–iron alloy layer may also be visible as a grayish color. Because (in theory) the alloy layer is more passive than tin and iron, it should dissolve only after all the tin and iron. However, because of its extreme thinness and its coupling effect with the tin, the potential of the can may be nobler than that of iron, which may also go into solution. In this situation, the enamel coating may serve as the cathode due to diffusion of protons through it.

Researchers[9] used tomato concentrate as an example of the above scenario. Although the overall rate of tin dissolution is reduced in the presence of an enamel coating, protection of the steel by the tin is also impaired due to reduction in the ratio of exposed tin to steel areas. Corrosion is concentrated in small areas and strong local currents may occur. The presence of corrosion accelerators such as nitrates aggravates the situation.

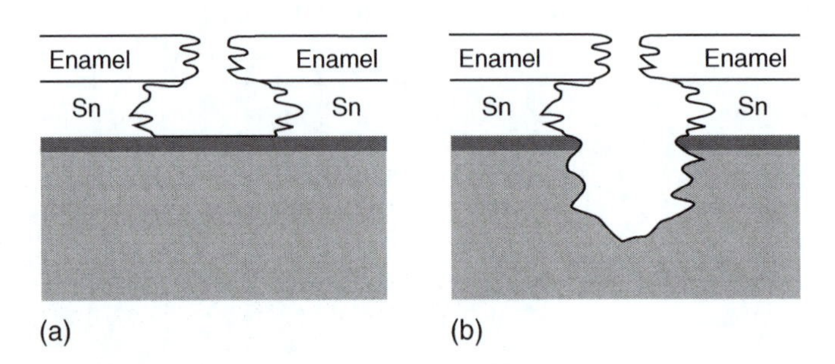

FIGURE 7.11 Schematic diagram illustrating two possible corrosion mechanisms of enameled tinplate: (a) anodic tin; (b) cathodic tin.

Where the alloy and iron are more anodic than the tin (as illustrated in Figure 7.11b), the alloy layer dissolves rapidly and the iron is attacked. However, because no tin is dissolved, the alloy layer is not laid bare and thus has no influence on the corrosion process. Failure eventually occurs due to pinhole formation (also known as *pitting corrosion*) but there is no undermining of the enamel, with corrosion usually starting at a point of discontinuity such as a scratch or pore in the coating. Aggressive products such as beets in acetic acid and berries are responsible for this type of corrosion. The cans do not appear to be corroded, and it is only on closer inspection (usually with a hands lens) that spots of corrosion (often with deep penetration into the steel) are visible.

Failure of enameled cans is often due to a reduction in the bond between the enamel and the metal surface, resulting in eventual lifting of the enamel coating. Thus, good adhesion is required to prevent anodic reactions, to counteract forces developed under the coating due to physical or chemical factors and to ensure an aesthetic appearance.

3. Corrosiveness of Foods

Food products and beverages are extremely complex chemical systems covering a wide range of pH and buffering properties, as well as a variable content of corrosion inhibitors or accelerators. Factors that influence the corrosiveness of food products and beverages can be divided into two groups: intensity and type of corrosive attack inherent in the food itself, and corrosiveness due to the processing and storage conditions. All these factors are interrelated, and may combine in a synergistic manner to accelerate corrosion.

The most important corrosion accelerators in foods include O_2, anthocyanins, nitrates, sulfur compounds and trimethylamines. Some typical corrosion reactions associated with these accelerators and their stoichiometric equivalents of dissolved tin are presented in Table 7.6.

From a corrosiveness point of view, it is convenient to divide foods into five classes:

1. Highly corrosive such as apple and grape juices, berries, cherries, prunes, pickles and sauerkraut.
2. Moderately corrosive such as apples, peaches, pears, citrus fruits and tomato juice.
3. Mildly corrosive such as peas, corn, meat and fish.
4. Strong detinners such as green beans, spinach, asparagus and tomato products.
5. Beverages are conveniently considered as a fifth class.

Although the above classification offers a broad guide, it is important to note that different varieties of the same food can exhibit as much variation in their corrosiveness as may exist between

TABLE 7.6
Some Corrosion Promoting Agents and their Mode of Reaction

Corrosion Accelerator	Reduction Product	Equivalent in Weight
Proton (H^+)	H_2	1 mL H_2 ≡ 5.3 mg Sn^{2+}
Oxygen (O_2)	H_2O	1 mL O_2 ≡ 10.6 mg Sn^{2+}
Sulfur dioxide (SO_2)	H_2S	1 mL SO_2 ≡ 5.5 mg Sn^{2+}
Sulfur (S)	H_2S	1 mg S ≡ 3.7 mg Sn^{2+}
Nitrate (NO_3)	NH_3	1 mg NO_3 ≡ 7.65 mg Sn^{2+}
Trimethylamine oxide (TMAO)	TMA	1 mg TMAO ≡ 1.57 mg Sn^{2+}

Source: From Mannheim, C. and Passy, N., *CRC Crit. Rev. Food Sci. Nutr.*, 17, 371–407, 1982. With permission.

different types of foods. Thus, for example, the same variety of fruit from different growing regions may vary significantly in terms of corrosiveness. The various factors that can influence the corrosiveness of food products and beverages are considered in more detail.

a. *Acidity*

No direct proportionality exists between the acidity of a product and the degree of corrosion of tinplate, where two products of the same acidity will not necessarily be equally corrosive. It also appears that pure solutions of organic acids are less corrosive than the fruit juices containing them, suggesting that fruit juices contain unidentified depolarizers that enhance the corrosive action of organic acids. It has been well established that the tendency of an acid to form a complex with dissolved tin has an important bearing on the relative polarity of tin and steel, and hence the degree of corrosion (see Section VII.B.1.a).

b. *pH*

As is the case with acidity, no direct proportionality exists between pH and the degree of corrosion of tinplate. This is not surprising given that the reaction product when a metal is dissolved is not always an ionic species but often a solid oxide or hydroxide. The pH of the system also determines the relative cathodic protection given to steel. In some cases, tin is cathodic to steel over a certain pH range (in the case of acetic acid the range is pH 2.0 to 4.5), while in others, it offers protection up to pH 4; above that level, it may accelerate corrosion.

c. *Sulfur Compounds*

Sulfur and sulfur compounds may be introduced into the can in a number of ways. They may be introduced in the form of spray residues from agricultural chemicals, residues from sulfur-containing preservatives or as components in sulfur-containing compounds such as proteins in meat, fish and certain vegetables. Proteins are degraded during heat processing, releasing free sulfide or hydrosulfide ions and evolving hydrogen sulfide gas into the headspace.

Trace amounts of sulfur compounds from agricultural chemicals (e.g., derivatives of thio- and dithiocarbamic acid fungicides) can lead to accelerated corrosion and failure of plain tinplate cans containing acid foods such as apricots and peaches. In addition, pitting corrosion can occur, and this has been attributed to inactivation of the tin coating by a protective film of sulfide having a more cathodic potential. Consequently, there is a significant reduction in the tin dissolution rate, and no electrochemical protection of the steel by the tin.

Sulfur dioxide may be directly reduced on the tin surface to sulfide or even to sulfur, with tin passing into solution and the development of unpleasant odors and flavors. Residual SO_2 accelerates corrosion through its action as a depolarizer, inducing a negative charge in the double layer. This repels the electrons from the electrode, thus shifting the potential in the positive direction.[9] Trace amounts of SO_2 as low as 1 mg kg^{-1} are sufficient to accelerate corrosion, but such corrosion problems may be overcome by the use of enameled cans.

There are two types of sulfide staining. One involves iron sulfide (sometimes called sulfide black) and the other involves tin sulfides. These two types of staining do not constitute a health hazard or lead to failure of the can. However, both types may cause adverse reactions from the consumer on aesthetic grounds.

Iron sulfide stains are characteristically black, and usually occur at isolated points on the can (mainly in the headspace region) during or immediately after heat processing. Iron sulfide is not formed at pH values below about 6. Thus, it is uncommon to find iron sulfide in the portion of the can that is in contact with the contents. However, the pH of condensed volatile matter in the headspace may be above 6. The problem may be overcome by using enameled cans or plain cans with enameled ends.

Tin sulfide staining is usually widespread throughout the can, and is blue-black or sometimes brown. Two stages are believed to be involved. The first stage is an oxidation of the tin, and the second is the deposition of an insoluble tin sulfide precipitate on the surface. It occurs during, or soon after, heat processing and shows little or no increase in intensity during storage. It may be prevented by using sulfur-resistant enamels into which quantities of zinc or aluminum compounds are added before being applied to the plate surface. These react with sulfur-bearing gases to form almost invisible white metal sulfides. However, this approach is not suitable for acid products because the acids may attack the coating to produce zinc or aluminum salts which could be harmful to health.

d. Nitrates

Nitrates are found in fruits and vegetables grown in heavily fertilized soils and may also occur in water supplies as a result of pollution by fertilizers. Vegetables such as green beans, spinach, turnips, lettuce, beets and radishes have often been shown to contain several thousand mg kg^{-1} of nitrates. Nitrates are very efficient cathode depolarizers because they are capable of being reduced all the way to ammonia. They have been responsible for serious economic and toxicological problems in some canned foods, notably tomato products. Although nitrates and nitrites are also present as intentional additives in processed meats, they present no problem because meat products are above the critical pH range (5.5) for detinning to occur via the nitrate-tin reduction system.

Nitrates act as electron acceptors, replacing the H_2 evolution reaction with the electron-nitrate two-step reduction reaction and shifting the reaction towards increased tin dissolution. The reduction reactions involved are thought to be as follows:

$$4Sn \rightarrow 4Sn^{2+} + 8e^- \tag{7.7}$$

$$NO_3^- + 2e^- + 2H^+ \rightarrow NO_2^- + H_2O \tag{7.8}$$

$$NO_2^- + 6e^- + 8H^+ \rightarrow NH_4^+ + 2H_2O \tag{7.9}$$

The first reduction reaction (Equation 7.8) is probably rate determining. The equations indicate that the rate of detinning depends on the nitrate concentration and pH. Nitrate does not immediately affect the corrosion rate, but begins to act after tin and iron ions have passed into solution. Although the ammonium ion is the major conversion product at pH 5 and less, above this pH other products are formed, including nitrous oxide (N_2O), nitric oxide (NO) and hydroxylamine. Oxygen present in the can at the time of processing triggers the nitrate-detinning reaction by increasing the initial rate of formation of Sn^{2+}(Equation 7.7).

Overcoming the problem by restricting the use of nitrate fertilizers has proved difficult, and instead efforts have been directed towards finding cultivars that do not accumulate high concentrations of nitrate. It is also possible to avoid the use of waters with high nitrate content for canning operations; a suggested maximum is 5 mg L^{-1} nitrate. The best solution to the problem at present is the use of enameled cans, although the development of corrosion inhibitors (the search for such substances is continuing) would offer an alternative solution.

e. Phosphates

Phosphates are naturally present in meat and are often intentionally added as polyphosphates to processed meat products such as hams to reduce the loss of water during processing. The presence of phosphates leads to increased discoloration due to iron phosphate and sulfide formation.

One measure used to counter the effect of polyphosphate in cans of processed meats is the introduction of a small area of aluminum, which acts as a sacrificial anode, protecting the tin surface. This is not always completely successful, and parchment liners may be used in addition or as an alternative.

f. Plant Pigments

The anthocyanins and related pigments are among the most important potential corrosion accelerators (cathodic depolarizers) as they are easily reduced. Anthocyanin pigments can also act as anodic depolarizers through their ability to form complexes with cations, particularly those of iron and tin salts. Analysis of samples of canned fruits after a period of storage usually shows a greater amount of tin in the drained fruit than in the syrup, indicating that at least part of the tin is combined in an insoluble form with some constituent within the fruit.

The nature of the anthocyanin pigment is also important. For example, raspberries contain cyanidin glucosides, which have *ortho*-dihydroxy groups in their structures. It is these groups that are involved in the formation of blue-tinted complexes with metals such as tin. The major pigment of strawberries is pelargonidin-3-glucoside, which does not possess the necessary *ortho*-dihydroxy groups for complex formation. Therefore, strawberries do not show the same shift to a blue color in the presence of tin salts as do raspberries.

Combination of metal ions with tannins has been observed in other fruits. For example, discoloration in canned cranberry has been attributed to the formation of a complex between tannins present in the fruit and tin salts. Darkening in canned maraschino cherries has been observed to be more severe when fruit of high tannin content is used, where a high tannin content is often found in unripe fruit and fruit that has been stored in wooden barrels. It should be noted that not all reactions between plant pigments and metal ions will produce undesirable colors, although this is usually the case with anthocyanin pigments.

g. Synthetic Colorings

The canned products that most commonly contain synthetic colorings are soft drinks, which basically consist of sugar-based syrups and carbonated water containing flavors, acidulants and colors. The behavior of soft drinks largely depends on the presence of azo dyes (e.g., amaranth) and the amount of residual O_2 in the filled can. Both of these components are capable of acting as corrosion accelerators and are potentially active corrosive agents. Tin dissolution may adversely affect the color of some products, and iron dissolution may lead to perforation and flavor defects. Thus, fully enameled cans are essential, and it is important to obtain near-perfect coverage by the enamel.

h. Copper

Foods containing dissolved copper will deposit it when put in a metallic container, and, in acid products, this can lead to accelerated corrosion, either by tending to strip the tin or by producing local attack on the steel. The central reason why copper accelerates the corrosion of steel is because copper catalyzes the reduction of O_2.

In the past, copper entered products by the solution of copper oxide from copper-bearing metal food contact equipment. However, with the replacement of copper equipment with stainless steel, the problem of copper in canned foods has been dramatically reduced. Another possible source of copper is from certain fungicides in cases where cannery water supplies have unacceptably high copper levels.

4. Effects of Processing and Storage

a. *Oxygen*

Oxygen may be dissolved in food products as it is naturally present in food tissues such as fruit and vegetable cells, and is inevitably entrained or absorbed by particulate products prior to their being filled into cans. Removal of as much of this O_2 as possible is an essential part of good cannery practice, and a variety of methods are used including hot filling, vacuum filling, exhausting, closure under vacuum, steam flow closure and vacuum syruping. In addition, positive control of the headspace volume is essential.

A larger headspace is likely to contain a higher residual O_2 concentration than a smaller one. However, the larger the headspace, the more room there is for accumulation of H_2 resulting from corrosion, and thus the greater the time required to form a H_2 swell. In practice, the cannery technologist has little room to maneuver the headspace if the declared weight on the can label is to be met.

Despite the above procedures, there will always be some O_2 present in the headspace of newly filled cans. The rate of O_2 consumption at this stage is quite rapid but decreases with time, the rate being a function of initial concentration, headspace volume, can vacuum, nature of the product and type of container. Oxygen acts as a depolarizer, accelerating corrosion by reacting with the H_2 formed in the can through a cathodic reaction, as shown in Equation 7.3.

b. *Thermal Processing*

Little is known about the effect of heat sterilization processes on corrosion rates, except that the quantity of metal dissolved during the process is very small. This is hardly surprising given the comparatively short processing times (typically 30 to 120 minutes) relative to the total shelf life of the canned product (typically 1 to 2 years). However, degradation products formed during thermal processing can become involved in corrosion (e.g., nonenzymic browning intermediates and sugar derivatives).

The importance of correct cooling has been emphasized, where failure to cool cans adequately (generally a target temperature of 40°C when the cans leave the retort) can lead to increased corrosion because the can center temperatures may take several days to return to ambient. On the other hand, cooling to too low a temperature may cause paneling of the can walls as well as external corrosion (see Section VII.B.5) because complete drying of the can exteriors will not occur.

c. *Storage Temperature*

The rate of a chemical reaction increases as the temperature is raised, and for many reactions, the rate doubles for each 10°C increase in temperature (i.e., the temperature quotient Q_{10} equals 2). Therefore, to minimize undesirable reactions such as nonenzymic browning in canned foods, it is preferable that storage temperatures are kept as low as practicable.

5. External Corrosion of Cans

Although tinplate is very durable in a dry atmosphere, it rusts readily in the presence of moisture, where rusting occurs more readily the thinner the tin coating. The presence of sulfur dioxide or oxides of nitrogen in the atmosphere accelerates the rate of corrosion since they dissolve to form acids. Chlorides (present in locations close to the sea) can also cause a rapid increase in the rate of corrosion.

The mechanism of external corrosion is complex. Under normal conditions on the outside of a can, tin is cathodic to iron, and when a galvanic cell is set up at pore sites, the attack on the iron is accelerated by the presence of tin. Rust forms as a result of corrosion of the iron. Three stages of exposure to the risk of external rusting may be differentiated: thermal processing; cooling of cans and storage.

a. *Thermal Processing*

The essential ingredient for rusting of cans during thermal processing is the presence of O_2 in the processing vessel (typically a steam retort). When cans are processed in water at atmospheric pressure (as is quite common for high acid, low pH products such as fruits), the risk reaches a maximum at a temperature of about 80°C. Above this temperature, the loss of dissolved O_2 counteracts the acceleration of the reaction rate at the higher temperatures. Therefore, when full boiling is not required, the water should be deoxygenated and preferably treated with corrosion restrainers.

When cans are processed in steam, it is important that proper venting of the retort occurs so that all the O_2 initially present is displaced from the retort by the steam. In addition, bleeder valves should be left open so that any residual or incoming O_2 may escape. As well as minimizing any external corrosion, the removal of air from the retort prevents pockets of air from surrounding individual cans and insulating them from the steam, the latter situation leading to underprocessing.

b. *Cooling of Cans*

Cans are typically cooled after thermal processing by passing water through the retort or placing the cans in cooling canals. However, in some canneries, hot cans are removed from the retort and left to cool in the air. If cans are cooled with water, then it is a regulatory requirement in virtually every country that such water contain a measurable amount of bactericide (typically chlorine) when it exits the retort. In the case of chlorine, this usually means that incoming water has a free chlorine content of around 5 to 10 mg L^{-1}.

When cans are water cooled, they should be around 40°C when they are removed from the retort. If they are warmer than this, then there is the risk of thermophilic spoilage, chemical degradation and significant internal corrosion. On the other hand, if they are cooler than this, insufficient heat will remain to evaporate any water adhering to the exterior of the can. Cans should be stacked in such a way so as to enable self-drying prior to labeling and packing.

The addition of a wetting agent to the cooling water, or the spraying of a dilute solution of a wetting agent onto the cans as they leave the cooling water, facilitates rapid drainage of the water and leaves it as a quickly evaporating thin layer rather than as isolated droplets. Accelerated drying methods such as the use of hot air or a steam spray, or rolling cans over a sterile absorbent surface, have advantages in minimizing external corrosion of cans.

c. *Storage*

If corrosion is to be prevented during storage, then the atmosphere surrounding the can must be free of corrosive vapors or chemicals, and not promote condensation of moisture. In addition, packaging materials in contact with the cans (generally paperboard cartons closed with adhesives) should be as free as possible from soluble chlorides, sulfates or other salts which may promote condensation of moisture and corrosion. Cartons usually have a water content of 10 to 12%, and when the air temperature rises, moisture evaporates from the cartons in the warmer outer zone and condenses in the cooler center of the stack. The time for rust to develop on wet cans held at various temperatures is presented in Table 7.7.

The shrink wrapping of cans, while protecting them from promoters of corrosion found in the atmosphere, can cause problems of condensation. If the air inside the shrink wrap contains a considerable quantity of water vapor and the package is later subjected to a drop in temperature, then condensation will occur.

Rusting can also be caused by unsuitable conditions of transport and storage where cycling of the humidity (described as "sweating") occurs.[16] This is especially so when cans are transported from temperate to tropical areas, or temperate to temperate areas via the tropics (e.g., from Australia to

TABLE 7.7
Time–Temperature Profile for Rust to Develop on Wet Cans of Vegetables Held in a Saturated Moisture Atmosphere

Temperature (°C)	Storage Limits
26.7–37.8	4 hours
15.6–21	1 day
10	2 days
4	4 days
0	6 days
−6.7 or below	≥ 90 days

Source: From Yang, T. C. S., Ambient storage, In *Food Storage Stability*, Taub, I. A. and Singh, R. P., Eds., CRC Press, Boca Raton, FL, 1998, chap. 17. With permission.

North America). Attempts to prevent condensation of moisture by free movement of air have usually been unsuccessful because the center of a stack of cartons filled with cans takes a long time to respond to the external temperature change, and may remain below the dew point for long periods.

C. Corrosion of ECCS

Although the chromium/chromium oxide layer on ECCS cans is only approximately 1/30 to 1/50 the thickness of a typical tinplate coating, it transforms the base steel into an excellent canmaking material by providing rust protection, outstanding enamel coating adhesion and good resistance to underfilm sulfide staining. However, ECCS cannot be used in food packaging unless it is enameled because of its lack of resistance to corrosion. Two-piece ECCS cans have significantly lower product iron contents compared with three-piece tinplate cans.

D. Corrosion of Aluminum

Aluminum rapidly forms a protective oxide film when exposed to air or water:

$$4Al + 3O_2 \rightarrow 2Al_2O_3 \quad (7.10)$$

The film is extremely thin (about 10 nm), but it renders the metal completely passive in the pH range 4 to 9. Because aluminum oxide is amphoteric, it will dissolve in acid or alkali to give the soluble aluminum cation or anion respectively:

$$Al_2O_3 + 6H^+ \rightarrow 2Al^{3+} + 3H_2O \quad (7.11)$$

$$Al_2O_3 + 2OH^- \rightarrow 2AlO_2^- + H_2O \quad (7.12)$$

Although, in practice, aluminum will corrode in acid solutions (pH below 4) and alkali (pH above 9), its corrosion resistance is excellent in the neutral range of pH 4 to 9. Having a strongly negative electrode potential ($E_H = -1.66$), aluminum is liable to undergo severe corrosion if brought into metallic contact with copper, iron or other more positive metals in the presence of an electrolyte (e.g., fruit juice). Thus, care must be taken to ensure that situations that could lead to such corrosion (e.g., processing equipment with food contact surfaces containing both aluminum and stainless steel parts) are avoided.

Aluminum levels are generally very low because of the very good enamel systems that prevent contact of the food with the metal. However, in certain products such as beer, even very low levels of aluminum can cause cloudiness or haze and render the product unacceptable.

Pitting corrosion of aluminum easy-open ends on cans of fruit juices has been experienced, with the corrosion taking place mainly along the depressed score line if the underside organic coating is torn or damaged by the scoring operation. Corrosion takes place with high chloride content products such as tomato and vegetable juices, and is accelerated when the end is used with a tinplated can body. This is because the aluminum end is anodic to the can body. This problem can be overcome by reapplying enamel to the end after scoring.

Products containing brine should not be packed in aluminum as they can produce rapid and dramatic corrosion, such corrosion becoming apparent in the form of container or end perforation within 24 hours.[14]

Although no foods have a pH greater than 8, cleaning solutions used in food processing plants frequently have pH values of 13. It is therefore important that these solutions do not come into contact with aluminum packaging materials, and that any package-contact surfaces cleaned with these solutions are thoroughly rinsed with water afterwards.

REFERENCES

1. Bishop, P. W., Who introduced the tin can?, *Food Technol.*, 32(4), 60–64, 1978, see also p. 67.
2. Blakistone, B., Retortable pouches, In *Encyclopedia of Agricultural, Food, and Biological Engineering*, Heldman, D. R., Ed., Marcel Dekker, New York, pp. 846–851, 2003.
3. Cowell, N. D., Who introduced the tin can? — A new candidate, *Food Technol.*, 49(12), 61–64, 1995.
4. Farrer, K. T. H., *A Settlement Amply Supplied*, Melbourne University Press, Victoria, Australia, 1980.
5. Good, R. H., Recent advances in metal can interior coatings, In *Food and Packaging Interactions, ACS Symposium Series #365*, Hotchkiss, J. H., Ed., American Chemical Society, Washington, DC, 1988, chap. 17.
6. Hernandez, R. J. and Giacin, J. R., Factors affecting permeation, sorption, and migration processes in package-product systems, In *Food Storage Stability*, Taub, I. A. and Singh, R. P., Eds., CRC Press, Boca Raton, FL, 1998, chap. 10.
7. Kraus, F. J. and Tarulis, G. J., Cans, steel, In *The Wiley Encyclopedia of Packaging Technology*, 2nd ed., Brody, A. L. and Marsh, K. S., Eds., Wiley, New York, pp. 144–155, 1997.
8. Lampi, R., Retort pouch, *Wiley Encyclopedia of Food Science and Technology*, 2nd edn., Vol. 3, Francis, F. J., Ed., Wiley, New York, pp. 2055–2062, 2000.
9. Mannheim, C. and Passy, N., Internal corrosion and shelf life of food cans and methods of evaluation, *CRC Crit. Rev. Food Sci. Nutr.*, 17, 371–407, 1982.
10. Morgan, E., *Tinplate and Modern Canmaking Technology*, Pergamon Press, Oxford, England, 1985.
11. Pilley, K. P., *Lacquers, Varnishes and Coatings for Food and Drink Cans and for the Metal Decorating Industry*, ICI Packaging Coatings, Birmingham, England, 1997.
12. Silbereis, J., Metal cans, fabrication, In *The Wiley Encyclopedia of Packaging Technology*, 2nd ed., Brody, A. L. and Marsh, K. S., Eds., Wiley, New York, pp. 615–629, 1997.
13. Thorne, S., *The History of Food Preservation*, Parthenon Publishing Group, London, England, 1986.
14. Turner, T. A., *Canmaking: The Technology of Metal Protection and Decoration*, Blackie Academic & Professional, London, 1998.
15. Turner, T. A., *Canmaking for Can Fillers*, CRC Press, Boca Raton, FL, 2001.
16. Yang, T. C. S., Ambient storage, In *Food Storage Stability*, Taub, I. A. and Singh, R. P., Eds., CRC Press, Boca Raton, FL, 1998, chap. 17.

8 Glass Packaging Materials

CONTENTS

I. INTRODUCTION

Glass has been defined by the American Society for Testing and Materials[1] as "an amorphous, inorganic product of fusion that has been cooled to a rigid condition without crystallizing." Although glass is often regarded as a synthetic material, it was formed naturally from common elements in the earth's crust long before the world was inhabited. Natural materials such as obsidian (from magma or molten igneous rock) and tektites (from meteors) have compositions and properties similar to those of synthetic glass; pumice is a naturally occurring foam glass.

Although the origin of the first synthetic glasses is lost in antiquity and legend, the first glass vessels were probably sculpted from solid blocks about 3000 BC. In about 1000 BC, the techniques of pouring molten glass or winding glass threads over a sand mold were developed, resulting in the formation of crude but useful glass objects. However, the real revolution in glassmaking came around 200 BC with the introduction of the blowing iron, a tube to which red-hot, highly malleable glass adheres. Blowing through one end of the iron causes the viscous liquid to balloon at the other end, leading to the production of hollow glass objects.

By 200 AD, articles of glass were in fairly common use in Roman households. During the following 1000 years, glassmaking techniques spread over Europe. However, glass remained expensive until improved techniques in the eighteenth and nineteenth centuries brought down the price of bottles and jars to a relatively affordable level.

Mechanization of glass container manufacture was introduced on a large scale in 1892, and several important developments occurred over the next few decades. These included the first fully automated machine for making bottles, which was designed and built in 1903 by Michael J. Owens at the Toledo, Ohio plant of Edward D. Libbey.

Added impetus was given to automatic production processes in 1923 with the development of the gob (mass or lump of molten glass) feeder, which ensured the rapid supply of more consistently sized gobs in bottle production. Soon afterwards, in 1925, the Hartford Empire Company developed their IS (now generally taken to mean "individual section," but actually named after its inventors, Ingersall and Smith) blow and blow (B&B) machine.[8] Used in conjunction with the gob feeders, IS machines allowed the simultaneous production of a number of bottles from one piece of equipment. The gob feeder-IS machine combination remains the basis of most automatic glass container production today.

Further developments have occurred, resulting in the production of a wide range of glass containers for packaging. The two main types of glass container used in food packaging are bottles (which have narrow necks) and jars (which have wide openings). About 75% of all glass food containers in the U.S. are bottles and approximately 85% of container glass is clear, the remainder being mainly amber. Generally, today's glass containers are lighter but stronger than their predecessors, with the weight of many bottles and jars having been reduced by 25 to 50% over the last 50 years. Through developments such as this, the glass container has remained competitive and continues to play a significant role in the packaging of food products.

II. COMPOSITION AND STRUCTURE

The basic raw materials for glassmaking come from mines or quarries and must be smelted or chemically reduced to their oxides at temperatures exceeding 1500°C. The principal ingredient of glass is silica derived from sand, flint or quartz. Silica can be melted at very high temperatures (1723°C) to form fused silica glass which, because it has a very high melting point, is used for specialized applications including some laboratory glass.

For most glass, silica is combined with other raw materials in various proportions. Alkali fluxes (commonly sodium and potassium carbonates) lower the fusion temperature and viscosity of silica. Calcium and magnesium carbonates (limestone and dolomite) act as stabilizers, preventing the glass from dissolving in water. Other ingredients are added to give glass certain physical properties. For example, lead gives clarity and brilliance although at the expense of softness of the glass; alumina increases hardness and durability. The addition of about 6% boron to form a borosilicate glass reduces the leaching of sodium (which is loosely combined with the silicon) from glass.

As a consequence of the sodium in glass being loosely combined in the silica matrix, the glass surface is subject to three forms of "corrosion": etching, leaching and weathering. Etching is characterized by alkaline attack, which slowly destroys the silica network, releasing other glass components. Leaching is characterized by acid attack in which hydrogen ions exchange for alkali or other positively charged mobile ions. The remaining glass (principally silica) usually retains its normal integrity. Although not fully understood, weathering is not a problem in commercial glass packaging applications since it may take centuries to become apparent. However, a mild form of weathering is commonly known as surface bloom and may occur under extended storage conditions.

The most aggressive solution on glass is double-distilled water at neutral pH 7. The effect of dilute acidic solutions is much less, the main action being the extraction of sodium ions, which are replaced by hydrogen ions. The result is a surface zone where the glass is depleted of sodium, this dealkalized layer forming a barrier to further ionic diffusion. It is worth remembering that the aqueous phase of almost all food products is acidic.

A typical formula for soda-lime glass is given in Table 8.1. In practice, however, the quantities vary slightly; for example, silica (SiO_2) 68 to 73%, calcia (CaO) 10 to 13%, soda (Na_2O) 12 to 15%, alumina (Al_2O_3) 1.5 to 2% and iron oxides (FeO) 0.05 to 0.25%, depending on the glassmaker and the raw materials being used. The loss on ignition or fusion loss (generally the oxides of carbon and sulfur) can vary from 7 to 15%, depending on the quantity of cullet used, there being less fusion loss the greater the quantity of cullet. Soda-lime glass accounts for nearly 90% of all glass produced and is used for the manufacture of containers where exceptional chemical durability and heat resistance are not required.[4] Replacement of alkali by boric oxide leads to the production of borosilicate glass which is used for glass ovenware.

Glass is neither a solid nor a liquid but exists in a vitreous or glassy state in which molecular units have a disordered arrangement but sufficient cohesion to produce mechanical rigidity. Although glass has many of the properties of a solid, it is really a highly viscous liquid. During cooling, glass undergoes a reversible change in viscosity, the final viscosity being so high as to make the glass rigid for all practical purposes. Although glass at ambient temperatures has the characteristics of a solid, it is a supercooled liquid and will flow even at ambient temperatures over long periods of time, albeit extremely slowly. Evidence for this can be obtained by examining very old window panes that are slightly thicker at the bottom than at the top.

Physically, glass has a random atomic structure in that the atoms are capable of arranging themselves in different orders. The basic structural unit is the silicon–oxygen tetrahedron in which a silicon atom is tetrahedrally coordinated to four surrounding oxygen atoms. However, although the silica atoms are always surrounded by four oxygen atoms, large groupings tend to be unordered. This amorphous structure, without slip planes formed by crystal boundaries that might allow deformation, is responsible for the stiffness and brittleness of glass.

TABLE 8.1
Typical Formula for a One-Tonne Batch of Soda-Lime Container Glass

Material	Weight (kg)	Oxides Supplied (kg)					LOI[a] (kg)
		SiO_2	Al_2O_3	CaO	Na_2O	FeO	
Sand SiO_2	300	299.3	0.2			0.3	0.5
Soda ash Na_2CO_3	100				58.3		41.7
Aragonite $CaCO_3$	90			49.0		0.02	40.7
Feldspar ($SiO_2 \cdot Al_2O_3$)	40	26.4	7.6	0.4	1.3	0.03	0.1
Salt cake NaCl	4				2.1		1.9
Cullet	460	333.7	9.2	48.8	67.2	1.03	0.1
Total	994	659.4	17.0	98.2	128.9	1.95	85.0
Yield of glass	909						
wt% oxides		72.6	1.9	10.8	14.2	0.1	

[a] Loss on ignition (also referred to as fusion loss).

Source: Adapted from Boyd, D. C., Danielson, P. S., and Thompson, D. A., Glass, In *Kirk–Othmer Encyclopedia of Chemical Technology*, 4th ed., Vol. 12, Kroschwitz, J., Ed., Wiley, New York, pp. 555–628, 1994. With permission.

III. PHYSICAL PROPERTIES

A. Mechanical Properties

Because of its amorphous structure, glass is brittle and usually breaks because of an applied tensile strength. It is now generally accepted that fracture of glass originates at small imperfections or flaws, the large majority of which are found at the surface. A bruise or contact with any hard body will produce on the glass surface very small cracks or checks that may be invisible to the naked eye. However, because of their extreme narrowness, they cause a concentration of stress that may be many times greater than the nominal stress at the section containing them. Because of their ductility, metals yield at such points and equalize stress before failure occurs. Given that glass cannot yield, the applied stress (when it is high enough) causes these flaws to propagate.[10] Thus, it is the ultimate tensile strength of a glass surface which determines when a container will break. The fracture formula is:

$$\text{Tensile Stress} + \text{Stress Concentrator} = \text{Fracture}$$

In practice, a stress concentrator may be a small crack or check induced in the manufacturing process, or a scratch resulting from careless container handling. Therefore, the major step taken to make glass more break resistant involves the elimination of surface flaws (e.g., microcracks) by careful handling during and after forming and annealing, since the condition of the surface has a great deal to do with its tensile properties.

The mechanical strength of a glass container is a measure of its ability to resist breaking when forces or impacts are applied. Glass deforms elastically until it breaks in direct proportion to the applied stress, the proportionality constant between the applied stress and the resulting strain being Young's modulus E. It is about 70 GPa for typical glass.[4]

The principles of fracture analysis or diagnosis of the cause(s) of glass container breakage have been described by Moody[11] in an excellent book that is regrettably now out of print. The following four aspects are important:

1. *Internal pressure resistance*. This is important for bottles produced for carbonated beverages, and when the glass container is likely to be processed in boiling water or in pressurized hot water. Internal pressure produces bending stresses at various points on the outer surface of the container, as shown in Figure 8.1(a).
2. *Vertical load strength*. While glass can resist severe compression, the design of the shoulder (see Figure 8.5 for details of glass container nomenclature) is important in minimizing breakage during high-speed filling and sealing operations.
3. *Resistance to impact*. Two forms of impact are important — a moving container contacting a stationary object (as when a bottle is dropped), and a moving object contacting a stationary bottle (as in a filling line). In the latter situation, design features are incorporated into the sidewall to strengthen contact points. The development of surface treatments (including energy absorbing coatings) to lessen the fragility of glass when it contacts a stationary object has been very successful. A cross-section of a round bottle illustrating the ways in which tensile stresses on the inside and outside surfaces vary at various points around the bottle circumference is shown in Figure 8.1(b).
4. *Resistance to scratches and abrasions*. The overall strength of glass can be significantly impaired by surface damage such as scratches and abrasions. This is especially important in the case of reduced wall thickness bottles such as "one-trip" bottles. Surface treatments involving tin compounds (in conjunction with other treatments) provide scuff resistance, thereby overcoming susceptibility to early failure during bottle life.

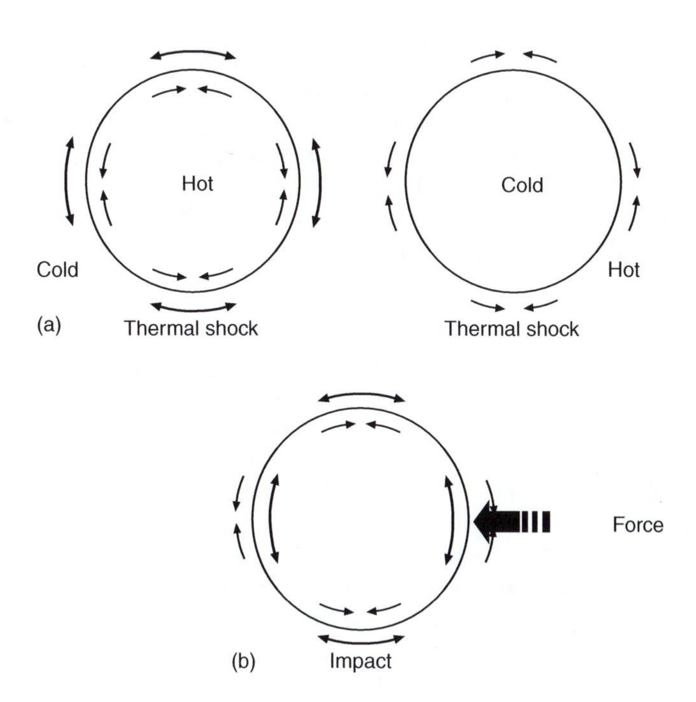

FIGURE 8.1 Cross-section of a round glass container illustrating various stresses on the inside and outside surfaces.

Although the mechanical strength of a bottle or jar can increase with glass weight, this is at the expense of thermal strength which decreases with increasing glass weight. Considerable expertise is required by the glassmaker to determine the most appropriate design to satisfy the mechanical strength requirements, and to balance the thermal strength demands of the finished product.

B. Thermal Properties

The thermal strength of a glass container is a measure of its ability to withstand sudden temperature change. In the food industry, the behavior of glass with respect to temperature is of major significance, because relative to other forms of food packaging, glass has the least resistance to temperature changes. The resistance to thermal failure depends on the type of glass employed, the shape of the container, and the wall thickness.

When a glass container is suddenly cooled (e.g., on removal from a hot oven), tensile stresses are set up on the outer surfaces, with compensating compressional stresses on the inner surface, as shown in Figure 8.1(a). Conversely, sudden heating leads to surface compression and internal tension. In both situations, the stresses are temporary and disappear when the equilibrium temperature has been reached. Because glass containers fracture only in tension, the temporary stresses from sudden cooling are much more damaging than those resulting from sudden heating, since the potentially damaged outside surface is in tension. In practice, it is found[11] that the amount of tension produced in one surface of a bottle by suddenly chilling it is about twice as great as the tension produced by suddenly heating the other surface, assuming the same temperature change in both cases.

Thermal shock resistance cannot be calculated directly because the strength of glass containers is greater under momentary stress than under prolonged load. Therefore, empirical testing procedures are used. ASTM C 149-86 (2000) covers the determination of the relative resistance of commercial glass containers (bottles and jars) to thermal shock, and is intended to apply to all types of glass containers that are required to withstand sudden temperature changes (thermal shock) in service such as in washing, pasteurization or "hot fill" processes, or in being transferred from a warmer to a colder medium or *vice versa*. Resistance to breaking is determined by transferring glass containers which have been totally immersed in a hot water bath (typically at 63°C) for 5 min to a cold water bath (typically at 21°C) and observing the number of breakages.

C. Optical Properties

Because glass has no crystalline structure, when it is homogeneous and free from any stresses, it is optically isotropic. The optical properties of glass relate to the degree of penetration of light and the subsequent effect of that transmission, transmission being a function of wavelength. The spectral transmission of glass is determined by reflection at the glass surface and the optical absorption within the glass. In silicate glasses, transmission is limited by the absorption of silica at approximately 150 nm in the UV region and at 6000 nm in the IR region. Iron impurities further reduce transmission in the UV and near-IR regions.[4]

Transmission may be controlled by the addition of coloring additives such as metallic oxides, sulfides or selenides, and the compounds that are frequently used are listed in Table 8.2. Most of the transition metal oxides (e.g., cobalt, nickel, chromium, iron, etc.) will give rise to absorption bands, not only in the visible but also in the UV and IR regions of the spectrum. The presence of iron oxide in glass produces a green color owing to the absorption bands in the UV and IR regions.[10]

The United States Pharmacopoeia[15] defines a light resistant container as one which passes no more than 10% of incident radiation at any wavelength between 290 and 450 nm through the average sidewall thickness. Amber glass provides this degree of light protection quite economically, as shown in Figure 8.2.

Glasses and other transparent materials tend to darken and lose much of their ability to transmit light when bombarded by high energy radiations such as those used in food irradiation. There are two principal causes of this coloration of glass. First, the impact of the radiations may displace electrons, which can become lodged in holes in the structure, forming color centers. Second, changes produced in the valence of bivalent or multivalent metal oxides may result in the increased absorption of light in the visible range. This second effect forms the basis of the process to protect glass from this coloration where a metal oxide (which will change its valence under bombardment more readily than the electrons which are displaced) is included in the composition of the glass. Provided that the oxide is free from serious light absorption bands in both valences, protection from

TABLE 8.2
Coloring Agents Used in Glass

Effect	Oxide
Colorless, UV absorbing	CeO_2, TiO_2
Blue	Co_3O_4, $Cu_2O + CuO$
Purple	Mn_2O_3, NiO
Green	Cr_2O_3, $Fe_2O_3 + Cr_2O_3 + CuO$, V_2O_3
Brown	MnO, $MnO + Fe_2O_3$, $TiO_2 + Fe_2O_3$, $MnO + CeO_2$
Amber	Na_2S
Yellow	CdS, $CeO_2 + TiO_2$
Orange	$CdS + Se$
Red	$CdS + Se$, Au, Cu, Sb_2S_3
Black	Co_3O_4 (+Mn, Ni, Fe, Cu, Cr oxides)

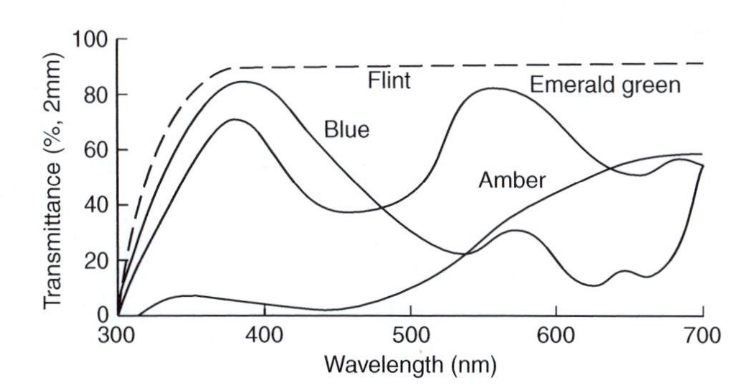

FIGURE 8.2 Typical radiation transmission of common container glasses.

discoloration may be obtained. The addition of CeO_2 (it is reduced to Ce_2O_3 by the radiations) in glasses in amounts up to 1.5% has proved to be an effective means of reducing coloration. Unfortunately, it is a very expensive oxide, so glass containers treated this way are significantly more costly than standard containers.

IV. MANUFACTURE

A. Mixing and Melting

The typical composition of a soda-lime glass is given in Table 8.1. The largest constituent (68 to 73%) is silica; the second largest constituent (15 to 50%) is cullet (i.e., scrap or recycled glass), originating as both glass scrap from the factory and recycled glass from consumers. Although the use of cullet can cause problems with the production of some types of glass unless there is good separation of colored glass and removal of associated material such as labels, the use of cullet is economically desirable since less energy is required to melt cullet than new raw materials. Cullet also reduces the amount of dust and other particulate matter that often accompanies a batch made exclusively from new raw materials.[4] Although the total primary energy use decreases as the percentage of cullet rises, the maximum energy saved is only about 13%.[7]

The raw materials are weighed, mixed and charged into a glass-melting furnace, which is maintained at a temperature of approximately 1500°C. Here, they are converted into molten glass that is chemically homogeneous and virtually free of gaseous inclusions (bubbles). The melting process consists of two phases: (1) changing the solids into a liquid, and (2) fining or "clearing up" of the liquid. During the refining process, gases (principally CO_2, SO_2 and water vapor) produced by the chemical reaction rise to the surface of the furnace and are removed. When the molten glass becomes free of gas (seed-free), it is then ready for forming into containers. It moves from the furnace into the working end of the furnace (mistakenly called the *refiner*) where thermal homogenization and cooling of the glass to the viscosity required for the particular operation begin. At this point, the temperature of the melt has been lowered from 1250–1350°C to approximately 1100°C.

The preferred energy source for glassmaking is natural gas, although alternate fuels such as oil and propane are used in some plants. With increasingly stringent environmental regulations limiting the emissions of NO_x from glass container furnaces, various systems have been introduced using natural gas and O_2 as the furnace fuel.[5] When air (78% N_2) is subjected to very high temperatures, various oxides of nitrogen are formed. By using natural gas and O_2 as the furnace fuel, there is no N_2 to be oxidized. In addition, there are improvements in energy efficiency given that only two volumes of O_2 are needed to burn one volume of natural gas compared to ten volumes when air is used. This reduces the total energy requirement by up to one third.[5]

B. Forming Processes

The glass is carried from the working end of the furnace to the forming machine in a channel-like structure called a *forehearth*, which is fired by a number of small burners, the aim being to ensure uniform temperature distribution throughout the depth of the glass. At the end of the forehearth is a gob-forming mechanism consisting of a rotating sleeve and vertical plunger. The glass exits in a continuous, viscous stream which is cut by rapidly moving, horizontal steel blades to form what is known as a "gob" (i.e., a mass or lump of molten glass).

Precise control of temperature and shape during the formation of the gob is required for the high-speed production of accurately formed glass containers. Temperatures in the vicinity of 1100°C varying by no more than ± 1°C are typical.

The process of converting a cylindrically shaped gob of glass into a bottle or jar is called forming, and it is essentially a controlled cooling process. While various types of forming machines are used throughout the world, the most predominant is the IS (individual section) machine. As its name implies, it consists of up to 16 sections, each one an individually functioning, hollow glass machine. It performs two basic functions: it shapes the gob into a hollow container, and simultaneously removes heat from the gob to prevent it from deforming significantly under its own weight.

Two basic types of processes are used to make containers on the IS machine: the blow and blow (B&B) and the press and blow (P&B). A closure size of approximately 35 mm is the dividing line between narrow-neck B&B containers (i.e., bottles) and wide-mouth P&B containers (i.e., jars).

1. Blow and Blow (B&B)

Bottles are normally produced by a two-step B&B process (Figure 8.3), whereby a gob of glass, accurately sheared in terms of weight and shape, is dropped into an externally air cooled, cast iron cavity to shape a preform (also known as a parison or body blank). Some of the glass flows over a plunger in the base of the mold, which is used to mold the finish of the container by means of ring molds. Compressed air is applied to force the glass down onto the plunger to form the neck ring. Sometimes, vacuum is applied from the bottom as an alternative or additional procedure.

When the finish molding is complete, the plunger is retracted and air blown in from the bottom, enlarging the size of the bubble until the glass is pressed out against the blank mold to form a hollow preform. This is then inverted and transferred to the blow mold where it elongates under its own weight. Air at about 200 kPa or vacuum is applied so that the glass is pressed against the metal surface of the blow mold, which is air cooled to ensure rapid removal of heat. The mold is then opened and the fully blown parison (now at approximately 650°C) is removed and briefly held

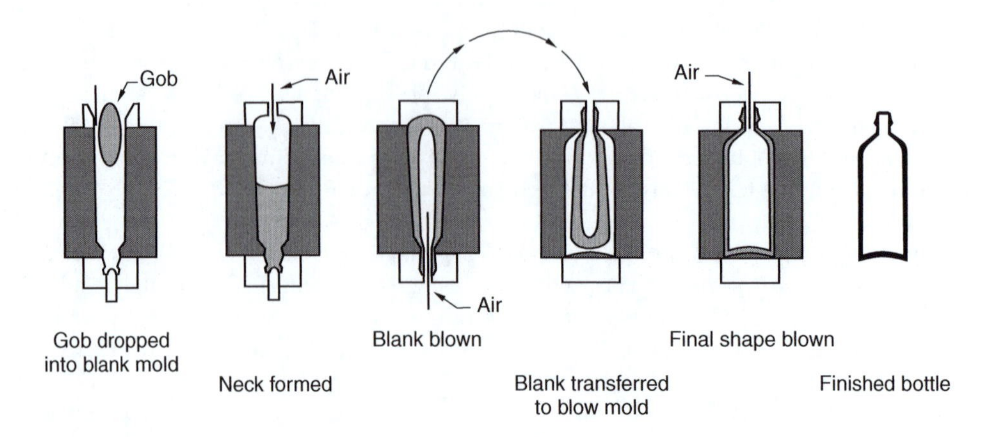

FIGURE 8.3 B&B process for glass container manufacture.

over a deadplate to allow air to flow up through the deadplate and around the container to further cool it. It is then transported to the annealing lehr.

2. Press and Blow (P&B)

In the case of jars, a two-step P&B process (Figure 8.4) is used. The body blank or parison is formed by pressing the gob of molten glass against the mold walls with a large plunger. When the cavity is filled, glass is then pushed down into the neck ring and the finish is formed. No baffle or counterblow air is used in the formation of the parison, the operation relying on the mechanical introduction of the plunger into the glass. The rest of the steps in the P&B process are identical to those in the B&B process.

3. Narrow Neck Press and Blow (NNPB)

NNPB is a more recent process for lightweight bottles, in which the gob is delivered into the blank mold and pressed by a metal plunger. The plunger and gob together have the same volume as the blank mold cavity. This enables the glassmaker to decide exactly how the glass is distributed in the parison and, hence, to be able to more accurately control the uniformity of glass distribution in the finished container. Indeed, weight savings of up to 30% can be made. The second stage is similar to the B&B process. The parison is blown to a finished container with a more uniform wall thickness and, as a result, higher strength.

C. Annealing

The term annealing generally refers to the removal of stress, the annealing temperature or point being defined (ASTM C 336-71) as the temperature at which stresses in the glass are relieved in a few minutes. The containers are transferred from the deadplate to a large oven, known as a lehr, which is equipped with a belt conveyer. The function of the annealing lehr is to produce a stable product by removing any residual stresses resulting from nonuniform cooling rates during forming and handling. This is achieved by raising the temperature of the container to approximately 540°C (almost the softening point of the glass), holding it there for a few minutes

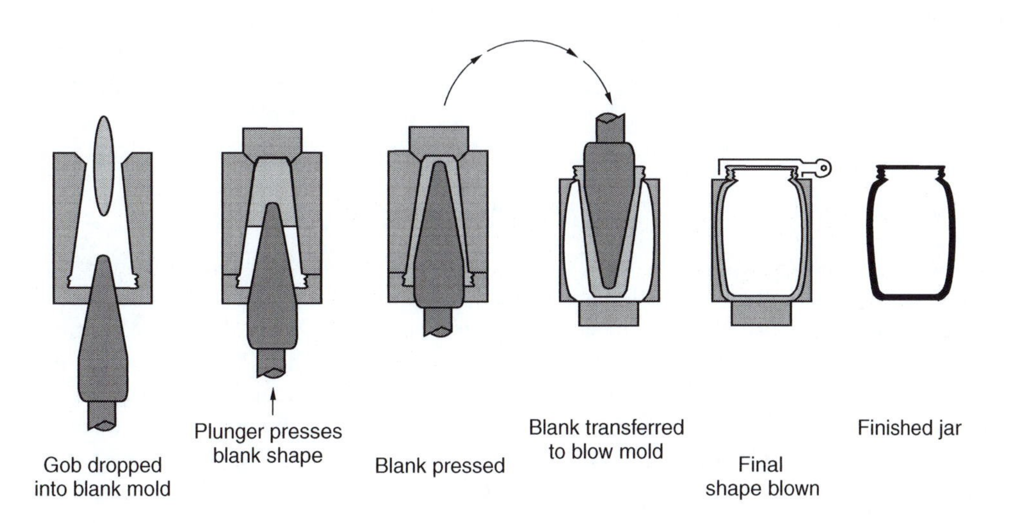

FIGURE 8.4 P&B process for glass container manufacture.

and then cooling at a rate that is consistent with the removal of stress from a predetermined wall thickness.

The critical area of temperature is between the upper annealing point (softening point) and the lower annealing point, after which they may be cooled at a rate which enables them to be handled as they emerge from the lehr. During cooling, the inside surface is hotter than the outside; this results in compression on the outer surface but tension at the inner surface. As mentioned above, glass fractures only in tension and usually at the surface. Sudden cooling introduces tensile stresses into the outer surfaces and compensating compressional stresses in the interior. Poorly annealed containers may be subject to breakage if the tension is high or the inner surface is bruised.[4]

D. Surface Treatments

The strength of a newly made glass container can be rapidly reduced by moisture or abrasion, and some form of surface treatment to increase the strength is essential, since glass is nonlubricious. Two general types of surface treatment are applied to glass containers to modify mechanical properties.[6]

1. Hot-End Treatment

In hot-end treatment (typically carried out while the glass container is at 550°C), vapor containing tin or titanium (generally in the form of a tetrachloride) is brought into contact with the outside of the container, forming a thin unimolecular film of metal oxide. This treatment prevents surface damage while the container is still hot, strengthens the surface and improves the adhesion of the subsequent cold-end coating.

2. Cold-End Treatment

Cold-end treatment (typically carried out while the glass container is at less than 100°C) is designed to protect the container surface and assist its flow through the filling line. Typically, it involves spraying an organic material in an aqueous base containing either waxes, stearates, silicones, oleic acid or polyethylene onto the outside of the container to increase its lubricity by providing a surface with a low coefficient of friction.[5] It is important to check the compatibility of the cold-end treatment with any adhesives used to attach labels. Sometimes, only the cold-end treatment is applied.

3. Shrink Sleeves

Although not strictly related to surface treatment of glass containers, shrink sleeves are mentioned here since they can have an important influence on the formation of imperfections that lead to container breakage due to surface contact. Most shrink sleeves are made of oriented plastic films that shrink around a glass container when heat is applied. Two types of protective labels are used on glass bottles in the form of a body sleeve: one is constructed from thin, foamed PS; the other is made from uniaxially oriented PVC or PS. The former offers some thermal insulation, while the latter (which can completely wrap the bottle from its neck to underneath the base if desired) contains the glass fragments and prevents shattered glass being scattered in all directions if the bottle is dropped. Shrink sleeves are discussed in more detail in Chapter 9 V D.

E. Defects in Glass Containers

Some 60 defects can occur in finished glass containers, ranging from critical defects such as "bird-swings" and "spikes" (long, thin strands inside the container that would probably break off when the container was filled) to minor defects such as "wavy appearance" (an irregular surface on the inside). Defects are classed as "critical" if they are hazardous to the user and render the container completely unusable, "major" if they reduce the usability of the container or its contents and "minor" if they detract from its appearance or acceptability to the consumer.[8]

Accurate classification of defects in glass packaging and their commercial significance are areas of specific expertise and no attempt will be made to describe or catalog them here.

V. GLASS CONTAINER DESIGN

One of the major advantages of glass as a packaging material is its capability to be formed into a wide range of shapes related to specific end uses, customer requirements and aesthetic appeal. The commercialization of computer-aided drafting (CAD) and computer-aided manufacture (CAM) techniques has made the task of designing new glass containers considerably easier and more rapid. This has led to greater flexibility in design and manufacture, and resulted in considerable efficiencies through a more thorough analysis of stresses and strength/weight factors and calculation of likely mechanical performance.

A. Glass Container Nomenclature

The basic nomenclature used for glass containers is shown in Figure 8.5. Usually, the shape of the container is determined by the nature of the product, each product group having a characteristic shape. Thus, liquid products generally have small diameter finishes for easier pouring; solid products require larger finishes for filling and removing the contents. As well as filling and emptying requirements, consideration must also be given to the nature and manner of labeling the container, and its compatibility with packaging and shipping systems.

The container *finish* (thus called because, in the early days of glass manufacturing, it was the part of the container to be fabricated last) is the part of the container that holds the cap or closure (i.e., the glass surrounding the opening in the container). It must be compatible with the cap or closure and can be broadly classified by size (i.e., diameter), sealing method (e.g., twist cap, cork, etc.) and special features (e.g., snap cap, pour-out, etc.).

The finish has several specific areas including the sealing surface, which may be on the top or side of the finish, or a combination of the two; the glass lug, which is one of several horizontal, tapering and protruding ridges of glass around the periphery of the finish on which the closure can be secured by twisting; the continuous thread, which is a spiral projecting glass ridge on the finish, intended to mesh with the thread of a screw-type closure; a transfer bead, which is a continuous horizontal ridge near the bottom of the finish, used in transferring the container from one part of the manufacturing operation to another; a vertical neck ring seam resulting from the joining of the two parts of the neck ring; and a neck ring parting line, which is a horizontal mark on the glass surface at the bottom of the neck ring or finish ring, resulting from the matching of the neck ring parts with the body mold parts. Not all glass containers have transfer beads or vertical neck ring seams.

Although there are literally hundreds of different finishes used on glass containers, glass finishes are standardized and a specific set of dimensions, specifications and tolerances has been established for every finish designation by the Glass Packaging Institute (GPI) in the U.S. and equivalent bodies in other parts of the world.

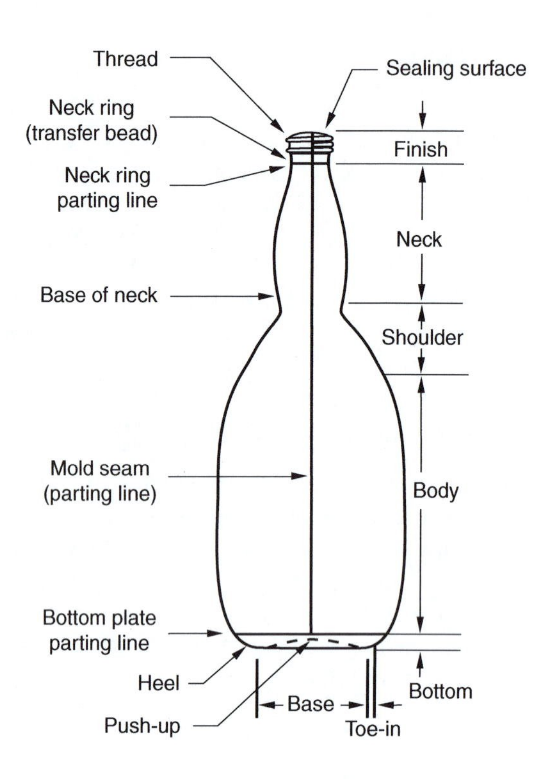

FIGURE 8.5 Glass container nomenclature. (*Source*: Adapted from Anon. Glass container design, In *The Wiley Encyclopedia of Packaging Technology*, 2nd ed., Brody, A. L. and Marsh, K. S., Eds., Wiley, New York, pp. 471–475, 1997. With permission.)

Once a design has been accepted, the molds used in the manufacturing process must be made. They are usually constructed of cast iron and consist of three parts: a bottom plate, a body mold (divided vertically into two halves) and a neck or finish mold, which is also usually split into two parts. Because of the high cost of mold manufacture, changes to container size and shape are usually made only if large quantities of the container are required. Generally, customers select their containers from the standard range provided by glass manufacturers unless they are extremely large users, in which case the extra expense of customized designs is justified.

The GPI has established limits that are generally accepted as reasonable tolerances by most manufacturers.[8] Allowances have been made for increases in container size as a consequence of mold wear, as well as expected process capabilities of the manufacturer. Although closer tolerances can be met, this often incurs a higher cost since molds must be replaced more frequently.

B. Glass Container Strength Factors

The shape, surface condition, applied stresses and glass weight all combine to determine the strength of a glass container.[3] Sharp transitions in container shape (e.g., a rectangular cross-section compared with a circular cross-section) lead to high stress concentrations. Small surface imperfections that form as a result of surface contact during the manufacturing process and subsequent handling operations can influence container strength. Good design will incorporate specific contact areas (e.g., knurls or small protrusions) that concentrate abrasions where they will have minimal effect on glass strength. Surface treatments (see Section IV.D) also assist in reducing surface abrasions.

The forces applied to a glass container during its intended use depend largely on the function of the container. Carbonated beverages and vacuum-packed foods develop internal pressure stresses, predominantly circumferential and longitudinal. In the cylindrical part of a typical glass bottle, the circumferential stress S depends on the bottle diameter d, the glass thickness x and the pressure p as follows[3]:

$$S = \frac{pd}{2x} \tag{8.1}$$

The longitudinal stress in this part of the bottle is one half of the circumferential value S; Equation 8.1 does not hold for the noncylindrical parts of a bottle.

Typical pressures inside a carbonated beverage bottle at ambient temperatures are 400 kPa (4 volumes of CO_2 gas per volume of beverage), rising to about 700 kPa at 40°C and 1000 kPa at pasteurization temperatures. Bottles for carbonated beverages have target bursting strengths well in excess of the equilibrium pressure of a carbonated beverage. The target bursting strength of a nonrefillable, one-way bottle is between 1240 and 1380 kPa, and for a returnable bottle, it is approximately 1720 kPa.[16] Willhoft[16] presented figures that indicated that the mean bursting pressure for a brand new, untouched glass bottle was 4054 kPa, falling to 2331 kPa on delivery to the bottler and to 1524 kPa after long use.

Vertical load stresses are generated by stacking containers on top of each other or by applying a closure; these compressive forces produce tensile stresses in the shoulder and heel region of up to 690 kPa. These stresses can be lowered by decreasing the diameter difference between the neck and the body, by increasing the shoulder radius and by reducing the diameter difference between the body and the bearing surface.[3]

During hot filling or pasteurizing of glass containers, the rapid temperature changes lead to the development of tension stresses on the cold surface and compression stresses on the hot surface, with additional bending stresses being generated by expansions and contractions of the container (see Section III.B). Thermal stresses can be reduced by minimizing the temperature gradient from the hot to the cold side, decreasing the glass thickness, and avoiding sharp corners, especially in the heel.

Stresses caused by steady-state thermal gradients may or may not cause failure, depending on the degree of constraint imposed by some parts of the container on others, or by the external mounting. Consequently, under minimum constraint and maximum uniformity of gradient through the thickness, very large temperature differences can be tolerated.[4]

VI. CLOSURES FOR GLASS CONTAINERS

A. Closure Functions

The final, critical aspect of glass packaging is the closure, which can consist of a cap, lid, cork or plug to seal the jar or bottle. Although glass is an excellent barrier to moisture vapor, gases and odors, an incorrectly designed or applied closure may negate the benefits that glass packaging offers in protecting food products from deterioration.

Closures are required to perform some or all of the following functions without affecting, or being affected by, the contents of the container:

1. Provide an effective hermetic (air-tight) seal to prevent the passage of solids, liquids or gases into or out of the container.
2. Provide easy opening and (when only part of the contents is used at one time) resealing of the container.
3. Provide evidence of inviolability (i.e., that unlawful access to the contents or their exposure to the atmosphere has not occurred). Roll-on pilfer proof (ROPP) closures, such as the aluminum roll-on closure used on beverage bottles, leave a tell-tale ring around the bottom of the neck of bottles after opening, and are designed to deter tampering with the contents by providing evidence of tampering.

B. Closure Terminology and Construction

Closures are frequently referred to by the following terms: the *panel* is the flat center area in the top of the closure; the *radius* or *shoulder* is the rounded area at the outer edge of the panel connecting the panel and the skirt; the *skirt* is the flat, almost vertical portion on the side of the closure, which serves as the gripping surface and may be smooth, knurled or fluted; and the *lug* is a horizontal inward protrusion from the curl, which seats under the thread or lug on the finish of the container, holding the closure in position.

Closures are made from one of two materials: metal or plastic. Metal closures are stamped out of sheets of tinplate, ECCS or aluminum, generally with a thickness of about 0.25 mm. The sheets are usually coated with enamels to prevent the metal from reacting with the contents of the container, and are frequently printed. Metal closures can take four forms: screw caps, crowns, lug caps and spin-on/roll-on closures.

Plastic closures are generally either compression or injection molded, the former being based on urea–formaldehyde or phenolic-formaldehyde resins, and the latter on a variety of thermoplastic polymers including PS, LDPE, HDPE, PP and PVC.

The actual sealing component of the closure is the *gasket* or *liner*, which must make intimate contact with the glass finish to form an effective seal. Gaskets are made from rubber or plastisols, the latter being suspensions of finely divided resin (often PVC) in a plasticizer, which can be flowed in or molded. Flowed-in compounds are almost invariably used in vacuum closures and closures for heat-processed foods.

Liners consist of a cushioning material (known as a *wad*) with a facing material, the purpose of the latter being to isolate the contents of the container from contact with the wad. Generally, the wad is made of composition cork (granules of cork bonded together with either a gelatin-type glue or a synthetic resin) or paperboard, and faced with either a coated paper, a paper faced with plastic film, a plain metal foil, a lacquered metal foil, or a metal foil faced with a plastic film or coated with a layer of a wax material. Thermoplastic resins have allowed the development of "linerless" plastic closures, which rely on a variety of styles of sealing rings molded into the closure to create a liquid or hermetic seal.

C. Food Container Closures

In general, closures used with food containers can be classified under four headings. These are now discussed in turn.

1. Closures to Retain Internal Pressure

The closures for this application are generally required to contain pressures from about 200 to over 800 kPa, as typically found in carbonated drinks and beer.

a. *Crown Cork*

The traditional pressure-retaining closure has been the crown cork, a crimp-on/pry-off friction-fitting closure made from tinplate with a fluted skirt (angled at 15° to maintain an efficient seal) and a cork or plastisol liner. The crown cork was invented in 1891 by William Painter, an American of British descent who lived in Baltimore, Maryland, who said of his invention (so it is claimed), "this closure gives a crowning and beautiful effect to the bottle" and the name crown stuck. It was simple and economical to produce and retained gas in the bottle. Although simple in concept, the crown cork provides a friction-fit sufficient to seal pressurized beverages. The flared cap skirt, in conjunction with the smoothness of the bottle neck, provides easy access through the prying motion of a bottle opener. The crown cork of today is still much the same, except that the teeth on the skirt have been reduced from 24 to 21, ECCS is used rather than tinplate, and PVC plastisol or some type of olefinic material linings have largely replaced cork.[12] A relatively recent improvement has been the introduction of twist-off crown corks.

The 26-mm crown cork is used worldwide, and is normally applied only to a glass finish. A crown applied to a 26-mm pry-off finish will require some type of opening device to remove the crown from the container, whereas a crown applied to a 26-mm twist finish can be removed by hand. The pry-off finish has a large, solid ring of glass around the outside. The corrugations of the crown can be crimped under the ring during application to hold the crown in place and form a solid seal. The twist-off finish has four noncontinuous threads and the corrugations of the crown are crimped into the threads to form the seal. The crown can be removed from the twist container by turning it 6 mm.

The crown cork is applied to a filled container using a crowner that exerts a straight downward force on the crown to crimp it onto the finish. The crowning head compresses the crown liner to form a tight seal, and bends the crown skirt downward and inward to lock it tightly under the locking ring on a pry-off finish. In the case of a twist-off finish, the knuckles between the corrugations are crimped into the threads in the bottle finish. This effectively puts threads into the inside side walls of the crown, allowing the crown to twist off.

Crowns are used in a wide range of beverage products as an inexpensive single use seal. They are also used for hot or cold filled, pasteurized and pressurized products, and for sterilized vacuum products.

b. *Roll-on Tamper-Evident*

A roll-on tamper-evident aluminum closure was first used in the 1920s as a closure on bottles of prescription drugs. It is now used where critical sealing requirements, such as carbonation retention, vacuum retention and hermetic sealing are to be met. The closure is produced as an unthreaded shell containing a liner, and is applied to the proper finish on a plastic or glass container. A thread is formed in the closure matching the bottle thread. The roll-on tamper-evident closure is produced in a wide range of sizes (from 18 to 38 mm), and is available in two different tamper-evident styles: the vertically scored and the standard band (both with and without venting). On opening, the vertically scored band ruptures along vertical score lines and is removed as part of the closure. The standard band separates from the closure during opening and remains on the bottle. Both standard and vertically scored closures can be reclosed after the original opening by turning the threaded closure back onto the container finish.

The roll-on tamper-evident closure is produced from coated, light gauge aluminum, and a range of lining materials (plastics, plastisols, pulpboard and facing, etc.) are available depending on the product to be packaged. The closure is coated on the inside surface for protective purposes, and can be printed and decorated in various colors on the outside surface to enhance the package's appeal.

Roll-on tamper-evident closures are applied to filled containers using roll-on capping machines. The capping head first applies a downward pressure that compresses the liner onto the sealing surface of the container. The capping head then forms a thread in the closure matching the container thread, while also tucking the tamper-evident band under a locking ring on the container finish.

Plastic versions of the closure are now available and used widely. The rolled-on or spun-on aluminum or plastic closure is especially popular for soft drinks in large containers where reuse is common. The same closures are applied to glass and plastic bottles.

2. Closures to Contain and Protect Contents

These closures are designed to contain and protect the contents with no internal pressure (e.g., wine in a bottle).

The most common closure for wine has been the traditional bark cork obtained from the holm oak tree *Quercus suber*, which grows mainly in Spain and Portugal. The cork stopper has been used as a closure on glass bottles for over 25 centuries,[2] and provides an incomparable friction-hold seal. With a high cellular density, cork is compressible, elastic, highly impervious to air and water and low in thermal conductivity. However, 5 to 10% of all wines sealed with natural cork suffer from TCA (2,4,6-trichloroanisole) taint, which can result in undesirable musty aromas and flavors, typically described as "wet cardboard" odor. TCA is a naturally occurring, odorous (albeit harmless) compound that certain micro-organisms produce when chlorine is present; it has a very low threshold of 6 parts per trillion (ppt). Various approaches have been adopted to overcome the problem of cork taint in wine.

The agglomerate cork, originally developed as a closure for sparkling wine, consists of small pieces or granules of clean, natural cork bound together with resin or a chemical binder into a single stopper, with one or more thin discs of intact natural cork stuck on the end intended to be in contact with the wine. The improvement results from an effective disinfecting or deodorizing process, which extracts volatile components from the cork material. This, it is claimed, eliminates the possibility of contaminants being retained inside the lenticels (pores through which gases are exchanged between the atmosphere and plant tissues).

In the early 1990s, many wineries tried synthetic corks as a way to avoid cork taint. Although some synthetics were criticized for leaching a plastic flavor into wines, their use is growing and new formulations are being introduced. In response to the increased use of synthetic corks, one of the world's largest cork manufacturers patented a process in 1999 that uses supercritical CO_2 for selective extraction of volatile compounds from cork, removing up to 97% of TCA.

A challenge to the cork was the development in France in the late 1960s of an aluminum rolled-on closure known as Stelvin®. Research suggested that it performed better than the bark cork. The wad consisted of three components: (1) an expanded LDPE foam substrate to provide controlled and uniform compressibility, (2) a layer of aluminum foil to provide a gas barrier, and (3) a PVdC copolymer facing, which isolated the aluminum foil from the product and provided an additional O_2 barrier. The closure was introduced to the Australian wine industry in 1976, and was used on approximately 20 million wine bottles from 1976 until the early 1980s. Although the closure eliminated the problem of oxidation and the risk of cork tainting,[13] consumers overwhelmingly rejected it, largely for aesthetic reasons. Most winemakers reverted to cork closures by the early 1980s. A newer version of the closure was released in 2000, and is gaining acceptance in many wine-producing countries. A recent book[14] contains references to many of the technical publications concerning the shelf life of wine in bottles sealed with Stelvin closures.

An Australian company has developed technology that reduces the likelihood of "cork taint" by applying a polymeric-based membrane to each end of the cork, which reduces chemicals entering the wine, regulates the passage of O_2 through the cork and retains cork moisture resulting in less cork breakage. It is claimed that 90 to 100% of the TCA in corks is prevented from entering the wine if corks with membranes are used. Another recent Australian innovation is a closure consisting of three components: (1) a cap which provides a tamper-evident clamp that locks onto the band of a cork mouth glass bottle; (2) a foil that provides an O_2 barrier, and (3) a hollow plunger filled with air at atmospheric pressure that creates the "pop" sound of a traditional cork on extraction, where the closure can be resealed after use. It is claimed to have an OTR $<$ 0.01 mL day^{-1}, and to be able to withstand an internal bottle pressure of up to 200 kPa.

3. Closures to Maintain Vacuum Inside Container

These closures are designed to maintain a vacuum inside glass containers that typically contain heat-processed food.[2] Three types of vacuum closures are used in food processing:

1. *Lug-type or twist cap.* This can be removed without a tool and forms a good reseal for storage. It has three, four or six lugs and a flowed-in plastisol liner.

2. *Press-on twist-off cap*. This has no lugs but the finish is threaded. The gasket is molded plastisol and covers the outer edge of the panel and the curl of the cap. It is widely used on baby foods. This type of closure is held on mainly by vacuum with some assistance from the thread impressions in the gasket wall.
3. *Pry-off (side seal) cap*. This is still widely used on retorted products. It consists of a cut rubber gasket held in place by being crimped under the curl. As with the preceding closure, the pry-off (side seal) cap is also mainly held in place by vacuum with slight assistance from the friction of the rubber gasket against the side of the container finish.

All of the above closures are made of metal (tinplate or ECCS) and are fitted either with flowed-in liners, gaskets or rubber rings.

Vacuum closures often have a safety button or flip panel consisting of a raised, circular area in the center of the panel that serves two purposes. First, it provides a visual indicator to the consumer that the package is properly sealed (containers with these types of closures usually have instructions on the closure warning consumers not to purchase or consume the product if the flip panel is up). Second, in the processing plant, the flip panel aids in automatic on-line detection (using a "dud" detector) of low vacuum or no vacuum containers.

4. Closures to Secure Contents Inside Container

Some closures are designed only to secure the contents inside the glass container (e.g., peanut butter). They are frequently screw caps with a minimum thread engagement of three quarters of a turn. They are made either from metal or plastic, and typically have either a plastic-coated paperboard wad inside, or they may be of the linerless plastic style.

REFERENCES

1. ASTM C 162-03, Standard Terminology of Glass and Glass Products, American Society for Testing and Materials, 2003. For the latest version of each standard, visit www.techstreet.com.
2. Anon. Closures, bottles and jars, In *The Wiley Encyclopedia of Packaging Technology*, 2nd ed., Brody, A. L. and Marsh, K. S., Eds., Wiley, New York, pp. 206–220, 1997.
3. Anon. Glass container design, In *The Wiley Encyclopedia of Packaging Technology*, 2nd ed., Brody, A. L. and Marsh, K. S., Eds., Wiley, New York, pp. 471–475, 1997.
4. Boyd, D. C., Danielson, P. S., and Thompson, D. A., Glass, In *Kirk–Othmer Encyclopedia of Chemical Technology*, 4th ed., Vol. 12, Kroschwitz, J., Ed., Wiley, New York, pp. 555–628, 1994.
5. Cavanagh, J., Glass container manufacturing, In *The Wiley Encyclopedia of Packaging Technology*, 2nd ed., Brody, A. L. and Marsh, K. S., Eds., Wiley, New York, pp. 475–484, 1997.
6. Doyle, P. J., Recent developments in the production of stronger glass containers, *Packag. Technol. Sci.*, 1, 47–53, 1988.
7. Gaines, L. L. and Mintz, M. M., *Energy Implications of Glass-Container Recycling*, Argonne National Laboratory, Argonne, IL, 1994, March.
8. Hanlon, J. F., Kelsey, R. J., and Forcino, H. E., *Handbook of Package Engineering*, 3rd ed., Technomic Publishing, Lancaster, PA, 1998.
9. Girling, P. J., Packaging in glass bottles, In *Handbook of Beverage Packaging*, Giles, G. A., Ed., CRC Press, Boca Raton, FL, 1999, chap. 3.
10. McLellan, G. W. and Shand, E. B., Glass technology, In *Glass Engineering Handbook*, 3rd ed., McGraw-Hill, New York, 1984, chap. 2.
11. Moody, B. E., *Packaging in Glass*, Hutchinson and Benham, London, England, 1977.
12. Pitman, K., Closures for beverage packaging, In *Handbook of Beverage Packaging*, Giles, G. A., Ed., CRC Press, Boca Raton, FL, 1999, chap. 11.

13. Rankine, B. C., Leyland, D. A., and Strain, J. J. G., Further studies on Stelvin and related wine bottle closures, *Australian Grapegrower & Winemaker Annual Technical Issue*, 196, 72, 1980, see also p. 74 and p. 76.
14. Stelzer, T., *Screwed for Good? The Case for Screw Caps on Red Wines*, Wine Press, Brisbane, Australia, 2003.
15. United States Pharmacopeia USP 27-NF 22. Rockville, Maryland: United States Pharmacopeial Convention, 948, 2004.
16. Willhoft, T., Victims of the pop bottle, *New Scientist*, 111(1522), 28–30, 1986.

9 Printing Processes, Inks, Adhesives and Labeling of Packaging Materials

CONTENTS

I. INTRODUCTION

Because most packaging materials are printed, a basic knowledge of the characteristics of the main printing processes is essential. In 594, the Chinese began to practice printing from a negative relief. Their method of rubbing off impressions from a wood block spread along the caravan routes to the Western world and, in 1400, the technique of printing with wooden blocks arrived in Europe from the Far East. The invention of paper, which was to provide the ideal surface for printing, also came from China.

In 1450, Johann Gutenberg invented the printing press in Germany by adapting the wine presses, which had been used in the Rhine Valley since the days of the Roman Empire. He used a recently perfected ink (based on linseed oil and soot) and also invented a functional metal alloy to mold the type. The printed word enabled information and knowledge, which was previously restricted to ecclesiastical establishments, to be widely disseminated and, by 1500, more than 9 million printed books were in circulation.

The basic principle of printing is that ink is deposited on an engraved plate and the inked image transferred to the substrate through contact. This can be performed in several ways when packaging material is to be printed. In direct printing, the inked plate makes direct contact with the packaging substrate. In indirect printing, the engraved plate transfers ink to an intermediate rubber blanket that then transfers the image to the packaging substrate. In stencil printing (e.g., screen printing), the ink is passed through a stencil to the substrate.

There are five different methods of printing in use today: relief (letterpress, flexography and flexo process); gravure or intaglio; lithography (offset) or planographic; screen or porous; and ink-jet or impactless.[8] In all but the latter method, ink is applied to a printing unit such as a cylinder or a plate, and is then transferred to the substrate by direct contact.

Plastics such as LDPE, HDPE, PP, PET and EVA copolymer cannot be satisfactorily printed unless their surfaces have been pretreated so as to obtain satisfactory adhesion between the ink and the plastic. This is because their inert, nonpolar surfaces do not permit any chemical or mechanical bonding between them and the ink. Various processes are used, all aimed at oxidizing the surface in some way. These include solvent treatment, chemical treatments, flame treatment and electrical treatment, and several of these processes were discussed in Chapter 5, Section III A. In addition to removing dust, oils, greases, processing aids, and so on, the surface of the plastic is activated by these processes and becomes more polar.

Photopolymers are light-sensitive plastics used to prepare letterpress, flexographic and offset printing plates. They have been available since 1974, and numerous systems have been developed for producing photopolymer plates. Basically, a photographic process is used in contrast to the photomechanical etching and molding system used to prepare most rubber plates. The photopolymer plate is formed by exposing the photopolymer to UV light through a film negative, which carries the image to be reproduced. Each photopolymer plate is an original derived directly from a negative.[4]

II. PRINTING PROCESSES

A. Relief

In this long-established process (commonly known as *letterpress*), the images or printing areas are raised above the nonprinting areas so that the ink rollers touch only the top surface of the raised areas. Originally, metal type was used, in which case, the process is called *letterpress*, but this has been largely replaced by synthetic rubber or photopolymer printing plates, in which case the process is called *flexographic*.

1. Letterpress

Letterpress printing presses are of three types: platen, flat-bed cylinder and rotary, the latter being by far the most common type for printing packaging materials. Plates must obviously be curved for mounting on rotary presses. They receive ink as they contact the inking rollers. An impression cylinder presses the substrate against the inked plate cylinder and transfers the image.

The inks used for letterpress printing are oil-based and slow drying, and have a pasty consistency, which makes the process a difficult one to apply to plastics films unless (like rigid PVC and RCF), they are not too susceptible to the high pressures which often have to be applied. A distinctive feature of letterpress printing is a "ghost-like" image around each character caused by the ink spreading slightly due to the pressure of the plate on the substrate. A slight embossing or denting sometimes appears on the reverse side of the surface, but the letterpress image is usually sharp and crisp.[6] Letterpress printing is still used for the printing of folding cartons, labels and all types of bags for dry goods.

2. Flexography

Flexography, a relief-printing technique and variation of letterpress printing, is a high-speed method that was developed primarily for printing packaging materials.[6] Introduced into the U.S. on a fairly broad scale from Germany in the early 1920s, it was known as *aniline* printing because, at that time, coal tar dyestuffs (derived from aniline oil) were used as the coloring ingredients for ink. In 1952, the name was changed to flexography. This is defined[4] as a method of direct rotary printing using resilient, raised image printing plates, affixable to plate cylinders of various repeat lengths, inked by a roll or doctor blade wiped metering roll, carrying fluid or paste type inks to virtually any substrate.

Generally, four rollers or cylinders are used. A rubber inking roller (fountain roller) revolves in an ink reservoir and transfers ink to a cavitated metering roller (often called an *anilox roller*), which is engraved such that it can hold ink in its recesses. An optional doctor blade removes excess ink from the surface of the anilox roller so that it transfers a controlled film of ink to the printing plates, which were made of rubber and referred to as *stereos* (an abbreviation for stereotypes), but are now generally made from photopolymer material. The printing plate then transfers this layer of ink to the substrate which is supported by the impression cylinder (Figure 9.1).

Flexographic printing can be carried out with a number of printing stations in sequence: one on top of the other (stack press), with the printing rollers arranged around a central large diameter drum (central impression) or with the printing stations arranged in a straight line (in-line). A stack press is used primarily for paper and laminated films; a central impression press for high-quality wide web films, and an in-line press for corrugated and folding cartons.[9] Because the costs of producing the plates are relatively low, flexographic printing is cost effective, especially for short runs.

Thin, fast-drying, solvent-based inks are used in flexography, permitting multicolored printing to be completed in one pass provided that oven drying is used. High speeds are possible. A major disadvantage was that very fine halftones were unable to be reproduced, because the inks had a low

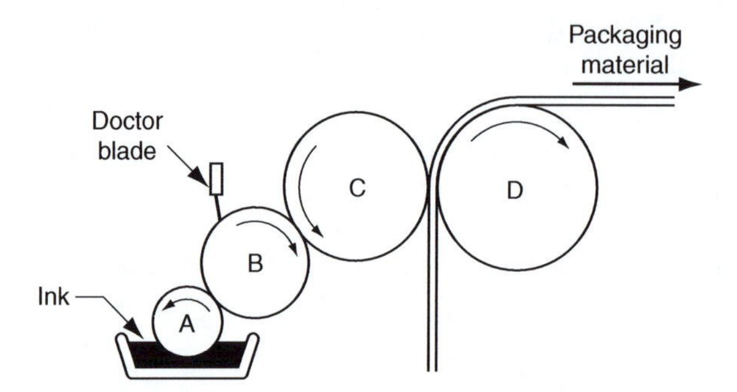

FIGURE 9.1 In flexographic printing, the fountain pan supplies ink to the rubber fountain roller, A, which supplies ink to the anilox roller, B, which transfers a uniform layer of ink to the printing cylinder, C. The impression cylinder, D, presses the packaging material against the printing cylinder from which ink is transferred to the packaging material. (*Source*: From Taggi, A. J., Printing: gravure and flexographic, In *The Wiley Encyclopedia of Packaging Technology*, 2nd ed., Brody, A. L. and Marsh, K. S., Eds., Wiley, New York, pp. 783–787, 1997. With permission.)

viscosity and it was difficult to get sharp images, with fine type having a tendency to fill in. However, with the introduction of photopolymer plates, this disadvantage has been overcome.

3. Flexo Process

For flexo process printing, a double-chambered doctor blade delivers the ink to the anilox roller in a closed system that ensures a constant flow of ink. A doctor blade removes excess ink from the anilox roller which is engraved with a very fine screen of 240 to 320 lines cm^{-1} (600 to 800 lines $in.^{-1}$). The engravings distribute a metered amount of ink to the print roller, which has printing plates mounted onto it. An impression roller presses the packaging material against the printing plate and ensures good transfer of ink to the paper (see Figure 9.2).

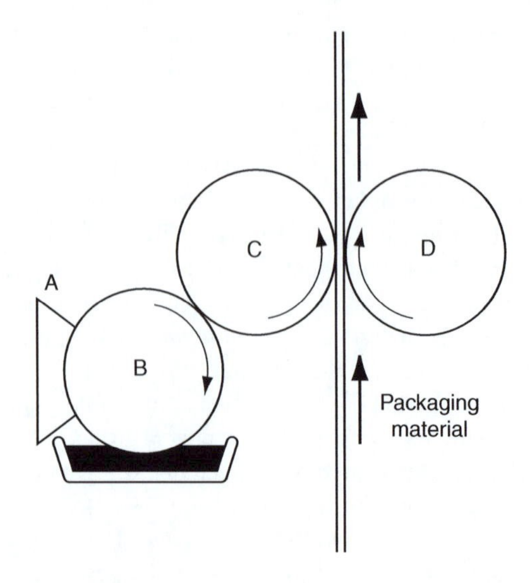

FIGURE 9.2 In flexo process printing, a chambered doctor blade, A, delivers ink to an anilox roller, B, which distributes a metered amount of ink to the printing cylinder C; the impression cylinder, D, presses the packaging material against the printing cylinder.

The quality of reproduction possible with flexo process printing has increased to the extent that it now approaches that of gravure printing. Photographic designs are able to be used.

B. Gravure

Gravure printing was known as *intaglio* from the Italian word for incising or engraving, reflecting the fact that this technique was first practiced in Florence in 1446 by the goldsmith and engraver Finiguerra. Gravure printing (more commonly known as *rotogravure printing* by the packaging industry because cylinders rather than plates are used) consists of a printing cylinder (image carrier), an impression cylinder and an inking system, as shown in Figure 9.3. The printing cylinder (usually made with a chrome-plated copper surface) has the image area etched to form a series of small cells of varying depth so that differing amounts of ink are picked up. During printing, the image carrier is immersed in fluid ink. As the image carrier rotates, ink fills the tiny cells and covers the surface of the cylinder. A doctor blade wipes away excess ink from the nonimage surface of the cylinder. As the cylinder comes into contact with the material to be printed, an impression cylinder (generally covered with a resilient rubber elastomer) presses the material into contact with the tiny cells of the printing cylinder, causing the ink in the cells to be transferred to the material through capillary action.

As with flexographic printing, gravure can print on a wide variety of materials and has found widespread use in the printing of cartons, foils, films, and papers for conversion into bags, and so on. The ideal substrates are generally smooth in finish (i.e., clay coated, supercalendered papers, films and foils) because effective ink transfer depends on thorough cell contact with the substrate.[4]

Gravure printing can easily be recognized because the entire image area is screened (usually 60 lines cm^{-1} or 150 lines $in.^{-1}$ for normal printing but more if a higher quality is required) to produce the tiny cells in the gravure cylinder. Thus, with a normal screen, there would be 3600 cells cm^{-2} or 22,500 cells $in.^{-2}$.

The preparation of the printing cylinders is costly and time-consuming, and gravure printing is therefore only economical for long production runs. Printing speeds are slower than those obtainable with flexographic printing processes. Despite these disadvantages, gravure printing produces high-quality, multicolor fine-detail printing.

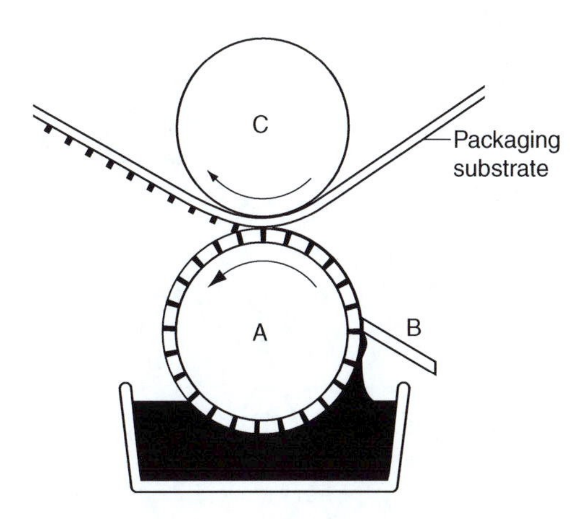

FIGURE 9.3 In rotogravure printing, an engraved cylinder, A, rotates through the ink, a doctor blade, B, removes ink from all but the etched recesses and the remaining ink leaves an imprint on the packaging material. C is a rubber impression cylinder.

C. Lithography

Lithographic printing (also known as *offset lithography* and occasionally as *planographic* printing) was invented by the Austrian artist Alois Senefelder in Munich in 1798 as a way to print musical scores more cheaply than by engraving. A lithograph is a print of a drawing done on a stone (usually a limestone). When Senefelder realized his stone-printing method could be used for many other things, he went into business with some partners to operate lithographic presses in the capitals of Europe. For 100 years, lithography was used to print packaging, cards, menus, book illustrations and labels — the first mass forms of advertising. It is still the major form of printmaking today.

Lithography involves printing from a flat surface, the image area being neither raised (as in letterpress and flexography) nor lowered (as in gravure). It is based on the principle that oil and water do not mix.

Oil-based ink is applied evenly through a series of rollers to the *offset* plate cylinder (usually made of aluminum). Water or fountain solution is simultaneously fed via rollers to the plate just before it contacts the inking rollers, as shown in Figure 9.4. Although very little moisture is required, the film of moisture is continuous on the nonimage areas of the plate and acts as a barrier, preventing adhesion of the ink. The plate accepts ink and repels water in the image areas. The image on the plate is transferred or offset to an intermediate or blanket cylinder covered with a rubber blanket. The material to be printed picks up the image as it passes between the blanket and the impression cylinder. The soft rubber blanket creates smooth, sharp images on a wide variety of materials, and is used extensively where illustrations are required on packaging materials.

Offset lithography produces quality printing on both rough and smooth papers, although coated papers are often used because they require less ink and give more brilliance to colors. It is used for printing labels and cartons, and for decorating metal containers, but is rarely used with plastic films. Offset printing plates are cheaper than those for rotogravure and can now deliver almost equivalent print quality.

A process that combines features of both letterpress and lithographic printing methods is known as *letterset* or *dry offset* printing. The term *dry* is used to differentiate it from the standard offset system that uses the incompatibility of water and inks to dampen the surface of the plate or substrate to prevent ink transfer.

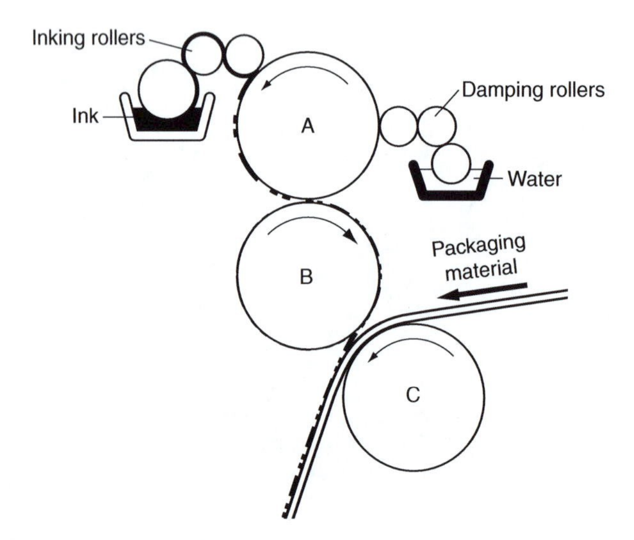

FIGURE 9.4 In offset lithography, the image is neither raised nor lowered. It is based on the principle that oil and water do not mix. A is the plate cylinder; B is the offset blanket cylinder and C is the impression cylinder.

In dry offset, ink is transferred by the raised surface of the relief plate to the rubber blanket, which then prints the entire multicolor copy taken from one to as many as six plate cylinders on the container in one operation. No chemical or water action is involved. This method has been especially developed to print round and tapered containers. Letterpress and lithographic inks can be used, the inks being either heat set or UV cured. However, special letterset inks provide very low odor printing for food cartons and confectionery wraps that require the use of water-miscible inks. Letterset is also used for printing on plastic bottles, cups and tubs, and on some metal packages, primarily tubes.[6]

D. Screen

Screen printing (also referred to as *serigraphy* or *porous* printing) is basically a stenciling process using a fine mesh screen made of silk, polyester or metal (fine wire) to support a stencil. The screen is completely coated on both sides with a light-sensitive emulsion, and a positive image of the graphics to be printed is placed on the outside of the emulsion screen. On exposure to intense light, the unshielded emulsion is cured (hardens) after which the shielded, uncured emulsion can be washed away. The nonblocked mesh area of the screen allows ink to pass through the fine mesh.

Ink is forced through the screen onto a substrate by a rubber blade or squeegee. Although screen printing was originally a manual process (and still is in some applications), power-screen presses with mechanical feed and delivery are in common use. Rotary screen presses are also available.

Screen printing inks can be solvent, oil, rubber or water-based, but are usually of the drying-oil type with the consistency of thick paints. Far greater amounts of ink are applied in screen printing compared with other methods of printing, and sometimes the texture of the screen can be recognized on the final image. Extra drying time is frequently needed.

Screen printing is a very versatile printing process, capable of printing on the widest variety of substrates including rounded and irregular surfaces (e.g., milk and soft drink bottles).[6] It is ideal for short production runs and is used for imaging glass, box wraps, folding cartons and plastic containers. However, it is relatively low speed and time is required for the ink to dry between colors.

E. Ink-Jet

Ink-jet or impactless printing is a noncontact, pressureless printing process, which produces an image with jets of colored ink. The ink is ejected through a nozzle under pressure, vibrated into uniform droplets that are charged electrostatically and directed by dot matrices housed in the printer's memory to form characters.

This printing process has been available since the mid-1960s and has found widespread application for the labeling, coding and dating of packaged food. The tiny droplets of ink can be projected into various recesses on the package, which, with the use of inks that adhere to nonabsorbent surfaces, makes it ideal for such applications.

III. INKS

A. Introduction

The major categories of printing processes are distinguished by whether the ink is held below, on or above, or passes through the surface that contacts the substrate to be printed. Recently, several *no surface* or *contactless* methods have been developed. These different geometries impose different requirements on the inks, which, in turn, largely determine the environmental impacts of each process.

1. Below the Surface

This includes the gravure process, using engraved plates, and the rotogravure process, using engraved cylinders. Ink is retained in depressions in the surface, and is transferred to the substrate when pressed against it. The ink needs to flow readily, and therefore has a relatively high solvent content.

2. On the Surface

This includes the lithographic processes that use differences in surface tension to create areas on the surface that are ink-repelling and ink-attracting. The type of ink used is invariably one that does not mix with water because the ink-repelling areas are water-attracting, and vice versa. A water-based "fountain solution" (often containing some alcohol or equivalent) is used to help keep the ink out of the ink-repelling areas. The ink needs to be somewhat higher in viscosity so that it sits on the surface without spreading. Both the hydrophobic ink solvent (often toluene) and the organics from the fountain solution can add to the environmental impacts. A relatively new process — waterless printing — is a type of lithographic printing that uses mechanical and temperature differences to accomplish the same purpose, with much lower solvent emissions.

3. Above the Surface

This consists of the archetypal printing processes associated with Gutenberg and the linotype machine. Raised letters and lines are inked and the ink is transferred to the substrate. The ink is low viscosity in order to spread uniformly on the ridges, and dries quickly once transferred to the substrate, thus requiring a volatile solvent. There are two major variations, depending on the nature of the plate or roller: letterpress and flexographic, as discussed above.

4. Through the Surface

This refers to the screen printing processes, in which the ink is forced through a polyester or wire mesh. A stencil is applied to the top of the mesh, and the ink passes through voids in the stencil, with the mesh regulating the flow of the ink.

5. No Surface

Modern electronics technology has made possible new approaches to transferring ink onto substrates in preset patterns with ink-jet printing, which is now a well-established technique.

B. Ink Components

Inks are designed for each of the five main printing processes. Flexography and gravure are known as the *liquid ink* processes, and are based on volatile solvents that evaporate readily at room temperatures. Lithography and letterpress are collectively known as the *paste ink* processes, using inks that are essentially nonvolatile at normal temperatures. Screen printing uses inks that fall between these two groups.

Printing inks are mixtures of three main types of ingredients: pigments, vehicles and additives. Pigments used in printing inks include both inorganic pigments (e.g., carbon black and titanium dioxide) and organic pigments, which are frequently dyes rendered insoluble by complexing with a metal ion. Most organic pigments are prepared from azo, anthraquinone and triarylmethane dyes, and phthalocyanines. Pigments produce color by selective absorption of light but, because they are solids, they also scatter light. Lead and other toxic pigments have been eliminated from inks for food packaging.

Vehicles generally consist of a resin or polymer with a liquid dispersant, which may be a solvent, oil or monomer. Choice of the vehicle for a printing ink depends on the printing process, how the ink will be dried and the substrate on which the image is to be printed. In lithography and letterpress, where inks are dried by absorption and oxidation, vehicles are generally mixtures of mineral and vegetable oils and resins. Flexographic inks, which are designed to dry quickly by evaporation, can be based on either water or organic solvents such as ethanol, ethyl acetate, *n*-propanol or isopropanol, with a wide variety of resins. Vehicles for gravure inks, which also dry by evaporation, may contain aromatic or aliphatic hydrocarbons and ketones as solvents. Inks for screen printing use organic solvents that are somewhat less volatile than those used for flexography or gravure (e.g., higher glycol ethers and aromatic and aliphatic hydrocarbons). Additives in inks include driers, waxes and plasticizers. UV radiation-cured inks, commonly based on acrylates, are used in all of the printing processes to varying degrees.[3]

The manufacture of inks consists of dissolving or dispersing resins in organic solvents or oils to produce the vehicle (varnish), mixing and dispersing the pigment or dye into the vehicle, and then introducing any additives.

Ink may be deposited on a substrate in various ways to obtain the required shade and depth of color. Solid deposition produces a continuous coating of ink, where the shade and depth of color depend on the type and amount of ink deposited on the substrate. Line printing produces a series of lines or crosshatches so that the ink is discontinuous. Shade and depth of color depend on the substrate, ink type, amount of ink and the relative concentration of lines or crosshatching. Dot printing (halftone) consists of a series of ink dots spread over the printed surface. The shade and depth of color depend on the substrate, ink type, amount of ink, dot spacing, dot size and dot density.

The inks used for the major packaging printing methods are discussed below.

C. Liquid Inks

1. Flexographic Ink

a. *Solvent-Based*

Typically, alcohols are the solvent of choice with addition of lower esters and small amounts of hydrocarbons to achieve solubility of the vehicle resin and proper drying of the ink film at press speed.[2] It is important that all the solvents used are screened to ensure that they do not interact with the printing plates and rollers. Since the 1990s, there has been a trend to phase out solvent-based inks so as to avoid the emission of VOCs, which are implicated in the formation of ground-level ozone and smog.

A typical formulation for a solvent-based flexographic ink used in the high-speed printing of LDPE film would be titanium dioxide pigment (40%), ethanol (26.7%), alcohol-soluble polyamide resin (20%), nitrocellulose varnish (5%), propyl acetate (5%), plasticizer (2%), wax (1%) and slip additive (0.3%).[2]

b. *Water-Based*

These are becoming increasingly common because of environmental concerns about the use of organic solvents in inks. Despite being termed *water-based* inks, the solvent used is not usually 100% water; up to 20% of an alcohol (typically ethanol) is added to increase drying speed, suppress foaming, increase resin compatibility and aid wetting of the plastic substrates. The water-based vehicles are generally emulsions or colloidal dispersions rather than true solutions, and include acrylic emulsion, maleic resin dispersion and styrene–maleic anhydride resins.[2]

2. Gravure Ink

a. *Solvent-Based*

A wider range of solvents can be used in gravure inks compared with flexographic inks because metal plates are used for gravure printing. Solvents used include aromatic hydrocarbons (toluene being the most common), aliphatic hydrocarbons, alcohols, esters, ketones, chlorinated solvents, nitroparaffins and glycol ethers, but their use is being phased out.

Technical developments have made it possible to almost completely prevent any toluene emissions in modern rotogravure printing processes. Nonetheless, the question is often asked as to whether ink containing toluene could be replaced by water-based rotogravure ink systems. However, in using water-based inks, there are considerable environmental and qualitative detriments that must be taken into consideration. Ink based on toluene is largely produced using regenerative raw material (resins), whereas water-based ink requires a production process which uses about three times as much energy. In addition, the energy required for drying the ink after the printing process is also higher, which results in increased emissions. In recycling, the color is harder to remove from paper that has been printed with water-based as opposed to solvent-based gravure inks. Finally, the quality of the colors does not meet today's market demands.

At present, toluene-based inks remain the standard for high-quality publication gravure printing, with almost all high-circulation magazines, catalogs and brochures being printed with this process. Although this solvent is listed as a hazardous air pollutant by the U.S. Environmental Protection Agency (EPA), publication gravure printers and their ink suppliers have concentrated on setting up systems for the recovery and recycling of toluene solvents. More than 90% of toluene in the inks is now returned to ink manufacturers. Many U.S. publication gravure printers have moved from aromatic to aliphatic solvent inks. Water-based gravure inks operate successfully in packaging printing, but they require a paper substrate with a minimum thickness of 250 gsm. There is a lack of scope for UV inks in gravure printing of packaging materials.

Regulations govern the use of toluene in ink, mainly because of the residues of the chemical left in the ink after printing, which could theoretically pose a threat to consumers. Toluene is considered to be a skin irritant, dangerous during long exposure through inhalation and a possible risk to the unborn child. It is also seen as a possible endocrine disrupter that could impair human fertility levels.

b. *Water-Based*

These are used for printing paper, paperboard substrates and increasingly for nonpaper substrates as solvent-based inks are phased out. To assist drying of water-based inks, the gravure cylinders are usually engraved or etched with shallower cells so that a thinner ink film is applied. However, to achieve the same relative printing density as solvent-based inks, a greater concentration of pigment must be used.

A typical formulation for a water-based packaging ink used to print paperboard cartons is acrylic emulsion (40%), water (26%), organic pigment (16.5%), isopropyl alcohol (7%), clay extender (5%), wax (2.5%), plasticizer (2%) and pH controller morpholine (1%).[2]

3. Screen Ink

Relatively little packaging material is screen printed. The two major types of inks used are solvent-based and plastisol, although other types are available.

D. Paste Inks

1. Offset Lithographic Inks

Lithographic ink is basically a concentrated dispersion of pigment in a viscous oil vehicle with various additives. Because the ink comes into contact with water during printing, it must be free from any tendency to bleed or to form an ink-in-water emulsion. The formation of a water-in-ink emulsion is unavoidable, but this does no harm unless the working consistency of the ink is affected.

These inks must resist the chemicals contained in the dampening solution, which is used to keep the plate constantly wetted. A typical ink for metal decoration would consist of organic pigment (15%), acrylate oligomer (40%), acrylate monomer (30%), photoinitiator and sensitizer (8%), tack reducer (4%) and wax (3%), and usually dried by simple radiation curing.[2] The inks used can also be dried by an oxidative process, which may be accelerated by the use of IR or UV radiation, and, in the case of metal packaging materials, by the use of heat (>150°C) or UV radiation.

2. Letterset Inks

The inks used generally have the viscosity and body of letterpress inks, and are used for the decoration of metal cans and plastic preformed tubs and containers; they are generally dried using heat (metal substrates) or UV radiation.

3. Letterpress Inks

These inks are used primarily for printing corrugated boxes and folding cartons, and drying is primarily oxidative or absorptive. This printing method is being increasingly replaced by water-based flexography.[2]

E. Thermochromic Inks

Thermochromic inks change color in response to fluctuations in temperature. They are reversible and will change color time after time with the appropriate exposure. These are highly specialized inks that combine standard ink components with one of several color-changing agents described below. Since these inks are used on a wide variety of substrates, they are offered in the typical solvent-based, water-based, plastisol, and UV formulations. Depending on the application, thermochromic inks can be applied with a number of printing processes, including offset lithography, flexography, gravure and screen printing.

The two types of thermochromic inks are liquid crystals and leuco (from the Greek word for white) dyes. Liquid crystal thermochromics are very difficult to work with and require highly specialized printing and handling techniques. For packaging applications, leucodye thermochromic inks are used in a wide range of applications including product labels. In its cool state, a leucodye exhibits color, and when warmed, it turns clear or translucent. It takes a 3 to 6°C shift to bring about a change in color.

Alkaline reduction of a dye produces the water soluble alkali metal salt or leuco form, which, on subsequent oxidation, reforms the original insoluble dye. A leucodye is a chemically reduced form of a dye, which, in most cases, is colorless or minimally colored and becomes colored by an oxidation step.

Some products printed with leucodye thermochromic inks change from one color to another, rather than transitioning from colored to clear. This is achieved by using an ink that combines a leucodye with a permanent-colored ink formulation. For example, a green ink may be formulated by adding a blue leucodye to a yellow ink. In its cool state, the printed ink layer is green and, once warmed, reverts to yellow as the leucodye becomes clear or translucent. Leucodyes can be designed to change color at various temperature ranges, from as low as −25 up to 66°C. A wide range of colors is also available.

In order to function, a leucodye requires a combination of chemicals working together in a special system of materials, which is microencapsulated to provide protection from the components of the ink. The microcapsules at 3 to 5 μm are at least ten times larger than the average pigment particle. Special considerations are usually involved in printing inks with these relatively large particles because the microencapsulation process cannot completely protect the leucodye system.

Under normal conditions, thermochromic leucodye inks have a shelf life of 6 months or more. After they are printed, they function, or continue to change color, for years. The postprint functionality can, however, be adversely affected by UV light, temperatures in excess of 121°C and aggressive solvents.

As the price of thermochromic inks decreases, their use will become more widespread on food packaging. One obvious area is to indicate to the consumer when the product is at the ideal temperature for consumption. In addition, a thermochromic ink that goes from colored to transparent can hide a warning message that will only be legible once a certain temperature is achieved. Fast-drying, irreversible thermochromic inks have also been formulated for the canning industry. Applied to the end of a can prior to retorting, they can provide a visual verification that the can has been retorted, changing color under the specific conditions of wet heating typical of the retort process. However, they do not indicate that an adequate F_o process has been achieved.

IV. ADHESIVES

Adhesives are widely used in the packaging industry. The principal uses include the forming and sealing of corrugated cases and folding cartons, the winding of tubes for cores and composite cans, the labeling of bottles, jars and other packages, the lamination of paper to paper, paperboard and foil, and the lamination of plastic films.

Adhesion is the process of bonding two surfaces together, the surfaces being referred to as *adherends*. An adhesive is any substance applied as a thin intermediate layer between two adherends that holds or bonds them together. Adsorption theory states that adhesion results from intimate intermolecular contact between two materials, and involves surface forces that develop between the atoms in the two surfaces. The most common surface forces that form at the adhesive-adherend interface are van der Waals forces. In addition, acid–base interactions and hydrogen bonds may also contribute to intrinsic adhesion forces.

Mechanical adhesion is where two porous materials are bonded by adhesive entering the mechanical structure (e.g., paper to paper). Porosity of the adherend will affect the degree of penetration — too little or too much porosity results in a weak bond. Materials such as paper can be coated to control penetration of adhesive or ink.

In contrast, chemical adhesion is the formation of chemical bonds between an adherent and the adhesive. Many adhesives are strongly polar (e.g., starch, PVA and casein). Paper is a polar adherend, whereas LDPE, PP, glass and metals are nonpolar adherends. Many adhesives work using both mechanisms.

The process of establishing intimate contact between an adhesive and adherend is known as *wetting*. To be effective, the adhesive must wet the surface of the adherend. For effective wetting, the surface must be clean and the surface tension of the solvent in the adhesive should be such that the adhesive will wet the surface. Wetting agents are sometimes added to improve this aspect. Polarity will also affect wetting with polar liquids wetting polar surfaces and vice versa. A continuous adhesive film across the surface improves adhesion, which is related to the surface tension and viscosity of the adhesive.

Tack is the ability to form a bond of measurable strength immediately. The elusive nature of the word (and of the concept) stems from the fact that tack results from a composite of several physical observations, parameters and concepts. Tack is the property of an adhesive that allows it to adhere to another surface on immediate contact. It is the "stickiness" of the adhesive while in a fluid or semifluid state.

Tack is not a true physical property of an adhesive, but rather a composite property that has a broad and somewhat qualitative meaning that is very useful in practice. Although there are many different ways to measure tack depending on the application, tack is simply the resistance to separation. Separation is rate and temperature sensitive and involves viscoelastic deformation of the bulk adhesive. The appropriate measurement is the work expended in separation rather than the force used.

Water-borne adhesives are the oldest and still the largest volume class of adhesive used in packaging.[5] Water-borne adhesives are slower drying than solvent-borne adhesives, requiring about three times more heat to dry; they also need more time to achieve steady-state performance during production runs. Water-borne adhesives generally do not provide the shear or peel strength that solvent-borne systems provide and once cured, water-borne adhesives usually do not have the moisture resistance of solvent-borne adhesives. However, water-borne adhesives can withstand wide temperature ranges, are easy and safe to handle and are low in cost.

A. Natural Materials

Until the 1940s, only naturally-derived materials were used as packaging adhesives. However, they have now been replaced by synthetic adhesives in many applications.

1. Starch

Starches in the form of amylose and amylopectin are obtained from plants such as wheat, potatoes and maize, and are then subjected to acid hydrolysis to produce smaller chain segments including dextrins. Plasticizers are often added, together with fillers such as kaolin clay and calcium carbonate, which modify the viscosity and reduce cost. Borax can act as a viscosity modifier, increase tack and prevent microbial growth.

The single largest use of starch adhesives is in the manufacture of corrugated board for shipping containers. Modified starches and dextrins are used in the sealing of cartons, winding of spiral tubes, seaming and forming of bags, and attaching labels to metal cans. For the labeling of glass bottles, an alkaline-treated starch adhesive with a special tacky, cohesive consistency is used.

Starch-based adhesives are easy to handle and inexpensive, but suffer from a relatively slow rate of bond formation, poor water resistance and limited adhesion to coatings and plastics.

2. Protein

Natural adhesives based on animal protein were once widely used, but are now used only in specific narrow areas where synthetics have been unable to match their performance. Protein-based adhesives are all polyamino acids derived from casein, soybeans or animal hides, bones and connective tissue (collagen).

a. Casein

This is produced by the acidification of skim milk, which results in the precipitation of the milk protein casein. It has high bond strength and good water resistance, and is the preferred adhesive for labeling glass beer bottles because it provides resistance to cold water immersion and can be removed by alkaline washing. It is also used as an ingredient in adhesives used to laminate aluminum foil to paper.[5]

b. Animal Glue

This is a water-based solution derived from collagen extracted from animal skin and bone by alkaline hydrolysis. It has a high level of hot tack and a long, gummy tack range. It is the standard adhesive used in forming rigid setup boxes.[5]

3. Natural Rubber Latex

This is extracted from the rubber tree *Hevea brasiliensis* and is used in a variety of self-seal applications because it is the only adhesive system that will form bonds only to itself with pressure. This property is used in self-seal candy wraps where it is called *cold-seal*.[5]

B. Synthetic Materials

1. Water-Borne Adhesives

These are the most broadly used class of adhesives in packaging, and are mainly resin emulsions consisting of PVA emulsions and stable suspensions of PVA particles in water. Recently, EVA copolymers or acrylic esters have greatly improved adhesion capabilities.[5] They are used to form, seal or label cartons, tubes, bags and bottles, and can be formulated to adhere to paper, glass, and most plastics and metals. Water-borne adhesives can be formulated to be very water-insensitive (for immersion resistance) or water-sensitive, depending on need.

2. Hot-Melt Adhesives

Hot-melt adhesives are 100% nonvolatile thermoplastic materials that can be melted by heat and then applied as a liquid to an adherend. The bond is formed when the adhesive resolidifies. Because of their extremely rapid rate of bond formation, they can be used successfully on high-speed packaging lines. They can be formulated to adhere to almost any surface due to the wide range of polymers and modifiers used. Their major weakness is the rapid falloff in strength at elevated temperatures which is not generally a problem in food packaging applications.

A typical hot-melt sealant is composed of three primary components: polymers (30 to 40%); tackifying resin (30 to 40%); petroleum wax (20 to 30%) plus antioxidants, fillers, plasticizers and blowing agents to enhance other properties. The tackifying resins control viscosity as well as wetting and adhesion. The function of the wax is to lower viscosity and control set speed. Fillers are added to opacify or modify the adhesive's flow characteristics as well as to reduce cost.[7]

The most widely used polymer is EVA copolymer which is normally run at 180°C. However, the availability of very low MW EVA copolymers has resulted in EVA hot-melt that can be run as low as 120°C, leading to energy savings and safer running conditions.[5] Other polymers used include PP, PETs, PAs and polyurethanes (the latter used for adhesive lamination of films such as PP).

3. Solvent-Borne Adhesives

Organic solvents have been used as adhesive carrier fluids and diluents, as well as for surface preparation and cleanup. Since the 1990s, environmental and workplace safety regulations on solvents have become increasing stringent, and there has been a trend to replace solvents and solvent-borne adhesives to avoid the emission of VOCs that takes place during formulation, application, drying and curing. As a consequence, the use of water-borne and hot-melt adhesives has grown. Today, only a small quantity of solvent-borne adhesives is used in specialized applications where water-borne or hot-melt systems do not meet the technical requirements.

V. LABELING

Labeling is a means of performing the communication function of packaging, informing the consumer about nutritional content, net weight, product use, and so on. Labeling acts as a silent salesperson through distinctive branding, facilitating identification at check-outs through the Universal Product Code (UPC). While almost all paper-based packaging (and increasingly metal and plastics packaging) is preprinted, many glass, plastic and metal packages still require labeling.

A. Glued-On Labels

These are the simplest type and consist of sheet material (typically paper), which has been printed and cut to size. They are attached to the package with adhesive, which is applied either at the time of application, or at the time of manufacture, in which case, the adhesive is activated with moisture immediately prior to application. This type of label is widely used for large volume items such as beer, soft drinks, wines and canned foods where high-speed application is required.

For returnable glass and plastic bottles, it is important that the wet strength of the paper is sufficient to ensure that the label can be removed in the bottle washer without repulping.

B. Self-Adhesive (Pressure-Sensitive) Labels

These can be made from paper, plastic or aluminum foil laminated to paper or plastic, and can be produced to adhere to a wide range of materials. They are supplied with an adhesive coated on the unprinted side and mounted on release paper, which is removed immediately before application to expose the adhesive.[1]

C. In-Mold Labels

In-mold labeling is a decorating technique used worldwide for blow molded bottles, as well as injection molded and thermoformed containers. It was pioneered in Europe for injection molding in the early 1970s, and in the U.S. for blow molding later in that decade. Printed labels can be applied to containers and lids during thermoforming, blow molding and injection molding.

The first in-mold labels (IMLs) consisted of paper, clay-coated on both sides. One side had a heat seal adhesive, and the other had an inked surface with an overprint coating to provide protection. Today, such paper labels comprise approximately half the IML market. Labels made with a plastic film form the other portion of the market.

A plastic IML is typically made from HDPE/LDPE blended material and is compatible with a host of plastic containers. What differentiates IMLs from conventional glue-on labels is the heat seal coating that is applied to the back side of the IML stock during the manufacturing process.

IMLs made from film offer better heat, moisture and chemical resistance than those labels made from paper. There are also recycling advantages with film labels. However, the greatest advantage with the use of film is a decorative consideration — the "no-label" look. This means it is possible to prepare an IML container that actually appears to have no label at all because the unprinted areas of the label blend into the container wall.

IML materials must be able to withstand the container manufacturing process. The heat generated during blow molding presents a challenge to most inks because pigments can change. Varnishes also face special challenges in IML use. The combination of heat and flexing as the container is shaped, followed by sudden cooling, can produce an "orange peel" effect (i.e., a deformation of the smooth surface).

During the in-mold labeling process, a label is placed in the open mold and held in place by vacuum ports, electrostatic attraction or other appropriate means. The mold closes and molten plastic resin is extruded into the mold where it conforms to the shape of the object. The hot plastic envelops the label, making it an integral part of the molded object. The difference between glue applied labels and IMLs is that glue applied labels are on the surface of the object while IMLs are in the wall of the object.

D. Sleeve Labels

A wide range of containers can be sleeve labeled including glass bottles, plastic bottles (extrusion blow molded PP and HDPE bottles as well as stretch blow molded PET) and metal cans. There is almost no restriction regarding the shape of the container. Sleeve labels shrink into or stretch around contours, penetrate variable geometries (such as hourglass shapes) and conform to irregular features (grips or slender necks). Preformed, printed sleeves are slipped over the container (normally a glass or plastic bottle) on-line, and are then either shrunk in a heat tunnel or (provided that the container is of simple shape) the actual label relaxes onto the container itself. A complex container shape will generally require the sleeve to be shrunk to the contours of the container.

Most shrink sleeves are made of oriented plastic films that shrink around a container when heat is applied. They are used as labels, tamper-evident neckbands and safety shields. Most shrink sleeves are made from PVC or PS film that has been uniaxially oriented to give the desired degree of shrink. The inherent property of uniaxially oriented film — to conform to almost any shape when heat is applied — vastly increases the range and variety of container designs that can use shrink sleeves.

Roll-fed stretch sleeves work well with straight sided single-serve bottles, and are often used with PET bottles. Roll-fed labels are relatively easy to apply, and no heat tunnel is needed.

Shrink-sleeve labels can decorate the complete surface of the container, including the closure if necessary, to provide tamper-evident neck and cap seals while offering maximum graphic space. One limitation as to the shape of the sleeved container is the amount of possible shrink; a diameter of 100 mm can be shrunk to a maximum of 40 to 35 mm. Shrink-sleeve material also provides a UV block.

Sleeve manufacture begins with the extrusion of the film material. This can be PVC, OPP, OPS or PET, and is typically 50-μm thick. Depending on the material, up to 75% shrinkage can be achieved. Although it does not have a very environmentally friendly image in some countries, PVC offers a high transparency with a high shrink of 65% at a low temperature. In contrast, OPP (which does not suffer from a negative environmental image) offers minimal distortion of the decoration but is less transparent, features a low rigidity and can shrink only 50%. OPP also goes through natural shrinkage, and the sleeve can relax after application. OPS has comparable mechanical and thermal properties. However, OPS is more transparent than OPP and has a minimum vertical shrinkage. At 75%, OPS offers the highest shrinkage, but is not very transparent and has low rigidity. PET has good mechanical properties, is highly transparent, has a high tensile strength and offers shrinkage of 70%. However, the raw material is quite expensive and shrinkage is less controllable, making it more difficult to control than PVC. After film extrusion, the material is gravure- or flexo printed in up to 10 colors. The sleeve is positioned before the containers enter the shrink tunnel. Shrinking is effected with hot air, infrared radiant heat, steam or a combination of these.

Thin PS foams coextruded with a surface layer for printability are used on beverage bottles. A typical construction would be a foamed PS layer of 300 μm with a 20 μm surface layer. The foamed PS reduces the noise level on filling lines, eliminates partitions in secondary containers and protects glass bottles in vending machines. It also insulates the container. Surface printed, shrinkable foam labels are used for prelabeling glass, plastic and microwavable containers.

REFERENCES

1. Anon, Labels and labeling machinery, In *The Wiley Encyclopedia of Packaging Technology*, 2nd ed., Brody, A. L. and Marsh, K. S., Eds., Wiley, New York, pp. 536–541, 1997.
2. Bassemir, R. W. and Bean, A. J., Inks, In *The Wiley Encyclopedia of Packaging Technology*, 2nd ed., Brody, A. L. and Marsh, K. S., Eds., Wiley, New York, pp. 511–514, 1997.
3. Bassemir, R. W., Bean, A., Wasilewski, O., Kline, D., Hillis, W., Su, C., Steel, I. R., and Rusterholz, W. E., Inks, In *Kirk–Othmer Encyclopedia of Chemical Technology*, 4th ed., Vol. 14, Kroschwitz, J., Ed., Wiley, New York, pp. 483–503, 1995.

4. Cotton, J. W., Ed., *Flexography Principles and Practices*, 3rd ed., Flexographic Technical Association, Huntington Station, New York, 1980.
5. Kaye, I., Adhesives, In *The Wiley Encyclopedia of Packaging Technology*, 2nd ed., Brody, A. L. and Marsh, K. S., Eds., Wiley, New York, pp. 23–25, 1997.
6. Lentz, J., Printing, In *The Wiley Encyclopedia of Packaging Technology*, Bakker, M. and Eckroth, D., Eds., Wiley, New York, pp. 554–559, 1986.
7. Pocius, A. V., Adhesives, In *Kirk–Othmer Encyclopedia of Chemical Technology*, 4th ed., Vol. 1, Kroschwitz, J., Ed., Wiley, New York, pp. 445–466, 1991.
8. Taggi, A. J. and Walker, P., Printing processes, In *Kirk–Othmer Encyclopedia of Chemical Technology*, 4th ed., Vol. 20, Kroschwitz, J., Ed., Wiley, New York, pp. 62–128, 1996.
9. Taggi, A. J., Printing: gravure and flexographic, In *The Wiley Encyclopedia of Packaging Technology*, 2nd ed., Brody, A. L. and Marsh, K. S., Eds., Wiley, New York, pp. 783–787, 1997.

10 Deteriorative Reactions in Foods

CONTENTS

I. INTRODUCTION

The principal aim of this chapter is to provide a brief overview of the major biochemical, chemical, physical and biological changes that occur in foods during processing and storage, and show how these combine to affect food quality. Knowledge of such changes is essential before a sensible choice of packaging materials can be made, because the rate and/or magnitude of such changes can often be minimized by selection of the correct packaging materials.

The deterioration of packaged foods — which includes virtually all foods because very few foods are currently sold without some form of packaging — depends largely on transfers that may occur between the internal environment inside the package, and the external environment, which is exposed to the hazards of storage and distribution. For example, there may be transfer of moisture vapor from a humid atmosphere into a dried product, or transfer of an undesirable odor from the external atmosphere into a high fat product. In addition to the ability of packaging materials to protect and preserve foods by minimizing or preventing such transfers, packaging materials must also protect the product from mechanical damage, and prevent or minimize misuse by consumers (including tampering).

Although certain types of deterioration will occur even if there is no transfer of mass (or heat, because some packaging materials can act as efficient insulators against fluctuations in ambient temperatures) between the package and its environment, it is often possible to prolong the shelf life of the food through the use of packaging.

It is important that food packaging is not considered in isolation from food processing and preservation, or indeed from food marketing and distribution. All of these factors interact in a complex way, and concentrating on only one aspect at the expense of the others is a surefire recipe for commercial failure.

The development of an analytical approach to food packaging is strongly recommended, and to achieve this successfully, a good understanding of food safety and quality is required. Without question, the more important of these is food safety, which is the freedom from harmful chemical or microbial contaminants at the time of consumption. Packaging is directly related to food safety in two ways.

Firstly, if the packaging material does not provide a suitable barrier around the food, then micro-organisms can contaminate the food and make it unsafe. However, microbial contamination can also arise if the packaging material permits the transfer of, for example, moisture or O_2 from the atmosphere into the package. In this situation, micro-organisms that are present in the food, but present no risk because of the initial absence of moisture or O_2, may subsequently be able to grow and present a risk to the consumer. Secondly, the migration of potentially toxic compounds from some packaging materials to the food is a possibility in certain situations, which gives rise to food safety concerns. In addition, migration of other components from packaging materials, while not harmful to human health, may adversely affect the quality of the product.

The major quality attributes of foods are texture, flavor, color, appearance and nutritive value, and these attributes can all undergo undesirable changes during processing and storage; a summary of such changes is given in Table 10.1. With the exception of nutritive value, the changes that can

TABLE 10.1
Classification of Undesirable Changes that Can Occur in Foods

Attribute	Undesirable Change
Texture	(a) Loss of solubility
	(b) Loss of water-holding capacity
	(c) Toughening
	(d) Softening
Flavor	Development of
	(e) Rancidity (hydrolytic or oxidative)
	(f) Cooked or caramel flavors
	(g) Other off-flavors
Color	(h) Darkening
	(i) Bleaching
	(j) Development of other off-colors
Appearance	(k) Increase in particle size
	(l) Decrease in particle size
	(m) Nonuniformity of particle size
Nutritive value	Loss or degradation of
	(n) Vitamins
	(o) Minerals
	(p) Proteins
	(q) Lipids

Source: From Fennema, O. R. and Tannenbaum, S. R., Introduction to food chemistry, In *Food Chemistry*, 3rd ed., Fennema, O. R., Ed., Marcel Dekker, New York, 1996, chap. 1. With permission.

occur in these attributes are readily apparent to the consumer, either prior to or during consumption. Packaging can affect the rate and magnitude of many of the quality changes shown in Table 10.1. For example, development of oxidative rancidity can often be minimized if the package is an effective O_2 barrier, flavor compounds can be absorbed by some types of packaging material and the particle size of many food powders can increase (i.e., clump) if the package is a poor moisture barrier. This chapter outlines the major biochemical, chemical, physical and biological changes that occur in foods during processing and storage, and shows how these combine to affect food quality. The following chapter addresses the issue of shelf life, which is very clearly related to food quality.

II. DETERIORATIVE REACTIONS IN FOODS

Knowledge of the kinds of deteriorative reactions that influence food quality is the first step in developing food packaging that will minimize undesirable changes in quality, and maximize the development and maintenance of desirable properties. Once the nature of the reactions is understood, knowledge of the factors that control the rates of these reactions is necessary in order to fully control the changes occurring in foods during storage (i.e., while packaged). The nature of the deteriorative reactions in foods is reviewed in this section, and the factors that control the rates of these reactions are discussed in the following section.

A. Enzymic Reactions

Enzymes are complex globular proteins that can act as catalysts, accelerating the rate of chemical reactions by factors of 10^{12} to 10^{20} over that of uncatalyzed reactions. An understanding of the biological mechanisms for controlling enzymic activities, and the biochemical mechanisms of enzyme action, can provide the food technologist with the means of effectively exploiting enzymes in food processing. From a food packaging point of view, knowledge of enzyme action is fundamental to fuller understanding of the implications of one form of packaging over another. The importance of enzymes to the food processor is often determined by the conditions prevailing within and outside the food. Control of these conditions is necessary to control enzymic activity during food processing and storage. The major factors that are useful in controlling enzyme activity are temperature, a_w (water activity), pH, chemicals that can inhibit enzyme action, alteration of substrates, alteration of products and preprocessing control.

Three of these factors are particularly relevant in a packaging context. The first is temperature, where the ability of a package to maintain a low product temperature and thus retard enzyme action will often increase product shelf life. The second important factor is a_w because the rate of enzyme activity is dependent on the amount of water available; low levels of water can severely restrict enzymic activities and even alter their pattern of activity. Finally, alteration of substrate (in particular, the ingress of O_2 into a package) is important in many oxygen-dependent reactions that are catalyzed by enzymes.

B. Chemical Reactions

Many of the chemical reactions occurring in foods can lead to a deterioration in food quality (both nutritional and sensory) or the impairment of food safety. The more important classes of these reactions are listed in Table 10.2 and are discussed fully in the standard textbook on food chemistry.[7] In the present context, it is noteworthy that such reaction classes can involve different reactants or substrates depending on the food and the particular conditions for processing or storage in question. The rates of these chemical reactions are dependent on a variety of factors amenable to control by packaging including light, O_2 concentration, temperature and a_w. Therefore, in certain circumstances, the package can play a major role in controlling these factors, and thus indirectly the rate of the deteriorative chemical reactions.

1. Sensory Quality

The two major chemical changes that occur during the processing and storage of foods, leading to a deterioration in sensory quality, are lipid oxidation and nonenzymic browning (NEB). Chemical reactions are also responsible for changes in the color and flavor of foods during processing and storage.

TABLE 10.2
Chemical Reactions that Can Lead to Deterioration of Food Quality or Impairment of Safety

Nonenzymic browning
Lipid hydrolysis
Lipid oxidation
Protein denaturation
Protein cross-linking
Oligo- and polysaccharide hydrolysis
Protein hydrolysis
Polysaccharide synthesis
Degradation of specific natural pigments
Glycolytic changes

Source: From Fennema, O. R. and Tannenbaum, S. R., Introduction to food chemistry, In *Food Chemistry*, 3rd ed., Fennema, O. R., Ed., Marcel Dekker, New York, 1996, chap. 1. With permission.

a. Lipid Oxidation

Autoxidation is the reaction, by a free radical mechanism, of molecular O_2, with hydrocarbons and other compounds. The reaction of free radicals with O_2 is extremely rapid, and many mechanisms for initiation of free radical reactions have been described. Autoxidation is a major cause of food deterioration; the crucial role that this reaction plays in the development of undesirable flavors and aromas in foods is well documented.

As well as being responsible for the development of off-flavors in foods, the products of lipid oxidation may also react with other food constituents such as proteins, resulting in extensive cross-linking of the protein chains through either protein–protein or protein–lipid cross-links.

Factors that influence the rate and course of oxidation of lipids are well known and include light, local O_2 concentration, high temperature, the presence of catalysts (generally transition metals such as iron and copper, but also heme pigments in muscle foods) and a_w. Control of these factors can significantly reduce the extent of lipid oxidation in foods.

b. Nonenzymic Browning

NEB is one of the major deteriorative chemical reactions that occur during storage of dried and concentrated foods. The NEB or Maillard reaction can be divided into three stages: (1) early Maillard reactions, which are chemically well-defined steps without browning, (2) advanced Maillard reactions, which lead to the formation of volatile or soluble substances, and (3) final Maillard reactions, leading to insoluble brown polymers.

The initial reaction involves a simple condensation between an aldehyde (usually a reducing sugar) and an amine (usually a protein or amino acid) to give a glycosylamine. The glycosylamine then undergoes an Amadori rearrangement to form an Amadori derivative. The formation of Amadori compounds accounts for the observed loss of both reducing sugar and amine during the Maillard reaction. Although the early Maillard reactions forming Amadori compounds do not cause browning, they do reduce nutritive value.

The final step of the advanced Maillard reaction is the formation of many heterocyclic compounds such as pyrazines and pyrroles, as well as brown melanoidin pigments. These pigments are formed by polymerization of the reactive compounds produced during the advanced Maillard reactions. The polymers are relatively inert and have a molecular weight greater than 1000.

c. Color Changes

Acceptability of color in a given food is influenced by many diverse factors, including cultural, geographical and sociological aspects of the population. However, regardless of these factors, certain food groups are only acceptable if they fall within a certain color range. The color of many foods is due to the presence of natural pigments. The major changes that these can undergo are briefly described below; a more detailed discussion can be found elsewhere.[17]

i. Chlorophylls

The name *chlorophyll* describes the green pigments involved in the photosynthesis of higher plants. The major change that chlorophylls can undergo is *pheophytinization* — the replacement of the central magnesium atom by hydrogen and the subsequent formation of a dull olive-brown pheophytin. Because this reaction is accelerated by heat and is acid catalyzed, it is unlikely to be influenced by the choice of packaging.

Almost any type of food processing or storage causes some deterioration of the chlorophyll pigments. Although pheophytinization is the major change, other reactions are possible. For example, dehydrated products such as green peas and beans packed in clear glass containers undergo photooxidation and loss of desirable color.

ii. Heme Pigments

Meat is an important part of many diets, and the color of red meat is due to the presence of the heme pigment *myoglobin*. Myoglobin is a complex muscle protein contained within the cells of the tissues where it acts as a temporary storehouse for the O_2 brought by the hemoglobin in the blood. The protein moiety is known as *globin* and the nonpeptide portion is called *heme*.

The color cycle in fresh meats is reversible and dynamic, with the three pigments oxymyoglobin, myoglobin and metmyoglobin constantly interconverted. In cured meat products, nitrite reacts with these pigments to form additional heme-based compounds. These reactions are discussed in more detail in Chapter 16. At this stage, it is sufficient to note that packaging has an extremely important influence on meat pigments. For example, at low partial pressures of O_2 (i.e., an almost impermeable package), the formation of brown metmyoglobin is favored. If the package is completely impermeable to O_2, then the heme pigments are fully reduced to the purple myoglobin.

iii. Anthocyanins

Anthocyanins are a group of more than 150 reddish water-soluble pigments that are very widespread in the plant kingdom. An anthocyanin pigment is composed of an aglycone (an anthocyanidin) esterified to one or more of five sugars (in order of relative abundance glucose, rhamnose, galactose, xylose and arabinose). The rate of anthocyanin destruction is pH dependent, being greater at higher pH values.

Of interest from a packaging point of view is the ability of some anthocyanins to form complexes with metals such as Al, Fe, Cu and Sn. These complexes generally result in a change in the color of the pigment (e.g., red sour cherries react with tin to form a purple complex) and are therefore undesirable. Because metal packaging materials such as cans could be sources of these metals, they are usually coated with special organic linings (enamels) to avoid these undesirable reactions (see Chapter 7).

iv. Carotenoids

The *carotenoids* are a group of mainly lipid-soluble compounds responsible for many of the yellow and red colors of plant and animal products. Carotenoids include a class of hydrocarbons called *carotenes* and their oxygenated derivatives called *xanthophylls*. Carotenoids can exist in the free state in plant tissue, or in solution in lipid media such as animal fatty tissue.

The main cause of carotenoid degradation in foods is oxidation. The mechanism of oxidation in processed foods is complex and depends on many factors. The pigments may autoxidize by reaction with atmospheric O_2 at rates dependent on light, heat and the presence of pro- and antioxidants.

v. Miscellaneous Natural Pigments

There are a number of other groups of compounds that are responsible for some of the colors in foods. These include *flavonoids* (yellow compounds with chemical structures similar to the anthocyanins), *proanthocyanins* (these are colorless but contribute to enzymic browning reactions in fruits and vegetables) and *tannins* (which contribute to enzymic browning reactions but their mechanisms of action are not well understood).

d. Flavor Changes

The term *flavor* has evolved to a usage that implies an overall integrated perception of the contributing senses of smell and taste at the time of food consumption. Specialized cells of the olfactory epithelium of the nasal cavity are able to detect trace amounts of volatile odorants. Taste buds located on the tongue and back of the oral cavity enable humans to sense sweetness, sourness, saltiness and bitterness, and these sensations contribute to the taste component of flavor.

In fruits and vegetables, enzymically generated compounds derived from long-chain fatty acids play an extremely important role in the formation of characteristic flavors. In addition, these types of reactions can lead to important off-flavors. Enzyme-induced oxidative breakdown of unsaturated fatty acids occurs extensively in plant tissues, and this yields characteristic aromas associated with some ripening fruits and disrupted tissues.[13]

Fats and oils are notorious for their role in the development of off-flavors through autoxidation. Aldehydes and ketones are the main volatiles from autoxidation, and these compounds can cause painty, fatty, metallic, papery and candle-like flavors in foods when their concentrations are sufficiently high. However, many of the desirable flavors of cooked and processed foods derive from modest concentrations of these compounds.[13] The permeability of packaging materials is important for retaining desirable volatile components within packages and for preventing undesirable components entering the package from the ambient atmosphere.

Many flavor compounds found in cooked or processed foods occur as the result of reactions common to all types of foods regardless of whether they are of animal, plant or microbial origin. These reactions take place when suitable reactants are present and appropriate conditions such as heat, pH and light exist. Packaging can play an important role in these reactions.

2. Nutritional Quality

As well as the chemical changes described above, which may have a deleterious effect on the sensory properties of foods, there are other chemical changes that can affect the nutritive value of foods. These reactions are discussed fully in standard textbooks.[1,2,12] Therefore, only a brief review of some of these reactions is presented here to illustrate the potential role of packaging in minimizing nutrient degradation in foods.

The four major factors that impact on nutrient degradation and can be controlled to varying extents by packaging are light, O_2 concentration, temperature and a_w. However, because of the diverse nature of the various nutrients as well as the chemical heterogeneity within each class of compounds and the complex interactions of the above variables, generalizations about nutrient degradation in foods are necessarily broad.

a. *Vitamins*

The chemical conversion of vitamins to biologically inactive products during the storage of foods has been the subject of extensive research. A generalized summary of vitamin stability is presented in Table 10.3, although it is important to note that exceptions exist and invalid conclusions could be reached on the basis of these generalizations.

Ascorbic acid is the most sensitive vitamin in foods; its stability varies markedly as a function of environmental conditions such as pH and the concentration of trace metal ions and O_2. The nature of the packaging material can significantly affect the stability of ascorbic acid in foods. The effectiveness of the material as a barrier to moisture and O_2, as well as the chemical nature of the surface exposed to the food, are important factors. For example, problems of ascorbic acid instability in aseptically packaged fruit juices have been encountered because of O_2 permeability and the O_2 dependence of the ascorbic acid degradation reaction. In addition, because of the preferential oxidation of metallic tin, citrus juices packaged in cans with a tin contact surface exhibit greater stability of ascorbic acid than those in enameled cans or glass containers. The aerobic and anaerobic degradation reactions of ascorbic acid in reduced-moisture foods are highly sensitive to a_w, the reaction rate increasing in an exponential fashion over the a_w range of 0.1 to 0.8.

b. *Proteins*

The nutritive value (and sometimes the wholesomeness) of proteins can be modified by heating and oxidation. Oxidation of proteins results in the formation of degradation products, which are known to detract from protein nutritive value. Proteins can also react with lipids to form complexes that

TABLE 10.3
General Stability of Vitamins to Environmental Effects

Nutrient	Oxygen	Light	Temperature
Vitamin A	U	U	U
Vitamin B_6	S	U	U
Vitamin B_{12}	U	U	S
Biotin	S	S	U
Vitamin C	U	U	U
Carotenes	U	U	U
Choline	U	S	S
Vitamin D	U	U	U
Folic acid	U	U	U
Inositol	S	S	U
Vitamin K	S	U	S
Niacin	S	S	S
Pantothenic acid	S	S	U
Riboflavin B_2	S	U	U
Thiamin B_1	U	S	U
Tocopherols	U	U	U

U, unstable; S, stable.

can affect food texture and, to a minor extent, protein nutritive value. In addition, the Maillard reaction can result in loss of nutritional properties, primarily from losses in the amino acid lysine.

c. Lipids

Lipids, especially when unsaturated, undergo many kinds of chemical changes during processing, and some of these changes can affect their nutritional value and wholesomeness. Peroxidizing lipids exert negative effects on the nutritive value of foods by their chemical interaction with proteins and vitamins. Oxygen often plays an important role in lipid degradation and packaging can play an important role in limiting or preventing O_2 ingress.

C. Physical Changes

The physical properties of foods can be defined as those properties that lend themselves to description and quantification by physical rather than chemical means. Their importance stretches from product handling, through processing, packaging and storage, to consumer acceptance. Physical properties include geometrical, thermal, optical, mechanical, rheological, electrical and hydrodynamic properties. Geometrical properties encompass the parameters of size, shape, volume, density and surface area as related to homogeneous food units, as well as geometrical texture characteristics. The latter can be subdivided into two classes: those referring to particle size and shape (e.g., gritty, grainy), and those referring to particle shape and orientation (e.g., fibrous, cellular).

Although many of these physical properties are important and must be considered in the design and operation of a successful packaging system, the focus here is on undesirable physical changes in packaged foods. The way that some of these changes can be affected by the nature of the packaging is now outlined.

Food powders are a diverse group that can be categorized in a number of ways.[14] On the basis of major chemical components, powders may be classified as starchy (e.g., wheat flour), proteinaceous (e.g., soy isolate), crystalline sugar (e.g., sucrose), amorphous sugars (e.g., dehydrated fruit juices) and fatty (e.g., soup mix). Powders may also be classified according to their particle size, although many food powders exhibit a range of several orders of magnitude in this parameter. They may also be classified according to their moisture sorption pattern, ranging from extremely hygroscopic (in the case of dehydrated fruit juices) through hygroscopic (in the case of spray dried coffee) to moderately hygroscopic (in the case of flours). Finally, powders can be classified as free flowing (e.g., granular sugar), moderately cohesive (e.g., flour) and very cohesive (most food powders after absorbing moisture).[14]

The major undesirable change in food powders is the sorption of moisture as a consequence of an inadequate barrier provided by the package, resulting in caking. This can occur either as a result of poor selection of packaging material in the first place, or failure of the package integrity during storage. Caking or spontaneous agglomeration of food powders (especially those containing soluble components or fats) occurs when they are exposed to moist atmospheres or elevated storage temperatures. The phenomenon can result in anything from small soft aggregates that break easily to rock hard lumps of variable size or solidification of the whole powder. In most cases, the process is initiated by the formation of liquid bridges between the particles that later solidify by drying or cooling.[14]

Anticaking agents are very fine powders of an inert chemical substance that are added to powders with much larger particle size in order to inhibit caking and improve flowability.[14] Studies on sucrose and onion powders showed that at ambient temperature, caking does not occur at a_ws of less than about 0.4. However, at higher activities ($a_w > 0.45$), the observed time to caking is inversely proportional to a_w, and at these levels, anticaking agents are completely ineffective. It appears that although they reduce interparticle attraction and interfere with the continuity of liquid bridges, they are unable to cover moisture sorption sites.

For foods containing solid carbohydrates, the largest effect in physical properties results from sorption of water, especially for the recrystallization of amorphous carbohydrate. Such changes can occur in boiled sweets (leading to stickiness or graining) and milk powders (leading to caking and lumpiness).

D. Biological Changes

1. Microbiological

Microorganisms can make both desirable and undesirable changes to the quality of foods, depending on whether or not they are introduced as an essential part of the food preservation process (e.g., as inocula in food fermentations) or arise adventitiously and subsequently grow to produce food spoilage. In the latter case, they only reach readily observable proportions when they are present in the food in large numbers. Because the initial population or microbial load is usually small, observable levels are only reached after extensive multiplication of the micro-organism(s) in the food.

The two major groups of micro-organisms found in foods are bacteria and fungi, the latter consisting of yeasts and molds. Bacteria are generally the fastest growing, so that in conditions favorable to both micro-organisms, bacteria will usually outgrow fungi. The phases through which the two groups pass are broadly similar: a period of adjustment or adaptation (known as the *lag* phase) is followed by accelerating growth until a steady, rapid rate (known as the *logarithmic* phase because growth is exponential) is achieved. After a time, the growth rate slows until growth and death are balanced and the population remains constant (known as the *stationary* phase). Eventually, death exceeds growth and the organisms enter the phase of decline.

Foods are frequently classified on the basis of their stability as nonperishable, semiperishable and perishable. An example of the first classification is sugar; provided it is kept dry, at ambient temperature and free from contamination, it should have a very long shelf life. However, few foods are truly nonperishable, and an important factor in determining their perishability is packaging.

For example, hermetically sealed and heat processed (e.g., canned) foods are generally regarded as nonperishable. However, they may become perishable under certain circumstances when an opportunity for recontamination is afforded following processing. Such an opportunity may arise if the can seams are faulty, or if there is excessive corrosion resulting in internal gas formation and eventual bursting of the can. Spoilage may also take place when the canned food is stored at unusually high temperatures where thermophilic spore-forming bacteria may multiply, causing undesirable changes such as flat sour spoilage.

Low moisture content foods such as flour, dried fruits and vegetables, and baked goods are classified as semiperishable. Frozen foods, although basically perishable, may be classified as semiperishable provided that they are properly stored at freezer temperatures.

The majority of foods (e.g., flesh foods such as meat and fish; milk, eggs and most fruits and vegetables) are classified as perishable unless they have been processed in some way. Often, the only form of processing that such foods receive is being packaged and kept under controlled temperature conditions.

The species of micro-organisms that cause the spoilage of particular foods are influenced by two factors: the nature of the foods and their surroundings. These factors are referred to as intrinsic and extrinsic parameters.

a. *Intrinsic Parameters*

These parameters are an inherent part of the food and are listed in Table 10.4. Most micro-organisms grow best at pH values around 7.0, while few grow below pH 4.0. Bacteria tend to be more fastidious in their relationships to pH than molds and yeasts, with the pathogenic bacteria being the most fastidious.

The minimum a_w values reported for growth of some micro-organisms in foods are presented in Table 10.5 and are discussed further in Section IV.B.3. It is noteworthy that yeasts and molds grow over a wider a_w range than bacteria.

Certain relationships have been shown to exist between a_w, temperature and nutrition.[10] Firstly, at any temperature, the ability of micro-organisms to grow is reduced as a_w is lowered. Secondly, the range of a_w over which growth occurs is greatest at the optimum temperature for growth. Thirdly, the presence of nutrients increases the range of a_w over which the organisms can survive. Therefore, the values given in Table 10.5 should be taken only as a guide.

The oxidation–reduction potential (E_h) of a substrate may be defined as the ease with which the substrate loses or gains electrons. The more highly oxidized a substance, the more positive will be its electrical potential. Aerobic micro-organisms require positive E_h values (oxidized) for growth while anaerobes require negative E_h values.[10] Among the substances in foods that help to maintain

TABLE 10.4
Intrinsic and Extrinsic Parameters Influencing Microbial Growth in Foods

Intrinsic Factors	Extrinsic Factors
pH	Storage temperature
a_w	Relative humidity of environment
E_h	
Nutrient content	Presence and concentration of gases in the environment
Antimicrobial constituents	
Biological structures	

TABLE 10.5
Approximate Minimum a_w Values for the Growth of Micro-Organisms of Importance in Foods

Organism	Minimum a_w
Most spoilage bacteria	0.91
Most spoilage yeasts	0.88
Most spoilage molds	0.80
Halophilic bacteria	0.75
Xerophilic molds	0.65
Osmophilic yeasts	0.60

reducing conditions are the sulfhydryl groups on proteins, and ascorbic acid and reducing sugars in fruits and vegetables. Deteriorative chemical reactions can alter the E_h value of foods during storage.

In order to grow and function normally, micro-organisms require several nutrients including water, a source of energy, a source of nitrogen, vitamins and related growth factors, and minerals. The availability of water is related directly to the a_w of the food. The primary source of nitrogen utilized by micro-organisms is amino acids. Growth factors and vitamin requirements tend to be specific to individual groups of micro-organisms. Some foods contain certain naturally occurring substances that have been shown to have antimicrobial activities, thus preventing or retarding the growth of specific micro-organisms in those foods.

b. *Extrinsic Parameters*

The extrinsic parameters of foods are those properties of the storage environment that affect both the foods and their micro-organisms. The growth rate of the micro-organisms responsible for spoilage primarily depends on such extrinsic parameters as storage temperature, relative humidity and gas composition of the surrounding atmosphere. These extrinsic parameters are discussed in more detail in Section IV.

The temperature of storage is particularly important, and several food preservation techniques (e.g., chilling) rely on reducing the temperature of the food to extend its shelf life. Although there is a very wide range of temperatures over which the growth of micro-organisms has been reported (−34 to 90°C), specific micro-organisms have relatively narrow temperature ranges over which growth is possible. Those that grow well below 20°C and have their optimum between 20 and 30°C are referred to as *psychrophiles* or *psychrotrophs.* Those that grow well between 20 and 45°C with optima between 30 and 40°C are referred to as *mesophiles*, and those that grow well at and above 45°C with optima between 55 and 65°C are referred to as *thermophiles*. Molds are able to grow over a wider range of temperature than bacteria, with many molds being capable of growth at refrigerator temperatures. Yeasts grow over the psychrophilic and mesophilic temperature ranges but generally not within the thermophilic range.

The relative humidity (RH) of the ambient environment can influence the a_w of the food unless the package provides a barrier. Many flexible plastic packaging materials provide good moisture barriers but none are completely impermeable, thus limiting the shelf life of low a_w foods.

The presence and concentration of gases in the environment has a considerable influence on the growth of micro-organisms. Increased concentrations of gases such as CO_2 are used to retard microbial growth and thus extend the shelf life of foods (see later chapters for a full discussion of this topic). Moreover, vacuum packaging (i.e., removal of air, and thus O_2, from a package prior to sealing) can also have a beneficial effect by preventing the growth of aerobic micro-organisms. This type of packaging (known as *modified atmosphere packaging* [MAP]) raises certain safety issues, which are discussed in Chapter 15. Most food pathogens do not grow at refrigerator temperatures, and CO_2 is not highly effective at nonrefrigeration temperatures. Therefore, most MAP food is usually held under refrigeration. Temperature abuse of the product (i.e., holding at nonrefrigerated temperatures) could allow the growth of organisms (including pathogens) that had been inhibited by CO_2 during storage at lower temperatures. For these reasons, it is difficult to evaluate MAP safety solely on the growth of certain pathogens at abusive temperatures.

c. *Influence of Packaging Material*

The protection of packaged food from contamination or attack by micro-organisms depends on the mechanical integrity of the package (e.g., the absence of breaks and seal imperfections), and on the resistance of the package to penetration by micro-organisms. Metal cans that are retorted after filling can leak during cooling, admitting any micro-organisms which may be present in the cooling water, even when the double seam is of a high quality. This fact is widely known in the canning industry and is the reason for the mandatory chlorination of cannery cooling water.

Extensive studies on a variety of plastic films and metal foils have shown that micro-organisms (including molds, yeasts and bacteria) cannot penetrate these materials in the absence of pinholes; in practice, thin sheets of packaging materials such as aluminum and plastic do contain pinholes. However, because of surface tension effects, micro-organisms cannot pass through very small pinholes, unless the micro-organisms are suspended in solutions containing wetting agents and the pressure outside the package is greater than that within.

2. Macrobiological

a. *Insect Pests*

The common insect pests of fresh food are flies (from the order Diptera), and cockroaches. They are attracted by food odors regardless of whether the food is fresh or beginning to decay. Any insect in food is a pest because not only does the food become contaminated with their bodies and excreta, but they are also capable of transmitting pathogens, including food poisoning organisms.[16]

In contrast, the main insect species important as pests of stored foods are entirely from the orders Lepidoptera (moths) and Coleoptera (beetles). They regularly damage and destroy large quantities of stored foods around the world every year. The number of species involved is not large and includes weevils, various other beetles and the larvae of several moths. Most stored product insects are cosmopolitan in that any given species is, for the most part, found worldwide in areas with similar climatic conditions. Warm humid environments promote insect growth, although most insects will not breed if the temperature exceeds about 35°C or falls below 10°C. Also, many insects cannot reproduce satisfactorily unless the moisture content of their food is greater than about 11%.

Moths and beetles are generally found in dry storage areas. They are able to survive on very small amounts of food, and thus can persist on food residues in improperly cleaned premises or equipment. Good ventilation, the use of cool storage areas and rotation of stock assists in keeping these pests at bay. Fundamental to the effective control of insect pests is an understanding of their life cycles and feeding habits.

In common with many insects, moths pass through four stages during their development: the egg, the larva (caterpillar), the pupa (chrysalis) and the adult moth. Food is consumed only in the larval stage. The presence of larvae can be recognized by a characteristic mixture of silken threads and frass (droppings) that they produce. Beetles have the same four life stages as moths, but they differ in that the adult beetle is a considerably harder-bodied insect than the moth, and may live and feed for months or even years. Cockroaches are larger, more robust insects, which are highly mobile. Their young are like small versions of the adult and, unlike beetles and moths, cockroaches do not inhabit packaged foods. Ants are a highly specialized group of insects which form nests, normally outside buildings. Worker ants may travel considerable distances and collect almost any type of food (especially sweet or high-protein foods) and take it back to the nest.

Mites that sometimes occur in stored foods are not insects but are closely related to spiders, having eight legs. They are minute in size, requiring a lens to see them, and are primarily pests of cereals and other foods with a moisture content of at least 12%. Mites are so small that the presence of a few would pass unnoticed; they produce a sour odor in the food.

The main categories of foods subject to pest attack are cereal grains and products derived from cereal grains, other seeds used as food (especially legumes), dairy products such as cheese and milk powders, dried fruits, dried and smoked meats, and nuts. As well as their possible health significance, the presence of insects and insect excreta in packaged foods may render products unsalable, causing considerable economic loss, as well as reduction in nutritional quality, production of off-flavors, and acceleration of decay processes due to the creation of higher temperatures and moisture levels.

Unlike micro-organisms, some insect species (penetrators) have the ability to bore through one or more of the flexible packaging materials in use today and take up residence inside. Other species (invaders) usually do not enter packages unless there is an existing opening. However, such openings need not be very large; for example, the adult saw-toothed grain beetle can enter an opening less than 1 mm in diameter.[9] Newly hatched larvae can enter much smaller openings; holes only 0.1 mm in diameter are sufficient to admit the larvae of some insects. Thus, package seal quality is critical in protecting foods from insect infestations.

Unless plastic films are laminated with foil or paper, insects are able to penetrate most of them quite easily, where the rate of penetration is usually directly related to film thickness. In general, thicker films are more resistant than thinner films, and oriented films tend to be more effective than cast films. The looseness of the film has also been reported to be an important factor, with loose films being more easily penetrated than tightly fitted films.

Generally, the penetration varies depending on the basic resin from which the film is made, on the combination of materials, on the package structure, and on the species and stage of insects involved. The relative resistance to insect penetration of common flexible packaging materials is given in Table 10.6. Where no thicknesses are given, the estimations are based on thicknesses commonly used in food packaging. Absolute values are difficult to determine because resistance to penetration is influenced by factors such as package configuration and the presence or absence of folds, tucks and other harborage sites. Therefore, after appropriate packaging materials have been selected, they must be evaluated *in situ* for insect resistance.[9]

b. Rodents

The rodents rats and mice are among humanity's most cunning and capable enemies. They have highly developed senses of touch, smell and hearing, and can identify new or unfamiliar objects in their environment. Rats can wriggle through openings the size of a quarter; a mouse needs a hole only as large as a nickel to gain access. Rats and mice carry disease-producing organisms on their feet or in their intestinal tracts and are known to harbor salmonellae of serotypes frequently associated with foodborne infections in humans. In addition to the public health consequences of rodent populations in close proximity to humans, these animals also compete intensively with humans for food.

Rats and mice gnaw to reach sources of food and drink and to keep their teeth short. Their incisor teeth are so strong that rats have been known to gnaw through lead pipes and

TABLE 10.6
Resistance of Various Materials to Insect Penetration

	Excellent	Good	Fair	Poor
Polycarbonate	x			
Poly(ethylene terephthalate)	x			
Cellulose acetate		x		
Polyamide		x		
Polyethylene (0.254 mm)		x		
Polypropylene (biaxially oriented)		x		
Poly(vinyl chloride) (unplasticized)		x		
Acrylonitrile			x	
Poly(tetrafluoroethylene)			x	
Polyethylene (0.123 mm)			x	
Regenerated cellulose film				x
Corrugated paperboard				x
Ethylene vinyl acetate copolymer				x
Ionomer				x
Kraft paper				x
Paper/foil/polyethylene laminate pouch				x
Polyethylene (0.0254-0.100 mm)				x
Poly(vinyl chloride) (plasticized)				x
Vinyl chloride/vinylidene chloride copolymer				x

Source: From Highland, H. A., Insect infestation of packages, In *Insect Management for Food Storage and Processing*, Baur, F. J., Ed., American Association of Cereal Chemists, St. Paul, Minnesota, 1984, chap. 23. With permission.

unhardened concrete, as well as sacks, wood and flexible packaging materials. Obviously, proper sanitation in food processing and storage areas is the most effective weapon in the fight against rodents, because all packaging materials apart from metal and glass containers can be attacked by rats and mice.

III. RATES OF DETERIORATIVE REACTIONS

As discussed in the preceding section, a number of deteriorative chemical, biochemical and microbiological reactions can occur in foods. The rates of these reactions depend on both intrinsic (compositional) and extrinsic (environmental) factors. As well as understanding the nature of these reactions, it is also important to have an appreciation of their rates, so that they can be controlled. Control of deteriorative reactions requires a quantitative analysis based on knowledge of the kinetics of food deterioration. Fortunately, simple chemical kinetics can be applied to such reactions.

Quantitative analysis of the deteriorative reactions that occur in a food during processing and storage requires the existence of a measurable index of deterioration; that is, a chemical, physical or sensory measurement or set of measurements that may be used reproducibly to assess the changes occurring. An increase or decrease in the index of deterioration must correlate with changes in food quality. For quantitative analysis of quality changes, the index must be expressed as a function of the conditions existing during processing and storage so that the changes can be predicted or simulated. Thus, calculation of quality losses requires a mathematical model that expresses the effect of intrinsic and extrinsic factors on the deterioration index.

The general equation describing quality loss may be written as:

$$-\mathrm{d}C/\mathrm{d}\theta = f(I_i, E_j) \tag{10.1}$$

where

$-\mathrm{d}C/\mathrm{d}\theta$ = rate of change of some index of deterioration C with time θ; a negative sign is used if the concentration of C decreases with time;
I_i = intrinsic factors ($i = 1\ldots m$);
E_j = extrinsic factors ($j = 1\ldots n$).

Because the quality of foods and the rate of quality changes during processing and storage depend on intrinsic factors, it is possible in many cases to correlate quality losses with the loss of a particular component such as a vitamin or pigment. The conversion of a single component or quality factor A to an end product B (e.g., conversion of chlorophyll to pheophytin, or conversion of ascorbic acid to brown pigments), may be written as:

$$A \rightarrow \text{intermediate products} \rightarrow B \tag{10.2}$$

The absolute concentrations of A or B need not be measured. For example, the production of brown pigments in foods is often measured as the increase in absorbance at 420 nm of an alcoholic extract of the food, and the change in absorbance used as an indicator of the extent of the reaction. Such quality loss can be represented as being proportional to the power of the concentration of the reactant or product:

$$-\mathrm{d}A/\mathrm{d}\theta = kA^n \tag{10.3}$$

or

$$\mathrm{d}B/\mathrm{d}\theta = kB^n \tag{10.4}$$

where

A and B = concentration of quality factor measured;
θ = time;
k = rate constant (dependent on extrinsic factors);
n = a power factor called the order of the reaction which defines whether or nor the rate is dependent on the concentration of A. The value of n can be a fraction or a whole number;
$\mathrm{d}A/\mathrm{d}\theta$ and $\mathrm{d}B/\mathrm{d}\theta$ = change in concentration of A or B with time.

Equation 10.4 implies that extrinsic parameters such as temperature, a_w and light intensity are held constant; if they are not, then their influence on the rate constant k must be taken into account in evaluating the equation. For most quality changes in foods, the reaction order n has generally been shown to be either 0 or 1.

From a packaging point of view, it is often useful to know the concentration of A or B at which the product is no longer acceptable, for example, when the concentration of a vitamin or pigment has fallen below some level (e.g., 50% reduction in concentration), or the concentration of some undesirable brown color has risen above some level. In these situations, the shelf life of the food (θ_s) is the time for the concentration of A (or B) to reach an undesirable level (A_e or B_e).

A. Zero-Order Reactions

When $n = 0$, the reaction is said to be pseudo zero-order with respect to A. Equation 10.4 can then be simplified to:

$$-\mathrm{d}A/\mathrm{d}\theta = k \tag{10.5}$$

Equation 10.5 implies that the rate of loss of A is constant with time and independent of the concentration of A. Rearranging and integrating Equation 10.5 between A_o, the concentration of A at $\theta = 0$, and A, the concentration of A at time θ:

$$\int_{A_o}^{A} \mathrm{d}A = -k \int_{0}^{\theta} \mathrm{d}\theta \tag{10.6}$$

yields

$$A = A_o - k\theta \tag{10.7}$$

or

$$A_o - A = k\theta \tag{10.8}$$

or

$$A_e = A_o - k\theta_s \tag{10.9}$$

where

A_e = value of A at end of shelf life;
θ_s = shelf life in days, months, years, and so on.

For a zero-order reaction, a plot of the amount of A remaining versus time yields a straight line (Figure 10.1) with the slope equal to the rate constant k in units of [concentration] [time^{-1}]. In other words, the loss of quality per day is constant when all extrinsic factors are held constant.

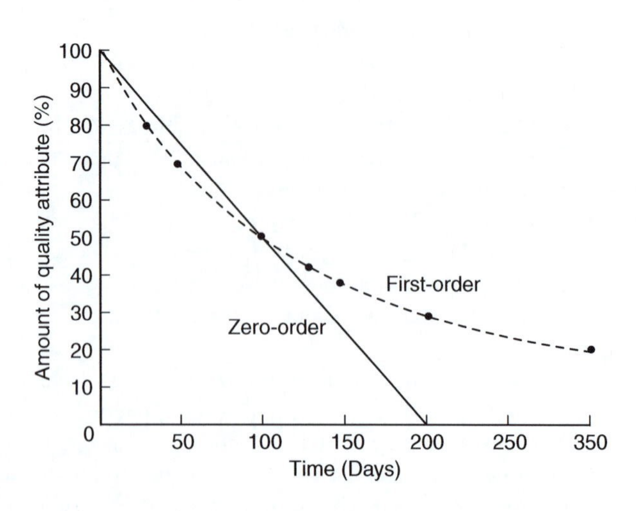

FIGURE 10.1 Change in quality versus time showing the effect of order of the reaction on extent of change.

Typical pseudo zero-order deteriorative reactions include NEB (e.g., in dry cereals and powdered dairy products), lipid oxidation (e.g., development of rancidity in snack foods, dry foods and frozen foods) and enzymic degradation (e.g., in fresh fruits and vegetables, some frozen foods and some refrigerated doughs).

It is important to appreciate that the order of the reaction (i.e., n) is strictly an empirical concept. Thus, a pseudo zero-order reaction does not imply that the mechanism is a monomolecular breakdown independent of the concentration of the reacting species. On the contrary, pseudo zero-order reactions are always an indication that a complex reaction is occurring involving a number of steps. All that a pseudo zero-order reaction suggests is that there is a high correlation between A and time.

Example 10.1. Orange juice was aseptically filled into hermetically sealed glass jars and laminated (polyethylene/paper/aluminum foil/polyethylene) cartons, and held at 25°C. The extent of browning (expressed as optical density [OD] at 420 nm) measured over a period of weeks gave the following results:

	Browning (optical density at 420 nm)	
Time (days)	**Carton**	**Jar**
0	0.100	0.100
10	0.123	0.114
20	0.147	0.127
30	0.171	0.141
40	0.195	0.155

Analyze the data to see if the browning reaction follows a pseudo zero-order reaction, and calculate the shelf life of juice in the two containers if the juice is unacceptable when browning exceeded 0.250 OD.

This problem involves an increase in product (browning) with time, so Equation 10.4 is appropriate; this can be integrated for $n = 0$ to give:

$$k = \frac{B - B_o}{\theta}$$

Carton	Jar
At $\theta = 40$ days,	
$k = (0.195 - 0.100)/40 = 2.38 \times 10^{-3}$ OD days^{-1}	$k = (0.155 - 0.100)/40 = 1.38 \times 10^{-3}$ OD days^{-1}
At $\theta = 20$ days,	
$k = (0.147 - 0.100)/20 = 2.35 \times 10^{-3}$ OD days^{-1}	$k = (0.127 - 0.100)/20 = 1.35 \times 10^{-3}$ OD days^{-1}

Because the rate constants for juice in each type of container agree closely after two time periods, there is some justification in treating the reaction as pseudo zero order.
To calculate the shelf life of the juice, the following form of the above equation is appropriate:

$$\theta = \frac{B - B_o}{k}$$

Carton	Jar
$\theta = (0.25 - 0.10)/2.375 \times 10^{-3} = 63$ days	$\theta = (0.25 - 0.10)/1.375 \times 10^{-3} = 109$ days

Thus, using this data, the shelf life of the orange juice packaged in a glass jar is 46 days longer than juice packaged in a laminated carton.

B. First-Order Reactions

In general, foods that do not follow a pseudo zero-order reaction deteriorate according to a pseudo first-order ($n = 1$) reaction in which the rate of loss is dependent on the amount left. In this case the solution of Equation 10.3 for $n = 1$ is:

$$\int_{A_o}^{A} \frac{dA}{A} = -k \int_{0}^{\theta} d\theta \tag{10.10}$$

and

$$\ln \frac{A}{A_o} = -k\theta \tag{10.11}$$

(where ln = natural logarithm) or

$$\ln A = \ln A_o - k\theta \tag{10.12}$$

or

$$A = A_o e^{-k\theta} \tag{10.13}$$

or

$$A_e = A_o e^{-k\theta s} \tag{10.14}$$

A plot of first-order data as the concentration of A versus time gives a curved line as shown in Figure 10.1. However, if the data are plotted as the base ten logarithm of A versus time a straight line is obtained, the slope of which is equal to $-k/2.303$. The units of k for a first-order reaction are [time^{-1}].

Typical pseudo first-order deteriorative reactions include NEB (e.g., loss of protein quality in dry foods), lipid oxidation (e.g., development of rancidity in salad oils and dry vegetables), vitamin loss in canned and dry foods, and microbial production of off-flavors and slime in flesh foods.

Example 10.2. Lemon juice at a concentration of 9°Brix is stored at 10°C and the concentration of ascorbic acid measured over a period of weeks to give the following results:

Time (weeks)	Ascorbic Acid (mg 100 mL^{-1})
0	52.9
4	45.1
8	38.3
12	32.9
16	26.7

Determine the rate constant for the loss of ascorbic acid assuming that the reaction is pseudo first order, and calculate the time for the ascorbic acid concentration in the juice to reach 20 mg 100 mL^{-1}.

From Equation 10.12:

$$k = (\ln A_o - \ln A)/\theta$$

After 16 weeks:

$$k = (\ln 52.9 - \ln 26.7)/16 = 0.043\ \text{weeks}^{-1}$$

After 8 weeks:

$$k = (\ln 52.9 - \ln 38.3)/8 = 0.040\ \text{weeks}^{-1}$$

To calculate the shelf life of the lemon juice:

$$\theta_s = (\ln A_o - \ln A_e)/k = (\ln 52.9 - \ln 20)/0.0415 = 23\ \text{weeks}$$

Thus, the concentration of ascorbic acid will have fallen to 20 mg 100 mL^{-1} after 23 weeks at 10°C.

C. Microbial Growth and Destruction

1. Microbial Growth

For the case of microbial growth, Equation 10.11 can be rewritten in the form:

$$\ln \frac{A}{A_o} = k\theta_{doub} \tag{10.15}$$

where

A_o = initial number of micro-organisms when $\theta = 0$;
$A = 2A_o$ (i.e., the number of organisms has doubled);
θ_{doub} = time for number of organisms to double (i.e., the generation or doubling time).

On substituting for A:

$$\ln 2 = k\theta_{doub} \tag{10.16}$$

and

$$k = \frac{0.693}{\theta_{doub}} \tag{10.17}$$

or

$$\theta_{doub} = \frac{0.693}{k} \tag{10.18}$$

This enables calculation of the generation or doubling time if the growth rate is known, or *vice versa*.

Example 10.3. Beef is to be packaged in plastic film and stored at chill temperatures. The initial level of contamination of the beef immediately after packaging is 10^3 micro-organisms per cm^2, and the maximum permitted level of micro-organisms is 10^8. Assuming that the micro-organisms are solely *Pseudomonas fluorescens*, which has a generation or doubling time of 8.5 h at 5°C, calculate the time for which the beef can be stored before the maximum permissible level of micro-organisms is reached.

From Equation 10.17:

$$k = \frac{0.693}{8.5} = 0.0815\ \mathrm{h}^{-1}$$

Substituting into Equation 10.15:

$$\ln\frac{10^8}{10^3} = 0.0815\theta_s$$

$$\therefore \theta_s = 141.5\ \mathrm{h}$$

If this shelf life were insufficient, then the storage temperature could be lowered. Given that the generation time at −2°C is 19 h, calculate the shelf life of the beef:

$$k = \frac{0.693}{19} = 0.0365\ \mathrm{h}^{-1}$$

and

$$\theta_s = \ln\frac{10^8}{10^3}\frac{1}{0.0365} = 315.4\ \mathrm{h}$$

If further extension of the shelf life were required, then the package could be flushed with CO_2 and the new shelf life calculated, provided the generation time for *P. fluorescens* at −2°C in a CO_2 atmosphere was known.

2. Microbial Destruction

For the kinetics of microbial destruction by heat and irradiation, the food industry uses a modified time term — the decimal reduction time or D value. This is defined as the time at constant temperature to reduce the population of micro-organisms by 90%. Mathematically, at time $\theta = D$, $A = 0.1\ A_o$.

Substituting into Equation 10.15:

$$\ln\frac{A_o}{0.1\ A_o} = kD$$

Because ln 10 = 2.303:

$$D = \frac{2.303}{k}$$

If the logarithm of the D value is plotted against the corresponding temperature, then a straight line is obtained, the slope of which is designated z. This can be defined as the temperature change necessary for a tenfold change in the D value or reaction rate. Mathematically,

$$\frac{k_T}{k_{T-z}} = 10 \qquad (10.19)$$

IV. EXTRINSIC FACTORS CONTROLLING THE RATES OF DETERIORATIVE REACTIONS

A. Effect of Temperature

Temperature is a key factor in determining the rate of deteriorative reactions, and in certain situations, the packaging material can affect the temperature of the food. This is particularly so with packaging materials which have insulating properties, and these types of packages are typically used for chilled and frozen foods. For packages that are stored in refrigerated display cabinets, most of the cooling takes place by conduction and convection. Simultaneously, there is a heat input by radiation from the fluorescent lamps used for lighting. Under these conditions, aluminum foil offers real advantages because of its high reflectivity and high conductivity. However, such advantages are seldom used in the packaging of frozen and chilled foods.

1. Linear Model

Early studies on the thermal processing of foods obtained a straight line when the thermal death times (now D values or the time for a 90% reduction in numbers) of micro-organisms were plotted against temperature on a linear scale. The equation of such a curve is:

$$\log(D_1/D_2) = (T_1 - T_2)/z \tag{10.20}$$

where

z = temperature change required to change the D value by a factor of 10.

For many micro-organisms of interest in food canning, $z = 10°C$, whereas for degradation of quality factors during thermal processing, $z = 32°C$. Reactions that have small z values are highly temperature dependent, whereas reactions with large z values are less influenced by temperature. This model has been found to be satisfactory for thermal processes[5] and is still in use today. In its more general form, it can be written as shown below because $k \propto D$:

$$k = k_r 10^{(T-T_r)/z} \tag{10.21}$$

A similar expression relating the rate of reactions and temperature has also been used for many years, especially in relation to shelf life plots (see Chapter 11):

$$k = k_o e^{b(T-T_o)} \tag{10.22}$$

where

k_o = rate at temperature T_o (°C);
k = rate at temperature T (°C);
b = a constant characteristic of the reaction;
e = 2.7183.

2. Arrhenius Relationship

The most common and generally valid relationship for the effect of temperature on the rate of deterioration is that of Arrhenius. The relationship is correctly expressed in the differential form:

$$\frac{d(\ln k)}{dT} = \frac{E_A}{RT^2} \tag{10.23}$$

For practical reasons, the integrated form is used:

$$k = k_o e^{-E_A/RT} \tag{10.24}$$

where

k = rate constant for deteriorative reaction;
k_o = constant, independent of temperature (also known as the Arrhenius, pre-exponential, collision or frequency factor);
E_A = activation energy (J mol^{-1});
R = ideal gas constant (8.314 J K^{-1} mol^{-1} = 1.987 cal K^{-1} mol^{-1});
T = absolute temperature (K).

The integrated relationship contains the inherent assumption that the activation energy and the pre-exponential factor do not change with temperature. This assumption is generally, but not universally, true. Therefore, predictions based on this model sometimes fail when applied over a temperature span of greater than about 40°C. Furthermore, when the reaction mechanism changes with temperature, the activation energy may vary substantially. The value of E_A is a measure of the temperature sensitivity of the reaction; that is, how much faster the reaction will proceed if the temperature is raised. Typical activation energies for reactions important in food deterioration are listed in Table 10.7.

The activation energy is generally derived from the slope of the plot of ln k versus $1/T$ and depends on factors such as a_w, moisture content, solids concentration and pH.

3. Temperature Quotient

Another term used to describe the response of biological systems to temperature change is the Q value, a quotient indicating how much more rapidly the reaction proceeds at temperature T_2 than at a lower temperature T_1. If Q reflects the change in rate for a 10°C rise in temperature, it is then called Q_{10}. Mathematically,

$$Q_{10} = \frac{k_{T+10}}{k_T} \tag{10.25}$$

When the Fahrenheit temperature scale is used instead of the Celsius scale, the symbol q_{10} is used. The relationship between Q_{10} and q_{10} is:

$$Q_{10} = (q_{10})^{1.8} \tag{10.26}$$

TABLE 10.7
Typical Activation Energies for Reactions Important in Food Deterioration

Reaction (kJ mol^{-1})	Activation Energy (E_A)
Diffusion-controlled reaction	8–40
Lipid oxidation	40–105
Flavor degradation in dried vegetables	40–105
Enzymic reactions	40–130
Hydrolysis	60–110
Vitamin degradation	85–130
Color degradation in dried vegetables	65–150
Nonenzymic browning	105–210
Protein denaturation	350–700

It can be shown that the rate of a deteriorative reaction at two temperatures is related to the shelf life at those two temperatures:

$$k_T \theta_{s_T} = k_{T+10} \theta_{s_{T+10}}$$

where

θ_{s_T} = shelf life at temperature T°C;
$\theta_{s_{T+10}}$ = shelf life at temperature $(T + 10)$°C.

Therefore,

$$Q_{10} = \frac{\theta_{s_T}}{\theta_{s_{T+10}}} \tag{10.28}$$

For any temperature difference Δ, which is not 10°C:

$$Q_{10}^{\Delta/10} = \frac{\theta_{s_{T_2}}}{\theta_{s_{T_1}}} \tag{10.29}$$

It can be shown when the Arrhenius model is used that:

$$\ln Q_{10} \approx \frac{10 E_A}{R T^2} \tag{10.30}$$

and when the linear model is used:

$$\ln Q_{10} = 10b \tag{10.31}$$

or

$$Q_{10} = e^{10b} \tag{10.32}$$

Note that Q_{10} is not constant but depends on both the E_A and the temperature, whereas E_A is assumed to be independent of temperature.

Example 10.4. The pseudo zero-order rate constant for the degradation of ascorbic acid in dried vegetables packaged in a polyester/polyethylene laminate pouch is 0.0745 mg 100 g^{-1} week^{-1} when stored at 30°C, and 0.0255 mg 100 g^{-1} week^{-1} when stored at 20°C. What is the Q_{10} and activation energy for the reaction?

From Equation 10.25:

$$Q_{10} = \frac{k_{T+10}}{k_r} = 0.0745/0.0255 = 2.92$$

From Equation 10.30:

$$\ln 2.92 = \frac{10 E_A}{8.314 \times 293 \times 303}$$

$$\therefore E_A = 79.1 \text{ kJ mol}^{-1}\text{K}^{-1}$$

It can also be shown that:

$$\frac{E_A}{RT^2} = \frac{\ln 10}{z} = \frac{2.3}{z} \tag{10.33}$$

By combining Equation 10.30 and Equation 10.33, it can be shown that:

$$z = \frac{10}{\log Q_{10}} \tag{10.34}$$

B. Effect of Water Activity (a_w)

1. Definitions

The parameter a_w is defined as the ratio of the water vapor pressure of a material to the vapor pressure of pure water at the same temperature. Mathematically,

$$a_w = p/p_o \tag{10.35}$$

where p = vapor pressure of water exerted by the food, and p_o = saturated vapor pressure of pure water at the same temperature. This concept is related to equilibrium relative humidity (ERH) in that ERH = 100 × a_w. However, while a_w is an intrinsic property of the food, ERH is a property of the atmosphere in equilibrium with the food. The a_w of most fresh foods is above 0.99. At subfreezing temperatures, a_w is defined as the vapor pressure of ice divided by the vapor pressure of supercooled water at the same temperature. Thus, the a_w of frozen foods depends only on their temperature; at −15°C, $a_w = 0.864$; at −10°C, 0.907; and at −5°C, 0.953.

As mentioned above in the definition of a_w, the temperature must be specified because a_w values are temperature dependent. The temperature dependence of a_w can be described by a modified form of the Clausius–Clapeyron equation:

$$\frac{d \ln a_w}{d(1/T)} = \frac{-\Delta H}{R} \tag{10.36}$$

or

$$\ln \frac{a_{w_2}}{a_{w_1}} = \frac{\Delta H}{R}\left[\frac{1}{T_1} - \frac{1}{T_2}\right] \tag{10.37}$$

where

a_{w_1} = water activity at temperature T_1 (K);
a_{w_2} = water activity at temperature T_2 (K);
ΔH = isosteric net heat of sorption at the moisture content of the food (J mol^{-1});
R = gas constant (8.314 J mol^{-1} K^{-1}).

Thus, from Equation 10.36, a plot of ln a_w versus $1/T$ at constant moisture content should be linear. Such plots are not always linear over wide temperature ranges, and they exhibit sharp breaks with the onset of ice formation.[6]

2. Isotherms

When a food is placed in an environment at a constant temperature and relative humidity, it will eventually come to equilibrium with that environment. The corresponding moisture content at steady-state is referred to as the equilibrium moisture content. When this moisture content (expressed as mass of water per unit mass of dry matter) is plotted against the corresponding relative humidity or a_w at constant temperature, a moisture sorption isotherm results (see Figure 10.2). Such plots are very useful in assessing the stability of foods and selecting effective packaging.

One complication is that a moisture sorption isotherm prepared by the addition of water to a dry sample (resorption) will not necessarily be superimposable with an isotherm prepared by removal of water from a wet sample (desorption). This lack of superimposibility is referred to as *hysteresis*. Typically, at any given a_w, the water content of the food will be greater during desorption than during resorption. This has important implications with respect to food stability, in that foods adjusted to the desired a_w by desorption, rather than resorption, may deteriorate more rapidly because of their higher moisture content.

Another complication occurs with some sugars. Crystalline sugars generally have completely different sorption isotherms from amorphous sugars, where the equilibrium moisture content is much lower for the crystalline form at any particular a_w. The amorphous form is often present when the food has been dried quickly (e.g., spray drying of milk often results in the formation of amorphous lactose); during storage it may slowly revert to the crystalline form. This results in a distinct break in the sorption isotherm, because the sugar releases moisture at constant a_w (see Figure 18.2).

Because a_w is temperature dependent, it follows that moisture sorption isotherms must also exhibit temperature dependence. Thus, at any given moisture content, a_w increases with increasing temperature, in agreement with the Clausius–Clapeyron equation (Equation 10.37) and shown in Figure 10.2.

Many relationships have been derived relating the a_w of a food to its moisture content. For many years, the most used relationship was the Brunauer–Emmett–Teller (BET) isotherm,

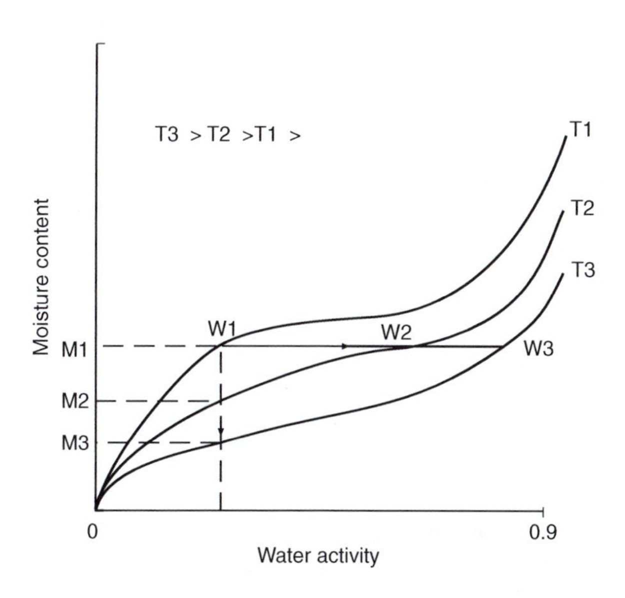

FIGURE 10.2 Schematic of a typical moisture sorption isotherm showing effect of temperature on water activity and moisture content.

which took the form:

$$\frac{a_w}{m(1 - a_w)} = \frac{1}{m_1 C} + \frac{C - 1}{m_1 C} a_w \tag{10.38}$$

where

m = equilibrium moisture content (dry weight basis) at water activity a_w;
m_1 = moisture content of the monolayer (dry weight basis);
C = a constant.

The BET equation has been used to estimate the monolayer value, which is equivalent to the amount of water held adsorbed on specific sites. For many foods, the monolayer value corresponds to an a_w of 0.2 to 0.4. As has been pointed out,[7] monolayer does not mean coverage of all dry matter with a closely packed, single layer of water molecules. Rather, the monolayer value should be regarded as the maximum amount of water that can be strongly bound to the dry matter; monolayer water constitutes a very small fraction of the total water in a high moisture food.

The Guggenheim–Anderson–de Boer (GAB) model has been widely used by European food researchers since the late 1970s, and has now gained worldwide acceptance. It is a three-parameter model with physically meaningful coefficients, which usually fits data very well up to 0.9 a_w. The model is:

$$\frac{m}{m_o} = \frac{Cka_w}{(1 - ka_w)(1 - ka_w + Cka_w)} \tag{10.39}$$

where

m = equilibrium moisture content (dry weight basis) at water activity a_w;
m_o = moisture content corresponding to saturation of all primary adsorption sites by one water molecule (equivalent to the BET monolayer);
C = Guggenheim constant;
k = factor correcting properties of the multilayer molecules with respect to the bulk liquid;

and

$C = c' \exp(H_1 - H_m)/RT$ where
H_1 = heat of condensation of pure water vapor;
H_m = total heat of sorption of the first layer on primary sites;

and

$k = k' \exp(H_1 - H_m)/RT$;
H_q = total heat of sorption of the multilayers;
k' and c' are adjusted constants for the temperature effect.

This model can be considered as an extension of the BET model, taking into account the modified properties of the sorbed water in the multilayer region. The real advantage of the GAB model is that it offers an objective method for fitting sorption isotherm data for a majority of foods up to 0.9 a_w, describing the temperature effect over a range of at least 40°. The final choice of a model will depend on a compromise between the desired closeness of fit and convenience with regard to the number of parameters involved.

3. Water Activity and Food Stability

Water may influence chemical reactivity in different ways. It may act as a reactant (e.g., in the case of sucrose hydrolysis) or as a solvent, where it many exert a dilution effect on the substrates, thus decreasing the reaction rate. Water may also change the mobility of the reactants by affecting the viscosity of the food systems, and may form hydrogen bonds or complexes with the reacting species. Thus, a very important practical aspect of a_w is to control undesirable chemical and enzymic reactions that reduce the shelf life of foods. It is a well-known generality that rates of changes in food properties can be minimized or accelerated over widely different values of a_w, as shown in Figure 10.3. Small changes in a_w can result in large changes in reaction rates.

a. Lipids

The influence of a_w on lipid oxidation has been studied extensively, mainly with the use of model systems. The general effect of a_w on lipid oxidation is shown in Figure 10.3. At very low a_w levels, foods containing unsaturated fats and exposed to atmospheric O_2 are highly susceptible to the development of oxidative rancidity. This high oxidative activity occurs at a_w levels below the monolayer level, and as a_w increases, both the rate and extent of autoxidation increase until an a_w in the range of 0.3 to 0.5 is reached. Above this point, the rate of oxidation increases until a steady state is reached, normally at a_w levels in excess of 0.75. At a_ws below the monolayer value, the oxidation rate decreases with increasing a_w. The rate reaches a minimum around the monolayer value and increases with further increases in a_w. Water may influence lipid oxidation by influencing the concentrations of initiating radicals present, the degree of contact and mobility of reactants and the relative importance of radical transfer versus recombination reactions.

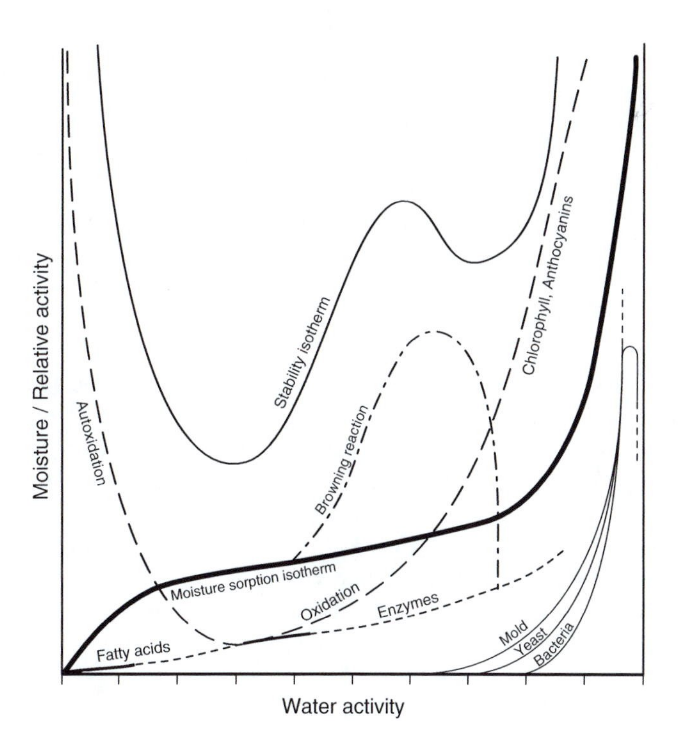

FIGURE 10.3 Rates of reactions as a function of water activity. (*Source*: From Rockland, L. B. and Beuchat, L. R., Eds., *Water Activity: Theory and Applications to Food*, Marcel Dekker, New York, 1987. With permission.)

b. Browning

Water may accelerate browning by imparting mobility to the substrates, or it may decrease the browning rate by diluting the reactive species. In the low a_w range, the mobility factor predominates, whereas the dilution factor predominates in the high a_w range. As a consequence, browning rate generally increases with increasing a_w at low moisture content, reaches a maximum at a_ws of 0.4 to 0.8, and decreases with further increase in a_w (see Figure 10.3).

c. Vitamins

The rate of degradation of vitamins A, B_1, B_2 and C increases as a_w increases over the range of 0.24 to 0.65. Generally, the rate of ascorbic acid degradation increases exponentially with increase in a_w. The photodegradation of riboflavin has been shown to increase with increasing a_w.

d. Enzymes

Near or below the monolayer a_w value, enzyme activities are generally minimized or cease. Above the monolayer value, enzyme activity increases with increasing a_w or increased substrate mobility, as illustrated in Figure 10.3. Substrates of high MW such as protein and starch are less mobile than low MW substrates such as glucose, and generally the latter have a lower a_w threshold for enzyme activity. At subfreezing temperatures, the reaction rate generally decreases with decreasing temperature and a_w, owing partly to the lower temperature, partly to the increase in viscosity of the partially frozen system, and partly to enzyme denaturation.

e. Pigments

Carotenoids are subject to changes due to heating and both enzymic and nonenzymic oxidation, all of which are influenced by water. Water appears to have a protective effect on oxidative degradation, apparently by reducing the free radical content. Values of a_w above the monolayer (up to a_w of 0.41) give almost complete protection against oxidative degradation.

Water activity has a definite influence on the rate of chlorophyll degradation, the rate decreasing with a decrease in a_w. The color of anthocyanins increases in intensity as a_w is lowered.

f. Texture

In the case of sugars, changes in a_w may either inhibit or promote a physical change in the nature of the sugar, which, in turn, affects the texture. A low a_w maintains the sugar in the form of a glassy, amorphous, free-flowing powder, whereas an increase in a_w promotes crystallization resulting in a sticky, caking powder, which has a much lower moisture content than the amorphous sugar.

The sensory crispness of several starch-based snack foods decreases with increasing a_w and most products become unacceptable in the a_w range of 0.35 to 0.50. There is no doubt that a_w has a major effect on textural properties of foods, but data relating textural changes to a_w is relatively sparse.

g. Microbial

Every micro-organism has a limiting a_w value below which it will not grow, form spores or produce toxic metabolites. Table 10.8 lists the range of a_ws that permit the growth of various common micro-organisms, together with common foods categorized according to their a_w.

TABLE 10.8
Water Activity and Growth of Micro-Organisms in Food

Range	Microorganisms generally inhibited by lowest a_w in this range	Foods generally within this range
1.00–0.95	*Pseudomonas, Escherichia, Proteus, Shigella, Klebsiella, Bacillus, Clostrium perfringens*, some yeasts	Highly perishable (fresh) foods and canned fruits, vegetables, meat, fish and milk; cooked sausages and breads; foods containing up to approximately 40% (w/w) sucrose or 7% NaCl
0.95–0.91	*Salmonella, Vibrio parahemolyticus, C. botulinum, Serratia, Lactobacillus, Pediococcus*, some molds, yeasts	Some cheeses (cheddar, Swiss, Munster, Provolone), cured meat (ham), some fruit juice concentrates; foods containing 55% (w/w) sucrose or 12% NaCl
0.91–0.87	Many yeasts (*Candida, Torulopsis, Hansenula*), *Micrococcus*	Fermented sausage (salami), sponge cakes, dry cheese, margarine; foods containing 65% (w/w) sucrose (saturated) or 15% NaCl
0.87–0.80	Most molds (mycotoxigenic penicillia), *Staphylococcus aureus*, most *Saccharomyces* (*bailii*) spp., *Debaryomyces*	Most fruit juice concentrates, sweetened condensed milk, chocolate, syrup, maple and fruit syrups; flour, rice, pulses containing 15–17% moisture; fruitcake, country-style ham, fondants, high-ratio cakes
0.80–0.75	Most halophilic bacteria, mycotoxigenic aspergilli	Jam, marmalade, marzipan, glacéd fruits, some marshmallows
0.75–0.65	Xerophilic molds (*Aspergillus chevalieri, A. candidus, Wallemia sebi*), *Saccharomyces bisporus*	Rolled oats containing approximately 10% moisture, grained nougats, fudge, marshmallows, jelly, molasses, raw cane sugar, some dried fruits, nuts
0.65–0.60	Osmophilic yeasts (*Saccharomyces rouxii*), few molds (*Aspergillus echinulatus, Monascus bisporus*)	Dried fruits containing 15–20% moisture, some toffees and caramels; honey
0.50	No microbial proliferation	Pasta containing approximately 12% moisture; spices containing approximately 10% moisture
0.40	No microbial proliferation	Whole egg powder containing approximately 5% moisture
0.30	No microbial proliferation	Cookies, crackers, bread crusts, and so on containing 3–5% moisture
0.20	No microbial proliferation	Whole milk powder containing 2–3% moisture, dried vegetables, cornflakes containing approximately 5% moisture

Source: From Beuchat, L., *Cereal Foods World*, 26, 345–349, 1981. With permission.

Water activity can influence each of the four main growth cycle phases by its effect on the germination time, the length of the lag phase, the growth rate phase, the size of the stationary population and the subsequent death rate. Generally, reducing the a_w of a given food increases the lag period and decreases the growth rate during the logarithmic phase, the maximum of which becomes lower.

Sporulation may occur at slightly below the minimum a_w for growth. In contrast, germination of spores of some micro-organisms may occur at a_w values below that required for growth. The minimum a_w for growth of micro-organisms is, without exception, less than or equal to the minimum a_w for toxin production. Optimal conditions of temperature, pH, O_2 tension and nutrient availability are necessary to permit sporulation, germination and toxin production at reduced a_w.

Whether a micro-organism survives or dies in a low a_w environment is influenced by intrinsic factors that are also responsible for its growth at higher a_w. These factors include water-binding properties, nutritive potential, pH, E_h and the presence of antimicrobial compounds. The influences exerted by these factors interact with a_w both singularly and in combination. Microbial growth and survival are not entirely ascribed to reduced a_w, but also to the nature of the solute. However, the exact nature of the role that water plays in the mechanism of cell survival is not clearly understood.

The main extrinsic factors relative to a_w that influence microbial deterioration in foods include temperature, O_2 and chemical treatments. These factors can all combine in a complex way to either encourage or discourage microbial growth.

C. Effect of Gas Atmosphere

Atmospheric O_2 generally has a detrimental effect on the nutritive quality of foods, and it is therefore desirable to maintain many types of foods at a low O_2 tension, or at least prevent a continuous supply of O_2 into the package. Lipid oxidation results in the formation of hydroperoxides, peroxides and epoxides, which, in turn, will oxidize or otherwise react with carotenoids, tocopherols and ascorbic acid to cause loss of vitamin activity. The decomposition of hydroperoxides to reactive carbonyl compounds could lead to losses of other vitamins, particularly thiamine, some forms of B_6 and pantothenic acid.[8] The destruction of other oxidizable vitamins such as folic acid, B_{12}, biotin and vitamin D is also likely.

With the exception of respiring fruits and vegetables and some flesh foods, changes in the gas atmosphere of packaged foods depend largely on the nature of the package. Adequately sealed metal and glass containers effectively prevent the interchange of gases between the food and the atmosphere. With flexible packaging, however, the diffusion of gases depends not only on the effectiveness of the closure, but also on the permeability of the packaging material which depends primarily on the physicochemical structure of the barrier. The gas permeabilities of the common thermoplastic packaging materials were given in Table 4.2.

The gas atmosphere inside food packages is often modified prior to closing by pulling a vacuum and removing most of the gases present, or by flushing the headspace area inside the package with an inert gas such as N_2 or CO_2. These procedures are generally referred to as MAP and are becoming increasing important, especially with the packaging of fresh fruits and vegetables, flesh foods and bakery products. Details of the actual procedures used and their effect on product shelf lives are discussed in Chapter 15. The associated technique of actively maintaining a desired gas atmosphere inside the package (referred to as *controlled atmosphere packaging* [CAP]) is restricted largely to cool stores and shipping containers.

D. Effect of Light

Many of the deteriorative changes in the nutritional quality of foods are initiated or accelerated by light. Light is essentially an electromagnetic vibration in the range between 4000 and 7000 Å. Each color is represented by a specific wavelength: violet is in the area of 4000 Å; blue and green are in the middle of the visible spectrum; and red is in the area of 7000 Å. The wavelength of UV light ranges between 2000 and 4000 Å. The catalytic effects of light are most pronounced in the lower wavelengths of the visible spectrum and in the ultraviolet spectrum. The intensity of light and the length of exposure are significant factors in the production of discoloration and flavor defects in packaged foods.

The total amount of light absorbed by a packaged food can be calculated using the following formula[11]:

$$I_a = I_i T_p \frac{1 - R_f}{(1 - R_f)R_p} \tag{10.40}$$

where
I_a = intensity of light absorbed by the food;
I_i = intensity of incident light;
T_p = fractional transmission by the packaging material;
R_p = fraction reflected by the packaging material;
R_f = fraction reflected by the food.

The fraction of the incident light transmitted by any given material can be considered to follow the Beer-Lambert law:

$$I_t = I_i e^{-\alpha X} \tag{10.41}$$

where
I_t = intensity of light transmitted by the packaging material;
α = the characteristic absorbance of the packaging material;
X = thickness of the packaging material.

The absorbance α varies not only with the nature of the packaging material, but also with the wavelength. Thus the amount of light transmitted through a given package will be dependent on the incident light and the properties of the packaging material. Some materials (e.g., LDPE) transmit both visible and UV light to a similar extent, whereas others (e.g., PVC) transmit visible light but absorb UV light.

Modification of plastic materials may be achieved by incorporation of dyes or application of coatings which absorb light at specific wavelengths. Glass is frequently modified by inclusion of color-producing agents or by application of coatings. In this way, a wide range of light transmission characteristics can be achieved in packages made of the same basic material.

There have been many studies demonstrating the effect of packaging materials with different light screening properties on the rate of deteriorative reactions in foods. One of the most commonly studied foods has been fluid milk, the extent of off-flavor development being related to the exposure interval, strength of light and amount of milk surface exposed. An excellent review on the influence of light transmittance of packaging materials on the shelf life of dairy products should be consulted for more details.[4]

The catalytic effect of light on the free radical reactions involved in fat oxidation is well established; such oxidation is effective not only in terms of lowering the nutritional value of the fat, but also in terms of producing toxic compounds from the fats and oils and destroying fat-soluble vitamins, in particular vitamins A and E.

In summary, light plays an important role in the deterioration of nutrients. Suitable packaging can offer direct protection by absorption or reflection of all or part of the incident light, depending on the light transmission characteristics of the packaging materials. Several plastic packaging materials, while transmitting similar amounts of light in the visible range, give varying degrees of protection against damaging UV wavelengths. These materials are often characterized by a cutoff wavelength below which transmission of light becomes negligible.

REFERENCES

1. Bender, D. A., *An Introduction to Nutrition and Metabolism*, 3rd ed., CRC Press, Boca Raton, Florida, 2002.
2. Berdanier, C. D., *Handbook of Nutrition and Food*, CRC Press, Boca Raton, Florida, 2002.
3. Beuchat, L., Microbial stability as affected by water activity, *Cereal Foods World*, 26, 345–349, 1981.
4. Bosset, J. O., Gallmann, P. U., and Sieber, R., Influence of light transmittance of packaging materials on the shelf life of milk and dairy products — a review, In *Food Packaging and Preservation*, Mathlouthi, M., Ed., Blackie Academic & Professional, Glasgow, 1994, chap. 9.
5. Cleland, A. C. and Robertson, G. L., Determination of thermal processes to ensure commercial sterility of food in cans, In *Developments in Food Preservation — 3*, Thorne, S., Ed., Elsevier Applied Science Publishers, Essex, England, 1985.
6. Fennema, O. R., Water and ice, In *Food Chemistry*, 3rd ed., Fennema, R., Ed., Marcel Dekker, New York, 1996, chap. 2.
7. Fennema, O. R. and Tannenbaum, S. R., Introduction to food chemistry, In *Food Chemistry*, 3rd ed., Fennema, O. R., Ed., Marcel Dekker, New York, 1996, chap. 1.
8. Gregory, J. F., Vitamins, In *Food Chemistry*, 3rd ed., Fennema, O. R., Ed., Marcel Dekker, New York, 1996, chap. 8.
9. Highland, H. A., Insect infestation of packages, In *Insect Management for Food Storage and Processing*, Baur, F. J., Ed., American Association of Cereal Chemists, St. Paul, Minnesota, 1984, chap. 23.
10. Jay, J. M., *Modern Food Microbiology*, 6th ed., Van Nostrand Company, New York, 2000, chap. 3.
11. Karel, M. and Lund, D. B., *Protective packaging, Physical Principles of Food Preservation*, 2nd ed., Marcel Dekker, New York, 2003, chap. 12.
12. Karmas, E. and Harris, R. S., *Nutritional Evaluation of Food Processing*, 3rd ed., Van Nostrand Reinhold Company, New York, 1988.
13. Lindsay, R. C., Flavors, In *Food Chemistry*, 3rd ed., Fennema, O. R., Ed., Marcel Dekker, New York, 1996, chap. 11.
14. Peleg, M., Physical characteristics of food powders, In *Physical Properties of Foods*, Peleg, M. and Bagley, E. B., Eds., Avi Publishing, Westport, Connecticut, 1983, chap. 10.
15. Rockland, L. B. and Beuchat, L. R., Eds., *Water Activity: Theory and Applications to Food*, Marcel Dekker, New York, 1987.
16. Sprenger, R. A., *Pest control, Hygiene for Management*, 2nd ed., Highfield Publications, Rotherham, England, 1985, chap. 13.
17. Von Elbe, J. H. and Schwartz, S. J., Colorants, In *Food Chemistry*, 3rd ed., Fennema, O. R., Ed., Marcel Dekker, New York, 1996, chap. 10.

11 Shelf Life of Foods

CONTENTS

I. DEFINITIONS

The term *food quality* has a variety of meanings to professionals in the food industry, but the ultimate arbiters of food quality must be consumers. This notion is embodied in the frequently cited definition of food quality as "the combination of attributes or characteristics of a product that have significance in determining the degree of acceptability of the product to a user."[33] Another definition of food quality is "the acceptance of the perceived characteristics of a product by consumers who are regular users of the product category or those who comprise the market segment."[7] The phrase "perceived characteristics" includes not only the food's sensory attributes, but also the perception of the food's safety, convenience, cost, value, and so on.[8]

The quality of most foods and beverages decreases with storage or holding time. Exceptions include distilled spirits (particularly whiskeys and brandies), which develop desirable flavor components during storage in wooden barrels, some wines, which undergo increases in flavor complexity during storage in glass bottles, and many cheese varieties, where enzymic degradation of proteins and carbohydrates, together with hydrolysis of fat and secondary chemical reactions, lead to desirable flavors and textures in aged cheeses.

For the majority of foods and beverages in which quality decreases with time, it follows that there will be a finite length of time before the product becomes unacceptable. This time from production to unacceptability is referred to as *shelf life*. Although the shelf lives of foods vary, they are generally determined routinely for each particular product by the manufacturer or processor. Quality loss during storage may be regarded as a form of processing at relatively low temperatures that goes on for rather a long time.[26] It is therefore not surprising that many of the concepts developed in connection with food processing find application in shelf life studies. Shelf life studies are an essential part of food product development, with the processor attempting to provide the longest practicable shelf life consistent with costs and the pattern of handling and use by distributors, retailers, and consumers.

Inadequate shelf life will often lead to consumer dissatisfaction and complaints. At best, such dissatisfaction will eventually affect the acceptance and sales of brand-name products, and, at worst, it can lead to malnutrition or even illness. Therefore, food processors pay considerable attention to determining the shelf lives of their products.

Despite its importance, there is no simple, generally accepted definition of shelf life in the food technology literature. The Institute of Food Technologists (IFT) in the U.S. has defined shelf life as "the period between the manufacture and the retail purchase of a food product, during which time the product is in a state of satisfactory quality in terms of nutritional value, taste, texture, and appearance".[1] This definition overlooks the fact that the consumer may store the product at home for some time before consuming it yet will still want the product to be of acceptable quality.

The Institute of Food Science and Technology (IFST) in the U.K. has defined shelf life as "the period of time during which the food product will remain safe; be certain to retain desired sensory, chemical, physical, microbiological, and functional characteristics; and comply with any label declaration of nutritional data when stored under the recommended conditions."[2]

Another definition is that "shelf life is the duration of that period between the packing of a product and the end of consumer quality as determined by the percentage of consumers who are displeased by the product".[18] This definition accounts for the variation in consumer perception of quality (i.e., not all consumers will find a product unacceptable at the same time) and has an economic element in that, because it is not possible to please all consumers all of the time, a baseline of consumer dissatisfaction must be established.[19] In the branch of statistics known as survival analysis, consumer dissatisfaction can be related to the survival function, defined as "the probability of a consumer accepting a product beyond a certain storage time."[13] Models permitting the application of survival analysis to the sensory shelf life of foods have been published.[13]

Put simply, shelf life is the time during which all of the primary characteristics of the food remain acceptable for consumption. Thus, the shelf life refers to the time for which a food can remain on both the retailer's and consumer's shelf before it becomes unacceptable.

The European Union (EU) has no definition of shelf life or legislation on how shelf life should be determined. The consolidated EU directive on food labeling (2000/13/EEC) requires prepackaged foods to bear a date of "minimum durability" or, in the case of foods which, from a microbiological point of view, are highly perishable, the "use by" date. The date of minimum durability is defined as the "date until which a foodstuff retains its specific properties when properly stored," and any special storage conditions (e.g., temperature not to exceed 7°C) must be specified. This concept (essentially equivalent to the "best before" date defined below) allows the processor to set the quality standard of the food, because the product would still be acceptable to many consumers after the "best before" date has passed.

It has been standard practice in almost all food manufacturing and processing establishments to put a "closed" code (so-called because only those with knowledge of the coding system can interpret the code) onto the packaged product. Generally this code indicates the time of processing and packaging, e.g., day and year, or shift, day and year, or hour, day and year. For many canned foods, it has long been mandatory to include such information on the end (lid) of the container in an embossed form, together with a code for the product itself to aid identification of unlabeled cans in the factory.

Dating of food products has been known to exist in the U.S. dairy industry since 1917, and in the 1930s U.S. consumers expressed a desire for an open dating regulation to indicate the freshness of their foods.[19] Since the advent of the consumer movement in the early 1970s, many different types of open dating systems have been proposed as part of the consumer's "right to know." An open date on a food product is a legible, easily read date which is displayed on the package with the purpose of informing the consumer about the shelf life of the product. The most common is the "best before" or "best if used by" date (the last date of maximum high quality), followed by the "use by" or "expiration" date (the date after which the food is no longer at an acceptable level of quality). However, because quality changes normally occur slowly, it is generally not possible to state that a food will be acceptable one day and unacceptable the next. Exceptions are fresh and minimally processed foods and baked goods such as bread. In the U.S., 29 states mandate some sort of open dating policies for food, and these policies are regulated by various state departments.[19]

Given the variety of definitions, it is not surprising that there is no uniform or universally accepted open dating system for packaged foods. In some countries, mandatory open dating of all perishable (and sometimes semiperishable) foods is required, while in other countries, such requirements are voluntary. Arguments can be advanced both for and against the open dating of foods. However, there is an increasing quantity of open-dated food on sale throughout the world, and this trend is likely to continue.

II. FACTORS CONTROLLING SHELF LIFE

The shelf life of a food is controlled by three factors:

1. The product characteristics including formulation and processing parameters (intrinsic factors).
2. The properties of the package.
3. The environment to which the product is exposed during distribution and storage (extrinsic factors).

Intrinsic factors include pH, water activity, enzymes, micro-organisms, and concentration of reactive compounds. Many of these factors can be controlled by selection of raw materials and ingredients, as well as the choice of processing parameters.

Extrinsic factors include temperature, relative humidity, light, total pressure and partial pressure of different gases, as well as mechanical stresses including consumer handling. Many of these factors can affect the rates of deteriorative reactions which occur during the shelf life of a product.

The properties of the package can have a significant effect on many of the extrinsic factors and thus indirectly on the rates of the deteriorative reactions. Thus, the shelf life of a food can be altered by changing its composition and formulation, processing parameters, packaging system or the environment to which it is exposed.[29]

A. Product Characteristics

1. Perishability

Based on the nature of the changes that can occur during storage, foods may be divided into three categories — perishable, semiperishable, and nonperishable or shelf stable, which translate into very short shelf life products, short to medium shelf life products and medium to long shelf life products.

Perishable foods are those that must be held at chill or freezer temperatures (i.e., 0 to 7°C or −12 to −18°C, respectively) if they are to be kept for more than short periods. Examples of such foods include milk, fresh flesh foods such as meat, poultry, and fish, minimally processed foods and many fresh fruits and vegetables.

Semiperishable foods are those which contain natural inhibitors (e.g., some cheeses, root vegetables, and eggs) or those that have received some type of mild preservation treatment (e.g., pasteurization of milk, smoking of hams, and pickling of vegetables), which produce greater tolerance to environmental conditions and abuse during distribution and handling.

Shelf stable foods are considered *nonperishable* at room temperatures. Many unprocessed foods fall into this category, and are unaffected by micro-organisms because of their low moisture content (e.g., cereal grains and nuts, and some confectionery products). Processed food products can be shelf stable if they are preserved by heat sterilization (e.g., canned foods), contain preservatives (e.g., soft drinks), are formulated as dry mixes (e.g., cake mixes) or processed to reduce their water content (e.g., raisins or crackers). However, shelf stable foods only retain this status if the integrity of the package containing them remains intact. Even then, their shelf life is finite owing to deteriorative chemical reactions which proceed at room temperature independently of the nature of the package, and the permeation through the package of gases, odors, and water vapor.

2. Bulk Density

The free space volume of a package (V) is directly related to the bulk density (ρ_b) and the true density (ρ_p) of the product as follows:

$$V = V_t - V_p = \frac{W}{\rho_b} - \frac{W}{\rho_p} \tag{11.1}$$

where
V_t = total volume of the package;
V_p = volume of the product;
W = weight of the product.

Thus, for packages of similar shape, equal weights of products of different bulk densities will have different free space volumes, and therefore package areas and package behavior will differ. This has important implications when changes are made in package size for the same product, or alterations are made to the process, resulting in changes to the product bulk density.

Although the true density of a food depends largely on its composition and cannot be changed significantly, the bulk density of food powders can be affected by processing and packaging. Some food powders (e.g., milk and coffee) are instantized by treating individual particles so that they form free-flowing agglomerates or aggregates in which there are relatively few points of contact;

the surface of each particle is thus more easily wetted when the powder is rehydrated. Instantization results in a reduction of bulk density — for example, for skim milk powder, from 0.64 to 0.55 g mL^{-1}. A wide range of bulk densities is encountered in food products, from around 0.056 g mL^{-1} for potato chips to 0.96 for granulated salt.

The free space volume has an important influence on the rate of oxidation of foods; if a food is packaged in air, then a large free space volume is undesirable because it constitutes a large O_2 reservoir. Conversely, if the product is packaged in an inert gas, then a large free space volume acts as a huge "sink" to minimize the effects of O_2 transferring through the package. It follows that a large package surface area and a low food bulk density result in greater O_2 transmission.

3. Concentration Effects

In Chapter 10, the major types of deteriorative reactions likely to be encountered in packaged foods were described, with the factors affecting the rates of these reactions quantified with the aid of simple chemical kinetic expressions. Thus, the progress of a deteriorative reaction can be monitored by following the change in concentration of some key component.

However, in many foods such as those containing whole tissue components, or where the reacting species are partially bound as in membranes, structural proteins or carbohydrates, the concentration varies from one point to another, even at zero time. Furthermore, because most of these compounds will have little opportunity to move, the concentration differences will get greater as the reactions proceed out from isolated initial foci. This has been described as the "brush-fire" effect and is especially important in chain reactions such as oxidation.

In addition, there may be several different stages of the deteriorative reaction proceeding at once, and the different stages may have different dependence on concentration and temperature, giving disguised kinetics. Such a situation is frequently the case for chain reactions and microbial growth, which have both a lag and a log phase with very different rate constants.

The point to be taken from this is that for many foods, it may be difficult to obtain kinetic data of use for predictive purposes. In such situations, use of sensory panels to determine the acceptability of the food is the recommended procedure.

B. Package Properties

Foods can be classified according to the degree of protection required, as shown in Table 11.1. The advantage of this sort of analysis is that attention can be focused on the key requirements of the package, such as maximum moisture gain or O_2 uptake. This then enables calculations to be made to determine whether a particular packaging material would provide the necessary barrier to give the desired product shelf life. In the case of metal cans and glass containers, these can be regarded as essentially impermeable to the passage of gases, odors, and water vapor, whereas paper-based packaging materials can be regarded as permeable. This then leaves plastics-based packaging materials, which provide varying degrees of protection, depending largely on the nature of the polymers used in their manufacture.

In Chapter 4, the permeability of thermoplastic polymers was discussed. The way in which this information can be utilized to select the most appropriate polymer for a particular product is discussed below.

The expression for the steady-state permeation of a gas or vapor through a thermoplastic material was derived earlier (see Equation 4.13), and can be written as:

$$\frac{\delta w}{\delta t} = \frac{P}{X} A(p_1 - p_2) \tag{11.2}$$

where P/X is the permeance (the permeability constant P divided by the thickness of the film X), A is the surface area of the package, p_1 and p_2 the partial pressures of water vapor outside and

TABLE 11.1
Degree of Protection Required by Various Foods and Beverages (Assuming 1 Year Shelf Life at 25°C)

Food/Beverage	Maximum Amount of O_2 gain (ppm)	Other Gas Protection Needed	Maximum Water Gain or Loss	Requires High Oil Resistance	Requires Good Barrier to Volatile Organics
Canned milk and flesh foods	1–5	No	3% loss	Yes	No
Baby foods	1–5	No	3% loss	Yes	Yes
Beers and wine	1–5	$<20\%$ CO_2 (or SO_2) loss	3% loss	No	Yes
Instant coffee	1–5	No	2% gain	Yes	Yes
Canned soups, vegetables and sauces	1–5	No	3% loss	No	No
Canned fruits	5–15	No	3% loss	No	Yes
Nuts, snacks	5–15	No	5% gain	Yes	No
Dried foods	5–15	No	1% gain	No	No
Fruit juices and drinks	10–40	No	3% loss	No	Yes
Carbonated soft drinks	10–40	$<20\%$ CO_2 loss	3% loss	No	Yes
Oils and shortenings	50–200	No	10% gain	Yes	No
Salad dressings	50–200	No	10% gain	Yes	Yes
Jams, jellies, syrups, pickles, olives and vinegars	50–200	No	10% gain	Yes	No
Liquors	50–200	No	3% loss	No	Yes
Condiments	50–200	No	1% gain	No	Yes
Peanut butter	50–200	No	10% gain	Yes	No

Source: Adapted from Salame, M., The use of low permeation thermoplastics in food and beverage packaging, In *Permeability of Plastic Films and Coatings*, Hopfenberg, H. B., Ed., Plenum, New York, p. 275, 1974. With permission.

inside the package, and $\delta w/\delta t$ the rate of gas or vapor transport across the film, where the latter term corresponds to Q/t in the integrated form of the expression.

1. Water Vapor Transfer

The prediction of moisture transfer either to or from a packaged food requires analysis of the above equation given certain boundary conditions. The simplest analysis requires the assumptions that P/X is constant, that the external environment is at constant temperature and humidity, and that p_2, the vapor pressure of the water in the food, follows some simple function of the moisture content.

External conditions will not remain constant during storage, distribution and retailing of a packaged food. Therefore, P/X will not be constant. However, using WVTRs determined at 38°C and 90% RH gives a "worst case" analysis; but if the food is being sold in markets in temperate climates, use of WVTRs determined at 25°C and 75% RH would be more appropriate. As was noted in Chapter 4, WVTRs can be converted to permeances by dividing by Δp.

A further assumption is that the moisture gradient inside the package is negligible; that is, the package should be the major resistance to water vapor transport. This is the case whenever P/X is less than about 10 g m^{-2} day^{-1} $(cm\ Hg)^{-1}$, which is the case for most films but not paperboard under high humidity conditions.

The critical point about Equation 11.2 is that the internal vapor pressure is not constant but varies with the moisture content of the food at any time. Thus, the rate of gain or loss of moisture is not constant but falls as Δp gets smaller. Therefore, some function of p_2 (the internal vapor pressure) as a function of the moisture content, must be inserted into the equation to be able to make proper predictions. If a constant rate is assumed, then the product will be overprotected.

In low and intermediate moisture foods, the internal vapor pressure is determined solely by the moisture sorption isotherm of the food.[3] As discussed in Chapter 10 (Section IV.B.2), several functions can be applied to describe a sorption isotherm, although the preferred one is the GAB model. If a linear model is used, then the result can be integrated directly, but if the GAB model is used, then it must be numerically evaluated.

In the simplest case when the isotherm is treated as a linear function,

$$m = ba_{\mathrm{w}} + c \tag{11.3}$$

where

m = moisture content in g H_2O g^{-1} solids;
a_{w} = water activity;
b = slope of curve;
c = constant.

The moisture content can be substituted for water gain using the relationship:

$$m = \frac{W(\text{weight of water transported})}{W_{\mathrm{s}}(\text{weight of dry solids enclosed})} \tag{11.4}$$

$$\therefore W = mW_{\mathrm{s}} \tag{11.5}$$

and

$$\delta W = \delta m W_{\mathrm{s}} \tag{11.6}$$

By substitution,

$$\frac{\delta W}{\delta t} = \frac{\delta m W_s}{\delta t} = \frac{P}{X} A \left[\frac{p_0 m_e}{b} - \frac{p_0 m}{b} \right] \tag{11.7}$$

which, on rearranging, gives:

$$\frac{\delta m}{m_e - m} = \frac{P}{X} \frac{A}{W_s} \frac{p_0}{b} \delta t \tag{11.8}$$

and on integrating

$$\ln \frac{m_e - m_i}{m_e - m} = \left[\frac{P}{X} \frac{A}{W_s} \frac{p_0}{b} \right] t \tag{11.9}$$

where

m_e = equilibrium moisture content of the food if exposed to external package RH;
m_i = initial moisture content of the food;
m = moisture content of the food at time t;
p_0 = vapor pressure of pure water at the storage temperature (*not* the actual vapor pressure outside the package).

A plot of the log of the unaccomplished moisture change — the term on the left-hand side of Equation 11.9 — versus time is a straight line with a slope equivalent to the bracketed term on the right-hand side of the equation.

The end of product shelf life is reached when $m = m_c$, the critical moisture content, at which time $t = \theta_s$, the shelf life. Thus, Equation 11.9 can be rewritten as:

$$\ln \frac{m_e - m_i}{m_e - m_c} = \frac{P}{X} \frac{A}{W_s} \frac{p_0}{b} \theta_s \tag{11.10}$$

The relationship between the initial, critical, and equilibrium moisture contents is illustrated in Figure 11.1.

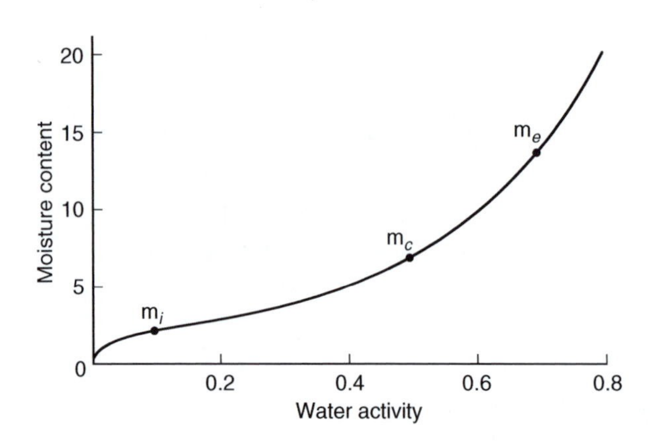

FIGURE 11.1 Typical moisture sorption isotherm for a snack bar, where m_i = initial moisture content; m_c = critical moisture content of product; m_e = equilibrium moisture content.

To simplify matters, the packaging parameters can be combined into one constant as:

$$\Omega = \frac{P}{X}\frac{A}{W_s} \tag{11.11}$$

Using Equation 11.11, one can calculate a minimum Ω, given a critical moisture content and maximum desired shelf life. Then, from Equation 11.10, for a given package size and weight of product, the permeance can be calculated and a packaging film(s) to satisfy this condition selected.

Equation 11.9 and the corresponding equation for moisture loss

$$\ln\frac{m_i - m_e}{m - m_e} = \frac{P}{X}\frac{A}{W_s}\frac{p_0}{b}t \tag{11.12}$$

have been extensively tested for foods and found to give excellent predictions of actual weight gain or loss. These equations are also useful when calculating the effect of changes in the external conditions (e.g., temperature and humidity), the surface area: volume ratio of the package, and variations in the initial moisture content of the product.

Given specific external conditions and a critical a_w for moisture gain, the shelf life is:

$$\theta_s = \Phi\frac{W_s}{A} = \Phi'\frac{V}{A} = \Phi''r \tag{11.13}$$

where Φ, Φ' and Φ'' are constants proportional to:

$$\left[\ln\frac{m_e - m_i}{m_e - m}\right] \div \left[\frac{P}{X}\frac{p_0}{b}\right] \tag{11.14}$$

where

W_s = weight of food solids = $\rho \times V$;
ρ = density of food;
V = volume of food;
r = characteristic package thickness;
θ_s = time to end of shelf life.

Because the $V : A$ ratio decreases as package size gets smaller by a factor equivalent to the characteristic thickness of the package, the shelf life using the same film will decrease directly by this thickness. Thus, to ensure adequate shelf life for a food in multiple-sized packages, shelf life tests should be based on the smallest package.

Example 11.1. The following data has been obtained for a snack bar, which is considered to be unacceptable once it has lost its crispness; it is stored at 30°C:

$m_i = 0.02$ g H_2O g^{-1} of solids
$m_e = 0.08$ g H_2O g^{-1} of solids
$m_c = 0.06$ g H_2O g^{-1} of solids
$P/X = 0.3$ g H_2O day^{-1} m^{-2} (mm Hg)$^{-1}$
Surface area of package = 0.150 m^2
Weight of dry solids = 500 g
Slope of the moisture sorption isotherm (b) = 0.06 g H_2O g^{-1} solids per unit a_w
Vapor pressure of pure water at 30°C = 31.8 mm Hg

The shelf life of the product can be calculated as follows using Equation 11.10:

$$\ln\frac{m_e - m_i}{m_e - m_c} = \frac{P}{X}\frac{A}{W_s}\frac{p_0}{b}\theta_s$$

$$\theta_s = \frac{\ln\dfrac{m_e - m_i}{m_e - m_c}}{\dfrac{P}{X}\dfrac{A}{W_s}\dfrac{p_0}{b}} = \frac{\ln[(0.08 - 0.02)/(0.08 - 0.06)]}{0.3\dfrac{0.150}{500}\dfrac{31.8}{0.06}} = 23 \text{ days}$$

If different packaging films were available with the following permeances, then the corresponding shelf lives could be calculated using the same formula:

Film A: $P/X = 0.05$ g H_2O day^{-1} m^{-2} (mm Hg)$^{-1}$; shelf life = 138 days
Film B: $P/X = 0.1$ g H_2O day^{-1} m^{-2} (mm Hg)$^{-1}$; shelf life = 69 days
Film C: $P/X = 0.2$ g H_2O day^{-1} m^{-2} (mm Hg)$^{-1}$; shelf life = 35 days

2. Gas and Odor Transfer

The gas of major importance in packaged foods is O_2 because it plays a crucial role in many reactions which affect the shelf life of foods (e.g., microbial growth, color changes in fresh and cured meats, oxidation of lipids and consequent rancidity, and senescence of fruits and vegetables).

The transfer of gases and odors through packaging materials can be analyzed in an analogous manner to that described for water vapor transfer, provided that values are known for the permeance of the packaging material to the appropriate gas, and the partial pressure of the gas inside and outside the package. Regrettably, the latter data are scarce for all but the common gases.

Packaging can control two variables with respect to O_2, and these can have different effects on the rates of oxidation reactions in foods:

1. Total amount of O_2 present. This influences the extent of the reaction, and in impermeable packages (e.g., hermetically sealed metal and glass containers), where the total amount of O_2 available to react with the food is finite, the extent of the reaction cannot exceed the amount corresponding to the complete exhaustion of the O_2 present inside the package at the time of sealing. This may or may not be sufficient to result in an unacceptable product quality after a certain period of time, dependent on the rate of the oxidation reaction. Of course, such a rate will be temperature dependent. With permeable packages (e.g., plastic packages), where ingress of O_2 will occur during storage, two factors are important: there may be sufficient O_2 inside the package to cause product unacceptability when it has all reacted with the food; or there may be sufficient transfer of O_2 through the package over time to result in product unacceptability through oxidation.
2. Concentration of O_2 in the food. In many cases, relationships between the O_2 partial pressure in the space surrounding the food and the rates of oxidation reactions can be established. If the food itself is very resistant to diffusion of O_2 (e.g., very dense products such as butter), then it will probably be very difficult to establish a relationship between the O_2 partial pressure in the space surrounding the food and the concentration of O_2 in the food.

The principal difference between predominantly water vapor-sensitive and O_2-sensitive foods is that the latter are generally more sensitive by two to four orders of magnitude. Thus, the amount of O_2 present in the air-filled headspace of O_2-sensitive foods must not be neglected when predicting their shelf life. This amount is actually 32 times higher per unit volume of air than per unit volume of O_2-saturated water. A further complicating factor with O_2-sensitive foods is that, in these foods, a concentration gradient occurs much more frequently than in moisture-sensitive foods. In the latter, it is practically limited to hard-boiled candies and freezer burn in frozen foods.

Example 11.2. Suppose that the feasibility of packaging fine wine in an oriented PET bottle with O_2 and SO_2 permeabilities of 0.30 and 3.0 P $\times 10^{-11}$ [mL(STP) cm cm^{-2} sec^{-1} (cm Hg)$^{-1}$], respectively, (all permeabilities calculated in air at 25°C and 50% RH on one side and 100% RH on the other) is to be investigated. If each bottle has a surface area of 720 cm^2, a thickness of 0.046 cm and holds 1 L of wine, calculate the shelf life of the wine in the bottle, assuming that the bottles are perfectly sealed with gas-impermeable closures.

(a) Oxygen Ingress

Because the atmosphere has an O_2 concentration of 21%, the O_2 vapor pressure outside the bottle will be $0.21 \times 76 = 16.0$ cm Hg. Assume that the O_2 vapor pressure inside the bottle is zero.

From Table 11.1, the maximum quantity of O_2 able to be absorbed by the wine and still retain acceptable quality is 5 ppm = 5 mg L^{-1} = 5×10^{-3} g L^{-1}, and because the bottle holds 1 L, 5×10^{-3} g of O_2 can be absorbed.

Using a value for the density of oxygen of 1.43×10^{-3} g mL^{-1}, the maximum quantity of oxygen (Q) permissible = $(5 \times 10^{-3})/(1.43 \times 10^{-3}) = 3.5$ mL.

Equation 10.6 can then be used:

$$\frac{Q}{t} = \frac{P}{X}A(p_1 - p_2) \tag{11.15}$$

which, on rearrangement and letting $t = \theta_s$ (the shelf life):

$$\theta_s = \frac{QX}{PA\Delta p} \tag{11.16}$$

$$\theta_s = \frac{3.5 \times 0.046}{0.3 \times 10^{-11} \times 720 \times 16} = 4.654 \times 10^6 \text{ sec} = 54 \text{ days}$$

(b) Sulfur Dioxide Egress

Assume that the initial concentration of SO_2 in the wine is 100 ppm, and that 50% of this is in the free form. The vapor pressure of 50 ppm SO_2 in the wine has been estimated to be 1.73×10^{-3} cm Hg.

If the units for ρ are g mL^{-1} and for V_w g, then Q (the quantity of SO_2 permeating through the bottle wall) will be in g SO_2 g^{-1} of wine. The initial level of free SO_2 is 50 ppm or 50 mg L^{-1}; this corresponds to 50 mg kg^{-1} of wine (assuming that the wine has the same density as water) or 5×10^{-5} g g^{-1}. In this problem, the shelf life of the wine can be considered to be over when half of the free SO_2 has permeated through the bottle (i.e., when 2.5×10^{-5} g g^{-1} has been lost). Using a value for the density of SO_2 of 2.93×10^{-3} g mL^{-1}, the maximum quantity of SO_2 that can be lost is $(2.5 \times 10^{-5})/(2.93 \times 10^{-3}) = 8.5 \times 10^{-3}$ mL g^{-1} = 8.5 mL for a 1 L bottle.

Assume a value for the partial pressure difference of SO_2 of 1.726×10^{-3} cm Hg. Substituting into Equation 11.16 and solving gives

$$\theta_s = \frac{8.5 \times 0.046}{3.0 \times 10^{-11} \times 720 \times 1.726 \times 10^{-3}} = 1.05 \times 10^{10} \text{ sec} = 333 \text{ years}$$

(c) Conclusion

Ingress of O_2 from the atmosphere into the wine is likely to be the major mode of failure for wine packaged in a PET bottle. The calculated shelf life of 54 days is probably too short, and therefore the bottle would have to be coated with a barrier material. Permeation of SO_2 through the bottle walls is not going to be a limiting factor in the shelf life of the wine.

Prediction of the shelf life of food products which deteriorate by two or more mechanisms simultaneously (e.g., oxidation resulting from ingress of oxygen and loss of crispness owing to ingress of moisture) is more complex. Some general approaches that can be applied have been proposed.[24] However, the amount of data necessary to develop the equations required for predictive purposes are prohibitively costly for the food industry. Hence, accelerated shelf life testing (ASLT) procedures are a more cost-effective and simpler method for the determination of product shelf life.

3. Package/Product Interaction

With certain products packaged with certain materials, the end of shelf life arises when an unacceptable degree of interaction between the package and the product has occurred. Several examples will be given to illustrate the nature of the problem.

The first example is that of a tomato product processed under typical conditions and packaged in a three piece can with a plain tinplate body and enameled ECCS ends. Over a storage period of 24 months at ambient temperature, several degradative reactions occurred. The concentration of tin ions in the product increased rapidly during the first 3 months from approximately 20 to about 160 ppm, reaching 280 ppm after 24 months. Iron also dissolved, increasing slowly from 8 to 10 ppm after 18 months to reach 14 ppm after 24 months. The flavor score (as determined by sensory evaluation) declined as a result of the increasing quantities of dissolved tin and iron; the color value (as determined instrumentally) showed a decrease owing to an increase in brown pigments, but remained acceptable.

The limiting factor for this particular product is the deterioration in flavor resulting from the dissolution of tin and iron from the package into the product, giving an acceptable shelf life of 18 to 24 months. If a longer shelf life was required, then it would be necessary to use a full enamel-lined can.

A second example involves an orange juice packaged aseptically in LDPE-foil-paper laminate cartons and glass containers. After 2.5 months storage at 25°C, an experienced taste panel detected a significant ($p \leq .05$) difference between the orange juices in cartons and glass containers. Analysis of the *d*-limonene (one of the major components of the essential oils in citrus juices) content showed that it had decreased from 70 to 40 ppm in the cartons within 35 days. The limonene had been absorbed (scalped) by the LDPE surface in contact with the orange juice. Moreover, ascorbic acid degradation and subsequent browning was accelerated owing to contact with the LDPE film.

Thus, the shelf life of aseptically packaged citrus juices in cartons is limited (largely as a result of package/product interaction) to about 9 months, the end of shelf life being determined by flavor changes to the juices as a result of "scalping" of the flavor components by the package.

C. Distribution Environment

1. Climatic

The deterioration in product quality of packaged foods is often closely related to the transfer of mass and heat through the package. Packaged foods may lose or gain moisture; they will also reflect the temperature of their environment because very few food packages are good insulators. Thus, the climatic conditions (i.e., temperature and humidity) of the distribution environment have an important influence on the rate of deterioration of packaged foods.

a. *Mass Transfer*

With mass transfer, the exchange of vapors and gases with the surrounding atmosphere is of primary concern. Water vapor and O_2 are generally of most importance, although the exchange of volatile aromas from or to the product from the surroundings can be important. As well as O_2, transmission of N_2 and CO_2 may have to be taken into account in packages where the concentration of these gases inside the package has been modified from ambient to inhibit or slow down deteriorative reactions in the food.

Generally, the difference in partial pressure of the vapor or gas across the package barrier will control the rate and extent of permeation, although transfer can also occur as a result of the presence of pinholes in the material, channels in seals and closures, or cracks that result from flexing of the packaging material during filling and subsequent handling. In contrast to the common gases, the partial pressure of water vapor in the atmosphere varies continuously, although the variation is generally much less in controlled climate stores.

To summarize, mass transfer depends on the partial pressure difference across the package barrier of gases and water vapor, and on the nature of the barrier itself. These factors were discussed above in Section II.B.

b. *Heat Transfer*

One of the major determinants of product shelf life is the temperature to which the product is exposed during the time from production to consumption. Without exception, food products are exposed to fluctuating temperature environments during this time, and to accurately estimate shelf life, the nature and extent of these temperature fluctuations need to be known. There is little point in carefully controlling the processing conditions inside the factory and then releasing the product into the distribution and retail system without knowledge of the conditions it will experience in that system. Such knowledge is essential in the case of products containing a "best before" or "use by" date.

The detailed climatic statistics of global maximum and mean temperatures that are available in many countries are of great assistance. Despite meteorological and secular trends, the daily (and even the annual) cycle of temperatures can be normalized to a standard cycle with a standard frequency distribution derived from the mean and range at many places. This is because of the sinusoidal trend of diurnal (Earth's rotation) and seasonal (Earth's revolution) solar radiation intensity.

The storage climates inside buildings such as warehouses and supermarkets are only broadly related to the external climate as reported by weather stations; climatic variations in temperature and humidity can differ as much between different building constructions as between seasons on one site. Table 11.2 shows the influence of season and type of premises on quarterly mean temperatures and humidities in the U.K.

If the major deteriorative reaction causing end of shelf life is known, then simple expressions can be derived to predict the extent of deterioration as a function of available time–temperature storage conditions. The basic types of deteriorative reactions that foods undergo were discussed in Chapter 10, together with the rates of these reactions and the factors controlling these rates. These reactions and their rates are now analyzed in relation to product shelf life.

Fundamental to such an analysis is that the particular food under consideration follows the laws of additivity and commutativity. *Additivity* implies that the total extent of the degradation reaction in the food produced by a succession of exposures at various temperatures is the simple sum of the separate amounts of degradation, regardless of the number or spacing of each time–temperature combination. *Commutativity* means that the total extent of the degradation reaction in the food is independent of the order of presentation of the various time–temperature experiences.

TABLE 11.2
United Kingdom Climate in Typical Premises

	Supermarkets		Heated Warehouses		Unheated Buildings	
Period	**°C**	**RH (%)**	**°C**	**RH (%)**	**°C**	**RH (%)**
March–May	17.3	48.3	15.9	53.4	11.1	67.4
June–August	21.1	53.9	20.8	66.6	17.1	66.5
September–November	18.2	55.9	17.0	60.8	12.7	75.3
December–February	14.5	48.0	12.7	55.3	4.8	79.2

Source: From Cairns, J. A., Elson, C. R., Gordon, G. A., and Steiner, E. H., *Research Report No. 174*, British Food Manufacturing Industry Research Association, Leatherhead, England, 1971. With permission.

i. Shelf Life Plots

A useful approach to quantifying the effect of temperature on food quality (especially when little data are available to get rate constants, or when only the time to reach a certain level of quality change has been determined) is to construct shelf life plots. As discussed in Chapter 10 (Section IV.A), several models are used to represent the relationship between the rate of a reaction (or the reciprocal of rate, which can be time for a specified loss in quality or shelf life) and temperature. The two most common models are the Arrhenius and linear, and these are shown in Figure 11.2.

The equations for these two plots are:

$$\theta_s = \theta_0 \exp \frac{E_A}{R}\left[\frac{1}{T_s} - \frac{1}{T_0}\right] \tag{11.17}$$

and

$$\theta_s = \theta_0 e^{-b(T_s - T_0)} \tag{11.18}$$

where

θ_s = shelf life at temperature T_s;
θ_0 = shelf life at temperature T_0.

If only a small temperature range is used (less than ±20°C), then there is little error in using the linear plot rather than the Arrhenius plot.

Most deteriorative reactions in foods can be classified as either zero or first order, and the way in which these two reaction orders can be used to predict the extent of deterioration as a function of temperature is now outlined.

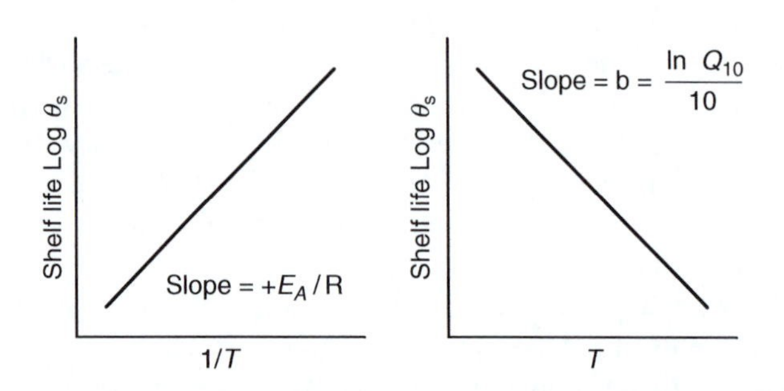

FIGURE 11.2 Arrhenius plot of log shelf life (θ_s) versus reciprocal of the absolute temperature (K) showing a slope of E_A/R, and (b) linear plot of log shelf life versus temperature (°C) showing a slope of b.

ii. Zero-Order Reaction Prediction

In Chapter 10, Equation 10.9 was derived for the change in a quality factor A when all extrinsic factors are held constant:

$$A_e = A_0 - k_z\theta_s \tag{11.19}$$

and

$$A_0 - A_e = k_z\theta_s \tag{11.20}$$

where

A_e = value of A at end of shelf life;
A_0 = value of A initially;
k_z = zero-order rate constant (time^{-1});
θ_s = shelf life in days, months, years, and so on.

For variable time–temperature storage conditions, Equation 11.19 can be modified as follows:

$$A_e = A_0 - \Sigma(k_i\theta_i) \tag{11.21}$$

where $\Sigma k_i\theta_i$ = the sum of the product of the rate constant, k_i, at each temperature, T_i, multiplied by the time interval, θ_i, at the average temperature, T_i, for the given time period, $\Delta\theta$.

To apply this method, the time–temperature history is broken up into suitable time periods and the average temperature in that time period determined. The rate constant for that period is then calculated from the shelf life plot using a zero-order reaction. The rate constant is multiplied by the time interval, θ_i, and the sum of the increments of $k_i\theta_i$ gives the total amount lost at any time.

Alternatively, instead of calculating actual rate constants, the time for the product to become unacceptable (i.e., for A to become A_e) can be measured, and Equation 11.21 modified to give:

f_c = fraction of shelf life consumed = change in A divided by total possible change in A

$$= \frac{A_0 - A}{A_0 - A_e} \tag{11.22}$$

$$= \frac{\Sigma(k_i\theta_i)}{\Sigma(k_i\theta_s)} \tag{11.23}$$

$$= \Sigma\left[\frac{\theta_i}{\theta_s}\right]T_i \tag{11.24}$$

A similar approach to that described earlier is employed. The temperature history is divided into suitable time periods and the average temperature, T_i, at each time period evaluated. The time held at that temperature, θ_i, is then divided by the shelf life, θ_s, for that particular temperature, and the fractional values summed up to give the fraction of shelf life consumed. Astute readers will recognize the similarity between this method and the graphical method used to determine the lethality of a thermal process. This is not surprising because both are concerned with summing the effects of various temperatures on the rates of reactions.

The shelf life can also be expressed in terms of the fraction of shelf life remaining, f_r:

$$f_r = 1 - f_c \tag{11.25}$$

Thus, for any temperature T_s,

$$f_r\theta_s = (1 - f_c)\theta_s = \text{shelf life left at temperature } T_s \qquad (11.26)$$

In other words, the shelf life left at any temperature is the fraction of shelf life remaining multiplied by the shelf life at that temperature.

The above method was initially developed by the U.S. Department of Agriculture in California during the 1950s for the determination of the shelf life of frozen foods. It is referred to as the *time–temperature-tolerance approach* (TTT).[34] In these and related studies, the period of time (designated as the *high quality life* [HQL]) for 70 to 80% of a trained taste panel to correctly identify the control samples (held at -29°C) from samples stored at various subzero temperatures using the triangle or duo–trio test was determined. The change in quality at this stage has been designated the *just noticeable difference* (JND) or *first noticeable difference* (FND). The HQL has no real commercial significance and is quite different from the *practical storage life* (PSL), which is of interest to food processors and consumers. The ratio between PSL and HQL is often referred to as the *acceptability factor* and can range from 2:1 up to 6:1.

The TTT work on frozen foods has generally demonstrated that the HQL varies exponentially with temperature. However, it has been subsequently shown that when overall quality is measured (rather than just one single quality factor), a semilogarithmic plot results in curved rather than straight lines. It was suggested that a semilogarithmic plot was convenient for products with very long keeping times, but for other products, a plot utilizing two linear scales is more convenient.

TTT relationships are not strict mathematical functions but empirical data subject to large variability, particularly because of variations in product, processing methods and packaging (the PPP factors). Therefore, any shelf life prediction made will be specific for a particular product (e.g., specific breed of animal slaughtered at a certain age or weight) that is processed, packaged, and stored under specific conditions. Failure to specify PPP factors leads to the vast plethora of seemingly contradictory shelf lives for frozen foods reported in the literature. For example, the frozen shelf life of cod stored at -18°C has been reported by various authors to be anywhere from 15 to 45 weeks, and it has been calculated that, on the basis of data in the literature, the 95% confidence interval for the HQL of frozen lean meat ranges from 8 months to 3 years. Thus, predictions cannot be made with any precision on the quality or quality change in a frozen food from knowledge of its time–temperature history and TTT literature data only. Therefore, in determining the shelf life of frozen foods, the PPP factors must be taken into account in addition to the TTT relationships.

Example 11.3. A frozen food (ground beef packaged in LDPE film) has a PSL at various temperatures as follows:

Temperature (°C)	PSL (days)
−8	120
−12	180
−15	230
−18	300
−20	350
−23	420
−25	480

Calculate the total loss of PSL along the freezer chain from processor to consumer given the time and temperature history shown below.

Links in the Freezer Chain	Average Temperature (°C)	Storage Time (days)	PSL (days)	PSL Loss (% day^{-1})	Loss (%)
Processor	−23	40	420	0.238	9.5
Transport	−20	2	350	0.286	0.6
Cold store	−25	190	480	0.208	39.9
Transport	−18	1	300	0.333	0.3
Wholesale	−23	30	420	0.238	7.2
Transport	−15	1	230	0.435	0.2
Display cabinet					
Center	−20	20	350	0.286	5.8
Upper layer	−12	6	180	0.556	3.4
Transport	−8	1/6	120	0.833	0.1
Consumer	−18	50	300	0.333	16.5
Total loss of PSL:		340			83.6

By dividing the PSL into 100, the product life loss per day as a percentage at that temperature is determined (e.g., at −23°C, 100/420 = 0.238% loss per day). When the storage time in Column 3 is multiplied by product life loss per day in Column 5, the product life loss expressed as a percentage of the PSL can be calculated.

The total loss of PSL for the ground beef at the end of the freezer chain (340 days) is 84%. Thus, 16% of its PSL is left; that is, the product could be kept by the consumer at −18°C for another 16 × 300/100 = 49 days before it exceeded its PSL.

iii. First-Order Reaction Prediction

The equivalent expression to Equation 11.19 for a first-order reaction was derived as Equation 10.14 for the case where all extrinsic factors are held constant:

$$A_e = A_0 \exp(-k\theta_s) \tag{11.27}$$

From this, an expression can be developed to predict the amount of shelf life used up as a function of variable temperature storage for a first-order reaction in the form:

$$A = A_0 \exp(-\Sigma k_i \theta_i) \tag{11.28}$$

where A = the amount of some quality factor remaining at the end of the time–temperature distribution, and $\Sigma k_i \theta_i$ has the same meaning as in Equation 11.21.

If the shelf life is based simply on some time to reach unacceptability, then Equation 11.28 can be modified to give an analogous expression to that derived for the TTT method. Note that because of the exponential loss of quality, A_e will never be zero. Thus,

$$\ln \frac{A}{A_0} = -\Sigma k_i \theta_i \tag{11.29}$$

and

$$k_i = \frac{\ln A_e/A_0}{\theta_s} \tag{11.30}$$

where

$\ln A/A_0$ = fraction of shelf life consumed at time θ;
$\ln A_e/A_0$ = fraction of shelf life consumed at time θ_s.

The fraction of shelf life remaining, f_r, is

$$f_r = 1 - \frac{\ln A_0/A}{\ln A_0/A_e} = 1 - \Sigma\left[\frac{\theta_i}{\theta_s}\right]_{T_i} \tag{11.31}$$

Example 11.4. The rate of protein quality loss (measured as lysine) in enriched pasta ($a_w = 0.49$) was determined at three temperatures assuming a first-order reaction with the following results:

Temperature (°C)	Rate Constant k (days^{-1} × 10^4)
30	61
37	95
45	156

Assuming that the pasta was packaged in a material that was totally impermeable to moisture vapor (i.e., there was no change in a_w), calculate the protein quality loss after the packaged pasta was exposed to the following time–temperature storage regime:

Temperature (°C)	Time (days)
25	60
32	30
41	10
38	8
29	22
23	100

The first step is to calculate the rate of protein quality loss at the various storage temperatures. Either the linear or the Arrhenius relationships can be assumed; the rate constants below were obtained graphically assuming each relationship. For the linear relationship, the logarithm of k was plotted against temperature (°C), whereas for the Arrhenius relationship, the logarithm of k was plotted against the reciprocal of the absolute temperature (K). It can be seen that use of either relationship gives essentially the same results.

Temperature (°C)	Rate Constant k (days^{-1} × 10^4)	
	Linear	Arrhenius
23	38	38
25	44	43
29	57	57
32	70	72
38	104	97
41	128	123

Using the rate constants obtained above and assuming a linear relationship, the following protein quality loss was calculated after each time–temperature storage period; a first-order reaction was assumed.

Temperature (°C)	Time (days)	Rate Constant k (days^{-1} × 10^{-4})	$k_i\theta_i$	Lysine Concentration (Assuming 100 at θ=0)
25	60	44	0.258	77.3
32	30	70	0.210	62.7
41	10	128	0.128	55.2
38	8	104	0.083	50.2
29	22	57	0.125	44.8
23	100	38	0.380	30.6

Thus, at the end of the storage regime, only 30.6% of the lysine present initially remains. In real life, the package is unlikely to be impermeable to moisture vapor, and therefore the a_w of the pasta would change with time. The change in a_w would, in turn, affect the rate constant, which depends on both temperature and a_w and therefore the calculated shelf life would not be the actual shelf life.

iv. Sequential Fluctuating Temperatures

Although the above analysis can be applied to any random time–temperature storage regime, in practice, many products are exposed to a sequential, regular, fluctuating, temperature profile, especially if held in trucks, rail-cars and uninsulated warehouses. This is because of the daily day–night pattern resulting from exposure to solar radiation. It has been found that many of these patterns can be assumed to follow either a square or sine wave form, as shown in Figure 11.3. For example, warehouse temperatures in various geographical locations in the U.S. are predictably cyclic, and nearly sinusoidal within cans stored in the warehouse.

Equations have been developed for both zero- and first-order reactions which enable calculation of the extent of a degradative reaction for a food subjected to either square or sine wave temperature functions. It can be shown that the extent of reaction after a period of time will be the same as it would have been if the food had been held at a certain steady "effective" temperature for the same length of time. This effective temperature will be higher than the arithmetic mean temperature. Comparisons for losses in a theoretical temperature distribution showed that for less than 50% degradation, the losses were about the same for zero and first order at any time, and thus determination of the reaction order is not critical. However, the temperature sensitivity (Q_{10}) of the reaction is very important in making predictions.[30]

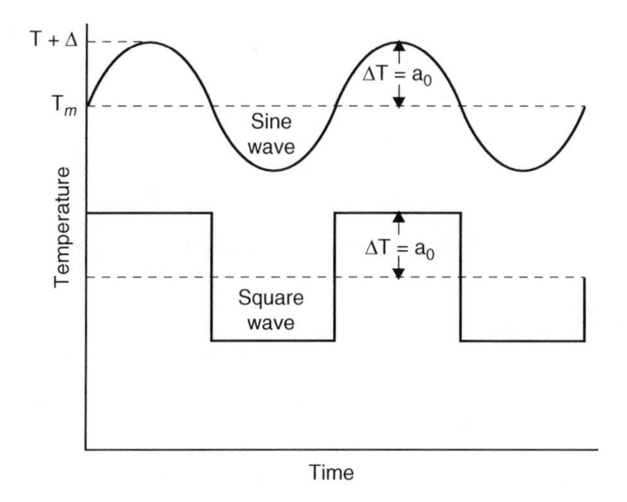

FIGURE 11.3 Square and sine wave temperature fluctuations of packaged foods where a_0 is the amplitude.

c. *Simultaneous Mass and Heat Transfer*

In the majority of distribution environments, many packaged foods undergo changes in both moisture content and temperature during storage as a result of variable temperature and humidity conditions in the environment. This has the effect of complicating the calculations for predicting the shelf life of packaged foods. For instance, the preceding example for the loss of protein quality in pasta assumed no change in a_w of the pasta as it underwent various changes in temperature. Clearly, it would be unlikely that the package would be totally impermeable to moisture vapor, and therefore a_w would change with time. This complicates the calculation of quality loss, because the rate is now dependent on both temperature and a_w.

A further complication is that data on the humidity distribution of environments where foods are stored are scarce and not as easily predicted as the external temperature distribution. Therefore, prediction of the actual shelf life loss of packaged foods will only be approximate. More complete data about the humidity distribution of food storage environments is required so that shelf life predictions can be further refined.

2. Physical

After processing and manufacture, the food product leaves the factory and usually moves into the company's warehouse. It is then transported by rail or truck to a distribution center. From the distribution center, the product may be taken to the retail outlet by truck or rail, or may be taken to a port for transport by sea or air to another distribution warehouse. Further transport by rail or truck then follows to the retail outlet.

Regardless of the distribution pattern, transportation damage may be incurred by the product. The extent of such damage will be a function of not only the packaging (primary, secondary, and tertiary), but also the nature of the distribution environment and the method of transportation.

Most secondary packages are stacked on pallets, which may or may not be shrink or stretch wrapped. Pallets are usually stacked two high in rail cars and one or two high in trucks; in warehouses, pallets may be stacked four high. Thus, the forces acting on the top tier of a four pallet stack will be quite different from those acting on the bottom tier of the same stack. The longer the time that the product spends in the distribution chain (this is usually directly related to the shelf life of the product), the more significant are the effects arising from pallet stacking height.

Ultimately, knowledge of the distribution environment (both climatic and physical) is essential before meaningful shelf life tests can be designed. Taking a product from the production line and conducting a shelf life test on it while ignoring possible distribution environment hazards will almost certainly lead to an overestimation of shelf life.

III. SHELF LIFE ESTIMATION

A. Introduction

There are at least three situations when shelf life estimation might be required:

1. To determine the shelf life of existing products.
2. To study the effect of specific factors and combinations of factors such as storage temperature, packaging materials, processing parameters or food additives on product shelf life.
3. To determine the shelf life of prototype or newly developed products.

Several established approaches are available for estimating the shelf life of foods[30]:

1. *Literature study*: the shelf life of an analogous product is obtained from the published literature or in-house company files. Examples can be found in recent books on the shelf life of foods.[11,14,21]
2. *Turnover time*: the average length of time that a product spends on the retail shelf is found by monitoring sales from retail outlets, and from this the required shelf life is estimated. This does not give the "true" shelf life of the product but rather the "required" shelf life, where it is implicitly assumed that the product is still acceptable for some time after the average period on the retail shelf.
3. *End point study*: random samples of the product are purchased from retail outlets and then tested in the laboratory to determine their quality. From this, a reasonable estimation of shelf life can be obtained because the product has been exposed to actual environmental stresses encountered during warehousing and retailing.
4. *ASLT*: laboratory studies are undertaken during which environmental conditions are accelerated by a known factor so that the product deteriorates at a faster than normal rate. This method requires that the effect of environmental conditions on product shelf life can be quantified.

B. Sensory Evaluation

Regardless of the method chosen or the reasons for its choice, sensory evaluation of the product is likely to be used either alone or in combination with instrumental or chemical analyses to determine the quality of the product. Not surprisingly, many food scientists and technologists in the industry attempt to replace human judgment with instrumental or chemical analyses because the latter are neither prone to fatigue nor subject to the physiological and psychological fluctuations that characterize human performance. However, because human judgment is the ultimate arbiter of food acceptability, it is essential that the results obtained from any instrumental or chemical analysis correlate closely with the sensory judgments for which they are to substitute. Correlation of values of individual chemical parameters with sensory data is often not straightforward because overall organoleptic quality is a composite of a number of changing factors. The relative contribution of each factor to the overall quality may vary at different levels of quality or at different storage conditions.[32] Other problems with sensory evaluation include the high cost of using large testing panels and the ethics of asking panelists to taste spoiled or potentially hazardous samples.

Three experimental designs are commonly used for the purpose of shelf life estimation: the paired comparison test; the duo–trio test; and the triangle test. Further details about these tests can be found in standard texts on sensory evaluation.[23] Descriptive methods are used to measure quantitative or qualitative characteristics of products and require specially trained panelists. Affective methods are used to evaluate preference, acceptance or opinions of products and do not require trained panelists.

The selection of a particular sensory evaluation procedure for evaluating products undergoing shelf life testing is dependent on the purpose of the test. Acceptability assessments by untrained panelists are essential to an open dating program, while discrimination testing with expert panels might be used to determine the effect of a new packaging material on product stability. However, an expert panel is not necessarily representative of consumers, much less different consumer segments. Even if that assumption can be made, a cutoff level of acceptability has to be decided. The time at which a large (but predetermined) percentage of panelists judge the food to be at or beyond that level is the end of shelf life.[32]

C. Shelf Life Failure

In shelf life testing, there can be one or more criteria that constitute sample failure. One criterion is an increase or decrease by a specified amount in the mean panel score. Another criterion is microbial deterioration of the sample to an extent that renders it unsuitable or unsafe for human consumption. Finally, changes in odor, color, texture, flavor, and so on, which render the sample unacceptable to either the panel or the consumer, are criteria for product failure. Thus, sample failure can be defined as the condition when the product exhibits either physical, chemical, microbiological or sensory characteristics that are unacceptable to the consumer, and the time required for the product to exhibit such conditions is the shelf life of the product.

However, a fundamental requirement in the analysis of data is knowledge of the statistical distribution of the observations, so that the mean time to failure and its standard deviation can be accurately estimated, and the probability of future failures predicted. The length of shelf life for food products is usually obtained from simple averages of time to failure on the assumption that the failure distribution is symmetrical. If the distribution is skewed, estimates of the mean time to failure and its standard deviation will be biased. Furthermore, when the experiment is terminated before all the samples have failed, the mean time to failure (based on simple averages) will be biased because of the inclusion of unfailed data.

In order to improve the methodology for estimating shelf life, knowledge of the statistical distribution of shelf life failures is required, together with an appropriate model for data analysis. Five statistical models — normal, lognormal, exponential, Weibull, and extreme-value distributions — have been fitted to failure data using the method of hazard plotting which provides information about the adequacy of fit of the observed data to the proposed model, the mean or median time to failure and the probability of future failures.[12] The Weibull distribution was suggested as the most appropriate shelf life model, and examples of its use to predict end of shelf life for food products have been presented.[6,10,18]

The nature of product failure over time is commonly represented by a "bath-tub" curve, which has many applications in the actuarial and engineering sciences; an example for a refrigerated food product is shown in Figure 11.4. At time X_0, the finished product leaves the processing plant and begins its journey to the many distribution outlets. During the time between X_0 and X_1, early failures may occur owing to faulty packaging (e.g., pinholes or poor seals) and product abuse. However, the early failures should not be taken as true failures relative to the shelf life of the product. From X_1 to X_2, no product failures (barring random fluctuation) would be expected.

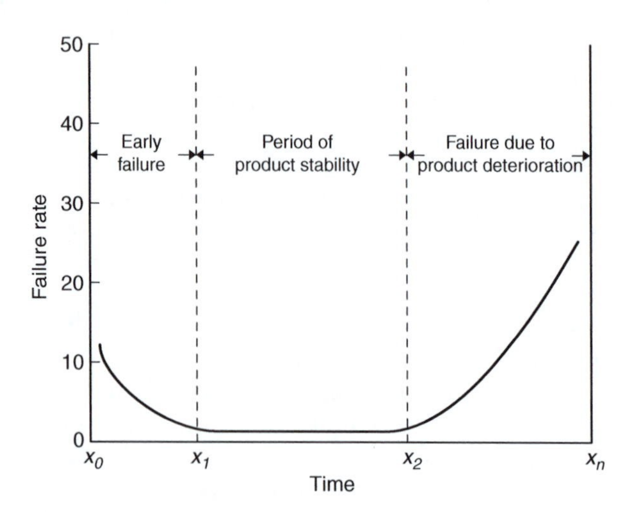

FIGURE 11.4 Bath-tub curve showing failure rate as a function of time.

From the time X_2 to the termination time X_n, the hazard (failure) rate increases, this time representing the true failure owing to deteriorative changes within the product. The length of shelf life is determined between the times X_0 and X_n, and the hazard function plays a central role in the analysis of failure data.

One challenge with shelf life testing is to develop experimental designs that minimize the number of samples required (thus minimizing the cost of the testing), while simultaneously providing reliable and statistically valid answers.

D. Accelerated Shelf Life Testing (ASLT)

1. Basic Principles

The basic assumption underlying ASLT is that the principles of chemical kinetics can be applied to quantify the effects that extrinsic factors such as temperature, humidity, gas atmosphere, and light have on the rate of deteriorative reactions.[24] These basic principles and the way in which they can be applied to foods are described in Chapter 10. By subjecting the food to controlled environments in which one or more of the extrinsic factors is maintained at a higher than normal level, the rates of deterioration will accelerate, resulting in a shorter than normal time to product failure. Because the effects of extrinsic factors on deterioration can be quantified, the magnitude of the acceleration can be calculated and the "true" shelf life of the product under normal conditions calculated. Thus, a shelf life test that would normally take a year can be completed in about a month if the storage temperature is raised by 20°C.[31]

The need for ASLT of food products is simple — because many foods have shelf lives of at least 1 year, evaluating the effect on shelf life of a change in the product (e.g., a new antioxidant or thickener), the process (e.g., a different time/temperature sterilization regime) or the packaging (e.g., a new polymeric film) would require shelf life trials lasting at least as long as the required shelf life of the product. Companies cannot afford to wait for such long periods to determine whether or not the new product/process/packaging will provide an adequate shelf life, because other decisions (e.g., to construct a new factory, order new equipment or arrange contracts for the supply of new packaging material) have lead times of months or years. Some way of speeding up the time required to determine the shelf life of a product is necessary, and ASLT has been developed for that reason. Such procedures have long been used in the pharmaceutical industry where shelf life and efficacy of drugs are closely related. However, the use of ASLT in the food industry is not as widespread as it might be, due in part to the lack of basic data on the effect of extrinsic factors on the rates of deteriorative reactions, in part to ignorance of the methodology required, and also due to skepticism of the advantages to be gained from using ASLT procedures.

As discussed in Chapter 10, quality loss for most foods follows either a zero- or first-order reaction. Figure 11.2 showed the logarithm of shelf life versus temperature and the inverse of absolute temperature. If only a small range of temperature is considered, then the former shelf life plot generally fits the data for food products.

For a given extent of deterioration and reaction order, the rate constant is inversely proportional to the time to reach some degree of quality loss. Thus, by taking the ratio of the shelf life between any two temperatures 10°C apart, the Q_{10} of the reaction can be found. This can be expressed by extension of Equation 10.25, assuming a linear shelf life plot:

$$Q_{10} = \frac{k_{T+10}}{k_T} = \frac{\theta_{s_T}}{\theta_{s_{T+10}}} \tag{11.32}$$

where

θ_{s_T} = shelf life at temperature T°C;
$\theta_{s_{T+10}}$ = shelf life at temperature $(T + 10)$°C.

TABLE 11.3
Effect of Q_{10} on Shelf Life

Temperature (°C)	Shelf Life (weeks)			
	$Q_{10} = 2$	$Q_{10} = 2.5$	$Q_{10} = 3$	$Q_{10} = 5$
50	2[a]	2[a]	2[a]	2[a]
40	4	5	6	10
30	8	12.5	18	50
20	16	31.3	54	4.8 (years)

[a] Arbitrarily set at 2 weeks at 50°C. Shelf lives at lower temperatures are calculated on this arbitrary assumption.

Source: From Labuza, T. P. and Kamman, J. F., Reaction kinetics and accelerated tests simulation as a function of temperature, In *Computer-Aided Techniques in Food Technology*, Saguy, I., Ed., Marcel Dekker, New York, 1983, chap. 4. With permission.

The effect of Q_{10} on shelf life is shown in Table 11.3, which illustrates the importance of accurate estimates of Q_{10} when making shelf life estimations. For example, if a product has a shelf life of 2 weeks at 50°C and a Q_{10} of 2, then it has a shelf life of 16 weeks at 20°C. However, if Q_{10} was 2.5 rather than 2, the shelf life at 20°C would be almost twice as long (31 weeks). Thus, a small error in Q_{10} can lead to huge differences in the estimated shelf life of the product. Typical Q_{10} values for foods are 1.1 to 4 for canned products, 1.5 to 10 for dehydrated products and 3 to 40 for frozen products.

A further use for Q_{10} values is illustrated in Figure 11.5, which depicts a shelf life plot for a product that has at least 18 months shelf life at 23°C. To determine the probable shelf life of the

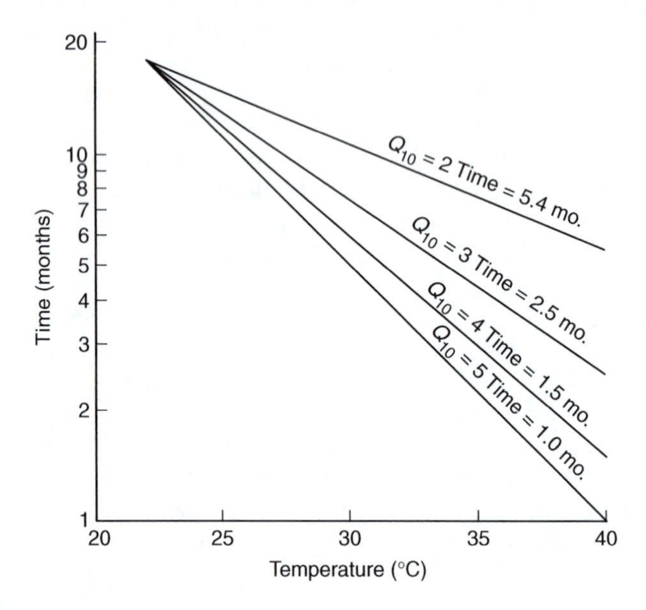

FIGURE 11.5 Hypothetical shelf life plot for various Q_{10}'s passing through a shelf life of 18 months at 23°C. 'Time' refers to the accelerated shelf life times required at 40°C for various Q_{10}'s. (*Source*: From Labuza, T. P. and Kamman, J. F., Reaction kinetics and accelerated tests simulation as a function of temperature, In *Computer-Aided Techniques in Food Technology*, Saguy, I., Ed., Marcel Dekker, New York, 1983, chap. 4. With permission.)

TABLE 11.4
Effect of E_A of the Key Deteriorative Reaction on the Time to Complete an ASLT Test for a Low Moisture Food Product with a Targeted Shelf Life of 2 Years at Ambient Storage

E_A (kJ mol^{-1})	Testing Time at 40°C (days)	Testing Time at 45°C (days)
45	224	171
85	78	47
125	28	13

Source: From Taoukis, P.S. and Giannakourou, M.C., in *Understanding and Measuring the Shelf-Life of Food*, Steele, R., Ed., CRC Press, Boca Raton, Florida, 2004, chap. 3. With permission.

product at 40°C, lines are drawn from the point corresponding to 18 months at 23°C to intersect a vertical line drawn at 40°C; the slope of each of the straight lines so drawn is dictated by the Q_{10} value. Thus, if the Q_{10} of the product was 5, then its shelf life at 40°C would be 1 month, increasing to 5.4 months if the Q_{10} was 2. Such a plot is helpful in deciding how long an ASLT is likely to run.

Instead of Q_{10} values, the E_A can be used to determine the duration of an ASLT. An example of the effect of E_A on the time to complete an ASLT is given in Table 11.4.

2. ASLT Procedures

The following procedure should be adopted in designing a shelf life test for a food product[32]:

1. Determine the microbiological safety and quality parameters for the product.
2. Select the key deteriorative reaction(s) that will cause quality loss and thus consumer unacceptability in the product, and decide what tests (sensory or instrumental) should be performed on the product during the trial.
3. Select the package to be used. Often, a range of packaging materials will be tested so that the most cost-effective material can be selected.
4. Select the extrinsic factors to be accelerated. Typical storage temperatures used for ASLT procedures are shown below, and it is usually necessary to select at least two.

Product	Test Temperatures (°C)	Control (°C)
Frozen	−7, −11, −15	<−40
Chilled	5, 10, 15, 20	0
Dry and IMF	25, 30, 35, 40, 45	−18
Canned	25, 30, 35, 40	4

5. Using a plot similar to that shown in Figure 11.5, determine how long the product must be held at each test temperature. If no Q_{10} values are known, then an open-ended ASLT will have to be conducted using a minimum of three test temperatures.
6. Determine the frequency of the tests. A good rule of thumb is that the time interval between tests at any temperature below the highest temperature should be no longer than

$$f_2 = f_1 Q_{10}^{\Delta T/10} \tag{11.33}$$

where
f_1 = the time between tests (e.g., days, weeks) at the highest test temperature T_1;
f_2 = the time between tests at any lower temperature T_2; and
ΔT = the difference in degrees Celsius between T_1 and T_2.
Thus, if a product is held at 40°C and tested once a month, then at 30°C with a Q_{10} of 3, the product should be tested at least every

$$f_2 = 1 \times 3^{(10/10)} = 3 \text{ months}$$

More frequent testing is desirable, especially if the Q_{10} is not accurately known, because at least six data points are needed to minimize statistical errors, otherwise the confidence in θ_s is significantly diminished.

7. Calculate the number of samples that must be stored at each test condition, including those samples which will be held as controls.
8. Begin the ASLTs, plotting the data as it comes to hand so that, if necessary, the frequency of sampling can be increased or decreased as appropriate.
9. From each test storage condition, estimate k or θ_s, and construct appropriate shelf life plots from which to estimate the potential shelf life of the product under normal storage conditions. Provided that the shelf life plots indicate that the product shelf life is at least as long as that desired by the company, then the product has a chance of performing satisfactorily in the marketplace.

3. Examples of ASLT Procedures

a. *Dehydrated Products*

In dehydrated vegetables, lipid and hydrolytic oxidation, together with nonenzymic browning and (in the case of green vegetables) chlorophyll degradation, are the major modes of deterioration. In dehydrated fruit, the major mode of deterioration is nonenzymic browning. Samaniego[28] used temperatures of 30 and 40°C to accelerate deterioration in sliced green beans and onion flakes and found that, for example, at 40°C, the shelf life of onions was 11 times shorter, and at 30°C, 3.5 times shorter than at 20°C when the a_w was 0.56.

b. *Frozen Foods*

Plots of HQL or PSL versus time were discussed in Section II.C.1.b; such curves suggest that accelerated tests could be used for predicting the shelf life of frozen foods with a considerable degree of accuracy. The shape of TTT curves for a wide range of products has been determined, and for any specific set of conditions (e.g., a particular product, process or package), a more detailed TTT curve could be determined. Then, accelerated tests could be carried out at temperatures as warm as −10 or even −8°C. As a result, the shelf life of a frozen food, which would normally be stored at −18°C, could be predicted in a few weeks or months at these higher temperatures.

Although mold growth has been recorded down to −17°C, no evidence of microbiological growth on meat products has been found at −8°C, and therefore −8°C is generally recommended as the warmest temperature for ASLT of meat. Due cognizance must be taken of those frozen products such as frozen bacon, which exhibit so-called "reverse stability" where the keeping quality is poorer at −25 than at −5°C.

In ASLT of frozen foods, the formation of ice has to be considered. As ice forms, the concentration of the unfrozen aqueous phase increases and influences reaction rates because they depend on both temperature and concentration. Below about −7°C, the relative change in concentration of the unfrozen aqueous phase is small, but storage at temperatures above −7°C should be avoided. In the temperature range between 0 and −7°C, the overall observed rate of reaction may increase, stay relatively constant or decrease depending on the specific system. Consequently, there is no generally applicable method to estimate low temperature shelf life from measurements made above −7°C.[25]

c. *Canned Foods*

It is generally assumed that if good manufacturing practices are followed, then microbial deterioration of canned foods will not be a problem. If there is thermophilic spoilage when canned foods are stored at elevated temperatures, then this is more than likely a result of inadequate cooling of the cans following thermal processing. Microbial spoilage at ambient temperatures is generally the result of "leaker" spoilage, thus called because the micro-organisms are drawn into the can during cooling; chlorination of cooling water according to good manufacturing practice will alleviate this problem. Thus, deteriorative reactions in canned foods will normally be limited to organoleptic changes such as loss of color, development of undesirable flavors and nutrient degradation.

Labuza[15] quotes a producer of canned meat products as stating that the major mode of deterioration is hydrogen gas production resulting from internal corrosion of the can. Samples were stored at 37.8°C to accelerate this deterioration; the shelf life at 37.8°C was considered to be 40% of the shelf life at 4.4°C, corresponding to a Q_{10} of 1.3.

d. *Oxygen-Sensitive Products*

In all the classical ASLT methods, temperature is the dominant acceleration factor used, and its effect on the rate of lipid oxidation is best analyzed in terms of the overall activation energy E_A for lipid oxidation. An inherent assumption in these tests is that E_A is the same in both the presence and absence of antioxidants, although indications are that it is in fact considerably lower in the latter case.

Other acceleration parameters used for shelf life are the O_2 partial pressure, reactant contact and the addition of catalysts. The effect of these factors is generally much less important than that of temperature. An exception is high fat products packaged in metal containers, where metal contamination of the product may be the most important factor in limiting shelf life.

4. Problems in the Use of ASLT Conditions

The potential problems and theoretical errors which can arise in the use of ASLT conditions have been described[17,18] as follows:

1. Errors in analytical or sensory evaluation. Generally, any analytical measure should be done with a variability of less than ±10% to minimize prediction errors.
2. As temperature rises, phase changes may occur (e.g., solid fat becomes liquid), which can accelerate certain reactions, with the result that at the lower temperature the actual shelf life will be longer than estimated.
3. Carbohydrates in the amorphous state may crystallize out at higher temperatures, with the result that the estimated shelf life is shorter than the actual shelf life at ambient conditions.
4. Freezing "control" samples can result in reactants being concentrated in the unfrozen liquid, creating a higher rate at the reduced temperature and thus confounding estimates.
5. If two reactions with different Q_{10} values cause quality loss in a food, then the reaction with the higher Q_{10} may predominate at higher temperatures while at normal storage temperatures the reaction with the lower Q_{10} may predominate, thus confounding the estimation.
6. The a_w of dry foods can increase with temperature, causing an increase in reaction rate for products of low a_w in sealed packages. This results in overprediction of "true" shelf life at the lower temperature.

7. The solubility of gases (especially O_2 in fat or water) decreases by almost 25% for each 10°C rise in temperature. Thus, an oxidative reaction such as loss of vitamin C or linoleic acid can decrease in rate if O_2 availability is the limiting factor. Therefore, at the higher temperature, the rate will be less than theoretical, which, in turn, will result in an underprediction of "true" shelf life at the normal storage temperature.
8. If the product is not placed in a totally impermeable pouch, then storage in high temperature/low humidity cabinets will generally enhance moisture loss, and this should decrease the rate of quality loss compared to no moisture change. This will result in a shorter estimated shelf life at the lower temperature.
9. If high enough temperatures are used, then proteins may become denatured, resulting in both increases or decreases in the reaction of certain amino acid side chains, leading to either under or overprediction of "true" shelf life.

Therefore, in light of the above points, the use of ASLT to estimate actual shelf life can be severely limited, except in the case of very simple chemical reactions. Consequently, food technologists should always confirm the ASLT results for a particular food product by conducting shelf life tests under actual environmental conditions. Once a relationship between ASLT and actual shelf life has been established for a particular product, then ASLT can be used for that product when process or package variables are to be evaluated.

Another point worth stressing is that ASLT is really only applicable in temperate climates. For those who produce food products in or for tropical climates, the ambient temperature in these countries is typically 30 to 40°C and even higher in warehouses, trucks, and so on. These temperatures correspond to those suggested for ASLTs in Section III.D.2 step 4. However, temperatures higher than these cannot be used for ASLT of food products in tropical countries because this would lead to reactions that were not representative of how the food products would deteriorate under tropical conditions.

E. Predicting Microbial Shelf Life

Microbial spoilage of food is an economically significant problem for food manufacturers, retailers, and consumers. Depending on the product, process, and storage conditions, the microbiological shelf life can be determined by either the growth of spoilage or pathogenic micro-organisms. In the case of spoilage micro-organisms, the traditional method for determining microbiological shelf life involved storing the product at different temperatures and determining spoilage by sensory evaluation or microbial count. Where the microbiological shelf life is determined by the growth of pathogenic micro-organisms, the traditional approach has been challenge testing of the product with the organism of concern, followed by storage at different temperatures and microbial analysis at certain intervals. For processes such as heat treatments, where the elimination of particular micro-organisms is required (e.g., canning), the use of inoculated packs is common.[4]

In recent years, the development and commercialization of predictive models has become relatively widespread. The use of such models can reduce the need for shelf life trials, challenge tests, product reformulations, and process modifications, thus saving both time and money. Although there are both mechanistic and empirical predictive models, the latter predominate. Empirical predictive models can be subdivided into probabilistic and kinetic models.[4] *Probabilistic* models describe the probability of a microbiological event occurring, and are used to predict whether certain micro-organisms will grow when they are close to their growth boundaries (e.g., whether *C. botulinum* will grow and produce toxin). The ultimate test for predictive models is whether they can be used to predict reliable outcomes in real situations. For a detailed discussion, the reader is referred to the standard text in this area.[22]

Predictive models have been used to determine the likely shelf life of perishable foods such as meat, fish, and milk, and recent publications in the area have been reviewed, together with a discussion of the limitations of such models.[4,9] Despite their increasing sophistication and widespread availability, models should not be relied on completely. Rather, models are best employed as tools to assist decision-making. Models do not completely negate the need for microbial testing, and do not replace the judgment of a trained and experienced food microbiologist.

IV. SHELF LIFE DEVICES

As mentioned at the outset of this chapter, the quality of most foods and beverages decreases over time. In other words, there is a continual loss of quality from the time they leave the food processor until they are consumed, even under ideal handling conditions. The goal of modern food distribution techniques is to minimize the extent of quality degradation so that the foods will reach the consumer's table as close to their original state as possible.

Of all the extrinsic factors that accelerate quality degradation, the one with the greatest influence is temperature. This fact is well known by food technologists, and most countries have codes of practice that specify optimum storage temperatures for many foods, particularly those classed as perishable. Despite these specifications, problems of storage temperature abuse arise all too frequently. One difficulty when storage temperature abuse is suspected lies in detecting the extent of the quality degradation without sampling the food (i.e., disturbing the integrity of the package). An associated difficulty is that many of those involved in the distribution chain are not trained to make reliable judgments about the quality of the food.

Thus, food processors require a simple way of indicating whether their products have been stored at undesirable temperatures, or better still, a means of indicating how much shelf life remains. Devices that provide the first category of information are time–temperature recorders; devices for the second category are time–temperature indicators (TTIs), and these are discussed further in Chapter 14 under the heading of "Intelligent Packaging."

REFERENCES

1. Anonymous, Shelf Life of Foods. Report by the Institute of Food Technologists' Expert Panel on Food Safety and Nutrition and the Committee on Public Information, Institute of Food Technologists, Chicago, Illinois, August 1974, *J. Food Sci.*, 39, 861–865, 1974.
2. Anonymous, *Shelf Life of Foods: Guidelines for Its Determination and Prediction*, Institute of Food Science and Technology, London, 1993.
3. Bell, L. N. and Labuza, T. P., *Moisture Sorption: Practical Aspects of Isotherm Measurement and Use*, American Association of Cereal Chemists, St Paul, MN, 2000.
4. Blackburn, CdeW, Modelling shelf life, In *The Stability and Shelf-Life of Food*, Kilcast, D. and Subramanian, P., Eds., CRC Press, Boca Raton, FL, 2000, chap. 3.
5. Cairns, J. A., Elson, C. R., Gordon, G. A., and Steiner, E. H., *Research Report No. 174*, British Food Manufacturing Industry Research Association, Leatherhead, England, 1971.
6. Cardelli, C. and Labuza, T. P., Application of Weibull hazard analysis to the determination of the shelf life of roasted coffee, *Lebensm Wiss u Technol.*, 34, 273–278, 2001.
7. Cardello, A. V., Food quality: relativity, context and consumer expectations, *Food Qual. Pref.*, 6, 163–170, 1995.
8. Cardello, A. V., Perception of food quality, In *Food Storage Stability*, Taub, I. A. and Singh, R. P., Eds., CRC Press, Boca Raton, FL, 1998, chap. 1.
9. Dens, E. J. and Van Impe, J. F., Modelling applied to foods: predictive microbiology for solid food systems, In *Food Preservation Techniques*, Zeuthen, P. and Bøgh-Sørensen, L., Eds., CRC Press, Boca Raton, FL, 2004, chap. 21.

10. Duyvesteyn, W. S., Shimoni, E., and Labuza, T. P., Determination of the end of shelf life for milk using Weibull hazard analysis, *Lebensm Wiss u Technol.*, 34, 143–148, 2001.
11. Eskin, N. A. M. and Robinson, D. S., Eds., *Food Shelf Life Stability: Chemical, Biochemical and Microbiological Changes*, CRC Press, Boca Raton, FL, 2001.
12. Gacula, M. C. and Singh, J., Shelf life testing experiments, *Statistical Methods in Food and Consumer Research*, Academic Press, New York, 1984, chap. 8.
13. Hough, G., Langohr, K., Gómez, G., and Curia, A., Survival analysis applied to sensory shelf life of foods, *J. Food Sci.*, 68, 359–366, 2003.
14. Kilcast, D. and Subramaniam, P., Eds., *The Stability and Shelf-Life of Food*, CRC Press, Boca Raton, FL, 2000.
15. Labuza, T. P., *Shelf-Life Dating of Foods*, Food and Nutrition Press, Westport, CT, 1982.
16. Labuza, T. P. and Kamman, J. F., Reaction kinetics and accelerated tests simulation as a function of temperature, In *Computer-Aided Techniques in Food Technology*, Saguy, I., Ed., Marcel Dekker, New York, 1983, chap. 4.
17. Labuza, T. P. and Schmidl, M. K., Accelerated shelf life testing of foods, *Food Technol.*, 39(9), 57–62, 64, 134, 1985.
18. Labuza, T. P. and Schmidl, M. K., Use of sensory data in the shelf life testing of foods: principles and graphical methods for evaluation, *Cereal Foods World*, 33, 193–205, 1988.
19. Labuza, T. P. and Szybist, L. M., Playing the open dating game, *Food Technol.*, 53(7), 70–85, 1999.
20. Man, C. M. D., Shelf-life testing, In *Understanding and Measuring the Shelf-Life of Food*, Steele, R., Ed., CRC Press, Boca Raton, FL, 2004, chap. 15.
21. Man, C. M. D. and Jones, A. A., Eds., *Shelf-Life Evaluation of Foods*, 2nd ed., Aspen Publishers, Gaithersburg, MD, 2000.
22. McMeekin, T. A., Olley, J. N., Ross, T., and Ratkowsky, D. A., *Predictive Microbiology: Theory and Application*, New York, Wiley, 1993.
23. Meilgaard, M. C., Civille, G. V., and Carr, B. T., *Sensory Evaluation Techniques*, 3rd ed., CRC Press, Boca Raton, FL, 1999.
24. Mizrahi, S., Accelerated shelf life tests, In *Understanding and Measuring the Shelf-Life of Food*, Steele, R., Ed., CRC Press, Boca Raton, FL, 2004, chap. 14.
25. Reid, D. S., Frozen foods shelf life, In *Encyclopedia of Agricultural, Food, and Biological Engineering*, Heldman, D. R., Ed., Marcel Dekker, New York, pp. 420–421, 2003.
26. Ross, E. W., Mathematical modeling of quality loss, In *Food Storage Stability*, Taub, I. A. and Singh, R. P., Eds., CRC Press, Boca Raton, FL, 1998, chap. 11.
27. Salame, M., The use of low permeation thermoplastics in food and beverage packaging, In *Permeability of Plastic Films and Coatings*, Hopfenberg, H. B., Ed., Plenum, New York, p. 275, 1974.
28. Samaniego-Esguerra, C. M. L., Boag, I. F., and Robertson, G. L., Kinetics of quality deterioration in dried onions and green beans as a function of temperature and water activity, *Lebensm Wiss u Technol.*, 24, 53–57, 1991.
29. Singh, T. K. and Cadwallader, K. R., The shelf life of foods: an overview, In *Freshness and Shelf Life of Foods, ACS Symposium Series #836*, Cadwallader, K. R. and Weenen, H., Eds., American Chemical Society, Washington, DC, 2003, chap. 1.
30. Singh, T. K. and Cadwallader, K. R., Ways of measuring shelf life and spoilage, In *Understanding and Measuring the Shelf-Life of Food*, Steele, R., Ed., CRC Press, Boca Raton, FL, 2004, chap. 9.
31. Taoukis, P. S. and Giannakourou, M. C., Temperature and food stability: analysis and control, In *Understanding and Measuring the Shelf-Life of Food*, Steele, R., Ed., CRC Press, Boca Raton, FL, 2004, chap. 3.
32. Taoukis, P. S., Labuza, T. P., and Saguy, I. S., Kinetics of food deterioration and shelf life prediction, In *Handbook of Food Engineering Practice*, Valenta, K. J., Rotstein, E. and Singh, R. P., Eds., CRC Press, Boca Raton, FL, 1997, chap. 9.
33. U.S. Department of Agriculture Marketing Workshop Report, 1951. In *Food Quality Assurance*, Gould W. A., Ed., AVI Publishing, Westport, CT, 1977.
34. Van Arsdel, W. B., Estimating quality change from a known temperature history, In *Quality and Stability of Frozen Foods*, Van Arsdel, W. B., Copley, M. J. and Olson, R. L., Eds., Wiley Interscience, New York, 1969, chap. 10.

12 Aseptic Packaging of Foods

CONTENTS

I. INTRODUCTION

Aseptic packaging is the filling of sterile containers with a commercially sterile product under aseptic conditions, and then sealing them so that reinfection is prevented; that is, so that they are hermetically sealed. Figure 12.1 illustrates the various aspects of aseptic packaging in diagrammatic form. The term *aseptic* implies the absence or exclusion of any unwanted organisms from the product, package or other specific areas, while the term *hermetic* (strictly *air tight*) is used to indicate suitable mechanical properties to exclude the entrance of micro-organisms into a package and gas or water vapor into (or from) the package. The term *commercially sterile* is generally taken to mean the absence of micro-organisms capable of reproducing in the food under nonrefrigerated conditions of storage and distribution, thus implying that the absolute absence of all micro-organisms need not be achieved.

Currently there are two specific fields of application for aseptic packaging: (1) packaging of presterilized and sterile products and (2) packaging of a nonsterile product to avoid infection by micro-organisms. Examples of the first application include milk and dairy products, puddings, desserts, fruit and vegetable juices, soups, sauces and products with particulates. Examples of the second application include fresh products such as fermented dairy products like yogurt.

The three major reasons for the use of aseptic packaging are: (1) to take advantage of high temperature–short time (HTST) sterilization processes, which are thermally efficient and generally give rise to products of a superior quality compared to those processed at lower temperatures for longer times, (2) to enable containers to be used that are unsuitable for in-package sterilization and (3) to extend the shelf life of products at normal temperatures by packaging them aseptically.

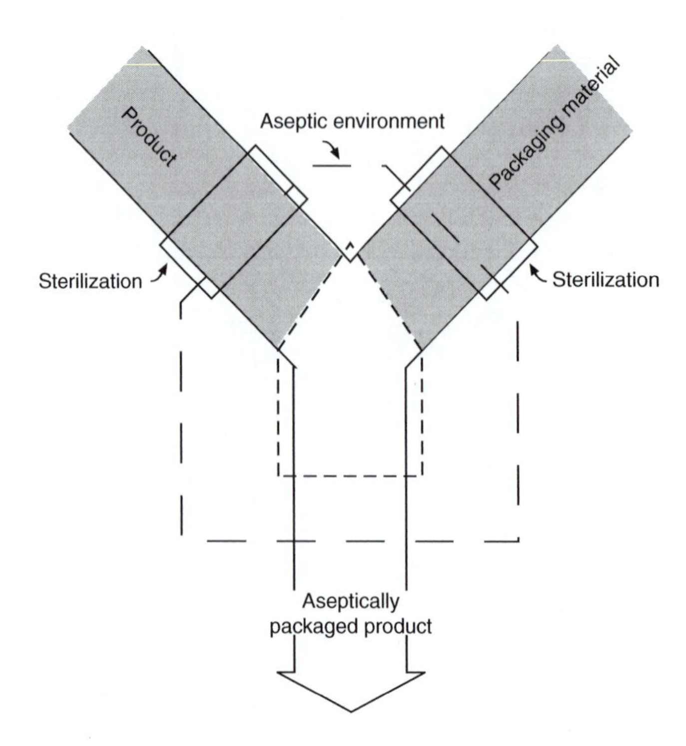

FIGURE 12.1 Diagrammatic representation of the aseptic packaging process.

A. Historical Development

The first aseptic packaging of food (specifically milk in metal cans) was carried out in Denmark by Nielsen before 1913 and a patent for this process (termed aseptic conservation) was granted in 1921. In 1917 in the U.S., Dunkley patented the method of sterilizing cans and lids with saturated steam; the cans were then filled with a presterilized product. In 1923, aseptically packaged milk from South Africa reached a trade fair in London in perfect condition. The American Can Company developed a filling machine in 1933 called the heat–cool fill (HCF) system, which used saturated steam under pressure to sterilize the cans and ends. The sterile cans were filled with sterile product and the ends sealed on in a closed chamber, which was kept pressurized with steam or a mixture of steam and air. Three commercial plants were built and operated on this principle until 1945.

In the 1940s in the U.S., W.M. Martin developed a process in which empty metal cans were sterilized by treatment with superheated steam at 210°C, before being filled with cold, sterile product. In 1950 the Dole Company bought the first commercial aseptic filling plant on the market.

At the end of the 1940s, a dairy enterprise (Alpura AG, Bern) and machinery manufacturer (Sulzer AG, Winterthur) in Switzerland combined their knowledge to develop ultrahigh temperature-sterilized, aseptically canned milk, which was subsequently marketed in Switzerland in 1953. However, this system was not economical, mainly because of the cost of the cans, and Alpura, in collaboration with Tetra Pak of Sweden, went on to develop an aseptic system based on paperboard laminate cartons. The first milk with a long shelf life to be packaged in this manner was sold in Switzerland in October 1961.[7]

B. Principles of Sterilization

The sterilization processes used in aseptic processing are variously described as high temperature–short time (HTST) and ultra heat treated or ultrahigh temperature (UHT). The HTST process is defined as sterilization by heat for times ranging from a few seconds to 6 min. The International Dairy Federation has suggested that UHT milk should be defined as "milk which has been subjected to a continuous flow heating process at a high temperature for a short time and which afterwards has been aseptically packaged. The heat treatment is to be at least 135°C for one or more seconds." More generally, the term UHT refers to in-line, continuous flow sterilization processes, which employ heat treatments within the temperature range 130 to 150°C with holding times of 2 to 8 sec. The upper end of the temperature range tends to be used for low viscosity products such as milk, and the lower end for more viscous products.

The quality advantage that accrues from the use of HTST and UHT processes can best be understood by comparing the z value of microbial destruction with the z value for the loss of desirable quality factors in the food such as nutrients. A common z value for the former is 10°C, and for the latter, 33°C. A C_o or cooking index has been proposed to describe the overall sensory quality deterioration which occurs during thermal processing. It is defined analogously to the F_o value, except that the reference temperature is 100°C, rather than 121.1°C:

$$C_o = 10^{(T-100)/z}T \tag{12.1}$$

The implications of the differences in z value are evident from a consideration of Figure 12.2. The regions which result in an F_o value of 10 are typical of the low temperature–long time (LTLT) processes used for sterilizing foods in conventional canning processes; the C_o values associated with such processes are in the order of 30 to 300. When a F_o value of 10 is achieved at higher temperatures (130 to 150°C), the corresponding C_o values are around 1 to 10. Thus, while there is the same level of microbial destruction in both LTLT and HTST processes, the latter results in considerably less degradation and explains the preference of consumers for HTST- rather than LTLT-processed foods. Many chemical reactions such as nonenzymic browning, chlorophyll

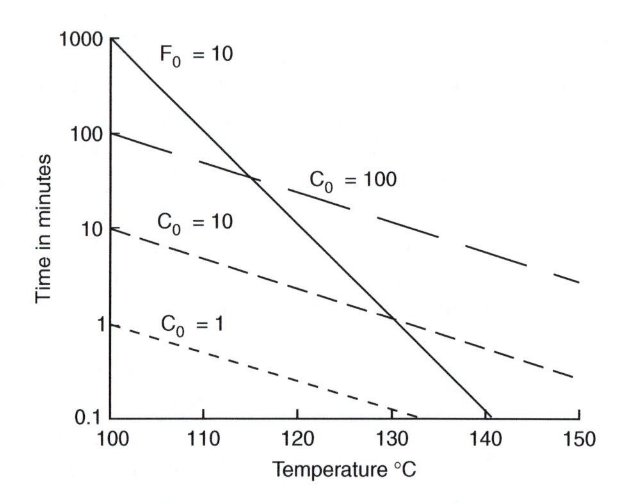

FIGURE 12.2 Comparative enzyme and bacterial spore inactivation curves.

degradation and nutrient loss have *z* values of 33 to 50, and therefore the extent of the reaction at HTST conditions will be significantly less than at LTLT conditions.

One problem associated with the use of HTST processes is that of adequate enzyme inactivation. This is a particular problem with vegetable enzymes (especially peroxidases), bacterial proteases and lipases produced by psychrotrophic bacteria, usually *Pseudomonas* spp. The heat resistance of bacterial enzymes is very high, and it has been calculated that one such enzyme is 4000 times more heat resistant than spores of *Bacillus stearothermophilus*, the most resistant of the spores to be destroyed by a heat treatment process. Bacterial enzymes that have shown extreme heat resistance have *z* values in the range of 20 to 60°C. Thus, as the processing temperature is raised an increasing percentage of the enzymes survive for the same sterilizing (F_o) performance. As a consequence, the probability of enzymic deterioration during storage of the processed product increases as the heat treatment temperature increases.[2] This can lead to quality defects during storage of UHT milk products; for example, the proteases can lead to the development of bitter flavors and age gelation (see Chapter 18 for more details).

From a commercial point of view, the aim of sterilizing food with the highest temperatures possible for the shortest possible times has lead to a number of different equipment designs and configurations. The heat exchange equipment can be classified into two groups[4]:

1. Indirect heat exchange where the product and the heat exchange fluid are separated by the heating surface. There are three types:
   (a) Tubular heat exchanger
   (b) Plate heat exchanger
   (c) Scraped surface heat exchanger
2. Direct heat exchange where steam is condensed in the product for heating, and evaporation removes vapor resulting in cooling. There are two means of heating:
   (a) Steam injection (steam into product)
   (b) Steam infusion (product into steam)

The objective in the design of the above equipment has been to improve the rate of heat transfer into and out of the food product in ways which minimize the required heating and cooling times. Generally, the direct systems give more rapid come-up and cool-down times than the indirect systems, and present fewer problems with formation of scale and burn-on. The indirect systems lend themselves to better heat recovery, tend to give a more stable output temperature and are not prone to contamination from condensables in the steam. In recent years, much work has been conducted on the use of microwave and dielectric heating, as well as resistive (ohmic) heating, for the sterilization of food. A few systems using such processes have become commercially available.

Regardless of the type of sterilizer used, the sterile product is cooled to an appropriate temperature, typically 20°C for low viscous food products like milk and fruit juices, and 40°C for products of higher viscosity such as puddings and desserts. A presterilized container is then filled with the cooled, sterile product.

An aseptic filling system must meet a series of requirements, each of which must be satisfied individually before the whole system can be considered satisfactory. These are[2]:

1. The container and method of closure must be suitable for aseptic filling, and must not allow the passage of micro-organisms into the sealed container during storage and distribution.
2. The container (or that part of it which comes into contact with the product) must be sterilized after it is formed and before being filled.
3. The container must be filled without contamination by micro-organisms either from the equipment surfaces or from the atmosphere surrounding the filler.
4. If any closure is needed, it must be sterilized immediately before it is applied.
5. The closure must be applied and sealed in place while the container is still within a sterile zone to prevent the passage of contaminating micro-organisms.

There are many possible ways of meeting the above requirements, and the remainder of this chapter will describe the major sterilization and packaging systems which have been developed and commercialized to meet the above requirements.

II. STERILIZATION OF PACKAGING MATERIAL FOOD CONTACT SURFACES

A. Required Count Reduction

The required count reduction (number of D values) for the sterilization of the food contact packaging material surface is determined by the type of product, its desired shelf life and the likely storage temperature. For nonsterile acidic products of pH < 4.5, a minimum of four decimal reductions in bacterial spores is required. For sterile, neutral, low acid products of pH > 4.5, a six decimal reduction is required. However, if there is the possibility that *Clostridium botulinum* is able to grow in the product, then a full 12 decimal reduction process should be given.

The nonsterility rate or error rate, E_r, is the number of nonsterile or faulty packages as a proportion of the total number of packages processed over a given period. It can be calculated using the following equation:

$$E_r = -A\frac{N}{R} \tag{12.2}$$

where N is the number of micro-organisms on, and A the area of, the food contact surface of the packaging material, and R is the number of decimal reductions obtained in the sterilization process. Smaller containers with a smaller food contact surface area will have correspondingly less initial contamination and a less severe sterilization process will be needed to give a certain E_r. Conversely, larger containers will require a more severe sterilization process. However, because container volume varies with the cube of the linear dimensions whereas surface area varies only with the square of the dimensions, variation in the sterilization process according to container size is less than might be expected.[2] Commercial requirements for less than one faulty package out of 10,000 produced are common.[5]

Because of the importance of the initial level of microbial contamination of the packaging material, steps should be taken to ensure that it is as low as possible. Thus, it should be produced, transported and stored under conditions which are as free from micro-organisms as possible.

Three main sterilization processes for packaging material are in common use, either individually or in combination: irradiation, heat and chemical treatments[3]; each of these is now considered in turn.

B. Irradiation

1. Ionizing Radiation

Particle irradiation techniques using gamma rays from cobalt-60 or cesium-137 have been used to sterilize the interior of sealed but empty containers, especially those made of materials which cannot withstand the temperatures needed for thermal sterilization or that, because of their shape, could not be conveniently sterilized by other means. A radiation dose of 25 kGy (2.5 Mrad) or more is generally accepted to be sufficient to ensure sterility. The packages are sealed into microbial-proof containers prior to the irradiation treatment. It is also possible to use low energy (100 keV), large area electron beams for the surface sterilization of packaging materials and containers.

2. Pulsed Light

By storing electrical energy in a capacitor and releasing it in short pulses, high peak power levels can be generated. The use of intense and short duration pulses of broad-spectrum "white" light to sterilize aseptic packaging material underwent considerable research and development in the 1990s. The duration of the pulses ranges from 1 μs to 0.1 sec, and the flashes are typically applied at a rate of 1 to 20 flashes per second. Approximately 25% of the emitted light, which has an intensity about 20,000 times that of sunlight at the Earth's surface, is UV, 45% is visible and 30% is infrared. Generally, a few flashes applied in a fraction of a second provide high levels of microbial inactivation. Despite a successful field trial, this system has not yet been commercialized.[1]

3. UV-C Radiation

UV radiation has a wavelength of 200 to 315 nm; it is most effective in terms of microbial destruction between 248 and 280 nm (the so-called *UV-C range*), with an optimum effectiveness at 253.7 nm. Mercury vapor lamps emit UV-C at 253.7 nm. UV-C irradiation is generally only used commercially in combination with hydrogen peroxide. Recently, UV (Excimer) lasers operating at 248 nm and using rare (noble) gas halides, such as krypton fluoride, have been commercialized.

C. Heat

Heat sterilization processes can involve either steam (moist heat) or dry heat. Steam is pure gaseous water with no air or other gases present. Dry heat is hot air in the absence of water molecules. The sterilization effect depends on time and temperature, and steam is much more efficient than dry heat. Steam sterilization at 121°C for 20 min is equivalent in effectiveness to dry heat sterilization at 170°C for 60 min.[6]

1. Saturated Steam

The most reliable sterilant is steam. However, in order to reach temperatures sufficiently high to achieve sterilization in a few seconds, the steam (and thus the packaging material with which it comes into contact) must be under pressure, necessitating the use of a pressure chamber. In addition, any air that enters the pressure chamber with the packaging material must be removed to prevent it interfering with the transfer of heat from the steam to the package surface.

Finally, condensation of steam during heating of the packaging material surface produces condensate which, if not removed, could remain in the container and dilute the product.

Despite these problems, saturated steam under pressure is being used to sterilize plastic containers. For example, immediately after deep-drawing, molded PS cups and foil lids are subjected to steam at 165°C and 600 kPa for 1.4 sec (cups) and 1.8 sec (lids). In order to limit the heating effect to the internal surface of the cups, the exterior of the cups is simultaneously cooled. This process has been shown to achieve a five to six decimal reduction in *Bacillus subtilis* spores.

2. Superheated Steam

Superheated steam was the method used in the 1950s for the sterilization of tinplate and aluminum cans and lids in the Martin-Dole aseptic canning process. The cans are passed continuously at normal pressure under saturated steam at 220 to 226°C for times of 36 to 45 sec, depending on the construction material given that aluminum cans have a shorter heating time because of their higher thermal conductivity.

3. Hot Air

Dry heat in the form of hot air has the advantage that high temperatures can be reached at atmospheric pressure, thus simplifying the mechanical design problems for a container sterilization system.[2] Hot air at a temperature of 315°C has been used to sterilize paperboard laminate cartons where a surface temperature of 145°C for 180 sec is reached. However, such a system is apparently only suitable for acidic products with a pH < 4.5.

4. Hot Air and Steam

A mixture of hot air and steam has been used to sterilize the inner surfaces of cups and lids made from PP, which is thermally stable up to 160°C. In this process, hot air is blown into the cups through a nozzle in such a way that the base and walls of the cup are uniformly heated.

5. Extrusion

During the extrusion of plastic granules prior to blow molding of plastic containers, temperatures of 180 to 230°C are reached for up to 3 min. However, because the temperature distribution inside the extruder is not uniform and the residence time of the plastic granules varies considerably, it is not possible to guarantee that all particles will achieve the minimum temperature and residence time necessary to result in sterility. Extrusion results in a three to four decimal reduction in microbial spores, and therefore extruded containers should only be aseptically filled with acidic products with a pH < 4.5. For products with a pH > 4.5, it is recommended that extruded containers be poststerilized with hydrogen peroxide or peracetic acid (PAA).

D. Chemical Treatments

1. Hydrogen Peroxide

The lethal effect of hydrogen peroxide (H_2O_2) on micro-organisms (including resistant spores) has been known for many years, with the first commercial aseptic filling system being devised in 1961. It used a combination of peroxide and heat to sterilize the surface of paperboard laminate packaging material. Although there have been many studies into the death of resistant spores in suspension in peroxide solutions, the actual mechanism of death is not fully understood. Because even concentrated peroxide solutions at room temperature have hardly any destructive effect, a minimum temperature of 70°C and a minimum concentration of 30% are necessary to achieve destruction of the most resistant spores on packaging materials within seconds.

Because there are many uncertainties in the use of peroxide for surface sterilization, it is difficult to predict the sterilizing effect that any specific combination of peroxide concentration and temperature is likely to have. Therefore, in most situations where peroxide is used to sterilize packaging materials prior to aseptic filling, the sterilization conditions have been determined empirically. These conditions include the peroxide solution concentration, the quantity applied to the packaging material per unit area, the intensity of the radiant or irradiant heat (see below) or the temperature and quantity of the drying air and the time for which it is applied.

The U.S. Food and Drug Administration has ruled that the concentration of H_2O_2 present in food products packaged in material sterilized by H_2O_2 must be no greater than 500 ppb (500 parts in 10^9) at the time of filling and must fall to approximately 1 ppb within 24 h. Because peroxide cannot be measured accurately in food products because of the presence of reducing compounds which rapidly eliminate it, checks of the initial level must be made on packs filled with water.

Because peroxide solution is unable to sterilize packaging material by itself, a number of systems have evolved which increase the efficacy of the peroxide treatment by combining it with heat and radiant or irradiant energy. These processes are briefly described below.

a. *Dipping Process*

In one process, the packaging material is unwound from a reel and passed through a bath of 30 to 33% H_2O_2 solution containing a wetting agent to ensure uniform wetting of plastic surfaces that tend to be hydrophobic. The liquid H_2O_2 solution is reduced to a thin film, either mechanically by means of a squeeze roll or with jets of sterile air; the adhering liquid film is then dried with hot air.

b. *Spraying Process*

In this process, H_2O_2 is sprayed through nozzles onto prefabricated packages. The peroxide is then dried using hot air. The death rate is dependent on the volume of sprayed H_2O_2 (larger volumes require longer drying times) and the temperature of the hot air. The trend now is to completely avoid spraying liquid droplets and use a mixture of hot air (130°C) and vaporized peroxide instead.

c. *Rinsing Process*

When the prefabricated container is of an intricate shape such that the spraying process is unsuitable, it can be rinsed with peroxide or a mixture of peroxide and PAA. After spraying, the container is drained and then dried with hot air. This process has been used to sterilize glass containers, metal cans and blow molded plastic bottles.

d. *Combined with UV Irradiation and Heat*

When UV irradiation and H_2O_2 are used together they act synergistically, with the UV irradiation promoting the breakdown of the peroxide into hydroxyl radicals. The overall lethal effect is greater than the sum of the individual effects of peroxide and irradiation, the optimum effect being at a relatively low peroxide concentration of between 0.5 and 5%.

It is usual to use heat as well as UV irradiation and peroxide in the sterilization of packaging materials. The advantage of such a combination is that much lower concentrations of peroxide can be used (less than 5% compared with 30 to 35% for peroxide and heat together) and the problems of atmospheric contamination by peroxide and residual peroxide in the product are reduced. However, because too high a peroxide concentration reduces the effectiveness of sterilization, strict control of concentration is essential.[2]

2. Peracetic Acid

PAA is a liquid sterilant, which is particularly effective against spores. It is produced by the oxidation of acetic acid by H_2O_2 and a solution containing PAA and H_2O_2 is effective against resistant bacterial spores, even at 20°C. It is used for sterilizing filling machine surfaces as well as packaging materials such as polyethylene terephthalate (PET) bottles prior to aseptic filling, the PET bottles being rinsed with sterile water rather than hot air.

E. Verification of Sterilization Processes

Sterilization of the packaging material is verified by inoculation of the surface of the web, cup or lid stock with the proper concentration of the test organism and allowing this to dry. The system is usually run as for a commercial test batch, and the finished containers are filled with an appropriate growth medium, incubated and observed for growth. At least 100 containers should be tested for each time, temperature or sterilizing agent concentration.

III. ASEPTIC PACKAGING SYSTEMS

The aseptic packaging system must be capable of filling the sterile product in an aseptic manner and of sealing the container hermetically so that sterility is maintained throughout the handling and distribution processes. An aseptic packaging system should be capable of meeting four criteria:

1. Able to be connected to the processing system in a manner that enables aseptic transfer of product to take place
2. Able to be effectively sterilized before use
3. Able to carry out the filling, sealing and critical transfer operations in a sterile environment
4. Able to be cleaned properly after use

The type of packaging material used is influenced by the nature of the product, the cost of both the product and the package and the preferences of the consumer. Although the most widespread consumer package for aseptic products is the paperboard laminate carton, five major categories of aseptic packaging equipment are available and their major features and characteristics are described below.

A. Carton Systems

The carton material consists of layers of unbleached or bleached paperboard coated internally and externally with polyethylene, resulting in a carton that is impermeable to liquids and in which the internal and external surfaces may be heat sealed. There is also a thin (6.3 μm) layer of aluminum foil which acts as an O_2 and light barrier. The structure of a typical paperboard carton is shown in Figure 12.3. The functions of the various layers are:

1. The outer polyethylene (15 g m^{-2}) protects the ink layer and enables the package flaps to be sealed.
2. The bleached paperboard (186 g m^{-2}) serves as a carrier of the decor and gives the package the required mechanical rigidity.
3. The laminated polyethylene (25 g m^{-2}) binds the aluminum to the paperboard.
4. The aluminum foil (6.3 μm) acts as a gas barrier and provides protection of the product from light.
5. The two inner polyethylene layers (15 g m^{-2} and 25 g m^{-2}) provide a liquid barrier.

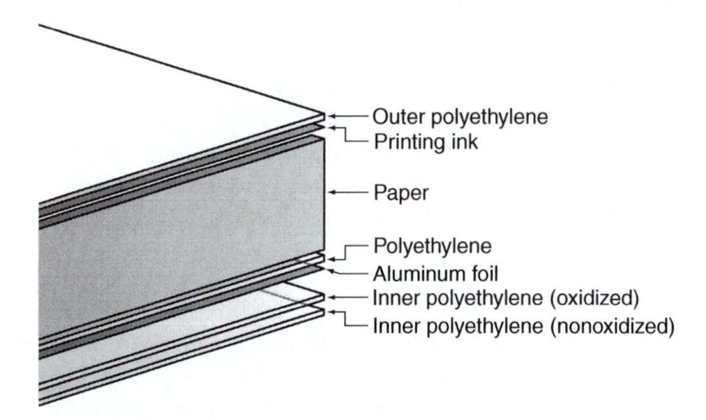

FIGURE 12.3 Typical structure of a paperboard laminate carton for aseptic filling.

1. Form-Fill-Seal Cartons

The packaging material is supplied in rolls that have been printed and creased, the latter being necessary to ease the forming process. A plastic strip is sealed to one edge (the reason for this is described later) and the packaging material sterilized using a wetting system or a deep bath system.

In the wetting system, a thin H_2O_2 film (15 to 35% concentration) containing a wetting agent to improve the formation of a liquid film is applied to the inner packaging material surface. The material then passes through a pair of rollers to remove excess liquid and under a tubular electric heater, which heats the inside surface to about 120°C and evaporates the H_2O_2.

In the deep bath system, the packaging material is fed through a deep bath containing H_2O_2 (35% concentration) at a minimum temperature of 70°C, the residence time being 6 sec. After squeezer rollers have removed much of the peroxide, both sides of the material are heated with air (directed through nozzles) at a temperature of 125°C to evaporate the peroxide.

The sterilized packaging material is fed into a machine where it is formed into a tube and closed at the longitudinal seal by a heat sealing element. In the process, the strip that was added prior to sterilization is heat sealed across the inner surface of the longitudinal seal to prevent contact between the outside and the inside of the carton. It also provides protection of the aluminum and paperboard layers from the product, which could corrode or swell the layers if such a strip were absent.

The tube is then filled with the product and a transverse seal made below the level of the product, thus ensuring that the package is completely filled. Alternatively, the packages may be produced with a headspace of up to 30% of the total filling volume by injection of either sterile air or other inert gases. The sterilization, filling and sealing processes are all performed inside a chamber maintained at an overpressure of 0.5 atm with sterile air.

One method of forming cartons from a continuous web is shown in Figure 12.4; also included in the diagram is a cross-section of the longitudinal seal. The sealed packages are then pressed by molds into rectangular blocks, after which the top and bottom flaps or wings are folded down and heat sealed to the body of the package using electrically heated air.

2. Prefabricated Cartons

In systems of this type, prefabricated carton blanks are used, with the cartons being die-cut, creased and the longitudinal seam completed at the factory of origin by skiving the inner layer of board and folding it back (see Figure 12.5). The cartons are delivered to the processor in lay-flat form ready to be finally shaped in the filler and the top seam formed and bonded. Although the above operations take place under nonsterile conditions, steps are taken to avoid recontamination.

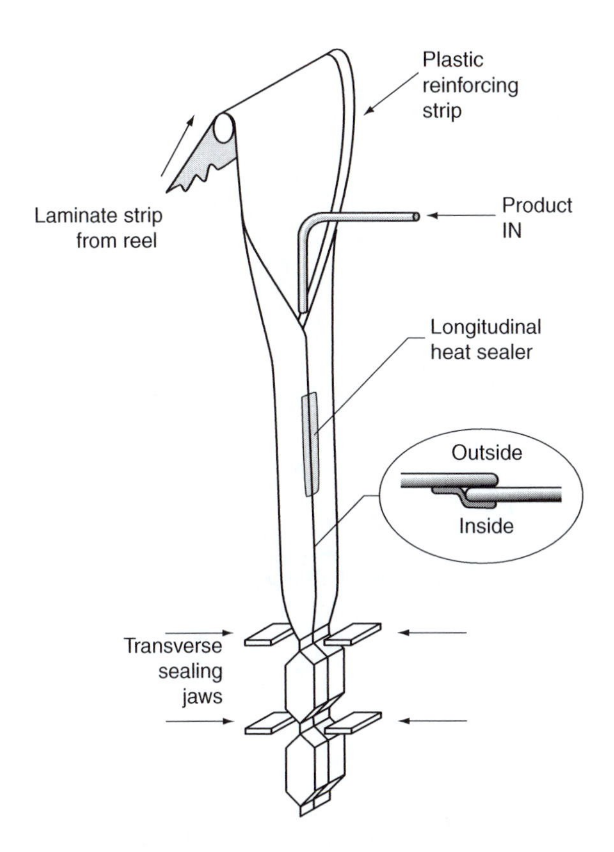

FIGURE 12.4 One method of forming cartons from a continuous web; the cross-section of the longitudinal seal is enlarged to show the polyethylene strip which protects the internal edge of the carton.

The aseptic area of the filling machine consists of several separate functional zones where operations are carried out in sequence. Sterility is maintained in each zone by a slight overpressure of sterile air. The inside surface of the carton is sterilized with a 35% solution of H_2O_2, delivered as either a fine spray or peroxide vapor in hot air so that the vapor condenses as liquid peroxide on the carton surface. The peroxide is removed by a jet of hot air at 170 to 200°C. Alternatively, the inside of the carton can be sprayed uniformly with a 1 to 2% solution of H_2O_2 and then irradiated for approximately 10 sec with high intensity UV radiation. The peroxide is then heated and removed by hot air jets.[2] The problems of residual peroxide in the carton and peroxide contamination in the surrounding atmosphere are more easily dealt with in this latter process because the total quantity of peroxide used is 20 to 30 times less than in the former.

The next stage of the process is filling. A certain amount of headspace is always advisable to ensure that the package can be opened and poured without spilling. A headspace is essential when the contents require shaking (as is the case with flavored milk drinks and pulpy fruit juices). It is advantageous to fill the area between the product and the top of the package with steam or an inert gas such as N_2 for products such as fruit juices. If steam is used, then the headspace volume is reduced as a result of condensation of the steam. A headspace is also crucial when it is not possible to seal the filled package through the product, such as when it contains particulate solids. A headspace ensures that sealing can occur above the product line, thus preventing solid material from getting caught up in the top seam where its presence would lead to loss of sterility.

The carton top is folded and closed after filling, where the seal is made by either induction heating or ultrasonic welding. Production and date codes are added afterwards with ink-jet printing or by burning into the top seam. The protruding flaps or "ears" on each side are folded down and

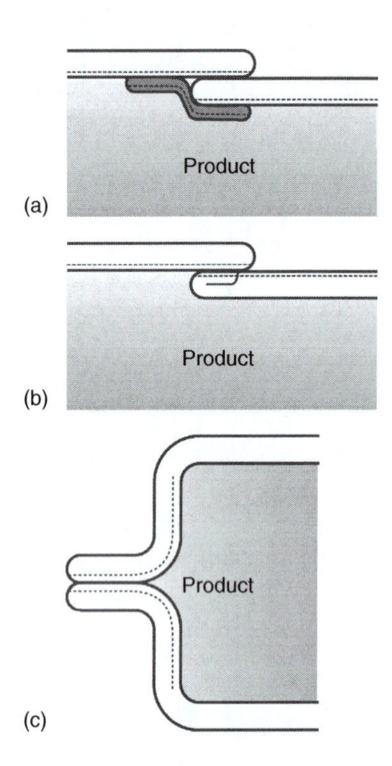

FIGURE 12.5 Three types of side seams used with aseptic paperboard laminate cartons: (a) plastic strip overlaps internal side of longitudinal seam (as used with form-fill-seal cartons); (b) inner layer of board is skived (pared down) and the reduced-caliper edge is folded back and sealed off from the product (as used with prefabricated cartons); (c) fin seal which avoids exposure of the product to any cut edges of paperboard.

sealed to the package with hot air. The finished cartons are then discharged onto a conveyor belt, ready for the final packing process.

B. Can Systems

The system for cans was pioneered by W.M. Martin in the late 1940s and in 1950 the first commercial aseptic filling machine was commissioned by the James Dole Corporation in California for soups. This system uses superheated steam at temperatures of up to 225°C for up to 40 sec to sterilize the cans and can ends. The three basic types of metal cans (tinplate, electrolytically chromium-coated steel [ECCS] and aluminum) can be used in this system and various can sizes, from 125 mL to 22 L, can be handled. Product quality is the same for large and small containers, an important feature with products which are difficult to process because of heat sensitivity or poor heat transfer properties.

After the cans are filled with the cold, sterile product they are sealed using a conventional can seamer which has been modified for aseptic operation. Superheated steam is used to maintain asepsis during the filling operation and this results in a high vacuum in the can. To prevent excessive vacuum in the can which could lead to leaker spoilage, either sterile air or N_2 is blown into the headspace of the can immediately prior to seaming; this results in a vacuum of about 275 mm Hg compared with 500 to 600 mm Hg without air injection.

An additional important packaging-related factor concerns the lining compound in the lid. At the seaming temperature of 220°C it is very plastic, and for this reason the seamed can must be transported in an upright position for at least 15 sec to allow the compound to settle down and hermetically seal the can. Only then may the can be rinsed to remove filling residues or transported by rolling.

Composite cans consisting of a spirally wound body made from laminations of foil, plastics and paper with metal ends are sterilized using hot air at 143°C for 3 min. This process renders the container and ends sterile in respect of acid-tolerant micro-organisms such as yeasts, molds and nonspore-forming bacteria, thus permitting the aseptic packaging of acidic products such as fruit juices and other beverages. The use of steam to sterilize composite cans is not practicable since swelling of the paper layers would result.

C. Bottle Systems

1. Glass

Aseptically filling glass bottles was attempted following the success of this method with metal cans. The bottles were sterilized with either saturated steam under pressure or dry heat. When the latter process was used, extended cooling with sterile air was required to minimize the risk of bottle breakage from thermal shock when bottles were filled with cool product. None of the prototypes for aseptically filling glass bottles reached commercial operation.[2]

However, there has recently been a revival of interest in aseptic packaging in glass containers and several new systems have been developed. One of these uses dry heat sterilization, while others use a H_2O_2 bath or spray followed by drying with hot air. None of these new glass bottle systems has found widespread acceptance.

2. Plastics

Blow molded plastic bottles have been used for many years as a cheaper alternative to glass for nonreturnable containers. High density polyethylene (HDPE) and PP are the two most common thermoplastics used, sometimes with pigments added so the contents are better protected from light. It is also possible to mold bottles from multilayered material, resulting in greatly improved barrier properties and, thus, longer shelf lives for the products packaged in them.

Three different types of systems are in use:

a. Nonsterile Bottles

After blowing (either immediately prior to the following steps or on a separate site) the plastic bottles are conveyed into a sterile chamber, the air pressure of which is kept at a slight overpressure with sterile air. The bottles are inverted and sprayed inside and outside with a solution of H_2O_2 or PAA. The H_2O_2 is evaporated by passing the bottles through a hot air tunnel prior to filling. Bottles sterilized with PAA are rinsed with sterile water and then filled. A chemically sterilized, heat sealable closure such as a plastic film or cap is then applied.

b. Sterile Blown Bottles

Bottles are extruded, blown with sterile air and sealed under conditions that ensure internal sterility. The sealed bottles are then introduced into a sterile chamber (maintained under a slight positive pressure) where the outside surfaces are sterilized with H_2O_2 sprays. The closed top of the bottle is cut away, the neck trimmed, the bottle filled and a foil cap or heat sealable closure (which has been sterilized outside the chamber) is applied.

c. Single Station Blowing, Filling and Sealing

In this mechanically complex system, the separate operations of parison extrusion, blow molding, bottle filling and sealing all take place in sequence in a single mold. Sterility of the inside surface of the containers is ensured by the high temperature of the plastics material during extrusion of the parison and the use of sterile air for blowing. The thermoplastics used for this type of container include HDPE, PP and PETG copolyester with the extrusion temperature ranging from 165 to 235°C, depending on the resin. After filling, the tube projecting from the bottle mold is vacuum-formed or sealed with jaws into a cap which closes the bottle. No special arrangements to ensure sterility are required since the filling and sealing are carried out within the closed mold.

D. Sachet and Pouch Systems

1. Form-Fill-Seal Systems

In this system a vertical form-fill-seal machine operates in a sterile chamber. The packaging material is passed through H_2O_2 and then drained and dried with hot air. The film used is typically a laminate of linear low density polyethylene with a center layer of ethylene vinyl alcohol (EVOH) copolymer and carbon black to give the pouch the required shelf life. An aseptic pouch, similar in structure to the laminate carton shown in Figure 12.3, was commercialized in 1997, the difference being that the paperboard was replaced by paper of 80 gsm. Similar filling machines and procedures are used as for aseptic cartons. Pouches typically have fin seals on all four sides.

2. Lay-Flat Tubing

This system uses a blown film polymer in the form of lay-flat tubing so that only a transverse seal is required to form the bag. The assumption is made that the inside of the tubing is sterile due to the temperature achieved during the extrusion process. Either a single film or a coextrusion can be used. The tubing is fed from the reel into a sterile chamber in which an overpressure of air is maintained. The sachets are sealed at the bottom, cut and moved to a filling section. They are sealed at the top after filling and exit the chamber through a water seal.

E. Cup Systems

1. Preformed Plastic Cups

The cups are usually made from high impact polystyrene (HIPS), PP or coextruded, multilayered polymers if improved barrier properties are required. A typical example of the latter cup would be an outer layer of HIPS, a laminating adhesive, a barrier layer of polyvinylidene chloride (PVdC) or EVOH copolymer, a laminating adhesive and finally low density polyethylene.

The cups are fed onto a conveyor which is inside a sterile tunnel supplied with sterile air. The cups are sprayed inside with 35% H_2O_2 solution and after about 3 sec the solution is removed with compressed hot air at a maximum temperature of 400°C, depending on the material from which the cups are made. The inside surface of the cups reaches a temperature of about 70°C which completes surface sterilization and reduces the peroxide residues to acceptable levels.

Cups can also be sterilized by carrying them through a 35% peroxide bath at 85 to 90°C before heating and passing through a water bath. The cups then enter a sterile chamber where sterilization is completed by spraying with sterile water and drying with hot air.

The sealing material (usually aluminum foil with a thin coating of a thermoplastic polymer) is typically sterilized with a 35% peroxide solution, which is then removed either by radiant heat, hot sterile air or by passing the material over a heated roller. In some systems, UV radiation is used, either alone or in conjunction with peroxide.[2]

2. Form-Fill-Seal Cups

Plastic material (commonly HIPS because it is easily thermoformed) in the form of a web is fed from a roll into a thermoformer to give multiple containers (still in web form). More complex coextruded multilayer materials that incorporate a barrier layer of either PVdC or EVOH copolymer can also be treated in this way. However, mechanical forming (rather than thermoforming) is used if an aluminum foil layer is incorporated into the laminate.[2]

The advantages in thermoforming cups from a reel compared to using premade cups include a favorable price ratio, simplified handling because the constant reloading of magazines is avoided, higher output from utilizing multiple tools, smaller storage requirements for the packaging materials and maximum sterility of the cups (both inside and outside surfaces) and lids from running the flat material through sterilizing baths.

Sterilization of the web is carried out prior to forming by passing it through a 35% H_2O_2 bath at room temperature, typical residence times being in the order of 15 sec. Air knives remove surplus liquid prior to the web passing through a sterile tunnel where it is prepared for thermoforming by heating it to 130 to 150°C. Alternatively, radiant heat can be used to heat the web after it has left the peroxide bath. The containers are then formed (usually by a combination of mechanical forming and compressed air) into a water-cooled mold below the web. Sterilization of the lidding material is achieved in a similar manner to the container web.

An alternative type of form-fill-seal system sterilizes the containers after they have been formed using saturated steam under pressure at 3 to 6 atm (135 to 165°C) for about 1.5 sec; the lidding material is again sterilized in a similar manner.

Using the high temperatures reached during the extrusion process to ensure sterility is the feature of another form-fill-seal system based on coextruded multilayer films which typically contain PVdC or EVOH copolymer as a barrier layer. The outer layer of the coextrusion is peeled away within a sterile chamber, exposing a sterile inner surface which is then heated by radiation and thermoformed into the desired container shape. It is important that uniform wall thickness is maintained during the thermoforming process, otherwise the thickness of the barrier layer will vary and dramatically affect the shelf life of the product. The lidding material also has a peelable outer layer which is removed to expose a sterile material which can be heat sealed in place.

A special feature of the system is the in-mold labeling process, wherein the label is positioned in the mold and applied and heat sealed to the external surface of the container during the thermoforming process; this imparts strength to the container as well as providing an economic labeling system.

IV. INTEGRITY TESTING OF ASEPTIC PACKAGES

Assessment of package integrity is one of the most critical issues in the aseptic packaging of foods and it is imperative that package integrity be maintained to ensure the safety and quality of the product. In addition to the performance tests described below, the effects that shipping containers, palletizing, packaging materials and the form of packaging have on the integrity of the aseptic package is also important and should be evaluated before a particular aseptic packaging system is used commercially.

Several performance tests are in use to assess the likelihood of an aseptic package maintaining its integrity during distribution and handling. Package and seal integrity have traditionally been verified using destructive methods such as biotesting, electrolytic testing, dye penetration or bubble testing. However, destructive test methods are often laborious and time-consuming to perform and clearly it is not possible to test and reject all defective packages.

There is growing interest in nondestructive (or noninvasive) package integrity testing, which allows the testing of every package produced while leaving both product and package intact. Such tests need to meet three criteria: nonspecificity, high sensitivity and rapidity. Most nondestructive leak inspection systems are based on a stimulus–response technique with stimuli including pressure, a trace gas such as CO_2 or helium and ultrasound. The package response can be package movement, pressure change, trace gas detection or sound attenuation.

Other nondestructive tests involving computer-aided video inspection or automatic profiling of the packages have been developed and are being improved upon all the time. Profile scanning with aseptic packages is ineffective because, even if a package leaks and air enters, the package profile does not change immediately.

Despite considerable research in recent years, the availability of commercially viable, nondestructive, package integrity testing equipment is still very limited. Continuing research is in progress and new possibilities will probably be commercialized in the coming years.

REFERENCES

1. Barbosa-Cánovas, G. V., Pothakamury, U. R., Palou, E., and Swanson, B. G., *Application of Light Pulses in the Sterilization of Foods and Packaging Materials*, *Nonthermal Preservation of Foods*, Marcel Dekker, New York, 1998, chap. 6.
2. Burton, H., *Ultra-High-Temperature Processing of Milk and Milk Products*, Elsevier Applied Science, Essex, England, 1988.
3. Holdsworth, S. D., *Aseptic Processing and Packaging of Food Products*, Elsevier Science, Barking, England, 1992.
4. Lewis, M. J. and Heppell, N. J., *Continuous Thermal Processing of Foods*, Aspen Publishers, Greenwood Village, Connecticut, 2000.
5. Moruzzi, M., Wallace, E. G., and Floros, J. D., Aseptic packaging machine pre-sterilisation and package sterilisation: statistical aspects of microbiological validation, *Food Control*, 11, 57–66, 2000.
6. Robertson, G. L., Ultra-high temperature treatment (UHT) (b) Aseptic packaging, In *Encyclopedia of Dairy Sciences*, Roginski, H., Fuquay, J. W. and Fox, P. F., Eds., Academic Press, London, Vol. 4, pp. 2637–2642, 2002.
7. Robertson, G. L., The paper beverage carton: past and future, *Food Technol.*, 56(7), 46–51, 2002.

13 Packaging of Microwavable Foods

CONTENTS

I. INTRODUCTION

The history of the microwave oven began in England in 1940 when two scientists working at Birmingham University (Professors Randall and Boot) devised an electronic tube (which they named a *cavity magnetron*) that generated large amounts of microwave energy very efficiently. The unique ability of the magnetron to transmit microwaves at very high power enabled radar equipment to be built that was much smaller, more powerful and more accurate than anything previously designed.

The magnetron was taken to the U.S. in September 1940, and Dr. Percy Spencer at Raytheon received a contract to make copies of it. In 1945, Spencer first popped some corn in front of a waveguide horn, and the idea of using microwaves for heating foods was borne. He filed a patent application that year, which was issued in 1950 (Method of treating foodstuffs, U.S. No. 2,495,429). In 1947, he filed another application that was issued in 1949 (Prepared food article and method of preparing, U.S. No. 2,480,679), which showed a pouch specifically designed for microwave heating of an entire cob of popcorn. The first commercial microwave oven was released in 1947 for institutional and restaurant use. When Spencer's basic patent expired in 1967, the microwave oven domestic market grew rapidly, achieving a penetration rate in U.S. households of over 100% (i.e., more than one per household on average) by the early 1990s.

TABLE 13.1
Milestones in the History of the Microwave Oven

1945	Dr. Percy Spencer from the Raytheon Company files a patent for a method of treating food by application of microwave energy for sufficient time to cook the food to a predetermined degree
1947	First commercial microwave oven introduced by Raytheon for restaurant and institutional use
1955	First consumer microwave oven introduced
1967	First countertop domestic microwave oven introduced
1975	First commercial use of susceptors for pizzas
1984	First use of susceptors in microwave popcorn applications

Consumer interest in the microwave created a growing demand for microwavable foods. Initially, many companies took advantage of the microwave mania by simply giving a new designation — either dual ovenable or microwavable — to an already established product, without changing either the formulation or the packaging. Today, the microwave foods segment has passed into a new phase with the development of microwave-only products. However, although much has been accomplished, more advanced technological research and development of microwavable foods and packaging is required to fully realize the potential of the microwave oven. Milestones in the history of the microwave oven are listed in Table 13.1.

II. BASIC PRINCIPLES

A. Microwave Oven Operation

Microwaves, a form of electromagnetic radiation, are characterized by wavelength and frequency. Virtually all microwave ovens used by consumers operate at a frequency of 2450 million cycles per sec (cps) (2450 MHz), and in free space (air is a good approximation), the wavelength associated with this frequency ($\lambda = c/\nu$, where λ = wavelength, c = speed of light = 3×10^8 m sec^{-1} and ν = frequency) is 122.4 mm.[3]

The energy delivered is in the form of sine waves, with the electrical and magnetic components orthogonal to each other and also in phase, as shown in Figure 13.1. The strength of the electric field in volts per unit distance oscillates from ground state to a maximum positive voltage, decaying back to

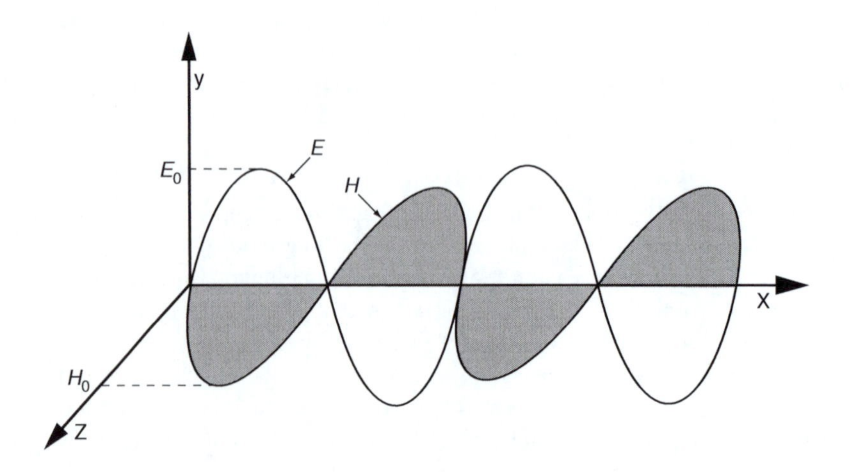

FIGURE 13.1 Diagrammatic view of a plane electromagnetic wave. E and H represent the electrical and magnetic components of the wave, respectively; E_0 and H_0 are their respective amplitudes.

zero, then building up to a maximum voltage of opposite polarity and decaying back to zero. This full cycle occurs 2450 million times per second, and the same phenomenon is seen in the magnetic field. The relationships between the components of any electromagnetic field are given by Maxwell's equations, a set of four equations first formulated by the Scottish physicist James Clerk Maxwell around 1860. They will not be given here but can be found in standard physics textbooks.

In the microwave oven, the oven cavity is three-dimensional and waves are scattered in all directions. For this reason, it is often convenient to consider the variations in electric field along the three principal directions in the oven. These are referred to as the *three orthogonal components* of the electric field.

Microwaves are generated in a device known as an *applicator* by means of a magnetron. The waves are transferred by means of waveguides to the food, which is placed in a cavity or oven. These waveguides are often aluminum tubes with rectangular or circular cross-sections along which the waves are internally reflected. In an attempt to provide an even distribution of the field, a metal fan, which reflects the radiation randomly, or a turntable, which rotates the food through a variable field, are incorporated into microwave ovens. The metallic walls of a microwave oven are coated with a material that almost totally reflects the microwaves, thus providing a set of boundary conditions. As a result, the walls of the oven do not rise in temperature. Because the walls are good conductors, an electric field parallel to the oven wall cannot exist at the wall.

The interface between the food and the air in the microwave oven (or between various packaging materials and the food) also imposes a boundary condition. At the air/food interface, there is a change in the way microwaves propagate (because of large differences in the dielectric properties between the food and the air), and this controls the ratio of the reflected (from the food surface) and transmitted (passing into the food) waves.

When a beam of microwaves is reflected, the incident beam and the reflected beam interact to form a standing wave pattern. The standing wave pattern is such that at certain distances from the reflecting surface, the field variations have maximum values which correspond to the added values of the incident and reflected waves at those points. At other distances between these maxima, the incident and reflected fields cancel out, resulting in a minimum value which is zero if the reflection is total. Although the incident and reflected beam are always in continuous motion, the resultant electric field pattern appears to be stationary.

The multiple reflections in three dimensions that occur in the microwave oven result in a number of possible three-dimensional standing wave patterns, where each pattern is referred to as a *mode*. In a typical microwave oven, the field can be described in terms of a mixture of about 20 or 30 modes, leading to the description of ovens as *multimode cavities*. Placing a container of food into an oven will change the reflections and therefore modify the mixture of modes in the oven.

B. Microwave Heating Mechanisms

Microwave heating depends on fundamental principles that must be considered in both product and package design. In conventional heating, foods are placed in a high temperature environment and absorb heat from the oven over time. In a microwave oven, foods are placed in an electromagnetic field at ambient (room) temperature and heat is generated by the food's own ingredients and sometimes the packaging. The temperature gradients that might be observed if the same food were heated in a conventional and a microwave oven are depicted in Figure 13.2.

The rapidly varying electric field of microwaves is responsible for most of the heating of food in a microwave oven. Although there is also an analogous set of magnetic field components, they are not normally important in determining the heating behavior because foods do not interact with magnetic fields. However, certain special types of packaging materials have been developed that do make use of magnetic interactions to generate heat; these are discussed in Section IV.B.

In developing products for microwave processing, it is important to recognize that microwaves are a form of energy, not a form of heat, and are only manifested as heat on interaction with a

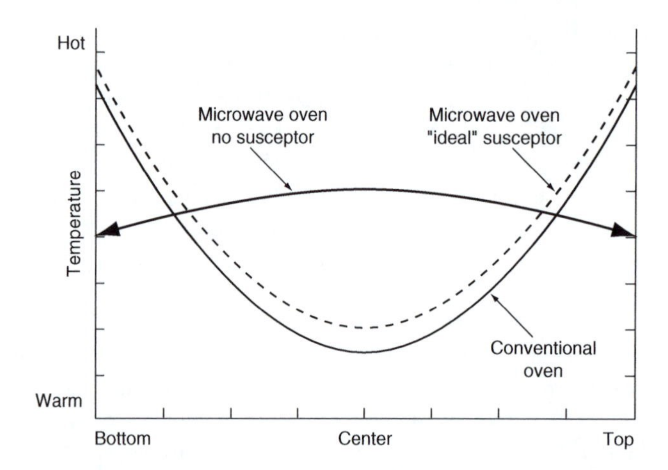

FIGURE 13.2 Temperature gradients in microwave and conventional heating of foods. (*Source*: From Turpin, C., *Microwave World* 10(6), 8–13, 1989. With permission.)

material as a result of one or more energy-transfer mechanisms. There are two main mechanisms by which microwaves produce heat in foods and these are now discussed in turn.

1. Dipole Rotation

Polar molecules (water is the most common polar material in foods) are charged, asymmetric molecules, which are randomly oriented under normal conditions. In the presence of an electric field, the polar molecules attempt to align themselves with the rapidly changing, alternating electric field. The polarity of the field is varied at the rate of the microwave frequency. In this field, the molecules act as miniature dipoles and, while oscillating around their axes in an attempt to go to the proper positive and negative poles, gain potential energy. As the field decays to zero, this gained energy is released as random kinetic energy or heat. In other words, the intermolecular friction which is created by the oscillation is manifested as a heating effect.[4]

2. Ionic Polarization

Ionic polarization occurs when ions in solution move in response to an electric field. Ions carry an electric charge and are accelerated by the electric field. Kinetic energy is given up by the field to the ions, which collide with other ions, converting kinetic energy into heat. The more concentrated or dense the solution, the greater the frequency of collisions and the more kinetic energy that is released. At microwave frequencies, numerous collisions occur and much heat is generated. However, it is a less important mechanism than dipole rotation.

Microwave heating is also dependent on the physical state of the material. As the temperature of a material increases, the molecules tend to line up more rapidly and return to the random state more rapidly. Liquid water absorbs microwave energy much more efficiently than ice because, in ice, the movement of water molecules is restricted. Therefore, ice is a poor microwave absorber. Thus, if ice and pockets of liquid water exist together (as in frozen foods), then the water pockets absorb energy much more rapidly than the adjacent frozen areas. The usual experience with microwave thawing is runaway heating, with one area of the product heating excessively while the remainder is still frozen.

With conventional heating, moisture is first flashed off the surface of the food and then water moves towards the surface as a result of a driving force due to a dry surface and a wet interior. This movement is primarily by capillarity and is limited by the diffusion rate. With microwave heating, the interior temperature may be higher than that on the surface, causing a higher internal vapor pressure, which effectively pumps water as vapor towards the surface. The result is often a much higher rate of water movement and evaporation than in a conventional oven, and frequently, the surface remains constantly wet. Because the air in the microwave oven is cool, the water vapor condenses. In addition, any evaporation will cause surface cooling.[13]

C. Dielectric Materials

The *dielectric constant* (ε') is the ratio of the capacity current through the material to the capacity current that would flow if the same field intensity were applied to free space. It is a measure of the speed of the electromagnetic wave through a material. The frequency of the waves in the microwave is constant (2450 MHz) regardless of the medium they are traversing. If the propagation speed changes and the frequency remains unchanged, then the wavelength will change. The larger the ε' for a given material, the shorter the wavelength, and thus the slower the wavespeed.[8] It is a measure of a material's ability to store electrical energy.

The *dielectric loss factor* (ε'') is an intrinsic property of the food and is a measure of a material's ability to dissipate electrical energy. It indicates the efficiency with which electromagnetic radiation is converted to heat. Thus, a large loss factor indicates that the food would heat readily. Materials with a high loss factor are termed *lossy materials* and are very suitable for microwave heating. The loss factor has been found to be dependent on the frequency of the radiation and the temperature.

The *loss tangent* (also known as the *dissipation factor*) is the ratio of the loss factor to the dielectric constant and represents the energy loss characteristic of the material:

$$\tan \delta = \frac{\varepsilon''}{\varepsilon'} \tag{13.1}$$

The terms loss factor and loss tangent are somewhat misleading terms, in that as these loss terms increase, the amount of power that the material possessing those properties will absorb actually increases. The degree of lossiness of a material (expressed by terms such as loss factor and loss tangent) varies irregularly with frequency, temperature and the nature of the material. The greater the lossiness of a material, the greater the absorption of microwave energy and the higher the rate of temperature increase within the material.

D. Energy Conversion

There are some popular misconceptions about microwave heating, and statements such as "the food heats up from the center" are not uncommon. Therefore, it is necessary to understand how microwave radiation is adsorbed by a food. For simplicity, consider microwave radiation that is incident on one surface only. As the wave passes through the food, it is attenuated (i.e., it loses energy). It is this energy that is converted to heat at the point where the energy is lost. Any temperature gradients that develop within the food are then eliminated by normal processes of conduction and/or convection.

The actual wavelength in different materials is considerably shorter, being reduced by the square root of the dielectric constant of that material:

$$\lambda_m = \lambda_0 \varepsilon'^{-2} \tag{13.2}$$

where

λ_m = wavelength in a material;
λ_0 = wavelength in free space;
ε' = dielectric constant of the material.

The fundamental equation for microwave power absorption by a food is:

$$P_v = kE^2 f\varepsilon'' \tag{13.3}$$

where

P_v = power developed in a volume of material;
k = a constant to determine units of power and volume;
E = electric field strength in volts per unit distance;
f = frequency of the microwave system;
ε'' = dielectric loss factor for the material.

The increase in temperature of a material per unit of time is given by:

$$\frac{dT}{d\theta} = \frac{k' E^2 f\varepsilon''}{\rho c} \tag{13.4}$$

where

ρ = density of the material;
c = specific heat;
k' = a constant to express heating in the desired temperature and time units.

Two of the parameters (field strength and frequency) are properties of the energy source (i.e., fixed by the microwave oven). The dielectric loss factor, density and specific heat are properties of the materials being heated. Increasing the value of any of these factors increases the amount of energy converted. The food product developer may be able to increase the latter two factors by altering the food composition.

As mentioned above, there are two main mechanisms by which materials may heat when exposed to a microwave field. Using ε' and ε'' to characterize heating in the oven eliminates the need to distinguish the particular mechanisms of the material under study, and they are therefore very useful characteristics (see Table 13.2).

E. Penetration of Microwaves

When microwaves strike the surface of a material, they arrive with some initial level of power P_0, which is a function of the power output of the magnetron, oven cavity size and uniformity of the fields. As the waves begin penetrating the material, some of the power is absorbed and the wavelength changes, depending on the dielectric constant of the material.

For practical use in industry, the dielectric properties are converted to an attenuation factor or its inverse, the penetration depth, which defines the depth into the material at which the energy has decreased to $1/e$ (about 37% because $e = 2.718282$). Penetration is also expressed as the half-power depth (HPD); that is, the depth from the surface of a material at which the power has been reduced to one half of the incident power (Table 13.2).

The penetration depth d can be calculated as follows:

$$d = \frac{\lambda_0}{2\pi}\left[\frac{2}{\varepsilon'[(1+\tan^2\delta)^{1/2} - 1]}\right]^{1/2} \tag{13.5}$$

The penetration depth can offer insight into the heating behavior of foods (see Table 13.2). For example, at a depth d into the product, 63% of the energy has been absorbed; at $2d$, only 14% of the incident power level remains, and after $3d$, just 5% is left. Thus, assuming a homogeneous material

TABLE 13.2
Dielectric Constant (ε'), Dielectric Loss Factor (ε'') and HPD of Various Materials at 2450 MHz

	ε'	ε''	HPD (cm)
Water (distilled at 25°C)	78	12	1.4
Water + 0.5 *M* NaCl (25°C)	42	67	0.2
Ice (− 12°C)	3.2	0.003	1160
Oil	2.5	0.15	
Milk	78	23	
Lean meat	40	12	
Fatty meat	53	16	
Cooked fish	46	12	
Ham (precooked, 20°C)	43	23	0.6
Potato			
Raw (25°C)	57	16	0.9
Mashed (30°C)	72	24	0.7
Butter (salted)	4.6	0.60	6.9
Butter (unsalted)	2.9	0.45	7.4
Bread dough	22.0	9	1.0
Bread	4.6	1.2	3.5
Paper	3	0.15	14.8
Polyethylene	2.3	0.003	700
Glass	5.2	0.02	

and constant penetration depth at different levels, it is apparent that the outer layer heats faster than the inner layers. As long as the product thickness is equivalent to or small compared with the penetration depth, this effect is not major, but for very thick products, the outer surface will get much hotter than the inner one. In addition, because microwaves strike materials from all sides, there could also be a "hot spot" in the center where the waves from all sides overlap, creating a superposition of power.

F. NONUNIFORM HEATING

Many of the more serious problems associated with foods heated in a microwave oven can be attributed to nonuniform heating.[9] For example, it is not uncommon for a frozen entree to thaw, heat and begin to boil around the edges, while the center remains frozen. If the heating time is chosen to ensure that the center temperature is acceptable (say 50°C), then the product around the edges is dried and charred. Alternatively, if the heating time is reduced to avoid deterioration around the edges, then the center of the product will be unacceptably cold. Because of such nonuniformity, it is impossible to achieve a result that approaches a satisfactory quality.

A number of factors contribute to the nonuniformity of microwave heating. The three principal effects are[9]:

1. Undesirable field distribution, with the microwave energy being significantly more intense around the periphery than in the center of the oven
2. Microwave energy entering through the sidewalls of the container, causing further heating of the food nearest the walls and very little heating in the central region
3. Differences in absorption of microwave energy for frozen and thawed foods, where there is a dramatic increase in lossiness as the food thaws

The energy required to melt frozen food can be determined using differential scanning calorimetry (DSC) techniques. For example, a frozen entree of pasta in sauce requires approximately 350 J g^{-1}.[9] From this, the average power absorption at the different positions in the product can be estimated using the following equation:

$$P = E/t \tag{13.6}$$

where
P = average power absorption (W g^{-1});
E = energy required to melt food (J g^{-1});
t = time to melt food (sec).

Near the edge of the product,

$$P = 350/90 = 3.9 \text{ W g}^{-1}$$

At the center of the product,

$$P = 350/600 = 0.6 \text{ W g}^{-1}$$

When the product begins to boil around the edges, moisture loss occurs and the rate of evaporation can be estimated as follows[9]:

measured enthalpy of evaporation = 1540 J g^{-1}
rate of energy absorption near the edge = $3.9 \times 60 = 234$ J g^{-1} min^{-1}

This corresponds to an evaporation rate of approximately 234/1540 or 15.2% by weight per minute (expressed as a percentage of the original weight). Because the temperature at the center of the product only reaches about 40°C, the moisture loss in that region is very low. It follows that 4 min of boiling would correspond to a weight loss of at least 60%. Obviously, weight loss around the edges far exceeds the average weight loss, which, for the frozen entree described here, has been measured as 22% after 12 min.[9]

Three distinct approaches are used to eliminate or minimize the effects of nonuniform heating. These are:

1. Food formulation, where by the careful selection of ingredients and food combinations, it is possible to prepare meals that are more tolerant to variations in heating.
2. Heating procedures, which include the use of intermediate standing times, low power settings and stirring operations (if appropriate) to allow the temperature of the overheated regions to be reduced. However, because of the low thermal conductivity of most foods and the short times involved, there is little significant transfer of heat. Thus, only minimal improvements in product quality are obtained using this approach.
3. Packaging systems, which interact with and modify the microwave field distributions by simultaneously increasing field intensities at the center of the food and reducing the field intensity at the edges. Such systems are discussed further in Section IV.

III. EFFECT OF FOOD PRODUCT

If the dielectric properties of the food materials are known, then the penetration depth can be calculated using Equation 13.5.[10,11] This offers insight into the heating behavior of foods and packaging. Determining dielectric properties is quite a sophisticated procedure, and most product developers rely on published data. The dielectric properties of foods at microwave frequencies are, for many practical purposes, determined by their moisture, solids and salts contents. Generally, it is found that both the dielectric constant and the loss factor are temperature dependent. The following example for water illustrates this:

At 1.5°C, $\varepsilon' = 80.5$, $\varepsilon'' = 24.96$ and $\tan\delta = 0.3100$. By calculation, $d = 14.2$ mm.
At 25°C, $\varepsilon' = 78$, $\varepsilon'' = 12.48$ and $\tan\delta = 0.16$. By calculation, $d = 27.6$ mm.
At 75°C, $\varepsilon' = 60.5$, $\varepsilon'' = 39.93$ and $\tan\delta = 0.660$. By calculation, $d = 76$ mm.

Compare these values with ice where $\varepsilon' = 3.20$, $\varepsilon'' = 0.0029$ and $\tan\delta = 0.0009$. In this situation, $d = 24.2$ m. This explains why it is difficult to rapidly thaw foods in a microwave oven; little energy is absorbed by ice, even when reasonably thin, giving rise to the expression that "ice is highly transparent to microwaves."

The physical geometry of the product exerts its influence in several ways. However, because the microwaves can be assumed to follow the laws of optics with reflections and refractions at interfaces between materials of different optical/dielectric properties, many of the peculiarities in microwave heating can be explained. The more regular the shape, the more uniform the heating. Sharp edges and corners should be avoided, because these will tend to overheat. Round is better than square, and a torus (dough nut) is an ideal shape. The sphere acts as a microwave resonator, the microwaves that enter the sphere being internally reflected because of their very limited angle of incidence and the effect of refraction.

IV. PACKAGING

Packaging materials can react in three ways to microwaves: they can reflect the radiation, absorb the radiation or transmit the radiation (so-called *RAT characteristics*). Transparent materials do not appreciably react to the microwave field of the oven or appreciably modify the power distribution in the oven, and are referred to as *passive packages*.[2] The various types of packaging materials used to package food in a microwave oven are described below.

A. Transparent Materials

In these types of packages, the microwaves penetrate the transparent material and are absorbed by the food. All the polymers currently used for food packaging, as well as paper products and glass, are microwave transparent. Because paper products contain some water and mobile ions in their structure, they will heat in the microwave, but the rate is generally slow and negligible.

Closure of these packages can enhance product performance and temperature uniformity by transforming the energy used to vaporize water that may otherwise escape into general heating of the product. Localized drying of products in corners and thin layers on the edges can lead to localized heating. As the product temperature rises, adjacent packaging materials will be heated by the product, and therefore temperature resistance of the packaging material needs to be evaluated.

The types of problems encountered in microwave ovens with foods packaged with transparent packaging materials include[2]:

1. Sogginess or lack of crisping development
2. Lack of browning or color development
3. Nonuniformity of moisture loss, leading to localized toughening
4. Nonuniform temperature distribution
5. Boil-out or run-off of sauces and toppings
6. Inappropriate heating rates for proper cooking of multiple component food products

B. Absorbent Materials

Packaging materials that absorb microwave energy and re-emit that energy as heat are commonly called *susceptors* but are also referred to as *receptors*, *absorbers* or *heater elements*. They have been used commercially in various forms since 1975, the key microwave susceptor patent being granted in 1986 to Oscar Seiferth (U.S. No. 4,641,005). Susceptors are used to achieve localized effects such as crisping (surface drying) and browning, and are an example of active packaging (see Chapter 14) because they perform functions beyond that of just containment and enhance the performance of the package.

An ideal susceptor would heat up very rapidly to a predetermined temperature, and remain at that temperature throughout the whole heating process. Unfortunately, thin film susceptors do not perform well in many applications; they either overheat locally and burn the product or become damaged during the heating process in the microwave. Such damage is characterized by a premature reduction in their heating capability and an unevenly heated product.[15]

To produce a thin film susceptor, particles of a metal such as aluminum are applied in a particular density (typically a thickness of less than 30 to 60 Å) to a heat resistant surface such as a polyester film, which is adhesive laminated to a low loss, temperature-stable substrate such as bleached kraft paperboard. Two techniques are used: thermal evaporation and sputtering.[5]

The basic principle of film deposition by thermal evaporation is that increasing the temperature of a material will also increase its equilibrium vapor pressure. When this vapor pressure is significantly higher than the residual pressure in the deposition chamber, a flux of atoms or molecules of the material will be projected in all directions from the heated evaporation source. A cold substrate in the path of this vapor flux provides an efficient nucleation site for these species. The process is carried out in a vacuum chamber, and the most common way to heat the source material is by the passage of a current.[5] The process is also known as *vacuum deposition* and was described in Chapter 5.

A comprehensive theoretical treatment of microwave interactive thin films has been published.[6] Although susceptors are referred to in the literature as *conductive coatings*, it is the resistance between the metal particles at their contact points that results in the generation of heat. Very thin layers of conductive material have significant electrical resistance and generate localized resistance heating when they are exposed to a rapidly oscillating electric field. When the susceptor is placed in a magnetic field, a current flows through the metallic coating. Provided that the surface resistivity is good, very rapid heating occurs by the conversion of the microwave energy into sensible heat, according to the equation:

$$P = I^2 R \tag{13.7}$$

where

P = power generated in watts;
I = current in amperes;
R = surface resistivity (the reciprocal of conductivity) in ohms per square.

The surface resistivity of a material is the ratio of the potential gradient parallel to the current along its surface to the current per unit width of surface, and is numerically equal to the surface resistance between two electrodes forming opposite sides of a square; the size of the square is immaterial, leading to the units of ohms per square.

The most important property of a metal film is its surface resistivity — a measure of a thin metal coating's ability to conduct electricity. For a thick metal, all of the energy of the incident microwave is reflected, and the surface resistivity is zero. As the metal thickness is gradually reduced, the surface resistivity starts to increase and some of the energy is now transmitted, some is absorbed and less is reflected. As the thickness of the metal is reduced further, the amount of energy

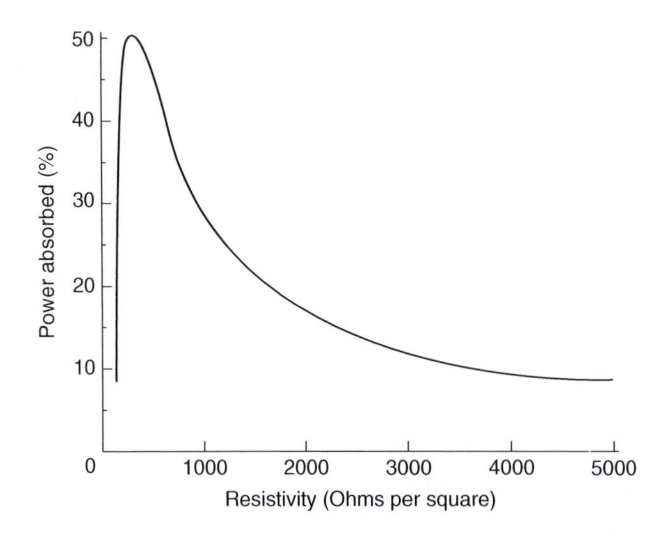

FIGURE 13.3 Percentage of microwave power absorbed as a function of surface resistivity.

absorbed increases rapidly until it reaches 50% of the incident energy, after which absorption gradually falls to zero while the surface resistivity increases to infinity.[14]

The relationship between the percentage of power absorbed and the surface resistivity is shown in Figure 13.3. Thin film susceptors with a surface resistivity between 50 and 250 ohm sq^{-1} would result in maximum power absorption. Such a susceptor is aluminum at a Macbeth optical density of 0.20 (63% transmission), which has a surface resistivity of 25 ohm sq^{-1}. Below an OD of 0.12 (70% transmission of light), the film does not function as a heater, while above an OD of 0.35 (45% light transmission), the film will cause arcing.

These films absorb about 40% of the incident energy, the remaining 60% being reflected or transmitted. Consequently, these films heat in the presence of microwave energy, and temperatures of around 250°C can be achieved in only a few seconds, as shown in Table 13.3. When this type of film is laminated to a paperboard surface, the result is a heater board.

All susceptors in current commercial use are powered by thin conductive coatings. Vacuum deposited aluminum is the most common material in general use. The four basic layers consist of the following:

1. A heating surface (often 12 μm of heat-set, biaxially orientated PET)
2. A thin metal layer (usually vacuum deposited aluminum at 50 to 250 ohm sq^{-1})
3. An adhesive
4. A substrate (usually paper or paperboard)

TABLE 13.3
Maximum Observed Interface Temperatures between Product and Susceptor

Product	Maximum Temperature (°C)	Heating Time (sec)	Comments
Susceptor alone	316	100	No food load
Popcorn	247	150	Normal
Popcorn	280	220	Extended heating
Fish fillet	222	290	Not turned
Pizza	223	290	Not turned

To improve performance, it is usual to have a deliberate pattern of metallization dependent on the nature of the food, rather than metallizing the whole surface. Patterned susceptors allow selective heating at different locations on the food. There are several methods that can be used to obtain pattern metallizations:

1. Pattern demetallization involves taking a full web of metallized substrate and coating the metal side with a clear lacquer in the pattern desired on the finished product. The coated substrate is then subjected to an acidic bath, which removes the unprotected metal, creating a susceptor pattern ready for lamination to paper or paperboard stock.
2. Hot stamp transfer combines conventional hot stamping technology with susceptor metallizing technology to create a transferable susceptor with standard hot stamping equipment. This type of pattern transfer best utilizes the susceptor film and provides more control over inventory by the converter who can change patterns simply by changing the die.
3. Hot nip transfer is similar to hot stamp transfer, the difference being that the final coating or adhesive coating is applied in a pattern using a cylinder similar to the demetallization process. This method allows the converter to produce a roll-to-roll transfer of susceptor directly to paper or paperboard stock.
4. Film to film transfer is done just like hot nip transfer, except that the susceptor is transferred to film rather than paper or paperboard.

Usually, the food is in direct contact with the heating surface, which may have a release coating. The metallized side of the susceptor is placed away from the food product to protect the metal from chemical or physical damage. Susceptor temperatures will usually not exceed 115 to 130°C when in close contact with the food, but may otherwise reach temperatures up to and above 250°C.[11]

The mass of the thin conductive coating in a susceptor is many thousands of times less than the food and, prior to exposure to microwaves, its lossiness is many thousands of times greater than any food. Consequently, its preferential power absorption in close proximity to the food is excellent. However, as soon as it starts absorbing microwave energy, the thin coating breaks up into numerous "islands," resulting in a sharp (and essentially irreversible) drop in lossiness until a tolerable power density is reached.

This break up of the thin coating provides a maximum temperature limit, although unfortunately this upper limit is very hard to control or predict because it is dependent on the balance between microwave power absorption and heat removal. The need for a heat sink to pull heat away from the susceptor coating in order to avoid an excessive decrease in lossiness is one reason why intimate contact with the food is important to effective thin film susceptor operation.[14] As with other packaging components used in the microwave, there are no guarantees on heating performance for the multitude of products where the susceptor would appear to be applicable. Relatively little has been published in the scientific literature about heating, and even less about browning, in microwave ovens.

C. Shielding

Shields are metallic structures that are thick enough so as not to heat, but can reflect microwaves in the same way as the oven walls. For example, a product encased in foil would not exhibit any microwave heating, because no energy is able to penetrate the foil to reach the product. Examples of shield materials include foil, foil laminated to a substrate or aluminum sheet converted into pans and trays.

Special precautions are necessary when using shields. Although they are good conductors of heat, a large electrical potential can build up, and if the potential comes close to a grounded surface or another piece of foil with a different potential, an electric discharge (spark) can occur. This arcing phenomenon can occur between two packages, between layers of foil insulated from each other, across tears, wrinkles or gaps in a single piece of foil and between foil packages and oven walls. Arcing can be controlled using common sense and some simple precautions, but it is important to exhaustively test the package/product combination to understand the hazards involved.

D. Field Modification

A microwave package has been developed that is claimed to modify the electromagnetic field to give uniform heating, selective heating and browning while preventing arcing. A standard aluminum pan (coated in a microwave beige color to minimize reflected energy) with a specially constructed lid is used. Basically, the microwave power is redistributed to obtain a dramatic improvement in uniformity, compared with the more typical cooler center and hotter outer regions. This ability to control power absorption can be applied to modify the behavior of each compartment of a multicavity container.

The lids are designed according to the characteristics of the foods they will be used with and the amount of heating required. This information, developed on specially adapted finite-element analysis software, determines the placement and the number of aluminum conductors to be incorporated within the dome. The conductors can range from a single 50-mm strip to a series of differently shaped aluminum pieces dispersed throughout the dome. The shape of the conductors and where they are positioned in the container are critical to their effectiveness. Applications include pot pies, chicken dinners and macaroni cheese.

E. Doneness Indicators

Doneness indicators are devices that detect and visually indicate the state of readiness of foods heated in a microwave oven. They are classed as *intelligent packaging* and are discussed in Chapter 14 IV B2.

F. Testing Methods and Safety

ASTM[1] has published several standards for evaluating susceptors. F 874-98(2003) details a method for measuring surface temperatures attained by microwave susceptors. It is useful for measuring susceptor/food interface temperatures during microwave preparation of foods with susceptor-based packaging, heating pads, crisping sleeves and so on. It may also be used to measure the temperature of a susceptor exposed to extractives testing or in a liquid extraction cell to be used for nonvolatile extractives testing. The latter procedures are performed to establish test conditions for conducting extraction and migration studies using temperature versus time profiles approximating those for actual microwave preparation of the product.

F 1308-98(2003) covers a procedure for quantitating volatile compounds whose identity has been established and which are evolved when a microwave susceptor sample is tested under simulated use conditions. F 1519-98(2003) covers a procedure for identifying volatile extractables that are released when a microwave susceptor sample is tested under simulated end use conditions, where the extractables identified using GC/MS. F 1349-98(2003) applies to laminate susceptors constructed of paperboard, adhesive and a layer of PET susceptor. The adhesive and PET-related compounds are quantitated using this test method. F 1500-98(2003) covers a procedure for quantitating non-UV-absorbing nonvolatile compounds, which are extractable when the microwave susceptor is tested under simulated-use conditions for a particular food product.

The high temperatures reached by packaging materials in microwave ovens led the U.S. FDA to issue an intent to publish new regulations in 1989. The food and packaging industries responded by conducting extensive research and submitting the results to the FDA.[12] To date, however, the FDA has chosen not to issue regulations regarding microwave susceptors, relying instead on voluntary guidelines to assure consumer safety.[2]

V. CONCLUSION

The development of microwavable foods has not lived up to the forecasts made for it in the early 1990s as "one of the big growth areas during this decade." Technological limitations in meeting consumer expectations were undoubtedly one reason; it has proved impossible to change the laws of physics simply to meet market demand. To avoid failure in the market place, product and package development must be thoroughly researched. An understanding of the fundamental principles involved in the microwave heating of foods is essential to successful development. Because the package is an integral part of a microwave product, development of microwavable foods must go hand-in-hand with development of suitable packaging. There have already been many exciting developments in packaging for microwavable foods and a listing of important U.S. patents from 1949 to 1999 has been compiled.[2] A holistic approach to product and package design should lead to further significant improvements in the quality and convenience of microwave-heated foods.[7]

REFERENCES

1. ASTM: American Society for Testing and Materials, Philadelphia, Pennsylvania. For the latest version of each standard, visit www.techstreet.com.
2. Bohrer, T. H. and Brown, R. K., Packaging techniques for microwaveable foods, In *Handbook of Microwave Technology for Food Applications*, Datta, A. K. and Anantheswaran, R. C., Eds., Marcel Dekker, New York, 2001, chap. 12.
3. Decareau, R. V., *Microwaves in the Food Processing Industry*, Academic Press, New York, 1985.
4. Decareau, R. V. and Peterson, R. A., *Microwave Processing and Engineering*, Ellis Horwood, Chichester, 1986.
5. Grovenor, C. R. M., Polycrystalline conducting thin films, *Microelectronic Materials*, IOP Publishing, Bristol, 1989, chap. 4.
6. Habeger, C. C., Microwave interactive thin films, *Microwave World*, 18(1), 8–22, 1997.
7. Lefeuvre, S. A. E. and Audhuy-Peaudecerf, M. B. M., Microwavability of packaged foods, In *Food Packaging and Preservation*, Mathlouthi, M., Ed., Blackie Academic, Glasgow, 1994, chap. 4.
8. Lewis, M. J., Microwave and dielectric heating, *Physical Properties of Foods and Food Processing Systems*, Ellis Horwood, Chichester, 1987, chap. 9.
9. Lorenson, C., Ball, M. D., Hewitt, B. C., and Bellavance, D., Nonuniform heating of foods packaged in microwaveable containers, *TAPPI J.*, 73, 265–267, 1990.
10. Mudgett, R. E., Microwave properties and heating characteristics of foods, *Food Technol.*, 40(6), 84–93, 1986.
11. Ohlsson, T. and Bengtsson, N., Microwave technology and foods, *Adv. Food Nutr. Res.*, 43, 81–140, 2001.
12. Risch, S., Safety assessment of microwave susceptors and other high temperature packaging materials, *Food Addit. Contam.*, 6, 655–661, 1993.
13. Schiffman, R. F., The technology of microwaveable coated foods, In *Batters and Breadings in Food Processing*, Kulp, K. and Loewe, R., Eds., American Association of Cereal Chemists, St. Paul, Minnesota, 1990, chap. 9.
14. Turpin, C., Browning and crisping. The functions, design and operation of susceptors, *Microwave World*, 10(6), 8–13, 1989.
15. Zuckerman, H. and Miltz, J., Changes in thin-layer susceptors during microwave heating, *Packag. Technol. Sci.*, 7, 21–26, 1994.

14 Active and Intelligent Packaging

CONTENTS

I. HISTORICAL DEVELOPMENT

Smart films were first mentioned in the literature in 1986,[54] the word "smart" being used in connection with selectively permeable films used for MAP. The selective permeability was created by strip lamination of HDPE and PET films, allowing CO_2 to exit the package and preventing excess O_2 from entering. Such films are no longer available. Smart packages were defined in 1989[65] as "doing more than just offer protection. They interact with the product, and in some cases, actually respond to changes." According to one authority,[2] "the term smart packaging was coined in the mid-1980s to describe package structures that allegedly sensed changes in the internal or surrounding environment and altered some of their relevant properties in response. Simultaneously, academics and true (sic) researchers, concerned that the term was too juvenile, invented the term 'interactive' packaging to describe the same entities and later shortened it to 'active' packaging."

However, the term smart packaging is enjoying a renaissance in the twenty-first century and a *Smart Packaging Journal*[25] has been published online since mid-2002. It defines smart packaging as "the use of features of high added value that enhance the functionality of a product, notably mechanical, electronic and chemical features that improve safety and efficiency." The journal urges readers to forget about all the numerous definitions — active, diagnostic, intelligent, smart, functional, enhancement — to describe smart packaging and accept that it is all one big continuum of functionality. It is suggested that one should think of smart as meaning clever, neat or "wow."

The first widely reported use of the term active packaging is generally attributed to the seminal review presented at an EU conference in Iceland in 1987 by Professor Labuza[31,33] from the University of Minnesota. He had just spent a sabbatical year at 3M working with the packaging systems division on a new time–temperature indicator (TTI) and, during a visit to Tokyo Pack in 1986, had seen many commercial examples of active packaging. He and Curt Larson from 3M conceived the term *active packaging* for his valedictory presentation to 800 scientists at 3M at the conclusion of his sabbatical year.

Various authors have attempted to identify the origins of what is now referred to as active packaging. For example, the use of starch-based edible films (rice paper) by the ancient Chinese has been claimed as the first attempt in active packaging;[16] most people would classify it as an early example of edible packaging. An expert in the area of active packaging[39] has claimed that the most obvious commencement of active packaging was the use of tinplate for cans in 1810, because the tin acts as a sacrificial anode, rapidly consuming any residual O_2 in the can and dissolving to protect the iron base. He lists as a subsequent development the introduction of zinc oxide into enamel coatings to impart sulfur resistance by forming white zinc sulfide rather than the blue-black or brown tin sulfide or black iron sulfide (see Chapter 7). In this application, zinc oxide is used for cosmetic reasons and does not impact on the shelf life or organoleptic quality of canned foods.

The patent literature abounds with ideas that could be considered active packaging. A 1938 patent[55] from a Finnish researcher described the use of iron, zinc or manganese powders to remove O_2 from the headspace of cans; a 1943 patent[26] from a researcher in England described removing O_2 from a container containing vacuum or gas packed food in which a metal such as iron absorbed O_2 to form an oxide. In the U.S., the removal of O_2 by the catalytic conversion of H_2 to H_2O was first described in 1955[28] and applied to spray dried milk powder initially in tinplate cans and later in laminate pouches. The package was flushed with a mixture of H_2 (7%) in N_2 and necessitated the use of a palladium catalyst.

The fungistatic effectiveness of a wrapper made of RCF impregnated with sorbic acid and used to package natural and processed cheeses was reported in 1954,[50] together with an alternative approach where sorbic acid was mixed into a wax layer for natural cheese.[35] These were probably the first antimicrobial packaging films.

The use of enzymes and in particular glucose oxidase to remove O_2 has been studied since the initial patent in 1956,[42] which described the impregnation into or on a moistureproof or fabric sheet of glucose oxidase and catalase (the latter to destroy the H_2O_2 formed by the former). The concept of incorporating enzymes into a package material was overtly described in the 1956 patent[42] and, in 1958,[43] the first publication on the use of packets or sachets of chemicals in packages to remove O_2 appeared followed by one in 1961.[44] A 1968 German patent[5] proposed a sodium carbonate powder to absorb O_2 in food packages.

In 1970, researchers in Australia[45] and the U.S.[33] published details on the use of potassium permanganate as an ethylene absorbent in LDPE bags to delay ripening in bananas, but the approach was not widely adopted commercially for many years. In 1973, researchers in India[18] published details of fungistatic wrappers to extend food shelf life, but these were not commercialized.

Although most of the above patents were never commercialized, they laid the groundwork for the subsequent development in Japan in 1976[20] of the iron-based O_2 scavengers which are widely used today. The use in Japan of sachets containing iron powder to absorb O_2 is generally regarded as the first widespread commercial application of active packaging. Intelligent packaging, of which time–temperature indicators (TTIs) represent the most well-known example, have been used commercially since at least 1971.

II. DEFINITIONS

> "When I use a word," Humpty Dumpty said in rather a scornful tone, "it means just what I choose it to mean — neither more nor less."

Through the Looking Glass by Lewis Carroll

A variety of terms can be found in the literature to describe the rapidly growing and important areas of active and intelligent packaging, including active, interactive, smart, clever and intelligent packaging. Often the terms are used without being defined, or if they are defined, the definitions are either so broad as to include many packages that are neither active nor intelligent, or so narrow as to exclude important examples.

Before examining the various definitions, it is worth recapping some basic facts. Foods are packaged using a variety of materials, the primary functions of such materials being to contain and protect the food. In many cases, there is a headspace or void inside the package and the composition of that headspace can have an important influence on the shelf life of the food. In addition, the packaging material, or a component of it such as a seal or closure, may interact with the package or permit the transfer of certain compounds through the package. In addition to the food, the package might also contain a sachet or pad, which might absorb or emit a specific gas in the case of a sachet or water in the case of a pad. These facts should be kept in mind when the definitions of active and intelligent packaging are considered.

A. Active Packaging

First, a summary of what active packaging can achieve[1,4,11,62]: it can be used to remove an unwanted component (e.g., C_2H_4 produced by respiring fruits or O_2 present inside the package); add a desirable ingredient (e.g., CO_2 to inhibit microbial growth); prevent microbial growth (e.g., an antimicrobial chemical incorporated into a film); change the permeability to gases as the temperature changes by several of orders of magnitude greater than normal polymeric films; or change the physical conditions inside the package (e.g., remove water vapor by absorption or change the temperature of the food). The definitions of active packaging presented below (in chronological order) will be considered to see how inclusive or exclusive they are:

1. *Active packaging is the inclusion of subsidiary constituents into the packaging material or the package headspace with the intent of enhancing the performance of the package system.*[21]

This definition includes all the examples given above and has the advantage of not limiting the enhanced performance to quality, safety or sensory properties but rather takes a systems approach.

2. *Active packaging senses environmental changes and responds by changing its properties.*[3] Most of the examples given above would not qualify under this definition because, for example, an O_2 absorber does not sense environmental changes; it simply absorbs any O_2 present in the package headspace. However, this definition would include so-called intelligent polymers — highly permeable membranes used to control the flow of gases through packages as temperatures change (see Section III.D).

3. *Active packaging materials change the condition of the packaged food product to extend shelf life and/or improve microbial food safety and/or improve sensorial properties.*[63] This definition would include all the examples given above and can therefore be considered a broad and useful definition.

4. *Packaging may be termed active when it performs some desired role in food quality or safety other than providing an inert barrier to external conditions.*[41] The key words in this definition are "desired" and "inert" because, with the exception of glass, almost all packaging materials can, under certain circumstances, contribute undesirable compounds to the food (e.g., dioxins from bleached paperboard; tin from tinplate cans; monomers from plastic films). This is a broad definition that covers all types of active packaging. It could include virtually all polymeric films given that they are not inert but are, to varying degrees, permeable to gases and organic vapors. However, the use of the word "desired" excludes those situations where, for example, orange juice is sealed in an HDPE bottle and certain key flavor compounds are absorbed by the bottle over time; clearly, this is not active packaging because loss of flavor compounds is undesirable.

5. *Active packaging is packaging that changes the condition of the packaged food to extend shelf life or to improve safety or sensory properties, while maintaining the quality of the food.*[15] This definition was drawn up by participants in an EU project on active packaging and encompasses all types of active and passive packaging. It could be abridged slightly given that, in its broadest sense, shelf life implies that the food is still of acceptable quality, and quality includes safety and sensory aspects. This is a very broad definition and could include, for example, the placing of a loaf of bread inside a plastic bag and sealing it, or the hot-filling of juice into a plain tinplate can, followed by the seaming on of the end. Neither of these two examples should be regarded as active packaging even although they fall within this definition.

6. *Active packaging refers to the incorporation of certain additives into packaging film or within packaging containers with the aim of maintaining and extending product shelf life.*[13] This definition is similar to the previous one, except that it focuses on the additives that make the package active. It assumes (correctly) that product shelf life includes safety, sensory and quality aspects. However, by using the word "additive" it excludes, for example, self-heating packages and widgets.

7. *Active packaging is a packaging technique that actively and constantly either:*

(a) *changes package permeation properties or the concentration of different volatiles and gases in the package headspace during storage, or*

(b) *actively adds antimicrobial, antioxidative or other quality improving agents, e.g. flavor enhancing substances, via packaging materials into the packed food in small amounts during storage.*

These kinds of packaging materials can also be defined as interactive, because they are in active interaction with food. The aim of these active or interactive methods is to maintain a product's desired shelf life throughout storage.[24]

Part (a) of this definition includes situations where an O_2 absorber is incorporated into the package structure; minerals are added to a polymer; microperforations are made in a film to increase its permeability, as well as so-called intelligent polymer films. Whether they can all be described as interactive is a moot point.

In this book, a modified version of Definition 1 will be used (Table 14.2). *Active packaging* is defined as *packaging in which subsidiary constituents have been deliberately included in or on either the packaging material or the package headspace to enhance the performance of the package system.* The two key words are "deliberately" and "enhance." Implicit in this definition is that performance of the package system includes maintaining the sensory, safety and quality aspects of the food. This definition excludes a standard polymeric film from being classified as active packaging but includes the same film if, for example, it has been impregnated with an antimicrobial compound.

In light of the above definition, it is possible to list processes which are not active packaging although some authors have described them as such. MAP as normally practiced (see Chapter 15) is passive not active packaging *per se*, unless there is some way in which the package (or a sachet added to the package) actively affects the internal gas atmosphere other than via normal permeation through standard polymeric films. The interaction between tin and food constituents inside a can is not active packaging, and neither is scalping of aromatic flavors by, for example, the plastic layer in contact with fruit juices on the inside of paperboard cartons. The bulging of the ends of a metal can owing to internal gas formation (either as a result of microbial growth or from corrosion resulting in the production of H_2) is neither active nor intelligent packaging, although it does qualify under some of the definitions that have been given. Selected examples of active packaging systems are given in Table 14.1.

B. Intelligent Packaging

First, a summary of what intelligent packaging can achieve: it can indicate the quality of the food by signaling when it is ripe or fresh or whether its shelf life has expired; it can indicate the temperature of the food through the use of thermochromic inks or microwave doneness indicators (MDIs); it can indicate whether a package has been tampered with; and it can signal the location of a package.

Intelligent is defined as "having or showing understanding; clever, quick of mind." The primary meaning intended when the word is applied to packaging is "showing understanding" although "clever" is preferred for nontechnical audiences. The definitions below will now be considered to see how inclusive or exclusive they are:

1. *Intelligent packaging is an integral component or inherent property of a pack, product or pack/product configuration which confers intelligence appropriate to the function and use of the product itself.*[53] This relatively early definition is all-encompassing through the use of the word "appropriate." By not defining intelligence in the packaging context, it is too broad to be of much help to those unfamiliar with intelligent packaging.

2. *Intelligent packaging is a packaging technique containing an internal or external indicator for the active product history and quality determination.*[29] This simple (albeit good) definition is from a patent application for a visual O_2 indicator which was never commercialized. It stresses the key feature of intelligent packaging: an indicator of product history and quality.

3. *Intelligent packaging is packaging material able to monitor the conditions to which food is packaged, thus providing information on its quality.*[10] The use of the word "packaging material" limits the usefulness of this definition given that a TTI is not packaging material. This definition would be useful if the word "packaging" was used instead of "packaging material."

4. *Intelligent packaging acts as an intelligent messenger or as an information link.*[69] This broad definition sums up the essential attributes of intelligent packaging, but its broadness limits its use because it does not specify what sorts of messages or information are being transferred.

5. *Intelligent or smart packaging refers to packaging that senses and informs.*[12] This brief definition sums up the key features of intelligent packaging but, like the preceding definition, is of limited use because it does not specify what is being sensed.

6. *Intelligent packaging measures a component and signals the result.*[3] This brief definition is a restatement of the preceding definition and suffers from the same limitation.

TABLE 14.1
Selected Examples of Active Packaging Systems

Active Packaging System	Mechanisms	Food Applications
Oxygen scavengers	Iron-based, metal/acid, metal (e.g., platinum) catalyst, ascorbate/metallic salts, enzyme-based	Bread, cakes, cooked rice, biscuits, pizza, pasta, cheese, cured meats and fish, coffee, snack foods, dried foods and beverages
Carbon dioxide scavengers/emitters	Iron oxide/calcium hydroxide, ferrous carbonate/metal halide, calcium oxide/activated charcoal, ascorbate/sodium bicarbonate	Coffee, fresh meats and fish, nuts and other snack food products and sponge cakes
Ethylene scavengers	Potassium permanganate, activated carbon, activated clays/zeolites	Fruits and vegetables
Preservative releasers	Organic acids, silver zeolite, spice and herb extracts, BHA/BHT antioxidants, vitamin E antioxidant	Cereals, meats, fish, bread, cheese, snack foods, fruits and vegetables
Ethanol emitters	Encapsulated ethanol	Pizza crusts, cakes, bread, biscuits, fish and bakery products
Moisture absorbers	Poly (vinyl acetate) blanket, activated clays and minerals, silica gel	Fish, meats, poultry, snack, foods, cereals, dried foods, sandwiches, fruits and vegetables
Flavor/odor adsorbers	Cellulose triacetate, acetylated paper, citric acid, ferrous salt/ascorbate, activated carbon/clays/zeolites	Fruit juices, fried snack foods, fish, cereals, fruits, dairy products and poultry

BHA = butylated hydroxyanisole; BHT = butylated hydroxytoluene.
Source: From Day, B. P. F., Active packaging, In *Food Packaging Technology*, Coles, R., McDowell, D. and Kirwan, M., Eds., CRC Press, Boca Raton, FL, 2003, chap. 9. With permission.

7. *Intelligent, smart or clever packaging is defined as a packaging technique containing an external or internal indicator for the active product history and quality determination. These kinds of indicators could be called interactive, smart indicators because they are in active interaction with compounds originating from food.*[24] The first sentence is the same as Definition 2, while the second sentence would exclude TTIs, thermochromic inks and MDIs, which all respond to the temperature of the food or the package but not compounds originating from food. This definition also (correctly) excludes intelligent polymers—highly permeable membranes used to control the flow of gases through packages as temperatures change.

8. *Intelligent, smart or clever packaging is an advanced packaging system that contains attributes beyond the basic barrier properties but also contains additional functions. Simple intelligent packaging material contains a sensor, indicator or integrator that senses changes in the environment and consequently signals those changes. Interactive or responsive intelligent packaging, in addition to the above characteristics, contains a response mechanism that responds to the signal mechanism and starts to neutralize or undo the negative changes occurring in the food.*[38] This is a very comprehensive definition and the only example given by the authors of an interactive or responsive intelligent packaging application is the highly permeable membranes used to control the flow of gases through packages as temperatures change; this is not intelligent packaging because it does not signal changes in temperature.

In this book, *intelligent packaging* is defined as *packaging that contains an external or internal indicator to provide information about aspects of the history of the package and/or the*

TABLE 14.2
Definitions of Active and Intelligent Packaging

Active packaging. Packaging in which subsidiary constituents have been deliberately included in or on either the packaging material or the package headspace to enhance the performance of the package system

Intelligent packaging. Packaging that contains an external or internal indicator to provide information about aspects of the history of the package and/or the quality of the food

quality of the food (Table 14.2). This definition includes all the indicators (whether for gases, ripeness, temperature or tampering) as well as radio frequency data tags. The key word in the definition is "indicator."

In light of the above definition, it is possible to list packaging that is not intelligent packaging, although some authors have described it as such. As mentioned above, the so-called intelligent polymers do not qualify as intelligent packaging because they do not provide any information about the package or the food. Microwave susceptor packaging (see Section IV.B in Chapter 13) is also not intelligent packaging for the same reasons. Devices that have been used for many years, such as the button on metal closures attached to glass containers that have been retorted, and the tamper-evident band located on the skirt of screw closures attached to glass and plastic bottles, could qualify as intelligent packaging according to the above definition. However, only the newer and more sophisticated tamper-evident devices will be classified as intelligent packaging in this book. The pressure-relief valve found on packs of roasted coffee is not an intelligent package because it does not provide information; it simply relieves the pressure caused by a build-up of CO_2 inside the package.

III. ACTIVE PACKAGING SYSTEMS

Despite intensive research and development work, numerous patents, many conferences and countless publications on active packaging over the last 20 years, there are only a few commercially significant systems on the market. Of these, the O_2 absorbers contained in small sachets, which are added separately to the package headspace, are the most widely used, followed by ethanol emitters/generators, ethylene absorbers, moisture absorbers, off-odor absorbers, and CO_2 emitters and absorbers.

There are several ways in which active packaging systems can be classified; typically, the classification is based on what the system actually does (e.g., absorbs O_2) rather than its impact on the food (e.g., prevents oxidation). A similar classification will be adopted here. In addition, active packaging systems will be divided into two categories: those in which the active compounds are filled into sachets or pads, which are then placed inside packages, and those in which the active compounds are incorporated directly into or on the packaging materials.

A. Sachets and Pads

Sachets and pads can be a highly efficient form of active packaging, but they suffer from two major drawbacks: they cannot be used with liquid foods, and in packages made from or containing a flexible film, the film may cling to the sachet and isolate it from areas where it is needed to perform its function. To overcome the latter problem, the sachet can be glued to the inner wall of the package or the active ingredients incorporated into a label, which can be affixed to the inside wall of the package. Despite these drawbacks, sachets and pads are the most widely used forms of active packaging and the various functions which they perform are discussed below.

1. O_2 Absorbers

O_2 absorbers (also referred to as *scavengers*) use oxidation of either powdered iron or ascorbic acid, the former being more common. Powdered iron is used to provide a large reaction surface area, the reaction proceeding as follows[64]:

$$Fe \rightarrow Fe^{2+} + 2e^- \tag{14.1}$$

$$\tfrac{1}{2}O_2 + H_2O + 2e^- \rightarrow 2OH^- \tag{14.2}$$

$$Fe^{2+} + 2OH^- \rightarrow Fe(OH)_2 \tag{14.3}$$

$$Fe(OH)_2 + \tfrac{1}{4}O_2 + \tfrac{1}{2}H_2O \rightarrow Fe(OH)_3 \tag{14.4}$$

Overall:

$$Fe + \tfrac{3}{4}O_2 + 1\tfrac{1}{2}H_2O \rightarrow Fe(OH)_3 \tag{14.5}$$

By using iron powder, it is possible to reduce the O_2 concentration in the headspace to less than 0.01%, which is much lower that the typical 0.3 to 3.0% residual O_2 levels achievable by vacuum or gas flushing. In general, 1 g of iron can react with 0.0136 mol of O_2 (STP), which is equal to approximately 300 mL.[21] Various sizes of O_2 absorbers are available commercially with the ability to consume 20 to 2000 mL of O_2 (an air volume of 100 to 10,000 mL), and several interrelated factors influence the choice of the type and size of absorbent required.[49] These include:

- Nature of the food (i.e., size, shape, weight)
- a_w of the food
- Amount of dissolved O_2 in the food
- Desired shelf life of the product
- Initial O_2 level in the package headspace
- O_2 permeability of the packaging material

The last factor is critically important for the overall performance of the absorbent and the product shelf life, and if a long shelf life is desired, films containing PVdC or EVOH copolymer as a barrier layer are necessary. Such films have an O_2 permeability of $<0.004 \times 10^{-11}$ [mL(STP) cm cm^{-2} s^{-1} (cm Hg)$^{-1}$] and the headspace O_2 should be reduced to 100 ppm within 1 to 2 days and remain at that level for the duration of the storage period, provided packaging integrity is maintained.[49]

The most well known O_2 scavengers take the form of small sachets containing various iron-based powders, together with an assortment of catalysts that scavenge O_2 within the food package and irreversibly convert it to a stable oxide. Water is essential for O_2 absorbents to function (see Equation 14.2 above) and, in some sachets, the water required is added during manufacture while, in others, moisture must be absorbed from the food before O_2 can be absorbed. The iron powder is separated from the food by keeping it in a small sachet (labeled *do not eat*) that is highly permeable to O_2 and, in some cases, to water vapor.

The absorption kinetics of six O_2 scavengers has been reported.[60] It was found that the rates of O_2 absorption varied by factors of up to two between individual O_2 scavengers of the same type. To obtain consistent and reproducible results, it was recommended that multiple scavengers be used in a packaging system, which is clearly not a viable commercial option.

One disadvantage of iron-based scavengers is that they normally cannot pass the metal detectors often installed on packaging lines. Nonmetallic O_2 scavengers include those that use organic reducing agents such as ascorbic acid, ascorbate salts or catechol. They also include enzymic O_2 scavenger systems using either glucose oxidase or ethanol oxidase, which can be incorporated into sachets, adhesive labels or immobilized onto package surfaces. However, their use is not widespread.

O_2 absorbers were first marketed in Japan in 1977 under the trade name Ageless™, but were not adopted by North America and Europe until the 1980s and then only slowly. It has been suggested[13] that the reasons for the success of O_2 absorbers in Japan include the acceptance by Japanese consumers of innovative packaging and the hot and humid climate in Japan during the summer months, which is conducive to mold spoilage of food products. However, the first reason is more likely because O_2 absorbers are not found in many other countries with similar climatic conditions to Japan.

The possible accidental ingestion of the sachet contents by the consumer has been suggested as a reason for their limited commercial success, particularly in North America and Europe. However, ingestion does not result in adverse health impacts because a sachet typically contains 7 g of iron, which is approximately 160 times less than the LD_{50} for a 70 kg adult.[16] The recent development of O_2-absorbing adhesive labels that can be attached to the inside of packages has helped overcome this perceived problem and has aided the commercial acceptance of this technology, although their O_2-absorbing capacity is limited to 100 mL.[49]

O_2 absorbers have been used for a range of foods including sliced, cooked and cured meat and poultry products, coffee, pizzas, specialty bakery goods, dried food ingredients, cakes, breads, biscuits, croissants, fresh pastas, cured fish, tea, powdered milk, dried egg, spices, herbs, confectionery and snack food.[13]

2. CO_2 Absorbers/Emitters

Sachets that absorb only CO_2 are rare. More common are absorbent sachets that contain $Ca(OH)_2$ in addition to iron powder and, as a result, absorb CO_2 as well as O_2, which have found a niche application inside packages of roasted or ground coffee. Fresh roasted coffee releases considerable amounts of CO_2 (formed by the Maillard reaction during roasting) and, unless it is removed, it can cause swelling or even bursting of the package.

There are other sachets available based on either ascorbic acid and ferrous carbonate or ascorbic acid with sodium bicarbonate that absorb O_2 and generate an equivalent volume of CO_2, thus avoiding package collapse or the development of a partial vacuum, which can be a problem in flexible packages if only O_2 is removed.

3. Ethylene Adsorbers

The plant hormone ethylene (C_2H_4) is produced during the ripening of fruits and vegetables and can have both positive and negative effects on fresh produce. Positive effects include catalyzing the ripening process, while negative effects include increasing the respiration rate (which leads to softening of fruit tissue and accelerated senescence), degrading chlorophyll and promoting a number of postharvest disorders.

Many C_2H_4-adsorbing substances have been described in the patent literature, but those that have been commercialized are based on potassium permanganate ($KMnO_4$), which oxidizes C_2H_4 in a series of reactions to acetaldehyde and then acetic acid, which, in turn, can be further oxidized to CO_2 and H_2O[64]:

$$3C_2H_4 + 2KMnO_4 + H_2O \rightarrow 2MnO_2 + 3CH_3CHO + 2KOH \qquad (14.6)$$

$$3CH_3CHO + 2KMnO_4 + H_2O \rightarrow 3CH_3COOH + 2MnO_2 + 2KOH \qquad (14.7)$$

$$3CH_3COOH + 8KMnO_4 \rightarrow 6CO_2 + 8MnO_2 + 2H_2O \qquad (14.8)$$

Overall:

$$3C_2H_4 + 12KMnO_4 \rightarrow 12MnO_2 + 12KOH + 6CO_2 \qquad (14.9)$$

Because $KMnO_4$ is toxic, it cannot be integrated into food contact packaging. Instead, about 4 to 6% of $KMnO_4$ is added to an inert substrate with a large surface area such as perlite, alumina, silica gel, vermiculite, activated carbon or celite and placed inside a sachet, which can be safely added to packages.[70]

Another type of C_2H_4 adsorber available in Japan is based on a palladium catalyst on activated carbon, which adsorbs C_2H_4 and catalytically breaks it down.[70] The product comes in woven sachets that can be placed in packages of produce, but given the high cost of palladium, it has not found widespread use.

4. Ethanol Emitters

Ethanol (ethyl alcohol) has been used as an antimicrobial agent for centuries, Arabs having used it over 1000 years ago to prevent mold spoilage of fruit.[49] Ethanol exhibits antimicrobial effects even at low concentrations, and a novel and innovative method of generating ethanol vapor has been developed in Japan using sachets. The sachets contain ethanol (55%) and water (10%), which are adsorbed onto SiO_2 powder (35%) and filled into a paper–EVA copolymer sachet. To mask the odor of alcohol, some sachets contain traces of vanilla or other flavors. The sachet contents absorb moisture from the food and release ethanol vapor, so the a_w of the food is an important factor in the vaporization of ethanol into, and the absorption of ethanol from, the package headspace.[49]

Ethanol emitters are used mainly in Japan to extend the shelf life of high-moisture bakery products by up to 20 times. The main disadvantages of using ethanol vapors (apart from the cost) are the formation of off-flavors and off-odors in the food, and absorption of ethanol from the headspace by the food, which can lead to ethanol concentrations of up to 2% in the food and result in regulatory problems. This is not a problem if the product is heated in an oven prior to consumption because the ethanol will evaporate.[24]

5. Moisture/Water Absorbers

Liquid water can accumulate in packages as a result of temperature fluctuations in high-moisture packages, drip of tissue fluid from flesh foods and transpiration of horticultural produce. If this water is allowed to build up in the package, then it can lead to the growth of molds and bacteria as well as fogging of films.

Drip-absorbent pads (consisting of granules of a superabsorbent polymer sandwiched between two layers of a microporous or nonwoven polymer, which is sealed at the edges) have been used in the packaging of flesh foods to absorb liquid water. The polymers most frequently used for absorbing water are polyacrylate salts and graft copolymers of starch; such polymers are capable of absorbing 100 to 500 times their own weight of liquid water.[24]

B. Active Packaging Materials

1. O_2 Absorbing Materials

The most common O_2 absorbers used in the food industry are the porous sachets containing iron powder discussed above. However, there is still consumer resistance to sachets in many markets and they cannot be used in liquid foods. Furthermore, in close-fitting packages such as vacuum packs for cheese and meats where O_2 permeation is a prime cause of quality loss, sachets cannot be used. Therefore, a more attractive alternative is the incorporation of O_2-absorbing materials into the plastics components of the packaging material, and this has grown from a research curiosity to a small but rapidly developing commercial field over the last 25 years.[41]

O_2-absorbing polymers are versatile in that their layer thickness or blend composition can be varied to match the amount of O_2 to be removed. However, one major limitation to the large-scale adoption of the presently available O_2-absorbing materials is that the speed and capacity of O_2-absorbing films are considerably lower than the widely used O_2-absorbing sachets. An additional challenge is that the O_2-absorbing films must be stable in the O_2-rich environment of air and not start consuming O_2 until the food is packaged. Activation or triggering mechanisms based on exposure to light or elevated moisture have been the main approaches adopted, although the storage of O_2-absorbing PET preforms in barrier bags with a slight overpressure of N_2 to prevent absorption of O_2 until blown is also used.

The patent literature of the past 60 years contains many ideas for O_2-absorbing systems that can be incorporated into package structures, and various films with the required reactive ingredients dispersed within the polymer matrix or sandwiched between film layers have been trialed commercially. However, despite the enthusiastic write-ups in trade magazines and exciting presentations at conferences, very few have been successfully commercialized.

The first O_2-absorbing polymer that was useful as a blend with PET involved the cobalt-catalyzed oxidation of MXD6 polyamide. When used with 200 ppm of cobalt as the stearate salt, this polyblend allowed blowing of bottles with an O_2 permeability of essentially zero for 1 year.[41] More cost-effective ways of providing an O_2 barrier in PET bottles have limited the application of this approach. It should be noted that any O_2 absorber proposed as a middle layer in PET-based structures will not be a rapid headspace scavenger owing to the reasonable O_2 barrier provided by the inner PET layer.

In one approach currently being independently commercialized in both Australia and the U.S., a polymer-based absorber is coextruded in various packaging structures including bottles, films, coatings, sheet, adhesives, lacquers, can coatings and closure liners where it acts as both a headspace O_2 absorber and a barrier to O_2 permeation into the package. The O_2-absorbing capability is UV-activated, meaning it must be exposed to UV light before it can begin absorbing O_2. Another O_2-absorbing material is a copolymer that can serve as a clear layer in a bottle, jar or other rigid polyester container. A converter would laminate or extrude the material as a middle layer in a multilayer structure, which removes O_2 from within the package headspace, as well as any that enters the package by permeation or leakage.

In another approach developed to remove O_2 from beverages such as beer after they are capped, a multilayer barrier liner is fitted into crown corks, plastic and metal closures. The active ingredients in one liner consist of ascorbic acid, which is oxidized to dehydroascorbic acid while, in another, sodium sulfite is oxidized to sodium sulfate. Other proprietary systems are also available.

In Japan, a laminate has been used commercially for cooked rice and fruit jellies; it contains an iron-based O_2 absorber that can be thermoformed into a tray. Because it is activated by water, it is only applicable to foods with a high a_w but second and third generation versions that can react at lower a_ws are being developed.

An enzyme-based approach has been commercialized in Finland[24] involving glucose oxidase and catalase. The enzymes can be easily applied to the surface of polyolefins, which are very good substrates for immobilizing enzymes.

2. Ethylene Adsorbers

In the 1980s, a number of packaging films were released commercially based on the reputed ability of certain finely dispersed minerals to adsorb C_2H_4. These minerals were typically local kinds of clay such as pumice, zeolite, cristobalite (SiO_2) or clinoptilolite (hydrated NaKCaAl silicate) that were sintered together with a small amount of metal oxide before being dispersed in a plastic film.[70] The resulting films are translucent and have increased permeability to gases, which may, by itself, (regardless of any adsorption of C_2H_4) increase the shelf life of fresh fruits and vegetables. Although the minerals may have C_2H_4-adsorbing capacity, this capacity is often lacking after they have been incorporated into plastic films. There have been no peer-reviewed publications demonstrating the efficacy of these films in adsorbing C_2H_4 and extending shelf life of fruits and vegetables.

3. Flavor/Odor Absorbers

The sorption of flavors by polymeric packaging materials is known as scalping and can result in the loss of important flavors, the most researched area being fruit juices in contact with polyolefins. Although scalping is detrimental to food quality, it has been used in a positive way to selectively absorb unwanted odors or flavors.[64]

Volatile amines are a common taint formed by the breakdown of protein in fish muscle; they can be neutralized by various acidic compounds. Commercial bags made from a film containing a ferrous salt and an organic acid such as citric or ascorbic are available in Japan and are claimed to oxidize amines and other oxidizable odor-causing compounds as they are absorbed by the polymer.[40]

Odors that result from aldehydes such as hexanal and heptanal, which are formed from the breakdown of peroxides created during the initial stages of auto-oxidation of fats and oils, can be removed from package headspaces by active packaging. The technology is based on a molecular sieve with pore sizes of around 5 nm, and applications for snack foods, cereals, dairy products, poultry and fish have been claimed. Synthetic aluminosilicate zeolites, which have a highly porous structure, have been incorporated into packaging materials to adsorb odorous aldehydes.[13]

Removal of the bitter compounds limonin and naringin from citrus juices has also been demonstrated using a plastic bottle coated internally with cellulose acetate–butyrate to absorb limonin, and immobilization of naringinase in cellulose triacetate film to hydrolyze naringin. Neither process has been commercialized.

4. Antioxidant Release

Antioxidants can be incorporated in plastic films (particularly polyolefins) to stabilize the polymer and protect it from oxidative degradation. The potential for evaporative migration of antioxidants from plastics packaging into foods has been known for many years. The challenge lies in matching the rate of diffusion with the needs of the food. In the U.S., release of BHA and BHT from the inner plastic liner has been applied to breakfast cereals and snack products.[15] Recently, there has been interest in the use of vitamin E as a replacement for BHA and BHT.[13]

5. Antimicrobial Release

Antimicrobial agents incorporated into packaging materials could be used to prevent the growth of micro-organisms on the food surface and thus lead to an extension in shelf life or improved microbial safety of the food.[9,54] The burgeoning interest in antimicrobial food packaging is driven by increasing consumer demand for minimally processed, preservative-free foods. The use of antimicrobial films ensures that only low levels of preservative come into contact with the food compared to the direct addition of preservatives to the food.[63]

There are two mechanisms for the antimicrobial action of packaging films, which may have the antimicrobial agent either incorporated within the film or coated on the surface.[63] In one, the antimicrobial agent migrates partly or completely into the food or in the space surrounding the food and exercises its preservative action. In the other nonmigrating mechanism, the antimicrobial agent acts when the target micro-organisms come into contact with the surface of the film. Regardless of the mechanism, there must be physical contact between the packaging material and the food surface, such as that which occurs in vacuum packaged or shrink-wrapped foods. Another possibility is to incorporate the antimicrobial agent into an edible film or coating which can be applied by dipping or spraying onto the food.[6] In food contact applications, the substances incorporated into the film should be safe and slow to migrate.

Antimicrobial packaging materials have to extend the lag period and reduce the growth rate of micro-organisms to prolong shelf life and maintain food safety.[19] Substances that have been tested or proposed as antimicrobials include ethanol (discussed above) and other alcohols, organic acids and their salts (such as benzoates, propionates and sorbates[61]), fungicides (such as imazalil and benomyl), hexamethylenetetramine, enzymes (such as glucose oxidase and lysozyme), extracts from spices and herbs, SO_2 and ClO_2 and bacteriocins.

A *bacteriocin* is defined as a proteinaceous compound (usually a peptide) that has bactericidal action against a limited range of organisms, which are usually closely related to the producer organism.[7] Bacteriocins are commonly found in foods owing to their production by the lactic acid bacteria used to ferment dairy, vegetable and meat products. Nisin from *Lactococcus lactis* was one of the earliest bacteriocins to be described and is the only one that has been recognized as a safe biological food preservative: by FAO/WHO commission on food additives in 1968 and 20 years later by the FDA as GRAS. It is commercially exploited, particularly in processed cheeses and cold-pack cheese spreads, as it is effective against out-growth of and toxin production by *C. botulinum*.

Despite the large number of experimental studies on antimicrobials in packaging materials, there have been few commercial applications. This is because the inhibitory activity is lost when the antimicrobial is combined with the polymeric material resulting from incompatibility or heat lability during extrusion.[63] The legislative status of antimicrobials is also a limiting factor in their commercialization.

Most commercial antimicrobial films have been introduced in Japan. One is a synthetic zeolite that has had a portion of its sodium ions replaced with silver ions — silver combining a high antimicrobial activity with a very low human toxicity. The Ag-zeolite is extrusion coated as a thin (3 to 6 μm) layer directly onto a food contact film at levels of 1 to 3% w/w. It continuously releases a small quantity (ca. 10 ppb) of silver ions, resulting in long-term, broad spectrum, antimicrobial activity that is not harmful to tissue cells.[34] Because amino acids can react with silver ions, they are relatively ineffective in nutrient-rich foods, but highly effective in nutrient-poor drinks such as water or tea.[27]

Edible films and coatings have also been studied as carriers for antimicrobial agents such as benzoic acid, sorbic acid, propionic acid, lactic acid, nisin and lysozyme, and a recent review[6] should be consulted for further details.

6. Microwave Susceptors

Packaging materials that absorb microwave energy and convert it to heat are called *susceptors* and are described in Section IV.B of Chapter 13. They qualify as active packaging because they enhance the performance of the package by achieving localized effects such as browning and crisping of the food.

C. Self-Heating and Self-Cooling Packages

The concept of a self-heating container is not new, although earlier versions were not without their hazards. The U.K. armed services introduced a self-heating can in 1939, which relied on the burning of cordite (a smokeless propellant consisting of 65% gun cotton, 30% nitroglycerin and 5% mineral jelly) to provide the thermal energy, a design that could hardly be considered safe in untrained hands. More recent designs have all relied on an exothermic chemical reaction to generate heat; in most cases; the reaction is between CaO (quicklime) and a water-based solution, although armed forces tend to use the more expensive MgO because it heats more quickly. Although the fundamental chemistry is well known, the difficult part is optimizing the reaction and the thermal design of the container to provide an efficient, safe and cost-effective package.

Self-heating cans have been commercially available for decades and are particularly popular in Japan for sake, coffee, tea and ready meals. Recently, several self-heating cans have been developed, which all use the same basic heating mechanism but in a more convenient and efficient form. One is a three-piece, welded, retortable, steel container coated on both sides with PET and with a conventional ring-pull top. The can has an internal volume of 330 mL, but to make room for the heat-generating chamber, the can holds just 210 mL of beverage. An unusual steel bottom, drawn to a depth of 70 mm, is seamed on to become the chamber that holds the heat-generating components, which are activated by pushing in a button on the base of the can. This produces a chemical reaction between water and CaO contained in separate compartments in the can base. Once the reaction is complete, the can is shaken, turned right side up and left for 3 minutes while the liquid heats up to 60°C. The can is insulated so consumers do not burn their fingers, and a PP lip protector is applied to the top of the can to protect the consumer's lips from getting burned while drinking. Whether such relatively expensive cans will be commercially successful has not yet been determined.

Not all self-heating packages are cans. Retortable plastic trays based on a six-layer PP/EVOH copolymer structure with an under layer containing CaO have also been commercialized. In another approach using an electrochemical principle,[30] the flameless ration heater (FRH) was developed for the U.S. armed forces to heat Meals, Ready-To-Eat (MREs) for soldiers in the field. The FRH is based on the reaction between Mg and H_2O:

$$Mg + 2H_2O \rightarrow Mg(OH)_2 + H_2 + \text{heat} \quad (14.10)$$

Theoretically, 24 g of Mg releases 355 kJ of heat, sufficient to boil a liter of water. In practice, Mg has a protective oxide surface coating that prevents further oxidation, as well as the reaction shown in Equation 14.10. By mixing the Mg with NaCl and iron, the reaction proceeds, although the role that iron plays is unclear. The food is retorted in an alufoil/plastic pouch and packed in a cardboard carton. To heat the pouch, it is placed inside a plastic sleeve containing an FRH in which the chemicals are contained in a perforated fiberboard box. After water is added, the sleeve is placed inside the carton and as the reaction proceeds, the temperature of a 227-g food pouch increases by up to 55°C in 12 minutes.

Self-cooling cans have long been commercialized in Japan, utilizing an endothermic reaction based on the dissolution of ammonium nitrate and ammonium chloride in water. In the U.K., a self-chilling beverage can uses the latent heat of evaporating water to produce the cooling effect. The can consists of two parts: one is a standard aluminum can that contains an *evaporator*, which has a layer of gel sealed inside a vacuum on its inner surface. The other part, attached underneath the can, is an *absorber*, so-called because it contains a clay desiccant under vacuum that draws heat from the beverage through the evaporator into an insulated heat-sink container. Twisting the two parts of the can against each other breaks a seal and opens a connecting path between the gel and the desiccant. Water vapor then flows from the gel to the desiccant where it is condensed and absorbed. The unit has been designed to meet a target specification set by major beverage customers of cooling 300 mL of beverage in a 355-mL can by 17°C in 3 minutes, the unit continuing to extract heat until the drink reaches a temperature just above 0°C. The innovative technology of the self-chilling can will be a challenge to mass produce because it is one of the most complex types of packaging ever created.

D. Changing Gas Permeability Properties

Fresh fruits and vegetables when harvested consume O_2 and emit CO_2. When the fruits and vegetables are in a sealed package, the atmosphere will reach equilibrium levels of O_2 and CO_2, depending on the weight of the produce, its respiration rate and the permeability of the package. A specific beneficial atmosphere exists for each fruit and vegetable that helps preserve the quality and freshness of the produce with good temperature control. If the temperature increases above the chill range, then the O_2 consumption may increase beyond the rate at which O_2 can permeate through the packaging film and high levels of CO_2 will accumulate inside the package. This is because the respiration rates of fruits and vegetables increase more with temperature than do gas permeabilities of films. Therefore, higher than optimum temperatures can lead to anoxia and seriously damage the produce. As discussed in Chapter 17 on the MAP of horticultural products, none of the commercially available common polymer films has the requisite gas permeability required in this situation. Therefore, when packaging fresh produce, there is a need to provide greater package permeabilities and different selectivities to O_2 and CO_2 so as to maintain the desired atmospheres as the temperature changes.

This need has been met using Intellipac® technology which is based on an unusual "side chain crystallizable" (SCC) polymer having an internal temperature switch.[2,8] When elevated to the switch temperature, SCC polymers become molten fluids, which have inherently high gas-permeability coefficients. SCC polymers are unique because of their sharp melting transition and the ease with which it is possible to produce melting points in a specific temperature range. In SCC polymers, the side chain crystallizes independently from the main chain. Examples of such polymers are siloxanes or acrylic polymers in which the side chain has eight or more carbon atoms. Varying the chain length of the side chain can change the melting point of the polymer. Preparation of the acrylic polymers occurs in a solution using conventional free radical initiators. By making the appropriate copolymers, it is possible to produce any melting point from 0 to 68°C.

The SCC polymers are intrinsically highly permeable, but the polymer properties can also be modified by the inclusion of other monomers in order to change the relative permeability of CO_2 to O_2, to alter the temperature switch or other physical properties such as water vapor transmission rate.[8] The polymers are applied as a coating to a porous substrate, which is then cut into small patches and applied to a bag by a hot-stamping system incorporated into the bag-making process. The patch covers several holes that are cut into the bag material through which gases enter and exit.

The use of these highly permeable membranes essentially controls the flow of gases into and from the package. Altering the properties of the polymer provides specific O_2 permeabilities, specific CO_2:O_2 permeability ratios and changes in permeability with temperature.

The membrane is capable of being more than 1000 times more permeable to O_2 and CO_2 than a 50 μm LDPE film while still maintaining the same CO_2:O_2 permeability ratio. Unlike semipermeable films for respiring produce, which typically allow CO_2 out and O_2 in at a 6:1 ratio, by altering the polymer composition used to coat the membrane, it is possible to obtain CO_2:O_2 ratios from 2:1 to 18:1, depending on the requirements of the produce in question. The gases enter and exit at a predetermined ratio that maintains an optimal atmosphere. At elevated temperatures, when respiring produce needs more O_2, the polymer becomes more permeable, but at lower temperatures, permeability automatically decreases. In addition to the common gases, the membrane is also highly permeable to volatiles such as ethanol and ethyl acetate, which are generated from anaerobic respiration when produce is packaged in suboptimal atmospheres.

This new polymer technology is not intended as a substitute for good temperature control but rather as a solution to temporary and unexpected breaks in the cold chain during distribution or display. Despite their relatively high cost, these materials were commercialized during the 1990s, and membranes coated with SCC polymers are currently being employed for many fresh produce items including mixed cut vegetables, cut broccoli, cauliflower, asparagus, bananas and strawberries.

E. WIDGETS

A somewhat unusual example of active packaging is the highly successful foam-producing *widget*, originally developed for stout beer packaged in metal cans. A widget is generally defined as a device that is very useful for a particular job, and its synonyms are *gadget* and *gizmo*. Many British and Irish drinkers like their beer to have a low dissolved CO_2 content when poured (i.e., not gassy) but to have a good head, which results from large numbers of gas bubbles produced when pouring. With draught beer, this is easy to achieve by passing the beer through a venturi nozzle to introduce lots of bubbles when pouring. However, canned beer needs to have a higher internal CO_2 pressure to ensure that the cans have sufficient strength for stacking and also to provide some form of head when poured. The challenge was to develop a way of getting the dissolved gas out of solution rapidly when the can was opened to give the creamy head and low gas content preferred by consumers.

In 1985, a system was developed and patented[17] to release more of the dissolved CO_2 when the can is opened to produce the traditional creamy head. The widget itself is a small, plastic (or, in one case, aluminum), N_2-filled sphere with a tiny hole in it. The widget is inserted into the base of the can before filling; newer designs float in the beer and are oriented such that the hole is slightly below the level of the beer.[37]

At filling, a small quantity of liquid N_2 is injected with the beer, and the can is rapidly seamed before the N_2 evaporates. After the can is seamed, this liquid N_2 vaporizes and pressurizes the can, forcing a small quantity of beer through the tiny hole of the widget and compressing the N_2 inside. When the can is opened, the pressure inside drops suddenly, the compressed N_2 inside the widget expands and beer is pushed out through the tiny hole, causing the widget to spin and jet the gas and beer down through the can. This expelled beer nucleates a host of very small bubbles from the N_2/CO_2 supersaturated beer, causing the bubbles to rise to the surface and form the desired smooth, creamy head.

Other brewers have since come up with their own widget designs, which were introduced to draught beer in 1992, lager in 1994 and cider in 1997. In 2002, canned milk coffee containing a widget was marketed.

IV. INTELLIGENT PACKAGING

Intelligent packaging is defined as *packaging that contains an external or internal indicator to provide information about aspects of the history of the package or the quality of the food.*

Three types of intelligent packaging systems[38] are used to:

1. Improve product quality and product value, for example, quality indicators, temperature and TTIs, and gas concentration indicators
2. Provide more convenience, for example, quality, distribution and preparation/cooking methods
3. Provide protection against theft, counterfeiting and tampering

Each of these types is now discussed in more detail.

A. Improving Product Quality and Product Value

1. Quality Indicators

In this application of intelligent packaging, quality or freshness indicators are used to indicate if the quality of the product has become unacceptable during storage, transport, retailing and in consumers' homes. Intelligent indicators undergo a color change that remains permanent and is easy to read and interpret by consumers.[51]

Despite many attempts and several innovative approaches, no quality indicators are currently in widespread use by the food industry. However, two recent indicators that appear to have real commercial possibilities are described below. An indicator label that reacted with volatile amines from fish and changed color accordingly to indicate freshness had a short commercial life before being withdrawn in 2003.

a. *Kimchi Freshness Indicator*

Kimchi — a group of traditional fermented vegetable foods widely consumed as a side dish at every meal in Korea — is a product of natural mixed fermentation owing principally to lactic acid bacteria. It also contains various nutrients and functional components. Flavor, taste and texture are optimal when kimchi is properly fermented to pH 4.2 and titratable acidity of 0.6 to 0.8%. After optimum fermentation, kimchi quality deteriorates from formation of excessive organic acids and loss of texture. Changes in CO_2 concentration correlate highly with pH and titratable acidity.

A color indicator has been developed to monitor the ripeness of commercial kimchi products during distribution and sale.[22] The indicator ingredients consist of $Ca(OH)_2$ as a CO_2 absorbent and bromocresol purple or methyl red as a chemical dye. Using a mixture of polyurethane and PET polymers dissolved in organic solvents as a binding medium, the ingredients have been gravure printed onto polyamide films and then laminated with LDPE films to form an indicator. Color changes in both types of indicator correlated well with titratable acidity values of kimchi and commercialization is likely.

b. RipeSense

The RipeSense™ sensor, recently developed by scientists in New Zealand, enables consumers to choose fruit that best appeals to their taste. It works by detecting aroma compounds given off by the fruit as it ripens, changing a label on the package through a range of colors from red (firm) through orange to yellow (juicy). There is a good correlation between the amount of aroma that is produced and the actual softening of the fruit, so as the fruit softens, it produces more aroma and the sensor changes color. It has initially been trialed on pears, a fruit whose ripeness consumers have great difficulty in assessing.

The label is attached to the inside of a four-piece PET clamshell punnet with a tamper-evident seal. The package also has the advantage of protecting the fruit from the damage that often occurs as consumers handle produce prior to making their selection. Although the sensor has been developed only for pears, scientists hope to develop similar sensor labels for stone fruit, kiwifruit, avocado and melons, fruits that do not significantly change their skin color as they ripen.

2. Time–Temperature Indicators

TTIs are devices that integrate the exposure to temperature over time by accumulating the effect of such exposures and exhibiting a change of color (or other physical characteristic). Many devices that can be attached to food packages to integrate the time and temperature to which the package is exposed have been developed, and the patent literature contains designs of more than 200 such devices. Initially, the majority of these devices were developed specifically for frozen foods, but there is now widespread interest in shelf life devices for most categories of food, especially those where the rate of quality deterioration is highly temperature sensitive. Overviews of the major types of TTIs have been presented[46,56] together with their application to food quality monitoring.[66,67]

TTIs can be divided into two categories: *partial history indicators*, which do not respond unless some predetermined threshold temperature is exceeded, and *full history indicators*, which respond continuously to all temperatures. Partial history indicators are intended to identify abusive temperature conditions, and thus there is no direct correlation between food quality change and the response of this class of indicator. The difference in the way the full and partial history indicators respond to the same temperature history is shown in Figure 14.1.

The ASTM F 1416-96 (2003) *Standard Guide for Selection of Time–Temperature Indicators* covers information on the selection of commercially available TTIs for noninvasive, external package use on perishable products, such as food and pharmaceuticals. The guide stresses that it is the responsibility of the processor of the perishable product to determine the shelf life of a product at the appropriate temperatures, and to consult with the indicator manufacturer to select the available indicator that most closely matches the quality of the product as a function of time and temperature.

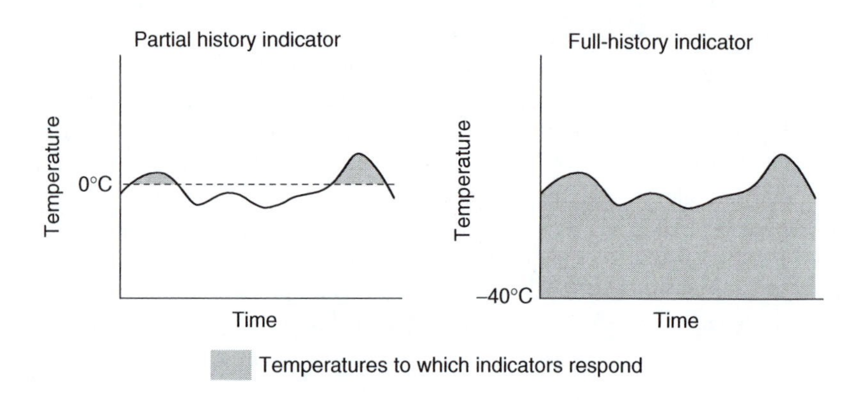

FIGURE 14.1 Types of time–temperature indicators.

Three of the most popular TTIs for which a large body of peer-reviewed scientific research has been published are discussed below.

a. *VITSAB*

The VITSAB™ (Visual Indicator Tag Systems AB) TTM (time–temperature monitor) was previously known as I-POINT™ and is a full history indicator consisting of an inner transparent plastic pouch with two compartments and an outer rectangular casing (62 × 25 mm) clear on one side with an adhesive backing on the other.[66] One compartment of the inner pouch contains a proprietary lipase enzyme and pH indicator dye and the other contains a lipase substrate (glycerol trihexanoate) in fluid suspension. The indicator is activated when the barrier separating the two compartments is broken by external pressure, and the enzyme initiates hydrolysis of the substrate. As the hydrolysis reaction proceeds, the solution pH irreversibly changes and, in turn, the pH indicator dye gradually changes color. The color change is compared visually to a reference color scale printed on a paper overlay, which partially covers the otherwise transparent packet. The VITSAB system can also be used in automatic optical reading systems where the results can be registered by a computer.

There are two basic types of VITSAB TTMs: the *master carton version* and the *consumer version*. The former, which is designated as an early warning indicator, is applied to the carton or pallet in the factory and activated by the pressure of the labeling machine. The color changes gradually from green to red, and four categories of quality are designated: green for excellent where 80% or less of the product's time–temperature tolerance (TTT) is used up; yellow for good when 80% is used up; brown or uncertain when 100% is used up; and red or overexposed when 130% or more is used up.

The consumer TTM is designed to be placed on individual consumer packages and consists of a single ampoule, which is also activated at the time of packaging. Its function is simpler in that only two color signals (green and yellow) are used to designate "fit for consumption" or "not fit for consumption." Such models are referred to as *go/no go* models.

b. *TEMPTIME*

The TEMPTIME™ (formerly LifeLines™) Fresh-Scan® labels provide a full history TTI, showing a response independently of a temperature threshold. The indicator is an adhesive-backed paper label (102 × 37 mm) consisting of three distinctive regions: an eight-digit number unique to each indicator; a two-digit code that identifies the indicator model; and a strip of material known as the indication band that changes color as a result of accumulated temperature exposure. The response of the TEMPTIME indicator can be monitored only with a hand-held microcomputer and an optical wand supplied by the manufacturer, together with software for data analysis.

The indication band contains diacetylene monomers (R–C≡C–C≡C–R), which appear colorless because they absorb light only in the UV portion of the spectrum. They undergo a time–temperature dependent polymerization to form a polymer with a conjugated backbone on which electrons are delocalized. These delocalized electrons absorb light in the visible portion of the spectrum and the polymer appears colored. A change of the side group R causes a dramatic change in the solid-state reaction kinetics, and these compounds have been found to follow Arrhenius-type kinetics over a wide temperature range. A variety of reaction schemes are possible with acetylenic materials.

A range of standard indicator labels is available that can be used to monitor products in ambient, chilled and frozen storage. The color change and bar code are monitored using a specially programmed, hand-held microcomputer with an optical wand, which records the decrease in reflectance as the indication band darkens. The labels contain standard bar codes alongside the indicator panel to enable product identification. The remaining product shelf life can be calculated from the change in color, based on prior product time–temperature experience which is programmed into the computer. The indicators have no means for in-field activation, and are shipped from the manufacturer already activated and responding to storage temperature. To minimize indicator response prior to use, they are stored at −24°C.

The TEMPTIME Fresh-Check® indicator has been developed for the consumer and consists of a small circle of polymer surrounded by a printed reference ring. The polymer gradually deepens in color to reflect cumulative temperature exposure. Consumers are advised on the package not to consume the product if the polymer center is darker than the reference ring, regardless of the use-by date. Red and yellow films containing a system of dyes cover the indicators and form an effective filter for preventing certain wavelengths of light from reaching the color-changing diacetylene coatings. This enables the use of the Fresh-Check TTI on consumer packages exposed to ambient lighting including retail displays. If not covered with one of these dyed films, then the TTI could darken due to light exposure, thus reducing the accuracy of its time–temperature monitoring.

c. MonitorMark

The 3M MonitorMark™ TTI is a partial history indicator consisting of five parts layered from the bottom as follows[66]: an 88 × 19 mm cardboard rectangle; a 28 × 12 mm pad containing a blue dye within a carrier substance; a plastic slip-tab for isolating the dye; a 7 mm blotter paper wick; and another 88 × 19 mm cardboard rectangle with five window cut-outs. The bottom piece has a pressure-sensitive adhesive backing, and the remaining surfaces are encased in transparent plastic film. Removal of the slip-tab brings the pad and wick into contact, the dye remaining within the pad until the carrier substance undergoes a phase change owing to temperature exposure above the response temperature. Typical esters used as carrier substances include octyl octanoate [melting point (m.p.) −17°C], dimethyl phthalate (m.p. −1.1°C) and butyl stearate (m.p. 12°C). Indicator response is measured by reading the distance the dye front has migrated past the windows in the indicator.

The indicator is designed as an abuse indicator, yielding no response unless a predetermined temperature is exceeded. It has a scale to indicate the length of accumulated exposure time above a predetermined temperature, and can be activated by pulling out a tab. Actual run-out times are cumulative, not necessarily continuous, and indicate how many hours the product has been exposed above the tag's stated temperature response (see Figure 14.1). If the temperature falls below the designated temperature, then the run-out time stops.

A range of tags is available depending on the particular abuse temperature for a specific product. Recommended storage conditions for the MonitorMark are 22°C and 20 to 60% relative humidity where shelf life is 2 years from date of manufacture. However, for ease of use, they can be stored below the stopping temperature but above −40°C. Response temperatures can vary from −17 to 48°C.

d. Applications

In developing an application of TTIs to shelf life monitoring, it is necessary to be able to define and measure quantitatively the key deteriorative reactions that contribute to quality loss in the particular food, and to integrate the expression for quality loss with the TTI response.[36] Paramount to the success of any TTI is recognition of the fact that unless it has the same or similar activation energy (E_A) as the quality-determining deteriorative reaction in the food it is monitoring, then it will overestimate or underestimate the loss of shelf life.[48]

Simple mathematical models for characterizing the three TTIs described above have been developed.[57,58] Results showed that the MonitorMark had an activation energy (E_A) of 41 kJ mol^{-1}, the TEMPTIME device about 86 kJ mol^{-1} and the VITSAB about 141 kJ mol^{-1}. Apparent E_As of spoilage indices for frozen foods in the temperature range of −12 to −23°C varied from 44 to 49 kJ mol^{-1} for fish, 46 to 64 kJ mol^{-1} for vegetables, 48 to 61 kJ mol^{-1} for fruit and 54 to 74 kJ mol^{-1} for meat. More recently, the E_As for two types of VITSAB devices were reported as 68.7 and 102.1 kJ mol^{-1} and for the TEMPTIME FreshCheck, 83.6 kJ mol^{-1}; the growth of specific spoilage bacteria on chilled fish was also modeled and the E_A of the spoilage indices was 82 kJ mol^{-1}.[59] It has been shown that a TTI with an E_A within ±20 kJ mol^{-1} of the E_A of the spoilage index can be used to satisfactorily monitor its shelf life.[59]

In a study[47] of cottage cheese stored under three isothermal conditions, changes in a VITSAB TTM correlated well with certain microbial and chemical changes, the response closely following a first order reaction. The study concluded that the TTM could be used as a tool for detecting and estimating thermal abuse during storage or transportation of perishable products.

With increasing incentives for suppliers to deliver high-quality food products, better quality control procedures during product transport and improved management of inventory storage will be needed to minimize quality deterioration during distribution.[68] TTIs are likely to play an increasingly important role in monitoring the shelf life of perishable and semiperishable foods, and modifications to and improvements in performance of the indicators discussed here are to be expected.

3. Gas Concentration Indicators

Indicators (either a small, individually packaged tablet or label) are available commercially which indicate the presence or absence of gases. For example, the most common O_2 indicator is pink when the ambient O_2 concentration is ≤0.1%, turning blue when the O_2 concentration is ≥0.5%. Such indicators can be included in anoxic packages to indicate the effective absorption of all O_2, and to warn if there is a breakdown in the O_2 barrier. The presence of O_2 will be indicated in 5 minutes or less, while the change from blue to pink may take 3 hours or more. When refrigerated and maintained in an oxygen-free state, the manufacturer claims a 6-month shelf life, during which time the indicator can cycle between low and high O_2 levels. Many other types of O_2 indicators are available commercially.[24]

Despite many patents and market tests, no indicators are commercially available at present for reliably indicating the CO_2 level inside packages.

B. Providing More Convenience

Packaging manufacturers have always strived to increase the convenience to the consumer of the package because improved convenience is a value-added function that customers are likely to pay extra for as lifestyles change. Three examples of intelligent packages that increase convenience are described below.

1. Thermochromic Inks

Temperature-sensitive thermochromic inks are available and were discussed in Section III.D of Chapter 9. They can be printed onto labels or containers that are to be heated or cooled prior to consumption to indicate the ideal drinking temperature for the product. Depending on their composition, the inks will change color at specific temperatures, and if appropriate colors are chosen, then hidden messages such as "drink now" or "too hot" become visible. Thermochromic technology for beverages first became popular with wine labeling, but has not yet been widely adopted.

Under normal conditions, thermochromic leucodye inks have a shelf life of 6 months or more. After they are printed, they function, or continue to change color, for years. The postprint functionality can, however, be adversely affected by UV light, temperatures in excess of 121°C and aggressive solvents including chlorine bleach.

2. Microwave Doneness Indicators (MDIs)

MDIs are devices that detect and visually indicate the state of readiness of foods heated in a microwave oven. The utility of such indicators has been recognized for many years because they obviate the need for complicated heating procedures and instructions to the consumer. However, a prerequisite for the use of doneness indicators is a product that heats uniformly enough so that there is a well-defined stage in the microwave heating sequence at which all the regions within the product simultaneously satisfy the two criteria. In products that heat nonuniformly, the hottest regions (usually around the edges) will trigger a doneness indicator long before the cooler regions have achieved an acceptable temperature.

The requirements for a visual indication of doneness mean that the preferred location for an indicator is on the lid or dome of a container directly above the food. As the food heats in the microwave oven, the space above the food will be heated and, in turn, the heat will be transferred to the lid. The relationship between the temperature of the food and the temperature of the lid forms the basis of an indicator system.

Although there are many temperature indicating papers and labels available that would give a visual indication when the target temperature is reached, most of these devices would be heated by the microwaves leading to false indications. A solution to this problem is the so-called *shield doneness indicator* discussed below.

Because field distributions in microwave ovens are complicated (see Chapter 13), the relationship between the field experienced by the food and the field experienced by the sensor on the lid will generally be different and will vary from oven to oven. As a result, field-sensitive indicator systems work satisfactorily in some ovens but give false indications in others.

Because the lid temperature generally reaches its equilibrium value before a target temperature has been achieved throughout the product, the sensor must incorporate a time-dependent mechanism. In essence, the detector is activated at a particular temperature and a progressive change in color occurs over the desired time period. Plastic containers of syrup for pancakes can be purchased in the U.S. that are labeled with a thermochromic ink dot to indicate that the syrup is at the right temperature following microwave heating.

If a temperature sensor is placed in contact with a metallic surface in a microwave oven, then it experiences greatly reduced field intensity because the electric fields parallel to the metal surface are very small at or near the surface of the metal. As a result, such a sensor responds primarily to temperature as it has little or no opportunity to absorb microwave energy. Such shielded doneness indicators based on this principle, which incorporate an aluminum foil label on a plastic lid, have been devised but have not been successfully commercialized.

The major limitation of doneness indicators is the difficulty in observing whether or not a color change has occurred without opening the microwave oven. To overcome this limitation, an innovative heating sensor that emitted an audible signal was developed in Japan, consisting of a whistling device on a portion of the lid that generated a sound when gas passed quickly through the device. This indicator worked best when heating foods with a high moisture content but it has not been commercialized.

3. Radio Frequency Identification (RFID)

RFID is the use of radio frequencies to read information on a small device known as a *tag*. Currently, these tags may take many forms: for example, the microchip may be the tag because it has a tiny antenna on it, or the tag may look like a banknote security ribbon. Other RFID tags can be applied to products and packaging in the form of a label. Some RFID tags have a microchip in them and some do not.

RFID is a term used for any device that can be sensed at a distance by radio frequencies with few problems from obstruction or disorientation. The origins of the term lie in the invention of tags that reflect or retransmit a radio frequency signal. The term tag is used to describe any small device (shapes vary from pendants to beads, nails, labels or microwires and fibers) that can be incorporated into paper, and even special printed inks on, for example, paper.

Almost all conventional RFID devices contain a transistor circuit employing a microchip. By contrast, the potential in low-cost RFID is split between chip-based technologies and "chipless" tags, which can still be interrogated through a brick wall and hold data, but are cheaper and more primitive in electronic performance than chip tags.

To date, RFIDs have been used to increase convenience and efficiency in supply chain management and tracability, being normally applied to secondary or tertiary packaging. However, if costs can be reduced significantly, then they could find application on individual consumer packages although it is unclear what benefits they might provide to the consumer.

C. Providing Protection against Theft, Counterfeiting and Tampering

Protection against theft and counterfeiting is a highly developed area for high value goods such as electronics and clothes. However, it has not found widespread application in the food industry because of the comparatively low unit value of packaged foods. To reduce the incidence of theft and counterfeiting, holograms, special inks and dyes, laser labels and electronic tags have been introduced but their use in food packaging is minimal, again largely for cost reasons.

Tampering has been a concern for food manufacturers for many years, and there are occasional well-publicized cases where major branded food products have been maliciously contaminated with a poisonous substance and the company held to ransom. However, there is no tamper-proof or tamper-evident package that will stop a determined person from contaminating a product. As mentioned earlier, devices such as the button on metal closures attached to glass containers that have been retorted, and the tamper-evident band located on the skirt of screw closures attached to glass and plastic bottles have been used for many years, but do not qualify as intelligent packaging.

However, intelligent tamper-evident technologies are being developed based on labels or seals that are transparent until the package is opened or tampered with, at which time they change color permanently or a word such as "stop" or "opened" becomes visible. Labels or seals that release dye on being ruptured are also in development but their widespread use on food packages is unlikely owing to reasons of cost.

V. SAFETY AND REGULATORY ISSUES

Despite the intensive R&D and increasing commercialization of active and intelligent packaging systems, no specific methods exist in national and international legislation to determine their suitability in direct contact with foods.[14] The result is that legislation applying to traditional packaging materials has been applied to active and intelligent packaging systems.[24]

The use of some types of active and intelligent packaging raises safety issues because of the potential effect of such packaging on the microbial ecology of the food.[23] For example, packaging that absorbs O_2 from inside the package will affect both the types and growth rate of the micro-organisms in foods, and could give rise to the growth of anaerobic pathogenic bacteria such as *C. botulinum*. The inclusion of antimicrobial agents in the contact layer of a packaging material may result in a change in the microbial ecology of the food, and the types of micro-organisms present on a food will be different from the same food packaged in a conventional manner. Antimicrobial films that only inhibit spoilage micro-organisms without affecting the growth of pathogenic bacteria will also raise food safety concerns.[13]

The key regulatory issue is food contact approval, because substances may migrate into the food from active packaging. Such migrants may be intentional or unintentional. Intentional migrants include antioxidants, ethanol and antimicrobial preservatives, which require regulatory approval in terms of their identity, concentration and possible toxicological effects.[13] Unintentional migrants include various metal compounds or other system components that could enter the food. In most countries, there are regulations limiting or prohibiting the quantities of such compounds in the food. However, no specific regulations exist on testing the suitability of active and intelligent packaging systems in direct contact with foods and, in many cases, the testing protocols used are not necessarily appropriate, being based on those developed for plastics packaging materials.

When considering migration from a regulatory point of view, active and intelligent packaging systems can be divided into three groups[24]:

Group 1. Systems in which no chemical substances are deliberately transferred into the packaged food but compounds are ad/absorbed from the atmosphere inside the package (e.g., O_2 and C_2H_4 absorbers). These systems are not intended to come into contact with the food, being either placed in sachets, attached to the inner surface of the package or incorporated into the packaging material. However, in practice, most of these systems are more or less in direct contact with the food.

Group 2. Systems that emit quality preserving agents such as CO_2 or ethanol vapor. They can only function if the agent comes into direct contact with the surface of the food where it influences microbial growth.

Group 3. Systems from which preserving agents are deliberately transferred onto the surface of the food (e.g., antimicrobial films that transfer sorbic acid or metal ions). In order to function, there has to be direct contact between the active ingredient and the food.

There are possible regulatory issues with all three groups. Firstly, there is the possibility that the components making up the ad/absorber or emitter in Group 1 and Group 2 will migrate to the food. Secondly, there are the inherent toxicological properties of the agents in Group 3 which deliberately contact the food. Unless they are common food additives (e.g., sorbic acid), then potential toxicological concerns will need to be addressed. Any unintentional additives that could be transferred from active and intelligent packaging systems into food will be treated in the same way as, for example, residual monomers in plastic films. A unique regulatory issue arises with ethanol emitters used on foods intended to be consumed without further cooking or processing: the residual ethanol inside the package could reach as high as 2% and may need to be declared on the label if it exceeds a threshold level. Such products may not then be legally sold to children in some countries.

VI. CONCLUSIONS

Active and intelligent packaging concepts are already in commercial use in many countries, particularly the U.S. and Japan. In Europe, legislative restrictions, fear of consumer resistance, lack of knowledge about effectiveness, and economic and environmental impacts have limited the application of many types of active and intelligent packaging.[13] However, there have been numerous attempts and many highly innovative developments in active and intelligent packaging that have shown great initial promise but have failed to be successfully commercialized. Additionally, attempts are regularly made to commercialize new types of active and intelligent packaging before rigorous scientific evaluations have been undertaken. Even with those systems that have been adopted commercially, the peer-reviewed literature is relatively sparse, making it difficult for food packaging technologists to separate the marketing hype from the technical reality.

Because current bar-coding technology cannot develop further, new technologies are taking the functionality of the traditional bar-code and integrating it with RFID and electronic article surveillance (EAS) technologies to provide added benefits at reasonable price. Eventually, this new technology will be affordable even for relatively low value goods such as individual food packages. The most easily quantifiable benefits of such packaging will be seen in the global retail supply chain where significant cost savings will be made and increases in efficiency realized. On a global scale, such devices will help reduce billions of dollars of lost revenue from inventory irregularities, shrinkage and theft throughout the supply chain.

New applications under active development in Japan, the U.S. and elsewhere are disposable, low-cost extensions of digital paper, smart ink technologies, paper batteries and laminar electronics. Self-powered, laminar polymeric electronics, which derive their power from ink-based anodic and cathodic materials that are directly ink-jet or screen-printed on polymeric substrates, are also on the horizon. With the development of new processes to print low-cost transistors and antennas directly onto polymer paper for packaging applications, images and sounds could be broadcast from packages. However, whether this kind of development is a commercially viable prospect remains to be seen. Such technologies could also have a dramatic effect in assisting those with impaired sight, as well as conveniently providing consumers with useful audio information such as nutritional data.

Other new smart packaging technologies that will soon be available include disposable, moving color displays that glow. These organic light emitting diodes (OLEDs) have already appeared in 2003 on cell phone displays, as have transparent laminar loudspeakers that give superb sound quality. Both are presently deposited on glass but the race is on to deposit them on common packaging materials at a cost where they can be disposable. Although they do not contribute to product shelf life or safety, these new technologies will provide novel opportunities for the package to communicate with the consumer.

Many companies are developing thin film transistor circuits (TFTCs) that can be deposited on paper or plastic film, and are safe enough to be eaten or reconfigured after use to become something different that is useful or amusing. There are also opportunities in packaging for more sophisticated TTIs, such as a self-adjusting use-by date that senses when the package was opened, how long and at what temperatures it was stored, speaks clearly and flashes if there is a danger from food poisoning. There have also been reports of a milk carton that says in a deep voice, "Put me back in the fridge" if it is left out too long. This is just one example of how companies are seeking to bring a much higher level of safety and ease of use to consumers.

While some of the examples outlined in the preceding paragraphs might seem ephemeral, there is no doubt that the use of active and intelligent packaging systems for food will become increasingly popular and new, innovative applications that deliver enhanced shelf life and greater assurance of safety will eventually become commonplace.

REFERENCES

1. Ahvenainen, R., Active and intelligent packaging: an introduction, In *Novel Food Packaging Techniques*, Ahvenainen, R., Ed., CRC Press, Boca Raton, FL, 2003, chap. 2.
2. Brody, A. L., Smart packaging becomes Intellipac™, *Food Technol.*, 54(6), 104–106, 2000.
3. Brody, A. L., What's the hottest food packaging technology today?, *Food Technol.*, 55(1), 82–84, 2001.
4. Brody, A. L., Strupinsky, E. R., and Kline, L. R., *Active Packaging for Food Applications*, Technomic Publishing, Lancaster, Pennsylvania, 2001.
5. Buchner, N., Oxygen-absorbing inclusions in food packaging, West German Patent 1,267,525, 1968.
6. Cagri, C., Ustunol, Z., and Ryser, E. T., Antimicrobial edible films and coatings, *J. Food Prot.*, 67, 833–848, 2004.
7. Chen, H. and Hoover, D. G., Bacteriocins and their food applications, *Compr. Rev. Food Sci. Food Saf.*, 2, 82–100, 2003.
8. Clarke, R., Intelligent packaging for safeguarding product quality, *Proceedings of the Second International Conference on Active and Intelligent Packaging*, September 7–8, 2000, Campden & Chorleywood Food Research Association, Chipping Campden, Gloucestershire, 2000.
9. Collins-Thompson, D. and Hwang, C.-A., Packaging with antimicrobial properties, In *Encyclopedia of Food Microbiology*, Robinson, R. K., Batt, C. A. and Patel, P. D., Eds., Academic Press, New York, pp. 416–418, 2000.
10. Dainelli, D., Regulatory aspects of active and intelligent packaging, *Proceedings of the International Conference on Active and Intelligent Packaging*, September 7–8, 2000, Campden & Chorleywood Food Research Association, Chipping Campden, Gloucestershire, 2000.
11. Day, B. P. F., Underlying principles of active packaging technology, *Food Cosmet. Drug Packag.*, 23, 134–139, 2000.
12. Day, B. P. F., Intelligent packaging for foodstuffs, *Food Cosmet. Drug Packag.*, 23, 233–239, 2000.
13. Day, B. P. F., Active packaging, In *Food Packaging Technology*, Coles, R., McDowell, D. and Kirwan, M., Eds., CRC Press, Boca Raton, FL, 2003, chap. 9.
14. de Kruijf, N. and Rijk, R., Legislative issues relating to active and intelligent packaging, In *Novel Food Packaging Techniques*, Ahvenainen, R., Ed., CRC Press, Boca Raton, FL, 2003, chap. 22.
15. de Kruijf, N. and van Beest, M. D., Active packaging, In *Encyclopedia of Agricultural, Food, and Biological Engineering*, Heldman, D. R., Ed., Marcel Dekker, New York, pp. 5–9, 2003.
16. Floros, J. D., Dock, L. L., and Han, J. H., Active packaging technologies and applications, *Food Cosmet. Drug Packag.*, 20(1), 10–17, 1997.
17. Forage, A. J. and Byrne, W. J., Beverage package and a method of packaging a beverage containing gas in solution, U.S. Patent 4,832,968, 1989.
18. Ghosh, K. G., Srivatsava, A. N., Nirmala, N., and Sharma, T. R., Development and application of fungistatic wrappers in food preservation. Part I. Wrappers obtained by impregnation method, *J. Food Sci. Technol.*, 10, 105–109, 1973.
19. Han, J. H., Antimicrobial food packaging, *Food Technol.*, 54(3), 56–65, 2000.
20. Harima, Y., Free oxygen scavenging packaging, In *Food Packaging*, Kadoya, T., Ed., Academic Press, San Diego, CA, 1990, chap. 13.
21. Hernandez, R. J. and Giacin, J. R., Factors affecting permeation, sorption, and migration processes in package-product systems, In *Food Storage Stability*, Taub, I. A. and Singh, R. P., Eds., CRC Press, Boca Raton, FL, 1998, chap. 10.
22. Hong, S.-I., Gravure-printed colour indicators for monitoring kimchi fermentation as a novel intelligent packaging, *Packag. Technol. Sci.*, 15, 155–160, 2002.
23. Hotchkiss, J. H., Safety considerations, In *Active Food Packaging*, Rooney, M. L., Ed., Chapman and Hall, London, 1995, chap. 11.
24. Hurme, E., Sipiläinen-Malm, T., and Ahvenainen, R., Active and intelligent packaging, In *Minimal Processing Technologies in the Food Industry*, Ohlsson, T. and Bengtsson, N., Eds., CRC Press, Boca Raton, FL, 2002, chap. 5.
25. www.idtechex.com.

26. Isherwood, F. A., Removing oxygen from a container containing vacuum or gas packed food in which a metal (ex Fe) absorbs oxygen to form an oxide, British Patent 553,991, 1943.
27. Ishitani, T., Active packaging for food quality preservation in Japan, In *Foods and Packaging Materials — Chemical Interactions*, Ackermann, P., Jägerstad, M. and Ohlsson, T., Eds., Royal Chemical Society, Cambridge, pp. 177–188, 1995.
28. King, J., Catalytic removal of oxygen from food containers, *Food Manuf.*, 30, 441–444, 1955.
29. Krumhar, K. C. and Karel, M., Visual indicator system, US Patent 5,096,813, 1992.
30. Kuhn, W. E., Hu, K. H., and Black, S. A., Flexible electrochemical heater, US Patent 4,522,190, 1985.
31. Labuza, T. P., Applications of 'active packaging' for improvement of shelf life and nutritional quality of fresh and extended shelf life foods, Icelandic Conference on Nutritional Impact of Food Processing, Reykjavik, Iceland, 1987.
32. Labuza, T. P. and Breene, W. M., Applications of 'active packaging' for improvement of shelf life and nutritional quality of fresh and extended shelf life foods, *J Food Process. Preserv.*, 13, 1–69, 1989.
33. Liu, F. W., Storage of bananas in polyethylene bags with an ethylene absorbent, *HortScience*, 5, 25–29, 1970.
34. Matsuura, T., Abe, Y., Sato, K., Okamoto, K., Ueshiger, M., and Akagawa, Y., Prolonged antimicrobial effect of tissue conditioners containing silver-zeolite, *J. Dentistry*, 25, 373–378, 1997.
35. Melnick, D. and Luckmann, F. H., Sorbic acid as a fungistatic agent for foods. IV. Migration of sorbic acid from wrapper into cheese, *Food Res.*, 19, 28–32, 1954.
36. Mendoza, T. F., Welt, B. A., Otwell, S., Teixeira, A. A., Kristonsson, H., and Balaban, M. O., Kinetic parameter estimation of time–temperature integrators intended for use with packaged fresh seafood, *J. Food Sci.*, 69, FMS90–FMS96, 2004.
37. Miles, R., Packaging beverages in cans, In *Handbook of Beverage Packaging*, Giles, G. A., Ed., CRC Press, Boca Raton, FL, 1999, chap. 2.
38. Rodrigues, E. T. and Han, J. H., Intelligent packaging, In *Encyclopedia of Agricultural, Food, and Biological Engineering*, Heldman, D. R., Ed., New York, Marcel Dekker, pp. 528–535, 2003.
39. Rooney, M. L., Overview of active food packaging, In *Active Food Packaging*, Rooney, M. L., Ed., Chapman and Hall, London, 1995, chap. 1.
40. Rooney, M. L., Active packaging in polymer films, In *Active Food Packaging*, Rooney, M. L., Ed., Chapman and Hall, London, 1995, chap. 4.
41. Rooney, M. L., Active packaging: science and application, In *Engineering and Food for the 21st Century*, Welti-Chanes, J., Barbosa-Canovas, G. V. and Aguilera, J. M., Eds., CRC, Boca Raton, FL, 2002, chap. 32.
42. Sarett, B. L. and Scott, D., Enzyme treated sheet product and article wrapped therewith, U.S. Patent 2,765,233, 1956.
43. Scott, D., Enzymatic oxygen removal from packaged foods, *Food Technol.*, 12(7), 7–11, 1958.
44. Scott, D. and Hammer, F., Oxygen-scavenging packet for in-package deoxygenation, *Food Technol.*, 15(2), 99–103, 1961.
45. Scott, K. J., McGlasson, W. B., and Roberts, E. A., Potassium permanganate as an ethylene absorbent in polyethylene bags to delay ripening of bananas during storage, *Aust. J. Exp. Agric. Anim. Husb.*, 10, 237–241, 1970.
46. Selman, J. D., Time–temperature indicators, In *Active Food Packaging*, Rooney, M. L., Ed., Chapman and Hall, London, 1995, chap. 10.
47. Shellhammer, T. H. and Singh, R. P., Monitoring chemical and microbial changes of cottage cheese using a full-history time–temperature indicator, *J. Food Sci.*, 56, 402–405, 1991, see also p. 410.
48. Shimoni, E., Anderson, E. M., and Labuza, T. P., Reliability of time–temperature indicators under temperature abuse, *J. Food Sci.*, 66, 1337–1340, 2001.
49. Smith, J. P., Hoshino, J., and Abe, Y., Interactive packaging involving sachet technology, In *Active Food Packaging*, Rooney, M. L., Ed., Chapman and Hall, London, 1995, chap. 6.
50. Smith, D. P. and Rollin, N. J., Sorbic acid as a fungistatic agent for foods. VIII. Need and efficacy in protecting packaged cheese, *Food Technol.*, 8(3), 133–135, 1954.
51. Smolander, M., The use of freshness indicators, In *Novel Food Packaging Techniques*, Ahvenainen, R., Ed., CRC Press, Boca Raton, FL, 2003, chap. 6.

52. Sneller, J. A., Smart films give big lift to controlled atmosphere packaging, *Mod. Plast. Int.*, 16(9), 58–59, 1986.
53. Summers, J., Intelligent packaging for quality, *Soft Drinks Manag. Int.*, May, 32–33, 1992, see also p. 36.
54. Suppakul, P., Miltz, J., Sonneveld, K., and Bigger, S. W., Active packaging technologies with an emphasis on antimicrobial packaging and its applications, *J. Food Sci.*, 68, 408–420, 2003.
55. Tallgren, H., Keeping food in closed containers with water carrier and oxidizable agents such as Zn dust, Fe powder, Mn dust, etc., British Patent 496,935, 1938.
56. Taoukis, P. S., Time–temperature indicators, In *Novel Food Packaging Techniques*, Ahvenainen, R., Ed., CRC Press, Boca Raton, FL, 2003, chap. 6.
57. Taoukis, P. S. and Labuza, T. P., Applicability of time–temperature indicators as shelf life monitors of food products, *J. Food Sci.*, 54, 783–788, 1989.
58. Taoukis, P. S., Fu, B., and Labuza, T. P., Time–temperature indicators, *Food Technol.*, 45(10), 70–82, 1991.
59. Taoukis, P. S., Koutsoumanis, K., and Nychas, G. J. E., Use of time–temperature integrators and predictive modelling for shelf life control of chilled fish under dynamic storage conditions, *Int. J. Food Microbiol.*, 53, 21–31, 1999.
60. Tewari, G., Jayas, D. S., Jeremiah, L. E., and Holley, R. A., Absorption kinetics of oxygen scavengers, *Int. J. Food Sci. Technol.*, 37, 209–217, 2002.
61. Vartianinen, J., Skytta, E., Enqvist, J., and Ahvenainen, R., Properties of antimicrobial plastics containing traditional food preservatives, *Packag. Technol. Sci.*, 16, 223–229, 2003.
62. Vermeiren, L., Devlieghere, F., van Beest, M., de Kruijf, N., and Debevere, J., Developments in the active packaging of foods, *Trends Food Sci. Technol.*, 10, 77–86, 1999.
63. Vermeiren, L., Devlieghere, F., and Debevere, J., Effectiveness of some recent antimicrobial packaging concepts, *Food Addit. Contam.*, 19(Suppl.), 163–171, 2002.
64. Vermeiren, L., Heirlings, L., Devlieghere, F., and Debevere, J., Oxygen, ethylene and other scavengers, In *Novel Food Packaging Techniques*, Ahvenainen, R., Ed., CRC Press, Boca Raton, FL, 2003, chap. 3.
65. Wagner, J., The advent of smart packaging, *Food Eng. Intl.*, 14(10), 11, 1989.
66. Wells, J. H. and Singh, R. P., The application of time–temperature indicator technology to food quality monitoring and perishable inventory management, In *Mathematical Modelling of Food Processing Operations*, Thorne, S., Ed., Elsevier Applied Science, London, 1998, chap. 7.
67. Welt, B. A., Sage, D. S., and Sage Berger, K. L., Performance specification of time–temperature integrators designed to protect against botulism in refrigerated fresh foods, *J. Food Sci.*, 68, 2–9, 2003.
68. Wright, B. B. and Taub, I. A., Stored product quality: open dating and temperature monitoring, In *Food Storage Stability*, Taub, I. A. and Singh, R. P., Eds., CRC Press, Boca Raton, FL, 1998, chap. 12.
69. Yam, K. L., Intelligent packaging for a future smart kitchen, *Packag. Technol. Sci.*, 13, 83–85, 2000.
70. Zagory, D., Ethylene-removing packaging, In *Active Food Packaging*, Rooney, M. L., Ed., Chapman and Hall, London, 1995, chap. 2.

15 Modified Atmosphere Packaging

CONTENTS

I. INTRODUCTION

A. Definitions

There is a continuous search for improved methods of transporting food products from producers to consumers. It has long been known (see Section IB) that the preservative effect of chilling can be greatly enhanced when it is combined with control or modification of the gas atmosphere during storage. Such methods have been used commercially for over 100 years for the bulk storage and transport of fresh meat and fruits and are referred to as *controlled atmosphere storage* (CAS). Since the 1970s and the widespread availability of polymeric packages, this approach has been applied to consumer packs and given the name *modified atmosphere packaging* (MAP) because the atmosphere surrounding the food is modified but not controlled.

MAP can be defined as the enclosure of food in a package in which the atmosphere inside the package is modified or altered to provide an optimum atmosphere for increasing shelf life and maintaining quality of the food. Modification of the atmosphere may be achieved either actively or passively. Active modification involves displacing the air with a controlled, desired mixture of gases, a procedure generally referred to as *gas flushing*. Passive modification occurs as a consequence of the food's respiration or the metabolism of micro-organisms associated with the food; the package structure normally incorporates a polymeric film, and so the permeation of gases through the film (which varies depending on the nature of the film and the storage temperature) also influences the composition of the atmosphere that develops.

Vacuum packaging of respiring foods or those containing viable micro-organisms such as flesh foods is clearly a form of MAP because after initial modification of the atmosphere by removal of most of the air, biological action continues to alter or modify the atmosphere inside the package. In vacuum packaging, elevated levels of CO_2 can be produced by micro-organisms or by respiring fruits and vegetables.

Two terms are in widespread use concerning procedures that involve changes in the gas atmosphere in bulk storage facilities. In CAS, the gas composition inside a food storage room is continually monitored and adjusted to maintain the optimum concentration within quite close tolerances. In contrast, the less common *modified atmosphere storage* (MAS) typically involves some initial modification of the atmospheric composition in an airtight storage room, which changes further with time as a result of the respiratory activity of the fresh food and the growth of micro-organisms. Because CAS is capital-intensive and expensive to operate, it is more appropriate for those foods that are amenable to long-term storage such as apples, kiwifruit, pears and meat.

Controlled atmosphere packaging (CAP) is, strictly speaking, the enclosure of food in a gas-impermeable package inside which the gaseous environment with respect to CO_2, O_2, N_2, water vapor and trace gases has been changed, and is selectively controlled to increase shelf life. Using this definition, there are no CAP systems in commercial use. However, the combination of in-package or in-film O_2 and C_2H_4 absorbers together with CO_2 release agents (i.e., active packaging as discussed in Chapter 14) could be classed as CAP, at least during the early stages of the storage life of the packaged product.

An associated technique is *hypobaric storage*, which consists of placing the food in an environment in which pressure, air temperature and humidity are precisely controlled, and the rate at which air in the storage environment is changed is closely regulated.[9] Unlike CAS and MAS, no gases other than air are required. The total pressure within the hypobaric chamber is important because the O_2 concentration is directly proportional to that pressure. Although much research has been carried out into the use of hypobaric conditions for refrigerated storage of flesh foods and horticultural products,[9] it has not been employed commercially to any great extent for the storage or transportation of foods. However, it is used commercially by growers of cut flowers.

B. History of MAP

The first recorded scientific investigation into the effect of modified atmospheres on fruit ripening appears to have been conducted by Jacques Etienne Berard, a professor at the School of Pharmacy at Montpellier in France, who published his findings in 1821. Berard recognized that harvested fruits utilize O_2 and give off CO_2, and that fruits placed in an atmosphere deprived of O_2 did not ripen as rapidly. There is no record of commercial use of this information for nearly 100 years.

However, a remarkable application of the principles of CAS took place in 1865 in Cleveland, Ohio when Benjamin Nyce built a reasonably airtight store that used ice for cooling and a special paste for filtering the atmosphere to remove CO_2. He operated this store for a few years but refused to permit others to use his patented procedures; there is no record of expanded use of his system.

The first American scientists to investigate CA storage were Thatcher and Booth of Washington State University. Around 1903, they performed 2 years of testing that proved promising but the work was discontinued. In the period between 1907 and 1915, research personnel at the U.S. Department of Agriculture and Cornell University studied the response of several fruits to both lower O_2 and higher CO_2 levels in storage atmospheres. This work was reported in various scientific journals but did not result in commercial applications.

The first intensive and systematic research on CAS of fruits was initiated in England in 1918 by Franklin Kidd (later knighted for his efforts) and Cyril West at the Low Temperature Research Station at Cambridge. Various temperatures and atmospheres were used with apples, pears, plums, strawberries, gooseberries and raspberries. The atmospheres were generated by fruit respiration, and were dependent on the O_2 consumed and the CO_2 evolved by the fruit within a gas-tight building. The first commercial CA store appears to have been constructed by a grower near Canterbury in Kent in 1929 and, by 1938, there were over 200 commercial CAS facilities in England. Historical details on the development of CAS for fresh produce have been provided by Dilley.[17]

The knowledge that CO_2 inhibits bacterial growth is not new; in 1877, Pasteur and Joubert observed that *Bacillus anthracis* could be killed by using CO_2, and 5 years later, the first paper on the preservative effect of CO_2 on extending the shelf life of beef was published in Germany by Kolbe. During the period from 1880 to late 1920, about 100 reports were published on the inhibitory effects of CO_2 on micro-organisms.[29] In 1930, Killefer in England demonstrated that lamb, pork and fish remained fresh twice as long in 100% CO_2 compared with storage in air at chill temperatures, and similar improvements were reported by other English researchers for bacon and beef. In 1933, Haines found that the doubling time of some common bacteria on meat stored in 10% CO_2 at 0°C was twice that in air at the same temperature. Practical application of these results was made in the shipment of chilled beef carcasses from Australasia to England from the early 1930s, with an atmosphere of 10% CO_2 and a temperature of -1°C providing a storage life of 40 to 50 days without spoilage.

Coyne from the U.K. reported in 1932 that fillets and whole fish at ice temperature could be kept twice as long if stored in an atmosphere containing a minimum of 25% CO_2 but that undesirable textural and visual changes occurred if the CO_2 concentration exceeded 80%. Although his results were taken to a semicommercial stage, the technique was never adopted by industry.

A comprehensive study into the use of CO_2-enriched atmospheres for extending the shelf life of poultry meat (chicken portions) was carried out in the U.S. by Ogilvy and Ayres in 1951. The maximum usable CO_2 concentration was 25% because, above this, the meat became discolored; even at 15%, a loss of bloom was sometimes noted.

In the 1950s, the U.S. Whirlpool Corporation, makers of household washers, driers and refrigerators, tried to develop small gas generators for household food preservation of fresh meats and produce using controlled atmospheres together with refrigeration. Their efforts did not prove successful but resulted in the Whirlpool Corporation building larger generators (Tectrol® units) for CA warehouses and, later, truck transports for apples, lettuces and a host of fresh foods. In the 1960s, this technology was spun off to a joint venture company called Transfresh (now Fresh Express), the world's leading producer of fresh cut vegetables.[7]

The first patent for MAP of red meat was issued to two workers at Unilever in 1969.[20] It described an atmosphere containing $\geq$70% O_2 and $\geq$10% CO_2, the balance being an inert gas. Under such a MA in a gas-impermeable container, beef was still in fresh condition after 15 days at 4°C.

In 1931, Skovholt and Bailey showed that storage of bread in atmospheres containing at least 17% CO_2 delayed appearance of mold, concentrations of 50% doubling the mold-free shelf life. Aulund from Norway confirmed these findings in 1961, achieving a mold-free shelf life of 16 days for rye bread packaged in an atmosphere of CO_2. During the 1960s, more extensive research was undertaken in the U.K. at the Chorleywood Flour, Milling and Baking Research Association into the gas packaging of bakery products using elevated levels of CO_2 to retard mold growth. Unlike flesh foods and fruits and vegetables, baked goods such as bread, pastries and cakes do not benefit from storage at chill temperatures because the rate of staling increases as the temperature is lowered. A large U.K. bakery used MAP in the late 1960s for cake and achieved shelf life extensions of 4 to 5 days. However, MAP of baked goods did not become significant until the late 1970s when new labeling regulations in Europe required a listing of all preservatives on the label. Adoption of MAP avoided the need to use, and thus list, preservatives and also gave a longer

TABLE 15.1
Advantages and Disadvantages of Modified Atmosphere Packaging

Advantages	Disadvantages
Shelf life increase by possibly 50 to 400%	Added costs for gases, packaging materials and machinery
Reduced economic losses due to longer shelf life	Temperature control necessary
Decreased distribution costs, longer distribution distances and fewer deliveries required	Different gas formulations for each product type
Provides a high quality product	Special equipment and training required
Easier separation of sliced products	Potential growth of food-borne pathogens due to temperature abuse by retailers and consumers
Centralized packaging and portion control	Increased pack volume adversely affects transport costs and retail display space
Improved presentation — clear view of product and all-around visibility	Loss of benefits once the pack is opened or leaks
Little or no need for chemical preservatives	CO_2 dissolving into the food could lead to pack collapse and increased drip
Sealed packages are barriers against product recontamination and drip from package	
Odorless and convenient packages	

Source: From Sivertsvik, M., Rosnes, J. T., and Bergslien, H., *Minimal Processing Technologies in the Food Industry*, CRC Press, Boca Raton, FL, 2002, chap. 4. With permission.

shelf life. Today, there is very little MAP of soft bakery goods in the U.S., despite the relatively widespread use of MAP for baked goods in Europe.[4,5]

In the U.S., vacuum packaging of poultry was introduced by Cryovac®, followed by the "boxed beef" concept in 1967, both involving vacuum packaging in low O_2 barrier materials. In these circumstances, the atmosphere around the meat becomes depleted in O_2 (often $<1\%$ v/v) and enriched in CO_2 ($>20\%$ v/v), resulting in microbial changes quite different from those observed during aerobic storage. Vacuum-packaged boxed beef is then distributed to retail outlets where it is converted into consumer units; vacuum-packaged, boxed pork and lamb followed in the 1970s.

In summary, the successful commercialization of MAP in the late 1970s was preceded by over 150 years of scientific research on the inhibitory effects of CO_2 on microbial growth, as well as the effect of gaseous atmospheres on respiring produce. It required the convergence of scientific knowledge, polymeric films, gas flushing/vacuum packaging equipment and cold distribution chains to achieve the commercial success it currently enjoys. Surprising to many is that MAP, in its many manifestations, is now well ahead of the more widely publicized canning, freezing, aseptic packaging, and retort pouch and tray packaging in terms of volume of food preserved.[7] Although extension of shelf life is the most apparent advantage of MAP, there are also several other advantages (as well as disadvantages) as shown in Table 15.1.

II. PRINCIPLES

MAP is used to delay deterioration of foods that are not sterile and whose enzymic systems may still be operative. With the exception of baked goods, MAP is always used in association with chill temperatures. Chill temperatures are those close to, but above, the freezing point of fresh foods, and are usually taken as −1 to +7°C. Holding food at chill temperatures is widely used as an effective short-term preservation method, which has the effect of retarding the following occurrences:

1. Growth of micro-organisms
2. Postharvest metabolic activities of intact plant tissues, and postslaughter metabolic activities of animal tissues
3. Deteriorative chemical reactions, including enzyme-catalyzed oxidative browning, oxidation of lipids, chemical changes associated with color degradation, autolysis of fish and loss of nutritive value of foods in general
4. Moisture loss

The effect of chilling on the microflora in a particular food depends on the temperature characteristics of the organisms, as well as the temperature and time of storage. As the temperature is lowered from the optimum, growth slows and eventually stops. Micro-organisms that can grow within the 0 to 7°C range are defined as *psychrotrophs*. The most important psychotropic bacteria as far as chill temperature preservation of food is concerned are from the genus *Pseudomonas*, but the pathogens *Clostridium botulinum* type E, *Yersinia enterocolitica*, *Listeria monocytogenes*, enterotoxigenic *Escherichia coli* and *Aeromonas hydrophila* are also able to grow at or below 6°C. Thus, chill temperatures cannot be relied on with certainty to keep foods safe because of the possible survival and growth of these pathogens at chill temperatures.

If cooling is too fast or if the temperature is reduced too near to the freezing point of the food, then chilling injury can result. This can manifest itself in various ways; for example, cold shortening of muscle and physiological disorders of fruits and vegetables. In general, each food has a minimum temperature below which it cannot be held without some undesirable changes occurring in that food.

The preservative effect of chilling can be greatly enhanced when it is combined with modification of the gas atmosphere. This is because many deteriorative reactions involve aerobic respiration in which the food or micro-organism consumes O_2 and produces CO_2 and water. By reducing O_2 concentration, aerobic respiration can be slowed. By increasing CO_2 concentration, microbial growth can be slowed or inhibited (see below).

In addition to the benefits resulting from modification of the atmosphere inside the package, other benefits from MAP for fresh foods can include maintenance of high relative humidity and reduction in water loss, as well as improved hygiene by reducing contamination during handling. In the case of fresh produce, surface abrasions are minimized by avoiding contact between the produce and the shipping container and there is a reduced spread of decay from one item to another. In many cases, the benefits of using MAP relate more to one or more of these positive effects than to changes in the O_2 and CO_2 concentrations inside the package. Negative effects of MAP of fresh produce include a slowing down in the cooling of the packaged products and increased potential for water condensation within the package, which may encourage fungal growth.[26]

Although there is considerable information available regarding suitable gas mixtures for different foods, there is still a lack of scientific detail regarding many aspects relating to MAP.[15] These include:

1. Mechanisms of action of CO_2 on micro-organisms
2. Safety of MAP packaged food products
3. Interactive effects of MAP and other preservation methods
4. Influence of CO_2 on microbial ecology of a food
5. Effect of MAP on nutritional quality of packaged foods

III. GASES USED IN MAP

The normal composition of air by volume is 78.08% nitrogen, 20.95% oxygen, 0.93% argon, 0.03% carbon dioxide and traces of nine other gases in very low concentrations. The three main gases used in MAP are O_2, CO_2 and N_2, either singly or in combination. Noble or "inert" gases such as argon are being used commercially for a wide range of products, although the literature on their application and benefits is extremely limited.[38] Experimental use of carbon monoxide (CO) and sulfur dioxide (SO_2) has also been reported.

A. Carbon Dioxide

Carbon dioxide is the most important gas in the MAP of foods because of its bacteriostatic and fungistatic properties. It inhibits the growth of many spoilage bacteria, the degree of inhibition increasing with increasing concentration. It is particularly effective against Gram-negative, aerobic spoilage bacteria such as *Pseudomonas* species.

Carbon dioxide is a colorless gas with a slight pungent odor at very high concentrations. It dissolves readily in water (e.g., 1.69 g kg^{-1} at 20°C; 2.32 g kg^{-1} at 10°C; 2.77 g kg^{-1} at 5°C; 3.35 g kg^{-1} at 0°C) where a small amount is hydrated to carbonic acid (H_2CO_3). Between pH 1 and 5.5, a CO_2 solution contains about 2% of the CO_2 as H_2CO_3 and the remainder exists as dissolved CO_2.[12] When the pH of a CO_2 solution rises from 5.5 to 8.0, the H_2CO_3 dissociates to H^+ and HCO_3^-:

$$CO_2 + H_2O \leftrightarrow H_2CO_3 \leftrightarrow HCO_3^- + H^+$$

CO_2 is also soluble in lipids and some other organic compounds. As with all gases, the solubility of CO_2 increases with decreasing temperature and therefore the antimicrobial activity of CO_2 is markedly greater at lower temperatures. This has significant implications for MAP of foods. The high solubility of CO_2 in high moisture/high fat foods such as meat, poultry and seafood can result in package collapse owing to the reduction of headspace volume. High levels of CO_2 can also result in increased drip or exudate from flesh foods, and the addition of absorbent pads in the base of the package is used to compensate for this.

B. Oxygen

Oxygen is a colorless, odorless gas that is highly reactive and supports combustion. It has a low solubility in water (e.g., 0.009 g kg^{-1} at 20°C; 0.011 g kg^{-1} at 10°C; 0.013 g kg^{-1} at 5°C; 0.015 g kg^{-1} at 0°C). Oxygen promotes several types of deteriorative reactions in foods including fat oxidation, browning reactions and pigment oxidation. Most of the common spoilage bacteria and fungi require O_2 for growth. For these reasons, O_2 is either excluded or the level set as low as possible. Exceptions occur where O_2 is needed for fruit and vegetable respiration or the retention of color in red meat.

C. Nitrogen

Nitrogen is an inert gas with no odor or taste. It has a lower density than air and a low solubility in water (0.018 g kg^{-1} at 20°C) and other food constituents, making it a useful filler gas in MAP to counteract package collapse caused by CO_2 dissolving in the food. Nitrogen indirectly influences the micro-organisms in perishable foods by retarding the growth of aerobic spoilage microbes but it does not prevent the growth of anaerobic bacteria.

D. Carbon Monoxide

Carbon monoxide is a colorless, tasteless and odorless gas, which is highly reactive and very flammable. It has a low solubility in water but is relatively soluble in some organic solvents.

Carbon monoxide has been studied in the MAP of meat where it has the potential to retard metmyoglobin formation and fat oxidation. Carbon monoxide combines with myoglobin to form the bright red pigment carboxymyoglobin, which is much more stable than oxymyoglobin; a CO concentration of 0.4% in a MAP of meat is sufficient to give a bright red color (see Chapter 16). CO at 5 to 10% (combined with less than 5% O_2) is an effective fungistat, which can be used on commodities that do not tolerate high CO_2 levels.[25] Carbon monoxide has not been approved by regulatory authorities for commercial use in the EU, but it is used in Norway for retail packaging of red meat. It has been sanctioned for use in the U.S. to prevent browning in packaged lettuce and for pretreating meat in a master pack system where it is considered a processing aid. Commercial application has been limited because of its toxicity, its explosive nature at 12.5 to 74.2% in air and the fact that it has a limited effect on micro-organisms.

E. Noble Gases

The *noble gases* are a family of elements characterized by their lack of reactivity and include helium (He), argon (Ar), xenon (Xe) and neon (Ne). Ar is increasingly being used in place of N_2 to flush the neck of wine bottles immediately prior to corking, apparently because, as a heavier gas, it is more effective at removing air. Although the noble gases are chemically inert, it has been suggested that they are biologically active and several patents have been issued.

A 1969 research paper[3] showed that high pressure (340 atm) treatment with Ar significantly inhibited the activity of tyrosinase but not invertase, trypsin or chymotrypsin. However, no attempt was made to separate the effect of high pressure from the effect of gas. Subsequently, in 1990, a patent was issued to one of the authors[33] for the MAP of fresh fruit pieces, where O_2 was introduced as a mixture with inert gases such as N_2, Ar, He and H_2. According to the patent, the inert gases act as bulking agents to ensure uniform distribution of O_2 throughout the container, aid in the transport of O_2 to the centers of the fruit pieces and prevent O_2 toxicity in the tissues. Furthermore, it was claimed that the inert gases may act as blockers of enzyme deteriorative actions.

In 1992, a patent[36] was issued for a gas mixture for preserving fish and seafood products, comprising 50 to 80% CO_2, 5 to 20% O_2 and 27 to 45% Ar. The patent claimed that this gas mixture slows down enzymatic reactions and microbiological development at the surface and inside the fish or seafood product.

A patented method[39] for regulating enzyme activity was issued in 1998, which entailed contacting one or more enzymes with a gas containing one or more noble gases or mixtures thereof. The patent claimed that enzyme activities could be regulated in a controlled and predictable manner by contacting with a gas containing one or more noble gases, where the noble gases have significant effects upon enzymes even at low pressure, and over a wide range of temperature.

It has also been claimed that Ar slows down the rate of production of volatile amino bases in seafood, inhibits enzymic discoloration, delays the onset of textural softening, extends the microbial lag phase, inhibits microbial oxidases and enhances the effectiveness of CO_2 by weakening microbes, thereby enabling food suppliers to use less CO_2 in MAP. A possible reason for these effects could be the greater solubility of Ar compared with O_2 and N_2, and its similar atomic size to O_2.[37] It is also claimed that Ar removes O_2 from packages more efficiently than N_2, and is probably more effective at displacing O_2 from cellular sites and enzymic O_2 receptors.[14] Because Ar is denser than N_2 and four times more efficient at displacing air, the difference in cost (Ar costs approximately five times more than N_2) is negligible.

However, a noted authority in the postharvest field[26] has stated that there is no evidence supporting the use of Ar, He or other noble gases as a replacement for N_2 in MAP of fresh produce, and no peer-reviewed papers on the use of noble gases in MAP have been published.

Despite the lack of scientific publications, noble gases are being used in a number of food applications including potato chips, processed meats, nuts, beverages, fresh pasta, chilled prepared meals and lettuce in U.S. and U.K. supermarkets based upon claims of an average 25%

improvement in shelf life. Some products, such as fresh pizza, have shelf life improvements of 40 to 50%. At present, nearly 200 different argon-packaged foods can be found on grocery store shelves.[38]

F. Gas Mixtures

The gas mixtures used for MAP of different foods depends on the nature of the food and the likely spoilage mechanisms. Where spoilage is mainly microbial, the CO_2 levels in the gas mix should be as high as possible, limited only by the negative effects of CO_2 (e.g., package collapse) on the specific food. Typical gas compositions for this situation are 30 to 60% CO_2 and 40 to 70% N_2. For oxygen-sensitive products, where spoilage is mainly by oxidative rancidity, 100% N_2 or N_2/CO_2 mixtures (if microbial spoilage is also important) are used. For respiring products, it is important to avoid too high a CO_2 level or too low an O_2 level (to avoid anaerobic respiration).

IV. Methods of Creating MA Conditions

A. Passive MA

In this approach (also known as *commodity-generated MA*), an atmosphere high in CO_2 and low in O_2 passively evolves within a sealed package over time as a result of the respiration of the product. Ideally, the gas permeabilities of the packaging film are such that sufficient O_2 can enter the package to avoid anoxic conditions and anaerobic respiration occurring, while at the same time, excess CO_2 can diffuse from the package to avoid injuriously high levels. Passive modification is commonly used for MAP of fresh respiring fruits and vegetables. Given the simplicity of this approach, and the many interrelated variables which affect respiration rate, considerable research is required if appropriate passive MA systems are to be developed for all horticultural commodities.

B. Active MA

Several methods can be used to actively modify the gas atmosphere inside a packaged product. These include vacuum packaging where the air is removed under vacuum and the package sealed. This method finds widest application for the packaging of flesh foods, particularly red meat. A two-stage method involves first removing the air inside the package using a vacuum, followed by flushing with the desired gas mixture. This creates the desired MA immediately after packaging in contrast with the passive approach, which may require a week or longer before achieving the same gas composition. In a third active MA method, no vacuum is used, but a gas mixture is injected into the package and the air swept or flushed out immediately prior to sealing, resulting in residual O_2 levels of 2 to 5%. For O_2-sensitive products, the two-stage method is preferred.

Regardless of whether vacuum or gas flush packaging is used to create a MA, the package itself must provide a barrier to permeation over the expected shelf life, otherwise the beneficial effects of reducing O_2 will be lost. In the case of vacuum packaging, the barrier issue relates only to O_2 ingress but, in gas flushing, both O_2 ingress and N_2 egress must be considered. An elegant analysis of the situation has been presented[8] and is repeated here.

In vacuum packaging, both O_2 and N_2 are potential entrants into the package. O_2 is a faster permeant than N_2 by a factor of about 4 to 6 (see Table 4.2); however, air contains only a quarter as much O_2 as it does N_2. Therefore, O_2 and N_2 are equally as likely to enter the package.

On the other hand, gas flush packaging (assuming that only N_2 is used) imposes a N_2 atmosphere inside the package, working against a 79% N_2 atmosphere in the surrounding air. Thus, the driving force for N_2 exiting the package is $1.00 - 0.79 = 0.21$ atmospheres, while the driving force for O_2 to enter is the partial pressure of O_2 in the air ≈ 0.21 atmospheres. Because the driving forces (i.e., the partial pressure differences inside and outside the package) are equal, and the permeation rate for N_2 egress is about a quarter to a sixth that of O_2 ingress, the package will slowly increase in pressure. If the package is made from a flexible film or has a film lid sealed to a plastic tray, then a "pillowing" effect is likely to occur. It is important that the reasons for this pillowing effect are clearly understood as it is not uncommon for it to be ascribed by the uninitiated to microbial growth, resulting in perfectly safe product being removed from sale.

In addition to the mechanical methods described above, adsorbers can be used inside the package, for example, to delay the climacteric rise in respiration for some fruits by adsorbing C_2H_4, to prevent the build-up of CO_2 to injurious levels by adsorbing CO_2 as well as to lower the concentration of O_2 through the use of O_2 adsorbers. Obviously, the use of gas adsorbers adds considerably to the cost, and therefore their use is limited to those commodities for which it is cost-effective. An alternative method for quickly reducing the O_2 content and increasing the CO_2 content within a package involves the use of ferrous carbonate inside a gas-permeable sachet; in the presence of moist air, the amorphous material oxidizes with the release of CO_2. The quantity of ferrous carbonate used must be carefully calculated because if too much is present inside the package, then anoxic conditions will be established.

Recently, another approach to creating MAs inside a package has been suggested.[37] It involves dissolving CO_2 into the product prior to packaging using a method called *soluble gas stabilization* (SGS). The CO_2 is dissolved into the food at low temperature (ca. 0°C) and elevated pressures (>2 atm), resulting in packages with smaller gas:product ratios and thus decreased package sizes for a given weight of product. This method has been suggested for the MAP of fish products, either alone, combined with traditional gas flushing or with vacuum packaging.

V. EQUIPMENT FOR MAP

Equipment for MAP must generally be capable of removing air from the package and replacing it with a mixture of gases. There are basically three types of packaging equipment used for MAP: horizontal or vertical form-fill-seal (FFS) machines using pouches or trays; snorkel machines using preformed bags or pouches; and chamber machines using preformed pouches or trays.

A. Form-Fill-Seal Machines

FFS machines can either form pouches (vertically or horizontally), or thermoformed trays with a heat-sealed lid, from rollstock. In the pouch version, the desired gas mixture is introduced into the package in a continuous countercurrent flow to force out the air, after which the ends of the web are heat sealed and the packages cut from one another. In the tray version, product is placed in the tray and a vacuum drawn, after which the desired gas mixture is introduced and the top web of film heat sealed to the base tray.

B. Chamber Machines

Here, the filled package (either a preformed pouch or tray inside a bag) is loaded into a chamber, a vacuum is pulled and the package is then flushed with the gas mixture and heat sealed. This is a batch process, is relatively slow and most suitable for bulk or master packs.

C. Snorkel Machines

Snorkel machines operate without a chamber. The product is placed inside a large flexible pouch (or bag) and positioned in the machine. Snorkels or probes are inserted into the pouch and remove air, after which the vacuum is broken by the addition of the desired gas mix. The probes are then removed and the package is heat sealed. These machines are used mainly for bulk packaging and for so-called *master packs* in which individual retail packs are packaged in a large MA pouch or bag.

VI. PACKAGING FOR MAP APPLICATIONS

The main characteristics to be considered when selecting packaging materials for MAP are permeability of the package to gases and water vapor, mechanical properties, heat sealability and transparency. For nonrespiring products, all of the common high gas barrier structures have been used in MAP, including laminates and coextruded films containing PVdC, EVOH and PAs as a barrier layer. The inside layer is usually LDPE to provide a good heat seal and moisture vapor barrier.

For the MAP of respiring produce such as fruits and vegetables, the choice of suitable packaging materials is much more complex and no easy solutions are available given the dynamic nature of the product. Ideally, the packaging material should maintain a low O_2 concentration (3 to 5%) in the headspace and prevent CO_2 levels exceeding 10 to 20%. Polyolefin films are normally used, but to achieve the desired MAP, it has been necessary to perforate the film or (more recently) use a special patch; O_2 and CO_2 absorbers have also been used on a limited scale. MAP materials for fruits and vegetables are discussed in more detail in Chapter 16.

To prevent condensation of water vapor on the inside of the package as a result of temperature differentials between the package contents and the packaging material, antifogging agents are used. These amphiphilic additives function by decreasing the interfacial tension between the polymer and the condensed moisture vapor, enabling the water droplets to coalesce and spread as a thin transparent layer across the surface of the film. Typical antifogging agents include nonionic ethoxylates and hydrophilic fatty acid esters.

For the trays used in MAP, there are no special requirements other than good thermoformability. Typical tray materials are PS and PVC, although the latter are much less common now in Europe due to environmental concerns among consumers. Preformed plastic-coated paperboard trays have also been used. Regardless of the type of tray, it is essential that the lidding film can be adequately sealed onto the tray.

VII. MICROBIOLOGY OF MAP

The species of micro-organisms that cause spoilage of particular foods are influenced by two factors: the nature of the foods and their surroundings. These are known as *intrinsic* and *extrinsic parameters* and were discussed in Chapter 10 (see especially Table 10.4). The two extrinsic factors most relevant in MAP are the gaseous composition of the in-pack environment and the temperature. The specific microbiology of various foods is dealt with in subsequent chapters of this book. However, general comments on the effects of MAs on food spoilage and pathogenic micro-organisms will be outlined here.

Microbial food spoilage is characterized by undesirable sensory changes to the odor, color, flavor and sometimes the texture of the food, making it inedible or unsaleable. Spoilage is an important safeguard in preventing food poisoning because the deterioration in food quality normally (but not always) warns the consumer that the food may be unsafe.

Micro-organisms have different respiratory and metabolic needs and can be grouped according to their O_2 needs as shown in Table 15.2. The effect of CO_2 on microbial growth is shown in Table 15.3.

Concentrations of CO_2 in excess of 5% v/v inhibit the growth of most food spoilage bacteria, especially psychotrophic species such as *Pseudomonas*, which grow on a wide range of refrigerated foods. The effect of CO_2 on bacterial growth is complex and the growth inhibition of micro-organisms in MA is determined by the concentration of dissolved CO_2 in

TABLE 15.2
Oxygen Requirements of Some Micro-Organisms of Relevance in Modified Atmosphere Packaging

Group	Spoilage Organisms	Pathogens
Aerobes (require atmospheric oxygen for growth)	*Micrococcus* sp. Molds (e.g., *Botrytis cinerea*) *Pseudomonas* sp.	*Bacillus cereus* *Yersinia enterocolitica* *Vibrio parahemolyticus* *Campylobacter jejuni*
Microaerophiles (require low levels of oxygen for growth)	*Latobacillus* sp. *Bacillus* spp. *Enterobacteriaceae*	*Listeria monocytogenes* *Aeromonas hydrophila* *Escherichia coli*
Facultative anaerobes (grow in presence or absence of oxygen)	*Brocothrix thermosphacta* *Shewanella putrefaciens* Yeasts	*Salmonella* spp. *Staphlococcus* spp. *Vibrio* sp.
Anaerobes (inhibited or killed by oxygen)	*Clostridium sporogenes* *Clostridium tyrobutyricum*	*Clostridium perfringens* *Clostridium botulinum*

Source: From Parry, R.T, *Principles and Applications of Modified Atmosphere Packaging of Foods*, Blackie Academic & Professional, London, England, 1993. With permission.

TABLE 15.3
Effect of CO_2 Atmosphere on Growth

Micro-organism	Type of Growth	Effect on Growth in CO_2 atm
Aeromonas spp.	Facultative	Inhibited (weakly)
Bacillus cereus	Facultative	Inhibited
Campylobacter jejuni	Microaerophilic	Inhibited, survival[a]
Clostridium botulinum proteolytic (A,B,F)	Anaerobic	Unaffected[b]
C. botulinum nonproteolytic (B,E,F)	Anaerobic	Unaffected[b]
C. perfringens	Anaerobic	Inhibited
Escherichia coli	Facultative	Inhibited (weakly)
Listeria monocytogenes	Facultative	Unaffected/inhibited[c]
Plesiomonas spp.	Facultative	Inhibited
Salmonella	Facultative	Inhibited[b]
Staphylococcus aureus	Facultative	Inhibited (weakly)
Vibrio cholerae	Facultative	Inhibited
V. parahemolyticus	Facultative	Inhibited
Yersinia enterocolitica	Facultative	Inhibited

[a] The bacteria survive better in CO_2 as compared to air, but growth is (weakly) inhibited.
[b] One report of growth stimulation under CO_2.
[c] Unaffected by CO_2 atmosphere if at least 5% O_2 present; inhibited under 100% CO_2.

Source: From Sivertsvik, M., Rosnes, J. T., and Bergslien, H., Modified atmosphere packaging, In *Minimal Processing Technologies in the Food Industry*, Ohlsson, T. and Bengtsson, N., Eds., CRC Press, Boca Raton, FL, 2002, chap. 4. With permission.

the product. Several mechanisms for the action of CO_2 on micro-organisms have been identified:

1. Alteration of cell membrane function including effects on nutrient uptake and absorption
2. Direct inhibition of enzymes or decreases in the rates of enzymic reactions
3. Penetration of bacterial membranes leading to intracellular pH changes
4. Direct changes in the physicochemical properties of proteins

A probable combination of all these activities accounts for the bacteriostatic effect.[37] However, CO_2 does not retard the growth of all types of micro-organisms; the growth of lactic acid bacteria, for example, is enhanced in the presence of CO_2 and low O_2 concentration. Another example is *L. monocytogenes*, which although inhibited by 100% CO_2, is unaffected by CO_2 if at least 5% O_2 is present.

Much research has been carried out on the safety and the health hazards of MAP of foods, especially on those pathogens able to multiply at chill temperatures, and those able to multiply in anaerobic conditions. There are seven food-borne pathogenic bacteria known to be capable of growth at or below 5°C: *C. botulinum* Type E, *L. monocytogenes*, *Y. enterocolitica*, *Vibrio parahemolyticus*, enterotoxigenic *E. coli*, *Bacillus cereus* and *Aeromonas hydrophila*. Two others are capable of growth at temperatures just above 5°C: *Staphylococcus aureus* at 6°C (10°C for toxin production) and *Salmonella* species at 7°C. Thus, it is vitally important that the MAs inhibit the growth of these organisms in foods under refrigerated storage. Fortunately, most of these organisms do not compete well with harmless bacteria, such as the *Lactobacillus* species, which grow rapidly if temperature abuse occurs.[31]

It has been found[23] that spores of *C. botulinum* types B, E and F grow and produce toxin after 5 weeks at 3°C, 3 to 4 weeks at 4°C and 2 to 3 weeks at 5°C. Growth occurred more frequently from spores of type F strains than for types B and E. Prior to this report, it was believed that spores of *C. botulinum* could not grow below 3.3°C. It has been demonstrated[21] that 100% CO_2 can have an inhibitory effect on the growth of *C. botulinum* at chill temperatures, and an increased inhibitory effect was observed when combining 100% CO_2 with increased NaCl levels and decreased pH.

Of most concern in MAP is the possible growth of *C. botulinum* Type E, which is associated with fishery products.[13] It is tolerant of low temperatures, anaerobic and may grow and produce a potent neurotoxin on the food before spoilage is detectable by the consumer. Storage at temperatures below 3°C should provide an adequate safeguard for products susceptible to *C. botulinum*.

The concept of barrier or hurdle technology was introduced in the 1980s.[28] Basically, this concept refers to combinations of different preservation factors ("hurdles") that are used to achieve multitarget, mild preservation effects. Hurdles used in addition to chill temperatures to limit or prevent the growth of psychrophilic pathogens during low temperature storage or limited temperature abuse of the food include CO_2, thermal processing, pH reduction (<5), preservatives, reduced a_w (<0.97) and competitive microflora (Table 15.4).

TABLE 15.4
Perishability of Meat Products as a Function of a_w and pH

Perishability	Criteria	Storage Temperature (°C)
Storable	$a_w < 0.95$, pH < 5.2 or $a_w < 0.91$ or pH < 5.0	No refrigeration
Perishable	$a_w < 0.95$ or pH < 5.2	<10
Easily perishable	$a_w < 0.95$, pH > 5.2	<5

Source: From Leistner L., Rödel, W., and Krispien, K., Microbiology of meat and meat products in high- and intermediate-moisture ranges, In *Water Activity: Influences on Food Quality*, Rockland, L. B. and Stewart, G. F., Eds., Academic Press, New York, pp. 855–916, 1981. With permission.

VIII. SAFETY OF MAP

The shelf life and safety of any MAP food is influenced by a number of factors including the nature of the food, the gaseous environment inside the package, the nature of the package, the storage temperature and the packaging process and machinery.[35] In discussing the safety of MAP foods, it is useful to divide them into two categories: those products such as smoked salmon, cured meats, fruits and salad vegetables, which are eaten without any prior heat treatment, and those such as fresh fish, raw meats and poultry products, which are subjected to a sufficient heat treatment to kill all vegetative pathogens.[37] Clearly, the first category presents more risks from a microbiological point of view. An excellent series of articles concerning the processing and safety of ready-to-eat, cold-smoked fish has been published.[1]

Chilled foods have been subjected to detailed regulatory controls in many countries, particularly with respect to temperature requirements, and at the international level, HACCP-based approaches to hygiene have been established.[10,11] One of the major concerns of MAP foods is temperature abuse because the biostatic effects of CO_2 are temperature-dependent, and a rise in temperature during storage could permit the growth of micro-organisms that had been inhibited by CO_2 at lower temperatures. If O_2 were present in the package, then growth of aerobic spoilage organisms during periods when the food was at nonrefrigerated temperatures would alert consumers to temperature abuse due to the appearance of undesirable odors, colors or slime. However, the absence of O_2 will favor the growth of anaerobic micro-organisms (including *C. botulinum*) over aerobic spoilage organisms. It has been demonstrated that anaerobic pathogens can grow at temperatures as low as 3°C and produce toxin without any sensory manifestation of food deterioration.

An area of active research is edible films for use in MAP systems. However, these films can create a very low O_2 environment where anaerobic pathogens such as *C. botulinum* may thrive. Antimicrobial compounds that can be incorporated into the coating are also being investigated.

Successful control of both product respiration and ethylene production and preservation by MAP can result in a fruit or vegetable product of high sensory quality. However, control of these processes is dependent on maintaining optimum temperature control along the entire food chain continuum, from processing, storage, transportation and retailing to in the home. Maintaining proper storage temperatures is often most difficult at retail and domestic level.

Currently, there is concern with psychrotrophic foodborne pathogens such as *L. monocytogenes*, *Y. enterocolitica* and *A. hydrophila*, as well as nonproteolytic *C. botulinum*, although clearly a number of other micro-organisms, especially *Salmonella*, *E. coli* O157:H7 and *Shigella* spp., can be potential health risks when present on MAP produce.[16,19] Although only two MAP produce products (coleslaw mix and ready-to-eat salad vegetables) have been implicated in foodborne illness outbreaks (botulism and salmonellosis, respectively), the potential for growth of pathogens exists. The success and microbiological safety of MAP is dependent on controlled, low temperature storage and the product's characteristics.[24]

It is difficult to evaluate the safety of MAP foods solely on the basis of the growth of certain pathogens at abusive temperatures because (1) most food pathogens do not grow at chill temperatures, and (2) CO_2 is not highly effective at nonrefrigeration temperatures. To minimize problems with pathogens inside MAP foods, the foods should be of the highest microbiological quality at the time of packaging, they should be processed and packaged under high standards of hygiene and sanitation, their temperature should be reduced as rapidly as possible, and the temperatures during distribution should be rigidly maintained as low as required to avoid anaerobic pathogen growth. If the above requirements are compromised in any way, serious public health hazards could result from ingestion of the MAP food.[30]

Predictive microbiology is being increasingly used as a tool to provide rapid and reliable answers concerning the probable growth of specific organisms under defined conditions. Models can be used to predict the probability of growth, the time until growth occurs or the growth rate of micro-organisms.[40] A comprehensive modeling program (available free of charge) has been produced by the U.S. Department of Agriculture called Pathogen Modeling Program. In the U.K., the FoodMicroModel was developed by the Ministry of Agriculture, Fisheries and Food. There are many advantages in the use of predictive models with chilled foods, especially in the decision-making processes of HACCP and risk analysis.

IX. REPFEDs AND *SOUS VIDE*

There is an increased interest in "fresh" preservative-free food with extended durability. In addition to foods packaged in MAs, there are refrigerated, processed foods with extended durability (REPFEDs) such as *sous vide* and cook-chill foods that are produced by the following general process. Meals or components of meals (which may include both raw and cooked components) are sealed in a heat-stable pouch and the packaged product is cooked at temperatures ranging from 65 to 95°C. The product is then cooled and stored at refrigeration temperatures (1 to 8°C) and has a shelf life of up to 42 days, dependent on the heat treatment and the storage temperature.[22]

If the pasteurization process is carried out adequately, then all vegetative bacteria present are killed but bacterial spores can survive this heating process. Because REPFEDS are mostly packed under vacuum or in an anaerobic atmosphere, growth of aerobic micro-organisms is restricted, while growth of anaerobic bacteria is favored. The ecological niche found in REPFEDS favors colonization by micro-organisms that produce heat-resistant spores and grow in the absence of O_2 at refrigeration temperatures. Nonproteolytic *C. botulinum* is the principal microbiological safety concern in REPFEDs.[22] Proteolytic *C. botulinum* and *Clostridium perfringens* are of concern when the storage temperature exceeds 10°C for a prolonged time.

Hurdle technology has been applied to REPFEDs to control growth of spoilage and pathogenic micro-organisms, although each hurdle is insufficient on its own to achieve the same effect. Typical hurdles include pH $\leq$ 5, salt concentration $\geq$3.5% and $a_w \leq 0.97$ throughout the food.

A food processing technology known as *sous vide* (literally "under vacuum") has been developed to enhance the shelf life of refrigerated foods and is a particular type of cook–chill process. Though lauded as a revolution by some, it is really the result of evolution of the conventional cook–chill process. It uses vacuum packaging, heating and rapid cooling, followed (normally) by chilled storage although two of the largest producers in the U.S. specialize in frozen *sous vide* products.[18]

The precursor of *sous vide* was first discussed in 1960[27] and the name "Frigi-Canning" given to this new process, which involved thermal processing in hermetically sealed cans or jars, followed by rapid chilling given that only vegetative forms of micro-organisms were destroyed. The first formal application of *sous vide* technology was developed and tested at two hospitals in Sweden from 1960 to 1965, the process being termed the "Nacka system" after one of the hospitals. *Sous vide* technology appeared in France in 1972 for the processing of ham, and the first large-scale application of modern *sous vide* technology occurred in 1985 by the French national railway company SNCF.[2] The *sous vide* method was developed for restaurant use in the early 1970s by a French chef George Pralus using a packaging material and system patented in the U.S.[34] He was commissioned to find a way to reduce the shrinkage of *foie gras* and developed a vacuum packaging and cooking process that not only dramatically reduced shrinkage (from 46 to only 5%) but also enhanced flavor. Today, *sous vide* products range from institutional foodservice to in-home convenience foods to restaurant meals.

Low acid, refrigerated foods, which are precooked, free of microbial inhibitors and vacuum packaged, present some very serious challenges because the partial cooking results in the destruction of the vegetative microflora, leaving heat-resistant spores as survivors. However, *sous vide* may be produced safely if proper controls (including HACCP) are in place.[35]

X. APPLICATIONS OF MAP

MAP is being successfully used by many food processing companies around the world to extend the shelf life and retain the quality of a wide variety of foods. Table 15.5 gives examples of foods currently packaged in MAs, together with the gas mixtures typically used. A detailed discussion of the MAP of these various foods can be found in subsequent chapters.

Several novel technologies offer the potential for further improvements in the shelf life and safety of MAP foods. These include the use of active and intelligent packaging as discussed in Chapter 14. In particular, O_2 and C_2H_4 absorbers and CO_2 emitters, used either alone or in

TABLE 15.5
Examples of Gas Mixtures for Selected Food Products

Product	Temperature (°C)	O_2 (%)	CO_2 (%)	N_2 (%)
Meat products				
Fresh red meat	0–2	40–80	20	Balance
Cured meat	1–3	0	30	70
Pork	0–2	40–80	20	Balance
Offal	0–1	40	50	10
Poultry	0–2	0	20–100	Balance
Fish				
White fish	0–2	30	40	30
Oily fish	0–2	0	60	40
Salmon	0–2	20	60	20
Scampi	0–2	30	40	30
Shrimp	0–2	30	40	30
Plant products				
Apples	0–4	1–3	0–3	Balance
Broccoli	0–1	3–5	10–15	Balance
Celery	2–5	4–6	3–5	Balance
Lettuce	<5	2–3	5–6	Balance
Tomatoes	7–12	4	4	Balance
Baked products				
Bread	RT[a]		60	40
Cakes	RT		60	40
Crumpets	RT		60	40
Crepes	RT		60	40
Fruit pies	RT		60	40
Pita bread	RT		60	40
Pasta and ready meals				
Pasta	4		80	20
Lasagna	2–4		70	30
Pizza	5		52	50
Quiche	5		50	50
Sausage rolls	4		80	20

[a] Room temperature; staling is accelerated at refrigerated temperatures.

Source: From Brody, A. L., Packaging: part IV — controlled/modified atmosphere/vacuum food packaging, *Wiley Encyclopedia of Food Science and Technology*, 2nd ed., Vol. 3, Francis, F. J., Ed., Wiley, New York, pp. 1830–1839, 2000.[6] With permission.

combination with MAs, are likely to find wider application as more cost-effective and innovative designs are commercialized.

The use of TTIs (also discussed in Chapter 14) on individual MA packs could enable the temperature to be monitored throughout the supply chain and a warning given if the food has suffered temperature abuse. Although not currently practicable owing largely to cost, innovative developments enabling each individual package to have its own TTI could make this a commercial reality.

The use of hurdle technology[29] involving MAP in combination with deliberate manipulation of a_w, pH or redox potential, as well as the use of preservatives, bacteriocins, ultra high pressure and edible coatings, could lead to better control of potential pathogens and thus a safer product.

In conclusion, MAP is a simple concept and is increasingly applied to many different foods, providing substantial extensions in shelf life and significant economies in production and distribution. However, MAP is not a substitute for good manufacturing practices, HACCP programs and proper temperature control during storage and distribution. Used intelligently and responsibly, MAP will likely become the dominant form of food preservation in the twenty-first century.

REFERENCES

1. Anon., Processing parameters needed to control pathogens in cold-smoked fish, *J. Food Sci.*, 66(7), S1062–S1112, 2001.
2. Bailey, J. D., Sous vide: past, present, and future, In *Principles of Modified-Atmosphere and Sous Vide Product Packaging*, Farber, J. M. and Dodds, K. L., Eds., Technomic Publishing, Lancaster, PA, 1995, chap. 10.
3. Behnke, J. R., Fennema, O., and Powrie, W. D., Enzyme-catalyzed reactions as influenced by inert gases at high pressures, *J. Food Sci.*, 34, 370–375, 1969.
4. Blakistone, B. A., Ed., *Principles and Applications of Modified Atmosphere Packaging of Foods*, 2nd ed., Blackie Academic & Professional, London, England, 1998.
5. Brody, A. L., *Controlled/Modified Atmosphere/Vacuum Packaging of Foods*, Food & Nutrition Press, Trumbull, CT, 1989.
6. Brody, A. L., Packaging: part IV — controlled/modified atmosphere/vacuum food packaging, *Wiley Encyclopedia of Food Science and Technology*, 2nd ed., Vol. 3, Francis, F. J., Ed., Wiley, New York, pp. 1830–1839, 2000.
7. Brody, A. L., Modified atmosphere packaging, In *Encyclopedia of Agricultural, Food, and Biological Engineering*, Heldman, D. R., Ed., Marcel Dekker, New York, pp. 666–670, 2003.
8. Brown, W. E., *Plastics in Food Packaging: Properties, Design and Fabrication*, Marcel Dekker, New York, 1992.
9. Burg, S. P., *Postharvest Physiology and Hypobaric Storage of Fresh Produce*, CABI Publishing, Wallingford, Oxfordshire, England, 2004.
10. Codex. *Hazard Analysis and Critical Control (HACCP) System and Guidelines for its Application, CAC/RCP 101969, Rev. 3*, Codex Alimentarius Commission, FAO/WHO, Rome, 1997.
11. Codex, *Code of Hygienic Practice for Refrigerated Packaged Foods with Extended Shelf Life*, Alinorm 97/13, Codex Alimentarius Commission, FAO/WHO, Rome, 1997.
12. Daniels, J. A., Krishnamurthi, R., and Rizvi, S. S. H., A review of the effects of carbon dioxide on microbial growth and food quality, *J. Food Prot.*, 48, 532–537, 1985.
13. Davis, H. K., Fish and shellfish, In *Principles and Applications of Modified Atmosphere Packaging of Foods*, 2nd ed., Blakistone, B. A., Ed., Blackie Academic & Professional, London, England, 1998, chap. 9.
14. Day, B. P. F., Novel MAP: a brand new approach, *Food Manuf.*, 73(11) p. 22, 1998.

15. Devlieghere, F., Gil, M. I., and Debevere, J., Modified atmosphere packaging (MAP), In *The Nutrition Handbook for Food Processors*, Henry, C. J. K. and Chapman, C., Eds., CRC Press, Boca Raton, FL, 2002, chap. 16.
16. Devlieghere, F. and Debevere, J., MAP, product safety and nutritional quality, In *Novel Food Packaging Techniques*, Ahvenainen, R., Ed., CRC Press, Boca Raton, FL, 2003, chap. 11.
17. Dilley, D. R., Historical aspects and perspectives of controlled atmosphere storage, In *Food Preservation by Modified Atmospheres*, Calderon, M. and Barkai-Golan, R., Eds., CRC Press, Boca Raton, FL, 1990, chap. 10.
18. Farber, J. M. and Dodds, K. L., Eds., *Principles of Modified Atmosphere and Sous Vide Product Packaging*, Technomic Publishing, Lancaster, PA, 1995.
19. Farber, J. M., Harris, L. J., Parish, M. E., Beuchat, L. R., Suslow, T. V., Gorney, J. R., Garrett, E. H., and Busta, F. F., Chapter IV Microbiological safety of controlled and modified atmosphere packaging of fresh and fresh-cut produce, *Compr. Rev. Food Sci. Food Saf.*, 2(Suppl.), 142–159, 2003.
20. Georgala, D. and Davidson, C.M., Packages for perishable foodstuffs, *French Patent*, 6,909,728, 1969.
21. Gibson, A. M., Ellis-Brownlee, R.-C. L., Cahill, M. E., Szabo, E. A., Fletcher, G. C., and Bremer, P. J., The effect of 100% CO_2 on the growth of nonproteolytic *Clostridium botulinum* at chill temperatures, *Int. J. Food Microbiol.*, 54, 39–48, 2000.
22. Gorris, L. G. M. and Peck, M. W., Microbiological safety considerations when using hurdle technology with refrigerated processed foods of extended durability, In *Sous Vide and Cook-Chill Processing for the Food Industry*, Ghazala, S., Ed., Aspen Publishers, Gaithersburg, MD, 1998, chap. 9.
23. Graham, A. F., Mason, D. R., Maxwell, F. J., and Peck, M. W., Effect of pH and NaCl on growth from spores of non-proteolytic *Clostridium botulinum* at chill temperatures, *Lett. Appl. Microbiol.*, 24, 95–100, 1997.
24. Hui, Y. H. and Nip, W.-K., Safety of vegetables and vegetable products, In *Handbook of Vegetable Preservation and Processing*, Hui, Y. H., Ghazala, S., Graham, D. M., Murrell, K. D. and Nip, W.-K., Eds., Marcel Dekker, New York, 2004, chap. 29.
25. Kader, A. A., Zagory, D., and Kerbel, E. L., Modified atmosphere packaging of fruits and vegetables, *Crit. Rev. Food Sci. Technol.*, 28, 1–30, 1989.
26. Kader, A. A. and Watkins, C. B., Modified atmosphere packaging — towards 2000 and beyond, *Hort. Technol.*, 10, 483–486, 2000.
27. Kohman, E. F., "Frigi-Canning", *Food Technol.*, 14, 254–257, 1960.
28. Leistner, L., Rödel, W., and Krispien, K., Microbiology of meat and meat products in high- and intermediate-moisture ranges, In *Water Activity: Influences on Food Quality*, Rockland, L. B. and Stewart, G. F., Eds., Academic Press, New York, pp. 855–916, 1981.
29. Molin, G., Modified atmospheres, In *The Microbiological Safety and Quality of Food*, Vol. 1, Lund, B. M., Baird-Parker, T. C. and Gould, G. W., Eds., Aspen Publishers, Gaithersburg, MD, 2000, chap. 10.
30. O'Connor-Shaw, R. and Reyes, V., Use of modified atmosphere packaging, In *Encyclopedia of Food Microbiology*, Vol. 2, Robinson, R. K., Batt, C. A. and Patel, P. D., Eds., Academic Press, San Diego, pp. 410–416, 2000.
31. Parry, R. T., Ed., *Principles and Applications of Modified Atmosphere Packaging of Foods*, Blackie Academic & Professional, London, England, 1993.
32. Phillips, C. A., Review: modified atmosphere packaging and its effects on the microbiological quality and safety of produce, *Int. J. Food Sci. Technol.*, 31, 463–479, 1996.
33. Powrie, W.D., Chiu, R., Wu, H. and Sekura, B.J., Preservation of cut and segmented fresh fruit pieces, U.S. Patent, 4,895,729, 1990.
34. Ready, C.A., Method of preparing and preserving ready-to-eat foods, U.S. Patent, 3,607,312, 1971.
35. Rhodehamel, E. J., FDA's concerns with sous vide processing, *Food Technol.*, 46(12), 73–77, 1992.
36. Schvester, P. and Saunders, R., Method for preservation of fresh fish or sea-food, U.S. Patent, 5,108, 656, 1992.

37. Sivertsvik, M., Rosnes, J. T., and Bergslien, H., Modified atmosphere packaging, In *Minimal Processing Technologies in the Food Industry*, Ohlsson, T. and Bengtsson, N., Eds., CRC Press, Boca Raton, FL, 2002, chap. 4.
38. Spencer, K. C. and Humphreys, D. J., Argon packaging and processing preserves and enhances flavor, freshness and shelf life of foods, In *Freshness and Shelf Life of Foods, ACS Symposium Series #836*, Cadwallader, K. R. and Weenen, H., Eds., American Chemical Society, Washington, DC, 2003, chap. 20.
39. Spencer, K. C., Schvester, P., and Boisrobert, C. E., Method for regulating enzyme activities by noble gases, U.S. Patent 5,707,842, 1998.
40. Walker, S. J. and Betts, G., Chilled foods microbiology, In *Chilled Foods: A Comprehensive Guide*, 2nd ed., Stringer, M. and Dennis, C., Eds., CRC Press, Boca Raton, FL, 2000, chap. 7.

16 Packaging of Flesh Foods

CONTENTS

I. INTRODUCTION

There is a continuous search for improved methods of transporting food products from producers to consumers. With the increasing urbanization of society, the problems associated with the keeping quality of fresh flesh foods have become accentuated. Large livestock slaughter and processing facilities have developed in areas where livestock production is highly concentrated; not surprisingly, such areas are well away from the centers of population density. In addition, there is a major world trade in fresh and preserved flesh foods. In all of these situations, packaging has a key role to play in protecting the product from extrinsic environmental influences and providing the shelf life required in the particular market concerned.

II. RED MEAT

In the packaging of red meat, there are two factors of major importance: color and microbiology. Both of these factors will be considered in some detail before specific packaging materials and systems are discussed, because an understanding of red meat color and microbiology is a prerequisite for the development of successful packaging for red meat. Color is generally regarded as the first limiting factor in shelf life, but when inappropriate hygienic and packaging methods are used, microbial spoilage is the principal determinant of meat product acceptability.

A. Color of Red Meat

1. Introduction

The importance of color as a marketing attribute of red meat is well established, especially for self-service retailing. Used to seeing bright red meat prepared for sale, consumers associate this color with good eating quality, although there is little correlation between the two. This association of the color of red meat (both in the chilled and frozen form) with freshness has been the dominant factor underlying retail meat marketing. The loss of this bright red color, known as loss of "bloom" in the industry, is affected by many factors, although the consumer will usually associate color loss with bacterial growth. Because of the importance of meat color, various methods of transportation, distribution and packaging have evolved that optimize the maintenance of a desirable meat color.

The color of meat as perceived by the consumer is primarily determined by such factors as the concentration and chemical form of the meat pigment myoglobin, the morphology of the muscle structure and the ability of the muscle to absorb or scatter incident light.

2. Myoglobin Pigments

Myoglobin is the principal pigment of fresh meat and the form that it takes is of prime importance in determining the color of the meat. The myoglobin molecule consists of a heme nucleus attached to a globulin-type protein component. The heme group consists of a flat porphyrin ring with a central iron atom, which has six bonding points or coordination links. Four of these are linked to nitrogen atoms; one is attached to the globin molecule and the remaining linkage is free to bind to other substances, usually water or O_2. The color of myoglobin depends on at least three factors:

1. The oxidation state of the iron atom: it may be in the reduced ferrous form (Fe^{2+}) or the oxidized ferric form (Fe^{3+})
2. The nature of the group at the sixth bonding point of iron
3. The state of the globin: it may be native as in raw meat, or denatured as in cooked meat

The color of fresh meat depends chiefly on the relative amounts of the three pigment derivatives of myoglobin present at the surface: myoglobin (Mb) also referred to as *deoxymyoglobin*;

TABLE 16.1
Main Forms of Pigment Found in Fresh Red Meat

Pigment	State of Iron	Sixth Linkage	Color
Myoglobin (Mb)	Fe^{2+}	H_2O	Purple
Oxymyoglobin (O_2Mb)	Fe^{2+}	O_2	Bright red
Metmyoglobin (MetMb)	Fe^{3+}	H_2O	Brown

oxymyoglobin (O_2Mb) and *metmyoglobin* (MetMb). Myoglobin (Mb) is purple in color ($\lambda_{max} = 555$ nm) and predominates in the absence of O_2. Oxymyoglobin (O_2Mb) is bright red in color ($\lambda_{max} = 542$ and 580 nm) and results when Mb is oxygenated or exposed to O_2, producing the familiar bloom of fresh meat. Metmyoglobin (MetMb) is brown in color ($\lambda_{max} = 505$ and 635 nm) and exists when the O_2 concentration is between 0.5 and 1%, or when meat is exposed to air for long periods of time. The brown MetMb cannot bind O_2 and is physiologically inactive; the sixth coordination position is occupied by water. The main forms of pigment found in fresh red meat are listed in Table 16.1.

The color reactions of myoglobin are all reversible and dynamic with respect to the three primary forms of myoglobin: Mb, O_2Mb and MetMb. It is rare when all the myoglobin pigments in meat are in the same form; generally, two or more of the pigments will be present with the predominant pigment being most noticable. Although MetMb cannot take up O_2, enzymes present in fresh meat are capable of reducing MetMb to Mb, which can then take up O_2 to form O_2Mb. As meat ages, the substrate for these enzymes is gradually used up; MetMb can no longer be reduced and the meat appears brown. The situation is summarized in Figure 16.1.

In extreme conditions, the pigment may be decomposed; the heme portion becomes detached from the protein, the porphyrin ring is disrupted and, finally, the iron atom is lost from the heme structure. Green choleglobin and colorless bile pigments are formed.

During storage, the rate of MetMb accumulation on the surface of red meat is related to many intrinsic factors including pH, muscle type and the age, breed, sex and diet of the animal, as well as

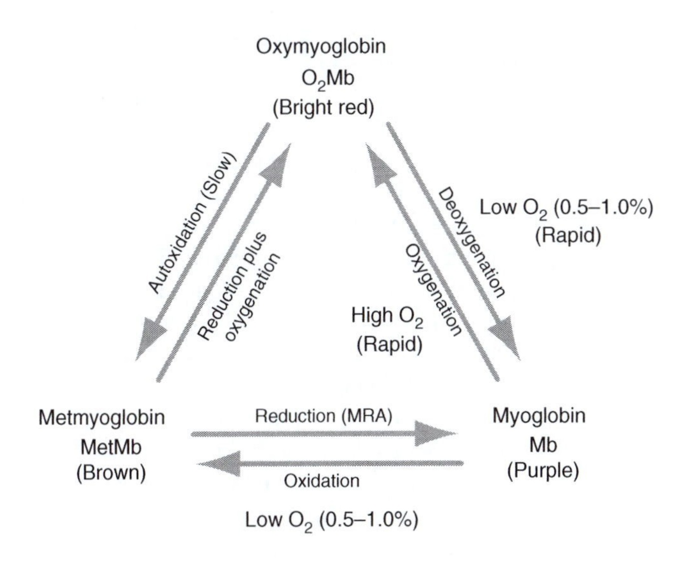

FIGURE 16.1 Interrelationships between the three major pigments of fresh red meat (MRA = metmyoglobin reduction activity).

TABLE 16.2
Summary of Major Factors Affecting Red Meat Color

Storage temperature	
High	Favors greater O_2 scavenging by residual respiratory enzymes, plus other oxygen-consuming processes such as fat oxidation
	Enhances dissociation of O_2 from O_2Mb, thereby increasing the tendency for autoxidation of the Mb produced
Low	Promotes increased penetration of O_2 into the surface
	Enhances O_2 solubility in tissue fluids
	Both above effects increase depth of O_2Mb at the surface
Oxygen partial pressure	
High	Favors formation of O_2Mb
Low	Favors formation of MetMb
Meat pH	
High	Accelerates respiratory activity of meat tissue resulting in thin layer of O_2Mb with underlying Mb more apparent
	Muscle fibers swell thereby reducing O_2 diffusion and therefore formation of O_2Mb
Low	Causes denaturation of the globin moiety and subsequent dissociation of O_2 from the heme
	Promotes oxidation of Mb

extrinsic factors including preslaughter conditions and the processing conditions used (e.g., electrical stimulation, hot boning, chilling mode). During retail display, physical factors such as temperature, O_2 availability, type and intensity of lighting, microbial growth and gas atmosphere surrounding the meat surface influence the shelf life of fresh red meats. A summary of the major factors affecting red meat color is presented in Table 16.2.

3. Role of Oxygen

a. *Oxygenation and Deoxygenation*

The reversible conversion of Mb to O_2Mb is described as *oxygenation* because O_2 becomes part of the pigment complex. This is not an oxidation reaction, however, because the iron component remains in the ferrous state (i.e., there is no loss of an electron). In the presence of O_2, Mb can exist either in the O_2Mb or MetMb state depending on the amount of O_2 present. Mb is the dominating form at 0% O_2, MetMb dominates at approximately 0.5 to 1%, depending on temperature and O_2Mb is the dominating form over approximately 4% O_2.[30] Owing to a lack of O_2, the color of the interior of fresh red meat is purple because of the presence of Mb. The outer most layer exposed to air is red owing to the presence of O_2Mb. A thin brown layer of MetMb can be observed between the red and purple regions resulting from a low concentration of O_2. This layer thickens within 1 or 2 days and becomes apparent first by the darkening of the translucent surface tissue and later by breaking through to the surface. At lower O_2 partial pressures, the MetMb layer will be nearer the surface until at the critical partial pressure it is at the surface. The rate of these changes is strongly influenced by temperature.

Oxygenation of purple Mb to red O_2Mb is rapid, with the surface of beef in air appearing red within half an hour at 5°C. In contrast, oxidation to MetMb is slow; it appears first as a thin brown layer at the limit of O_2 penetration. The red O_2Mb is stable as long as the heme remains oxygenated, although the O_2 is continually associating and dissociating from the heme.

The dissociation of O_2 from the heme (known as *deoxygenation*) is caused by conditions such as low pH (less than pH 5.4), high temperature, UV light, salts and especially low O_2 tensions. The deoxygenation of red O_2Mb results in Mb (which is very unstable), which then becomes oxidized to brown MetMb. Surface dessication increases the salt concentration and promotes the formation of MetMb.

b. Oxidation

Oxygen has sufficient oxidation potential to convert O_2Mb and Mb to MetMb. This conversion will occur as NADH and substrates from the glycolytic pathway are depleted. The conditions that cause deoxygenation of O_2Mb to Mb are also responsible for the conversion of unstable reduced Mb to brown MetMb.

The oxidation reaction of O_2Mb to form MetMb, and Mb to form MetMb, involves the loss of an electron from the iron of the heme ($Fe^{2+} \rightarrow Fe^{3+}$) and is termed *autoxidation*; it is a comparatively slow reaction.[60] The maximum rate of MetMb formation has been reported at O_2 partial pressures of 7.5 ± 3 mm Hg at 7°C and 6 ± 3 mm Hg at 0°C for beef (one atmosphere of pure O_2 is 760 mm Hg). Any value is very dependent on temperature, although independent of pH.[37] If meat is held under high O_2 partial pressures (i.e., above 30 mm Hg), then the rate is independent of the O_2 partial pressure and autoxidation of Mb is minimized.[44] These results have led to the development of systems allowing transport of meat at high ($>60\%$) O_2 partial pressures.

c. Reduction

Biochemical reduction is a process in which a chemical compound gains an electron; it works in opposition to oxidation. *MetMb reduction* refers to the conversion of Fe^{3+} heme iron to Fe^{2+} heme iron. When this occurs, the concentration of MetMb decreases and that of Mb or O_2Mb increases.[12] Although MetMb is stable, it is slowly reconverted to Mb in *post rigor* meat, primarily by enzymic reactions collectively termed *metmyoglobin reducing activity* (MRA). These enzymic reactions are believed to be independent of O_2 partial pressure.[37] Some muscles will remain bright red for longer periods of time than other muscles because any MetMb formed is reduced back to Mb and immediately oxygenated back to red O_2Mb; such muscles are said to have a higher MRA. Loss of reducing activity in meat during storage is a result of a combination of factors including fall in tissue pH, depletion of required substrates and cofactors and, ultimately, complete loss of structural integrity and functional properties of the mitochondria.[44]

Although the formation of MetMb in meat can be inhibited by packaging in vacuum or high O_2 atmospheres, other approaches have been attempted including feeding of α-tocopherol (vitamin E) to the animals or treatment of the meat with vitamin C.[37]

d. Bacterial Discoloration

Bacteria cause discoloration of meat in their logarithmic growth phase. This is attributed to the high O_2 demand of aerobic bacteria, which reduces the O_2 tension at the meat surface and causes the formation of brown MetMb.

Some bacteria also produce by-products that oxidize the iron molecule. The most common are hydrogen sulfide (H_2S) and hydrogen peroxide (H_2O_2), which react with unstable Mb to produce the green pigments sulfmyoglobin and choleglobin, respectively. H_2S production causes green discoloration on vacuum-packaged meat; it is generally found only on meat with a high pH ($pH > 6$). On opening vacuum packaged meat that has green discolorations on it, the green sulfmyoglobin becomes oxygenated to oxysulfmyoglobin, which is red in appearance. However, the color change does not cause any diminution in the H_2S off-odor resembling rotten eggs. H_2O_2 can either cause greening owing to oxidation of heme pigments to form choleglobin or degradation beyond porphyrins to yellow or colorless bile pigments.

4. Color Intensity

The color intensity of meat is determined by *ante mortem* factors such as species, stress, sex and age of the animal, the *post mortem* pH rate of decline and the ultimate pH of the meat. During *post mortem* glycolysis, the pH of normal tissue falls from an *in vivo* value of 7.2 to an ultimate value of

5.4 to 5.6. As well as influencing the color, the rate of fall of pH and the ultimate pH value also influence the *water-holding capacity* (WHC) and texture of the meat. Differences in color intensity between species are primarily caused by differing concentrations of Mb. Thus, beef, which has the highest concentration, is the darkest of the meat species. Lamb is intermediate in color and Mb concentration and pork has the lowest concentration of Mb and, as such, is the lightest in color.

The color intensity of meat is also influenced by pH and morphology of the muscle structure. The pH of muscle is largely influenced by the conditions that exist immediately prior to or just after slaughter. Short-term violent excitement immediately before slaughter or the slow cooling of carcasses after slaughter can result in meat having a low ultimate pH. This is caused by a buildup of lactic acid, which is formed as a metabolic end product in the anaerobic breakdown of glucose and glycogen.[45] This condition is known as PSE because it produces *p*ale, *s*oft and *e*xudative meat as a result of an abnormally rapid fall in pH immediately after slaughter, when the combined effects of low pH and high temperature in the muscle lead to denaturation of sarcoplasmic and myofibrillar proteins. Because muscle glycolysis is relatively more rapid in hogs than bovines, PSE is rarely observed in beef animals.

PSE meat causes problems in packaging; the color is abnormally pale due to denatured protein, the physical structure increasing light scattering, making the meat more opaque, and autoxidation increases, which causing color fading. In addition, because of its low WHC resulting from protein denaturation, PSE meat produces excessive drip.

Biochemical conditions directly opposite to those producing PSE meat give rise to another type of abnormal meat from all meat animals but more especially beef cattle and hogs. This condition has been described as even more troublesome than PSE from a packaging point of view.[26] The problem meat, known as dark-cutting or *d*ark, *f*irm *d*ry (DFD), is translucent and sticky to touch, and unacceptable for retail packaging because of its dark purple color. This meat has a high pH that results from a low residual glycogen level remaining in the muscle at slaughter, which is attributable to excessive stress (such as an extended transit haul) or exercise prior to slaughter. Such treatment causes depletion of the muscle glycogen before death. The low residual glycogen means that insufficient lactic acid is formed by glycolysis during the rigor process to lower the pH.

The muscle fibers of DRD meat are swollen and tightly packed together forming a barrier to the diffusion of O_2 and the absorption of light. In addition, the high pH of the meat accelerates respiratory activity of the meat tissue, resulting in a very thin layer of red O_2Mb, with the underlying purple Mb more visually apparent. Moreover, the high pH encourages the growth of putrefactive micro-organisms, thereby significantly reducing the keeping quality of this meat.[26] For example, vacuum-packaged beef of normal pH can be stored at chill temperatures for periods in excess of 10 weeks, whereas DRD beef held under the same conditions will generally spoil within about 6 weeks. However, it has been shown that both normal and high pH cuts of beef packaged under CO_2 can remain unspoiled at 15 weeks.[22]

5. Role of Carbon Dioxide and Carbon Monoxide

It was reported in the 1970s that meat stored in high concentrations of CO_2 often developed a grayish tinge, supposedly because of the lowering of the pH and subsequent precipitation of some sarcoplasmic proteins. It was therefore recommended that CO_2 not be used in concentrations exceeding 20%. The claim was also made that no MetMb formation occurred in atmospheres containing high concentrations of CO_2, provided that the O_2 partial pressure exceeded 5%. A later report indicating that 50 to 80% CO_2 was often found in residual air spaces in vacuum packages with no associated detrimental effect on meat color led to a re-evaluation of the effect of CO_2 on red meat color.

The realization that discoloration in high CO_2 atmospheres must result from O_2 led to the use of 100% CO_2 atmospheres on the grounds that meat color would not be adversely affected if O_2 was rigorously excluded from the packages. In the case of raw red meats, the color was actually improved by enzymic reduction of pigment that oxidized before the meat was packaged. However, the stability of raw meat color on exposure to air decreased as storage life increased, with the color stability at display reaching a minimum value after 12 weeks storage. After that time, acceptable meat color at display was maintained for about 40% of the time that fresh meat retained acceptable color. This was claimed to be an inevitable consequence of prolonged chill temperature storage and not an effect of the 100% CO_2 atmosphere.[18] Recently, 36 studies on the effect of CO_2 on red meat color have been reviewed[29] and the importance of excluding O_2 confirmed.

Carbon monoxide (CO) has the potential to retard MetMb formation and fat oxidation. CO combines with Mb to form the bright cherry-red pigment carboxymyoglobin (COMb), which is spectrally very similar to O_2Mb. Because COMb is much more stable toward oxidation than O_2Mb by virtue of the stronger association of CO to the iron-porphyrin site on the myoglobin molecule, the addition of CO at low levels can negate the detrimental color changes associated with the high levels of CO_2 required to maintain wholesomeness over prolonged transit times.[52]

Retail meat can be packaged in gas mixtures containing 60 to 70% CO_2, 30 to 40% N_2 and $<0.5\%$ CO. This gas mixture provides a unique combination of a long microbiological shelf life and a stable, cherry-red color of the meat. The shelf life of meat packaged in the CO mixture is longer than that of meat packaged in the commonly used high O_2 atmospheres of approximately 70% O_2 and 30% CO_2.[33,52] A recent study[28] using 0.4% CO with beef steaks and ground beef found that the COMb formed did not mask microbial spoilage.

Despite these advantages, CO has not (with one notable exception) been approved by regulatory agencies for retail use with meats, largely because of safety considerations. The notable exception is Norway where the packaging of red meat in retail packs with 0.4% CO has been permitted since 1985; it is reportedly also permitted in some Asian countries. In early 2002, the U.S. FDA cleared a system for pretreating meat with 0.4% CO in a master pack system, where the CO was considered as a processing aid and thus did not require labeling. However, the use of sealed retail packages containing CO is prohibited in the U.S.

6. Lighting

Lighting is an important factor when presenting meat and meat products for retail sale. A high percentage of red in the light gives a particularly intense red impression, and products that are already beginning to turn a grayish-red in daylight appear saturated with red when exposed to reddish lighting.

Incident light is a contributory factor in the discoloration of both fresh and frozen meat as, for example, in a retail supermarket display case. The extent of the effect depends on such factors as the wavelength and intensity of the light, temperature, oxygen partial pressure, meat pH and storage time.[26] Provided that UV wavelengths are avoided, the effect of light is small under usual refrigerated conditions.

7. Effect of Temperature

The optimum temperature for storing chilled meats is the minimum that can be maintained indefinitely without freezing the muscle tissue; in practice, this temperature is in the range -1.5 ± 0.5°C.[19] Discoloration (rather than microbial spoilage) is likely to limit the shelf life of red meats when they are stored in aerobic atmospheres, with the rate of discoloration increasing linearly with temperature. For meats stored in MAs rich in O_2, the effect of temperature on the rate of discoloration does not appear to be well identified in the literature, but it seems likely that it follows a similar pattern as for meat stored in air.[19]

When meat is stored anaerobically, the color stability of muscle tissue increases at first (probably related to the relatively rapid loss of respiratory activity) and then declines (probably reflecting the decay of MRA). The rate at which color stability degrades is twice as fast at 5°C and four times as fast at 10°C as at 0°C.[19]

8. Effect of Freezing

The color of frozen meat initially depends on the rate of freezing and the resultant size of ice crystals in the surface layer. Slow freezing produces large ice crystals with poor light-scattering properties, giving the meat a dark, translucent appearance. Fast freezing, on the other hand, results in the formation of small ice crystals, which scatter light and make the meat surface pale and opaque.

In frozen red meat, the principal deterioration during storage is photo-oxidation of the pigment. Whereas under direct illumination, chilled meat oxidation begins in the subsurface layer and progresses towards the surface, frozen meat oxidizes from the surface inwards. Loss of redness is detrimental to the marketing of frozen as well as chilled meat.

Considerable improvements (in the order of 5 to 10 times) in the color shelf life of prepackaged frozen beef are possible if the light intensity in a normal retail cabinet is reduced by 90 to 95%.

B. Microbiology of Red Meat

1. Introduction

The composition of the flora that is present on meat at the onset of spoilage is affected by the qualities of the tissue on which the bacteria are growing, the composition of the atmosphere around the product, and the numbers and composition of the microflora on the meat at the time of packaging.[20] Spoilage flora is generally dominated by those species that can grow most rapidly under the conditions at which the meat is held. At chill temperatures, species of *Pseudomonas* and *Lactobacillus* can outgrow competing species under aerobic and anaerobic conditions, respectively.

On meat stored aerobically (i.e., in air or in O_2-permeable packaging materials) at chill temperatures (-1 to $+5$°C), the flora is usually dominated by species of *Pseudomonas*. These organisms are strictly aerobic and can grow at their maximum rate with O_2 concentrations in the atmosphere of 1% or less. High CO_2 and low O_2 tension depress the growth of pseudomonads. Microbial degradation of refrigerated beef in an aerobic environment produces a gradual elevation in pH leading to the release of NH_3 following microbial attack on the amino acids. The critical factor is the availability of glucose to the bacteria.[20] Most spoilage species (including the dominant pseudomonads) utilize this substrate preferentially. When bacteria can no longer obtain sufficient glucose, they begin to degrade amino acids, giving rise to NH_3 as well as other odors and highly offensive end products.

Any O_2 remaining in vacuum packages immediately after sealing is converted to CO_2 by meat tissue respiration and bacterial activity. In an anaerobic environment such as that which exists in vacuum- and some MA-packaged meats, the absence of O_2 inhibits the growth of the putrefying aerobes, and anaerobic and facultative anaerobic strains such as *Lactobacillus*, *Brochothrix thermosphacta* and *Enterobacteriaceae* constitute the dominant microbial flora. However, in many cases, lactic acid bacteria (particularly leuconostocs species) become dominant in refrigerated, vacuum-packaged meat. However, these bacteria do not produce grossly offensive byproducts.[20] Inclusion of small amounts of CO in anaerobic atmospheres does not affect development of the spoilage flora.[53]

At least five factors are implicated in slowing or preventing growth of spoilage organisms in vacuum packages: (1) reduced O_2 concentration; (2) elevated CO_2 concentration; (3) normally low (5.5 to 5.8) pH of meat; (4) low storage temperature; (5) antimicrobial activities of *Lactobacilli* species.

Raw meat cannot normally be stored for extended periods under refrigeration (-1 to $+4$°C) because of the proliferation of psychrotrophic food spoilage micro-organisms and texture degradation caused by the presence of endogenous proteolytic enzymes.

2. Effect of Temperature

Temperature is of the utmost importance in controlling the types of micro-organisms that develop on meats, because these products are normally held at chill temperatures. Essentially, all studies on the spoilage of red meats, poultry and seafood carried out over the past 30 years have dealt with products stored at chill temperatures. The temperature not only determines whether microbial numbers increase or decrease, but also influences the nature of the flora that becomes dominant. Low temperatures select for cold-tolerant organisms (mainly psychrotrophs), and these become the most important component of the spoilage flora. The storage life of meat with or without packaging at 0, 2 and 5°C is about 70, 50 and 30%, respectively, of the storage life that would be obtained if the product were stored at −1.5°C. Storage at chill temperatures can delay, but not prevent, the ultimate onset of microbial spoilage.[19]

It must be emphasized that no matter how good the packaging is, the maximum storage life of packaged meat will not be achieved unless very close attention is paid to control of the storage temperature. Although this can be relatively easily achieved in laboratory trials, the temperature control of commercial quantities of packaged meats is often variable, with the result that product shelf lives are also variable. When the expected shelf life is not achieved, attention should be focused on the temperatures achieved during storage, distribution and retailing, rather than blaming the packaging or those who conducted the laboratory trial.

3. Effect of Gaseous Atmosphere

Any form of packaging changes the environment to which micro-organisms are exposed, with the main change being a modification of the gaseous atmosphere. It is the composition of this atmosphere that largely determines the extent and type of spoilage that develops during storage. Although a number of different gas compositions have been found to have preserving properties on flesh foods, pure CO_2 or combinations of CO_2 with O_2 or N_2 are by far the most important. CO_2 is highly soluble in both muscle and fat tissues, where the solubility in muscle tissue decreases with decreasing pH and increasing temperature, and the solubility in fat increases with increasing temperature within the chill temperature range.[19]

The solubility of CO_2 in muscle tissue of pH 5.5 at 0°C is approximately 960 mL kg^{-1} of tissue at STP. As the temperature increases, the solubility decreases by 19 mL kg^{-1} for each °C rise. As tissue pH increases, the solubility decreases by 360 mL kg^{-1} for each pH unit. With raw meat, CO_2 at ≈1.5 L kg^{-1} meat is required to saturate the product without the packaging collapsing tightly around the contents.[17] The large gas volume initially required can result in pouches at first overfilling the cartons in which they are contained. However, the gas dissolves sufficiently rapidly for the pack volume to fall to its final value during overnight storage.[18]

Models for CO_2 solubility as a function of packaging and storage parameters such as the ratio of product weight to headspace volume, temperature and initial CO_2 levels have been developed.[61] The amount of absorbed CO_2 ranges from 0 to 1.8 L CO_2 kg^{-1} of meat depending on the applied CO_2 partial pressure, temperature, pH of the meat and so on. Equilibrium with respect to gas absorption is obtained during the first 1 to 2 days, although microbial and meat metabolism can cause slight changes in gas composition by consuming O_2 and producing CO_2.[30]

A large number of papers has been published on the effects of CO_2 on flesh foods and on the inhibitory properties of CO_2 on specific micro-organisms. CO_2 is effective when the spoilage microflora is dominated by Gram-negative, aerobic, psychrotrophic bacteria, thus making the use of CO_2 with fresh meats, poultry and seafood efficacious.

CO_2 selectively inhibits the growth of Gram-negative bacteria such as pseudomonads and related psychrotrophs, but has less effect on lactic acid bacteria. The growth rate of pseudomonads decreases with increasing CO_2 concentrations up to about 20%, concentrations beyond that level doing little to reduce the growth rate provided the atmosphere is aerobic. The maximum reduction in the rates of growth of pseudomonads is about 50%, which is sufficient to allow lactic acid bacteria to outgrow pseudomonads and dominate the spoilage flora.[16]

There are four major effects of CO_2 on food spoilage:

1. The exclusion of O_2 by replacement with CO_2 may contribute slightly to the overall effect by slowing the growth rate of aerobic bacteria.
2. The ease with which CO_2 penetrates the cell may facilitate its chemical effects on the internal metabolic processes.
3. CO_2 is able to produce a rapid acidification of the internal pH of the cell with possible ramifications relating to metabolic activities.
4. CO_2 appears to exert an effect on certain enzyme systems.

With regard to the optimal concentration of CO_2, there is considerable ambiguity among the reported values of various researchers, as well as variations in methodologies. It is important to appreciate the influence of temperature on the growth-inhibitory effect of CO_2 on micro-organisms. Generally, the relative inhibitory effect decreases as the temperature increases. However, because the solubility of CO_2 decreases as the temperature increases, it is necessary to take this into account before interpreting any results. Many results were obtained in liquid media, but the temperature effect may not show the same characteristics in muscle foods where the microbial activity is located at the surface. In this situation, one is dealing with a solid–gas interface rather than a population dispersed in a liquid medium.

The high partial pressure of oxygen-free CO_2 extends the lag phase and the generation time and slows the growth rate of all spoilage organisms, the types that cause spoilage being either totally inhibited or subjected to a long lag time of at least 12 weeks at −1°C. The only potent spoilage organism not fully controlled is *Brochothrix thermosphacta*; however, its minimum growth temperature is raised in an oxygen-free CO_2 atmosphere from below the freezing point of meat to above 0°C. As a consequence, even heavily contaminated, high pH meats stored at −1°C will remain unspoiled for at least 3 months.[16]

Unlike CO_2, the solubility of O_2 in muscle and fat tissues is low. However, O_2 is converted to CO_2 by the respiratory activities of both muscle tissue and bacteria. When packaging fresh meat products, an elevated O_2 partial pressure needs to be maintained to ensure that the meat pigment is in the O_2Mb state. Using response surface methodology, it has been shown that the O_2 level has no adverse effect on the color shelf life in the interval between approximately 40 to 80%. Despite this finding, an O_2 level of 70 to 80% is commonly used commercially, even though a level of 40% is sufficient to ensure stability of the bright red color.[30]

The solubility of N_2 in tissues is low and the gas is metabolically inert. Therefore, the only function of N_2 in a package atmosphere is to buffer against changes in the volume of the atmosphere that could lead to package collapse and crushing of the contents.[19]

CO is a potent inhibitor of cytochrome oxidase and hence aerobic respiration of many organisms. However, the effect of low levels ($<1\%$) of CO on the microbiological condition of meat is negligible.[52] This is partly attributable to the fact that CO is removed quickly from the package atmosphere as it reacts rapidly and essentially irreversibly with Mb to form the cherry-red carboxymyoglobin.

C. Lipid Oxidation

Lipid oxidation is a leading cause of quality deterioration in muscle foods, resulting in rancidity, off-flavors, off-odors as well as color and texture deterioration.[30] It is believed to be initiated in the highly unsaturated phospholipids fraction in subcellular membranes. The major products of lipid oxidation are the hydroperoxides, which break down into secondary products such as aldehydes, alcohols, hydrocarbons and ketones. It is these secondary products that contribute to the off-flavors generated during storage.[59] A distinctive off-flavor that develops rapidly in meat that has been precooked, chill-stored and reheated is termed warmed-over flavor and autoxidation of membrane phospholipids is widely accepted as the cause.[30]

Mechanisms of myoglobin-induced oxidation/peroxidation of polyunsaturated fatty acids are still a matter of dispute.[3] At pH values of relevance for meat and meat products, both MetMb and O_2Mb have been shown to be major initiators of lipid oxidation and peroxidation. An in-depth knowledge of the relationship between lipid and protein oxidation might reveal important information regarding the oxidative stability of meat and meat products.

Lipid oxidation is not usually a limiting factor in conventional retail trays of meat overwrapped with a polyolefin film as such films are permeable to gases and allow volatile odors to escape. However, with MA packs the volatile products are retained within the package and can be clearly detected by consumers when opened.[59] Storage of meat in high O_2 atmospheres leads to a limited shelf life owing to lipid oxidation.[30]

D. Vacuum Packaging of Fresh Meat

In the U.S. prior to 1967, beef carcass shipments were traditionally made to retail outlets where carcasses were fabricated into retail items by butchers in individual stores. The "boxed beef" concept was developed by French scientists as early as 1932 to prolong the shelf life of frozen meat used as military provisions.[59] It was introduced in the U.S. in 1967, dramatically changing beef processing, distribution and retail fabrication without affecting retail presentation to the consumer.[5] The basis for the boxed beef concept was vacuum packaging into plastic bags with low permeabilities to gases.

The introduction of vacuum packaging for the distribution and storage of chilled beef was one of the greatest innovations in meat handling. Beef carcasses are broken down into primal and subprimal cuts, separated into boneless and bone-in cuts, and then vacuum packaged. Because only about two thirds of a beef carcass is usable meat, there are advantages in reduced refrigerated space for transportation and storage and less packaging material when boneless beef rather than bone-in or carcasses are stored and distributed. Another advantage is that the tenderness of the beef can be improved by aging without the evaporative weight loss incurred when carcasses are hung in the conventional manner.

It is now common practice to butcher beef carcasses at the slaughter house into primal joints, which are then vacuum packaged. In these circumstances, the atmosphere around the meat becomes depleted in O_2 (often $<1\%$ v/v) and enriched in CO_2 ($>20\%$ v/v), resulting in microbial changes quite different from those observed during aerobic storage.[6] Vacuum-packaged boxed beef is then distributed to retail outlets where these primals and subprimals are fabricated into consumer units, overwrapped in O_2-permeable film on PS foam or PVC trays and displayed for sale.[57] Most U.S. beef carcasses are fabricated and converted to vacuum-packaged, primal and subprimal cuts within 64 hours of entering the chill cooler, and normally appear in retail markets 14 to 28 days after fabrication.[50]

Because the a_w of chilled meat is very high, unprotected meat will lose weight by evaporation and its appearance will deteriorate. Further weight loss will occur when meat is cut, because the exposed surfaces exude liquid, which detracts from the appearance of packaged meat. This can be overcome by including an absorbent pad in the base of the package. Although efficient chilling can reduce the quantity of exudate, a certain amount will always be present when meat cuts are held for retailing. This unattractive bloody exudate found in vacuum packaged beef is referred to as "purge," "weep" and "drip"; 1 to 2% purge is considered acceptable, while 4% is considered excessive. Values of 2 to 4% drip can have substantial economic implications if not controlled.

Vacuum packaging achieves its preservative effect by maintaining the product in an O_2-deficient environment (nominally <500 ppm). Any residual O_2 is rapidly consumed by meat and muscle pigments and CO_2 is produced as the end product of tissue and microbial respiration.

In anoxic conditions, potent spoilage bacteria are severely or totally inhibited on low pH (<5.8) meat. However, their growth on high pH muscle tissue, or extensive fat cover of inevitably neutral pH, will result in relatively rapid spoilage in a vacuum pack. Vacuum packaging can therefore extend the shelf life of primal cuts composed largely of low (normal) pH muscle tissue such as beef and venison by about fivefold over that achieved in air. Typical shelf lives of normal pH meat are 12 to 14 weeks at 0°C. For other meats and small cuts, only a twofold extension of shelf life can be safely anticipated.[18]

The shelf life of retail packages of beef is determined by microbial contamination, microbial activity and the MRA of the muscles in the cuts.[21] Exhaustion of MRA usually antedates microbial activity and cuts turn brown (ending shelf life) before bacteria have caused deterioration. Between 2 and 20% of all beef cuts sold in U.S. retail stores are discounted or even discarded as a result of the loss of the desirable cherry-red color that consumers associate with freshness.[50]

1. Vacuum-Packaging Systems

Vacuum packaging involves enclosing large joints (typically 3 to 15 kg in weight) in flexible plastic containers (usually bags) to prevent moisture loss and exclude O_2 from the meat's surface. Packing under a vacuum reduces the volume of air sealed in with the meat. The plastic materials used for vacuum packaging must have low moisture and gas permeabilities and be strong enough to hold heavy beef joints. In packaging fresh meat primals, many cuts contain bones, which are often sharp and abrasive and readily puncture the flexible plastic materials used in vacuum packaging. To overcome bone puncture, a material consisting of a wax-impregnated and coated cotton scrim is employed.

Reported O_2 permeabilities of packaging films are usually measured at ambient temperatures and moderate humidities (typically 23°C and 75% RH), but both temperature and humidity can affect the rates at which gases are transmitted through films. Data on the O_2 permeabilities of packaging films at chill temperatures are sparse, and those that do exist often do not include a complete specification of the film under test or the test conditions. This is further complicated by the variety of test methods and units used. The O_2 permeabilities at subzero temperatures of two plastic films used for the vacuum packaging of meat have been reported.[35] One film was a PA–LDPE laminate, while the other was an EVA copolymer–PVdC copolymer laminate. Their OTRs at −1°C were reported as 2.0 and 0.6 mL m^{-2} day^{-1} atm^{-1}, respectively, approximately 1/50 of the values obtained at 23°C and 90% RH.

Once meat has been vacuum packaged in an O_2 barrier material with an OTR (at 23°C and 75% RH) less than 50 mL m^{-2} day^{-1} atm^{-1} and adequately sealed to prevent air re-entry, the shelf life of the meat is very much the same regardless of the packaging material. Therefore, the significant differences between high O_2 barrier packaging materials/systems are not so much in the structure as they are in the physical properties, the production speeds of the system and the abuse resistance of the package itself. The relationship between O_2 permeability and storage life of vacuum packaged beef is shown in Table 16.3 and clearly indicates the advantages in terms of shelf life of using a high O_2 barrier package.

The O_2 in the small volume of residual air inside a vacuum package is quickly consumed by meat respiration so that, within 2 days, the O_2 partial pressure at the surface of the meat drops below 10 mm Hg. At these very low partial pressures, the penetration limit of O_2 is very near the surface, and the thin brown layer of MetMb that develops cannot conceal the underlying Mb; therefore, the visible color of vacuum-packaged beef is purple.[55]

The efficacy of vacuum packaging depends on close contact between a film of low gas permeability and all product surfaces. If there are vacuities within the pack, then these will develop an O_2-containing atmosphere as gases permeate into the package during storage. Bacterial growth will accelerate and product color will deteriorate because of oxidation of Mb at meat surfaces exposed to such an atmosphere. In addition, when the meat surface is large relative to the meat

TABLE 16.3
Relationship between Film Oxygen Permeability of Vacuum Packs and the Storage Life of Normal-pH Beef at 0°C

Film Permeability (mL m^{-2} atm^{-1} day^{-1} at 25°C)	Storage Life (days)
0	105
190	105
290	77–105
532	42–63
818	24–42
920	14–28

Source: From Newton, K. G. and Rigg, W. J., *J. Appl. Bacteriol.*, 47, 433–441, 1979.[41] With permission.

mass, an extensive film surface is presented for O_2 permeation and color deterioration can also occur. Consequently, vacuum packaging is relatively ineffective for preserving products such as carcasses whose shapes preclude close application of the packaging film to all surfaces. Furthermore, because the anoxic conditions within the package result in the formation of purple Mb, vacuum packaging for retail display packages of red meat has not met with wide success because of consumers' negative perception of the color purple.

Four basic methods are available for vacuum packaging meats, which are now discussed in turn.

a. *Shrink Bag Method*

This system involves placing the meat into a heat shrinkable barrier bag (typically a triple-layer coextruded film constructed from EVA copolymer–PVdC copolymer–EVA copolymer, but sometimes PA is used the barrier layer with an ionomer as the inner or outer layer). The bag is then evacuated prior to sealing. In the past, this was achieved by applying a metal clip around the twisted neck of the bag, but today, heated jaws are used. The bag is then heat shrunk by placing in water at 90°C. After shrinking, the bag conforms closely to the meat and produces a tight vacuum pack. Very high vacuum levels are achieved on rotary single-chamber machines, which also heat seal shrink bags, and owing to their improved productivity and versatility, these machines have become the industry standard. Large primal cuts up to 205 mm deep and 460 mm wide can be vacuum packaged using this type of equipment.

b. *Nonshrink Bag Method*

In this technique, meat is placed into a preformed plastic bag, which is then put in an enclosed chamber that is evacuated. When a predetermined low pressure has been reached, heated jaws close and weld the mouth of the bag. Typical bag constructions consist of laminates or coextrusions, which include PET as the outside layer to provide strength, PA as the middle layer to provide a good O_2 barrier, and inner layers of LDPE, ionomer or EVA copolymer, which are good moisture barriers and can be easily heat sealed. A typical structure would be ionomer–PA–EVA copolymer.

Because the bags used in this system are not heat shrinkable, purge tends to accumulate in corners during storage. Secondary sealing (also referred to as *self-welding*) has been introduced to overcome this problem, and involves passing the vacuum packs through a heating tunnel. Because of the ease with which these films are heat sealed, the excess area of film around the meat is sealed. If performed perfectly, then the packaging material size can be reduced or sealed to a point that approximates the surface area of the meat inside. This prevents purge from spreading into the film wrinkles, folds and corners as well as providing a wider heat seal, which reportedly reduces the incidence of leaking bags.

c. *Thermoforming Method*

In this method, deep trays are thermoformed in-line from a base web of plastic. Meat is placed in the trays and an upper web of plastic is heat sealed under vacuum to form a lid. Generally, the materials used for thermoforming are laminates of PA, PET or PVC, sometimes with a PVdC copolymer coating and heat sealing layers such as LDPE, EVA copolymer or ionomer.

d. *Vacuum Skin Packaging*

Vacuum skin packaging (VSP) involves production of a skin package in which the product is the forming mold. It was first introduced using an ionomer film, which softens on heating to such an extent that it can be draped over sharp objects without puncturing. In this technique, meat portions are skin packed in a barrier film material, the top web of which is softened by heating before applying a vacuum and sealing. During this operation, the soft film molds itself to the shape of the meat to give a skintight package, the meat thus being held under anaerobic conditions. It forms closely around the meat and seals to the base film. Although this type of vacuum pack is an excellent method of presentation, the meat remains in an unoxygenated state, which is not accepted by most consumers, even although it will still oxygenate to the customary red color when exposed to air.

In a comparison with MAP meat,[56] it was found that VSP meat remained acceptable microbiologically for up to at least 2 weeks after packing, during which time the color, although not red, was unchanged. However, the meat retained its ability to develop a bright red color when exposed to air for at least 2 weeks during storage at 1°C. As is typical for vacuum-packaged meat, lactic acid bacteria predominated on VSP samples on which spoilage bacteria grew slowly, if at all, resulting in a long, odor-free shelf life. Off-odors developed much more rapidly in MAP packs than in VSP. In choosing between MAP and VSP it was concluded that the retailer must balance the attractive red color achieved over a relatively short period by MAP against the longer-term stability (albeit with an unaccustomed purple meat color) afforded by VSP. Because long-term stability can be achieved only at the expense of color, the authors were unable to determine a ready compromise between these two divergent packaging systems.

2. Shelf Life of Vacuum-Packaged Red Meats

In any discussion of the shelf life of packaged meats, it must be borne in mind that comparisons of published data are difficult. Moreover, making generalizations is foolhardy because there are a large number of variables that interact to determine the actual shelf life. The most important of these variables is temperature, and the statement that samples of vacuum-packaged meat were held at a particular temperature is not especially helpful, unless the range of temperatures encountered by the samples during storage is specified. Generally, it can be said that laboratory scale trials control temperatures over a much smaller range compared with commercial scale trials, and this is frequently the reason for the longer shelf life obtained in laboratory scale trials compared with that obtained in commercial production.

Other important variables include the microbiological status of the meat at the time of packing and the method used to determine the end of shelf life of the meat. These methods range from objective assessments of microbial counts to subjective assessments by taste panels consisting of trained or untrained members. Clearly, attempting to draw generally applicable conclusions from numerous published reports where the magnitude of these variables differs (or is not even specified) would be misleading.

a. Beef

Vacuum-packaged beef of normal pH can be stored at chiller temperatures for periods in excess of 10 weeks. However, high-pH, DFD beef held under the same conditions will generally spoil within about 6 weeks.[55] The degree of vacuum has been reported as having no effect on sensory panel ratings of vacuum-packaged beef cuts. There are also advantages in aging beef in vacuum packages, as opposed to carcass aging, including less loss owing to water evaporation, less necessity for trimming of exposed surfaces and more efficient use of refrigerated space.

b. Lamb and Pork

Although for many years, vacuum packaging was only applied to chilled beef, it is now also applied to lamb and pork. Because of their relatively small size, pork and lamb carcasses are only partially boned before packaging, and the presence of bone can lead to puncturing of the package unless precautions are taken.

In contrast to beef cuts, much of the surface of lamb cuts is adipose (rather than muscle) tissue. Adipose tissue has pH values close to neutrality and has no significant respiratory activity. Packaged lamb can therefore present a heterogeneous environment for microbial growth. This different microbial environment of high pH and, possibly, relatively high O_2 concentration, probably accounts for the shelf life of only 6 to 8 weeks reported for vacuum-packaged lamb, whereas 11 to 12 weeks is routinely attainable with beef. The longer shelf life of vacuum-packaged beef is a result of its flora being dominated by lactobacilli, which overgrow other spoilage organisms in the low temperature, low pH, low O_2, high CO_2 environment of the package.

Various conflicting reports have appeared in the literature concerning the chill temperature shelf life of pork joints. For example, figures of little more than 2 weeks at 1°C and 3 to 4 weeks at 2°C were reported in an extensive review.[55] However, it appears that limited use is being made of vacuum packaging for wholesale cuts of pork, where the attraction is likely to be convenience in butchery rather than savings in weight loss.

E. Modified Atmosphere Packaging of Fresh Meat

Vacuum packaging has the inherent disadvantage that both package and meat are subjected to mechanical strain. Mechanical pressure on the meat may increase drip loss, and if bone is present and not adequately covered with a suitable material, the pack may be ruptured. As an alternative to vacuum packaging, attempts have been made to store meat under various gaseous atmospheres, a process referred to as *modified atmosphere packaging* (MAP). The strain on the packaging material can be alleviated by introducing another gas or mixture of gases after evacuation and before sealing. Typical polymers used for the packaging of chilled meat (both vacuum and MAP) are presented in Table 16.4.

The first patent for MAP of red meat was issued to two workers at Unilever in 1969. The intention has generally been to preserve the fresh meat color (O_2Mb) and prevent anaerobic spoilage by using high concentrations of O_2 (50 to 100%) along with 15 to 50% CO_2 to restrict the growth of *Pseudomonas* and related bacteria. O_2 at concentrations of 75% penetrates almost twice as far into the surface of meat as it does at the 21% level in air. The resultant thick layer of oxygenated tissue allows red meat to retain its bright red color for up to 5 to 6 days in normal retail display cabinets. In practice, however, the storage life in packs containing high concentrations of O_2 is usually less than in conventional vacuum packs.

Master packaging in gas-impermeable bags (backflushed with CO_2 or N_2) of meat in retail trays overwrapped with high O_2-permeable film has the potential to provide sufficient shelf life to facilitate centralized meat cutting and packaging operations.[57] However, permanent discoloration can occur if residual O_2 is high enough to exhaust the MRA of the muscle. The use of O_2 scavengers

TABLE 16.4
Typical Materials Used for Packaging Chilled Meat

Pack Type	Bottom Web Materials	Top Web Materials (where applicable)
Flexible vacuum pack	PA–LDPE, coextruded as 5 layer film	
Flexible MAP pack	PA–LDPE	OPA–LDPE
	PA–EVOH–LDPE	
	PA–EVOH–PA–LDPE	PET–PVdC–LDPE
	PP–EVOH–LDPE	
	LDPE–EVOH–LDPE	
Rigid vacuum pack	APET	OPA–LDPE
	PVC or PVC–LDPE	PET–PVdC–LDPE
	PS–EVOH–LDPE	OPA–LDPE–EVOH–LDPE
		PET–LDPE–EVOH–LDPE
Rigid MAP pack	PVC	OPA–LDPE
	PVC–LDPE or PVC–EVOH–LDPE	PET–PVdC–LDPE
	APET	OPA–LDPE–EVOH–LDPE
	APET–LDPE or APET–EVOH–LDPE	PET–LDPE–EVOH–LDPE
	PS–EVOH–LDPE	
Skin packs	PVC–LDPE	Several combinations of up to seven
	PS–EVOH–LDPE	or more layers but incorporating
	APET	EVOH as gas barrier
	APET–LDPE	

Source: From Mondry, H., Packaging systems for processed meat, In *Meat Quality and Meat Packaging*, Taylor, A. A., Raimundo, A., Severini, M. and Smulders, F. J. M., Eds., ECCEAMST (European Consortium for Continuing Education in Advanced Meat Science and Technology), Utrecht, Holland, pp. 323–356, 1996.[40] With permission.

inside the retail trays has been shown to improve the shelf life of retail-ready meat cuts by achieving an O_2 concentration of $\leq$500 ppm in the pack atmosphere.[58]

Because containers for gas packaging are good gas barriers, the internal atmosphere will be modified by the meat during storage. The relative volumes of gas and meat are therefore important in determining the progress of the changes in concentration of gases during storage, and the high solubility of CO_2, compared with the relatively low solubility of O_2 and N_2, in meat must be taken into account. In addition, the safety aspects of MAP must be fully appreciated before use is made of this technology.[10,32]

MAP of fresh meat can be classified into three categories: high oxygen, low oxygen and ultra low oxygen.

1. High-Oxygen MAP

High-O_2 MAP systems with atmospheres of 20 to 30% CO_2, 60 to 80% O_2 and up to 20% N_2 are used to both extend the color stability and delay microbial spoilage of display-packaged meat. Although both the color stability and the time until spoilage are approximately doubled by high-O_2 MAP, this extension in shelf life is not wholly adequate for many commercial purposes.[18]

Studies[55] following the changes in gas volume and composition when beef was stored in completely impermeable containers with mixtures of CO_2 and relatively high concentrations of O_2 showed that an initial O_2 concentration of 80% was depleted to approximately 65% within 2 days in a typical commercial package where headspace volume was roughly equal to meat volume. The CO_2 level changed only slightly during this time because the gas dissolving in the meat was balanced by CO_2 from tissue respiration. If packages with O_2-rich atmospheres are to be stored for relatively long times, then the volume of the pack atmosphere should be about three times the volume of the product, to avoid excessive decreases in O_2 concentration.[19] In many retail situations, a package that meets these requirements is impractical.

The O_2 level during storage determines the effectiveness of color retention. The depth of oxygenation with 80%, 60% and 40% concentrations is shown in Table 16.5 and indicates the time by which surface color is no longer considered red. Many studies have concluded that mixtures of 75 to 80% O_2 with 25 to 15% CO_2 are the most effective, but there have been reports of off-odors and rancidity in meats stored in high O_2 concentrations.[55] High O_2 MAP provides a shelf life at chill temperatures of only 5 to 10 days.

For the above reasons, use of high-O_2 MAP for display packaging depends largely on commercial circumstances.

2. Low-Oxygen MAP

a. *Carbon Dioxide Gas Flush*

In these packages, the air is largely displaced by CO_2, either by itself or mixed with N_2 or air. In general, the shelf life extension is similar to that achieved with vacuum packaging, with higher concentrations of CO_2 leading to a longer shelf life. N_2 is inert and plays no active part.[55] Because of the low-O_2 concentrations, the formation of O_2Mb on the meat surface is precluded or hindered, and red meats rapidly discolor as a result of the formation of MetMb. Thus, a low-O_2 MAP system is unsuitable for use in retail packs of red meat.

b. *Nitrogen Gas Flush*

As an inert gas, N_2 is convenient for gas packaging; it is generally considered to be a neutral filler because it influences neither the color of the meat nor its microbiological quality. If the air in the package is removed prior to the addition of N_2, then the effect on meat is similar to that of vacuum packaging, except that residual O_2 is diluted and MetMb formation on the surface should be less pronounced than with a vacuum.[55] However, although the formation of MetMb on the surface is reduced, the N_2 also dilutes the CO_2 produced by tissue respiration, prolonging the time required for the concentration to accumulate to levels sufficient to inhibit growth of spoilage bacteria.

Although there is little commercial use of N_2 flushing with fresh meat, several studies have confirmed that 100% N_2 is as effective as vacuum for storing fresh meat joints; the only advantage is reduced exudate owing to less mechanical pressure on meat compared with vacuum packaging.

TABLE 16.5
Thickness of Oxygenated Layer in Beef Samples Stored in Different Mixtures of O_2 and CO_2

	Thickness of Oxygenated Layer (mm)				
Gas Mixture	1 day	2 days	5 days	8 days	12 days
80% O_2 + 20% CO_2	11	11	14	14	—
60% O_2 + 20% CO_2 + 20% N_2	9	9	8	—	—
40% O_2 + 20% CO_2 + 40% N_2	8	7	—	—	—

Source: From Taylor, A. A., Packaging fresh meat, In *Developments in Meat Science*, Vol. 3, Lawrie, R. A., Ed., Elsevier Applied Science Publishers, Essex, England, 1985, chap. 4. With permission.

3. Ultra Low-Oxygen MAP

This type of MAP could also be described as high or saturated CO_2 MAP. It has been identified by some experts[19] as controlled atmosphere packaging (CAP). However, CAP implies that an atmosphere of known, desired composition is established within a package and is controlled (i.e., remains unchanged) during the life of the package.[34] Because in this type of package no control is extended over the package atmosphere after sealing with respect to its volume or composition, it is misleading to describe it as CAP; ultra low-O_2 MAP is a preferable term.

CO_2 is highly soluble in both water and oils. Therefore, when CO_2 is applied to meat in a rigid pack, the gas will be absorbed by the muscle and fat tissue until equilibrium is attained. At equilibrium, the partial pressure of CO_2 will be less than that of the original gas mixture, and the total gas pressure will also be less than that at which the gas mixture was initially applied.[17]

Similar considerations apply in flexible packaging systems. If CO_2 alone is added, then the pack will collapse around the meat as gas is absorbed, unless CO_2 is added in excess of the quantity required to saturate the meat (1.5 L CO_2 kg^{-1} meat) at atmospheric pressure. In general, meat will absorb approximately its own volume of gas in an atmosphere of 100% CO_2.[19]

As noted earlier in this chapter, red meat color is not adversely affected by high-CO_2 atmospheres if O_2 is rigorously excluded from the package. To ensure that the package is impermeable to O_2, the package is constructed from a laminate containing aluminum foil or two webs of metalized film.[4,18] Such a packaging system incorporating specialized gassing equipment has been developed, initially for achieving a long shelf life, for chilled lamb. A filled pouch is evacuated in a chamber under reducing external pressure to maintain free pathways for the exhaustion of air. The pouch is then collapsed by mild external pressure to expel residual air and is subsequently inflated under increasing external pressure to minimize stress during gassing of the pack. With this type of relatively elaborate filling procedure, residual O_2 levels of $<0.05\%$ (500 ppm) can be reliably obtained provided that the CO_2 gas contains 0.03% or less O_2.[34] With O_2-free CO_2, residual O_2 concentrations after pack sealing of 100 ppm have been obtained. However, even 100 ppm O_2 can result in discoloration of product, although such discoloration is generally transient. MetMb is usually reduced to Mb within 4 days as anoxic conditions are established and maintained.[19]

This system (known as CAPTECH) was initially developed to allow convenient sea shipment of chilled lamb primal cuts and whole lamb carcasses from New Zealand to northern hemisphere markets. After 24 weeks storage for lamb carcasses and 20 weeks for lamb primal cuts at -1°C, the product was still acceptable, whereas vacuum-packaged controls began to spoil after 11 weeks. The end of shelf life of the CAPTECH products was not microbial spoilage but the development of excessive meat tenderness. Maintaining a storage temperature of -1°C under commercial conditions without allowing the meat to freeze presents a serious challenge. However, this system is increasingly being used commercially for the above applications, as well as for the prolonged storage of shelf-ready consumer packs of beef and lamb which have shelf lives at -1°C of 18 and 20 weeks, respectively. Trials also demonstrated that it is suitable for the prolonged storage of pork, poultry, venison, offal, fish, cooked meats, blanched vegetables and complete meals.[18]

F. Packaging of Frozen and Restructured Meats

Frozen meat is stored and displayed between -10 and -30°C to arrest microbiological growth.[55] Therefore, the changes in meat most influenced by packaging are those associated with appearance, with color and the absence of frost inside the package being the two most important features in this regard.

If a bright red color is required, then it must be produced by oxygenation of the meat surface before freezing, followed by packaging in a material that is relatively permeable to O_2. It has been claimed[55] that ionomer film will keep the bright red color for at least a year if the meat is stored in the dark at −20°C. However, when exposed to light, the red color begins to darken after about a week. This is caused by light-activated oxidation of the pigment at the meat surface and is inevitable in meat that has been frozen in the bright red state and subsequently exposed to light.[55]

When frozen meat and meat products are stored without an adequate moisture vapor barrier, an opaque dehydrated surface known as freezer burn is formed. Freezer burn is caused by the sublimation of ice on the surface of the product when the water vapor pressure of the ice is higher than the vapor pressure in the surrounding air inside the package. Histological studies have revealed that the spongy freezer burn area in frozen muscle and liver possesses microscopic cavities, which scattered light; these cavities were previously occupied by ice crystals.

The key to avoiding freezer burn and lessening oxidative deterioration during frozen storage is to eliminate or reduce the headspace in the package, which should also serve as an effective barrier to O_2 and water vapor. Freezer burn can occur even when using a packaging material that is an excellent barrier to moisture vapor if the package headspace has not been essentially eliminated. Oxidative changes are even more effectively reduced through exclusion of air by means of vacuum packaging.

Several packaging systems are available that satisfy the above requirements for frozen meat. Vacuum packing followed by heat shrinking of the package has been available for many years, as has VSP. Because there is no space between the meat and the packaging material, frost cannot develop to mask the attractive appearance. Some systems can handle either chilled or frozen cuts of meat, whereas others can only pack meat which is already frozen. When the meat is frozen prior to VSP, the final radiant heat sealing operation glazes the surface of the frozen meat and produces an appearance similar to that of chilled meat.[55]

Restructured meat products can be manufactured by three methods: (1) chunking and forming; (2) flaking and forming; (3) tearing and forming. Chunked and formed products are normally either cured and heat processed or fresh frozen. Cured products include sausages and loaves; fresh products are shaped by stuffing into casings or forming into logs. Flaking and forming processes are normally used only to make fresh products such as steaks, cutlets, chops and roasts. Tearing and forming produces products similar to those made by flaking and forming.[25]

There are four major problems related to the packaging of restructured meat products: (1) microbial spoilage; (2) change in color; (3) lipid oxidation and moisture loss.[25] Because most restructured meat products are distributed frozen, microbial spoilage is generally limited to cured meat products, which are discussed in the next section.

Frozen, restructured meat products have traditionally being packaged in bags, pouches, trays, overwraps and plastic-coated paperboard, with polyolefins being the most common material used. Of course, they must contain appropriate plasticizers so that their mechanical properties are not impaired at subzero temperatures. VSP is also used for frozen, restructured meat products. Typically, a heat-softened ionomer film is draped over the product, which is supported on a lower web of the same material. Air is withdrawn from between the two webs and the webs heat sealed together, resulting in a package that is sealed skintight to the edge of the product, regardless of its contour or size. Because there are no empty spaces for moisture condensation to occur, freezer burn is virtually eliminated during frozen storage.[25]

III. CURED AND COOKED MEATS

A significant volume of meat products are cured; therefore, an understanding of the changes meat pigments undergo during the curing process is important. A cured meat is one to which sodium chloride has been added and in which the native meat pigment Mb is, as a result of reaction with nitric oxide (NO), mainly in the nitroso form. The precursor of NO is either $NaNO_2$ or KNO_2, which may be added directly or result from nitrate reduction. Nitrite is also an oxidizing agent and rapidly converts Mb to MetMb. Nitric oxide then combines with MetMb to form nitrosylmetmyoglobin

(NOMetMb), which is reduced by various reducing agents such as added ascorbate to nitrosylmyoglobin (NOMb).[41] NO also reacts directly with Mb to form NOMb:

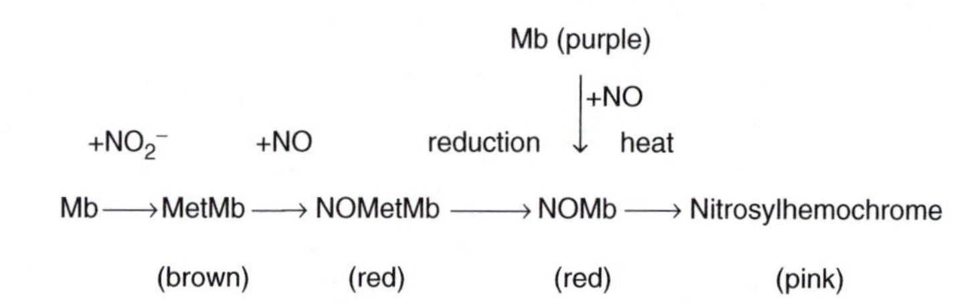

The attractive red color of cured meats before cooking is essentially that of NOMb; it has the red color characteristic of fresh meat O_2Mb. Heating converts NOMb to denatured globin nitrosylhemochrome, which is stable and characteristically red-pink in cooked, cured meats. Residual nitrite is required at the time of cooking and bacon with low or zero nitrite at the time of frying may be gray after cooking. In traditional processes, nitrate was the only source of nitrite, which was formed by the action of bacteria (mainly micrococci) on the nitrate. However, in modern rapid-cure processes, nitrite is added directly to the brine and a proportion of nitrate may also be included.

During storage, cured meats deteriorate firstly because of discoloration, secondly because of oxidative rancidity in the fat, and thirdly on account of microbial changes, the latter having become of greater importance since the advent of prepackaged methods of retailing.[36]

Although the pigment of cured meats (NOMb) is stable in the absence of O_2 or under vacuum, its oxidation to MetMb is very rapid when O_2 is present. The rate of NOMb oxidation increases directly with increasing O_2 tension, unlike Mb itself where the rate of oxidation is maximal at 4 mm Hg O_2 partial pressure. The most common and effective antioxidants used are ascorbate or erythrobate, which are either incorporated into the curing brine or sprayed onto the surface of the product after maturing.

Nitrosylmyoglobin and nitrosylhemochrome are much more susceptible to light than Mb and cured meats can fade after 1 h under retail display lighting conditions. Because light accelerates oxidative changes only in the presence of O_2, vacuum or inert gas packaging can eliminate the effect. Holding vacuum packaged cured meats in the dark for 1 to 2 days before exposing them to display lights allows residual surface O_2 to be depleted by micro-organisms and tissue activity, thus reducing subsequent color deterioration.[38] Several other techniques can reduce light-related color damage.[25] The most efficient way is environmental control; that is, changing the type of light or using light shields, but this is usually impractical. Modification of the package can also result in less light damage. This can be achieved by the addition of additives to the package such as coloring agents, UV absorbers and other materials that decrease transmission by increasing opaqueness of the film. Packages can be designed to reduce the total exposed area by adding labels, printing on the film and increasing film thickness. In addition, the product can be stored with the clear side facing down in the display case. However, all these techniques decrease the ability of customers to view the products inside the package.

To inhibit color changes in cured meat products, a lower level of available O_2 than that required to shift the microbial population from aerobic to anaerobic is required.[13,38] In cooked beef roasts, N_2 packaging significantly improves appearance by retarding greenish discoloration. Light increases the frequency of green discoloration of vacuum packaged samples after 28 days of storage.

Smoke, traditionally produced by the slow combustion of sawdust derived from hard woods, inhibits microbial growth, retards fat oxidation and imparts flavor to cured meats. However, today, smoke essences are frequently used instead of actual smoke, and they contribute largely only flavor to cured meats.

Cured hams undergo a different type of spoilage from that of fresh or smoked hams, primarily a result of the fact that curing solutions pumped into hams contain sugars, which are fermented by the

TABLE 16.6
Factors Influencing the Form of Microbial Spoilage of Cured Meats

Product Characteristic	Environmental Factors
Nature of tissue (fat, lean, etc.)	Storage atmosphere (O_2; CO_2)
pH of tissue	Storage temperature and time
Moisture content	Heat treatment
Levels of curing salts (NaCl, nitrite, nitrate)	
Smoke components	
Polyphosphates, sugars and other curing adjuncts	

Source: From Gardner, G. A., Microbial spoilage of cured meats, In *Food Microbiology: Advances and Prospects*, Roberts, T. A. and Skinner, F.A., Eds., Academic Press, London, England, 179–197, 1983. With permission.

natural flora of the ham, and also by those organisms such as lactobacilli, which are pumped into the product in the curing solution.[31]

Recently, a study[39] on the factors affecting light-induced oxidative discoloration of cured ham (CO_2:N_2 20:80) during 14 days storage at 5°C drew attention to the important effect of headspace volume, which directly influences the total amount of O_2 available for oxidation. As well as keeping the headspace O_2 level low, the headspace volume should also be small. A packaging film with an OTR of 0.5 mL m^{-2} day^{-1} atm^{-1} was recommended.

Currently, there is widespread production of sliced, cooked, vacuum-packaged cured meats including cooked ham, corned beef, emulsion-type sausages and luncheon meats. The salt contents are usually in the range of 2 to 4%, pH values are normally >6.0, and residual nitrite levels vary from 10 to 200 $\mu g\ g^{-1}$, but are mostly $<100\ \mu g\ g^{-1}$.[14] Bacterial spoilage of these products is largely influenced by the nature of the meat, together with environmental factors (see Table 16.6) and the degree and composition of the initial microflora.

Semipreserved cured meats include hams, bacons and sausages packaged either in metal containers or plastic films and cooked after packaging to an internal temperature of between 65 and 75°C. This destroys most vegetative micro-organisms and gives the product a keeping quality of 6 months if held below 5°C. There is concern over the safety of these types of products in environments where rigid temperature control cannot be guaranteed.

In contrast, shelf stable canned cured meats (SSCCM) have shelf lives comparable to canned meat products. However, compared with canned meat products whose safety depends on heat treatment alone (provided that there is no postprocess contamination), the thermal process for SSCCM is relatively mild and allows the survival of a significant number of bacterial spores. These must be adequately inhibited from outgrowth by the NaCl/nitrite combination.

IV. POULTRY

Raw poultry meat is a perishable commodity of relatively high pH (5.6 for breast muscle and up to 6.4 for leg muscle), which readily supports the growth of micro-organisms when stored under chill or ambient conditions. The shelf life of such meat depends on the combined effects of certain intrinsic and extrinsic factors, including the numbers and types of psychrotrophic spoilage organisms present initially, the storage temperature, muscle pH and type (red or white), as well as the kind of packaging material used and the gaseous environment of the product.

Poultry muscle generally has low concentrations of Mb and high rates of O_2 consumption, which means that little O_2Mb is formed when poultry muscle is exposed to air. Consumers are accustomed to the appearance of poultry meat where Mb and MetMb pigments are dominant. Therefore, while the color of poultry meat is not enhanced by storage under high-O_2 atmospheres, the appearance of the meat is not grossly degraded by its exposure to the anaerobic conditions or low concentrations of O_2 that would cause red meats to lose their attractive red color.[19]

The major factor limiting the shelf life of poultry is microbial spoilage, especially growth of *Pseudomonas* and *Achromobacter* species and enterobacteria, the latter causing putrid spoilage of the product after relatively short storage times. These Gram-negative aerobic spoilage organisms can be effectively inhibited by CO_2 concentrations of 20%.

The main pathogenic organisms associated with poultry and poultry products are *Salmonella* spp., *Staphylococcus aureus* and *Clostridium perfringens.* Most studies on the extension of shelf life using CO_2 in MAs have concentrated on the suppression of spoilage organisms rather than the survival and growth of pathogens.

The vacuum packaging of poultry carcasses, cuts and other manufactured products can extend shelf life, provided that the product is held under chill conditions. Vacuum packaging of poultry meat leads to the development of mainly lactic acid bacteria, sometimes accompanied by cold-tolerant coliforms. Although vacuum packaging may be used to encourage the development of an atmosphere around the product that delays microbial spoilage, other systems currently in use involve the addition of at least 20% CO_2 to either individually packaged items or bulk packs of varying size in an O_2-impermeable plastic film.

Based on research in the U.S. in the 1950s, it was long considered that the maximum usable CO_2 concentration for poultry was 25% because, it was claimed, above this level the meat became discolored with grayish tinges; even at 15%, a loss of bloom was sometimes noted. However, gas compositions of 25 to 50% CO_2 and 50 to 75% N_2 are used commercially, and it appears that the occasional discoloration problems may result from high residual levels of O_2 rather than the concentration of CO_2.[18]

Although the number and types of micro-organisms found on stored poultry is an important factor when determining the shelf life, the real determinant is the sensory quality of the raw and cooked product. Unfortunately, most published studies have not included sensory tests, although one study evaluated the quality of raw and cooked poultry stored under MA and refrigeration for up to 5 weeks.[27] Their data indicated that MAP (80% CO_2) poultry would be quite acceptable to consumers for up to 4 to 6 weeks depending on the temperature of storage. They noted that commercial poultry processors may not get as long a shelf life because of difficulties in controlling the packaging process and temperature under production conditions.

Recently, TTIs were used to monitor the quality of MAP broiler chicken cuts stored under different temperature conditions.[51] It was found that microbiological shelf life could be considerably improved when the cold chain was carefully maintained. Temperature had a critical effect on the numbers of *Enterobacteriaceae*, proteolytic bacteria, H_2S-producing bacteria and clostridia — the microbial groups likely to have an effect on the sensory quality.

In a study of broiler carcasses packaged under vacuum in film of low O_2 permeability, or under CO_2 in gas-impermeable packages,[23] shelf life was a function of storage temperature, packaging and O_2 availability. Putrid spoilage in gas-impermeable packages after 7 weeks storage at 3°C or 14 weeks storage at -1.5°C was attributed to *Enterobacteriaceae*. In vacuum packages with OTRs of 30 to 40 mL m^{-2} day^{-1}, putrid odors were detected after 2 weeks storage at 3°C and 3 weeks storage at -1.5°C.

Concerning the safety of MAP chicken, the possible problem organisms are *C. jejuni*, which may be able to survive better in a MAP product, and *L. monocytogenes* and *A. hydrophila*, which, because of the extended storage lives of the MAP products, may have additional time to grow to potentially high numbers. Although *C. perfringens* may be able to survive better in some MAs as compared with air, it would not be able to grow at the chill temperatures commonly used for MAP products. Thus, it is unlikely to be a health hazard in a MAP product unless the product is temperature abused, because high numbers of the organism must be ingested to cause illness.[11]

A wide range of manufactured poultry products has been developed including rolls, roasts, burgers and sausages. However, in most cases, very little information is available on either keeping quality or the influence on shelf life of particular packaging materials.[8]

Most studies on possible methods of extending shelf life of poultry meat have been on chicken with little attention paid to other poultry species. It cannot be assumed that the keeping quality and shelf life characteristics of turkeys and ducks are the same in all respects as those of chickens, particularly in relation to flavor changes under different storage conditions, which do not always correlate with off-odor development.

V. SEAFOOD

A. Types of Spoilage

Flesh foods such as fish and shellfish are highly perishable owing to their high a_w, relatively high pH and the presence of autolytic enzymes, which cause the rapid development of undesirable odors and flavors. The chemical composition and microbial flora of seafood vary considerably between species, different fishing grounds and seasons, but the pH of most fish is >6.0. The flesh of most marine fish and shellfish contain large amounts of nonprotein nitrogen (NPN) including the compound trimethylamine oxide (TMAO). After death, TMAO can serve as a terminal electron acceptor for some spoilage bacteria, enabling them to grow when O_2 levels are depleted; the ammoniacal susbstance trimethylamine (TMA) is released as a consequence.[7] Nonfatty fish such as cod and haddock have lipid contents of 1 to 2% in contrast to fatty fish such as herring and mackerel, which can have lipid contents of more than 30%. The high degree of unsaturation that gives fish oils their nutritional significance also makes them very vulnerable to oxidation.

It is generally accepted that the internal flesh of healthy, live fish is sterile; micro-organisms that exist on fresh fish are generally found in the gills, the outer slime and the intestines. The *post mortem* changes leading to spoilage depend principally on the chemical composition of the fish, its microbial flora, subsequent handling and processing and storage. The spoilage of salt and fresh water fish appears to occur in essentially the same manner. Spoilage results from changes brought about by (1) reactions that result from autolytic enzymes; (2) metabolic activities of micro-organisms; (3) chemical reactions such as oxidation.

Immediately *post mortem*, a whole series of tissue enzyme reactions begin the process of autolysis (basically self-digestion of the fish muscle), which eventually leads to spoilage. The autolytic enzyme reactions predominate for 4 to 6 days at 0°C, after which the products of bacterial activity become increasingly evident with the appearance of undesirable odors and flavors. The rates of autolytic changes are determined by many factors but the most important are temperature, pH, availability of O_2 and the physiological condition of the fish before death.

As spoilage proceeds, there is a gradual invasion of the flesh by bacteria from the outer surfaces. Because bacteria can generally use only very basic nutrients as food, bacterial spoilage does not normally commence in whole fish until autolysis is well advanced. Breakdown of the muscle structure occurs only after spoilage has proceeded well beyond the point of rejection. The development of objectionable slimes, odors and flavors results mainly from bacterial activity.

Fish from warm sea waters generally carry larger numbers of bacteria than fish from colder waters, with Gram-positive organisms forming a large or predominant fraction of the flora, along with Gram-negative organisms of the types found on fish from colder waters. The major spoilage organisms found on fish stored under aerobic conditions are Gram-negative species, particularly pseudomonads, *Shewanella putrefaciens* (formerly known both as *Pseudomonas putrefaciens* and

Alteromonas putrefaciens) and *Photobacterium phosphoreum*. As mentioned earlier, some of these organisms can utilize TMAO and continue with oxidative metabolism under anaerobic conditions. *P. phosphoreum* is responsible for spoilage of cod fish under anaerobic conditions but it is not known if this bacterium is the general specific spoilage organism (SSO) for all marine temperate fishes.[47] Lactic acid bacteria and *Brochothrix thermosphacta* have been identified as the typical SSOs of freshwater fish and fish from warmer waters.[46]

The third type of spoilage is chemical spoilage — primarily oxidation of the fatty compounds leading to the development of rancid flavors. The rate of rancidity development is closely related to the temperature of storage and reactions can still occur at freezer temperatures as low as −30°C. Some substances such as salt and some processes such as drying and smoking can aggravate the oxidation problem, and therefore frozen smoked fish has a shorter shelf life than unsmoked fish of the same species.

Shellfish are divided into two main groups: crustaceans (which include shrimp, prawn, lobster, crabs and crayfish) and mollusks (which include oysters, clams, squid, cockles, mussels and scallops). The microbial flora of shellfish reflect the waters from which they are caught, contaminants from the deck and handlers and the quality of the washing waters used. Mollusks differ from crustacean shellfish and nonfatty fish in having a significant content of carbohydrate material and a lower total quantity of nitrogen in their flesh. Because the carbohydrate is largely in the form of glycogen, the spoilage of molluscan shellfish is largely fermentative.[31] This results in a progressive fall in tissue pH as spoilage develops, from pH 5.9 to 6.2 in fresh mollusks to less than about pH 5.5 in spoiled mollusks.

B. Vacuum and Modified Atmosphere Packaging

Because of the high pH of most fish muscle, growth of *S. putrefaciens* is not prevented by vacuum packaging, and this organism may dominate the spoilage process in vacuum-packaged fish.[16] However, *P. phosphoreum* has been identified as the organism responsible for spoilage of fish such as cod in vacuum and MA packs.[46] The rates of growth of both these organisms are not greatly reduced by anaerobic conditions and, consequently, the extension of shelf life achieved by the vacuum packaging of fish is often small. In contrast to *S. putrefaciens*, which is greatly inhibited by high concentrations of CO_2, *P. phosphoreum* has been shown to be relatively insensitive to CO_2 and may spoil fish in MA packs after a minimally longer storage life than that of the same product stored in air.[16]

The first extensive research on seafood stored in CO_2 was reported in the early 1930s in the U.K., U.S. and Russia. In a 100% CO_2 atmosphere, fish kept fresh 2 to 3 times longer than control fish in air at the same temperature.[54] The absorption of CO_2 altered the pH of fish from 6.6 to 6.2, but this was reversed on subsequent exposure to air.[47]

Many of the reported studies on MAP of seafood do not specify the gas volume to product weight ratio. For a given gas mixture, a high gas-to-fish ratio will present a very different chemical balance from a low ratio. Moreover, changes in the CO_2 and O_2 levels inside the package headspace during storage are seldom measured, making comparison between different studies difficult. Although any chemical effects on fish tissues will be affected by the amount of CO_2 that dissolves, the bacteriostatic effect is more likely to be influenced by the residual atmosphere because most spoilage organisms are present on the surface.

The effects of MAP on seafood are similar to those described earlier in this chapter for meat and poultry. Vacuum and MAP (including flushing with N_2 and CO_2) suppress the normal spoilage bacteria that cause off-odors and -flavors, thereby extending the shelf life of seafood. Microorganisms that are not usually involved in aerobic spoilage eventually predominate. These microorganisms such as lactobacilli are less affected by the elevated CO_2 atmosphere and grow more slowly than the normal aerobic spoilage bacteria. Because the micro-organisms that predominate under MAP cause less noticeable and less offensive organoleptic changes, the net result is a significant extension of shelf life under MAP at refrigeration temperatures, compared with packaging under air. However, a recent review concluded that MAP confers little or no additional increase in shelf life compared with vacuum packaging. The best effect of MAP storage on shelf life has been obtained with fish from warm waters, the spoilage flora being dominated by Gram-positive micro-organisms.[47]

Recently, the use of MAP (CO_2:N_2 60:40) in conjunction with "superchilling" or "partial freezing" (reducing the temperature of the fish to 1 to 2°C below its freezing point) was compared with storage in air at normal chilled (4°C) and superchilled (−2°C) temperatures.[48] The shelf life of MAP salmon at −2°C (24 days) was 2.5 times that of MAP salmon at 4°C, and 3.5 times that of salmon at 4°C stored under air. It was suggested that the synergistic effect of MA and temperature may be caused by increased solubility of CO_2 at the superchilled temperature.

Effective gas compositions vary according to fish species, with low-O_2 concentrations being used with fatty fish, which are susceptible to oxidative rancidity. Generally, gas mixtures for nonfatty fish and shellfish are 25 to 35% O_2, 35 to 45% CO_2 and 25 to 35% N_2, and for smoked and fatty fish, 35 to 45% CO_2 and 55 to 65% N_2.[9] Problems caused by too high a CO_2 level include pack collapse, increased drip, CO_2 taint, which gives an acid flavor to certain species of fish, and clouding of the eyes, which consumers often use as an indicator of freshness.

It is evident from the published literature that MAP can extend the shelf life of a variety of fish and fish products. However, MAP is not equally effective for extending the shelf life of all fish products.[49] Most studies of MAP of seafood indicate a shelf life extension from a few days up to a week or more compared with air storage, depending on species and temperature. Differences in spoilage microflora and pH are mainly responsible for the observed differences in shelf life[47] provided similar gas-to-product ratios are used. Although some claims have been made of a shelf life of up to 3 to 4 weeks for the chill storage of MAP fish, this is generally considered to be unrealistic unless superchilled temperatures are used. General target shelf lives of these products are in the range of 10 to 14 days, but may reach 18 to 20 days if the temperature is controlled very tightly just above freezing point.

C. Safety Aspects of Packaged Seafood

Fish and shellfish are vehicles for transmission of foodborne diseases. There is a large body of evidence that indicates that fresh fish may be contaminated with the anaerobe *C. botulinum* either as a result of it being present in the microbiota of the fish ecosystem, or as a result of postcatching contamination during processing. Conditions for growth of *C. botulinum* occur only a few millimeters below the surface of fish flesh, where spores may arise through gaping, knife cuts and other punctures.[7] Foodborne botulism outbreaks caused by fish are predominantly a result of type E strains of *C. botulinum* in products that are consumed without further heat processing. Because many fish products are not cooked prior to consumption, the hazard from botulinal toxin is real.

It is now recognized that the growth of *C. botulinum* in foods does not depend on the total exclusion of O_2, nor does the inclusion of O_2 ensure that growth of *C. botulinum* is prevented.[15,47] A summary of recent results on the growth and toxin production by *C. botulinum* is presented in Table 16.7. While some studies have detected botulinal toxin in MAP fish prior to the products being considered spoiled, other studies have shown that MA or vacuum packaged fish spoil prior to, or in parallel with, toxin production. Despite these inconclusive results, there is a potential threat for a packaged fish product to become toxic prior to spoilage at storage temperatures of 8°C or above.[47]

Whereas the presence of CO_2 does not lead to an increase in growth of *C. botulinum*, replacing air with N_2 produces anaerobic conditions and an increased susceptibility to the growth of *C. botulinum* in advance of spoilage signals resulting from spoilage microflora. Vacuum packaging in high barrier films produces the same growth conditions as replacing air with N_2. However, if an

TABLE 16.7
Growth and Toxin Production by *C. Botulinum* in Packaged Fishery Products

Type of Fishery Product	Storage Temperature (°C)	Atmosphere CO_2:N_2:O_2	Toxin Detection (days)	Shelf Life (days)
Salmon fillet	16	Air	4	4
	16	75:25:0	4	5–6
	16	Vacuum	3	3
	8	Air	17	13–17
	8	75:25:0	24	20–24
	8	Vacuum	10	>6, <10
	4	Air	>66	24–27
	4	75:25:0	>80	55–62
	4	Vacuum	>66	34–38
Tilapia fillets	16	Air	4	3
	16	75:25:0	4	4
	16	Vacuum	3	3
	8	Air	20	6
	8	75:25:0	40	17
	8	Vacuum	17	10
	4	Air	>47	10
	4	75:25:0	>90	80
	4	Vacuum	>90	47
Catfish fillets	16	Air	3	3
	16	75:25:0	4	4
	16	Vacuum	3	3
	8	Air	9	6
	8	75:25:0	18	13
	8	Vacuum	6	6
	4	Air	>54	13
	4	75:25:0	>75	38–40
	4	Vacuum	46	20–24
Cod fillets	16	Air	>7	3–4
	16	75:25:0	7	6
	16	Vacuum	7	3–4
	8	Air	>41	13–17
	8	75:25:0	>60	24–27
	8	Vacuum	17	13
	4	Air	>60	20–24
	4	75:25:0	>90	55–60
	4	Vacuum	>5	24–27
Rainbow trout fillets	10	Vacuum skin packaging	6	3
	4	Vacuum skin packaging	<21	12
Channel catfish	10	Air	4	4–6
	10	80:20:0	4	2–6
	4	Air	9	9
	4	80:20:0	18	9–12

Source: From Sivertsvik, M., Jeksrud, W. K., and Rosnes, J. T., *Int. J. Food Sci. Technol.*, 37, 107–127, 2002. With permission.

O_2-permeable film is used for vacuum packaging, then distinct spoilage signals will be produced if the pack is held at abuse temperatures because of the ingress of O_2 into the package.[1]

High temperature abuse (21 to 27°C) for periods of 12 to 24 hours is a major concern because MAP fish generally do not become overtly spoiled under these conditions yet may be toxic, whereas fish held at the same temperatures under similar aerobic conditions start to become putrid before toxin production occurs. If MAP fish are held at very high refrigeration temperatures (greater than 10°C), then strong spoilage signals may not develop in advance of *C. botulinum* toxin production.

Vacuum and MAP alone are not capable of providing the safety required for extended storage of seafood with respect to outgrowth and toxin production by *C. botulinum* type E. A fail-safe mechanism by which storage temperature could be maintained at or below 3°C is required. In order to effectively utilize the extended shelf life aspect of vacuum and MAP, some intervention is needed to assure the delay of toxin production if even mild temperature abuse should occur.

In an extensive 5-year study involving 927 experiments and 18,700 samples, nonproteolytic *C. botulinum* types B, E and F in a variety of fresh fish stored between 4 and 30°C for up to 60 days were evaluated.[2] A general formula that provides the most conservative model for the prediction of lag times was developed, and its utility was demonstrated by its ability to predict the time before toxigenesis of *C. botulinum* in inoculated fish stored under different MAs. Models based on lag times are regarded as most appropriate for pathogens with zero growth tolerance, such as *C. botulinum*, *Salmonella* and *Listeria*.

Another study[24] found growth and toxin production from spores of *C. botulinum* types B, E and F after 5 weeks at 3°C, 3 to 4 weeks at 4°C and 2 to 3 weeks at 5°C. Growth occurred more frequently from spores of type F strains than for types B and E.

The use of active packaging technologies such as O_2 scavengers and CO_2 emitters does not improve microbial safety above that obtained by traditional MAP, and gives little or no additional shelf life to fresh seafood products compared with MAP and vacuum packaging.[46]

The only effective way to assure the safety of chilled vacuum-packaged or MAP fish products would be to either: (1) keep the product at or below 3°C at all times; (2) heat the product sufficiently to destroy spores of all strains; or (3) heat the product sufficiently to inactivate the nonproteolytic spores and then keep the product well below 10°C.[11] The latter two points may be effective from a theoretical standpoint but, in practice, it may be difficult in a fish processing environment to avoid postprocessing contamination with spores of *C. botulinum*.[42] As mentioned earlier, many seafood products are eaten in the raw state; this makes the application of heat as a method of ensuring their microbiological safety an unacceptable approach. When chill storage is the only controlling factor, storage at temperatures between 5 and 8°C should be limited to 5 days; shelf lives up to 10 days could be assigned for storage temperatures at 5°C or below.

REFERENCES

1. Ashie, I. N. A., Smith, J. P., and Simpson, B. K., Spoilage and shelf life extension of fresh fish and shellfish, *Crit. Rev. Food Sci. Technol.*, 36, 87–121, 1996.
2. Baker, D. A. and Genigeorgis, C., Predicting the safe storage of fresh fish under modified atmospheres with respect to *Clostridium botulinum* toxigenesis by modeling length of the lag phase of growth, *J. Food Prot.*, 53, 131–140, 1990, 153.
3. Baron, C. P. and Andersen, H. J., Myoglobin-induced lipid oxidation. A review, *J. Agric. Food Chem.*, 50, 3887–3897, 2002.
4. Bell, R. G., Meat packaging: protection, preservation and presentation, In *Meat Science and Applications*, Hui, Y. H., Nip, W.-K., Rogers, R. W. and Young, O. A., Eds., Marcel Dekker, New York, 2001, chap. 19.

5. Brody, A. L., Modified atmosphere/vacuum packaging of meat, In *Controlled/Modified Atmosphere/Vacuum Packaging of Foods*, Brody, A. L., Ed., Food & Nutrition Press, Trumbull, CT, 1989, chap. 2.
6. Dainty, R. H., Shaw, B. G., and Roberts, T. A., Microbial and chemical changes in chill-stored red meats, In *Food Microbiology: Advances and Prospects*, Roberts, T. A. and Skinner, F. A., Eds., Academic Press, London, England, pp. 151–178, 1983.
7. Davis, H., Fish and shellfish, In *Principles and Applications of Modified Atmosphere Packaging of Foods*, 2nd ed., Blakistone, B. A., Ed., Blackie Academic & Professional, London, England, 1998, chap. 9.
8. Dawson, L. E., Packaging of processed poultry, In *The Microbiology of Poultry Meat Products*, Cunningham, F. E. and Cox, N. A., Eds., Academic Press, London, England, 1987, chap. 7.
9. Day, B. P. F., Modified atmosphere packaging of chilled fish and seafood products, In *Farmed Fish Quality*, Kestin, S. C. and Warriss, P. D., Eds., Blackwell Science, Oxford, England, 2001, chap. 24.
10. Devlieghere, F. and Debevere, J., MAP, product safety and nutritional quality, In *Novel Food Packaging Techniques*, Ahvenainen, R., Ed., CRC Press, Boca Raton, FL, 2003, chap. 11.
11. Farber, J. M., Microbiological aspects of modified-atmosphere packaging technology — a review, *J. Food Prot.*, 54, 58–70, 1991.
12. Faustman, C., Postmortem changes in muscle foods, In *Muscle Foods: Meat, Poultry and Seafood Technology*, Kinsman, D. M., Kotula, A. W. and Breidenstein, B. C., Eds., Chapman & Hall, New York, 1994, chap. 3.
13. García-Esteban, M., Ansorena, D., and Astiasarán, I., Comparison of modified atmosphere packaging and vacuum packaging for long period storage of dry-cured ham: effects on color, texture and microbiological quality, *Meat Sci.*, 67, 57–63, 2004.
14. Gardner, G. A., Microbial spoilage of cured meats, In *Food Microbiology: Advances and Prospects*, Roberts, T. A. and Skinner, F. A., Eds., Academic Press, London, England, pp. 179–197, 1983.
15. Gibson, A. M., Ellis-Brownlee, R.-C. L., Cahill, M. E., Szabo, E. A., Fletcher, G. C., and Bremer, P. J., The effect of 100% CO_2 on the growth of nonproteolytic *Clostridium botulinum* at chill temperatures, *Int. J. Food Microbiol.*, 54, 39–48, 2000.
16. Gill, A. O. and Gill, C. O., MAP and quality of fresh meats, poultry and fin fish, In *Innovations in Food Packaging*, Han, J. H., Ed., Academic Press, San Diego, CA, 2005, chap. 16.
17. Gill, C. O., The solubility of carbon dioxide in meat, *Meat Sci.*, 22, 65–71, 1981.
18. Gill, C. O., Controlled atmosphere packaging of chilled meat, *Food Control*, 1, 74–78, 1990.
19. Gill, C. O., Active packaging in practice: meat, In *Novel Food Packaging Techniques*, Ahvenainen, R., Ed., CRC Press, Boca Raton, FL, 2003, chap. 17.
20. Gill, C. O., Spoilage, factors affecting (a) microbiological, In *Encyclopaedia of Meat Science*, Jensen, W. K., Devine, C. and Dikeman, M., Eds., Academic Press, London, 2004.
21. Gill, C. O., Visible contamination on animals and carcasses and the microbiological condition of meat, *J. Food Prot.*, 67, 413–419, 2004.
22. Gill, C. O. and Penney, N., Packaging conditions for extended storage of chilled, dark, firm, dry beef, *Meat Sci.*, 18, 41–53, 1986.
23. Gill, C. O., Harrison, C. L., and Penney, N., The storage life of chicken carcasses packaged under carbon dioxide, *Int. J. Food Micro.*, 11, 151–158, 1990.
24. Graham, A. F., Mason, D. R., Maxwell, F. J., and Peck, M. W., Effect of pH and NaCl on growth from spores of non-proteolytic *Clostridium botulinum* at chill temperature, *Lett. Appl. Microbiol.*, 24, 95–100, 1997.
25. Harte, B. R., Packaging of restructured meats, In *Advances in Meat Research*, Pearson, A. M. and Dutson, T. R., Eds., Elsevier Applied Science Publishers, Essex, England, 1987, chap. 13.
26. Hood, D. E., The chemistry of vacuum and gas packaging of meat, In *Recent Advances in the Chemistry of Meat*, Bailey, A. J., Ed., Royal Society of Chemistry, Cambridge, England, 1984, chap. 11.
27. Hotchkiss, J. H., Baker, R. C., and Qureshi, R. A., Elevated carbon dioxide atmospheres for packaging poultry. II. Effects of chicken quarters and bulk packages, *Poult. Sci.*, 64, 333–340, 1985.
28. Hunt, M. C., Mancini, R. A., Hachmeister, K. A., Kropf, D. H., Merriman, M., Delduca, G., and Milliken, G., Carbon monoxide in modified atmosphere packaging affects color, shelf life, and microorganisms of beef steaks and ground beef, *J. Food Sci.*, 69, FCT45–FCT52, 2004.

29. Jacobsen, M. and Bertelsen, G., The use of CO_2 in packaging of fresh red meat and its effects on chemical quality changes in the meat: a review, *J. Muscle Foods*, 13, 143–168, 2002.
30. Jacobsen, M. and Bertelsen, G., Active packaging and colour control: the case of meat, In *Novel Food Packaging Techniques*, Ahvenainen, R., Ed., CRC Press, Boca Raton, FL, 2003, chap. 19.
31. Jay, J. M., *Modern Food Microbiology*, 6th ed., Aspen Publishers, Gaithersburg, Maryland, 2000.
32. Jones, M. V., Modified atmospheres, In *Mechanisms of Action of Food Preservation Procedures*, Gould, G. W., Ed., Elsevier Science Publishers, Essex, England, 1989, chap. 10.
33. Krause, T. R., Sebranek, J. G., Rust, R. E., and Honeyman, M. S., Use of carbon monoxide packaging for improving the shelf life of pork, *J Food Sci.*, 68, 2596–2603, 2003.
34. Kropf, D. H., Modified Atmosphere Packaging, In *Wiley Encyclopedia of Food Science and Technology*, 2nd ed., Vol. 3, Francis, F. J., Ed., Wiley, New York, pp. 1561–1567, 2000.
35. Lambden, A. E., Chadwick, D., and Gill, C. O., Technical note: oxygen permeability at sub-zero temperatures of plastic film used for vacuum-packaging meat, *J Food Technol.*, 20, 281–283, 1985.
36. Lawrie, R. A., *Lawrie's Meat Science*, 6th ed., Woodhead Publishing, Cambridge, England, 1998.
37. Ledward, D. A., Colour of raw and cooked meat, In *The Chemistry of Muscle-based Foods*, Johnston, D. E., Knight, M. K. and Ledward, D. A., Eds., Royal Society of Chemistry, Cambridge, England, pp. 128–144, 1992.
38. Lundquist, B. R., Protective packaging of meat and meat products, In *The Science of Meat and Meat Products*, Price, J. F. and Schweigert, B. S., Eds., Food and Nutrition Press, Westport, CT, 1987, chap. 14.
39. Møller, J. K. S., Jakobsen, M., Weber, C. J., Martinussen, T., Skibsted, L. H., and Bertelsen, G., Optimisation of colour stability of cured ham during packaging and retail display by a multifactorial design, *Meat Sci.*, 63, 169–175, 2003.
40. Mondry, H., Packaging systems for processed meat, In *Meat Quality and Meat Packaging*, Taylor, A. A., Raimundo, A., Severini, M. and Smulders, F. J. M., Eds., ECCEAMST (European Consortium for Continuing Education in Advanced Meat Science and Technology), Utrecht, Holland, pp. 323–356, 1996.
41. Newton, K. G. and Rigg, W. J., The effect of film permeability on the storage life and microbiology of vacuum-packaged meat, *J. Appl. Bacteriol.*, 47, 433–441, 1979.
42. Notermans, S., Dufrenne, J., and Lund, B. M., Botulism risk of refrigerated, processed foods of extended durability, *J. Food Prot.*, 53, 1020–1024, 1990.
43. Pegg, R. B. and Shahidi, F., Unraveling the chemical identity of meat pigments, *Crit. Rev. Food Sci. Nutr.*, 37, 561–589, 1997.
44. Renerre, M., Oxidative processes and myoglobin, In *Antioxidants in Muscle Foods*, Decker, E., Faustman, C. and Lopez-Bote, C. J., Eds., Wiley-Interscience, New York, 2000, chap. 5.
45. Seideman, S. C., Cross, H. R., Smith, G. C., and Durland, P. R., Factors associated with fresh meat color: a review, *J. Food Qual.*, 6, 211–225, 1984.
46. Sivertsvik, M., Active packaging in practice: fish, In *Novel Food Packaging Techniques*, Ahvenainen, R., Ed., CRC Press, Boca Raton, FL, 2003, chap. 18.
47. Sivertsvik, M., Jeksrud, W. K., and Rosnes, J. T., A review of modified atmosphere packaging of fish and fishery products — significance of microbial growth, activities and safety, *Int. J. Food Sci. Technol.*, 37, 107–127, 2002.
48. Sivertsvik, M., Rosnes, J. T., and Kleiberg, G. H., Effect of modified atmosphere packaging and superchilled storage on the microbial and sensory quality of Atlantic salmon (*Salmo salar*) fillets, *J. Food Sci.*, 68, 1467–1472, 2003.
49. Skura, B. J., Modified atmosphere packaging of fish and fish products, In *Modified Atmosphere Packaging of Food*, Ooraikul, B. and Stiles, M. E., Eds., Ellis Horwood, Chichester, England, 1991, chap. 6.
50. Smith, G. C., Belk, K. E., Sofos, J. N., Tatum, J. D., and Williams, S. N., Economic implications of improved color stability in beef, In *Antioxidants in Muscle Foods*, Decker, E., Faustman, C. and Lopez-Bote, C. J., Eds., Wiley-Interscience, New York, 2000, chap. 15.
51. Smolander, M., Alakomi, H.-L., Ritvanen, T., Vainionpää, J., and Ahvenainen, R., Monitoring of the quality of modified atmosphere packaged broiler chicken cuts stored in different temperature conditions. A. Time-temperature indicators as quality-indicating tools, *Food Control*, 15, 217–229, 2004.

52. Sørheim, O., Aune, T., and Nesbakken, T., Technological, hygienic and toxicological aspects of carbon monoxide use in modified atmosphere packaging of meat, *Trends Food Sci. Technol.*, 8, 307–312, 1997.
53. Sørheim, O., Nissen, H., and Nesbakken, T., The storage life of beef and pork packaged in an atmosphere with low carbon monoxide and high carbon dioxide, *Meat Sci.*, 52, 157–164, 1999.
54. Stammen, K., Gerdes, D., and Caporaso, F., Modified atmosphere packaging of seafood, *Crit. Rev. Food Sci. Technol.*, 29, 301–331, 1990.
55. Taylor, A. A., Packaging fresh meat, In *Developments in Meat Science*, Vol. 3, Lawrie, R. A., Ed., Elsevier Applied Science Publishers, Essex, England, 1985, chap. 4.
56. Taylor, A. A., Down, N. F., and Shaw, B. G., A comparison of modified atmosphere and vacuum skin packaging for the storage of red meats, *Int. J. Food Sci. Technol.*, 25, 98–104, 1990.
57. Tewari, G., Jayes, D. S., and Holley, R. A., Centralized packaging of retail meat cuts: a review, *J. Food Prot.*, 62, 418–425, 1999.
58. Tewari, G., Jeremiah, L. E., Jayas, D. S., and Holley, R. A., Improved use of oxygen scavengers to stabilize the colour of retail-ready meat cuts stored in modified atmospheres, *Int. J. Food Sci. Technol.*, 37, 199–207, 2002.
59. Walsh, H. M. and Kerry, J. P., Meat packaging, In *Meat Processing — Improving Quality*, Kerry, J., Kerry, J. and Ledward, D., Eds., CRC Press, Boca Raton, FL, 2002, chap. 20.
60. Young, O. A. and West, J., Meat color, In *Meat Science and Applications*, Hui, Y. H., Nip, W.-K., Rogers, R. W. and Young, O. A., Eds., Marcel Dekker, New York, 2001, chap. 3.
61. Zhao, Y., Wells, J. H., and McMillin, K. W., Applications of dynamic modified atmosphere packaging systems for fresh red meats: a review, *J. Muscle Foods*, 5, 299–328, 1994.

17 Packaging of Horticultural Products

CONTENTS

I. INTRODUCTION

Fresh fruits and vegetables are essential components of the human diet as they contain a number of nutritionally important compounds such as vitamins which cannot be synthesized by the human body. A fruit or vegetable is a living, respiring, edible tissue which has been detached from the parent plant. Fruits and vegetables are perishable products with an active metabolism during the

TABLE 17.1
Classification of Horticultural Produce According to the Plant Organ Used

Class	Commodities (Examples)
Root vegetables	Carrots, celery, garlic, horseradish, onions, parsnips, radishes, turnips
Tubers	Potatoes, yams, Jerusalem artichokes
Leaf and stem vegetables	Brussel sprouts, cabbage, celery, chicory, Chinese cabbage, cress, green onions, kale, lettuce, spinach
Flower vegetables	Artichokes, broccoli, cauliflower
Immature fruit vegetables	Beans, cucumber, gherkins, okra, peas, peppers, squash, sweet corn
Mature fruit vegetables	Melons, tomatoes
Reproductive organs	Most fruits

postharvest period. In simple terms, the shelf life of fruits and vegetables can be extended by retarding the physiological, pathological and physical deteriorative processes (generally referred to as postharvest handling) or by inactivating the physiological processes (generally referred to as food preservation). Packaging has an important role to play in both the handling and preservation approaches to maximizing the shelf life; this chapter will primarily focus on the former because the packaging requirements of the preservation processes are not unique to horticultural products.

Difficulties arise when attempts are made to draw a clear line between fruits and vegetables. Fruits and vegetables cannot be clearly delineated botanically or morphologically as they encompass numerous organs in vegetative or reproductive stages and belong to a large number of botanical families. Fruits tend to be restricted to reproductive organs arising from the development of floral tissues, with or without fertilization, while vegetables consist simply of edible plant tissues. Thus, some fruits are included as vegetables and *vice versa*. Table 17.1 classifies horticultural produce according to the plant organ used, mainly based on the form in which the product is handled such as root, tuber, stem or leaf.

II. POSTHARVEST PHYSIOLOGY

Growth, maturation and senescence are the three important phases through which fruits and vegetables pass, with the first two terms often referred to as *fruit development*. *Ripening* (a term reserved for fruit) generally begins during the later stages of maturation and is considered the beginning of *senescence*, a term defined as the period when anabolic or synthetic biochemical processes give way to catabolic or degradative processes leading to aging and final death of the tissue.

A. Respiration

Respiration involves the oxidation of energy-rich organic substrates normally present in cells, such as starch, sugars and organic acids, to simpler molecules (CO_2 and H_2O) with the concurrent production of energy (ATP and heat) and other molecules which can be used by the cell for synthetic reactions. The principal carrier of free energy is ATP. The greatest yield of energy is obtained when the process takes place in the presence of molecular O_2. Respiration is then said to be aerobic. If hexose sugar is used as the substrate, then the overall equation can be written as follows:

$$C_6H_{12}O_6 + 6O_2 \rightarrow 6CO_2 + 6H_2O + \text{energy} \qquad (17.1)$$

This transformation actually takes place in a large number of individual stages with the participation of many different enzyme systems. The water produced remains within the tissue but the CO_2 escapes and accounts for part of the weight loss of harvested fruits and vegetables, typically in the range of 3 to 5%. When 1 mol of hexose sugar is oxidized, 36 mol of ATP (each possessing 32 kJ of useful energy) are formed. This represents about 40% of the total free energy change, the remainder being dissipated as heat.[36] Rapid removal of this heat is usually desirable and it is important that the packaging assists, rather than impedes, this process.

The rate of respiration is often a good index of the storage life of horticultural products: the higher the rate, the shorter the life; the lower the rate, the longer the life. Table 17.2 classifies vegetables according to their respiration intensities.

It is possible to evaluate the nature of the respiratory process from measurements of CO_2 and O_2. The ratio of the volume of CO_2 released to the volume of O_2 absorbed in respiration is termed the respiratory quotient (RQ). Values of RQ range from 0.7 to 1.3 for aerobic respiration, depending on the substrate being oxidized: $RQ = 1$ for carbohydrates, $RQ < 1$ for lipids and $RQ > 1$ for organic acids.[25]

Anaerobic respiration (sometimes called *fermentation*) involves the incomplete oxidation of compounds in the absence of O_2 and results in the accumulation of ethanol, acetaldehyde and CO_2. Much lower amounts of energy (2 mol of ATP) and CO_2 are produced from 1 mol of hexose sugar than that produced under aerobic conditions.[36] In addition, compared with aerobic respiration, very little heat energy (approximately 5%) is produced for a given amount of carbohydrate oxidation in anaerobic respiration. The O_2 concentration at which a shift from aerobic to anaerobic respiration occurs varies among tissues and is known as the extinction point.[25] Very high RQ values (>1.3) usually indicate anaerobic respiration.

1. Internal Factors Affecting Respiration

Variations in the rate of respiration occur during organ development; as fruits increase in size, the total amount of CO_2 emitted increases although the respiration rate calculated on a per unit weight basis decreases continually. A majority of studies on gas exchange in plant tissues indicate that the skin represents the main significant barrier to diffusion.[22] Clearly, the rate at which the three gases O_2, CO_2 and C_2H_4 diffuse through the tissue will have a significant effect on the rate of respiration.

Fruits may be divided into climacteric and nonclimacteric types. Climacteric fruits are those in which ripening is associated with a distinct increase in respiration and C_2H_4 production, the respiration rate rising up to the climacteric peak and then declining. Such an increase can occur while the fruit is attached to or separated from the plant. A further distinguishing feature is that treatment of climacteric fruits with C_2H_4 or propylene stimulates both respiration and autocatalytic

TABLE 17.2
Classification of Vegetables According to Respiration Intensity

Class	Respiration Intensity at 10°C ($mg\ CO_2\ kg^{-1}\ h^{-1}$)	Commodities
Very low	Below 10	Onions
Low	10–20	Cabbage, cucumber, melons, tomatoes, turnips
Moderate	20–40	Carrots, celery, gherkins, leeks, peppers, rhubarb
High	40–70	Asparagus (blanched), eggplant, fennel, lettuce, radishes
Very high	70–100	Beans, Brussel sprouts, mushrooms, savoy cabbage, spinach
Extremely high	Above 100	Broccoli, peas, sweet corn

Source: From Weichmann, J., Low oxygen effects, In *Postharvest Physiology of Vegetables*, Weichmann, J., Ed., Marcel Dekker, New York, 1987. chap. 10. With permission.

C_2H_4 production. Low temperatures greatly reduce the magnitude of the climacteric. The climacteric generally coincides with changes associated with ripening such as color changes, softening, increased tissue permeability and the development of characteristic aromas. Typical climacteric fruits include apples, pears, peaches, nectarines, bananas, mangoes, plums, tomatoes and avocados.

In nonclimacteric fruit, ripening is protracted and the attainment of the ripe state is not associated with a marked increase in respiration or C_2H_4 production. Treatment of nonclimacteric fruit with C_2H_4 stimulates respiration only; there is no increase in autocatalytic C_2H_4 production. Citrus fruits, strawberries and pineapples are examples of nonclimacteric fruits.

Generally, vegetables do not show a sudden increase in metabolic activity that parallels the onset of the climacteric in fruit unless sprouting and regrowth is initiated.

2. External Factors Affecting Respiration

a. *Temperature*

Temperature is the most important environmental factor in the postharvest life of horticultural products because of its dramatic effect on the rates of biological reactions, including respiration.[25] Typical Q_{10} (the ratio of the respiration rates for a 10°C interval) values for vegetables are 2.5 to 4.0 at 0 to 10°C, 2.0 to 2.5 at 10 to 20°C, 1.5 to 2.0 at 20 to 30°C and 1.0 to 1.5 at 30 to 40°C (see Chapter 10 IV A 3). Taking mean Q_{10} values, it can be calculated that the relative rate of respiration would increase from 1.0 at 0°C to 3.0 at 10°C, 7.5 at 20°C, 15.0 at 30°C and 22.5 at 40°C.[25] These figures dramatically illustrate the need to reduce the temperature of fresh fruits and vegetables as soon as possible after harvesting in order to maximize the shelf life.

The rate of increase in respiration rates declines with an increase in temperature up to 40°C, with the Q_{10} becoming less than one as the tissue nears its thermal death point (about 50 to 55°C) when enzyme proteins are denatured and metabolism becomes disorderly.[25]

The storage life of some fruits and vegetables, primarily those of tropical or subtropical origin, can be limited by chilling injury, a disorder induced in whole plants or susceptible tissues with low (0 to 12°C) but nonfreezing temperatures. The extent of the chilling injury is influenced by the temperature, the duration of the exposure to a given temperature and the chilling sensitivity of the particular fruit or vegetable. The symptoms of chilling injury may not be evident while the produce is held at chill temperatures, only becoming apparent after transfer to a higher temperature. Chilling injury prevents some fruits from ripening and increases their susceptibility to fungal spoilage. Chilling injury is generally associated with necrosis of groups of cells situated either externally, leading to the formation of depressed areas, pitting and external discoloration, or internally, leading to internal browning.

b. *Ethylene*

Ethylene is a natural plant hormone and plays a central role in the initiation of ripening. It is physiologically active in trace amounts (less than 0.1 ppm). The capacity to produce C_2H_4 varies greatly among fruits as shown in Table 17.3, with those fruits exhibiting moderate-to-very-high production rates generally classified as climacteric fruits. Vegetables also produce C_2H_4, although data are comparatively scarce. Some typical figures in $\mu L\ kg^{-1}\ h^{-1}$ at 20°C range from 0.04 for whole carrot roots, 0.02 for potato slices, 0.6 for intact cabbage head, 1.1 for summer squash, 1.7 for cauliflower head, 2.3 for Brussels sprouts to 3.3 to 27.0 for broccoli heads. Some fruit-vegetables such as tomatoes and melons are climacteric, while other fruit-vegetables such as cucumbers are not. The production of C_2H_4 by edible floral parts, such as cauliflower and broccoli, may be quite high and comparable with that of tomatoes.

Exposing climacteric fruits to C_2H_4 during their preclimacteric stage reduces the time to the start of the climacteric rise in respiration. The magnitude of the final respiratory rise is controlled by the fruit's endogenous C_2H_4 production and is not influenced by added C_2H_4. A reduction in C_2H_4

TABLE 17.3
Classification of Fruits According to their Maximum Ethylene Production Rate

Ethylene Production Rate $\mu L\ kg^{-1}\ h^{-1}$ at 20°C	Fruits
Very low: 0.01–0.1	Cherries, citrus, grapes, pomegranates, strawberries
Low: 0.1–1.0	Blueberries, kiwifruit, peppers, persimmon, pineapples, raspberries
Moderate: 1.0–10.0	Bananas, figs, honeydew melons, mango, tomatoes
High: 10.0–100.0	Apples, apricots, avocados, plums, cantaloupe, nectarines, papaya, peaches, pears
Very high: >100.0	Cherimoya, mamey apples, passion fruit, sapote

Source: From Kader, A. A., *Food Technol.*, 34(3), 51–54, 1980. With permission.

production and sensitivity associated with modified atmospheres can delay the onset of the climacteric and prolong the storage life of these fruits.

When nonclimacteric tissues are exposed to C_2H_4, a climacteric-like rise in respiration is induced proportional to the C_2H_4 concentration. Respiration rates return to their pretreatment level when the C_2H_4 is removed. Nonclimacteric fruits and vegetables can benefit from the reduced C_2H_4 sensitivity and a lower respiration rate of modified atmospheres.

The potent effects of C_2H_4 on plant growth, development and senescence mean that this gas, which is commonly found in the environment, can greatly reduce the storage life of perishable produce sensitive to it. Important effects of C_2H_4 in hastening the deterioration of perishable produce include accelerated senescence and loss of green color in leafy vegetables and some immature fruits (e.g., cucumbers and squash), accelerated ripening of fruits during handling and storage, russet spotting on lettuce, formation of a bitter compound (isocoumarin) in carrots, sprouting of potatoes, abscission of leaves (e.g., in cauliflower and cabbage) and toughening (lignification) of asparagus.

In addition to C_2H_4, several other hydrocarbons such as propylene and acetylene mimic C_2H_4 effects on the respiration rates of fruits and vegetables.[25] Ethylene production is reduced by low O_2, high CO_2 or both and the effects are additive.

c. Oxygen and Carbon Dioxide Concentration

A simple consideration of Equation 17.1 would suggest that if the CO_2 in the atmosphere were augmented (or the O_2 decreased), the respiration rate and the storage life would be extended. Such is the case and the application of this approach is considered further in Section III. Fresh fruits and vegetables vary greatly in their relative tolerance to low O_2 concentrations (see Table 17.4) and elevated CO_2 concentrations (see Table 17.5).

Reduction of the O_2 concentration to less than 10% provides a tool for controlling the respiration rate and slowing down senescence, although an adequate O_2 concentration must be available to maintain aerobic respiration. Vegetable crops usually require a minimum O_2 content of 1 to 3% in the storage atmosphere and, at O_2 contents below 2%, most vegetables react with a sudden increase in CO_2 production. Glycolysis results in the formation of acetaldehyde, CO_2 and finally ethanol. Both acetaldehyde and ethanol are toxic to plant cells.

Elevated CO_2 levels can inhibit, promote or have no effect on C_2H_4 production by fruits. However, CO_2 has been shown to be a competitive inhibitor of C_2H_4 action, delaying fruit ripening by displacing C_2H_4 from its receptor site. The respiration rate of most root- or bulb-type vegetables when stored under elevated CO_2 levels is stimulated, probably as an injury response that leads to further disorders and shortened life.[15] At concentrations above 20%, a significant increase in anaerobic respiration occurs which can irreversibly damage plant tissue.

TABLE 17.4
Classification of Fruits and Vegetables According to Their Tolerance to Low O_2 Concentrations

Minimum O_2 Concentration Tolerated (%)	Commodities
0.5	Tree nuts, dried fruits and vegetables
1.0	Some cultivars of apples and pears, broccoli, mushrooms, garlic, onions, most cut or sliced (minimally processed) fruits and vegetables
2.0	Most cultivars of apples and pears, kiwifruit, apricots, cherries, nectarines, peaches, plums, strawberries, papaya, pineapple, olives, cantaloupe, sweet corn, green beans, celery, lettuce, cabbage, cauliflower, Brussel sprouts
3.0	Avocados, persimmon, tomatoes, peppers, cucumber, artichoke
5.0	Citrus fruits, green peas, asparagus, potatoes, sweet potatoes

Source: From Kader, A. A., Zagory, D., and Kerbel, E. L., *CRC Crit. Rev. Food Sci. Nutr.*, 28, 1989. With permission.

Low O_2 and/or high CO_2 can both reduce the incidence and severity of certain physiological disorders such as those induced by C_2H_4 (scalding of apples and pears) and chilling injury of some fruits and vegetables (e.g., avocados, citrus fruits, chili peppers and okra). On the other hand, O_2 and CO_2 levels beyond those tolerated by the commodity can induce physiological disorders, such as brown stain on lettuce, internal browning and surface pitting of pome fruits and blackheart in potatoes.[22]

TABLE 17.5
Classification of Fruits and Vegetables According to Their Tolerance to Elevated CO_2 Concentrations

Minimum CO_2 Concentration Tolerated (%)	Commodities
2	Golden Delicious apples, Asian pears, European pears, apricots, grapes, olives, tomatoes, peppers (sweet), lettuce, endive, Chinese cabbage, celery, artichoke, sweet potatoes
5	Apples (most cultivars), peaches, nectarines, plums, oranges, avocados, bananas, mango, papaya, kiwifruit, cranberries, peas, peppers (chili), eggplant, cauliflower, cabbage, Brussels sprouts, radishes, carrots
10	Grapefruit, lemons, lime, persimmon, pineapple, cucumber, summer squash, snap beans, okra, asparagus, broccoli, parsley, leeks, green onions, dry onions, garlic, potatoes
15	Strawberries, raspberries, blackberries, blueberries, cherries, figs, cantaloupe, sweet corn, mushrooms, spinach, kale, Swiss chard

Source: From Kader, A. A., Zagory, D., and Kerbel, E. L., *CRC Crit. Rev. Food Sci. Nutr.*, 28, 1989. With permission.

d. Stresses

Physical damage such as surface injuries, impact bruising and vibration bruising can stimulate respiration and C_2H_4 production rates, depending on the variety of fruit or vegetable and the severity of the damage. The extent of the increase in respiration rate is usually proportional to the severity of bruising.[25] However, the commercial implications of the bruising (i.e., the fact that consumers prefer not to purchase bruised fruits and vegetables) are of far greater importance than its effect on the respiration rate.

Water stress resulting from loss of water to the surrounding atmosphere can stimulate the respiration rate, although the respiration rate may be reduced once the water loss exceeds 5%, coinciding with noticable wilting and shriveling of the tissue (see next section).

B. Transpiration

1. Introduction

All fruits and vegetables continue to lose water through transpiration after they are harvested and this loss of water is one of the main processes that affect their commercial and physiological deterioration. If transpiration is not retarded, it induces wilting, shrinkage and loss of firmness, crispness and succulence, with concomitant deterioration in appearance, texture and flavor. Most fruits and vegetables lose their freshness when the water loss is 3 to 10% of their initial weight. As well as loss of weight and freshness, transpiration induces water stress which has been shown to accelerate the senescence of fruits and vegetables.

2. Factors Influencing Transpiration

Several factors such as the surface area to volume ratio, nature of the surface coating, relative humidity, temperature, atmospheric pressure and extent of any mechanical damage influence the transpiration process in fruits and vegetables.

The surface area to volume ratio can range from 50 to 100 $cm^2\ mL^{-1}$ for individual edible leaves to 5 to 10 $cm^2\ mL^{-1}$ for small soft fruits such as currants, 2 to 5 $cm^2\ mL^{-1}$ for larger soft fruits such as strawberries, 0.5 to 1.5 $cm^2\ mL^{-1}$ for tubers, pome, stone and citrus fruits, bananas and onions, to 0.2 to 0.5 $cm^2\ mL^{-1}$ for densely packed cabbage.[7] Other factors being equal, a leaf will lose water and weight much faster than a fruit, and a small fruit or root or tuber will lose weight faster than a larger one.

While O_2, CO_2 and C_2H_4 diffuse mainly through air-filled stomata, lenticels, floral ends and stem scars, water vapor preferentially diffuses through the liquid aqueous phase of the cuticle.[47] In addition, there can be significant water vapor loss from cut tissues such as asparagus and celery ends.

Mechanical damage and physical injury, such as bruising, scratches and surface cuts, greatly accelerate the rate of water loss from fruit and vegetable tissue. Some tuber and root vegetables such as onions and potatoes retain the capacity to seal off wound areas after harvest when held at the appropriate temperatures and humidities (known as *curing*).

As mentioned earlier, respiration generates heat, which is dissipated through direct heat transfer to the environment and evaporation of water. The heat of respiration raises the tissue temperature and therefore increases transpiration.

The ambient RH is not a very reliable guide to likely water loss and it is more useful to calculate the water vapor pressure deficit (VPD) at a particular temperature and humidity. The VPD of the air is defined as the difference between the water vapor pressure of the ambient air and that of saturated air at the same temperature.[6] Thus, the drier the air, the greater the VPD and the more rapidly any produce held in that environment will transpire. To minimize transpiration, produce should be held at low temperature, high RH and as small a VPD as possible. Alternatively, packaging materials with very low permeability to water vapor can be used. However, the problem in these situations is that decay processes are favored.

C. Postharvest Decay

Once micro-organisms are established in plant tissue, they can proliferate and secrete enzymes to bring about quality deterioration. Frequently there is a lag period between infection and any manifestation of the presence of micro-organisms. The most common pathogens causing harvested vegetables to rot are fungi such as *Alternaria, Botrytis, Diplodia, Monilinia, Penicillium, Phomopsis, Rhizopus* and *Sclerotinia* and the bacteria *Erwinia* and *Pseudomonas*. The majority of these can only invade damaged tissue such as bruised structures and fractured cells. The development of postharvest decay is favored by high temperatures and high humidities, the latter often as a result of condensation which may affect spore germination and can cause tissue anaerobiosis.

Acidic fruit tissue is generally attacked by fungi while many vegetables with a tissue pH above 4.5 are more commonly attacked by bacteria. Initially only one or a few pathogens invade and break down the tissues, followed by a broad spectrum attack of several weak pathogens, which results in complete loss of the produce.

III. MODIFIED ATMOSPHERE PACKAGING OF FRESH HORTICULTURAL PRODUCE

A. Introduction

As stated earlier, a simple consideration of Equation 17.1 would suggest that if the CO_2 in the atmosphere were augmented (or the O_2 decreased), the respiration rate and storage life would be extended.[26] This is the basis of modified atmosphere packaging (MAP) as discussed in Chapter 15. The recommended concentrations of O_2 and CO_2 for fruits and vegetables can be found in the published literature[23,31] and will not be tabulated here. A very effective way of plotting this data has been presented with CO_2 concentration as the ordinate and O_2 concentration as the abscissa.[31] Figures 17.1 and 17.2 show such plots for some common fruits and vegetables. The windows represent the boundary of recommended gas concentrations; the smaller the window the more rigid the design requirement.

As shown in Table 17.6, the permeability ratio β (permeability coefficients of CO_2:O_2) for air is approximately 0.8 and this is represented as line A–C in Figures 17.1 and 17.2. An otherwise impermeable package with a few small holes can be used to create atmospheres along this line; no vegetable windows fall on this line but it does pass through the window for berry fruits and figs. Line A–B is plotted for a permeability ratio β of 5.0 which is approximately the value for LDPE and PVdC copolymer films. This line bisects several windows indicating that these films could be used successfully for some fruits and vegetables. However, PVdC copolymer is suitable only for produce with very low respiration rates because it has such low gas permeabilities (Table 17.6). The CO_2–O_2 atmospheres that lie between lines A–B and A–C may be created by using packages made from LDPE film with pinholes or microporous windows.[31]

Since the 1960s, attempts have been made to create and maintain modified atmospheres within plastic polymeric films. As discussed in Chapter 14, the availability of absorbers of O_2, CO_2, C_2H_4 and water provides additional tools for the packaging technologist to use to maintain a desired atmosphere within a package. Details of such systems are now discussed.

B. Factors Affecting MAP

The conditions created and maintained within a package are the net result of the interplay among several factors, including those related to the specific fruit or vegetable, and those related to the surrounding environment.[46]

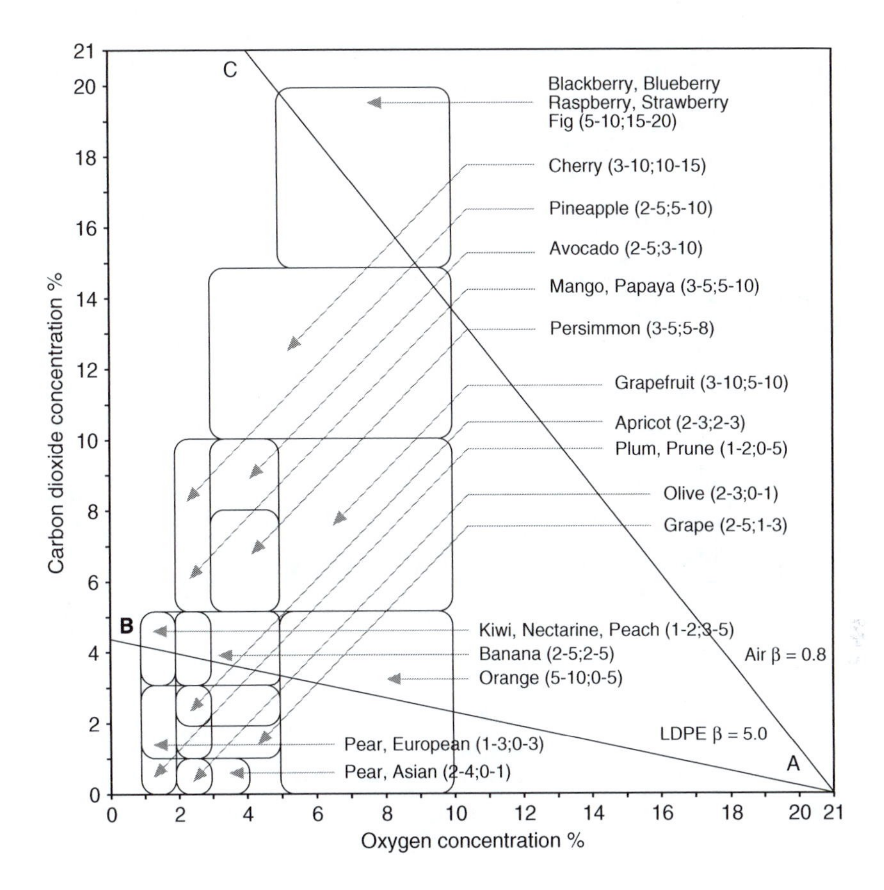

FIGURE 17.1 Recommended modified atmospheres for storage of fruits. (*Source*: From Kader, A. A., Singh, R. P., and Mannapperuma, J. D., Technologies to extend the refrigerated shelf life of fresh fruits, In *Food Storage Stability*, Taub, I. A. and Singh, R. P., Eds., CRC Press, Boca Raton, FL, 1998, chap. 16. With permission.)

1. Resistance to Diffusion

Movement of O_2, CO_2 and C_2H_4 in fruit tissue occurs by diffusion of the gas molecules under a concentration gradient. Oxygen in the environment immediately surrounding the external surface of the tissue diffuses in the gas phase through the dermal system and into the intercellular system; it then diffuses from the intercellular atmosphere into the cellular solution (cell sap) from where it diffuses in solution within the cell to centers of consumption. CO_2 and C_2H_4 produced in the cell sap diffuse outwards to the ambient environment under a concentration gradient.

The rate of gas movement depends on the properties of the gas molecule, the magnitude of the gradient and the physical properties of the intervening barriers. Both the solubility and diffusivity of each gas are important for its diffusion across barriers, with CO_2 moving more readily than O_2; CO_2 and C_2H_4 have similar diffusion rates.[25]

Different commodities have different total amounts of internal air space; for example, potatoes have only 1 to 2% while tomatoes have 15 to 20% and apples have 25 to 30%. A limited amount of air space could lead to a significant tissue resistance to gas diffusion. In addition, the cell wall appears to present some resistance to gas diffusion and thus a gradient between the cells and the intercellular space may be expected to develop.[22]

Internal concentrations of O_2 and CO_2 in plant tissues depend on the stage of maturity at harvest, respiration rate, temperature, composition of the external atmosphere and added barriers.[25]

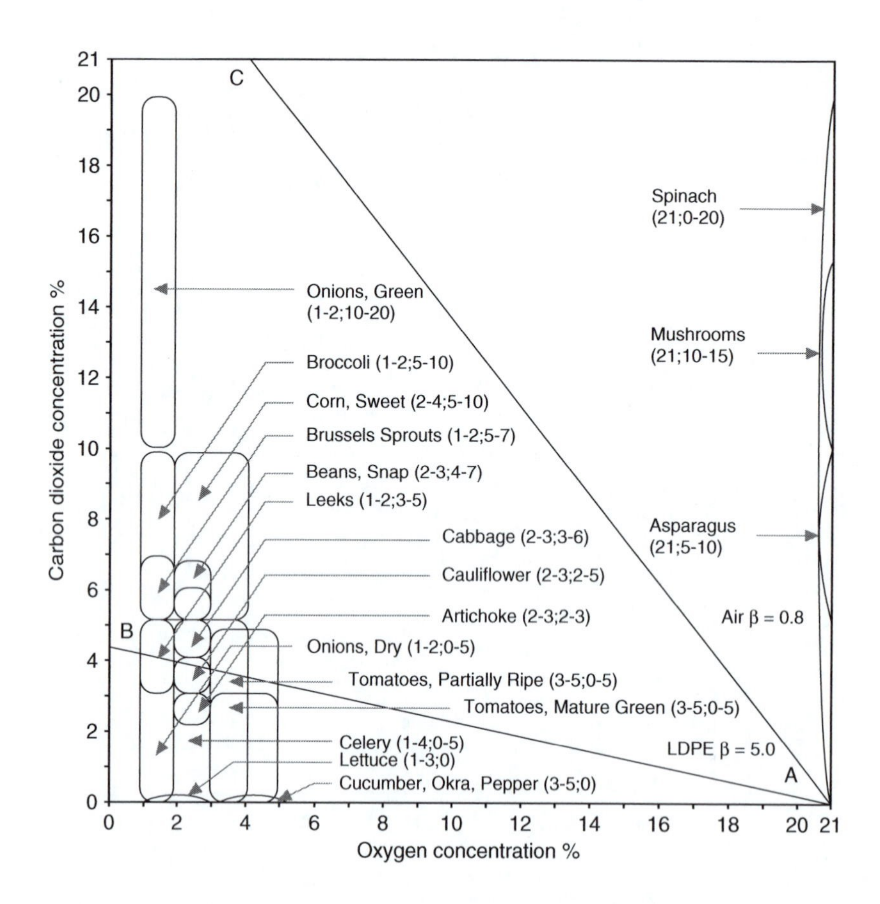

FIGURE 17.2 Recommended modified atmospheres for storage of vegetables. (*Source*: From Kader, A. A., Singh, R. P., and Mannapperuma, J. D., Technologies to extend the refrigerated shelf life of fresh fruits, In *Food Storage Stability*, Taub, I. A. and Singh, R. P., Eds., CRC Press, Boca Raton, FL, 1998, chap. 16. With permission.)

TABLE 17.6
Permeability Data of Some Polymeric Films, Air and Water

	Permeability × 10^{11} [mL(STP) cm cm^{-2} sec^{-1} (cm Hg)$^{-1}$]		Activation Energy (kJ mol^{-1})		Permeability Ratio (β)
	O_2	CO_2	O_2	CO_2	CO_2/O_2
Polyethylene (density 0.914)	30.0	131.6	42.4	38.9	4.39
Polypropylene	17.4	75.5	47.7	38.1	4.34
Poly(vinyl chloride)	0.47	1.64	55.6	56.9	3.49
Poly(vinylidene chloride)	0.055	0.31	66.5	51.4	5.64
Air	2.5×10^8	1.9×10^8	3.6	3.6	0.76
Water	9.0×10^2	2.1×10^4	15.8	15.8	23.33

Source: From Kader, A. A., Singh, R. P., and Mannapperuma, J. D., Technologies to extend the refrigerated shelf life of fresh fruits, In *Food Storage Stability*, Taub, I. A. and Singh, R. P., Eds., CRC Press, Boca Raton, FL, 1998, chap. 16. With permission.

A majority of studies on gas exchange in plant tissues indicate that the skin represents the main significant barrier to gas diffusion. Skin resistance to water vapor diffusion is much lower than resistance to O_2, CO_2 or C_2H_4 diffusion. As plant organs advance into the senescent stage, cell walls and membranes begin to break down, flooding some of the intercellular space with cell sap. In addition, the peel of some fruit may shrivel after prolonged exposure to lower water vapor concentrations. Consequently, the resistance of tissue to gas diffusion increases and may become significant.[22]

2. Respiration

The respiration rate of a commodity inside a polymeric film package depends on the kind of commodity, its stage of maturity and physical condition, the concentrations of O_2, CO_2 and C_2H_4 inside the package, the quantity of product within the package, temperature and possibly light.[22] These factors were discussed in Section II.A. Suffice it to say that the interplay of these factors leads to a complex situation about which it is difficult to make quantitative predictions (see Section III.D).

3. Temperature

It is impossible to make any specific predictions about the magnitude of the effect of a change in temperature on MAP because a number of key, interrelated factors are involved. Any change in temperature will affect the respiration rate; it will also affect gas diffusion between the cell sap and the intercellular spaces, the solubility of gases in liquids decreasing with increasing temperature. In addition, any change in temperature will affect the permeability of the plastic film surrounding the fruit or vegetable.

It is extremely unlikely that, as a result of any change in temperature, the change in respiration rate will match the change in gas permeability rates of the film. For a match to occur, the activation energy E_p of permeation would have to equal the activation energy of the respiration process. If the activation energies are not approximately equal, the equilibrium gas concentrations inside the package will change which in turn will affect the respiration rate. If the temperature stabilizes, new equilibrium conditions will be established; however, if the temperature continues to change, steady-state conditions will never be reached. It is the complex nature of this dynamic situation which makes modeling of MAP so difficult (see Section III.D).

C. Methods of Creating MA Conditions

The two methods for creating MA conditions are passive and active and they were discussed in Chapter 15. Passive MA creation relies on the respiration of the produce and the gas permeability properties of the film to achieve the desired MA. Active MA creation involves gas flushing and gas scavenging technology by adding, for example, CO_2 and N_2, or removing O_2 during packaging. Although the passive method was the traditional approach taken with horticultural produce, the active approach is increasingly common. The main reason for this is the time required to achieve the desired MA: a week or longer may be necessary using the passive approach as shown in Figure 17.3. Such a delay is particularly unacceptable for fresh cut (minimally processed) produce where, for example, enzymic browning reactions need to be inhibited as soon as possible.

D. Design of MAPs

1. General Concepts

The selection of suitable packaging materials and the evaluation of MAP for fresh fruits and vegetables has traditionally been a largely empirical, trial-and-error exercise that is

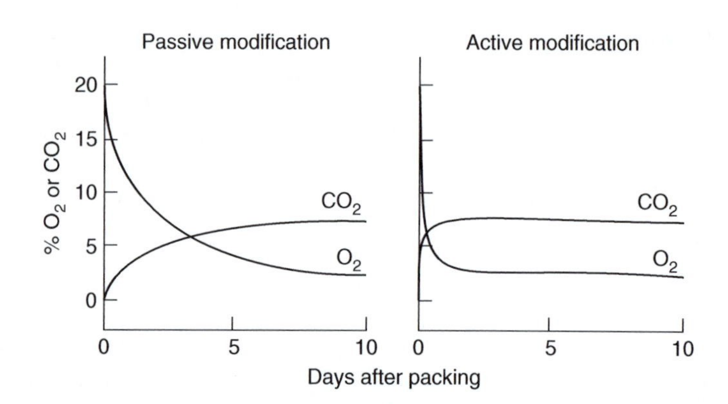

FIGURE 17.3 Relative changes in O_2 and CO_2 concentrations during passive modification and active modification of packaged horticultural products. (*Source*: From Zagory, D. and Kader, A. A., *Food Technol.*, 42(9), 70–77, 1988. With permission.)

time-consuming, subjective and often without unifying principles to guide the research and development efforts. This empirical approach can lead to long testing times, high development costs, costly overpackaging and the absence of a mechanism to fine-tune a packaging system once it has been developed. In an attempt to put the design of MAP on an analytical basis, a number of research groups have proposed and tested mathematical models but few of them are used commercially.[24,27]

In order to predict what the equilibrium gas concentrations will be and how long it will take to achieve equilibrium, a model should take into account the following factors[47]:

1. The effects of changing O_2 and CO_2 concentrations on the respiration rate
2. The possibility that the RQ is not equal to 1
3. The permeability of the film to O_2 and CO_2
4. The effect of temperature on film permeability
5. The surface area and headspace of the package
6. The resistance of the commodity to diffusion of gases through it
7. The optimal atmosphere for the commodity of interest
8. The gas concentrations which are likely to be deleterious to the commodity and if they are likely to be reached before or after equilibrium

The key part of any model is the relationship between the rate of respiration by the commodity and the permeation rate of respiratory gases through the package. As the O_2 supply within the plant cells begins to decrease as a result of respiration, the partial pressures of O_2 between the cells and the package interior and exterior atmospheres become unbalanced, causing some O_2 to diffuse into the package. When the rate of O_2 consumption exceeds the rate of permeation into the package, the O_2 concentration in the package decreases. This, in turn, further slows the respiration rate to the point where the rate of O_2 consumed equals the rate of O_2 permeation through the package film.

True equilibrium in the package would be reached when the gas flux due to respiration equaled the gas flux due to permeation; that is, O_2 was being consumed at the same rate at which it was entering the package, and CO_2 was being produced at the same rate at which it was leaving the package. However, unless the package is held under very closely controlled temperature conditions, it is doubtful if true equilibrium is ever achieved. Furthermore, because no polymeric film has the same permeability to O_2 as to CO_2, and because anoxic conditions and resultant anaerobic respiration are to be avoided at all costs, steady-state conditions inside a package are generally taken to be when the rate of O_2 consumption equals the rate of O_2 permeability through the film.

In such situations, the concentration of CO_2 inside the package will increase over time, because the partial pressure difference of CO_2 across the package film will, at least in the early stages, be so low as to result in negligible gas flux to the external atmosphere.

The real challenge in designing and developing MA packages for fresh produce is to determine the O_2 consumption and CO_2 evolution rates under the variable atmosphere conditions existing in the package.[16] These values can be found in the literature for many products under normal atmosphere but not under abnormal atmosphere conditions.

In examining the factors listed above, several important aspects must be remembered. The standard approach to measuring the gas permeabilities of films (see Chapter 4) uses Fick's law with some simplifying assumptions, the most salient in the present context being that there is a constant partial pressure difference across the film of the gas under consideration. Clearly, in MAP, this is not the case until (if ever) equilibrium is reached. Thus, the use of traditional film permeability data (if it is available at the temperatures of interest) to calculate likely gas fluxes will require accurate data on the pressure gradients across the package at any particular time.

The ratio of the permeability coefficients to CO_2 and O_2 of a polymeric film (β) can be expressed by the following equation:

$$\beta = \frac{P_C}{P_O} = \frac{p_{O_o} - p_{O_i}}{p_{C_i} - p_{C_o}} \quad (17.2)$$

where P_C and P_O are the permeability coefficients of CO_2 and O_2, respectively, p_{O_o} is the partial pressure of O_2 outside the package, p_{O_i} is the partial pressure of O_2 inside the package, p_{C_i} is the partial pressure of CO_2 inside the package and p_{C_o} is the partial pressure of CO_2 outside the package (essentially zero). If air surrounds the package, then p_{O_o} is 0.21 atm and, because p_{C_o} is essentially zero, Equation 17.2 can thus be simplified to:

$$\beta = \frac{0.21 - p_{O_i}}{p_{C_i}} \quad (17.3)$$

In the case where the film has similar permeability coefficients to both O_2 and CO_2 (i.e., $\beta = 1$), the sum of the partial pressures of O_2 and CO_2 inside the package will equal 0.21 atm. The final volume concentrations of O_2 and CO_2 inside the package will be approximately 21%, the exact value depending on the RQ of the product. Therefore products requiring optimum atmospheres of say 5% O_2 and 15% CO_2 could achieve this by using a perforated package. From Figure 17.1, it can be seen that blackberries, blueberries, figs, raspberries, strawberries and cherries are suitable for perforated packages.

However, because in reality the permeability coefficients of commercial films differ and P_C is typically 3 to 6 times P_O (see Table 17.6), the sum of the partial pressures will be less than 0.21 atm. Although the partial pressure of CO_2 inside the package increases and the partial pressure of O_2 decreases, the total pressure of these two gases will be less than 0.21 atm. Given that according to the gas laws, the total pressure times the volume is a constant, the volume of the package will decrease. As the volume of the package decreases, the partial pressure of N_2 inside the package will exceed the partial pressure of N_2 outside. This results in the permeation of N_2 through the package to the outside and, in certain situations, the reduction in package volume is such that the film adheres to the surface of the fruit or vegetable. In other situations, such as the onset of anaerobic respiration where there is a rapid build-up of CO_2, the total pressure inside the package may increase, causing the package to bulge. This would cause the partial pressure of both O_2 and N_2 inside the package to decrease and favor permeation of these gases into the package. The actual scenario depends on whether the package is rigid or flexible. In a rigid package, the free volume remains constant, while the total pressure may change, whereas in a perfectly flexible package, the total pressure remains constant while the free volume may change.[42]

Equation 17.2 can also be used to calculate the ratio of the CO_2 to O_2 permeability coefficients to give a desired gas atmosphere inside the package. For example, if it is desired to create an atmosphere containing 2% O_2 and 6% CO_2, then

$$\frac{P_C}{P_O} = \frac{0.21 - 0.02}{0.06} = 3.2 \tag{17.4}$$

Therefore, if a film was selected which had a CO_2 permeability coefficient 3.2 times its O_2 permeability at the temperature of storage (from Table 17.6, PVC is close to this at 3.5), an atmosphere containing 2% O_2 and 6% CO_2 could develop. However, it is not only the ratios of the permeability coefficients which are important, but the absolute values. If the permeability coefficients are very small, then the gas fluxes will be very low and there is the possibility of anoxic conditions or undesirably high concentrations of CO_2 developing inside the package.

It is worth noting that the ratio of the CO_2 to O_2 permeability coefficients for a particular film is not constant but depends on the temperature. In a study based on 122 permeability measurements, ratios at 0, 5, 10, 15 and 20°C were 5.08, 4.63, 4.16, 3.79 and 3.45; data from eight other film types (not specified) showed that, with one exception, the ratios increased as the temperatures decreased towards 0°C.[43] As many authors have stated, there is a dearth of actual film permeability values at the temperatures and humidities likely to be encountered during storage of MAP horticultural produce.

Another important variable can be the free or headspace volume of the package. In converting concentration measurements into volume of gases, it is usually assumed that the total pressure in the enclosed system is constant over time. However, this is true only when the respiratory quotient of the produce is unity. As was discussed earlier, this is certainly not always the case in MAP of fresh produce.

2. Developing a Predictive Model

Because the shelf life of fresh horticultural produce can be extended by confinement in an atmosphere of reduced O_2 and elevated CO_2, it is of practical importance to be able to design a package to maintain the desired concentrations of O_2 and CO_2 inside a single retail package of fresh produce. A predictive model makes the design process more focused and rapid.[17]

The influx of O_2 and efflux of CO_2 at a specific temperature are determined by the permeability of the package, the thickness and surface area of the film, and the partial pressure gradients of O_2 and CO_2 inside and outside the package. The first three parameters should be specified by the designer of the package, while the partial pressure gradients are dictated by the desired atmospheric composition inside the package and the ambient atmosphere.

The starting point for the development of a predictive model for MAP is the application of Fick's law to the package system. From Equation 4.13, the flux of O_2 through a polymeric film can be described as follows:

$$Q_O = \frac{P_O}{X} A(p_{O_o} - p_{O_i}) \tag{17.5}$$

where Q_O is the diffusive flux of O_2 through the package in unit time; P_O is the oxygen permeability coefficient of the package at the temperature of storage; A is the surface area of the package; X is the thickness of the film; p_{O_o} is the partial pressure of O_2 outside the package (0.21 atm); and p_{O_i} is the partial pressure of O_2 inside the package.

A similar expression can be written for the flux of CO_2 through the package:

$$Q_C = \frac{P_C}{X} A(p_{C_i} - p_{C_o}) \tag{17.6}$$

where Q_C is the diffusive flux of CO_2 through the package in unit time; P_C is the CO_2 permeability coefficient of the package at the temperature of storage; p_{C_i} is the partial pressure of CO_2 inside the package; and p_{C_o} is the partial pressure of CO_2 outside the package (essentially zero).

The flux of O_2 into the fruit is a function of the respiration rate (RR_O) which in turn is a function of the O_2 (and maybe CO_2) concentration within the package[12]:

$$Q_f = RR_O W \tag{17.7}$$

where Q_f is the flux of O_2 into the fruit per unit of time, RR_O is the respiration rate or O_2 uptake of the fruit (mL kg^{-1} h^{-1}) as a function of O_2 partial pressure inside the package and W is the weight of fruit in the package.

At equilibrium, the O_2 flux through the film and into the fruit should be equal; that is, $Q_O = Q_f$. On rearranging Equation 17.5 and Equation 17.7,

$$\frac{P_O A}{X} = \frac{RR_O W}{(p_{O_o} - p_{O_i})} \tag{17.8}$$

Thus, Equation 17.8 enables the prediction of the O_2 permeability requirements provided RR_O, the relationship between the respiration rate and O_2 concentration, is known. Inspection of Equation 17.8 indicates that the equilibrium concentration of O_2 (p_{O_i}) is independent of time; this is understandable because the time to equilibrium depends on the free or headspace volume of the package. The smaller the free volume, the faster equilibrium or steady-state conditions are reached.

A similar equation to Equation 17.8 can be derived for CO_2:

$$\frac{P_C A}{X} = \frac{RR_C W}{(p_{C_i} - p_{C_o})} \tag{17.9}$$

Equation 17.8 and Equation 17.9 can be rearranged to make AW/X the subject of each equation and then combined. If gas concentrations (c), rather than partial pressures (p), are used and the ratio of P_C to P_O is designated as β, then

$$c_{C_i} = c_{C_o} + \frac{1}{\beta}(c_{O_o} - c_{O_i})\frac{RR_C}{RR_O} \tag{17.10}$$

In the case of LDPE where $\beta \approx 5$, and assuming that the respiratory quotient $RR_C/RR_O = 1$, Equation 17.10 yields a straight line with a slope of 1/5, shown as line A–B in Figures 17.1 and 17.2. The implication is that only the modified atmospheres falling along the line A–B can be generated by LDPE film, for example, 2% CO_2 and 11% O_2, or 1% CO_2 and 16% O_2.[23] However, the ratio of CO_2 generation to O_2 consumption (the respiratory quotient RQ) also affects the slope of the line A–B; RQs greater than unity will result in slopes greater than $1/\beta$ and *vice versa*.

It is helpful to understand the importance of film thickness and surface area on MAP. Consideration of Equation 17.8 and Equation 17.9 indicates that doubling the thickness of a film will halve the amount of gas that can move across the film in a given amount of time; making the film thinner will have the opposite effect. Once the β of the film is known, the thickness of the film will help determine where on the β line the actual gas concentrations in the package will lie (Figures 17.1 and 17.2).[46] Increasing the thickness or decreasing the surface area of the film will move the gas equilibrium concentrations up the β line owing to reduced gas movement across the film. Increasing the weight of produce inside the package will have a similar effect because more produce will consume more O_2 and produce more CO_2.

The quantitative relationship between these factors can be expressed by rearranging Equation 17.8 and substituting concentrations for partial pressures:

$$P_O = \frac{RR_O WX}{A(c_{O_o} - c_{O_i})} \tag{17.11}$$

Equation 17.11 enables calculation of the desired permeability of a plastic film to supply sufficient O_2 to a product in order to prevent anaerobic conditions, provided that the product respiration rate at the desired O_2 concentration is known; such data can be found in the scientific literature.[46]

The above concepts were used in the development of a model to analyze the effect of the key parameters on the design of polymeric MA packages.[23] The design parameters selected were *W*, *A*, *X*, P_O and P_C, the last three parameters being fixed when a specific film of given thickness is selected, leaving only *W* and *A*.

If the ratio *W:A* is denoted by τ, then Equation 17.9 and Equation 17.10 can be arranged to yield:

$$c_{O_i} = c_{O_o} - \frac{XRR_O}{P_O}\tau \tag{17.12}$$

$$c_{C_i} = c_{C_o} - \frac{XRR_C}{P_C}\tau \tag{17.13}$$

If the respiration rates do not change with gas concentrations, then any changes in τ simply move the gas compositions inside the package along the line A–B. As τ increases, the O_2 concentration decreases and the CO_2 concentration increases. Thus, the package atmosphere moves towards B on the line A–B in direct proportion to the increase in τ.

However, the respiration rate usually decreases as the package atmosphere moves towards B, and therefore the actual change in package atmosphere is smaller than that dictated by the increase in τ alone. Similarly, the effect of any decrease in τ will also be moderated. Therefore, the effect of any errors made in the selection of τ will be moderated to a large extent by the response of respiration rate to the change in package atmosphere.[23]

For models such as that described above to be widely applicable, a great deal of data is required, including the permeances of potentially useful films at likely storage temperatures, and the relationship between the rate of respiration and O_2 concentration. It is also necessary to have similar data for CO_2 so that the potentially damaging effects of high CO_2 concentrations on produce quality can be avoided. Even with all this data, the reality is that few polymeric films are capable of maintaining the desired MAs[10,19,45,48] and that is why perforated films and, more recently, films with a breathable membrane that has a very high O_2 permeability, an adjustable selectivity for CO_2 and O_2, and permeability that increases dramatically as temperature increases,[28] have been developed.

IV. PACKAGING OF HORTICULTURAL PRODUCTS

A. Fresh and Minimally Processed Horticultural Produce

1. Introduction

The quality of fruits and vegetables comprises several parameters including flavor, aroma, texture, appearance, nutrition and safety, and the relative importance of each parameter depends on the particular commodity and its intended use. The quality of the produce is compromised whenever one of these parameters falls below a certain desirable level. The most important aspect of produce quality is freshness, typified by the quality of a fruit or vegetable when it is freshly harvested.[8]

Fresh fruits and vegetables are expected to be crisp, not tough, sweet (where appropriate), juicy, nutritious and free from defects.[46] The challenge of produce marketing is to maintain these properties of freshness during long transportation and marketing periods and packaging has as key role to play here.

In addition to maintaining freshness, modern retail packaging of fresh horticultural produce is expected to meet a wide range of requirements including prevention of mechanical damage resulting from handling, compression and impact; minimization of weight loss and shrinkage; and, if the produce is at ambient temperature at the time of packing, the ability to cool the produce rapidly after packing.

Minimally processed fruits and vegetables (MPFVs) are products that have the attributes of convenience and fresh-like quality, their forms varying widely depending on the nature of the unprocessed commodity and how it is normally consumed. They are sometimes referred to as *fresh-cut* produce, defined as any fresh fruit or vegetable or any combination thereof that has been physically altered from its original form but remains in a fresh state.[13,29] MPFVs have gained a significant proportion of the fresh produce market since their introduction to the U.S. market in the 1970s and to European markets in the 1980s. The purpose of minimal processing is to deliver to the consumer a like-fresh product with an extended shelf life, while simultaneously maintaining the nutritional and sensory quality and ensuring food safety.

MPFVs remain biologically and physiologically active, and this results in increased metabolic activity including increased respiration rates. In some cases, C_2H_4 production may also increase. The physiology of minimally processed products is essentially that of wounded tissues as a result of mechanical injury resulting from processes such as cutting, trimming and peeling. Wounding of tissues induces a number of physiological disorders that need to be minimized to get fresh-like quality products. The intensity of the wound response is affected by a number of factors, the most significant being species and variety, O_2 and CO_2 concentrations, water vapor pressure, and the presence of inhibitors.

Of key importance with MPFVs is the control of enzymes, either endogenous from the produce itself or exogenous from invading micro-organisms, to maintain the firm, crisp texture and bright, light color. The enzymes can be controlled either by inactivation or chemical or physical means. Methods that are currently used include temperature, sulfite solutions, modified atmospheres, ascorbic acid compounds, divalent ions, very high pressures, additives such as L-cysteine (which inactivates polyphenoloxidase), vanillin (which inhibits polygalacturonase), mannose (which reduces C_2H_4 production, respiration and softening in pears) and gases such as SO_2, CO and ethylene oxide. However, it is unlikely that the widespread use of the latter three gases would ever gain regulatory approval or meet with consumer acceptance given the current trend towards additive-free food.

Microbial deterioration of MPFV can be controlled by several methods including the use of chill temperatures, reduction of the total microbial population by the use of heat or irradiation, and the use of antagonistic organisms that control growth of undesirable micro-organisms.[34]

2. Packaging Materials

At the outset, it must be recognized that fresh produce is not a single commodity since there are hundreds of different fruits and vegetables, each with its own particular requirements for package performance. Very little has been published on the packaging of MPFVs, but generally the same packaging processes are used as for fresh produce. However, allowance has to be made for differences in the respiration rate of produce which has been processed in some way. There is an increasing trend away from heavily processed MPFVs, and although some research has been undertaken to develop the most effective packaging systems, more is required.

Normally, a shelf life for packaged fresh produce and MPFVs of at least 7 days at refrigerated conditions is required and to achieve this MAP is usually necessary. The primary spoilage mechanisms are the metabolism of the tissue and microbial growth; both will cause deterioration of the tissue and must be controlled to maintain tissue viability. Generally, the shelf life of MPFVs is inversely proportional to respiration rate.[37] For MPFVs, the greatest hurdle to commercial marketing is the limited shelf life, which is due to excessive tissue softening and cut surface browning.[39]

Chill temperatures during storage, dipping in antibrowning solutions and MAP are the most common approaches used to preserve the initial color of MPFVs. Ascorbic acid and its derivatives have been used in numerous studies in fruits in concentrations ranging from 0.5 to 4%.

The beneficial effects of MAP for MPFVs have been reviewed.[1,2,40] Depleted O_2 or enriched CO_2 levels reduce respiration and decrease C_2H_4 production, inhibit or delay enzymic reactions, alleviate physiological disorders and preserve the product from quality losses. However, exposure to O_2 or CO_2 levels outside the limits of tolerance may lead to anaerobic respiration with the production of undesirable metabolites and other physiological disorders. Nevertheless, the levels of O_2 and CO_2 required to avoid tissue damage or quality loss are unknown for most fruits and vegetables. Low O_2 atmospheres act synergistically with elevated CO_2 levels to reduce C_2H_4 production and respiration rates but do not completely stop senescence and tissue breakdown.

In what at first seems counterintuitive, high O_2 MAP has been found to be effective at inhibiting enzymic browning, preventing anaerobic fermentation and inhibiting aerobic and anaerobic microbial growth.[9,20] Target gas concentrations immediately after packaging are 80 to 95% O_2 and 5 to 20% N_2, the aim being to achieve O_2 levels $>40\%$ and CO_2 levels of 10 to 25% throughout chilled storage. High O_2 barrier films are not suitable, the recommended film being 30 μm OPP.

Polymeric films are the most common materials used for the packaging of horticultural products including MPFVs.[38] Early work in the area stressed the primary role of packaging to reduce transpiration, with many studies encouraging film perforation to avoid the development of injurious atmospheres inside the packages. In addition to enabling the creation of MA conditions, polymeric films provide other benefits including maintenance of high RH and reduction of water loss; improved sanitation by reducing contamination during handling; minimal surface abrasions by avoiding contact between the commodity and the shipping container; reduced spread of decay from one produce item to another; use of the film as a carrier of fungicides, scald inhibitors, C_2H_4 absorbers or other chemicals; facilitation of brand identification; and providing relevant information to consumers.[24]

In packages where it is not intended to create a MA, the main concern is to avoid anoxic conditions and condensation of water vapor inside the package. This is most easily achieved either by incomplete sealing or perforation of the plastic packaging as discussed in Chapter 5.

The relative effect of diffusive flow through holes on package atmosphere can be appreciated by comparing the permeability of gases in air with permeability of the gases in polymers as shown in Table 17.6. Air is much more permeable than polymeric films, so that even a very small hole in a polymeric package can affect the package atmosphere very significantly. This phenomenon is used to advantage with microporous films.

The effect of thin layers and droplets of water on the inside surface of films can also be appreciated by reference to Table 17.6, which shows that the permeability of gases is much higher in water than in polymers. As a result, thin layers and droplets of water forming inside polymeric packages do not significantly affect the gas atmosphere in the package.[23]

Although many polymeric films are available for packaging purposes, relatively few have been used to package fresh produce. This is not surprising because few have gas permeabilities that make them suitable for MAP. Those most likely to be suitable include laminated or coextruded materials consisting of blends of LLDPE, LDPE or OPP with EVA copolymer. The polyolefin resins provide excellent strength and are good moisture barriers, while EVA copolymer provides sealability and a higher O_2 permeability than the pure polyolefin resins. Table 17.7 gives details of packaging materials which have been suggested for MPFVs. The O_2 and CO_2 permeabilities of the above films are required at actual conditions of use, but regrettably such data are scant, making the development

TABLE 17.7
Packaging Materials for Vegetables

Vegetable	Packaging Materials and Thickness
Peeled potato (both whole and sliced)	LDPE, 50 μm; PA–LDPE, 70–100 μm or comparable
Grated carrot	OPP, 40 μm; microperforated OPP; LDPE–EVA–OPP, 30–40 μm
Sliced swede	LDPE, 50 μm
Grated swede	LDPE–EVA–OPP, 40 μm
Sliced beetroot	LDPE, 50 μm; PA–LDPE, 70–100 μm or comparable
Grated beetroot	OPP, 40 μm; microperforated OPP; PE–EVA–OPP, 30–40 μm
Shredded Chinese cabbage	OPP, 40 μm; LDPE–EVA–OPP, 30–40 μm
Shredded white cabbage	OPP, 40 μm; LDPE/EVA/OPP–O, 30–40 μm
Shredded onion	OPP, 40 μm; PA–LDPE, 70–100 μm or comparable
Shredded leek	LDPE, 50 μm; OPP 40 μm; PA–LDPE, 70–100 μm or comparable

Source: From Laurile, E. and Ahvenainen, R., Minimal processing in practice, In *Minimal Processing Technologies in the Food Industry*, Ohlsson, T. and Bengtsson, N., Eds., CRC Press, Boca Raton, FL, 2002, chap. 9. With permission.

of suitable packages a trial and error exercise in many cases. The use of such information was illustrated in Section III.D.

As illustrated in Figures 17.1 and 17.2, there are some fruits and vegetables for which LDPE film will never be suitable if a MAP is required. Thus, there is a need for modification of the permeability properties of the common polymeric films to make them more suitable for MAP and developments in this area have already been referred to in Chapter 5. Approaches adopted have included inorganic fillers, perforations and porous patches. The most exciting development has been the commercialization of temperature responsive films with a breathable membrane that has a very high O_2 permeability, an adjustable selectivity for CO_2 and O_2 and permeability that increases dramatically as temperature increases[28]; such films were described in Chapter 14 III.D.

The use of Ar in MAP was discussed in Chapter 15. Recommended MAP of salads using Ar is 70 to 90% Ar, 0 to 20% CO_2 and 1 to 15% O_2.[41] Over an average 5-day shelf life, O_2 levels declined from 10 to 15% toward 0%, while CO_2 levels rose from an initial 5 to 15% to 20 to 30% at end of life. Although increasing concentrations of CO_2 can extend the shelf life, there are undesirable side effects including bleaching of color, generation of off-tastes and deliquescence (dissolving of CO_2 in water), particularly in colored produce such as carrots and red cabbage. The use of Ar rather than N_2 slowed degradation by inhibition of oxidases from both product and microbial sources, with the total viable count being 40% less.[41]

3. Safety of MAP Produce

The general safety aspects of MAP were discussed in Chapter 15. In this section, specific comments relating to MAP of fruits and vegetables are discussed. For a more detailed discussion, the reader is referred to recent reviews.[11,18]

It is generally believed that with the use of permeable films, spoilage will occur before toxin production is an issue. However, MAP of produce should always incorporate packaging materials that will not lead to an anoxic package environment when the product is stored at the intended temperature; that is, high O_2 barrier films should not be used. Spoilage of fresh produce is mainly due to the background micro-organisms that can vary greatly for each product and storage conditions. The elimination or significant inhibition of spoilage micro-organisms should not be practiced, as their interaction with pathogens may play an integral role in product safety.[11]

The commonly encountered microflora of fruits and vegetables are *Pseudomonas* spp., *Erwinia herbicola, Flavobacterium, Xanthomonas, Enterobacter agglomerans*, lactic acid bacteria such as *Leuconostoc mesenteroides* and *Lactobacillus* spp., and molds and yeasts.[5,33] Although this microflora is largely responsible for the spoilage of fresh produce, it can vary greatly for each product and storage conditions. Temperature can play a large role in determining the outcome of the final microflora found on refrigerated fruits and vegetables, leading to a selection for psychrotrophs and a decrease in the number of mesophilic micro-organisms.

The effect of MAP on lactic acid bacteria can vary depending on the type of produce packaged. The increased CO_2 and decreased O_2 concentrations used in MAP generally favor the growth of lactic acid bacteria. This can expedite the spoilage of produce sensitive to lactic acid bacteria, such as lettuce, chicory leaves and carrots.[33] The effect of MAP on yeasts is negligible, but because molds are aerobic micro-organisms, CO_2 can cause growth inhibition at concentrations as low as 10%.[32] The antimicrobial properties of high CO_2 concentrations are mostly due to a reduction of pH and interference with the cellular metabolism.

The concern when using MAP for fruits and vegetables arises from the potential for foodborne pathogens which may be resistant to moderate to high levels of CO_2 ($\leq$50%) and outgrow spoilage micro-organisms, which may be sensitive to the MA. High levels of O_2 can inhibit the growth of both anaerobic and aerobic micro-organisms because the optimal O_2 level for growth (21% for aerobes, 0 to 2% for anaerobes) is surpassed. However, there have also been reports of high O_2 (that is, 80 to 90%) stimulating the growth of foodborne pathogens such as *Escherichia coli* and *Listeria monocytogenes*.[3]

Atmospheres with low O_2 levels inhibit the growth of most aerobic spoilage micro-organisms, which usually warn consumers of spoilage, while the growth of pathogens, especially the anaerobic psychrotrophic, nonproteolytic *Clostridia*, may be allowed or even stimulated. However, there have been few studies on the effect of MAP conditions on the microbial safety and stability of fresh-cut fruits.

At extremely low O_2 levels (<1%), anaerobic respiration can occur, resulting in tissue destruction and the production of substances that contribute to off-flavors and off-odors,[46] as well as the potential for growth of foodborne pathogens such as *Clostridium botulinum*.[4] The absence of outbreaks of botulism linked to MAP produce indicates that *C. botulinum* may be competitively inhibited under the packaging and resident flora conditions of these products. However, more research needs to be conducted to examine the potential for growth of *C. botulinum* in a wide variety of MAP produce stored at mildly abusive temperatures such as 7 to 12°C. In addition, it has been suggested that other hurdles besides temperature need to be examined to prevent botulinum toxin production.[11]

Concerns about possible pathogen contamination in MAP produce have focused on *L. monocytogenes* owing to its ability to grow at refrigeration temperatures, remaining largely unaffected by MAP while the normal microflora is inhibited.[3] Thus, although MAP produce can remain organoleptically acceptable, *L. monocytogenes* with a reduced microflora can grow at low temperatures (especially if low levels of lactic acid bacteria are present) to reach potentially harmful levels during the extended storage life of MAP produce.[11]

Edible biodegradable coatings are gaining in popularity, but a number of problems have also been associated with their use. For example, modification of the internal gas composition of the product due to high CO_2 and low O_2 can cause problems such as anaerobic fermentation of apples and bananas, rapid weight loss of tomatoes, elevated levels of core flush for apples, rapid decay in cucumbers and so on.[35] Edible films can create very low O_2 environments where anaerobic pathogens such as *C. botulinum* may thrive; however, antimicrobial compounds can be incorporated into the coating in this scenario.[14] Because the antimicrobial or antioxidant can be incorporated and applied directly to the surface of the product, only small quantities are required.

The efficacy of MAP to control the physiological decay of fresh-cut fruits warrants further investigation, and knowledge about the influence of MAP on the microbiological safety of these foods is still judged as inadequate by experts.[11] The emergence of psychrotrophic pathogens such as *L. monocytogenes*, *Aeromonas hydrophilia*, *Yersinia enterocolitica*, mesophiles such as *Salmonella* spp., *Staphylococus* spp., and the microaerophilic *Campylobacter jejuni* is of greatest concern and further investigation is warranted. Despite these concerns, MAP of horticultural produce is now widespread with an extremely low level food poisoning outbreaks.[18]

In some countries, temperature abuse is a widespread problem in the distribution chain, and can occur at any of the stages from storage through transportation to retail display and consumer handling. In these situations, it is advisable to restrict the shelf life so that psychrotrophic pathogens have insufficient time to multiply and produce toxin. Where the shelf life is greater than 10 days and the storage temperature is likely to exceed 3°C, it has been suggested[30] that the products should meet one or more of the following controlling factors:

- A minimum heat treatment such as 90°C for 10 min
- A pH of 5 or less throughout the food
- A salt level of 3.5% (aqueous) throughout the food
- a_w of 0.97 or less throughout the food
- Any combination of heat and preservative factors which has been shown to prevent growth of toxin production by *C. botulinum*

However, given that the aim is to maintain the freshness of MPFVs, the above treatments are not practicable and limiting the shelf life is the best solution.

B. Frozen

Packaging materials for frozen fruits and vegetables must protect the product from moisture loss, light and oxygen, of which the first is the most important. Freezer burn or sublimation of water vapor from the surface of frozen foods results in dehydration with a concomitant loss in weight, and their visual appearance deteriorates. All the common polymeric films have satisfactory water vapor transmission rates at freezer temperatures.

The earliest form of packaging material for frozen fruits and vegetables was waxed cartonboard, often with a moistureproof RCF overwrap. These were replaced with folding cartons with a hot-melt coating of PVdC copolymer and the ability for the flaps to be heat sealed. Although still used to a small extent for low production volumes, the majority of frozen fruits and vegetables today are packaged in polymeric films based on blends of polyolefins, the major component of which is LDPE. Sufficient plasticizer is added to ensure that the films retain their flexibility at low temperature. It is also common for the film to contain a white pigment to protect the contents from light which could oxidize the pigments. The film is usually supplied in roll form from which it is converted into a tube, then filled and sealed continuously in a form/fill/seal type of machine. Premade bags are used for low-volume packaging operations.

C. Canned

The thermal processes used for canned fruits and vegetables differ markedly depending on the pH of the product: low acid products; that is, products with a pH greater than 4.5 (which includes most vegetables) require a full 12D process, typically 60 to 90 min at 121°C. In contrast, those products with a pH less than 4.5 need only a mild heat treatment, typically 20 min in boiling water. Some products are acidified to lower the pH below 4.5 and thus avoid the more severe heat treatment.

The majority of "canned" fruits and vegetables are packaged either in tinplate or ECCS cans or glass jars. The cans must have the correct internal enamel applied to avoid corrosion of the tinplate.

It is important that all the air is removed from the product prior to packaging to minimize internal corrosion. For acid fruits such as raspberries, which contain red/blue anthocyanin pigments, the enamel coating must be particularly rigorous because the pigments act as depolarizers, accelerating the rate of corrosion. With some fruits, only the ends of the can are enameled, and for pineapple, a plain can is used so that as the tin dissolves from the tinplate, it reacts with certain constituents of the pineapple and a yellow color develops. White aluminum-pigmented epoxy resin enamels are used with fruits in some countries.

Many vegetables contain sulfur compounds that can break down during heat processing to release H_2S. This can react with the tin and iron of the metal can to form black metallic sulfides which cause an unsightly staining of the can and also of the contents. While this process is encouraged and, indeed, is essential for the production of the desired flavor and color in canned asparagus, it is avoided with other vegetables by the use of special enamels which contain zinc oxide. This reacts with the H_2S to produce barely detectable white zinc sulfide on the inner surface of the can. White aluminum-pigmented enamels based on epoxy resins are also used for cans containing vegetables.

Glass containers are still used for packaging some commercially processed fruits and vegetables, generally for products at the premium end of the market. This is largely because the production rates for glass containers are much lower than those possible for metal cans. Cylindrical, widemouth glass jars are commonly used with either a twist-off or pry-off cap made from mild steel or aluminum coated with a suitable lining material. Considerably greater operator skills are required to retort glass jars compared to metal cans, because failure to control the overpressure correctly can result in either shattered containers or the loss of pry-off caps.

Retortable pouches made from laminates of plastic film generally with an aluminum central layer can also be used for the packaging of fruits and vegetables which are preserved by the use of heat. In addition, the use of retortable paperboard laminate cartons has recently commenced for fruits and vegetables.

D. Dehydrated

The packaging of dehydrated fruits and vegetables requires the use of a package that will prevent or, at the very least, minimize the ingress of moisture and, in certain instances, O_2. For example, products which contain carotenoid pigments (e.g., carrots and apricots) can undergo oxidative deterioration, and dehydrated potatoes are liable to develop stale rancid flavors unless O_2 is excluded. Vacuum or inert gas packaging may be used if the product is particularly sensitive to oxidation.

Many vegetables, for example green beans, peas and cabbage, are treated with sulfur dioxide prior to drying to retard nonenzymic browning (the principal cause of deterioration in dehydrated vegetables) and increase the retention of ascorbic acid. Sulfur dioxide also has a useful antimicrobial effect during the initial stages of drying and, by varying the form in which it is introduced (sodium sulfite or metabisulfite), it can be used to control pH, which, in turn, influences the color and subsequent handling and drying characteristics of the product. Concentrations of sulfur dioxide in the dried product normally range from between 200 and 500 ppm for potatoes to between 2000 and 2500 ppm for cabbage.

For the packaging of dehydrated fruits and vegetables, the material normally used consists of one or more polymeric films having the desired barrier properties. This implies that the material must be a very good barrier to water vapor and, depending on the particular product, a good barrier to O_2 and maybe SO_2 and certain volatiles. For premium products, it is common to use a laminate where the center layer is aluminum foil coated on both sides with polymeric films.

REFERENCES

1. Ahvenainen, R., New approaches in improving the shelf life of minimally processed fruit and vegetables, *Trends Food Sci. Technol.*, 7, 179–187, 1996.
2. Ahvenainen, R., Minimal processing of fresh produce, In *Minimally Processed Fruits and Vegetables: Fundamental Aspects and Applications*, Alzamora, S. M., Tapia, M. S. and López-Malo, A., Eds., Aspen Publishers, Gaithersburg, MD, 2000, chap. 16.
3. Amanatidou, A., Smid, E. J., and Gorris, L. G. M., Effect of elevated oxygen and carbon dioxide on the surface growth of vegetable-associated microorganisms, *J. Appl. Microbiol.*, 86, 429–438, 1999.
4. Austin, J. W., Dodds, K. L., Blanchfield, B., and Farber, J. N., Growth and toxin production by *Clostridium botulinum* on inoculated fresh-cut packaged vegetables, *J. Food Prot.*, 61, 324–328, 1998.
5. Bennik, M. H. J., Vorstman, W., Smid, E. J., and Gorris, L. G. M., The influence of oxygen and carbon dioxide on the growth of prevalent *Enterobacteriaceae* and *Pseudomonas* species isolated from fresh and controlled-atmosphere-stored vegetables, *Food Microbiol.*, 15, 459–469, 1998.
6. Ben-Yohoshua, S. and Rodov, V., Transpiration and water stress, In *Postharvest Physiology and Pathology of Vegetables*, 2nd ed., Marcel Dekker, New York, 2002, chap. 5.
7. Burton, W. G., *Post-Harvest Physiology of Food Crops*, Longman, New York, 1982.
8. Cardello, A. V. and Schutz, H. G., The concept of food freshness: uncovering its meaning and importance to consumers, In *Freshness and Shelf Life of Foods, ACS Symposium Series #836*, Cadwallader, K. R. and Weenen, H., Eds., American Chemical Society, Washington, DC, 2003, chap. 20.
9. Day, B. P. F., Novel MAP applications for fresh-prepared produce, In *Novel Food Packaging Techniques*, Ahvenainen, R., Ed., CRC Press, Boca Raton, FL, 2003, chap. 10.
10. Exama, A., Arul, J., Lencki, R. W., Lee, L. Z., and Toupin, C., Suitability of plastic films for modified atmosphere packaging of fruits and vegetables, *J. Food Sci.*, 58, 1365–1370, 1993.
11. Farber, J. N., Harris, L. J., Parish, M. E., Beuchat, L. R., Suslow, T. V., Gorney, J. R., Garrett, E. H., and Busta, F. F., Chapter IV. Microbiological safety of controlled and modified atmosphere packaging of fresh and fresh-cut produce, *Compr. Rev. Food Sci. Food Saf.*, 2(Suppl.), 142–159, 2003.
12. Fonseca, S. C., Oliveira, F. A. R., and Brecht, J. K., Modelling respiration rate of fresh fruits and vegetables for modified atmosphere packages: a review, *J. Food Eng.*, 52, 99–119, 2002.
13. Garrett, E. H., Fresh-cut produce, In *Principles and Applications of Modified Atmosphere Packaging of Foods*, 2nd ed., Blakistone, B. A., Ed., Blackie Academic & Professional, London, England, 1998, chap. 6.
14. Guilbert, S., Gontard, N., and Gorris, L. G. M., Prolongation of the shelf life of perishable food products using biodegradable films and coatings, *Lebens Wiss Technol.*, 29, 10–17, 1996.
15. Herner, R. C., High CO_2 effects on plant organs, In *Postharvest Physiology of Vegetables*, Weichmann, J., Ed., Marcel Dekker, New York, 1987, chap. 11.
16. Hertog, M. L. A. T. M., MAP performance under dynamic temperature conditions, In *Novel Food Packaging Techniques*, Ahvenainen, R., Ed., CRC Press, Boca Raton, FL, 2003, chap. 27.
17. Hertog, M. L. A. T. M. and Banks, N. H., Improving MAP through conceptual models, In *Novel Food Packaging Techniques*, Ahvenainen, R., Ed., CRC Press, Boca Raton, FL, 2003, chap. 16.
18. Hui, Y. H. and Nip, W.-K., Safety of vegetables and vegetable products, In *Handbook of Vegetable Preservation and Processing*, Hui, Y. H., Ghazala, S., Graham, D. M., Murrell, K. D. and Nip, W.-K., Eds., Marcel Dekker, New York, 2004, chap. 29.
19. Jacxsens, L., Devlieghere, F., De Rudder, T., and Debevere, J., Designing equilibrium modified atmosphere packages for fresh-cut vegetables subjected to changes in temperature, *Lebensm-Wiss-Technol.*, 33, 178–187, 2000.
20. Jacxsens, L., Devlieghere, F., Van der Steen, C., and Debevere, J., Effect of high oxygen atmosphere packaging on microbial growth and sensorial qualities of fresh-cut produce, *Int. J. Food Microbiol.*, 71, 197–210, 2001.
21. Kader, A. A., Prevention of ripening in fruits by use of controlled atmospheres, *Food Technol.*, 34(3), 51–54, 1980.

22. Kader, A. A., Zagory, D., and Kerbel, E. L., Modified atmosphere packaging of fruits and vegetables, *CRC Crit. Rev. Food Sci. Nutr.*, 28, 1–30, 1989.
23. Kader, A. A., Singh, R. P., and Mannapperuma, J. D., Technologies to extend the refrigerated shelf life of fresh fruits, In *Food Storage Stability*, Taub, I. A. and Singh, R. P., Eds., CRC Press, Boca Raton, FL, 1998, chap. 16.
24. Kader, A. A. and Watkins, C. B., Modified atmosphere packaging — toward and beyond, *HortTechnology*, 10, 483–486, 2000.
25. Kader, A. A. and Saltveit, M. E., Respiration and gas exchange, In *Postharvest Physiology and Pathology of Vegetables*, 2nd ed., Marcel Dekker, New York, 2002, chap. 2.
26. Kader, A. A. and Saltveit, M. E., Atmosphere modification, In *Postharvest Physiology and Pathology of Vegetables*, 2nd ed., Marcel Dekker, New York, 2002, chap. 9.
27. Lakakul, R., Beaudry, R. M., and Hernandez, R. J., Modeling respiration of apple slices in modified-atmosphere packages, *J. Food Sci.*, 64, 105–110, 1999.
28. Lange, D. L., New film technologies for horticultural products, *HortTechnology*, 10, 487–490, 2000.
29. Lamkanra, O., Ed., *Fresh-Cut Fruits and Vegetables: Science, Technology and Market*, CRC Press, Boca Raton, FL, 2002.
30. Laurile, E. and Ahvenainen, R., Minimal processing in practice, In *Minimal Processing Technologies in the Food Industry*, Ohlsson, T. and Bengtsson, N., Eds., CRC Press, Boca Raton, FL, 2002, chap. 9.
31. Mannapperuma, J. D. and Singh, R. P., Modeling of gas exchange in polymeric packages of fresh fruits and vegetables, In *Minimal Processing of Foods and Process Optimization — An Interface*, Singh, R. P. and Oliveira, F. A. R., Eds., CRC Press, Boca Raton, FL, pp. 437–458, 1994.
32. Molin, G., Modified atmospheres, In *The Microbiological Safety and Quality of Food*, Lund, B. M., Baird-Parker, T. C. and Gould, G. W., Eds., Aspen Publishers, Gaithersburg, MD, pp. 214–234, 2000.
33. Nguyen-the, C. and Carlin, F., The microbiology of minimally processed fresh fruits and vegetables, *Crit. Rev. Food Sci. Nutr.*, 34, 371–401, 1994.
34. Novak, J. S., Sapers, G. M., and Juneja, V. K., *Microbial Safety of Minimally Processed Foods*, CRC Press, Boca Raton, FL, 2002.
35. Park, H. J., Chinnan, M. S., and Shewfelt, R. L., Edible coating effects on storage life and quality of tomatoes, *J. Food Sci.*, 59, 568–570, 1994.
36. Powrie, W. D. and Skura, B. J., Modified atmosphere packaging of fruits and vegetables, In *Modified Atmosphere Packaging of Food*, Ooraikul, B. and Stiles, M. E., Eds., Ellis Horwood Ltd., Chichester, England, 1991, chap. 7.
37. Reyes, V. G., Improved preservation systems for minimally processed vegetables, *Food Australia*, 48(2), 87–90, 1996.
38. Smith, J. P., Ramaswamy, H. S., and Raghavan, G. S. V., Packaging fruits and vegetables, In *Handbook of Postharvest Technology: Cereals, Fruits, Vegetables, Tea, Spices*, Chakraverty, A., Mujumdar, A. S., Raghavan, G. S. V. and Ramaswamy, H. S., Eds., Marcel Dekker, New York, 2003, chap. 19.
39. Soliva-Fortuny, R. C. and Martin-Belloso, O., New advances in extending the shelf life of fresh-cut fruits: a review, *Trends Food Sci. Technol.*, 14, 341–353, 2003.
40. Solomos, T., Principles underlying modified atmosphere packaging, In *Minimally Processed Refrigerated Fruits and Vegetables*, Wiley, R. C., Ed., Chapman & Hall, New York, pp. 183–225, 1997.
41. Spencer, K. C. and Humphreys, D. J., Argon packaging and processing preserves and enhances flavor, freshness and shelf life of foods, In *Freshness and Shelf Life of Foods, ACS Symposium Series #836*, Cadwallader, K. R. and Weenen, H., Eds., American Chemical Society, Washington, DC, 2003, chap. 20.
42. Talasila, P. C. and Cameron, A. C., Prediction equations for gases in flexible modified-atmosphere packages of respiring produce are different than those for rigid packages, *J. Food Sci.*, 62, 926–930, 1997.
43. Tolle, W. E., Variables affecting film permeability requirements for modified-atmosphere storage of apples, *US Dept Agric. Tech. Bull.*, 1422 1971.
44. Weichmann, J., Low oxygen effects, In *Postharvest Physiology of Vegetables*, Weichmann, J., Ed., Marcel Dekker, New York, 1987, chap. 10.

45. Yam, K. L. and Lee, D. S., Design of modified atmosphere packaging for fresh produce, In *Active Food Packaging*, Rooney, M. L., Ed., Chapman & Hall, London, England, 1995, chap. 3.
46. Zagory, D., Principles and practice of modified atmosphere packaging of horticultural commodities, In *Principles of Modified Atmosphere and Sous Vide Product Packaging*, Farber, J. M. and Dodds, K. L., Eds., Technomic Publishing, Lancaster, PA, pp. 175–206, 1995.
47. Zagory, D. and Kader, A. A., Modified atmosphere packaging of fresh produce, *Food Technol.*, 42(9), 70–77, 1988.
48. Zanderighi, L., How to design perforated polymeric films for modified atmosphere packs (MAP), *Packag. Technol. Sci.*, 14, 253–266, 2001.

18 Packaging of Dairy Products

CONTENTS

I. INTRODUCTION

Particularly in the Western world, milk from cattle (*Bos taurus*) accounts for nearly all the milk produced for human consumption. The composition of milk reflects the fact that it is the sole source of food for the very young mammal. Hence, it is composed of a complex mixture of lipids, proteins, carbohydrates, vitamins and minerals. The approximate composition of milk and the range of average compositions for milks of lowland breeds of cattle are given in Table 18.1. The water phase carries some of the constituents in suspension, while others are in solution. The fat is suspended in very small droplets as an oil in water emulsion and rises slowly to the surface on standing, a process often termed *creaming*.

Milk is processed into a variety of products, all having different and varying packaging requirements.[48] The simplest product is pasteurized milk where, after a mild heat treatment, the milk is filled into a variety of packaging media and distributed. The shelf life of such a product varies from 2 to 15 days depending largely on storage temperature and type of packaging. UHT milk is subjected to a more complex process and the packaging is also more sophisticated; its shelf life can be up to 8 months. Cream is typically processed and packaged in a similar way to fluid milk and has a similar shelf life. Fermented dairy products, although being subjected to more complex processing operations, are also typically packaged in an analogous manner to fluid milk.

The other dairy products (butter, cheese and powders) are quite different in nature to fluid milk and their packaging requirements are therefore also quite different. Each of these groups of dairy products will be discussed in turn.

TABLE 18.1
Approximate Composition of Bovine Milk from Lowland Breeds

Component	Average Content (% w/w)	Range (% w/w)
Water	87.1	85.3–88.7
Solids-not-fat	8.9	7.9–10.0
Fat in dry matter	31	22–38
Fat	4.0	2.5–5.5
Protein	3.25	2.3–4.4
Casein	2.6	1.7–3.5
Lactose	5.0	4.9–5.0
Mineral substances	0.7	0.57–0.83
Organic acids	0.17	0.12–0.21
Miscellaneous	0.15	

Source: From Walstra, P., Geurts, T. J., Noomen, A., Jellema, A., and van Boekel, M. A. J. S., *Dairy Technology: Principles of Milk Properties and Processes*, Marcel Dekker, New York, 1999. With permission.

II. FLUID MILK

A. Pasteurized Milk

Milk for liquid consumption is often standardized with respect to fat content, and homogenized to retard the natural tendency for the fat globules to coalesce and rise to the surface. In response to consumer needs, a range of nonstandard fluid milk products have been developed in recent years. These products have varying (reduced) fat levels and additives such as calcium or other nutrients (e.g., the fat-soluble vitamins). From a packaging point of view, these nonstandard milks can be treated analogously to standard milks.

1. Effect of Micro-Organisms

In virtually all countries, liquid milk for consumption must be pasteurized and cooled before it is packed. The primary purpose of pasteurization is to destroy any pathogenic micro-organisms present in the raw milk to make it safe for human consumption, while simultaneously prolonging the shelf life of the milk by destroying other micro-organisms and enzymes that might ruin the flavor.

In general, only 90 to 99% of all micro-organisms present in raw milk are destroyed by pasteurization, and pasteurized milk producers depend on refrigerated storage to achieve the required shelf life. Postheat treatment contamination (PHTC) with cold-tolerant Gram-negative spoilage bacteria is the limiting factor affecting the shelf life of commercial pasteurized milks at refrigerated storage temperatures. Although these bacteria are completely inactivated by pasteurization, they are regularly found in pasteurized products. Inefficient sanitation of milk contact surfaces and contamination from the dairy plant atmosphere are the major causes of PHTC, with most problems arising in the filling line where open containers permit ingress of contaminants.[35] Upgrading of pasteurized milk handling and packaging systems to an ultra clean standard is an effective method for extending shelf life at refrigeration temperatures. It is also essential to control the number of stoppages on high-speed lines.

In the absence of PHTC, the shelf life of pasteurized milk products depends on the activity of heat-resistant organisms that survive pasteurization, and their level of activity depends on the storage temperature. A shelf life of at least 8 to 10 days at 6 to 8°C is typical, while at 5°C, a shelf life of 18 to 20 days is realistic for pasteurized milk products processed using an ultra clean packaging system.

2. Effect of Temperature

a. Thermalization

Thermalization is a heat treatment of lower intensity than low-temperature–long-time (LTLT) pasteurization (see below), usually 60 to 69°C for 20 sec. The purpose is to kill bacteria, especially psychrotrophs, because several of these produce heat-resistant lipases and proteases that may eventually cause deterioration of milk products.[53]

b. Pasteurization

All fluid milk, except for a small quantity of "certified" raw milk, is pasteurized at either 63°C for 30 min (referred to as the LTLT method or sometimes the Holder process) or at 72°C for at least 15 sec (referred to as the high-temperature–short-time [HTST] method). The milk is then quickly cooled to 5°C or less. These heat treatments are designed to destroy micro-organisms that produce disease and to reduce the number of spoilage micro-organisms present. They do not sterilize the product.

Ultrapasteurization involves heating the milk at or above 138°C for at least 2 sec to destroy all pathogenic organisms. Ultrapasteurized products (also known as extended shelf life [ESL] or superpasteurized) are packaged using aseptic filling machines into presterilized containers and kept refrigerated. These milk products have an ESL at low refrigerated temperatures of up to 90 days with little effect on their nutritive value.[19] They also have superior sensory properties compared with UHT milk as a result of a milder heat treatment and chill storage. However, limited data are available in the scientific literature on the safety, sensory qualities and shelf life of ESL milk.[39]

The disease-producing organisms of chief concern in milk are *Mycobacterium tuberculosis*, a non-spore-forming bacterium that causes tuberculosis and is frequently found in the milk of infected animals; Brucella species that cause brucellosis in animals and humans; and *Coxiella burnetti* that causes a febrile disease in humans known as Q fever. The most resistant of these three organisms is *C. burnetti*, which is characterized by a $D_{65.6}$ of 30 to 36 sec and a z of 4 to 5°C. *M. tuberculosis* is characterized by a $D_{65.6}$ of 12 to 15 sec and a z of 4 to 5°C, with the Brucella species characterized by a $D_{65.6}$ of 6 to 12 sec and a z of 4 to 5°C. It is left as an exercise for the interested reader to calculate the number of decimal reductions achieved for each of the above micro-organisms by the LTLT and HTST pasteurization processes.

c. Shelf Life

One of the most critical factors affecting the shelf life of pasteurized dairy products is the temperature of storage. Attention has focused on ways of predicting the effect of temperature on the growth of bacteria in foods. The shelf life of pasteurized milk is determined mainly by the level of contamination with Gram-negative psychotrophic bacteria, although the microflora of pasteurized milk varies significantly with storage temperature. Thus, while spoilage at refrigeration temperatures is mainly a result of the growth of *Pseudomonas* spp., Enterobacteriaceae and Gram-positive bacteria assume greater importance in the spoilage of milks stored at temperatures above 10°C.

In a study seeking to determine the maximum shelf life of fat-free pasteurized milk,[10] no correlation was found between the microbial count at the end of shelf life and the sensory quality of the milk. The sensory shelf life of the milk stored in paperboard cartons at 2, 5, 7, 12 and 14°C was 15.8, 13.7, 12.3, 4.6 and 3.9 days, respectively.

Ultrapasteurized ESL milk keeps longer [up to 90 days under refrigerated (<4°C) conditions but typically 45 to 60 days], but poses greater challenges than ordinary pasteurized milk.[19] In pasteurized milk, spoilage organisms limit the shelf life. As a result, pasteurized milk spoils before becoming unsafe. With ESL milk, the thermal process of ultrapasteurization destroys spoilage organisms along with pathogens. If the milk is contaminated after ultrapasteurization, then the product could potentially develop high levels of pathogens without the usual signs of spoilage.

3. Effect of Light

During processing, distribution, storage and marketing, milk may be exposed to natural and artificial light. Flavor changes as well as loss of vitamins and other nutritional components are attributed to chemical reactions induced in milk by light, particularly in wavelengths ranging from 420 to 550 nm.[9,47] Taste panels have indicated that in the early stages of oxidation, milk loses its naturally fresh flavor and becomes quite flat in taste without being objectionable as in the more advanced stages of oxidation.

Riboflavin (vitamin B_2) plays a central role because it is not only destroyed by light but, in addition, catalyzes the development of oxidized flavor and ascorbic acid oxidation by generating excited state (singlet) oxygen. Because of its absorption by riboflavin, light of wavelengths of 350 to 550 nm are the most damaging, the maximum damage occurring at about 450 nm. These wavelengths are contained in the emission spectra of white fluorescent tubes.[3]

Riboflavin destruction by light is greater in low/nonfat milk than whole milk, because light of 400 to 500 nm wavelengths can penetrate 40 to 50% deeper into low/nonfat milk than whole milk.[1] Destruction of nutrients by light could present legal problems in connection with nutrient standards or labels, because after storage in supermarket cabinets, the milk may not meet regulatory requirements. U.S. federal regulations require low fat and nonfat milk to be fortified with vitamin A, which may degrade from light exposure in a retail dairy case.

A study was conducted on the relative destruction of vitamin A and riboflavin in low fat milk when exposed to fluorescent light at an intensity of 2000 $lm\ m^{-2}$ for 24 h.[44] It was found that more than 75% of the added vitamin A was destroyed in glass, clear polycarbonate and polyethylene containers. Paperboard containers provided the most protection, while gold-tinted polycarbonate, which blocks light of 400 to 480 nm, provided the second best protection. The presence of milk fat appears to protect against vitamin A degradation in fluid milk products, but adversely affects the flavor quality of milk after exposure to light. Even a brief, moderate light exposure (2 h at 2000 lx) can reduce the nutritional value and flavor quality of fluid milk products.[54]

Pasteurized milk in LDPE bags without a light barrier suffered vitamin C losses in excess of 50% after 12 h exposure to cool white light, whereas no loss was observed in milk packaged in paperboard cartons.[3]

The extent of nutrient loss in a supermarket depends on the proximity of the containers to the light source, the number and wattage of light bulbs in the display case, the size of the exposed surface area and the length of exposure. The detrimental effects of fluorescent light on milk can be alleviated by: (1) selecting packaging materials to minimize transmission of light; (2) reducing light intensities in display cases to 500 $lm\ m^{-2}$; (3) using yellow or yellow-green lamps or filters in display cases; (4) rotating packages at retail outlets to limit prolonged detrimental exposure to light.[44]

Because it is clear that exposure of milk to light in the 400 to 550 nm wavelength region can result in the development of off-flavors and destruction of nutrients, packaging materials used for milk should ideally not transmit more than 8% of incident light at 500 nm wavelength and not more than 2% at 400 nm. Exposure to direct sunlight should be avoided under all circumstances because this also tends to increase the temperature of the milk, thus accelerating microbial spoilage.

4. Effect of Gases

Oxygen plays an important role in the light-induced development of off-flavors in milk. Pasteurized milk at filling is generally saturated with O_2 (about 8 ppm), but if no additional O_2 can gain access, its content falls and the rate of adverse reactions slows or stops. However, additional O_2 from the headspace or entering through a permeable container will maintain the O_2 content and keep the rate of oxidative reactions high.

It has been known for many years that dissolved CO_2 inhibits certain spoilage microorganisms in foods, and in the early 1980s, the addition of CO_2 to preserve raw milk was reported.[25] Since that time, considerable research has been reported on the use of CO_2 to extend the shelf life of dairy products but it has not yet been adopted commercially, partly because, in most countries, food regulations would need to be changed to make it legal. Quantifying the effects of CO_2 on growth of spoilage organisms will be critical for optimizing its use and ensuring safe and wholesome products, and work in this area is just beginning.[21] The shelf life (defined as the time to reach 10^6 cfu mL^{-1}) of low fat milk packaged in high-barrier plastic pouches and held at 6.1°C increased from 9.6 days with no added CO_2 to 19.1 days when CO_2 was present at 21.5 m*M*.[27]

5. Packaging Materials

Milk for retail sale was traditionally packaged in refillable glass bottles. However, today, single-serve paperboard cartons and plastic containers of various compositions and constructions dominate. The packaging material is central to the protection of the flavor and nutritional qualities of fluid market milk.[46,52]

The total amount of light passing through the container wall depends on the material from which the container is made, and also on the color either incorporated into the material or used in printing it. The color determines the wavelength of the light reaching the milk.

Unpigmented HDPE milk bottles in the 350 to 800 nm spectral region have been found to transmit 58 to 79% of the incident light. Light transmission was reduced by pigmenting with titanium dioxide (1.6%), the bottle being opaque below 390 nm. The use of a colorant with the titanium dioxide pigment further reduces transmission of light with wavelengths less than 600 nm. The unprinted area of a paperboard carton had less than 1.5% transmission below 550 nm and was opaque to wavelengths below 430 nm.[36]

In a study on light induced quality deterioration of milk,[43] four common milk packaging materials were used: clear polyethylene pouch; coextruded laminate polyethylene pouch (outer white layer and an inner black pigmented layer); standard paperboard carton; and returnable plastic jug. Off-flavor was detected in all containers except the laminate pouch.

The effect of prolonged light exposure on the chemical changes in whole and 2% fat pasteurized milk stored at 4°C in clear PET bottles was compared with milk stored in green PET bottles, PET bottles incorporating a UV blocker, PET bottles with exterior labels, HDPE jugs and LDPE pouches.[5,51] The milk stored in the green PET bottles experienced less lipid oxidation and vitamin A loss than milk stored in clear PET bottles, HDPE jugs or LDPE pouches. The PET bottles with UV blockers slowed vitamin A degradation but had little effect on lipid oxidation. Blocking visible light with translucent labels helped to inhibit lipid oxidation and vitamin A degradation.

Recent studies on the shelf life of whole and low fat (1.5%) pasteurized milk stored at 4°C for 7 days in a variety of packages have been reported.[34,56] The five packages evaluated were: (1) multilayer pigmented [HDPE + 2% TiO_2–HDPE + 4% carbon black–HDPE + 2% TiO_2]; (2) monolayer pigmented [HDPE + 2% TiO_2]; (3) clear PET; (4) pigmented [PET + 2% TiO_2]; (5) LDPE-coated paperboard cartons. Chemical and microbiological parameters showed satisfactory protection of whole milk in all five packages. For the whole milk, vitamin A losses were 8.8, 10.5, 50.9, 29.8 and 14.0%, while riboflavin losses were 18.4, 20.6, 47.1, 30.9 and 19.8%. For the low fat milk, vitamin A losses were 11, 11, 31, 11 and 16%, while riboflavin losses were 28, 30, 40, 33 and 28%. The best overall protection was provided by the multilayer followed by the monolayer HDPE bottles.

B. UHT Milk

The International Dairy Federation has suggested that UHT milk should be defined as "milk which has been subjected to a continuous-flow heating process at a high temperature for a short time and which afterwards has been aseptically packaged. The heat treatment is to be at least 135°C for one or more seconds."

The development of UHT milk processing methods for sterilizing in a continuous flow has brought about the need for aseptic packaging of the product. It is only through the use of aseptic packaging that the benefits of UHT processing can be fully realized.

1. Process Description

a. *Sterilization*

UHT processes can be classified as either directly or indirectly heated according to the kind of heat exchangers used (see Ref. 8 for details). Most European regulations rely on spoilage data as a measure of how well an aseptic system works. However, the U.S. FDA requires microbiological (challenge) and chemical tests to document whether an aseptic system provides an adequate margin of safety.

b. *Packaging*

The various aseptic packaging systems available commercially have been described in Chapter 11 and Ref. 40. Although all the different systems can be used for UHT milk, the most widely used are the paperboard/foil/plastic carton and the plastic container.

2. Microbiology

Destruction of micro-organisms during UHT processing has been well documented.[4] The types of spores that have been investigated as of particular relevance to the UHT processing of milk and milk products are those of *B. stearothermophilus*, *B. subtilus*, *B. coagulans* and *B. cereus*. Spores of anaerobic organisms such as *Cl. botulinum* and *Cl. sporogenes* have also been studied albeit to a lesser extent. There is considerable variation in results between investigators; some of the most widely divergent results have been attributed to unsatisfactory experimentation.[4]

It has been recommended[4] that a z value of 9.0 to 9.3°C is generally applicable to resistant spores and more specifically applicable to the effect of UHT processes on spores in milk products. However, no generally applicable values can be given for the D_{121} values to be used in any estimate of the sterilizing effect of a UHT process because the D_{121} value will depend on the type of spore that is important in a particular application.

An additional problem in calculating sterilizing times and temperatures for UHT processes is the increasing evidence that, in many spore populations, there is a very small proportion of spores that are extremely heat resistant. Thus, instead of the thermal destruction curve being linear, the curve shows a "tail" at low proportions of survivors and, in severe cases, an almost constant level of survivors is reached. In such situations, a constant level of spoilage will be obtained (determined by the level of thermal death at which the constant level is reached and the initial concentration of the organisms concerned) regardless of how severe a heat treatment is given. Fortunately, the constant level of spoilage is normally very low in relation to the likely initial spore concentration. However, if the hygiene standard of the raw milk part of a processing plant is low so that large numbers of such spores can develop, then there is obviously a risk of spoilage of UHT products, which no increase in process severity can deal with.[4] Therefore, it is particularly important that close attention be paid to plant hygiene at all stages of the process, including those before heat treatment.

3. Nutrition

The nutritive value of UHT milk can be reduced at two stages: during the UHT treatment and during storage after packaging. The nutritive values of milk components such as fat, fat-soluble vitamins, carbohydrates and minerals are essentially unaffected, whereas values of other components such as water-soluble vitamins and proteins are adversely affected. Nutrient loss during storage is a function of the temperature of storage, the initial O_2 content of the milk and the nature of the packaging material (in particular, its opacity and permeability to O_2).

a. Vitamins

In general, it appears that vitamins are more stable under UHT processing conditions than under pasteurization or other low temperature heat treatments. Fat-soluble vitamins (A, D and E) as well as some water-soluble vitamins (riboflavin, nicotinic acid and biotin) are heat stable and are not adversely affected by UHT processing. Vitamins such as folic acid, C and B_{12} are lost to different extents.[41] Considerable variation (from 0 to 100%) has been observed in the losses of water-soluble vitamins during UHT processing and subsequent storage. If the amount of O_2 dissolved in milk is limited, then ascorbic acid losses are minimal. The nutritional value of UHT milk can deteriorate during storage to an extent that is highly dependent on the O_2 level in the product, temperature and exposure to light. A high nutritional value of UHT milk can be attained by achieving low O_2 levels, selecting packaging materials with effective O_2 barriers and preventing exposure to light by using opaque containers. Low O_2 levels can be achieved either by the use of a deaerator before heating or evaporative cooling after processing, the latter being an essential part of the direct heating process.

b. Proteins

The milk constituents that undergo the greatest change during UHT processing and storage are the proteins. Alterations in proteins are related to many technological problems with UHT products such as flavor, gelation, sediment formation, fouling of heat transfer surfaces, loss of nutritional value and browning.

Severe heat treatment causes considerable denaturation (up to 80%) of the serum proteins of milk, especially β-lactoglobulin. Direct heated UHT milk has less serum protein denaturation than indirect heated milk. Although available lysine levels in milk are reduced as a result of the Maillard reaction, the decrease is small and does not represent a significant loss in nutritional value. No significant changes in other amino acids occur, either during processing or storage.

4. Biochemical and Physical Aspects

Extensive research has reported the presence and characteristics of heat-resistant enzymes in milk and their effects on UHT products during storage.[4] Proteases and lipases are of greatest concern. Although phosphatase activity is always zero after milk has been sterilized, it may be reactivated after prolonged storage, where the extent of reactivation increases with storage time and temperature.

Age gelation is an irreversible phenomenon that occurs during storage of UHT processed milk products, ultimately transforming the product into a gel. It is considered the most important quality problem associated with this type of product, because once the product has gelled, it has reached the end of its shelf life. The severity of the heat treatment, both prior to and during the sterilization process, critically affects age gelation in UHT milk products, with gelation being less critical in UHT milk than that in UHT concentrated milk. Sterilized milk produced by the direct heat UHT process is more prone to gelation than that prepared using the indirect method, probably owing to the better control over the severity of heat treatment given in the latter.[41]

Researchers are still not sure whether gelation is attributable to enzymic action or chemical and physical processes. For many years, it was considered that coagulation was caused by the slow action of heat resistant proteases from psychrotrophs such as *Pseudomonad* spp. However, age gelation has occurred where proteolytic activity was not evident, and has not occurred on other occasions when proteolytic activity was evident. A mechanism consisting of an enzymic triggering stage followed by a nonenzymic aggregation phase has been suggested. Although proteolysis is involved, nonenzymic mechanisms play a major role in governing the phenomenon of age gelation, especially those affecting interactions between caseins and whey proteins.[41]

The best way of avoiding age gelation is to prevent the development of heat-resistant enzymes in the milk before processing. This can be achieved by preventing contamination by the causal micro-organisms, and particularly by keeping the storage time short and the storage temperature low (e.g., below 5°C) to prevent the growth of psychrotrophs.[4]

5. Flavor

The flavor of UHT milk is different from pasteurized milk, with the former generally having a flatter or "purer" taste owing to the removal of most of the feedy or barny odors. Fresh UHT milk is characterized by a poor flavor, described as a noticeable "heated" flavor, and by a sulfurous odor note. The initial sensorial characteristics of UHT milk disappear within a few days of storage and a characteristic UHT milk flavor develops with storage time.[41]

A U.S. committee on flavor nomenclature[45] has hypothesized that there are four kinds of heat-induced flavors: cooked or sulfurous; heated or rich; caramelized; and scorched. Milk which has been heated to 135 to 150°C for several seconds exhibits a strong sulfurous or cooked flavor. After several days of storage, this flavor disappears to leave a rich or heated note. Volatile sulfides are believed to contribute to the cooked flavor and it has been suggested that the Maillard nonenzymic browning reaction causes the caramelized flavors. The compounds responsible for the rich or heated note have not been clearly elucidated, and it is possible that what many researchers describe as "stale" is a combination of "rich or heated" and "caramelized."[29]

Oxidized or rancid off-flavors can also develop in UHT milk, the extent depending on the level of O_2 and the storage temperature.[42] There is a direct relationship between the level of dissolved O_2 in a product and the headspace volume of a sealed container.

6. Packaging Materials

The most common packaging material used for UHT milk is the paperboard laminate carton, although increasing quantities of plastic-based packages are now being used.

A comparison of UHT milk packaged in polyethylene-lined paperboard cartons with and without a layer of aluminum foil[14] showed that the cartons containing a layer of aluminum foil did not lose any weight during a 44-day storage period at temperatures ranging from 4 to 38°C, whereas the other cartons lost up to 1% at the highest temperature. The O_2 in the milk in the aluminum foil cartons remained almost unchanged at 1 ppm, whereas the milk in the other type of carton became saturated with O_2 (8 to 9 ppm) after a few days. Most of the oxidative changes in this type of carton occurred in the first 2 or 3 days after packing, and this milk was acceptable for up to 3 weeks when stored at 15°C. Milk packaged in the carton containing aluminum foil was organoleptically acceptable for up to 2 months, even when stored at 38°C.

The sorption of dairy flavor compounds (aldehydes and methyl ketones) by LDPE and PP films has been investigated quantitatively in an attempt to assist aseptic processors select appropriate packaging materials for maximum flavor stability. PP sorbed these compounds to a greater extent than LDPE. Headspace analysis of UHT processed milk packaged in aseptic paperboard cartons

TABLE 18.2
Characteristics of Coextruded HDPE Packaging Materials for UHT Milk

Packaging Material	Characteristics
A	Three-layer HDPE: black (2% carbon) middle layer and white (3% TiO_2) inside and outside layers
B	Three-layer HDPE, same as A with an additional PVdC copolymer coating (5–6 μm)
C	Five-layer: white HDPE outside layer (3% TiO_2); adhesive layer; black (2% carbon) EVOH layer; adhesive; white (3% TiO_2) HDPE layer

Source: From Mottar, J., The usefulness of co-extruded high density polyethylene for packaging UHT milk, IDF Dairy Packaging Newsletter #15, April, 1987, 4. With permission.

revealed a loss of higher MW flavor compounds after 12 weeks storage, owing to the interaction between the LDPE packaging material and the milk.[17]

Three different coextruded HDPE materials were tested for the packaging of aseptically processed full-cream milk; the characteristics of the materials are summarized in Table 18.2.[33] The packaged milk was stored at room temperature (20°C) behind normal window glass for 3 months. It was found that a layer of black (2% carbon) HDPE gave sufficient protection from light. Although the EVOH copolymer layer provided the most efficient gas barrier and resulted in less quality loss, the HDPE on its own and with a PVdC copolymer coating was quite suitable, provided a shelf life of 3 months at 20°C was not exceeded.

III. FERMENTED PRODUCTS

Fermented milks are products prepared from milks (whole, partially or fully skimmed milk, concentrated milk or milk reconstituted from partially or fully skimmed dried milk), homogenized or nonhomogenized, pasteurized or sterilized, and fermented by means of specific micro-organisms.

Yogurt, the most important of the fermented milk products, is a coagulated milk product obtained by lactic acid fermentation through the action of typically *Lactobacillus bulgaricus* and *Streptococcus thermophilus*. It may contain added fruit or fruit flavors as well as carbohydrate sweetening matter. Kefir has an alcohol content of 0.5 to 2% and koumiss of 2 to 3%, and both contain considerable quantities of CO_2, which is formed by the heterofermentative, aroma-forming lactic acid bacteria.

The protective effect of five different types of containers for packaging solid unflavored yogurt stored at 8°C in the dark and less than 2000 lx for 12 h day^{-1} has been evaluated.[3] The results are summarized in Table 18.3, with the rankings being based on the results of sensory analyses and the determinations of peroxide value, vitamins A and B_2, and instrumental color measurements. The transparent, brown-pigmented glass jar was found to offer superior protection against light and O_2. The authors recommended that at a storage temperature of 8°C, the shelf life of yogurt packaged nonaseptically in the brown glass container was 16 to 18 days.

Plastic tubs (usually made from HIPS) are commonly used to package yoghurt, covered with an aluminum foil that is heat sealed to the rim of the container. In addition, rectangular gable-topped polyethylene-coated paperboard packages are used, sometimes incorporating aluminum foil as a barrier in the laminate. Although the traditional glass bottle with a foil cap is still used in some parts

TABLE 18.3
Summary of Protective Effects of Different Packaging Materials for Yoghurt

Materials	Protection against: Light	Protection against: Oxygen	Rank in Order of Decreasing Total Protection
Transparent brown glass	Good	Perfect	1
Unpigmented glass	Moderate	Perfect	4
Transparent brown PS	Good	Moderate	3
Unpigmented PS	Bad	Moderate	5
Paperboard and PS	Excellent	Bad	2

Source: From Bosset, J. O., Daget, N., Desarzens, C., Dieffenbacher, A., Fluckiger, E., Lavanchy, P., Nick, B., Pauchard, J.-P., and Tagliaferri, E., Influence of light transmittance and gas permeability of various packing materials on the quality of whole natural yoghurt during storage, In *Food Packaging and Preservation*, Mathlouthi, M., Ed., Elsevier, London, 1986, chap. 18. With permission.

of the world, it is increasingly being replaced by blow-molded polyethylene containers of similar shape, sealed with a close fitting plastic cap.

Today, it is common to add probiotic bacteria such as *Lactobacillus acidophilus* and *Bifidobacterium* spp. to yoghurt to impart health benefits to consumers. These bacteria require a low O_2 environment for maximum viability, where dissolved O_2 has a negative effect on their viability. A recent study[30] found that HIPS was ineffective as an O_2 barrier in current packaging systems. During the normal shelf life, the mean dissolved O_2 content of commercial yoghurt increased from 20 to 50 ppm. The use of a packaging material with an added O_2 barrier layer (HIPS-tie-EVOH-tie-LDPE) had a lowering effect on the dissolved O_2 content in the commercial yoghurt, with an initial O_2 content of 20 ppm decreasing to 8 ppm over 42 days.

IV. BUTTER AND SPREADS

A. Composition

Butter is a fat product obtained from milk and has the following main characteristics: a minimum milk fat content of 80 to 82%; total fat-free dry milk solids of 2%; and a maximum moisture content of 16%. Approved additives include food colorants (annatto, β-carotene and curcumin), sodium chloride and cultures of harmless lactic acid forming bacteria. The pH can be adjusted if required by the addition of approved neutralizing salts to a maximum level of 2 mg kg^{-1}.

During the production process, the milk fat is concentrated by two successive physical steps: separation of the milk and churning of the cream, which is subject to phase inversion by physical disruption of the natural milk fat globule membrane. After washing with clean water to remove milk solids, the granules are physically worked into a uniform mass that is called *butter*. Thus, butter is basically a high fat-content product in which the fat phase is the continuous one. However, the water phase (consisting mainly of small droplets) can migrate slowly. Butter is normally classified into four groups:

1. Unsalted butter (i.e., butter not flavored with salt) prepared from sweet cream
2. Salted butter prepared from sweet cream (salt content in the range of 0.2 to 2.0%)
3. Unsalted butter prepared from sour cream
4. Salted butter prepared from sour cream

The most common type of butter in the U.S. is salted prepared from sweet cream. It typically has a moisture content of 15.8% w/w and a fat content of 82% w/w.

The shelf life of butter is influenced by microbial, enzymic and chemical reactions. In addition, the susceptibility of butter to readily absorb odors from the surrounding environment can also limit its shelf life. Therefore, the choice of packaging material is usually made with the intention of delaying or preventing the undesirable reactions.

The chemical composition of butterfat plays an important role in oxidation. Heavy metals or their salts (particularly copper) have a very strong catalytic effect, with low pH promoting oxidation. Free fatty acids, fat-soluble amino acids and carotene also promote oxidation, although under the influence of light, carotene acts as an antioxidant. A natural antioxidant in butter is tocopherol (vitamin E). Butter is best stored at −25°C and sweet cream salted butter keeps satisfactorily for several years at this temperature. Although slightly oxidized flavors are expected by many consumers and are disguised by salt addition, the shelf life of butter can be usefully prolonged by exclusion of O_2 during packaging and subsequent storage.[35]

The chemical composition of the water phase varies depending on the type of butter with respect to pH and salt, protein and lactose content. Consequently, different microfloras can develop in, or are inhibited by, the various butters. For example, salted butter made from sour cream inhibits the development of certain micro-organisms.

Dairy-based spreads are manufactured using similar technology to that used to produce margarine and may have fat contents of 37.5 to 76.3%. In contrast to butter, the amount of fat present is generally low but high levels of milk protein may be incorporated to stabilize the product. Because of the high water content, the water in oil emulsion may have limited stability and this limits shelf life, especially when the product is subject to temperature cycling. A further problem is the potential for bacterial growth and spoilage as a result of the higher water content, limiting the shelf life of spreads especially when stored above 4°C.[35]

B. Packaging Requirements

1. Oxidation

Butter is very susceptible to light-induced flavors, which are mostly accompanied by oxidation defects. Light-induced oxidation of butter may occur when it is inadequately protected under illumination. The degree of deterioration depends on factors such as light source, wavelength of light, exposure time, distance of the butter from the light source and β-carotene content of the butter.[18] Butter is generally exposed to fluorescent light during storage in retail display cabinets.

The packaging material has a marked effect on the intensity of the oxidized flavor that develops, depending on the amount of light transmitted. Butter that is packaged in various types of conventional parchment papers develops objectionable levels of oxidized off-flavor after a few hours exposure in a supermarket display cabinet. However, aluminum foil laminates are satisfactory even after 48 days of continuous exposure. In a study[11] that investigated the influence of light on butter packaged with various materials (five mica-filled HDPE-based plastics, some containing yellow pigments; two parchment papers; two aluminum metallized papers, and aluminum foil glue laminated to bleached sulfite paper), oxidation occurred on the light-exposed butter surface with all the materials except the aluminum foil laminate when exposed to the three strongest light intensities.

Various packaging materials differ markedly in their transmission of light as shown in Figure 18.1. The transmission of parchment (A) ranges from 46 to 64%, while the presence of yellow pigments in B markedly reduces transmission to 1 to 17% in the 300 to 500 nm range and to less than 50% in the higher range. However, the reduced transmission below 500 nm does not appreciably reduce oxidation, indicating that longer wavelengths are important. Metallized papers

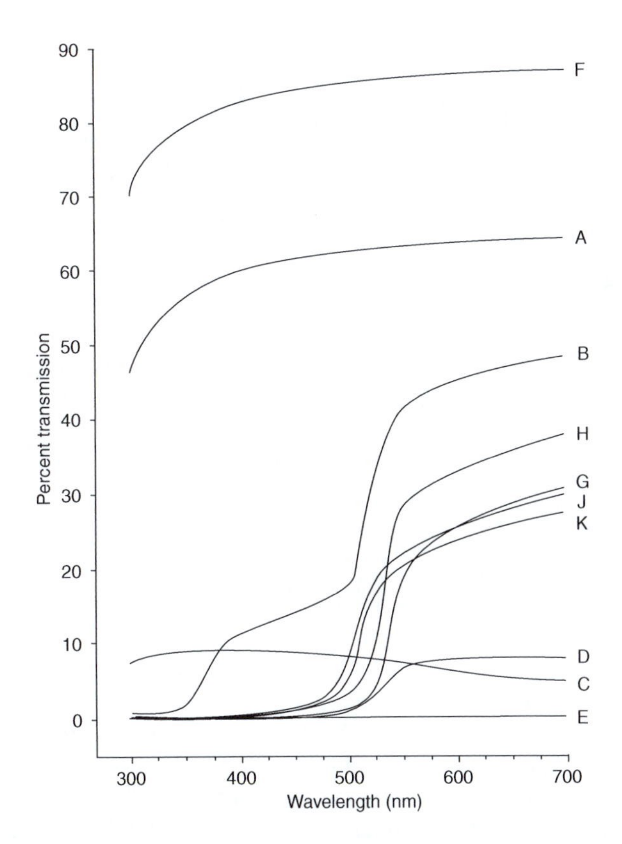

FIGURE 18.1 Spectra of light transmission by paper-based (A, B, C and D); foil laminate (E); and polyethylene-based packaging materials (F, G, H, J and K). (*Source*: From Emmons, D. B., Froehlich, D. A., Paquette, G. J., Butler, G., Beckett, D. C., Modler, H. W., Brackenridge. P. and Daniels, G., *J. Dairy Sci.* 69, 2248–2267, 1986. With permission.)

transmit less than 10% of light and the foil paper laminate transmits no measurable light. Oxidation is greater at lower intensities of light and at shorter times of exposure than expected from that observed at higher intensities and longer exposure times.

Metallic contamination of butter from dairy processing equipment is now minimal owing to the widespread use of stainless steel. In light of the fact that surface oxidation has been documented as the major flavor defect in butter at retail level,[12] butter should be packaged in laminates containing aluminum foil, which effectively block out all the light. This type of packaging material is common in Europe and is becoming increasingly so in North America.

2. Water Vapor Permeability

The surface desiccation of butter results in discoloration, a defect known as *primrosing*. However, if the packaging material is completely impermeable to water vapor, then there might be an increased risk of surface mold growth in areas where pockets of moisture have developed. The surface drying that occurs in butter packaged in parchment would thus be an advantage if the butter were susceptible to mold growth.

The permeability requirements of a satisfactory packaging material for butter vary from country to country, but a maximum of 3 gsm per day is generally agreed upon.

3. Odor Permeability

The susceptibility of fats to absorb odor compounds means that a satisfactory packaging material for butter should protect the butter from such odors. One simple but common test is to store the packaged butter for 24 h at 18°C over clove oil, after which time the butter should have no odor or taste of cloves.

4. Packaging in Current Use

Retail packaging of butter is commonly in aluminum foil (0.009 mm thick), laminated either to 40-gsm greaseproof paper or vegetable parchment, or sometimes just the paper or parchment alone. Other types of packaging for butter and dairy spreads include plastic tubs thermoformed from white-pigmented HIPS or PVC with a tight-fitting lid of the same material. To fill tubs with butter, it must be packed directly from the churn or reworked immediately prior to packing so that it will flow into the package and fill it efficiently.

V. CHEESE

A. Classification

Cheese is the generic name for a group of fermented milk-based food products produced in at least 500 varieties throughout the world. Although some soft cheese varieties are consumed fresh (i.e., without a ripening period), production of the vast majority of cheese varieties can be subdivided into two well-defined phases: manufacture and ripening. Despite differences in detail in the manufacturing processes used for individual varieties, the conversion of milk into cheese generally comprises four stages:

1. Coagulation — physicochemical changes in the casein micelles owing to the action of proteolytic enzymes or lactic acid lead to the formation of a protein network called coagulum or gel
2. Drainage — separation of the whey, after mechanical rupture of the coagulum, by molding and in certain cases by pressure, to obtain a curd
3. Salting — incorporation of salt by deposition on the surface or within the body of the cheese, or by immersion in brine
4. Ripening — biochemical changes in the constituents of the curd brought about by the action of enzymes, mostly of microbial origin

Cheese manufacture can be viewed as essentially a dehydration process in which the fat and casein in milk are concentrated between six- and twelvefold, depending on the variety. The degree of hydration is regulated by the extent and combination of the first three steps listed above, in addition to the chemical composition of the milk. In turn, the level of moisture in the cheese, the salt content and the cheese microflora regulate the biochemical changes that occur during ripening and thus determine the flavor, aroma and texture of the finished cheese. Although the nature and quality of the finished cheese are largely determined by the manufacturing steps, it is during the ripening phase that the characteristic flavor and texture of the individual cheese varieties develop.

With the exception of some soft cheeses, most cheese varieties are not ready for consumption at the end of manufacture but undergo a period of ripening (also referred to as *curing* or *maturation*), which varies from about 4 weeks to more than 2 years. The duration of ripening is generally inversely related to the moisture content of the cheese, although many varieties may be consumed at any of several stages of maturity depending on the flavor preferences of consumers.

B. Microbiology

A great variety of microbial species are involved in the ripening of cheese, the total population generally exceeding 10^9 organisms per gram. The principal bacterial groups involved in ripening are the lactic acid *Streptococci*, *Leuconostoc*, *Lactobacilli* and *Propionibacteria* species. *Micrococci* and *Corynebacteria* species may also be involved; they are aerobic and salt tolerant and thus grow especially on the surface of cheeses.

Yeasts are widely distributed in nature and are found in raw milk and some cheeses, the basic flora in the majority of cheeses being species of the genus *Kluyveromyces*. Yeasts produce enzymes capable of degrading the constituents of the curd and so can contribute to modifying the texture of the cheese and to the development of flavor and aroma. They are capable of converting lactose into CO_2 and may also take part in lipid degradation.

Among the fungal flora found in or on mold-ripened cheeses, species of the genus *Penicillium* are of particular importance. *Penicillium camemberti* is the original mold of Camembert and Brie and has a single habitat — the surface of a few cheeses and the environment of cheese factories. By contrast, *Penicillium roqueforti*, a microaerophilic mold, is widely distributed and is the mold of mold-ripened cheese; it grows very well at quite low O_2 levels (5%). *Geotrichum candidum* is present on certain soft cheeses where it forms a characteristic grayish-white crust on the surface. The role played by molds in ripening is a major one for soft and mold-ripened cheeses.

Fermentation of lactose to lactic acid during manufacture and the metabolism of residual lactose during the initial stages of ripening reduce the pH of cheese to around 5 depending on variety; at this pH the growth of many pathogenic bacteria is inhibited. During ripening the pH of cheese rises owing to the formation of alkaline N-containing compounds or the catabolism of lactic acid.[13]

The influence of pH on microbial growth and enzyme activity is particularly decisive. Only lactic acid bacteria, yeasts and molds can grow at pH values below 5. Enzymes are also very sensitive to variations in pH, most microbial proteases having greatest activity in the pH range of 5.0 to 7.5, and lipases in the range of 7.5 to 9.0. Curd at the end of drainage has a pH of less than 5.5 and is the site of active lactic acid fermentation. This acidic character of the curd is a necessity in that it slows down enzyme action and impedes the development of a harmful bacterial flora.

Soft cheese curd has a pH below 5, the pH of Camembert increasing from 4.5 to over 6.0 in the interior and over 7.0 on the surface by the end of ripening (about 30 days). Hard cheeses have pH values around 5.2 to 5.3 and neutralization of the cheese is undesirable. The pH tends to remain at about this level during ripening, partly because of the high degree of mineralization of the cheese, which gives it a high buffering capacity.

C. Packaging Requirements

The two key parameters contributing to the stability of cheeses are pH and a_w. However, neither of these parameters is low enough to ensure complete stabilization of the product, with the result that cheeses as a class lie between perishable foods on the one side and intermediate moisture foods on the other. While the packaging will have no influence on the pH of the cheese, the a_w of the surface (and ultimately the interior) of the cheese will be affected by the water vapor permeability of the packaging material.

Two other key factors that must be considered in the packaging of cheese (and indeed in the packaging of all dairy products) are the effect of light and O_2. As discussed earlier in this chapter, light initiates the oxidation of fats, even at temperatures found in refrigerated display cabinets. In unripened cheeses, this gives rise to off-flavors, which have been described as "cardboardy" or "metallic." Photooxidation of cheeses may be reduced by (1) minimizing light exposure; (2) optimizing the package barrier to light; (3) minimizing both the headspace volume and its residual O_2 levels.[32]

A recent review of light-induced changes in packaging cheeses[32] concluded that integrating results and conclusions on the effects of light on the sensory characteristics of packaged cheeses is very complicated because published reports contain various levels of detail about the experimental setups and methodologies. It was suggested that there is an urgent need for a more systematic approach to shelf life testing with respect to time, temperature, light sources and their intensity.

The oxidation reactions initiated by light may continue even if the cheese is subsequently protected from light,[20] and any metallic ions contained in the packaging material will catalyze the reactions. In addition, the ingress of O_2 through the packaging film is undesirable as it will contribute to the oxidation of fats and the growth of undesirable micro-organisms.

For the purposes of providing a framework to discuss packaging requirements, the following classification scheme for cheeses will be used: very hard and hard; semisoft and soft; fresh; and processed.

1. Very Hard and Hard

The very hard class of cheese (sometimes referred to as *grating grade*) is ripened by bacteria and is characterized by a moisture content (on a fat free basis) of $<51\%$, (e.g., Parmesan 42%; Romano 31%; Mozzarella 45%). The hard class of cheese is also ripened by bacteria and is characterized by a moisture content of 49 to 63%, where the range of 49 to 56% is categorized as hard, and the range of 54 to 63% being categorized as semihard. Cheeses in this class include Cheddar (still the cheese produced in the greatest quantities worldwide), Edam, Gouda, Cheshire, Gloucester, Derby and Leicester, as well as those with eyes such as Emmental and Gruyere. Also included in this class are Provolone, Mozzarella and Kasseri.

a. *Rindless Ripening*

Rindless cheese may be defined as cheese that has been ripened under a plastic film that allows little or no evaporation into the atmosphere to occur. Cheeses to which the technique of rindless ripening can be applied are obviously those in which the rind does not play an essential role in ripening. Such cheeses include the cooked, hard cheeses such as Emmental, as well as the uncooked hard cheeses such as Cheddar, Edam, Gouda and Saint Paulin. A feature of all these cheeses is that they do not have surfaces covered with molds, bacteria or yeasts producing enzymes responsible for ripening.[16]

The development of the technique of ripening under film (also known as *rindless ripening*) was motivated by various factors. One was the need to produce blocks better suited to a high degree of mechanization. Another was to increase cost-effectiveness by replacing earlier techniques of coating, thus reducing handling during ripening and losses due to drying-out. The first techniques (which appeared around 1930 and were used for Cheddar) involved coating the blocks with mineral oil. In 1950, Stine of Kraft Foods filed two patents concerning the use of a plastic packaging material for blocks of Swiss cheese; these patents have since passed into the public domain. A further factor was the development of prepacked, self-service, consumer portions of cheese for or by supermarkets, where the losses resulting from removal of the rind were considered unacceptable. This led to the idea of rindless cheese.[16]

There are three variables that may be used to adjust the process of ripening rindless cheese: the permeability of the packaging material; the ripening temperature; and the ripening time. Here, discussion is limited to the first of these variables. The general requirements for a plastic film for packaging rindless cheese have been outlined[16] and the nature of the films proposed for the ripening of cooked and uncooked hard cheeses is presented in Table 18.4 and their transmission rates are presented in Table 18.5.

For packaging sliced cheeses (9 slices each measuring $12 \times 12 \times 0.3$ cm per package), OPA–LLDPE is commonly used. This material has an O_2 transmission rate at 50% RH and

TABLE 18.4
Composition of Films Proposed for Packaging Rindless Cheeses

Material	Outside First Layer	Second Layer	Inside Third Layer
1	Copolymer of ethylene and vinyl acetate Thickness: 30 μm	Copolymer of vinylidene chloride and vinyl chloride Thickness: 8 μm	Irradiated copolymer of ethylene and vinyl acetate Thickness: 12 μm
2	Copolymer of ethylene and vinyl acetate Thickness: 15 μm	Copolymer of vinylidene chloride and vinyl chloride Thickness: 25 μm	Irradiated copolymer of ethylene and vinyl acetate Thickness: 12 μm
3	Nylon 6 (m.pt. 208°C) Thickness: 30 μm	Surlyn ionomer (sodium form) Thickness: 30 μm	
4	Mixture of 30% nylon 6 and 70% Surlyn ionomer (zinc form) Thickness: 49 μm	Copolymer of ethylene and 4.5% vinyl acetate Thickness: 89 μm	
5	Surlyn ionomer (sodium form) Thickness: 29 μm	Copolymer of ethylene and 3.5% vinyl acetate colored pale yellow by an inorganic pigment Thickness: 106 μm	
6	Copolymer of ethylene and 3.5% vinyl acetate colored pale yellow by an inorganic pigment Thickness: 26.5 μm	Surlyn ionomer (sodium form) Thickness: 108 μm	

Source: From Fradin, M., Ripening under film, In *Cheesemaking Science and Technology*, Eck, A., Ed., Lavoisier Publishing, New York, 1987, chap. 13. With permission.

23°C of 40 mL m^{-2} day^{-1} atm^{-1}; the package is sealed with a MA containing 25% CO_2 and 75% N_2.

Analysis of the gas found in the holes of, for example, an Emmental cheese revealed that it was composed of 95% CO_2 and 5% N_2. This atmosphere protects the entire cheese against the development of molds.[49]

TABLE 18.5
Transmission Rates of Films Proposed for Packaging Rindless Cheeses (See Table 18.4)

	Oxygen (mL m^{-2} day^{-1} atm^{-1})		Carbon Dioxide (mL m^{-2} day^{-1} atm^{-1})		Water Vapor (g m^{-2} day^{-1} at 38°C)
Material	0% RH	80% RH	0% RH	80% RH	90% RH
1	155	155	765	765	17.0
2	250	250	1500	1500	29.5
3	25	125	85	455	6.2
4	600	595	2120	2325	4.2
5	955	865	3685	3295	2.3
6	440	640	1470	1565	4.4

Source: From Fradin, M., Ripening under film, In *Cheesemaking Science and Technology*, Eck, A., Ed., Lavoisier Publishing, New York, 1987, chap. 13. With permission.

b. Role of Oxygen

The fate of O_2 in cheese is still not completely understood. What is certain is that the gas present in the space between the packaging material and the cheese (either produced as a product of enzymic action in the cheese, left in the package after sealing or diffusing through the packaging material) determines whether or not microbial growth will occur on the surface of the cheese.

The extent to which oxidation occurs depends largely on the storage temperature and time; also important are the surface area:volume ratio of the cheese and the O_2 permeability of the packaging material. The latter will vary with temperature and may vary with RH, depending on the chemical nature of the material.

Packaging films for Cheddar cheese must be sufficiently impermeable to O_2 to prevent fat oxidation and mold growth. Data on the minimum partial pressure of O_2 necessary for the development of molds are scarce. In the absence of CO_2, molds will grow at O_2 partial pressures below 1 mm Hg. In the presence of CO_2 at partial pressures of 65 to 110 mm Hg (8.5 to 15%), microbial growth is inhibited up to an O_2 partial pressure of 15 to 20 mm Hg.

It is likely that problems of mold growth on hard cheeses packaged in the films currently in common use are a function of hygienic conditions in the packing room, the degree of vacuum inside the package and the integrity of the heat seal, rather than the O_2 permeability of the packaging material used.

c. Gas Flush and Vacuum Packaging

The introduction of pure CO_2 into the package often produces the appearance of vacuum packaging after a certain period of storage. This is the result of absorption of CO_2 by the cheese, as well as some loss by diffusion through the packaging material. Because the permeability of polymer films to CO_2 is approximately 4 to 6 times that of O_2 or N_2, the rate of loss of CO_2 is greater than the rate at which these other gases can permeate in from the surrounding atmosphere. As a result, there is a decrease in volume of the package and the packaging material collapses around the cheese. A similar phenomenon was described in Chapter 17 in connection with MAP of fruits and vegetables. The more CO_2 that the cheese has "lost" at the time of cutting/slicing and packaging, the greater the absorption of CO_2 from gas flushing, and the greater the contraction of the package volume. Thus, in the gas flush packaging of cheese slices it is recommended that a mixture of gases (CO_2 and N_2 in the ratio 80:20 or 70:30) be used to avoid the slices being pressed against each other by atmospheric pressure.[2,49]

A study[6] into the MAP of Cheddar cheese shreds stored at 4°C under fluorescent light for 6 weeks found that 100% CO_2 atmospheres potentiated light-induced oxidation and color bleaching. However, residual O_2 concentration in the CO_2 gas was 2.69%, which may explain the photooxidation. Alterations in volatiles indicated that shreds packaged under 100% N_2 were highly susceptible to molding, even though visible signs of mold growth were not detected. Although high concentrations of either CO_2 or N_2 resulted in undesirable changes during storage, no recommendations as to the optimum gas mixture were given.

2. Semisoft and Soft

This class of cheese is also ripened by bacteria and is characterized by a moisture content of 61 to 69% on a fat-free basis and 43 to 55% on a total basis. Soft cheeses are cheeses which, independently of lactic acid fermentation, have undergone other ripening processes. The body is neither cooked nor pressed and may contain internal mold.

Examples of semisoft and soft cheeses include those which have been ripened by bacteria such as Brick and Munster, those ripened by bacteria and surface micro-organisms such as Limburger and Trappist, those ripened by surface mold such as Brie and Camembert, and those ripened by internal (blue) mold such as Roquefort, Gorgonzola and Stilton. Other semihard cheeses include Harvati, Samsø and Edam.

a. *Light*

Sliced, rindless Harvarti cheese packaged in an atmosphere of 25% CO_2 and 75% N_2 and stored under light for up to 21 days at 5°C showed a decrease in yellowness, an increase in redness and no significant changes in lightness.[26,31] Sliced, rindless Samsø cheese packaged in atmospheres of CO_2 and N_2 (0:100; 20:80 and 100:0 with residual O_2 at time of packing $<0.24\%$) and stored under varying light conditions for up to 21 days at 5°C also showed significantly decreased yellowness and increased redness.[23] Cheese stored in 100% CO_2 had significantly lower lightness and was described by the sensory panel as having a rancid taste and odor as well as a dry/crumbly texture.

In contrast to fresh cheeses, soft cheeses only require limited protection against light to maintain quality. When the cheese is ripe enough to be packaged, the influence of light on soft cheeses with surface mold is of little importance. Although light can slow down or hinder the germination of conidia, there is no such hindering effect with thick mycelium layers at an advanced stage of ripening. The same applies to red smear, where light does not even reach the internal mold of blue veined cheeses unless they are sliced. Light-induced fat oxidation in soft cheeses has not been observed, and mold mycelium and layers of bacteria, as well as the rind which forms during ripening, also provide protection against light.

b. *Gases*

The consumption of O_2 and the production of CO_2 are closely connected with the total bacterial count and the ripening conditions of temperature, humidity, pH, a_w, and so on. In the case of Camembert, for example, there is an increase in the starter bacterial count to about $10^{10}\ g^{-1}$ after the first turning, declining more or less rapidly until ripe for packaging to stabilize at about 10^7–$10^{-9}\ g^{-1}$. For a Camembert weighing 100 g, the peak O_2 requirement is about 15 mL h^{-1} with a CO_2 release of 10 mL h^{-1}. On further ripening prior to packaging, this drops to about 4.5 mL h^{-1} for O_2 and about 3.6 mL h^{-1} for CO_2.

Measurements taken of unpackaged and packaged Camembert have shown that the unpackaged cheese loses about 0.04% of its weight per hour during ripening, whereas packaged cheese loses about 0.006% per hour as water vapor.[22] These figures were obtained at 10°C and temperature clearly has an important influence. Thus, at 20°C the O_2 requirement is 11 mL h^{-1} and at 30°C it is 17.5 mL h^{-1}. The surface area:volume ratio is also important in determining the level of CO_2 release, the greater the surface area for a given cheese weight, the higher the CO_2 release.

The effects of MAP on the growth of *L. monocytogenes* in mold-ripened Stilton cheese during refrigerated storage over a 6-week period have been studied.[55] When samples were inoculated with *L. monocytogenes* and stored under MAP at N_2:CO_2:O_2 ratios of 80:10:10, 100:0:0 and 80:20:0, a significant decrease in count was found in samples stored in the 80:10:10 atmosphere. A greater inhibitory effect was achieved when CO_2 concentration was increased to 20% than by reducing the O_2 content. Results indicated that an 80:10:10 ratio is not suitable for use with blue Stilton cheese when *L. monocytogenes* may be present.

c. *Humidity*

The growth of micro-organisms in and on soft cheese is dependent on the a_w of the cheese, which is clearly influenced by the water vapor permeability of the packaging material. If the humidity of the air under the package becomes too high, then the fungus mycelium begins to turn yellow and exude liquid in the form of droplets, finally showing autolytic symptoms. This results in a clear shift of the flora composition towards hydrotrophic bacteria such as *B. linens*. Conversely, if the permeability of the packaging material is too low, the growth of mold or surface smear may be stopped, resulting in the appearance of facultative anaerobic bacteria with strong proteolytic activity.

Most polymeric packaging materials are too impermeable to be used for soft cheeses and must therefore be perforated before use. No definite rules can be laid down with regard to the number and size of the perforations as these must be found experimentally for the particular type of cheese. It should also be borne in mind that soft cheeses of the same type (e.g., Camembert) may behave differently depending on the method of production.[22]

In some countries, it is permissible to treat the packaging materials with fungicide to prevent subsequent mold growth on the surface after packaging. Typically, the packaging material is impregnated with sorbic acid or its salts, where the concentration needs to be sufficiently high (and the packaging material sufficiently close to the cheese surface) in order to prove successful.

d. *Ripened by Internal Mold*

The packaging material for internal molded cheese should allow for the passage of O_2 to promote mold development in the curing channels of the cheese. The package should also allow for a certain permeability of CO_2 and water vapor.[37] However, blue veined cheeses seem to be less dependent on the gas permeability of the packaging material than other soft and semisoft cheeses. For example, *P. roquefortii* grows in O_2 concentrations as low as 5%, as found in the curd holes of the cheese. The presence of CO_2 seems to stimulate growth, and therefore more impermeable packaging materials such as aluminum foil, PP film or thermoformed packages made from rigid PVC or PS with a transparent multilayer film lid for portions have proved effective.[22]

e. *Ripened by Surface Mold*

In surface mold-ripened cheeses such as Camembert and Brie, lactate is metabolized to CO_2 and H_2O by the metabolic activity of *P. camemberti*. For these cheeses, it is important that packing does not take place until the mold has grown to a certain extent. The packaging material must have a limited permeability to O_2 to minimize the risk of development of anaerobic proteolytic bacteria, which can also develop if the permeability to water vapor is too low resulting in condensation inside the package. The material should not adhere to the surface mold of the cheese.[37]

A suitable material for the packaging of these cheeses is perforated OPP, where the perforations are necessary to allow the passage of controlled quantities of water vapor. Paper is also used for packaging but, if it is in direct contact with the surface of the cheese, it must be coated with wax or laminated to perforated film to avoid decomposition of the paper by cellulase enzymes produced by certain molds. Aluminum foil is also used at a thickness of 7 to 9 μm if laminated to other materials and 12 μm if used alone. It must be given a protective coating of lacquer otherwise ammonia (a metabolite in the cheese ripening process) many corrode the metal, turning it black. It is preferable not to perforate the inner layer in direct contact with the surface of the cheese in order to avoid mycelium growing through the perforation and becoming apparent on the outside of the package.

Thermoformed packages made from PS and PVC are used, sometimes extrusion coated with LDPE to reduce water vapor transmission rates. Special packages for the sterilization of Camembert are made from combinations which consist of OPP–PVdC copolymer–PP, PET–PVdC copolymer–PP or OPA–PP–PVdC copolymer for deeper containers with higher puncture resistance.[49]

With few exceptions, soft cheeses are placed in an additional outer package prior to marketing. Traditionally, these consisted of wood, but the use of paperboard or a combination of a plastic base and a paperboard cover has become the norm. The plastic bases (PVC or HIPS) are designed in such a way that corrugated channels in the base ensure that sufficient gas and water vapor exchange occur.

f. *Ripened by Smear Coat*

This category, which includes Limburger, Münster and Brick, has a bacterium (typically *Brevibacterium linens*) smeared on the surface. These cheeses are sometimes wrapped in clear or orange pigmented OPP combined with greaseproof paper. Vegetable parchment is also recommended because of its mechanical strength when wet and its consequent moistening function for the surface floras of the cheese, which require a high a_w for their growth. Although vegetable parchment is sometimes used alone, it is more often used in a combination laminated to aluminum foil or an aluminum/tissue paper laminate.

3. Fresh

Fresh cheeses are slow drainage cheeses, which have been subjected to lactic acid fermentation; they are characterized by a moisture content of >80%. The three major types of fresh cheeses are Cottage, Quark and Petit Suisse.

a. *Manufacturing Processes*

i. *Cottage*

Cottage cheese is normally prepared from pasteurized skim milk by the *in situ* production of lactic acid by starters consisting of *Streptococcus lactis*, *Streptococcus cremoris* and *Leuconostoc citrovorum* (for flavor). The desired pH is ≈4.6, which may be reached in 5 to 16 h depending on the level (0.5 to 5%) of starter addition and the set temperature (22 to 32°C). The coagulum is cut and cooked to 50 to 55°C over a 1.5 h-period during which time considerable syneresis occurs and the curd assumes a firm, meaty texture. After removable of the whey and washing of the curd, salt (≈1% NaCl) is added and the curd mixed with a creaming mix to give a level of ≈4% fat in the finished cheese. A good-quality product has a shelf life of 1 to 2 weeks at chill temperatures, which may be extended by packaging in an atmosphere of CO_2.[15]

ii. *Quark*

Quark is produced in a very similar manner to Cottage cheese, with *Streptococcus diacetylactis* replacing *Leuconostoc citrovorum* for flavor development. After setting, the coagulum is broken and the whey removed from the uncooked curd/whey mixture by filtration or centrifugation. Compared with Cottage cheese, quark has a smooth consistency and a significant lactose content because it is not washed. Quark typically has a moisture content of 82%.[15]

iii. *Cream and Petit Suisse*

These are produced by *in situ* production of acid in cream and typically have a moisture content of 54%. After coagulation, the curds are separated from the whey by filtration or centrifugation. Gums may be added to cream cheeses to improve texture and consistency, and they may also be heat treated and homogenized to produce a product with a longer shelf life.[15] Cream cheese has a shelf life of 3 to 6 months, while Petit Suisse has a shelf life of 3 to 4 weeks.[22]

b. *Packaging Requirements*

The packaging requirements for fresh and cream cheeses are similar to those for the other types of cheeses, namely protection against light, O_2 and loss of moisture. Although fresh cheeses do not seem to be as sensitive to the influence of light as milk, cream or butter, the cream cheeses with their greater fat content (≈34%) are. For all the cheeses in this class, the packaging must provide protection against light transmission. The O_2 in fresh cheeses may either be present in the cheese as a result of the processing techniques used (e.g., centrifugation), in the headspace inside the package or permeate through the package over time. Because of their high water activites, the adsorption of moisture from the atmosphere is of little importance. However, loss of water through evaporation, particularly from the surface, must be avoided.

c. *Packaging Materials*

Genuine vegetable parchment or greaseproof paper was frequently used in the past to package fresh cheese and is still used today for Petit Suisse. It is usual for the paper to have a basis weight of 40 to 60 g m^{-2}. Paper coated with paraffin or PVdC copolymer is still sometimes used in the form of a banderole; for example, for packaging an unripened cheese intended for consumption within a short time.[49]

While a number of plastics have been used over the years, the standard material is HIPS which is thermoformed on form-fill-seal machines. It is also coextruded or extrusion coated with PVC or PVdC copolymer to improve its barrier properties, and pigmented with TiO_2 to provide a better barrier to light. Thermoformed containers made of HIPS are used for ladled fresh cheese and the base is separated from the cheese by a perforated PVC disk for whey drainage; no further outer packaging is required.[22] The use of PVC (both for the complete container and as a disk in the previous example) is favored because of its inertness towards the product, its impermeability to water vapor and gas, as well as its extraordinary resistance to fats.[49]

Injection molded containers made of HDPE with slits in the side to allow drainage of whey are also used, where the fresh cheese is ladled directly into the containers. Outer packaging in the form of a thermoformed PA–LDPE combination makes these containers tight and ready for use.[22]

Aluminum foil with a thickness of 7 to 20 μm can be used, with the thicker foils (15 to 20 μm) formed into containers of either rectangular shape with straight walls, or cylindrical section with corrugated or pleated sides. In all cases, the aluminum must be protected against corrosion, either by applying a suitable enamel or by laminating with LDPE or PP. If this is not done, then the whey coming into contact with the aluminum will cause the formation of aluminum lactate and attack the walls of the container, sometimes perforating them.[49] MAP has been suggested to maintain the quality of Cottage cheese.[28]

4. Processed

a. *Manufacture*

Attempts at the end of the nineteenth century to export hard cheeses from Europe to tropical countries were largely unsuccessful. This led to the development by the Swiss company Gerber in 1911 of “processed” cheese. This consisted of removing the rind from Gruyère or Emmental cheese and heating it to about 80°C while stirring in a solution of sodium citrate. The cheese formed a “sol,” which could be packaged in a metal foil while hot. On cooling, this gave a gel that was pleasant to eat, had a taste reminiscent of the original cheese and (if the acidity was properly controlled) had good keeping qualities. In 1917, processed cheese was manufactured in the U.S. from Cheddar cheese using a mixture of citrates and orthophosphates.[38]

Today, a wide range of processed cheeses is available containing a variety of flavoring compounds and fruits, vegetables or nuts. Preservatives such as sorbic and propionic acids and their salts are sometimes added as preservatives; their main function is to prevent mold growth. Nisin may be added to prevent the growth of anaerobic spore-formers such as *Clostridia* spp. The addition of antioxidants is also permitted in many countries.

b. *Packaging*

Traditional packaging of processed cheese consisted of triangular portions (usually weighing 20 to 30 g) packaged in tinfoil because of its resistance to corrosion by processing salts. For economic and technical reasons, tinfoil was replaced by heat sealable, lacquered aluminum foil. The thickness of the aluminum varies from 12 to 15 μm. To facilitate opening a portion of the cheese, an opening device is provided. These devices consist of narrow strips of PET film sealed onto the inner side of the aluminum foil before this is formed. They extend several millimeters beyond the packaging material so that they can be grasped between two fingers. Their point of exit from the packaging must itself be sealed to avoid any possibility of leakage or contamination. The strips are normally colored red to attract the consumer's attention.[49] The triangular portions are often assembled into a circular paperboard carton or plastic container complete with lid.

Processed cheeses are also very often packaged in plastic, either in film or in a thermoformed container. The following materials are used[49]:

1. PVC for containers filled below 80°C
2. PVC coextruded or laminated in combinations such as PVC (450 μm)–PVdC or EVOH copolymer (20 μm)–LDPE or PP (50 to 90 μm)
3. PP with a melt range from 165 to 170°C
4. PET

Slices of processed cheese were first marketed in the U.S. in 1950, and were manufactured either by forming strips of cheese, which were then cut up and packaged, or by molding the cheese in the form of a tube around which a web of plastic was wrapped, the whole assembly then being flattened and cooled. The films most often used are laminates of PET–LDPE, or if a more impermeable package is required, PET–PVdC copolymer–LDPE or OPP–EVOH copolymer–LDPE.[49]

Some processed cheese is packaged in either tinplate or aluminum cans, both of which must be adequately enameled inside to prevent corrosion occurring. A small quantity of spreadable processed cheese is packaged in tubes, which used to be manufactured from aluminum but are now made of five-layer laminates containing aluminum foil as the central core. The tubes are filled through the unsealed base opening, which is then welded by a high-frequency current.[49]

VI. MILK POWDERS

A. Manufacture

Milk and milk products are dried mainly by spray drying. This involves converting concentrated milk into a fog-like mist (atomizing) where it is given a large surface area, and exposing this mist to a flow of hot air in a drying chamber. When the atomized product is in contact with the hot air, the moisture evaporates quickly and the solids are recovered as a powder consisting of fine, hollow, spherical particles with some occluded air.

An important quality attribute of milk powder is the bulk density. It is obviously of considerable interest from an economic point of view because it influences the cost of storage, packaging and transport. The bulk density is governed chiefly by the total solids of the feed to the atomizer, but also by the temperature of the drying air. Agglomeration of dried particles (part of the instantizing process) also has a marked affect on the density, a heavily agglomerated particle being very light.

Instant milk powder is powder that has been manufactured in such a way that it has better reconstitution properties than normal powder. The instantizing process for skim milk powder (SMP) consists of agglomeration of the particles into porous aggregates of sizes up to 2 to 3 mm. As a result of agglomeration (achieved by wetting the powder, which causes the particles to agglomerate and then redrying them), the amount of interstitial air (i.e., the air between the particles) is increased. Reconstitution commences when the interstitial air is replaced by water.

Instantization results in a reduction of bulk density (e.g., for SMP) from 0.64 to 0.55 g mL^{-1}, although the bulk density of agglomerated particles can be as low as 0.35 to 0.40 g mL^{-1}. The free space volume has an important influence on the rate of oxidation of foods, where if a food is packaged in air, a large free space volume is undesirable because it constitutes a large O_2 reservoir. Conversely, if the product is packaged in an inert gas, then a large free space volume acts as a large "sink" to minimize the effect of O_2 transferred through the package. It follows that a large package surface area and a low bulk density result in greater O_2 transmission.

The instantizing process for whole milk powder (WMP) is more complicated because of the hydrophobic nature of the fat. Reconstitution of the agglomerates will not take place in water at temperatures below 45°C unless the powder particles have been coated with a surface-active or wetting agent. The use of lecithin as a surface-active agent is accepted world-wide, and it is mixed with butter oil and sprayed at 70°C onto the powder (which should be approximately 50°C) to give a final concentration of 0.2% in the powder. The best results are obtained when the powder is packaged at this temperature, but because the hot powder is very vulnerable to oxidation of the fat, it must be packaged with inert gas to reduce the O_2 level in the package to a maximum of 2%.

SMP typically has a maximum fat content of 1.25% and a moisture content of 4%, whereas WMP has a minimum fat content of 26% and a moisture content of 2.5%. Buttermilk (the liquid remaining after the manufacture of butter) can be evaporated and spray dried, as can whey, which results from the manufacture of casein, quarg and cheese.

B. Packaging Requirements

1. Oxidation

The detrimental effect of O_2 on the flavor of milk products (especially those high in fat) has already been discussed earlier in this chapter. It is therefore not surprising that the most effective method of extending the shelf life of milk powder is to package it in an airtight package from which air has been removed and replaced by a chemically inert gas such as N_2 — a process commonly known as gas packing (see Section D.2). This is particularly true for WMP where the shelf life is governed to a large extent by the rate of oxidation of the unsaturated fats and the consequent development of objectionable flavors. The advantage of gas packing SMP is much less but it has been found to be worthwhile in preventing the development of stale flavors, especially where the storage period might be prolonged or the storage temperature high. An alternative to gas packing is vacuum packaging and this is discussed in Section D.1.

2. Water Vapor Permeability

The rate of development of oxidized flavor or oxidative rancidity during the storage of WMP depends on both the O_2 concentration and the a_w. Therefore, if the maximum shelf life possible is to be obtained from milk powders (especially WMPs), it is important that the moisture content corresponds to the water activity (a_w) at which the rate of lipid oxidation is a minimum. This is usually taken to be the a_w which corresponds to the monolayer value.

A generalized moisture sorption isotherm for skim and whole milk powders is presented in Figure 18.2 and shows a break at $a_w = 0.5$ owing to lactose crystallization.[50] In practice, differences in equilibrium moisture content are observed as a result of differences in the pretreatment of the milk as well as its specific composition. For example, the presence of fat in the powder lowers the moisture content at constant a_w.[24] The lactose content of WMP is 36 to 39%, that of SMP 50 to 55% and that of whey powder 70 to 80%. At low a_ws, lactose generally occurs in the anhydrous form, the less hygroscopic α-lactose monohydrate containing about 5% water as water of hydration. The change in crystal structure from the anhydrous to the hydrated form can therefore only take place when the moisture content exceeds 5%. The proportion of hydrate increases as a_w rises above 0.5.

In selecting a suitable packaging material for milk powders, three factors must be taken into account: the initial moisture content of the powder; the final acceptable (critical) moisture content of the powder, and the shelf life required. Assuming that the powder equilibrates rapidly whenever

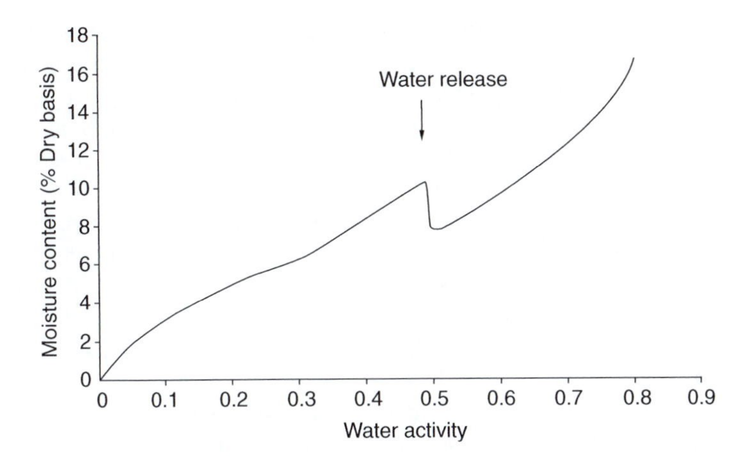

FIGURE 18.2 Generalized moisture sorption isotherm for milk powders showing a break at $a_w = 0.5$ due to lactose crystallization. (*Source*: From Thomas, M. E. C., Scher, J., Desobry-Banon, S., and Desobry, S., *Crit. Rev. Food Sci. Nutr.*, 44, 297–322, 2004. With permission.)

moisture enters the package, the maximum quantity of moisture that can enter the package can then be calculated and the maximum water vapor permeability of the packaging material specified. This will vary depending on the surface area of the package and the weight of dry solids in the package, a point often overlooked when pack sizes are changed without any consideration of the resultant effect on product moisture content. These factors were discussed earlier in Chapter 11.

3. Light

It is necessary to protect milk powders (especially those made from whole milk) from light, otherwise oxidative reactions will be accelerated. The nature of these reactions has been discussed earlier in this chapter.

C. Packaging Materials

1. Metal Cans

The traditional method for packaging milk powders for consumers uses three-piece tinplate cans where the atmospheric air is withdrawn from the powder and replaced with an inert gas such as N_2 prior to seaming the base onto the can. When correctly seamed, the can is essentially impermeable to oxygen, water vapor and light, and can be filled at high speeds. Its mechanical strength facilitates transport and handling, and the reuse possibilities of the empty can contribute to its popularity in many parts of the developing world. It is usual for the top of the can to have a lid that can be levered off and, in order to provide a gas-tight seal under the lid, an aluminum foil diaphragm is sealed to the rim of the can. This is punctured by the consumer immediately prior to use. The use of an easy-open lid incorporating a ring-pull made from scored aluminum is now quite common; a plastic overclosure is supplied to provide a limited degree of protection once the metal end has been removed.

2. Laminates

In recent years, aluminum foil/plastic laminates (sometimes containing a paper layer as well) have been introduced as a replacement for the tinplate can. The laminates can be formed, filled, gas flushed and sealed on a single machine from reel stock. Gas flushing is achieved by saturating the powder with inert gas but does not include the evacuation step used with cans. The main advantages associated with the laminated type of material are lower material cost and lighter material weight. The disadvantages are that such laminate packs do not have the mechanical strength and durability of rigid containers, and there can be difficulty in obtaining a satisfactory heat seal because of contamination of the heat seal area by powder during filling.

The physical strength of a laminate depends on the materials of construction. Aluminum foil of 9 μm thickness laminated to paper weighing 45 g m^{-2} and 25 μm thick LDPE will give a burst strength of 179 kPa. However, elimination of the paper and the introduction of a PET outer layer will change the physical properties considerably without greatly affecting the chemical properties. For example, 9 μm aluminum foil laminated with 12.5 μm PET and 64 μm LDPE increases the burst strength to 290 kPa.[7]

FFS machines are usually designed to produce gusseted or ungusseted pouches. As there is less handling of the material with FFS machines during the production sequence, it is possible to omit the paper layer normally incorporated into premade bags. A typical material construction for milk powder on FFS machines is PET (17 μm)–LDPE (9 μm)–foil (9 μm)–LDPE (70 μm).[7]

3. Fiber Cans

Fiber cans or composites manufactured by spiral winding of paperboard strip are obtainable with a wide variety of liners. They can give a similar degree of protection to that obtained from aluminum foil–LDPE–paper bags and strength similar to a metal can. They have the added advantages of being lighter than metal cans and not corroding, a problem that can occur with metal cans under high humidity conditions. A typical specification for fiber cans for use with either whole or SMP is 0.9 mm board and a foil coating of 5 μm with a nitrocellulose lacquer to protect the powder from the aluminum foil. An outer decorative label incorporating a fiber seal material gives increased protection against moisture penetration.[7]

Occluded air trapped in vacuoles within powder particles is not removed during gas flushing of cans or laminate packs. Depending on the volume of occluded air in the powder, the equilibrium residual O_2 content of the pack could exceed the 0.02 to 0.03 mL g^{-1} normally tolerated. This problem can be alleviated by conditioning the powder under vacuum for 24 to 48 h prior to gas flushing to remove occluded air.

D. Packaging Techniques

1. Vacuum Packing

An alternative method of reducing the O_2 content of packaged WMP is by compression of the powder. However, although such a procedure reduces the air content, it results in a significant increase in bulk density and (usually) a decrease in performance as measured by ease of solubility. Vacuum packing is used to reduce the O_2 content and it achieves this by evacuation of the interstitial air and some compression as a consequence of the vacuumization. The main problem associated with vacuum packing of milk powder is that of removing air from the package without removing powder fines, which could damage the vacuum pump and contaminate the sealing area of the laminate bag. This problem can be avoided by applying the vacuum at a slow rate so that powder particles are not disturbed.

2. Gas Packing

The technique of gas packing is simple and, provided that the package is airtight, the O_2 content can easily be reduced from the 21% present in air to 1% or less immediately after the operation is finished. However, with spray dried powder this diminution in O_2 content is limited mainly to the atmosphere surrounding the hollow powder particles. Up to 28 days may need to elapse before equilibrium is established between the gases within and around the particles, although most of the change will usually be complete after 7 to 10 days. During this desorption period, the O_2 content within the can may rise to as much as 5%; this is considerably above the level of 1 to 1.5% required to ensure optimal keeping quality. When O_2 levels of the order of 1% are needed, there is no alternative in the usual gas packing technique but to store the powder after the initial gas packing long enough for desorption to take place, and then to gas pack the cans a second time. This is costly and time consuming, requires considerable storage space, increases the handling of the cans and is unsuitable for normal factory routine. Thus, any process that will remove the desorbed O_2 as it is liberated is likely to receive close scrutiny.

REFERENCES

1. Allen, C. and Parks, O. W., Photodegradation of riboflavin in milks exposed to fluorescent light, *J. Dairy Sci.*, 62, 1377–1379, 1979.
2. Alves, R. M. V., Sarantopoulos, C. I. F. de L., Dender, A. G. F. van, and Faria, J. de A. F., Stability of sliced mozeralla cheese in modified atmosphere packaging, *J. Food Prot.*, 59, 838–844, 1996.
3. Bosset, J. O., Daget, N., Desarzens, C., Dieffenbacher, A., Fluckiger, E., Lavanchy, P., Nick, B., Pauchard, J.-P., and Tagliaferri, E., Influence of light transmittance and gas permeability of various packing materials on the quality of whole natural yoghurt during storage, In *Food Packaging and Preservation*, Mathlouthi, M., Ed., Elsevier, London, 1986, chap. 18.
4. Burton, H., *Ultra-High-Temperature Processing of Milk and Milk Products*, Elsevier, Essex, 1988.
5. Cladman, W., Scheffer, S., Goodrich, N., and Griffiths, M. W., Shelf-life of milk packaged in plastic containers with and without treatment to reduce light transmission, *Int. Dairy J.*, 8, 629–636, 1998.
6. Colchin, L. M., Owens, S. L., Lyubachevskaya, G., Boyle-Roden, E., Russek-Cohen, E., and Rankin, S. A., Modified atmosphere packaged Cheddar cheese shreds: influence of fluorescent light exposure and gas type on color and production of volatile compounds, *J. Agric. Food Chem.*, 49, 2277–2282, 2001.
7. Cummins, N., Milk powder, *Technical Guide to the Packaging of Milk and Milk Products., Bulletin, #143*, 2nd ed., International Dairy Federation, Brussels, 1982, chap. 19.
8. Deeth, H. C. and Datta, N., Ultra-High Temperature Treatment (UHT)/Heating Systems, *Encyclopedia of Dairy Sciences*, Vol. 4, Roginski, H., Fuquay, J. W. and Fox, P. F., Eds., Academic Press, London, pp. 2642–2652, 2002.
9. Dimick, P. S., Review. Photochemical effects on flavor and nutrients of fluid milk, *Can. Inst. Food Sci. Technol. J.*, 15, 247–256, 1982.
10. Duyvesteyn, W. S., Shimoni, E., and Labuza, T. P., Determination of the end of shelf life for milk using Weibull hazard method, *Lebensm.-Wiss. Technol.*, 34, 143–148, 2001.
11. Emmons, D. B., Froehlich, D. A., Paquette, G. J., Butler, G., Beckett, D. C., Modler, H. W., Brackenridge, P., and Daniels, G., Light transmission characteristics of wrapping materials and oxidation of butter by fluorescent light, *J. Dairy Sci.*, 69, 2248–2267, 1986.
12. Emmons, D. B., Froelich, D. A., Paquette, G. J., Beckett, D. C., Modler, H. W., Butler, G., Brackenbridge, P., and Daniels, G., Flavor stability of butter prints during frozen and refrigerated storage, *J. Dairy Sci.*, 69, 2451–2457, 1986.
13. Farkye, N. Y. and Fox, P. F., Objective indices of cheese ripening, *Trends Food Sci. Technol.*, 1, 37–40, 1990.
14. Farrer, K. T. H., *Light Damage in Milk*, Farrer Consultants, Blackburn, Victoria, 1983.
15. Fox, P. F., Cheese: an overview, In *Cheese: Chemistry, Physics and Microbiology*, Vol. 1, Fox, P. F., Ed., Elsevier, Essex, 1987, chap. 1.
16. Fradin, M., Ripening under film, In *Cheesemaking Science and Technology*, Eck, A., Ed., Lavoisier Publishing, New York, 1987, chap. 13.
17. Hansen, A. P. and Arora, D. K., Loss of flavor compounds from aseptically processed food products packaged in aseptic cartons, In *Barrier Polymers and Structures, ACS Symposium Series #423*, Koros, W. J., Ed., American Chemical Society, Washington, DC, 1990, chap. 17.

18. Hansen, E. and Skibsted, L. H., Light induced oxidative changes in a model dairy spread. Wavelength dependence of quantum yields and inner-filter protection by β-carotene, *J. Agric. Food Chem.*, 48, 3090–3094, 2000.
19. Henyon, D. K., Extended shelf life milks in North America: a perspective, *Int. J. Dairy Technol.*, 52, 95–101, 1999.
20. Hong, C. M., Wendorff, W. L., and Bradley, R. L., Effects of packaging and lighting on pink discoloration and lipid oxidation of annatto-colored cheeses, *J. Dairy Sci.*, 78, 1896–1902, 1995.
21. Hotchkiss, J. H., Chen, J. H., and Lawless, H. T., Combined effects of carbon dioxide addition and barrier films on microbial and sensory changes in pasteurized milk, *J. Dairy Sci.*, 82, 690–695, 1999.
22. International Dairy Federation. *Packaging of Butter, Soft Cheese and Fresh Cheese, Bulletin #214*, International Dairy Federation, Brussels, 1987.
23. Juric, M., Bertelsen, G., Mortensen, G., and Petersen, M. A., Light-induced colour and aroma changes in sliced, modified atmosphere packaged semi-hard cheeses, *Int. Dairy J.*, 13, 239–249, 2003.
24. Kessler, H. G., Dry products — sorption properties and keeping quality, *Food Engineering and Dairy Technology*, Verlag A. Kessler, Freising, 1981, chap. 9.
25. King, J. S. and Mabbit, L. A., Preservation of raw milk by addition of carbon dioxide, *J. Dairy Res.*, 49, 439–447, 1982.
26. Kristensen, D., Orlien, V., Mortensen, G., Brockhoff, P., and Skibsted, L. H., Light-induced oxidation in sliced Harvarti cheese packaged in modified atmosphere, *Int. Dairy J.*, 10, 95–103, 2000.
27. Loss, C. R. and Hotchkiss, J. H., The use of dissolved carbon dioxide to extend the shelf life of dairy products, In *Dairy Processing: Improving Quality*, Smit, G., Ed., CRC Press, Boca Raton, Florida, 2003, chap. 18.
28. Maniar, A. B., Marcy, J. E., Bishop, J. R., and Duncan, S. E., Modified atmosphere packaging to maintain direct set cottage cheese quality, *J. Food Sci.*, 59, 1305–1308, 1994, see also 1327.
29. Mehta, R. S., Milk processed at ultra-high-temperatures — a review, *J. Food Prot.*, 43, 212–225, 1980.
30. Miller, C. R., Nguyen, M. H., Rooney, M. L., and Kailasapathy, K., The control of dissolved oxygen content in probiotic yoghurts by alternative packaging materials, *Packag. Technol. Sci.*, 16, 61–67, 2003.
31. Mortensen, G., Sørensen, J., and Stapelfeldt, H., Effect of light and oxygen transmission characteristics of packaging materials on photo-oxidative quality changes in Harvarti cheeses, *Packag. Technol. Sci.*, 15, 121–127, 2002.
32. Mortensen, G., Bertelsen, G., Mortensen, B. K., and Stapelfeldt, H., Light-induced changes in packaged cheeses — a review, *Int. Dairy J.*, 14, 85–102, 2004.
33. Mottar, J., The usefulness of co-extruded high density polyethylene for packaging UHT milk, IDF Dairy Packaging Newsletter #15, April, 1987, 4.
34. Moyssiadi, T., Badeka, A., Kondyli, E., Vakirtzi, T., Savvaidis, I., and Kontominas, M. G., Effect of light transmittance and oxygen permeability of various packaging materials on keeping quality of low fat pasteurized milk: chemical and sensorial aspects, *Int. Dairy J.*, 14, 429–436, 2004.
35. Muir, D. D. and Banks, J. M., Factors affecting the shelf life of milk and milk products, In *Dairy Processing: Improving Quality*, Smit, G., Ed., CRC Press, Boca Raton, Florida, 2003, chap. 9.
36. Nelson, K. H. and Cathcart, W. M., Transmission of light through pigmented polyethylene milk bottles, *J. Food Prot.*, 47, 346–348, 1984.
37. Odet, G. and Zachrison, C., Cheese, In *Technical Guide to the Packaging of Milk and Milk Products. Bulletin #143*, 2nd ed., International Dairy Federation, Brussels, 1982, chap. 23.
38. Patart, J. P., Process cheeses, In *Cheesemaking Science and Technology*, 2nd ed., Eck, A., Ed., Lavoisier Publishing, New York, 1987, chap. 19.
39. Rankin, S. A., Liquid milk products/super-pasteurized milk, In *Encyclopedia of Dairy Sciences*, Vol. 2, Roginski, H., Fuquay, J. W. and Fox, P. F., Eds., Academic Press, London, pp. 1633–1637, 2002.
40. Robertson, G. L., Ultra-high temperature treatment (UHT)/aseptic packaging, In *Encyclopedia of Dairy Sciences*, Vol. 4, Roginski, H., Fuquay, J. W. and Fox, P. F., Eds., Academic Press, London, pp. 2637–2642, 2002.
41. Rosenberg, M., Liquid milk products/sterilized milk, In *Encyclopedia of Dairy Sciences*, Vol. 2, Roginski, H., Fuquay, J. W. and Fox, P. F., Eds., Academic Press, London, pp. 1637–1646, 2002.

42. Rysstad, G., Ebbesey, A., and Eggestad, J., Sensory and chemical quality of UHT milk stored in paperboard cartons with different oxygen and light barriers, *Food Addit. Contam.*, 15, 112–122, 1998.
43. Satter, A. and deMan, J. M., Effect of packaging materials on light-induced quality deterioration of milk, *Can. Inst. Food Sci. Technol. J.*, 6, 170–177, 1973.
44. Senyk, G. F. and Shipe, W. F., Protecting your milk from nutrient losses, *Dairy Field*, 164(3), 81–85, 1981.
45. Shipe, W. F., Bassette, R., Deane, D. D., Dunkley, W. L., Hammond, E. G., Harper, W. J., Kleyn, D. H., Morgan, M. E., Nelson, J. H., and Scanlan, R. A., Off flavor of milk: nomenclature, standards and bibliography, *J. Dairy Sci.*, 61, 855–869, 1978.
46. Simon, M. and Hansen, A. P., Effect of various dairy packaging materials on the shelf life and flavor of pasteurized milks, *J. Dairy Sci.*, 84, 767–773, 2001.
47. Skibsted, L.H., Light induced changes in dairy products, In *Packaging of Milk Products*. International Dairy Federation. Bulletin of the International Dairy Federation 346, Brussels, 2000, pp. 3–9.
48. Smit, G., Ed., *Dairy Processing: Improving Quality*, CRC Press, Boca Raton, Florida, 2003.
49. Stehle, G., Materials for packaging, In *Cheesemaking Science and Technology*, 2nd ed., Eck, A., Ed., Lavoisier Publishing, New York, 1987, chap. 18.
50. Thomas, M. E. C., Scher, J., Desobry-Banon, S., and Desobry, S., Milk powders ageing: effect on physical and functional properties, *Crit. Rev. Food Sci. Nutr.*, 44, 297–322, 2004.
51. van Aardt, M., Duncan, J. S. E., Marcy, E., Long, T. E., and Hackey, C. R., Effectiveness of poly(ethylene terephthalate) and high density polyethylene in protection of milk flavor, *J. Dairy Sci.*, 84, 1341–1347, 2001.
52. Vassila, E., Badeka, A., Kondyli, E., Savvaidis, I., and Kontominas, M. G., Chemical and microbiological changes in fluid milk as affected by packaging conditions, *Int. Dairy J.*, 12, 715–722, 2002.
53. Walstra, P., Geurts, T. J., Noomen, A., Jellema, A., and van Boekel, M. A. J. S., *Dairy Technology: Principles of Milk Properties and Processes*, Marcel Dekker, New York, 1999.
54. Whited, L. J., Hammond, B. H., Chapman, K. W., and Boor, K. J., Vitamin A degradation and light-oxidized flavor defects in milk, *J. Dairy Sci.*, 85, 351–354, 2002.
55. Whitley, E., Muir, D., and Waites, W. M., The growth of Listeria monocytogenes in cheese packaged under a modified atmosphere, *J. Appl. Micribiol.*, 88, 52–57, 2000.
56. Zygoura, P., Moyssiadi, T., Badeka, A., Kondyli, E., Savvaidis, I., and Kontominas, M. G., Shelf life of whole pasteurized milk in Greece: effect of packaging material, *Food Chem.*, 87, 1–9, 2004.

19 Packaging of Cereals, Snack Foods and Confectionery

CONTENTS

I. INTRODUCTION

Cereals are the fruits of cultivated grasses, members of the monocotyledonous family *Gramineae*. The principal cereal crops are wheat, barley, oats, rye, rice, maize, sorghum and the millets. Cereals have been important crops for thousands of years, and the successful production, storage and use of cereals has contributed significantly to the development of modern civilization. Today, cereals and cereal-based products are an important part of the diet in most countries, and new products based on cereals are developed and marketed to increasingly sophisticated consumers each year.

II. GRAINS

The cereals of commerce and industry are harvested, transported and stored in the form of grain. The anatomical structure of all cereal grains is basically similar, differing from one cereal to another in detail only.

The mature grain of the common cereals consists of carbohydrates, nitrogenous compounds (mainly proteins), lipids, mineral matter and water, together with small quantities of vitamins, enzymes and other substances, some of which are important nutrients in the human diet. Carbohydrates are quantitatively the most important constituents, forming 77 to 87% of the total dry matter. The lipids in milled cereal products are liable to undergo two types of deterioration: hydrolysis from endogenous lipases; and oxidation from endogenous lipoxygenases or molecular O_2. The products of lipid hydrolysis are glycerol and free fatty acids, which give rise to unpleasant odors. The products of lipid oxidation cause the odor and flavor of rancidity. Damage to the grain and the fragmentation that occurs in milling promote deterioration by bringing the lipid and the enzyme together.[33]

The hazards to grain in storage are moisture, temperature, fungi, bacteria, insects and other pests. If the grain moisture content can be controlled, then the hazards resulting from temperature rise, fungi and insects can be largely avoided. However, it is not the moisture content *per se* that is important for long-term storage, but rather the water activity a_w, a “safe” moisture content for long-term storage of grain and oil-seeds usually being accepted as one in equilibrium with 70% RH (i.e., 0.70 a_w). Above 0.75 a_w, molds will develop rapidly during storage, and heating will occur with subsequent deterioration in, and loss of, product. The moisture content corresponding to 0.70 a_w varies quite considerably among the different grains, thus emphasizing the utility of expressing stability in terms of a_w rather than moisture content.

Storage facilities for grains take many forms, ranging from piles of unprotected grain on the ground, underground pits or containers and piles of bagged grain, to storage bins of many sizes, shapes and types of construction. It is essential that the grain has been dried to a moisture content corresponding to 0.70 a_w or less prior to packaging and storage. Consumer packages for grain commonly consist of heat-sealed pouches made from LDPE film; these provide a satisfactory moisture barrier and result in the required shelf lives for the grains.

A. Wheat

The wheat grain is a living, respiring organism, which usually carries endemic fungi. Respiration is slow at 14% moisture content at 20°C, but rises as moisture content and temperature rise. The process of respiration generates heat (which is difficult to remove because wheat is a poor conductor) as well as CO_2 and water vapor, resulting in a loss in weight. Unless the grain is turned over to allow evaporation of the water, it will sweat and become caked in the bin.

Wheat at moisture contents between 16% and 30% can support fungal growth and there is thus the risk of mycotoxin production. Above 30% moisture content, wheat is susceptible to bacterial attack, leading to spoilage, excessive heat production and possibly charring. Insect life also becomes more active as the temperature rises and, because of their respiration, live insects in grain also raise the grain temperature. Deterioration during storage is aggravated by mechanical damage during harvesting because micro-organisms attack damaged grains more readily than intact grains.[33]

The shelf life for wheat as a function of moisture content and temperature is presented in Table 19.1. This table shows the importance of drying grain in that a drop of 3% in moisture content increases the shelf life by a factor of four. The major problems at higher moisture contents include accelerated wheat enzyme activities and microbial spoilage.

B. Flour

Wheat flour is the product prepared from grain by grinding or milling processes in which the bran and germ are partly removed and the remainder is comminuted to a suitable degree of fineness.[33] It has been recommended that for long storage periods, flour should be stored in a closed atmosphere. Under these conditions, flour acidity increases owing to the accumulation of linoleic and linolenic acids.

Flour is stored commercially in bags or bulk bins. The storage hazards for flour are similar to those for wheat in storage, which include mold and bacterial attack, insect infestation, oxidative rancidity and eventual deterioration of baking quality. The expected shelf life of plain white

TABLE 19.1
Safe Storage Life (Days) for Grains as a Function of Moisture Content and Temperature

Grain Temperature (°C)	Grain Moisture (%)		
	14	15.5	17
10.0	256	128	64
15.5	128	64	32
21.1	64	32	16
26.7	32	16	8
32.2	16	8	4
37.8	8	4	2

Source: From Bailey, J. E., Whole grain storage, In *Storage of Grains and Their Products*, Christensen, C. M., Ed., American Association of Cereal Chemists, St. Paul, MN, 1974, chap. 8.[5] With permission.

flour packaged in paper bags and stored in cool, dry conditions and protected from infestation is 2 to 3 years. The rate of increase in acidity increases with storage temperature and decreasing flour grade (i.e., as the ash residue increases). Thus, the shelf life of brown and wholemeal flours is shorter than that of white flour. Freedom from insect infestation during storage can be ensured only if the flour is free from insect life at the time of packing and if the storage area is free from infestation.

The optimum moisture content for the storage of flour is related to the intended shelf life, the barrier properties of the packaging material, as well as the ambient temperature and humidity. For use within a few weeks, flour can be packaged at 14% moisture content, but at moisture contents higher than 13%, mustiness resulting from mold growth may develop over time. At moisture contents lower than 12%, the risk of lipid oxidation and the development of rancidity increases.[33] A moisture content of 12% corresponds to approximately 0.5 a_w.

As with wheat and other grains, the moisture content is an unreliable guide to stability. To determine the moisture content that corresponds to the maximum a_w for stability at a particular storage temperature, moisture sorption isotherms at various temperatures are required. This will vary depending on the type of grain as well as the variety. Data on both the shelf life of flour at various a_ws and values for critical moisture contents are lacking, and it is therefore difficult to specify the type of moisture vapor barrier required in a package.

Notwithstanding the lack of information about critical moisture contents and shelf lives as a function of a_w, bags made from cotton twill or paper have been used successfully for decades for consumer packs of flour. Kraft paper bags with an LDPE liner would provide additional protection and therefore a longer shelf life, but this does not seem to be justified. A detailed discussion of the various packaging materials that have been used for packaging cereals and their products has been presented elsewhere.[61]

C. Rice

Cultivated on every continent except Antarctica, rice is a crop that feeds half of the world's population and has fed more people over a longer period of time than any other crop. Brown (unmilled) rice is more nutritious than milled rice, but storage stability problems and a traditional consumer preference for whole (milled) rice have limited the quantities of brown rice packaged and sold for direct consumption. A major deterrent to greater user of brown rice is the accumulation of free fatty acids in rice stored under warm and humid conditions. Fatty acids can be released by lipases present in the rice aleurone (bran) layer of damaged grains and by high lipase-containing bacteria and fungi adhering to rice.

In a study aiming to evaluate the effects of CO_2 gas flushing on the shelf life of brown rice,[51] greater stability of the rice was obtained at refrigerated (4°C) storage temperatures when it was packaged in a laminate film (PA–EVA copolymer) bag, rather than a regular (unspecified) plastic bag; gas flushing with CO_2 of the samples in the laminate bag improved stability even further. However, no differences between the three types of bags were found when stored at room temperature (24°C). The gas-flushed bags formed a hard, rigid pack similar to a vacuum-packaged product and maintained this appearance for at least 3 years. It was suggested that if bulk rice in warehouses could be stored as brown rice (minus hulls) instead of as rough rice (with hulls) in laminate gas-flushed bags, savings of at least 20% by weight and 30 to 35% by volume should result.

III. BREAKFAST CEREALS

Breakfast cereal foods can be classified according to the amount of domestic cooking required, the form of the product and the cereal used as raw material. All cereals contain a large proportion of starch, which, in its natural form, is insoluble, tasteless and unsuited for human consumption. It must be cooked to make it digestible and acceptable. In the case of hot cereals, the cooking is carried out in the home, while ready-to-eat cereals are cooked during manufacture.[37]

If the cereal is cooked with excess water and moderate heat as in boiling, then the starch gelatinizes and becomes susceptible to starch-hydrolyzing enzymes in the human digestive system. If the cereal is cooked with a minimum of water (or without water) but at higher temperatures as in toasting, then nonenzymic browning between protein and reducing sugars may occur and there may be some depolymerization of the starch.

A. Manufacture

Ready-to-eat cereals probably owe their origin to the Seventh Day Adventist Church whose members, preferring an entirely vegetable diet, experimented with the processing of cereals in the mid-nineteenth century. A granulated product called "Granula" and made by Jackson in 1863 may have been the first commercially available ready-to-eat breakfast cereal. A similar product called "Granola" was made by Kellogg by grinding biscuits made from wheatmeal, oatmeal and maize meal.[33]

Ready-to-eat cereals comprise flaked, puffed, shredded and granulated products, generally made from wheat, maize or rice, although oats and barley are also used. The basic cereal may be enriched with sugar, honey or malt extract. All types are prepared by processes that tend to cause hydrolysis (dextrinization) rather than gelatinization of the starch.

Flaked products are made from wheat, corn, oats or rice. After cooking (often at elevated pressure) and the addition of flavorings such as malt, sugar and salt, the cereal is dried to 15 to 20% moisture content and conditioned for 1 to 3 days. It is then flaked, toasted, cooled and packaged.

Puffed products are prepared from conditioned whole grain wheat, rice, oats or pearl barley, or a dough made from corn meal or oat flour with the addition of sugar, salt and sometimes oil. It is cooked for 20 min under pressure, dried to 14 to 16% moisture content and pelleted by extrusion through a die. A batch of the conditioned grain or pelleted dough is fed into a heated pressure chamber, which is injected with steam. The starch becomes gelatinized and expansion of water vapor on release of the pressure causes a several-fold increase in volume. The puffed product is dried to 3% moisture content by toasting and then cooled and packaged.

Shredded products are made from whole wheat grains which are cooked to gelatinize the starch. After cooling and conditioning, the grain is fed through shredders. The shreds are baked for 20 min at 260°C, dried to 1% moisture content, cooled and packaged.

Granulated products are made from yeast dough consisting of wheat flour and salt. The dough is baked as large loaves, which are then broken up, dried and ground to a standard degree of fineness.

Flaked or puffed cereals are sometimes coated with sugar or candy to provide a hard, transparent coating that does not become sticky even under humid conditions. The sugar content of cornflakes increases from 7 to 43% as a result of the coating process, and that of puffed wheat from 2 to 51%.[33]

B. Deterioration

There are five modes of deterioration to be considered when selecting suitable packaging materials for breakfast cereals. They are:

1. Moisture gain resulting in loss of crispness
2. Lipid oxidation resulting in rancidity and off-flavors
3. Loss of vitamins
4. Breakage, resulting in an aesthetically undesirable product
5. Loss of aroma from flavored product

The shelf life of breakfast cereals depends to a large extent on the content and quality of the oil contained in them. Thus, products made from cereals with a low oil content such as wheat, barley, rice and maize grits (oil content: 1.5 to 2.0%) have a longer shelf life than products made from oats (oil content: 4 to 11%, average 7%). Although whole corn has a high oil content (4.4%), most of the oil is contained in the germ, which is removed in making grits.[33]

C. Packaging

The materials used for the packaging of ready-to-eat breakfast cereals are discussed below in relation to the major indices of failure.

1. Loss of Crispness

Data on the permissible increase in moisture content or a_w before loss of crispness occurs are still relatively scant and, without this data, it is difficult to specify precisely the type of moisture vapor barrier required to achieve a given shelf life.

Packaging of breakfast cereals has traditionally been in fiberboard boxes with a super-calendered, waxed, glassine liner. As well as providing a barrier to moisture vapor, the liner must also confine cereal aromas within the packaged product and simultaneously prevent foreign odors entering. It should also be reclosable to protect the cereal remaining in the package.[47] The glassine liner has been largely replaced by various plastic materials, in particular thin gauge HDPE, which is usually folded rather than heat sealed. HDPE coextruded with a thin layer of EVA copolymer is also used, where the EVA copolymer permits a lower heat seal temperature and offers the consumer an appealing and peelable seal.[47]

In a few cases where the cereal product is not hygroscopic or retains a satisfactory texture when in equilibrium with the ambient atmosphere, a liner may not be needed for moisture protection and may even serve to entrap rancid aromas. Where this is the case, either no liner or one which is vapor permeable may be used.[47] Some shredded wheat products are in this category and are discussed below.

2. Lipid Oxidation

The primary mode of chemical deterioration in dry cereals is lipid oxidation and two reasons have been advanced for this.[38] First, the a_w of dry cereals is at or below the monolayer, which essentially stops all other types of deteriorative reactions. Second, unsaturated fats are required in lipid oxidation, and the grains used in breakfast cereals have a high ratio of unsaturated to saturated fats.

To minimize oxidative rancidity, it is important that the package excludes light. Excluding O_2 may be of limited assistance in extending shelf life although O_2 is almost never rate limiting.[38] For this reason, most companies do not bother to use packaging that is a good O_2 barrier. In a study of the storage stability of a flaked oat cereal product packaged in materials of different O_2 barrier properties with and without the addition of an iron-based O_2 absorber,[55] the absorber retarded or delayed lipid oxidation provided that it was used with a packaging material that was a good O_2 barrier such as PVdC copolymer–coated PP–LDPE.

Although the use of an antioxidant in the package liner has been shown to be successful in extending shelf life, it is not generally permitted in most countries.

3. Loss of Vitamins

The vitamin and mineral fortification of cereals is widely practiced in many countries and there are usually associated nutritional labeling requirements. The major factor influencing vitamin loss in packaged cereals is the temperature of storage. In a study on the effects of processing and storage on micronutrients in breakfast cereals, it was concluded that micronutrient loss would not be a major factor in determining the shelf life of dry cereals. There were no substantial losses of added vitamins during normal shelf lives, with the possible exception of vitamin A and, to a slight extent, vitamin C. Vitamin A survived for 6 months (the average distribution time) at room temperature with no measurable loss.[38]

4. Mechanical Damage

The rigidity of the carton stock and the compression resistance of the finished carton must together provide the necessary resistance to product breakage throughout production line operations, warehouse storage and distribution from the manufacturer to the retailer and consumer. Rigidity also prevents the bulging of the carton.[47] Protecting breakfast cereals from breakage does not appear to be a problem using currently available carton stock and carton designs.

5. Loss of Flavor

This can be a problem with certain cereal products to which fruit flavors have been added prior to packaging. In these situations, loss of flavor results in the product being considered to be at the end of its shelf life by the consumer. A study[46] evaluating two typical cereal liner materials (HDPE and glassine) found that the permeability coefficients of *d*-limonene (a common flavor component in citrus products) in the HDPE liner were three to four orders of magnitude higher than those in glassine. It was also found that the solubility of *d*-limonene in the glassine liner was substantially lower than in the HDPE liner for the same vapor pressures. Hence, equilibrium distribution of the limonene vapor between a product such as a fruit-flavored cereal and the respective liners will result in a much lower limonene concentration within the glassine liner, and "scalping" of the cereal flavor can be assumed to be much more significant in the HDPE liner.

IV. PASTAS

Although the word "pasta" is traditionally associated with Italy and with durum wheat semolina, Italy cannot claim to have invented this popular food, and semolina is not the original raw material. The Chinese invented pasta that was produced as noodles from rice and legume flours several thousand years BCE. Today, pasta consumption is increasing in many countries to the extent that pasta can be classed as a truly international food.

Part of the appeal of pasta products (macaroni, spaghetti, vermicelli, noodles) is that they may be prepared from several raw materials according to countless formulations, and can be cooked and served in numerous ways to various tastes. Durum wheat semolina is considered to be the best raw material for pasta making because of the functional characteristics of its proteins and its high pigment content.[24] The limited availability and high cost of durum wheat compared with other cereals has led to the use of flours and starches from rice, maize, barley, soft wheat, cassava and potato in pasta formulations in various parts of the world.[34]

The original composition of pasta (water, durum wheat semolina and egg) has altered considerably to include vitamin supplements, iron salts, powdered vegetables, tomato concentrate, milk protein, other cereal flours, and meat and cheese in filled pasta in order to satisfy the tastes and food habits of different populations. The introduction of MAP has enabled certain types of pasta (particularly the fresh and filled varieties) to progress from being a small-scale manufacturing operation to an established position in the food industry where, for example, fresh product is distributed across the U.S.

Pasta can be subdivided into two categories. The first, macaroni, has come to represent a generic family of over 140 items in the U.S. and includes spaghetti, macaroni and vermicelli. This class of product is made from semolina, water and, in most cases, added vitamins and minerals, and contains about 1.5% fat.

The second broad category of pasta is noodles, which contain the same ingredients as pasta as well as egg solids at a minimum level of 5.5%. The egg solids are usually in the form of yolks to enhance color, and the lipid content of the noodles is about 4.6%. Noodles may also contain additional protein from sources such as soy flour.

A. Dried Pasta

Dried pasta is produced by the reduction of dough moisture content from 30 to about 11% by means of a dehydration process, the length of which is determined by the temperature used. In recent years, the drying of pasta products at temperatures above 60°C has become widely accepted by pasta manufacturers, with benefits of this approach including control of bacteria in egg products and shorter drying cycles. If the drying stage of pasta manufacture is not properly controlled, then extensive growth of micro-organisms such as *Salmonella* spp. and *Staphylococcus aureus* can occur, resulting in a potential hazard to public health.[3]

Two modes have been identified for pasta product failure: moisture gain or loss and color loss. The major mode for pasta failure is moisture gain, where the optimum moisture content for pasta storage appears to be 10 to 11%. If the pasta moisture content increases to 13 to 16%, then mold growth (which makes the pasta unfit for consumption) and starch recrystallization or retrogradation (which makes the pasta unacceptably tough when cooked) occur.[38] As with other dried products, the optimum moisture content for stability is derived from the maximum "safe" a_w; that is, a moisture content that is in equilibrium with 70% RH or an a_w of 0.70.[15]

For macaroni, a moisture content of 12.8% corresponds to an a_w of 0.70 at 25°C, while for egg noodles, the corresponding moisture content is 14.7% at 25°C and 13.3% at 27°C for vermicelli. The a_w corresponding to a moisture content of 10 to 11% is about 0.56 at 25°C for macaroni, 0.44 for egg noodles at 25°C and 0.45 at 27°C for vermicelli, thus providing a margin for moisture increase during storage before the critical moisture content is reached. If the moisture content of pasta is allowed to fall to less than a certain level (9.5% in the case of vermicelli, which corresponds to 0.32 a_w), it becomes too fragile and unacceptable in quality.

A second mode of deterioration (especially for noodle products) is color loss through oxidation of carotene pigments by lipoxidase enzymes found naturally in semolina flour. These enzymes oxidize lipids and the peroxides formed attack the pigments. Light accelerates the oxidation process. An associated mode of deterioration is staling, which results from the oxidation of the lipids in the product. An analysis of published shelf life data[38] suggests that the basic mode of failure in dried pasta is lipid oxidation rather than moisture gain.

Manufacturers of pasta claim shelf lives of macaroni and spaghetti products ranging from 6 months to indefinite, and noodle shelf life claims range from 1 to 6 months. The exact shelf life depends on the storage temperature, RH and packaging material, none of which is well defined. The traditional packaging material for dried pasta was the paperboard carton, which frequently contained a plastic window so that the customer could view the contents. Today, most pasta products are packaged in plastic films such as OPP or LDPE–PET laminate.

B. Fresh Pasta

The production of pasta involves kneading of the dough, followed by extrusion or lamination and then drawing. In the case of special pasta, this latter stage is accompanied by filling, using a cooked meat or vegetable–cheese mixture, thus resulting in a variety of potential microbiological flora. The physicochemical characteristics of the main types of fresh pasta on the Italian market are presented in Table 19.2. The Barilla style pasta contains salt, which lowers its a_w.

Fresh pasta products are not usually subjected to pasteurization but are refrigerated for retail distribution. The microbiological quality of fresh pasta will thus depend on the quality of the raw materials, the cleanliness and hygiene of the processing environment and equipment and the handling of the product during production and packing. In addition, the temperature of the product

TABLE 19.2
Physicochemical Characteristics and Shelf Life of Main Types of Fresh Pasta

Fresh Pasta as	Water Content (Range % w/w)	Water Activity a_w	Shelf Life (days) MAP	Shelf Life (days) ATM[a]
Loose or nonhermetic package (very fresh)	30–40	0.96–0.98	–	1–5
Hermetic package	26–30	0.91–0.95	20–30	<15
Hermetic package and final pasteurization	26–30	0.92–0.95	30–90	30–50
Hermetic package and final pasteurization (Barilla style)	22–24	0.88–0.89	>90	–

[a] ATM, normal atmosphere.

Source: From Castelvetri, F., Microbiological implications of modified atmosphere packaging of fresh pasta products, In *Proceedings of the Sixth International Conference on Controlled/Modified Atmosphere/Vacuum Packaging*, Schotland Business Research Inc., San Diego, CA, pp. 253–265, 1991. With permission.

during storage, distribution and retailing is crucial for microbiological quality. However, some companies are now pasteurizing the pasta after packing using hot air or microwaves.[16]

These pasteurized fresh pasta products fall into the category of refrigerated, processed foods with extended durability (REPFEDs). If the pasteurization process is carried out adequately, then all vegetative bacteria present are killed. However, spores of many bacteria (including *C. botulinum*) can survive the heating process. If these foods are stored at temperatures between 3 and 10°C, then there may be a botulism risk if spores of group II (nonproteolytic strains) survived the heating process during preparation.[3] A study[49] of such foods indicated that heating before consumption was not always sufficient to inactivate botulism toxin completely. In order to ensure that the risk of botulism from these foods is controlled adequately, REPFEDS must be stored at a temperature <3.0°C. If this temperature cannot be guaranteed, then the storage time has to be limited.

Today, the use of MAP has become widespread for fresh pasta products.[58] Typical gas compositions used are 100% N_2 or 70 to 80% CO_2 and 20 to 30% N_2. A survey[52] for *Staphylococci* spp. and their enterotoxins in wet pasta packaged under a CO_2:N_2 (20:80) mixture from five processors showed that 12% of fresh products were contaminated with *S. aureus*. The pasta had a recommended shelf life of 4 weeks when stored at 4°C. The results showed that proper refrigeration was essential to ensure safety of MAP wet pastas.

The actual packaging materials used for fresh pasta products depend on whether or not the product is pasteurized (in which case, the package must be able to withstand the pasteurization process without deforming) and whether or not the product is to be heated in its package in a microwave by the consumer (in which case, the package must be able to withstand domestic microwave temperatures). For products that are not pasteurized or are not intended to be heated in their package, a rigid tray of PVC–LDPE onto which is sealed a PA–LDPE film is common. However, if microwave heating is used, then the rigid tray is usually made from CPET and the film may be based on PVdC copolymer-coated PET or PP.

V. BAKERY PRODUCTS

A. Manufacture

Bakery products have been an important part of a balanced diet for thousands of years. Flour and its principal baked product, bread, are the cheapest and most important staple foods for many nations of the world. The function of baking is to present cereal flours in an attractive, palatable

TABLE 19.3
Water Activity and pH of Typical Bakery Products

Water Activity (a_w)	pH	Products
0.99		Creams, custards
0.97	6.0	Crumpets
0.95–0.99	5.6	Breads, fermented products
0.90–0.95		Moist cakes (e.g., carrot cake)
0.91	6.3	Yeasted pastries (e.g., danish, croissant)
0.84	4.2	Fruit pies
0.82–0.83	6.3	Chocolate-coated doughnuts
0.80–0.89		Plain cakes (e.g., madeira, sponge) cakes
0.70–0.79		Fruit cakes
0.65–0.66	5.6	Bread crumbs, biscuit crumbs
0.60–0.69		Some dried fruits or fruit cakes
0.61		Biscuits, chocolate, some dried fruits
0.30		Pastries

Source: From Cauvain, S.P. and Young, L.S., Eds., *Bakery Food Manufacture and Quality: Water Control and Effects*, Blackwell Science, Oxford, England, 2000; Smith, J. P., Daifas, D. P., El-Khoury, W., Koukoutsis, J., and El-Khoury, A., *Crit. Rev. Food Sci. Nutr.*, 44, 19–55, 2004. With permission.

and digestible form. Bread is made by baking a dough, which consists largely of wheat flour, water, yeast and salt. Other ingredients that may be added include flours of other cereals (e.g., malt flour and soy flour), fat, yeast foods, emulsifiers, milk and milk products, fruit and gluten.[33]

Currently, a wide variety of bakery products can be found on supermarket shelves including breads, unsweetened rolls and buns, doughnuts, sweet and savory pies, pizza, quiche, cakes, pastries, biscuits, crackers and cookies.[8] It is useful to classify bakery products on the basis of their a_w and pH, because these parameters are a good indication of the spoilage problems likely to be encountered. The pH and a_w of a range of bakery products is presented in Table 19.3. Biscuits, crackers and cookies will be discussed in Section VI.

B. Deterioration

Although many modes of deterioration are possible in bakery products, the three most important are now discussed in turn.

1. Microbiological Spoilage

Microbial growth, particularly mold growth, is the major factor limiting the shelf life of bakery products. Spoilage types for typical bakery products are shown in Table 19.4. Chemical preservatives are used by the bakery industry to prevent or retard microbiological spoilage; chemicals used include calcium and sodium propionate, sorbic acid, potassium sorbate, sodium diacetate, methylparaben, propylparaben, sodium benzoate and acetic acid at levels of 0.005 to 0.5% w/w.[60] The extension of possible shelf life through the use of preservatives is limited by the development of off-odors and flavors or effects on product quality.[56] In addition, the trend towards foods free of preservatives is driving the development of alternative methods to overcome the problem. Principal among these is MAP, using mainly CO_2.

It has been mentioned earlier in this book (see Chapter 15 in particular) that CO_2 has an inhibitory effect on the growth of certain micro-organisms, where its effectiveness increases as the

TABLE 19.4
Spoilage Types for Typical Bakery Products

Water Activity (a_w)	Products	Spoilage Types
0.99	Creams, custards	Bacterial spoilage (e.g., "rope" mold growth and "chalk molds")
0.90–0.97	Breads, crumpets, part-baked yeasted products	Bacterial spoilage (e.g., "rope" mold growth and "chalk molds")
0.90–0.95	Moist cakes (e.g., carrot cake)	Mold and yeast, bacterial spoilage (e.g., "rope")
0.8–0.89	Plain cakes	Molds and yeasts
0.7–0.79	Fruit cakes	Xerophilic molds and osmophilic yeasts
0.6–0.69	Some dried fruits or fruit cakes	Specialized xerophilic molds and osmophilic yeasts, sugar-tolerant yeasts
<0.6	Biscuits, chocolate, some dried fruits	No microbial spoilage

Source: From Cauvain, S.P. and Young, L.S., Eds., *Bakery Food Manufacture and Quality: Water Control and Effects*, Blackwell Science, Oxford, England, 2000. With permission.

product storage temperature is reduced. Micro-organisms tend to vary in their tolerance of CO_2, with molds generally being more affected than bacteria or yeasts.

For bakery products with an a_w of 0.86 or above, the *Penicillium* group of molds tend to govern mold-free shelf life, but as the a_w falls below this level, the *Eurotium* (*Aspergillus*) glaucus group of molds predominate. These latter organisms are more susceptible to the effects of CO_2 than *Penicillium* species, although some of the latter species (in particular *P. roquefortii*, a common contaminant of rye bread) are much more CO_2 resistant than others. The type of molds present is a more important factor affecting the antimold activity of CO_2 than the a_w of the products. Bakery products such as cakes, which have an a_w of 0.85 or below, can be expected to show a large increase in mold-free shelf life by packaging in CO_2.[60]

With high a_w bakery products, the shelf life is sometimes limited by the growth of yeasts or lactic acid bacteria rather than molds. These micro-organisms are resistant to the effects of CO_2 and cause spoilage either in the form of visible growth or by the generation of quantities of CO_2, which cause the package to expand. Of particular importance is the group of filamentous yeasts known as *chalk molds* (usually *Pichia burtonii*). Chalk molds produce a white powdery spreading growth, which tends to be more obvious on the surfaces of dark-colored breads. Lactic acid bacteria (particularly *Leuconostoc mesenteroides*) have been found to be responsible for spoilage of gas-packaged crumpets. Fortunately, postbaking contamination with yeasts and lactic acid bacteria is uncommon and can be controlled by the adoption of appropriate hygienic precautions.[56]

2. Physical Spoilage

a. Staling

Staling is the common description of the decreasing consumer acceptance of bakery products as they age, caused by changes in crumb other than those resulting from the action of spoilage organisms; it is the major mode of deterioration. It has been defined as "all the physicochemical changes that occur after baking" and "almost any change, short of microbiological spoilage, that occurs in bread or other products during the post baking period, making it less acceptable to the consumer."[63] Typical changes are hardening of the crumb, softening of the crust and loss of the flavor associated with fresh bread. The most widely used indicator of staling is measurement of the increase in crumb firmness.[25] Most white bread in the U.S. has a commercial shelf life of

2 days after which it is no longer considered fresh because of the staling process. It has been estimated that 3% of bread is returned to the manufacturer as a result of staling, posing an economic burden to both the baking industry and consumers.[63]

Bread staling is a complex phenomenon in which multiple mechanisms operate, and neither the bread system nor the staling process is understood well at the molecular level. The most plausible hypothesis is that retrogradation of amylopectin occurs, and because water molecules are incorporated into the crystallites, the distribution of water is shifted from gluten to starch/amylopectin, thereby changing the nature of the gluten network.[25] The formation of complexes between starch polymers, lipids and flour proteins is thought to inhibit the aggregation of amylose and amylopectin, so that, for example, cookies and biscuits (which have a higher lipid content than bread) tend to stale more slowly. Staling can be prevented if bread is stored above 55°C or below − 18°C.[32] An antistaling enzyme, which works by hydrolyzing the amylopectin fraction and thereby preventing recrystallization and hence staling, is commercially available.[12]

An interesting aspect of the staling of bread is that it has a negative temperature coefficient, in that as the temperature increases up to 55°C, the rate of staling decreases. The staling rate passes through a maximum close to 4°C and decreases as the temperature declines below this point. From a packaging point of view, nothing can generally be done through the selection of different packaging materials to either accelerate or impede the rate of staling, although an exception is pita bread, which is discussed in the section below on gas packaging.

Evidence for the effect of CO_2 on the rate of staling is conflicting. Some studies[35] have shown that the staling rate of white and wholemeal breads and biscuits is significantly reduced when packaged in CO_2 as compared with N_2 or air, and that toast bread and rye bread packaged in CO_2 is less stale and has a better aroma and taste after storage than comparable products packaged in air. In contrast, no difference was found[21] on the rate of crumb firming between white pan bread, pound cake or sponge cake stored in CO_2, N_2 or air, and, in a further study,[53] no significant difference in the staling rate of white bread when packaged in these gases was observed. There is a need for further research to determine the role, if any, of CO_2-enriched atmospheres in delaying staling.

b. Moisture Loss/Gain

The cooling of bread can create problems, particularly when the bread is to be sliced and packaged before sale. Bread leaves the oven with the crumb at a temperature of about 98°C and a moisture content at the center of about 45%. The crust is hotter (about 150°C) but much drier (1 to 2% moisture content) and cools rapidly. During cooling, moisture moves outwards from the interior towards the crust and then to the atmosphere. Excessive drying during cooling results in weight loss and poor crumb characteristics. If the moisture content of the crust rises considerably during cooling, then the texture of the crust becomes leathery and tough and the attractive crispness of freshly baked bread is lost.[33]

After cooling, the moisture content of white bread is 36% and the a_w is 0.96. Packaging has a major influence over whether or not a bakery product will gain or lose moisture during storage, although the difference between the a_w of the product and the RH of the ambient atmosphere will be the primary determinant. Softness or resistance to deformation and recovery from deformation are important crumb characteristics and are directly affected by the level of moisture remaining in the product, coupled with a fully developed and resilient crumb structure.[17]

The loss of crust crispness is an important and beneficial change in sandwich bread types because it adds to the perception of freshness by the consumer. When purchasing bread, consumers can only assess freshness by squeezing the loaf, knowing that fresh bread has little resistance to squeezing and will rapidly spring back to its original shape.[18]

Although loss of crust crispness as a result of moisture gain is, at least in theory, a possible cause of end of shelf life for bread, such an occurrence is rare and can generally be overcome by selecting a more permeable package. An associated problem with moisture gain is an increase in crust a_w and an increased likelihood of mold growth. Conversely, excessive moisture loss can also be easily controlled by selecting a less permeable packaging material.

C. Packaging

The objective in packaging bread is to maintain the bread in a fresh condition by preventing too rapid drying out without providing too good a moisture barrier, which would promote mold growth on a soggy crust. The most commonly used material is an LDPE bag in which the end is twisted and sealed with a PS tag. This form of packaging helps retard one mode of deterioration in bread; namely, moisture loss. However, the moisture that tends to migrate from the crumb to the crust is prevented by the package from passing freely into the atmosphere, and this results in a crust with a tough, leathery consistency. Specialty breads such as French and Italian are packaged in OPP bags perforated with small holes which allow moisture to escape and thus retain a crisp crust. Several hole sizes and densities are available, depending on the particular product and its surface area:volume ratio. Recently, an assessment of the risk of physical contamination of bread packaged in perforated OPP films and sold in self-service retail outlets found a correlation between the risk and the geometrical characteristics of the film including the size of the holes and the area of the holes as a percentage of the total surface area.[54]

Vacuum packaging is not a suitable technology to extend the mold-free shelf life of most soft bakery products because the product is crushed under a vacuum. However, it has been used to prevent mold problems in flat breads such as naan and pita, and pizza crusts.

An alternative to vacuum packaging is to modify the atmosphere inside the package and the three approaches that have been investigated are discussed below.

1. Gas Packaging

The idea of modifying the gas phase inside a package of bread in order to extend the product shelf life is not new. In fact, it was shown as early as 1933[57] that the storage of bread in atmospheres containing at least 17% CO_2 delayed the appearance of mold. At a concentration of 50% CO_2, the mold-free shelf life of bread was doubled under storage conditions favoring mold development.

Extensive research on the use of CO_2 for extending the shelf life of bakery products was undertaken by Seiler in the U.K. in the 1960s.[56] In detailed studies with bread and cake stored at 21 and 27°C and CO_2 concentrations of 0 to 60%, it was shown that the mold-free shelf life increased with increasing CO_2 concentrations, with the effect being greater at lower temperatures. Subsequent studies with mixtures of CO_2 and N_2 and with 100% CO_2 confirmed the need for CO_2 in the package headspace, where simply displacing headspace O_2 with N_2 alone was insufficient to prevent mold growth.

English-style crumpets are a bakery product made from wheat flour, water, salt, vinegar and yeast or leavening agents. Traditionally, they are packaged in LDPE pouches after baking and cooling, and consumed within a few days or frozen if a longer shelf life is required. A major problem with this type of product is mold growth, which is visible within a few days, and spoilage is aggravated by bacterial growth and activities, producing off-odors and discoloration. Crumpets processed with 0.07% potassium sorbate and packaged in LDPE-coated PA film containing various combinations of CO_2 and N_2 were found to have a shelf life of 14 days at room temperature when the gas mixture was 1:1 CO_2 and N_2. After 14 days, the packages swelled as a result of CO_2 production by *Bacillus* species and lactic acid bacteria. Use of a 100% CO_2 atmosphere inside the package resulted in excessive shrinkage of the package within about 7 days owing to absorption of CO_2 by the crumpets.[50] To attain the desired shelf life of 1 month, the temperature at all stages of distribution and retailing must be kept below 24°C. Above this temperature, CO_2 production by micro-organisms exceeded the rate at which CO_2 could be absorbed by the product. The gas composition used was 3:2 CO_2 and N_2. The packaging materials used in this study were a PET–LDPE top web and a PA–LDPE bottom web.

Pita bread (also called Arabic, Egyptian baladi, flat or pocket bread) is made from flour, water and yeast, and has a shelf life of only a few hours, mainly because of its large surface area to volume ratio. Hardening caused by staling and drying is the main factor limiting shelf life, the moisture content and a_w of unpackaged pita bread reducing from 55.5% and 0.95 to 40.3% and 0.92, respectively, in 6 hours. Using a laminate film containing EVOH copolymer as a barrier layer, a gas atmosphere of 99.5% CO_2 or a mixture of CO_2 and N_2 in the ratio 73:27 enabled the shelf life of pita bread to be extended to 14 days at which time yeast growth terminated shelf life. Staling, as determined by means of a penetrometer, was delayed in MAP pita bread.[4] It has also been shown[9] that microbial spoilage can be prevented for up to 28 days in packages containing EVOH copolymer and flushed with 100% CO_2. Nonetheless, stale flavors developed after 21 days and no clear pattern of firming over time or between treatments was found.

A range of gas mixtures has been used to extend the shelf life of bakery products, from 100% CO_2 to 50% CO_2 and 50% N_2. The optimum blend of gases for a specific product cannot be determined by trial and error but only through a detailed, systematic study of the variables influencing product shelf life. Extensions of 3 weeks to 3 months at room temperature are achievable using appropriate mixtures of CO_2 and N_2.[59]

In a recent review on the use of MAs for the packaging of bakery products,[60] it was concluded that MAP may not be suitable for all types of bakery products, and that knowledge of a product's physical, chemical and microbiological characteristics is critical to the success of this technology. Furthermore, the importance of combining technologies such as O_2 absorbers and ethanol vapor generators to increase the shelf life of bakery products was emphasized.

Although the first commercial application of MAP for bakery products was in the U.K., the commercialization of MAP is most widespread in Europe. This is attributed[56] to the implementation of new labeling regulations requiring that the presence of a preservative be declared, the possibility that the conventional preservatives would no longer be permitted, and the longer increases in shelf life from MAP than were possible using propionates or sorbates. Safety concerns have limited the use of MAP technology in the North American marketplace.[60]

2. Alcohol Vapor

It has long been known that ethanol is a powerful bactericide. Indeed, ethanol is still used in this application for the sterilization of surgical instruments and working surfaces. Ethanol is also a very effective antifungal agent and, as such, it has the potential to extend the shelf life of bakery products.

In 1976, a U.S. patent was granted to cover the use of ethanol for retarding mold growth in partially baked pizza crust. The data demonstrated that where pizza bases were sprayed on all surfaces with 95% ethanol to give a concentration of 2% based on product weight, the shelf life was increased by up to fivefold. Subsequently, the U.S. FDA affirmed the GRAS status of ethanol as a direct human food ingredient and permitted its use for spraying prebaked pizza bases at concentrations up to 2% by weight.[53]

Extensive tests were carried out in the U.K. to determine the effectiveness of treatment with 95% ethanol in increasing the mold-free shelf life of a range of bakery products.[56] At a given level of treatment, the antimold activity of ethanol was greater when the products were tightly packaged in film than when loosely packaged, and when gas impermeable films were used. The ability of ethanol to act as a vapor phase inhibitor has been confirmed by the finding that similar increases in mold-free shelf life were obtained when the same amount of ethanol was sprayed over all surfaces of the product prior to packaging and sealing to when ethanol was merely added to the base of the same size bag before adding the product and sealing. Extensions in mold-free shelf life varied according to the type of product, tightness of the package, gas permeability of the packaging material and seal integrity. Treatment with 0.5% by product weight of food grade ethanol (95%) was found, under optimum conditions, to at least double mold-free shelf life, while treatment with 1.0% was found to at least treble shelf life. Ethanol was found to retard the rate of staling of both bread and cake in addition to inhibiting mold growth. The flavor of ethanol could be detected by sensory panels in cake treated with 1.0% but not 0.5% ethanol by product weight.[56]

Another way of adding ethanol (besides the injection or deposition of a small amount into the package immediately prior to sealing) is to use sachets of paper–EVA copolymer containing powdered silica gel (35% w/w) onto which food grade ethanol (55% w/w) has been absorbed. Vanilla is also added to mask the smell of ethanol. These sachets are available commercially and can be placed inside the package prior to sealing. They allow the slow release of alcohol vapor, which exerts the preservative effect. The extension in shelf life has been shown by the supplier of the sachets to depend on the ethanol permeability of the packaging material, the integrity of the seals, the a_w of the food and the type of micro-organisms present. Data on the permeability of plastic films to ethanol are scant; the permeability coefficient for ethanol through LDPE has been reported as 0.28 g mm m^{-2} day^{-1} atm^{-1}.

Ethanol vapor generators have been shown to be effective in controlling at least 10 species of molds, including *Aspergillus* and *Penicillium* species; 15 species of bacteria including *Staphylococcus*, *Salmonella* and *E. coli* spp; and three species of spoilage yeast.[60] For products with a_ws of less than 0.90, long increases in mold-free shelf life can be obtained using 0.34 to 0.69% of ethanol by weight of the product. For higher a_w products, up to 4% ethanol by weight may be required.

3. Oxygen Absorbers

It is very difficult to reduce the O_2 content to a very low level in packages of bakery products. The porous interiors of these products tend to trap O_2 in such a way that it does not readily interchange with gas that is flowing through the package as occurs in a simple flushing operation carried out as part of MAP. Repeated vacuumizing followed by release of vacuum with an anaerobic gas system would probably solve this problem but would also tend to collapse products such as bread and rolls.[43] One approach to overcome this problem is to place an O_2-absorbent material inside the package after it has been flushed with N_2 or CO_2. Sachets containing forms of iron, which rapidly react with O_2, have been evaluated for this purpose.

In tests reported by Seiler,[56] slices of bread and madeira cake that had been artificially inoculated with molds were placed in bags of O_2 impermeable film with an O_2-absorbent sachet and carefully heat sealed; no mold growth appeared even after prolonged storage at 27°C. In further tests in which films with higher O_2 permeability and leaking seals were used, the extensions in mold-free shelf life were greatly reduced. Permeability of the film was found to be more important than leakage sites in the seals. It was concluded from these tests that the use of O_2-absorbing sachets can result in commercially worthwhile increases in shelf life of baked products, provided that the packaging material used is sufficiently oxygen impermeable and the packages are well sealed.

Further tests have been reported[1] in which chocolate cake was packaged with an O_2-absorbent sachet inside a high gas barrier film (PVdC copolymer-coated oriented PA–LDPE) and stored at 20°C. Although a blue mold colony was formed in the control group on the 14th day, no mold colony was found in the package containing the sachet after 30 days. The chocolate flavor lessened in the control group by the 7th day and the chocolate changed in quality from the 14th day as a result of mold growth. Almost no reduction in flavor was found in the package containing the sachet after 30 days.

VI. BISCUITS, COOKIES AND CRACKERS

A. Manufacture

The three basic ingredients used to manufacture products in this category are wheat flour, fat and sugar, which are combined in different combinations, together with salt and other ingredients in lesser quantities, to produce a full range of products. The mixing process achieves two purposes: intermingling of ingredients, and certain chemical and physical changes that depend on the type of product being produced.

Doughs fall into two categories: hard and short. Hard doughs (crackers and "tea and coffee" biscuits) have low fat, high water and receive a single-stage mix. Short doughs (pastry products, shortbread and American-style cookies) have high fat, low water and usually high sugar levels, and are usually mixed in two stages. Baking is carried out in tunnel ovens where the formed dough pieces are conveyed through a series of heated oven sections. From the oven, the product is taken onto a series of conveyors where it cools and loses the last traces of moisture. After cooling, some products are directly packaged, but many types require additional processing to add nonbaked enrichments such as cream and chocolate.[2,42]

B. Deterioration

Three modes of deterioration are usually associated with biscuits: loss of crispness; development of rancidity; and development of fat bloom. The possible role that packaging may play in controlling the rate of deterioration is now discussed for each mode.

1. Loss of Crispness

Freshly baked biscuits usually have moisture contents within the range of 1 to 5% and a_ws of about 0.1.[62] If biscuits are completely sealed in moistureproof packaging, the small amount of moisture in the atmosphere within the package will rapidly come into equilibrium with that in the product and no further change will take place. However, if the packaging material is permeable to water vapor or if the package seals are not perfect, moisture from the air will enter the package and ultimately lead to a loss of crispness.

The moisture content and a_w at which biscuits lose their crisp texture have not been accurately determined for most biscuits. Using a published a_w value for crackers[39] of 0.43, corresponding moisture contents for various types of biscuits at 25°C can be interpolated from the moisture sorption isotherms presented by Wade[62]; they range from about 5% for sugar-snap cookies to 8.5% for water biscuits.

A report[62] indicated that in the case of semisweet and short biscuits, increased moisture content can promote the development of stale flavors without affecting the texture of the products. It was found that freshly baked product with no added water (moisture contents 1.9 to 2.4%) kept unchanged for 12 months (the maximum period of the trial), whereas samples at moisture contents of 3.9 to 4.1% developed a stale flavor after 6 months storage. Samples at moisture contents of 5.2 to 5.8% were stale after only 6 weeks of storage. Although the stale flavor was not identified, it did not result from rancidity because the fat used in the biscuits contained antioxidants.

Obviously, it is useful to know the initial and critical moisture contents or a_ws of biscuits so that a satisfactory package can be selected in order to give the desired shelf life in a particular environment. Such data can be used to determine the shelf life of the crackers if the maximum permissible moisture content or a_w before they become unacceptable is known.

2. Development of Rancidity

It has been suggested[62] that the permeability of the packaging material to O_2 may be less important as a contributor to the development of oxidative rancidity than is frequently assumed. Unless the package seals are very efficient, atmospheric O_2 as well as moisture will penetrate into the product during storage, and, in any case, sufficient O_2 will normally be present in the package at the time of sealing to initiate reactions leading to rancidity if other conditions in the product are favorable.

3. Development of Fat Bloom

Fat bloom is a gray discoloration that can occur on the surface of biscuits during storage. Its identity is easily confirmed by gently warming the product when the discoloration will disappear.[62] The formation of bloom is accelerated by cyclic variations in temperature during storage of the products, and is associated with the use of certain fats and fat blends. This mode of deterioration is unlikely to be affected by different packaging materials.

C. Packaging

A detailed description of the packaging options and types of packages used for biscuits, cookies and crackers has been presented elsewhere.[2] Therefore, the following discussion focuses on ways in which the choice of packaging material can retard the rate of deterioration in biscuits.

The traditional material used for the packaging of biscuits has been RCF coated with either LDPE or PVdC copolymer, and often with a layer of glassine in direct contact with the product if it contained fat. However, this combination of material has been largely replaced by OPP, either as plain or pearlized OPP film, coextruded OPP film or acrylic-coated on both sides. Plain OPP films are economical but generally require a heat seal coating to improve sealability. Coextruded OPP films provide superior seal strength. If a superior O_2 barrier is required, then acrylic-coated OPP is used, and one side is sometimes coated with PVdC copolymer rather than acrylic. In addition, acrylic and PVdC copolymer-coated OPP films provide a superior flavor and aroma barrier compared with that of uncoated OPP.

Mechanical protection is generally provided either by placing the product in a protective rigid container such as a carton of appropriate caliper, or by packaging the product tightly together, the choice between the two options depending on a number of factors. If the product is particularly moisture sensitive, then the carton will need to be overwrapped with a film that can provide a good barrier to moisture vapor. A further option is to place the biscuits inside a tray (usually made from thermoformed PVC or HIPS) and then overwrapping the tray with a film to provide suitable protection from moisture vapor and O_2.

VII. SNACK FOODS

Dictionary definitions of a snack include "a slight or casual or hurried meal, a small portion of food or drink, or a very light meal." Until the 1970s, commercial snack foods could be said to be potato chips or crisps, nuts, cookies and confectionery. Snack foods now include a very wide range of products, including potato and corn chips, alkali-cooked corn tortilla chips, pretzels, popcorn, extruder puffed and baked/fried products, half-products, meat snacks and rice-based snacks.[11,41,44]

Today, the snack food industry increasingly relies on extrusion cooking processes. A major distinction can be made between direct expanded snack foods and the expanded pellet forms.[20] The former are usually very light structures, which emerge from the cooker/extruder and require only adjustment of moisture content before enrobing and flavoring, whereas the latter are typically compact and dense and require rather specialized drying before expanding. This is achieved by means of a number of techniques including frying in oil, rotating in sand or salt roasters, microwave heating or fluid bed toasters.

A. Fried Snack Foods

1. Manufacture

Fried snack foods can consist of many different ingredients. Although the most popular have been based on potatoes and nuts, large quantities are also made from cereal ingredients, the most widely used cereal being corn. Common to all these snacks is fat which is used as a processing agent to dehydrate the product (as in the case of potato chips) or puff it (as in the case of some extruded products) and develop characteristic flavors. As a consequence, the end of shelf life of many fried snack foods is closely related to the development of rancidity by the fat.

The manufacture of potato chips (also referred to as crisps in some countries) is quite straightforward. After washing, peeling and trimming, potatoes are thinly sliced, washed to remove adhering starch granules, blanched and then dried before passing on a conveyor through hot oil in which they are rapidly dehydrated and cooked. Excess oil is drained or centrifuged off and the chips cooled, salted, flavored (usually by powder adhesion to the residual fat on the chips) and packaged. During the processing of potatoes to chips, the moisture content of the potato is reduced from about 79% to 5%, and of the final 95% of dry matter in chips, 35–40% is fat.

2. Deterioration

There are two major modes of deterioration of fried snack foods: development of fat rancidity and loss of crispness.

All fats are subject to deterioration by oxidative and hydrolytic rancidity which leads to the formation of objectionable odors and flavors. Hydrolytic rancidity is responsible for the development of "soapy" flavors and for facilitating deterioration by direct oxidation. Oxidative rancidity results in food spoilage associated with fat deterioration, i.e., the presence of pungent or acrid odors, and this is the more important of the two mechanisms with respect to food acceptability.[38]

The susceptibility of fried snack foods to oxidative rancidity depends on the type of fat used and the number of unsaturated bonds in the fatty acid moiety. Oxidation of oils is mainly responsible for volatile compound changes in potato chips during storage. To minimize the development of rancidity, the product must be protected from O_2, light and trace quantities of metal ions. The addition of phenolic-type antioxidants such as butylated hydroxyanisole (BHA), butylated hydroxytoluene (BHT) and tertiarybutylhydroquinine (TBHQ) is very helpful but is not always permitted by legislation.

Several approaches for improving the storage stability of potato chips are available. Because the bulk density of chips is typically 0.056 g mL^{-1}, they have a very large headspace volume per unit weight of product. If the product is packaged at atmospheric O_2 concentration, then the headspace O_2 is sufficient to cause O_2 uptake in excess of 3 mL O_2 (STP) g^{-1}. Consequently, inert gas packaging results in a very significant increase in the storage life of potato chips, provided that the headspace oxygen concentrations attained are below 1% and the package permeability to O_2 is very low. The package should be designed to avoid light penetration.

Crispness is a salient textural characteristic for fried snack foods, and its loss owing to absorption of moisture is a major cause of snack food rejection by consumers. Water affects the texture of snack foods by plasticizing and softening the starch/protein matrix, which alters the mechanical strength of the product. Although several investigators have discussed a water content limit at which the textural quality of a snack food product becomes organoleptically unacceptable

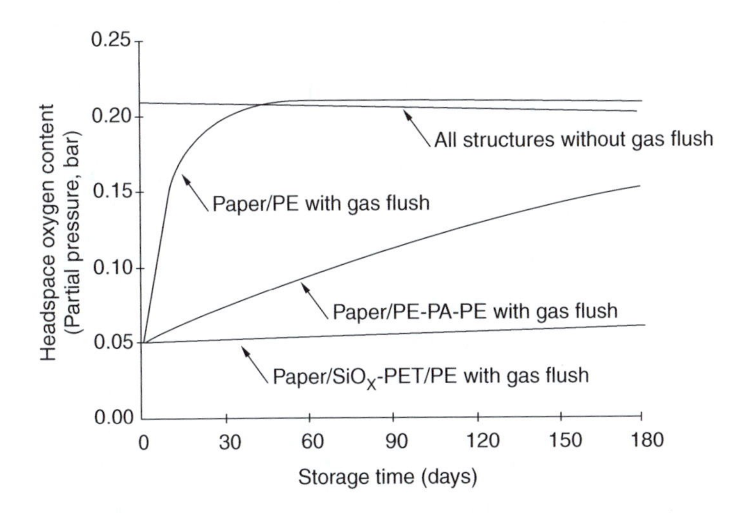

FIGURE 19.1 Influence of different packaging materials and of gas flushing on the oxygen content of potato chip packages at 25°C. (*Source*: From Pfeiffer, C., d'Aujourd'hui, M., Nuessli, J., and Escher, F., *Food Technol.*, 53(6), 52–59, 1999.[53] With permission.)

(typically 3 to 3.5%), water content limits are strongly dependent on the method used for moisture determination. A more reliable approach to establishing moisture conditions for textural acceptance is as a function of a_w. The critical a_w for potato chips and corn is in the range of 0.40 to 0.50. The initial a_w for potato chips has been reported as 0.076, which corresponds to a moisture content of 0.65 g H_2O per 100 g solids, well below the monolayer value of 2.9 g H_2O per 100 g solids.[38] The influence of different packaging materials and of gas flushing on the O_2 content of potato chip packages is shown in Figure 19.1. The critical a_w value for crispness in nuts has been reported as <0.65.[10]

3. Packaging

From the modes of deterioration discussed above, it is clear that a satisfactory package for fried snack foods would need to provide a good barrier to O_2, light and moisture.[22]

Fried snack foods are typically packaged in multilayer structures, although spiral-wound, paperboard cans lined with aluminum foil or a barrier polymer and sealed under vacuum with an LDPE–foil end are used for some specialty products that also require mechanical protection. In addition, the use of metal cans for fried nuts is popular for premium products, where the container is usually gas flushed with N_2 immediately prior to seaming.

Limited information is available on the effects of packaging materials on the stability of snack foods during ambient storage. Because these products are frequently displayed for sale under fluorescent lights, flexible packages are usually pigmented or (occasionally) placed inside paperboard cartons. The use of metallized films is widespread and although they are reasonably efficient light barriers, they do permit some light to penetrate into the package.

Potato chips packaged in OPP–LDPE–PVC, HDPE–EVA copolymer plus a UV light-absorbing compound, or HDPE–EVA copolymer plus a titanium dioxide light barrier developed distinct oxidized flavors within 7 days when stored at 21°C, 55% RH and under 140 to 230 ft candles of continuous fluorescent light. Potato chips stored under the same conditions, but packaged in HDPE plus titanium dioxide and a brown light-absorbing pigment construction or an aluminum foil–LDPE construction, were stable throughout 10 weeks of storage.[11] Oxygen-barrier film

TABLE 19.5
Construction and Properties of Flexible Packaging Materials Used for Storage of Potato Chips

Film Construction	WVTR[a] (mL m^{-2} day^{-1})	OTR[b] (mL m^{-2} day^{-1})
OPP–LDPE–OPP–PVdC copolymer	3.5	6
HDPE–HDPE + UVab–EVA	3.0	1500
HDPE–HDPE + TiO_2–EVA	3.0	1500
HDPE + TiO_2–HDPE + Br[c]–HDPE	3.0	1500
OPP–LDPE–Foil–LDPE–HDPE–EVA	0	0

[a] Measured at 37.8°C and 90% RH.
[b] Measured at 22.8°C and 0% RH.
[c] Brown-pigmented light barrier.

Source: From Kubiak, C. L., Austin, J. A., and Lindsay, R. C., *J. Food Prot.*, 45, 801–805, 1982.[36] With permission.

characteristics did not influence the oxidative stability of the air-packaged potato chips. The barrier properties of the various films used are presented in Table 19.5.

An economical method for packaging peanuts and pecans for long-term storage utilizes the CO_2-adsorption properties of these commodities and involves placing them in plastic pouches impervious to air and CO_2, flushing with CO_2 and then heat sealing the pouches. CO_2 is adsorbed into the pores of the commodities resulting in the formation of a vacuum inside the pouches. Both shelled raw peanuts and shelled roasted and blanched peanuts are protected from any significant deterioration of flavor and other quality factors for up to 12 months.

The formation of a collapsed, tight package a few weeks after vacuum-packaging of cereals, grains and nuts owing to the adsorption of CO_2 is well documented. If these products are packaged in CO_2 in a plastic bag of low gas permeability and the bag is sealed, then the package volume gradually decreases as CO_2 gas is absorbed and finally a skintight package is obtained.

B. Extruded and Puffed Snacks

1. Manufacture

Extrusion has provided a means of manufacturing new and novel food products and has revolutionized many conventional snack manufacturing processes. The most popular and successful extruders used in the production of snack foods have been single-screw extruders, although twin-screw extruders are also used. The extruder must exercise a number of functions in a short time under controlled, continuous or steady-state operating conditions. These functions may include heating, cooling, conveying, feeding, compressing, reacting, mixing, melting, cooking, texturing and shaping.[30]

The majority of extruded snacks on the market fall into the category of expanded snacks. They are usually light with a low bulk density and are seasoned with an array of flavors, oils and salt. A typical manufacturing process would consist of blending of the ingredients with water prior to being fed into the extruder. As the mix passes through the extruder, it is compressed, with the work performed on the mix during extrusion being transformed into heat. The combination of pressure and heat causes the mix to become very viscous and as it passes through the extruder heads, the superheated moisture instantaneously vaporizes, resulting in puffing of the product.

The moisture content of the extruded product is normally between 8 to 10% on a wet basis and it must be reduced to 1 to 2% to give the desired product crispness. Additional drying at temperatures up to 150°C for 4 to 6 min is used. Sometimes, the product is subsequently fried in oil to remove moisture and develop a desirable flavor. The product is then cooled and any fines are removed prior to coating with flavors. This is achieved by first spraying the product with vegetable oil and then dusting with a variety of dry flavors or seasonings. Alternatively, the oils, flavors and seasonings may be mixed together and applied to the extruded product as it is tumbled in a flavor-application reel.[30]

Puffed snack foods such as popcorn were originally made by placing grains of corn onto very hot plates. This caused the moisture in the grains to suddenly expand into steam, thus causing the grain to be puffed and simultaneously cooked. This method was refined by heating the grain in a quick-release but hermetically sealed cylinder where the sudden release of pressure caused the grain to puff or expand. A similar principle is used in extrusion. The efficiency of the process has been improved so that many cereals can now be puffed up to four to eight times their original size; these expanded, original-texture grains are used in many snack products.

2. Deterioration

The major mode of deterioration for extruded and puffed snacks is loss of crispness. The critical a_w for puffed corn curl has been reported[32] as 0.36, which corresponds to a moisture content of 4.2 g H_2O per 100 g solids. The initial a_w of this product was 0.082 with a corresponding moisture content of 1.83 g H_2O per 100 g solids. For popcorn, initial and critical a_ws were 0.062 and 0.49, with corresponding moisture contents of 1.70 and 6.1 g H_2O per 100 g of solids. The critical a_w for extruded rice snacks has been reported to be 0.43, which corresponds to 6.5% moisture content.[19] The moisture sorption isotherm for puffed corn curls is shown in Figure 19.2, and the sensory crispness intensity of popcorn as a function of a_w (with the critical a_w marked as a_c) is shown in

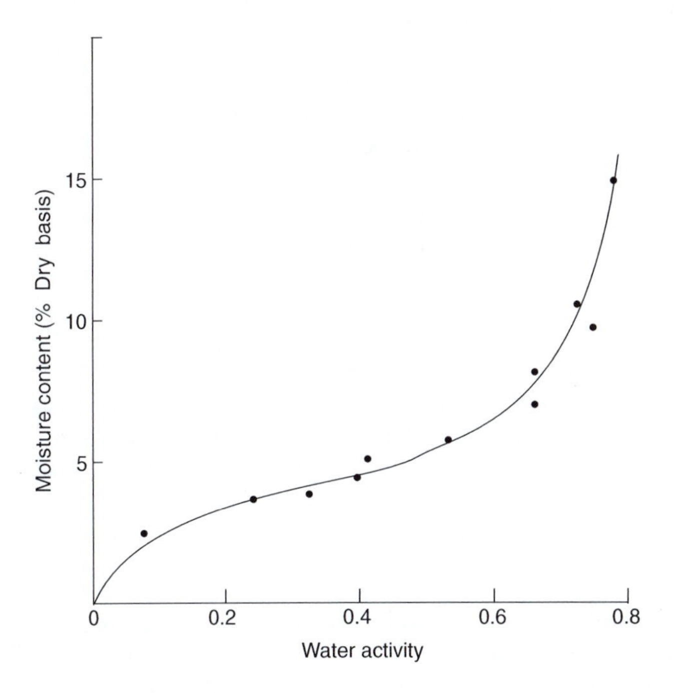

FIGURE 19.2 Moisture sorption isotherm for puffed corn curls at 20°C. (*Source*: From Katz, E. E. and Labuza, T. P., *J. Food Sci.*, 46, 403–409, 1981. With permission.)

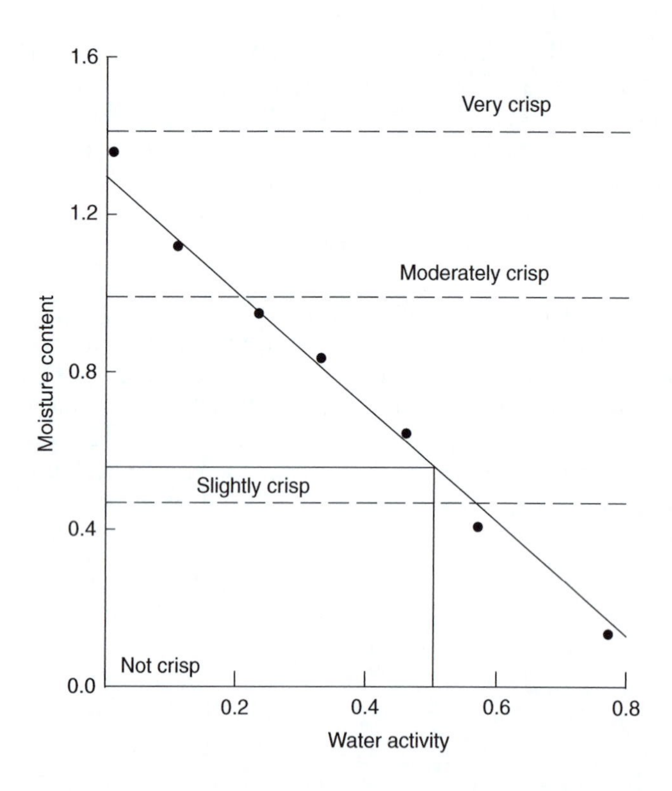

FIGURE 19.3 Sensory crispness intensity of popcorn as a function of water activity; the critical a_w is shown as a_c. (*Source*: From Katz, E. E. and Labuza, T. P., *J. Food Sci.*, 46, 403–409, 1981. With permission.)

Figure 19.3. A report on the effects of a_w on the textural characteristics of puffed rice cakes[29] indicated that rice cake lost its crispness and became tough as the a_w increased above 0.44, a critical point with respect to texture. Rice cakes with a_w between 0.23 and 0.44 were crisp and low in hardness. Data such as these are essential before selection of a suitable packaging material can be made.[7]

The development of stale or oxidized and rancid flavors and odors can also be a problem limiting the shelf life of certain extruded and puffed snacks.

3. Packaging

Many extruded and puffed snack foods are packaged in identical material to that discussed above for fried snack foods. However, because the major mode of deterioration is loss of crispness, a package that provides a good barrier to water vapor is the primary requirement. Some extruded and puffed snacks are comparatively less sensitive to O_2 than fried snack foods and the O_2 barrier requirements of the packages are consequently less stringent.

C. Fruit-Based Snacks

A large variety of dried fruits, usually with added nuts and honey, sugar or syrup, are used in the manufacture of this type of snack food. Their composition is infinitely variable, leading to the development of bars with a wide variety of flavors and textures. Generally, these products are in the intermediate moisture category and have a_ws in the range of 0.20 to 0.40. This is because of their relatively high sugar content (25 to 30%) mainly in the form of glucose, sucrose and fructose from several ingredients such as corn syrup, glucose syrup, sweetened condensed (evaporated) skimmed milk, raw cane sugar and invert sugar syrup.

Also in this category are snacks based largely on flaked cereals such as oatflakes/wheatflakes or puffed cereals such as rice, to which dried fruits, nuts and carob chips are also added to produce variations in flavor and texture. Again, many of these snacks have a chewy texture and an a_w which places them in the intermediate moisture category. The fat content of these chewy bars can be as high as 24%.

Packaging materials for these fruit-based snacks must provide a barrier to water vapor ingress to avoid the development of stickiness as a result of moisture uptake by the sugar in the product. The a_w to maintain chewiness has been given as <0.50 a_w[10] although no specific product was mentioned. In very dry atmospheres, moisture loss could be a problem. Before the packaging requirements can be specified with any degree of precision, the moisture sorption isotherm for this type of product would be required. A useful simulation of moisture transfer in packaged cereal–fruit systems has been published.[40]

VIII. CONFECTIONERY

A. Sugar Confectionery (Candy)

1. Manufacture

The words "candy" and "confectionery" are often used interchangeably to mean a sweet sugar- or chocolate-based snack or desert food. Strictly defined, the word *candy* refers to products containing sugar as the dominant component and does not include chocolate products. *Confection* is a more general term referring to any sweet product manufactured from sugar and other ingredients. Products falling within the latter definition will be discussed in this section.

Confectionery has four major basic ingredients: sucrose, glucose, fructose and water. These can be combined with flavorings and colorings to make confections such as fondants and boiled sweets. When milk solids, fat, proteins and modified starches are added as well, products such as jellies, pastilles, toffees and caramels can be produced. The primary difference between the various confectionery products is in the amounts and types of sugars and in the amount of water. In one group (which includes the hard candies and soft, chewy products), uncrystallized sugars are present in a very viscous solution, which is handled like a solid at room temperature. In the other group (which includes fondant crèmes and other crystallized or grained products), sucrose is present in crystalline form, usually as microscopic crystals dispersed in syrup.[26]

Hard candies (also known as *boiled sweets*) are made almost entirely of sugar, typically combinations of sucrose, glucose and fructose. The mixture of sugars is heated under vacuum to reduce the moisture content to about 1%, after which flavors, colors and acid (normally citric) are mixed in, and the plastic mass is formed into the desired shape. After cooling to set, the sweets are packaged as rapidly as possible, preferably while still warm.

Confections such as fondant and fudge are composed of sucrose in the form of crystals suspended in a saturated sugar solution. A typical composition of base fondant is about 75% sucrose, 15% glucose and 10% water. The mixture is boiled at 120°C and cooled rapidly to about 50°C when it is subjected to violent agitation to cause rapid nucleation and crystallization. Fudge is basically a toffee or caramel with a high sugar content, which has been deliberately crystallized during processing. Typically, caramel is cooked to about 120°C, cooled to about 50°C and then fondant added to initiate crystallization; it is then cooled to allow crystallization to occur.[23]

The solid crystals in fondant and fudge have no effect on a_w, and the phases are stable provided that the system is protected from atmospheric moisture by suitable packaging material. The a_w of the syrup phase is related to both its composition and to any water loss or gain that may have occurred; a_ws of 0.78 and 0.65 have been quoted for fondant and fudge, respectively.[48]

Toffees and caramels are made by boiling sucrose, glucose syrup, condensed milk, vegetable fats and salt. Other ingredients may include cream, butter and various flavorings, although the inherent flavor results from the Maillard nonenzymic browning reaction between reducing sugars and milk proteins.[31] After cooling to about 45 to 50°C, the toffee is cut and packaged. Because caramel is a dispersion of fat globules in a high solids, highly supersaturated, viscous sugar matrix that cannot crystallize, all the component sucrose molecules, rather than just those in the syrup phase, influence the a_w.[14]

2. Deterioration

An important difference in the mode of deterioration between the various candies is that crystallized products tend to dry out under normal storage conditions, while uncrystallized products tend to pick up moisture from the atmosphere. The a_ws of various types of sugar confectionery are presented in Table 19.6.

One of the major defects of confectionery is crystallization of the sugar. Hard candies rely on their viscosity to prevent sucrose crystallization. Their a_w is usually lower than the surrounding atmosphere, and if they absorb moisture, the surface viscosity decreases very rapidly to the point where the solution flows under finger pressure, giving a sticky candy. If the candies are not individually wrapped, then they may stick together. Also, the diluted solution allows more rapid mass transfer of sucrose and it becomes possible for crystals to grow. As sucrose begins to crystallize, the sugar concentration in the remaining solution becomes still lower, making crystal growth even more rapid. Within hours, the entire surface may be covered with crystalline material, a defect known as graining.[26]

Fondants generally have an a_w that is higher than the humidity of the surrounding atmosphere and thus they tend to dry out. This results in increased crystallization. One significant change arising from the loss of moisture is a decrease in volume. If the fondant is coated with chocolate, then this volume decrease would cause either collapse of the coating or an air space inside the coating. Both consequences are undesirable, and although a collapsed coating is more apparent, an

TABLE 19.6
Water Activities of Various Kinds of Confectionery

Type of Product	Average a_w	Usual Limits
Boiled sweets	0.28	<0.30
Toffee	0.47	<0.48
Caramels	0.50	0.45–0.55
Nougat	0.55	0.50–0.60
Gums and pastilles	0.60	0.51–0.64
Liquorice	0.64	0.53–0.66
Fruit jellies	0.65	0.60–0.70
Fudge	0.65	0.60–0.70
Turkish delight	0.66	0.60–0.70
Marshmallow	0.72	0.63–0.73
Fondant	0.78	0.75–0.84

Source: From Nelson, C., Wrapping, packaging and shelf life evaluation, In *Sugar Confectionary Manufacture*, 2nd ed., Jackson, E. B., Ed., Blackie Academic & Professional, London, 1995. With permission.

air space can permit condensation of moisture, which would lead to a localized area of high a_w where microbial growth could occur.[26]

Two types of flavor changes can occur in confectionery: loss of desired flavor components and development of off-flavors. Packaging has an important role to play in preventing loss of desirable flavors from the product and the entry of O_2, which can oxidize flavors.

Many confections, notably the hard candies, are brightly colored and therefore color stability is important. The main cause of color fading during storage is bleaching by light. Use of a package that is a complete barrier to light is generally undesirable because the attractiveness of brightly colored confectionery is an important selling point. The use of synthetic, light-stable colors in hard candies is widespread.

3. Packaging

The type of packaging required to protect the confection from moisture uptake will depend on the a_w of the confection and the RH of the ambient atmosphere. The packaging type will also depend on whether or not the confection deteriorates by gain or loss of moisture under these conditions.

It has been customary to wrap boiled sweets, toffees and caramels individually, partly as a hygienic measure, partly to protect them from atmospheric moisture, partly to prevent them from sticking together and partly to avoid the intermingling of the flavors of assortments. Because the wrapping machines operate at very high speeds, the mechanical and electrostatic properties also largely govern the choice of packaging materials to be used.[27] Both ends of the wrapper sleeve are twisted to effect a closure.

Fibrous materials are generally unsuitable because they promote adhesion of the wrapper to the sweet. Materials such as waxed paper, waxed glassine and moistureproof, plasticized RCF have been used successfully for many years, but the use of cast PP (which holds the twist better than RCF) is now widespread. Although these materials offer some barrier to moisture vapor, they provide little protection in this situation because the overlap is not sealed. Thus, an outer package that provides a barrier to moisture is required, and this is usually a heat-sealed bag made from coated RCF or a polyolefin, although metal containers, glass jars and foil or metallized laminates are also used. In some situations, cartons made from paperboard are used. The board may be coated with LDPE, or the carton may have a plain or waxed glassine liner or one made from thin gauge HDPE.

B. Chocolate

1. Manufacture

Chocolate is a suspension of finely ground, roasted cocoa beans or cocoa mass and sugar particles in cocoa butter (the lipid fraction of the cocoa mass). Milk chocolate is similar but with the addition of whole milk powder. Cocoa beans are roasted to develop flavor and then ground to liberate some fat. Other ingredients are mixed in and the particles further reduced in size to a maximum of about 30 μm. The final stage is *conching* — agitation under heat (75 to 80°C for plain and 50 to 60°C for milk chocolate) to remove water and unwanted volatile substances, improve flavor and texture, and to reduce viscosity. During this process, an enormous increase in surface area of the solid particles occurs, and the resulting chocolate mass becomes quite dry. The sugar particles are not easily wetted by the cocoa butter and, although small, are compacted into aggregates.[6]

Although cocoa butter can crystallize in a number of polymorphic forms, only one such form (B or form V) is stable. Chocolate is tempered by heating to 50°C to remove all crystal nuclei and then cooled gradually to about 27°C where crystallization commences. The temperature is then raised to 32°C when only a finely dispersed seed crystal of form V remains. The chocolate is then cooled at a controlled rate when stable crystals form, and the surface of the chocolate takes on a good gloss.

If chocolate is cooled rapidly, then unstable polymorphs can form and these will ultimately give rise to fat bloom — a coarse crystal growth on the surface of the product.[13]

2. Deterioration

Chocolate is very sensitive to temperature, the melting point of fat (which forms the stable crystal form) being about 37°C. If melting and subsequent solidification occur, then the surface texture becomes rough. If the temperature oscillates just below the melting point, then fat will move from the body of the chocolate to the surface (a defect known as *fat bloom*), resulting in a grayish discoloration. Bloom can also arise as a result of using fats in chocolate that are incompatible with cocoa butter.[28] Generally, there is little that can be done by packaging to prevent such defects occurring.

Sugar bloom is another defect that affects the appearance of chocolate. While on cursory examination, it appears similar to fat bloom, sugar bloom consists of a layer of sugar crystals on the surface. It is caused by exposing the surface of chocolate to air of high humidity or by the use of refined sugar having a high moisture content.[28]

Oxidative and lipolytic rancidity are flavor defects, the former coming from oxidation of unsaturated fats and the latter from enzymic hydrolysis of short- and medium-chain triglycerides. Cocoa butter contains tocopherols (liposoluble vitamin E compounds), which act as antioxidants and therefore confer a natural protection against oxidation during storage.[28]

Being high in fat, chocolate is very likely to absorb any foreign odors from the surrounding atmosphere unless adequately protected by suitable packaging materials.

3. Packaging

Chocolate is sold in many forms where the product often contains not only chocolate but also other ingredients such as fruits, nuts and caramel, which influence the likelihood of occurrence of different storage defects.

Suitable packaging for chocolate must provide a good barrier to light, O_2, moisture vapor and foreign odors. The most common material used to package blocks of chocolate used to be unsealed aluminum foil of 0.009 mm thickness. At this thickness it exhibits "dead wrap" characteristics (i.e., takes on the shape of the product around which it is placed). No sealing was possible but folding and overlapping of the foil provided adequate protection. In warmer climates, a layer of waxed tissue paper was placed inside the foil to prevent fat staining of the outer package; it was also claimed to offer protection against odors.[45]

Today, it is more common to package chocolate blocks in a laminate consisting of aluminum foil and LDPE, making it possible to heat seal the package. Such a package is a better moisture and odor barrier than the foil and the foil/paper packages. Such packages sometimes contain a layer of paper, either between the foil and the LDPE or on the outside of the foil when the latter is laminated directly to the LDPE. In some packages, the foil is replaced by PVdC copolymer, the latter being applied as a thin coating on the LDPE.

Chocolate-coated confectionery such as coated caramel bars are typically packaged in pearlized OPP, which is cold sealed longitudinally and at each end, where the cold sealants avoid the risk of melting the chocolate.

Packages of individual chocolates are frequently purchased for gifts and special occasions, and this is reflected in the sometimes very elaborate packaging used. Molded PVC or HIPS trays with individual cavities for each piece are common, with many of the individual chocolates being wrapped in colored, thin gauge, aluminum foil. The trays are placed inside paperboard boxes and overwrapped with RCF or polyolefin film. Sometimes, metal or glass containers are used, which provide an excellent barrier to moisture.

REFERENCES

1. Abe, S. and Kondoh, Y., Oxygen absorbers, In *Controlled/Modified Atmosphere/Vacuum Packaging of Foods*, Brody, A. L., Ed., Food & Nutrition Press, Trumbull, CT, 1989, chap. 9.
2. Almond, W. N. V., *Biscuits, Cookies and Crackers, The Biscuit Making Process*, Vol. 2, Elsevier Science Publishers, Essex, England, 1989, chap. 8.
3. Aureli, P., Fenicia, L., Gianfranceschi, M., and Pasolini, B., *Microbiological Aspects of Fresh and Dried Pasta*, Mercier, Ch. and Cantarelli, C., Eds., Elsevier Applied Science Publishers, London, pp. 109–121, 1986.
4. Avital, Y. and Mannheim, C. H., Modified atmosphere packaging of pita (pocket) bread, *Pack. Technol. Sci.*, 1, 17–23, 1988.
5. Bailey, J. E., Whole grain storage, In *Storage of Grains and Their Products*, Christensen, C. M., Ed., American Association of Cereal Chemists, St. Paul, MN, 1974, chap. 8.
6. Beckett, S. T., *Industrial Chocolate Manufacture and Use*, 3rd ed., Blackwell Publishing, Oxford, England, 1999.
7. Bell, L. N. and Labuza, T. P., *Moisture Sorption: Practical Aspects of Isotherm Measurement and Use*, 2nd ed., American Association of Cereal Chemists, St Paul, MN, 2000.
8. Bent, A. J., Ed., *Technology of Cakemaking*, Blackie Academic & Professional, London, 1998.
9. Black, R. G., Quail, K. J., Reyes, V., Kuzyk, M., and Ruddick, L., Shelf-life extension of pita bread by modified atmosphere packaging, *Food Aust.*, 45, 387–391, 1993.
10. Bone, D. P., Practical applications of water activity and moisture relations in foods, In *Water Activity: Theory and Applications to Food*, Rockland, L. B. and Beuchat, L. R., Eds., Marcel Dekker, New York, 1987, chap. 15.
11. Booth, R. G., *Snack Food*, Van Nostrand Reinhold, New York, 1990.
12. Boyd, P. J. and Hebeda, R. E., Anti-staling enzyme for baked goods, *Food Technol.*, 44, 129–132, 1990.
13. Bralsford, R. and Le Fort, J., Chocolate confectionery, In *Snack Food*, Booth, R. G., Ed., Van Nostrand Reinhold, New York, 1990, chap. 4.
14. Brockway, B., Applications to confectionery products, In *Water and Food Quality*, Hardman, T. M., Ed., Elsevier Science Publishers, Essex, England, 1989, chap. 9.
15. Cardoso, G. and Labuza, T. P., Effect of temperature and humidity on moisture transport for pasta packaging material, *J. Food Technol.*, 18, 587–593, 1983.
16. Castelvetri, F., Microbiological implications of modified atmosphere packaging of fresh pasta products, In *Proceedings of the Sixth International Conference on Controlled/Modified Atmosphere/Vacuum Packaging*, Schotland Business Research Inc., San Diego, CA, pp. 253–265, 1991.
17. Cauvain, S. P. and Young, L. S., Eds., *Technology of Breadmaking*, Blackie Academic & Professional, London, 1998.
18. Cauvain, S. P. and Young, L. S., *Bakery Food Manufacture and Quality: Water Control and Effects*, Blackwell Science, Oxford, England, 2000.
19. Chauhan, G. S. and Bains, G. S., Equilibrium moisture content, BET monolayer and crispness of extruded rice-legume snacks, *Int. J. Food Sci. Technol.*, 25, 360–363, 1990.
20. Cosgriff, M., Papotto, G., and Stefani, L., *Extruded Snack Food Technologies*, Mercier, Ch. and Cantarelli, C., Eds., Elsevier Applied Science Publishers, London, pp. 174–189, 1986.
21. Doerry, W. T., Packaging bakery products in controlled atmospheres, *Am. Inst. Baking Tech. Bull.*, 7, 1–21, 1985.
22. Dunn, T., Product protection and packaging materials, In *Snack Foods Processing*, Lusas, E. W. and Rooney, L. W., Eds., CRC Press, Boca Raton, FL, 2001, chap. 22.
23. Edwards, W. P., *The Science of Sugar Confectionery*, Royal Society of Chemistry, Cambridge, England, 2000.
24. Fabriani, G. and Lintas, C., Eds., *Durum Wheat: Chemistry and Technology*, American Association of Cereal Chemists, St. Paul, MN, 1988.
25. Gray, J. A. and Bemiller, J. N., Bread staling: molecular basis and control, *Compr. Rev. Food Sci. Food Saf.*, 2, 1–21, 2003.

26. Hansen, T. J., Candy and sugar confectionery, In *Handbook of Food and Beverage Stability*, Charalambous, G., Ed., Academic Press, Orlando, FL, 1986, chap. 7.
27. Hooper, J. H., *Confectionary Packaging Equipment*, Aspen Publishers, Gaithersburg, MD, 1999.
28. Horman, I., Chocolate, In *Handbook of Food and Beverage Stability*, Charalambous, G., Ed., Academic Press, Orlando, FL, 1986, chap. 6.
29. Hsieh, F. L., Hu, L., Huff, H. E., and Peng, I. C., Effects of water activity on textural characteristics of puffed rice cake, *Lebensm-Wiss u-Technol.*, 23, 471–473, 1990.
30. Huber, G. R. and Rokey, G. J., Extruded snacks, In *Snack Food*, Booth, R. G., Ed., Van Nostrand Reinhold, New York, 1990, chap. 7.
31. Jackson, E. B., Ed., *Sugar Confectionary Manufacture*, 2nd ed., Blackie Academic & Professional, London, 1995.
32. Katz, E. E. and Labuza, T. P., Effect of water activity on the sensory crispness and mechanical deformation of snack food products, *J. Food Sci.*, 46, 403–409, 1981.
33. Kent, N. L. and Evers, A. D., *Technology of Cereals: an Introduction for Students of Food Science and Agriculture*, 4th ed., Pergamon Press, Oxford, England, 1994.
34. Kill, R. C. and Turnbull, K., Eds., *Pasta and Semolina Technology*, Blackwell Science, Oxford, England, 2001.
35. Knorr, D. and Tompkins, R. I., Effect of carbon dioxide modified atmosphere on the compressibility of stored baked goods, *J. Food Sci.*, 50, 1172–1177, 1985.
36. Kubiak, C. L., Austin, J. A., and Lindsay, R. C., Influence of package construction on stability of potato chips exposed to fluorescent light, *J. Food Prot.*, 45, 801–805, 1982.
37. Kulp, K. and Ponte, J. G., Eds., *Handbook of Cereal Science and Technology*, 2nd ed., Marcel Dekker, New York, 2000.
38. Labuza, T. P., *Shelf-life Dating of Foods*, Food and Nutrition Press, Westport, CT, 1982.
39. Labuza, T. P. and Contreras-Medellin, R., Prediction of moisture protection requirements for foods, *Cereal Foods World*, 26, 335–343, 1981.
40. Labuza, T. P. and Sapru, V., Moisture transfer simulation in packaged cereal-fruit systems, *J. Food Eng.*, 27, 45–61, 1996.
41. Lusas, E. W. and Rooney, L. W., Eds., *Snack Foods Processing*, CRC Press, Boca Raton, FL, 2001.
42. Manley, D. J. R., *Technology of Biscuits, Crackers and Cookies*, 3rd ed., CRC Press, Boca Raton, FL, 2000.
43. Matz, S. A., *Bakery Technology: Packaging, Nutrition, Product Development, Quality Assurance*, Elsevier Science Publishers Ltd, Essex, England, 1989.
44. Matz, S. A., *Snack Food Technology*, 3rd ed., Van Nostrand Reinhold, New York, 1993.
45. Minifie, B. W., *Chocolate, Cocoa and Confectionary: Science and Technology*, 3rd ed., Van Nostrand Reinhold, New York, 1989, chap. 22.
46. Mohney, S. M., Hernandez, R. J., Giacin, J. R., Harte, B. R., and Miltz, J., Permeability and solubility of d-limonene vapor in cereal package liners, *J. Food Sci.*, 53, 253–257, 1988.
47. Monahan, E. J., Packaging of ready-to-eat breakfast cereals, *Cereal Foods World*, 33, 215–221, 1988.
48. Nelson, C., Wrapping, packaging and shelf life evaluation, In *Sugar Confectionary Manufacture*, 2nd ed., Jackson, E. B., Ed., Blackie Academic & Professional, London, 1995, chap. 17.
49. Notermans, S. H. W., Dufrenne, J., and Lund, B. M., Botulism risk of refrigerated, processed foods of extended durability, *J. Food Prot.*, 53, 1020–1024, 1990.
50. Ooraikul, B., Modified atmosphere packaging of bakery products, In *Modified Atmosphere Packaging of Food*, Ooraikul, B. and Stiles, M. E., Eds., Ellis Horwood, Chichester, England, 1991, chap. 4.
51. Ory, R. L., DeLucca, A. J., St Angelo, A. J., and Dupuy, H. P., Storage quality of brown rice as affected by packaging with and without carbon dioxide, *J. Food Prot.*, 43, 929–932, 1980.
52. Park, C. E., Szabo, R., and Jean, A., A survey of wet pasta packaged under a CO_2:N_2 (20:80) mixture for staphylococci and their enterotoxins, *Can. Inst. Food Sci. Technol. J.*, 21, 109–115, 1988.
53. Pfeiffer, C., d'Aujourd'hui, M., Nuessli, J., and Escher, F., Optimizing food packaging and shelf life, *Food Technol.*, 53(6), 52–59, 1999.

54. Piergiovanni, L., Limbo, S., Riva, M., and Fava, P., Assessment of the risk of physical contamination of bread packaged in perforated oriented polypropylene films: measurements, procedures and results, *Food Addit. Contam.*, 20, 186–195, 2003.
55. Sakamaki, C., Gray, J. I., and Harte, B. R., The influence of selected barriers and oxygen absorbers on the stability of oat cereal during storage, *J. Pack. Technol.*, 2, 98–103, 1988.
56. Seiler, D. A. L., Bakery products, In *Principles and Applications of Modified Atmosphere Packaging of Food*, 2nd ed., Blakistone, B. A., Ed., Blackie Academic & Professional, London, 1998, chap. 7.
57. Skovholt, O. and Bailey, C., The influence of humidity and carbon dioxide upon the development of moulds on bread, *Cereal Chem.*, 10, 446–451, 1933.
58. Smith, J. P. and Simpson, B. K., Modified atmosphere packaging of bakery and pasta products, In *Principles of Modified Atmosphere and Sous Vide Product Packaging*, Farber, J. M. and Dodds, K. L., Eds., Technomic Publishing, Lancaster, 1995, chap. 9.
59. Smith, J. P. and Simpson, B. K., Modified atmosphere packaging, In *Baked Goods Freshness: Technology, Evaluation and Inhibition of Staling*, Hebeda, R. E. and Zobel, H. F., Eds., Marcel Dekker, New York, 1996, chap. 8.
60. Smith, J. P., Daifas, D. P., El-Khoury, W., Koukoutsis, J., and El-Khoury, A., Shelf life and safety concerns of bakery products — a review, *Crit. Rev. Food Sci. Nutr.*, 44, 19–55, 2004.
61. Southwick, C. A., Packaging of cereal products, In *Storage of Grains and Their Products*, 3rd ed., Christensen, C. M., Ed., American Association of Cereal Chemists, St. Paul, MN, 1982, chap. 14.
62. Wade, P., *Biscuits, Cookies and Crackers, The Principles of the Craft*, Vol. 1, Elsevier Science Publishers, Essex, England, 1988, chap. 4.
63. Zobel, H. F. and Kulp, K., The staling mechanism, In *Baked Goods Freshness: Technology, Evaluation and Inhibition of Staling*, Hebeda, R. E. and Zobel, H. F., Eds., Marcel Dekker, New York, 1996, chap. 1.

20 Packaging of Beverages

CONTENTS

I. INTRODUCTION

Beverages are an important part of the diet of all humans and have been since the earliest times. Although the origins of many beverages are unknown, there is no doubt that the range and sophistication of beverages has increased dramatically over recent decades. Much of this growth can be attributed to developments in packaging, which have made it possible for a large national and international trade in beverages to flourish. Today, a wide variety of quite different beverages are consumed in the home, at work and at a myriad of sporting, leisure and entertainment activities, and the full range of packaging media is used (either alone or in varying combinations) to bring these beverages to consumers. This chapter discusses the major categories of beverages, including their manufacture, deterioration and packaging.

II. WATER

A. INTRODUCTION

In Europe and China, spring and mineral waters have been consumed for centuries by local inhabitants, immigrants and invaders. For example, Evian natural mineral water began its present-day history in 1789 when a French marquis began bottling the "miraculous" waters on his estate bordering Lake Geneva (Lac Léman). Similarly, Perrier water was first bottled as Perrier sparkling mineral water in 1863.[35]

Bottled water is now widely available for sale and its consumption has risen dramatically over the past 15 years. The growth in bottled water is influenced by three public concerns or fears: a declining quality from often overworked municipal water supplies; possible toxic contamination of ground water sources; and a general increased interest in personal health. It has also become a "must have" fashion accessory for many consumers. There is a public perception that bottled water is safe, natural and free from additives such as fluoride and chlorine.

Bottled water can be divided into nonsparkling or still, and sparkling. It can also be divided into either natural or processed categories, where natural water is bottled directly from underground sources, while processed water is tap or well water that is highly filtered or distilled.

Natural bottled waters are sold with the understanding (and in Europe, the legal requirement) that they have not been subjected to any treatment that would remove natural indigenous bacteria, which are believed to have medicinal and therapeutic qualities. It has never been proven that the ingested levels of indigenous micro-organisms in bottled water have an adverse effect on health. Despite this, much controversy surrounds the question of the potential pathogenicity of indigenous micro-organisms in mineral waters.[34]

Bottled water is defined by the U.S. FDA as "water that is intended for human consumption and that is sealed in bottles or other containers with no added ingredients except that it may contain safe and suitable antimicrobial agents".[35] However, antimicrobial agents are not permitted in many other countries. The water may be subjected to a number of treatments including distillation, carbonation, ozonation, chlorination, filtration and so on, depending on the quality of the source water, the type of bottled water being manufactured and where it is being manufactured.

Based on available information, bottled water sold worldwide by members of the bottled water associations has generally been found to be of good microbiological quality and is not considered to pose any microbiological threat. However, not all bottled water manufacturers belong to these organizations, and thus not all bottled water is produced in compliance with their code of practice.[34]

B. Deterioration

The major deteriorative reaction in bottled water is microbial growth. To avoid this, the water is usually treated prior to bottling with chlorine or ozone, the latter being preferred because it is much faster acting than the former. However, when the water is packaged in plastic bottles, residual ozone in the water can attack the plastic causing an odor and taste problem, although the presence of antioxidants in the plastic minimizes the problem.[36]

The source of water is never sterile and contains sufficient trace nutrients for microbial growth. Various saprophytic bacteria, yeasts and molds are commonly found in water, which may also be contaminated with pathogenic bacteria and parasites.[16] In addition, water may become contaminated during transport and processing, with bottles and closures also being a possible source of contamination.

The most obvious manifestation of spoilage in bottled water is the appearance of floating pieces of mold mycelium, where *Penicillium* species, along with *Cladosporium* and *Phaeoramularia*, are the most commonly isolated fungi.[16] *Pseudomonas aeruginosa*, an organism associated with soil contaminated with human and animal faeces, is not generally associated with spoilage of bottled water but its presence can affect water color, clarity and taste. It is capable of growing to high numbers in minimal nutrient environments such as deionized and demineralized water and has been implicated in food- and water-borne disease. Most *P. aeruginosa* strains are resistant to commonly used antibiotics and sanitizers/disinfectants but should be inactivated by pasteurization. It has been isolated from bottled waters from Brazil, Canada, France, Germany, Indonesia, Spain and the U.S.[16]

Biofilm proliferation has been reported on the walls of PVC containers and may be partly responsible for the long-term survival of bacteria in bottled water, where several authors have found higher counts and faster growth in bottled water stored in plastic bottles than in glass bottles.[34]

The use of returnable containers for bottled water is common in many countries. Only a limited number of studies have been performed on the contamination risk by micro-organisms when classical caustic cleaning is applied to rinse returnable containers. It has been found that under optimal conditions, the following classification could be made in decreasing order of microbial rinsability: glass $>$ PET $>$ PC $>$ PP $=$ PVC $>$ HDPE. Even at optimal rinsing conditions, it was not possible to totally remove all bacteria from the sides of the containers, leading to the recommendation that bottled water should be disinfected by ozonation.[35]

The presence of dissolved O_2 in many still waters is a major contributor to the desirable taste and mouth feel of the water. Loss of much of this dissolved O_2 leads to the water being judged unacceptable by consumers. A further deterioration in bottled water has been the development of a "plastic" taste, which is sometimes found in water which has been packaged in unsuitable plastic containers.[28]

Sensory evaluation of six commercial bottled mineral waters and bottled tap water, which had either been filtered with activated carbon or electrolyzed after the addition of calcium salts, and then stored in polyethylene bottles at room temperature, was undertaken in Japan.[18] Results showed that carbon-filtered water rated higher than electrolyzed water which rated higher than the commercial samples.

C. Packaging

Glass bottles were long considered the container of choice for sparkling waters and carbonated soft drinks (see Section VI.C.1) but, in recent years, PET bottles have gained an increasing share of this sector. The majority of still waters are now packaged in plastic containers, with five principal resin alternatives—HDPE, PP, PVC, PC and PET. Antioxidant additives are used as processing aids in the plastic resins and it is important that they not migrate into the water and cause undesirable odors and taste.

PVC has been a very common bottle resin, particularly in Europe but less so in the U.S., for which special low taste and odor compounds are used. Adverse publicity in 1974 about VCM residues in PVC led to a steady decline in the use of PVC as a food packaging material, especially in Europe. PET has gained ground with the popularity of on-site *hole-through-the-wall* (HTW) bottle manufacturing facilities, helping it make further inroads into the PVC bottle market share. Problems with VCM residues have long since been addressed, and PVC is generally regarded as the most suitable polymer from a technical point of view. A PVC bottle is crystal clear with a high gloss surface finish and, because it can be extrusion blow molded, cut-through handles can be featured.[13] In addition, it provides a relatively high O_2 barrier that assists the retention of dissolved O_2 found in many bottled waters, thus allowing for a more consistent taste and longer shelf life.

A major problem with HDPE is its translucent white color that does little to enhance the appeal of water. Although PP has a somewhat hazy appearance, this virtually disappears when filled with water. PC is a very suitable resin for bottle manufacture but is not used in the nonreturnable, one-way water market because of its high cost relative to other resins. PET is widely used for the packaging of both carbonated soft drinks and waters. However, careful control of melt temperature is required during processing to minimize the production of acetaldehyde, a substance with a fruity odor that can alter water taste. Paperboard–alufoil–plastic cartons are also used but on a limited scale because many consumers prefer to see the product.

A study[36] into the effects of ozonation on the taste of water packaged in HDPE and glass bottles found that there was a taste preference for the water stored in glass. The development of off-taste occurred least in glass containers and most often in HDPE containers with antioxidant BHT; off-taste was noted with increasing frequency as levels of BHT were decreased in HDPE. Moreover, softer water appeared to be less sensitive to development of off-taste than hard water.

III. COFFEE

A. Manufacture

The fruit (called a *cherry*) from trees of the genus *Coffea* (the two most important species are *C. arabica* and *C. robusta*) is a small pod that contains two coffee beans. The outer skin and pulp layers are removed using one of two processes: a "dry" method where the cherries are dried and then passed through a hulling machine, which removes the outer skin, dry fruit and parchment, and the "wet" method in which the beans are softened in water, depulped mechanically, fermented in large tanks to remove a mucilaginous parchment layer that encases the beans and then dried. Milling removes the parchment shell leaving a gray-green colored bean.

Three operations are needed to convert green coffee beans into a consumable beverage—roasting, grinding and brewing. Roasting develops the characteristic flavor and headspace aroma of coffee, while grinding is necessary so that both the soluble solids and volatile flavor substances can be sufficiently extracted by infusion or brewing with hot water to provide a beverage of required strength, either for immediate consumption or an extract for subsequent drying to make instant coffee.

Roasting of coffee beans involves rapid heating to raise the bean temperature to about 180°C. Roasting is normally carried out under atmospheric conditions with hot combustion gases and excess air and results in both chemical and physical changes in the green beans. The chemical changes give rise to the characteristic flavor and aroma of roasted coffee, and in addition, CO_2 is formed from decomposition of carbohydrates and other chemical reactions. These latter processes have important implications for the packaging stage. The roasted whole beans may be packaged directly or else ground first prior to packaging. To maximize the retention of flavor volatiles within the roasted and ground coffee, cryogenic grinding (i.e., at subzero temperatures with the use of CO_2 or liquid N_2) is recommended.

The bulk density of ground coffee is assessed in two ways—free fall or packed measurement. Values will differ according to the blend, degree of roasting (decreasing with increasing severity), degree of grinding (increasing with increasing fineness) and moisture content. A fine grind coffee may have a bulk density of 0.39 to 0.47 g mL^{-1} in free fall.[10] Bulk density is clearly of importance from a packaging point of view, because coffee is sold by weight and if the bulk density varies, then the headspace in the package will also vary.

Instant coffee is produced by extracting both soluble solids and volatile aroma/flavor compounds from roasted and ground coffee beans with water. Extracts of around 25% w/w soluble solids concentration are obtained, and these extracts are further concentrated using evaporation or freeze concentration and then dried using either spray drying or freeze drying. The latter is more common and is typically followed by agglomeration to produce granules of around 1400 μm on average. The agglomerates are then dried in a fluidized bed to the desired final moisture content. Freeze-dried coffee has a longer shelf life than spray-dried coffee, although the reason for this difference is not clear.

B. Deterioration

The major deteriorative reaction in coffee is *staling*, thought to result from a loss of flavor volatiles or chemical changes in the volatile components caused by moisture and O_2 absorption. The aroma degeneration during staling has been described[29] as changing from flat to old to sharply rancid, with a cocoa odor appearing in the advanced stage. Concurrently, the taste of the coffee changes from flat to bitter, old and rancid, at which stage it should no longer be sold. The extent of staling is increased by increased moisture content and holding at higher temperatures.

O_2 is believed to be absorbed by roast and ground coffee, which provides a reservoir for O_2 to cause subsequent deterioration. It has been observed that in the vacuum packing of roasted coffee, the in-package O_2 value may in fact rise after several days as a result of desorption into the vacuum space, followed by its fall again during storage. There is little information on the precise relationship between the in-package percentage O_2 content and stability for roast whole beans (RWB).[10] However, for roast and ground coffee, there is some evidence of a linear correlation between the logarithm of the O_2 content and the time to end of shelf life for a constant quality level. For example, for a closed package at 0.5% initial O_2 content by volume, stored at 21°C, the corresponding shelf lives were 6, 12 to 17 and 20 to 25 months for high, medium and low flavor quality criteria, respectively; at 1% initial O_2 content, the shelf lives were 4, 9 to 17 and 14 to 20 months, respectively.[27] Extrapolation to 21% O_2 gave a shelf life of 10 to 15 days for medium flavor quality. Differences would be expected for coffees of different grind and roast degree, and blend. Storage temperature would also have an important effect on shelf life.

Another study seeking to determine the shelf life of roast and ground coffee found that shelf life decreased with increase in O_2 partial pressure, a_w and temperature.[8] Oxygen had the greatest effect with an approximately 20-fold reduction in shelf life when the pressure was increased from 0.5 to 21.3 kPa. Increasing a_w by 0.1 led to a 60% decrease in shelf life, while a temperature increase of 10°C decreased shelf life from 15 to 23%. Storing the coffee in air at 4°C gave a 44% increase in shelf life compared with that at 22°C.

Roast coffee will also gradually release its volatiles to the atmosphere, and after a certain proportion of the volatiles have been released, the coffee is regarded as unacceptable.

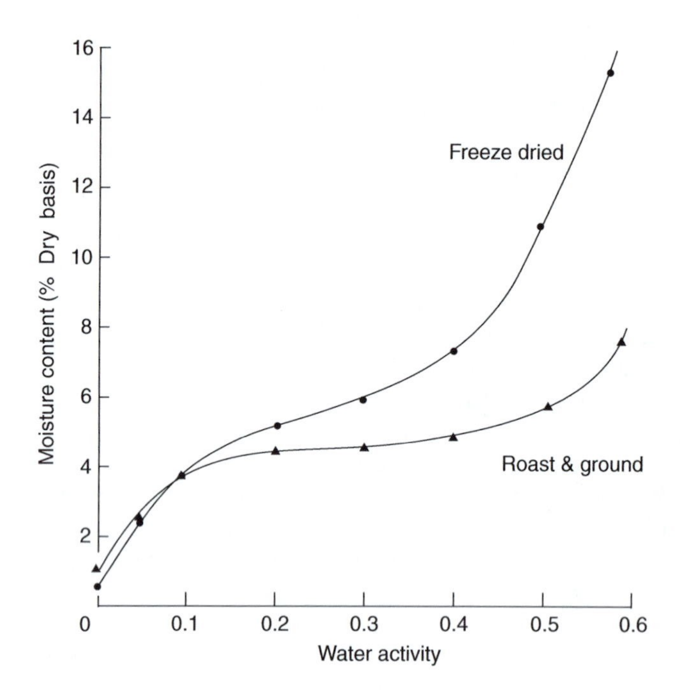

FIGURE 20.1 Typical moisture sorption isotherms for coffee.

Another reaction that is not strictly deterioration is the evolution of CO_2. During the roasting process, a large quantity of CO_2 is emitted both from the pyrolysis of the beans as well as the heating gases. The beans absorb a large amount of the CO_2 and part of this CO_2 is freed when the cells are crushed by grinding. During the first 24 hours after grinding, CO_2 is given off quite rapidly.[29] Fine grind coffee (500 to 600 μm) degasses more rapidly than coarse grind (800 to 1000 μm) coffee, and thus the latter presents more of a problem when vacuum packed.

The major deteriorative reaction in instant coffee is ingress of moisture vapor, resulting in caking when the moisture content reaches 7 to 8%. Instant coffee typically has a moisture content of 2 to 4% at the time of packing, which corresponds to a very low a_w (Figure 20.1). Some instant coffees have a surface application of coffee oils carrying aromatics, which enhance the headspace aroma, and these are susceptible to flavor deterioration from O_2 and moisture pickup.

C. Packaging

1. Roasted Whole Beans

Depending on the shelf life required, the choice of packaging material has to be considered with regard to moisture vapor ingress, O_2 permeability, CO_2 and volatile component egress, and grease resistance if oily, dark-roasted beans are being packaged. The major problem with the packaging of RWB is the evolution of CO_2.

The quantity of CO_2 entrapped in RWB is of the order of 2 to 5 mL g^{-1} (measured at STP) and probably higher, depending on the method of measurement.[10] Thus, when RWB are placed within a closed package, the CO_2 released will have to be contained within the available headspace within the package or, depending on the nature of the packaging material, permeate through the package.

Some useful calculations have been presented[10] to illustrate the magnitude of the packaging problem and assist in evaluating alternative packages. Assuming that 100 g of RWB are placed in a package, they will occupy about 330 mL, where about 150 mL is occupied by the beans themselves but only 75 mL by the solid material of the beans because their internal porosity is 50%. If it is further assumed that the internal volume of the package is 10% greater than that occupied by the beans, then the total package volume will be 366 mL, of which 366 − 150 = 216 mL will be available for takeup of released CO_2. Every 100 mL of CO_2 introduced inside the package will increase the total internal pressure by 100/216 = 0.46 atm. From 100 g of RWB, up to 500 mL of CO_2 could be released, generating a total internal pressure of 2.3 atm.

Five possible solutions to the problem created by CO_2 release from RWB have been suggested[10]:

1. Use a sufficiently rigid container to withstand the increase in internal pressure — this was considered to be impractical.
2. Use a packaging material either sufficiently permeable to CO_2 or fitted with a one-way valve.
3. Use a package under vacuum — this would not solve the problem because even if a full vacuum were obtained in the package, it might well be inadequate to cope with the quantity of CO_2 released.
4. Use a package with a very large headspace — this was judged to be commercially unacceptable.
5. Allow a substantial release of CO_2 by holding the beans before packing.

In Table 4.2, the permeability constants of plastic films was presented, and in all cases the permeability constant for CO_2 was greater than that for O_2 by, on average, a factor of 6. Thus, if plastic laminates are used to package RWB, then a satisfactory O_2 barrier can be obtained, which will permit some CO_2 egress. However, it is still necessary to reduce the initial O_2 content inside the package immediately prior to sealing, and this is best achieved by flushing of the package and contents with an inert gas, either CO_2 or more typically N_2.

For longer shelf life, nonreturn valves are popular. These consist of a plastic valve fitted to the outside of an otherwise impermeable package. At a preset pressure, the valve opens to release excess CO_2. Vacuum packaging is not often used, although there are such packs, also fitted with one-way valves, for RWB that have been degassed over a period of several days.[27] CO_2-absorbing sachets are also used.

2. Roasted and Ground Coffee

Substantial quantities of CO_2 are released on grinding, especially with the finer grinds. For example, estimates have been given that 45% is released within 5 min of grinding for a fine ground coffee while others have determined that 30% is lost within 5 min of grinding to an average particle size of 1000 μm and 70% at 500 μm.

O_2 is a prime determinant of shelf life and there are three main ways of lowering its concentration inside a package. The first method is to apply a high vacuum immediately after filling into the package and then sealing. The second is to flush the roast and ground coffee and package with an inert gas immediately prior to sealing. The third is to place an O_2-absorbing sachet inside the package.

When used by consumers, packages of roast and ground coffee will be opened and closed frequently; in such situations, the shelf life of the product is essentially that of an air pack.

The use of in-package percentage O_2 content figures shortly after packing is widespread, as is the use of in-package vacuum measurements by special gauges, although measurement of the weight of O_2 per unit weight of coffee is regarded as more satisfactory.[10] Such a maximum figure for satisfactory shelf life in a commercial pack might be 20 μg of O_2 g^{-1} of coffee (including that which has been absorbed), which corresponds to 1% O_2 content in a high-vacuum pack immediately after desorption. Another often quoted figure for the amount of O_2 required to "stale" roast coffee such that it is no longer acceptable is 14 mL lb^{-1} of coffee, which corresponds to 40 $\mu g\ g^{-1}$.[10] However, such figures do not indicate the rate of staling, which will obviously be strongly influenced by the storage temperature.

a. Metal Cans

The oldest type of commercial package for roast and ground coffee is the vacuum-packed tinplate can. It provides impermeability to moisture vapor, gases and volatiles, and can be made with a scored, detachable aluminum end over which a threaded screw cap is placed. After filling, a high vacuum is pulled on the can; this should be at least −95 kPa (713 mm Hg) to give an O_2 content of less than 1%. After sealing, evolution of CO_2 gas will reduce the vacuum until atmospheric pressure is restored. It is important that pressure does not build up inside the container during storage; to prevent this occurring, the roast and ground coffee must be degassed in bulk to reduce its CO_2 content to an acceptable level. The type of calculation required to ensure that pressure will not develop during storage of the packaged coffee has been presented elsewhere.[10]

Inert gas packing of roast and ground coffee in metal containers is also practiced. The usual procedure is first to apply a vacuum and then release it with an inert gas, usually N_2. In order to prevent excessive pressures developing in the can, it is necessary to degas the coffee to a much lower level than for high vacuum packing. Alternatively, a relatively low vacuum can be applied to the gas-purged can before finally sealing.[10]

b. Hard Packs

As an alternative to the metal can, so-called "hard" or "shape-retentive" packages of flexible laminated materials shaped into bags have been developed and widely used. These packs are called "hard" because on application of a high vacuum after filling and sealing, the material collapses onto the coffee to form a "brick," which is hard to the touch. However, if significant quantities of CO_2 are evolved or air enters from the atmosphere, then the bag will become "soft." This is an undesirable condition because consumers may erroneously perceive that the coffee has deteriorated, although this is not necessarily the case.

Many flexible laminates used for this type of package used to contain a central layer of aluminum foil but this has been replaced in many situations by a metallized layer. A typical early construction was 12 μm PET–12 μm Al foil–70 μm LDPE, while contemporary structures are metallized PET laminated to LDPE. Most U.S. coffee roasters employ a more durable four-ply structure, which utilizes BOPP or BON (polyamide) in addition to OPET, alufoil and LDPE.[3]

The internal pressure at which the packs will become noticeably soft is of the order of 0.5 to 0.75 atm.[10] This means that coffee that has been satisfactorily packaged in metal containers will need to be degassed to a greater extent before packing in laminate packs if a similar shelf life is desired. Degassing times range from 1 to 30 h, depending on the degree of roasting and the fineness of the grinding.

A recent development to avoid the problem of softening of the package or permit the packaging of freshly ground coffee is to include inside the package a sachet containing a sorbent for CO_2. The use of such sachets has now become quite widespread.

c. Soft Packs

These packs or pouches (sometimes referred to as *pillow packs*) are similar to those described above for the packaging of whole roast beans, and similar considerations apply. No vacuum is used and the O_2 content is reduced by flushing with an inert gas immediately prior to sealing. To provide an adequate shelf life, the packaging material must provide a good barrier to gases and water vapor.

Again, adequate degassing of the ground coffee must be carried out prior to packing to avoid pressure building up inside the package.

The use of sachets containing a sorbent for CO_2, as described above for hard packs, is also possible for soft packs. However, it has become more common to incorporate a one-way valve into the side of the package which opens and releases CO_2 when a certain internal pressure is exceeded. Once the internal pressure decreases to a set level, the valve closes. Three types of valve construction are available: a valve with a fully protruding cap, a valve with the cap half-protruding and a valve with the cap completely sunk into the base plate.[3]

3. Instant Coffee

In a review of the limited shelf life data and accompanying information on instant coffee, it was concluded[9] that, provided the moisture content was maintained at less than 4 to 5% w/w, coffee will retain its original quality for at least 2 years at ambient (temperate) conditions. With more sophisticated products (i.e., with higher retained levels of volatiles), a shelf life of at least 18 months is possible, provided that the initial O_2 headspace contents are lowered to less than 4.0%.

For many years, instant coffee for the retail market was packaged in either tinplate cans or glass jars of various shapes. A laminated paper or metal foil diaphragm sealed to the rim of the container provided an effective barrier to moisture vapor and O_2, over which a tight-fitting metal lid was placed in the case of tinplate containers, and a screw cap of plastic or metal in the case of glass jars. More recently, "refill" packs of instant coffee packaged in flexible laminates of PET–Al foil–LDPE or metallized PET–LDPE have reduced the quantity of coffee packaged in metal containers and glass jars. The contents of such packs after opening in the home are refilled into glass or metal containers, and adequate shelf lives are generally obtained.

IV. TEA

The appealing characteristics of tea as a beverage are its taste, aroma and color. Polyphenols (such as catechins) and amino acids (such as theanine) are the main contributors to the unique taste and color of tea. The components of essential oil in fresh tea leaves and volatile compounds developed during the manufacturing process form the characteristic tea flavor.[38]

Although many types of tea are available around the world, teas may be classified into three general categories according to the manufacturing process used: fermented (black tea), semifermented (oolong and pouchong) and nonfermented (green tea). The different manufacturing conditions result in differences in taste, aroma and color as well as in storage stability and shelf life.[38]

A. Manufacture

1. Black Tea

Black tea goes through four processes: withering, rolling, fermentation and firing. After picking, the leaves are placed on trays to wither or dry for about 18 h. Heat may be applied to reduce the time to 6 to 8 h. The leaves are then rolled, which distorts the shape and crushes the cell walls, releasing oxidizing enzymes. The leaves are then spread out and held at ambient temperature and saturated humidity for about 3 h. This is the fermentation stage and results in the leaves changing from green to copper-red, as well as developments in the flavor. The final stage is heating to 82 to 93°C to denature the enzymes and stop fermentation. The moisture content is reduced to about 3%.

Oolong and pouchong teas are semifermented teas, with the fermentation process typically lasting 5 to 6 h, after which the leaves are heated with agitation in a pan at 200°C until the desired final moisture content is reached.

2. Green Tea

After picking, the leaves are steamed (Japanese type) or panned (Chinese type) to inactivate polyphenoloxidase and other enzymes responsible for fermentation. The leaves are then rolled and dried by a process similar to that used for black tea.

B. Deterioration

1. Black Tea

During storage, black tea consumes O_2 and evolves CO_2, the latter process being mainly anaerobic. Deterioration of black tea is caused by losses of volatile components; changes in catechins, amino acids, theaflavins and other pigments; and increases in undesirable taints arising from oxidative reaction products from fatty acids and oxidation and condensation products from soluble polyphenols such as catechins and theaflavins. All of these reactions are accelerated by increases in tea moisture content, elevated temperature and exposure to light.[38] Although lipid oxidation is insignificant except under hot, dry conditions, oxidation of free fatty acids released during storage occurs during brewing and has a profound influence on the quality of the liquor.

It has been reported that at around 32% RH and 20°C under conditions excluding light, black tea could be stored for a period of 300 days without loss of tea character. Exclusion of light is clearly important because photooxidation of lipids and nonenzymic browning reactions, both of which contribute to quality loss in black tea, are accelerated by light.

2. Green Tea

Deterioration in the quality of green tea during storage is recognized by the following phenomena[38]: (1) reduction in ascorbic acid content; (2) change in color from bright green to olive green and then brownish green; (3) change in characteristic leafy and refreshing odor to dull and heavy odor; (4) changes from a well-balanced, complex taste to a flat taste lacking in characteristic briskness. All of these reactions are accelerated by moisture, O_2, elevated temperatures and exposure to light, much as in the case of black tea.

C. Packaging

Loose tea is packaged in a multitude of different shapes, sizes and types of materials, the most common being a paperboard carton with either an aluminum foil liner or an overwrap of PP or RCF. Metal containers with snap-on lids are also used for some premium products. Tea bags have now become the most popular form of retail packaging, and considerable development has gone into improving the tissue paper used for this type of package, with porous wet-strength paper being required. Once filled, the tea bags must be placed inside a package that provides an adequate barrier to moisture vapor. Paperboard cartons overwrapped with PP or RCF are most common.

Given that the initial moisture content of tea is 3 to 4%, a good moisture vapor barrier is required to prevent a relatively rapid increase in moisture content. However, even when tea has gained a large amount of moisture, the market value is similar to tea protected from moisture, leading to the conclusion that moisture content alone does not lead to deterioration of tea. A reduction in the levels of various chemical constituents is more likely to contribute to the end of shelf life of tea than an increase in moisture content. The volatile fraction of black tea shows an overall decline during storage that is accelerated by moisture uptake and, to some extent, storage at elevated temperatures.

The storage stability of green tea is the lowest among various teas including black tea, oolong tea and pouchong tea. To most effectively protect the quality of green tea during storage, it is

TABLE 20.1
Changes in Moisture Content and Ascorbic Acid during Storage at 25°C and 80% RH of Green Tea in Pouches Made of Various Materials

Material of Pouch	% Moisture Content			% Loss of Ascorbic		
	1	2 (in months)	3	1	2 (in months)	3
RCF–9 μm Al–40 μm LDPE	3.5	3.9	4.0	7.4	6.7	8.8
RCF–112 μm Al–40 μm LDPE	3.6	3.2	3.5	1.4	5.0	8.6
RCF–15 μm Al–40 μm LDPE	3.2	3.2	3.2	4.6	5.6	8.7
20 μm OPP–2 μm PVdC copolymer–60 μm LDPE	4.4	5.4	6.0	8.1	20.5	32.1
RCF–13 μm LDPE–40 gsm paper–20 μm LDPE	6.9	8.8	10.8	27.9	42.1	69.4

necessary to use nitrogen gas flushing or vacuum packaging.[38] Changes in the moisture content and ascorbic acid content of green tea during storage in pouches made of various materials are shown in Table 20.1.

V. JUICES

A. Manufacture

Although fruit juices were originally developed to use up the surplus fresh fruit production, fruit (in particular citrus and apple) is now specially grown for juicing. A variety of different types of juice is available. These include clear clarified juices such as grape, apple and blackcurrant; light cloud juices such as pineapple; heavy cloud juices containing cellular material in suspension such as orange and grapefruit juices; pulpy juices such as tomato-based products; and nectars made by pulping whole fruits like peaches, apricots and comminuted citrus products.

The quality of a juice depends essentially on the species and maturity of the fresh fruit. The main factors that influence the quality are the sugar-to-acid ratio, the aroma volatiles, the phenolic components and the ascorbic acid content. Satisfactory fruit juice production depends on sound judgment of the raw materials and blending procedures adopted. Details of the juice extraction and subsequent processing operations used are outside the scope of this book because they tend to be specific for the different types of juices; the reader is referred to standard texts on the subject for more specific information.[5,6]

A key step in the processing of fruit juices from the packaging point of view is the *deaeration* step. This is important both to minimize oxidative reactions in the juice (e.g., oxidation of ascorbic acid and flavor compounds) and reduce corrosion if the juice is subsequently packaged in a metal container.

B. Deterioration

The four key deteriorative reactions in juices are microbiological spoilage, nonenzymic browning, oxidation resulting in loss or degradation of flavor components and nutrients (essentially ascorbic acid), and absorption of flavor compounds by the package. Although preservatives were commonly added to fruit juices to overcome microbiological problems, the recent consumer preference for preservative-free foods has seen their use diminish. Instead, attention to good manufacturing practice in the plant coupled, in many cases, with aseptic processing and packaging has obviated the need for them.

The rate of browning and nutrient degradation in fruit juices is largely a function of storage temperature, although the rate is in part dependent on the packaging material. For example, a study[22] comparing the quality of citrus juices aseptically packaged into laminated cartons with packaging in glass containers found that the extent of browning and loss of ascorbic acid was greater in cartons than in glass, presumably because of O_2 permeation into the carton.

Pasteurization was originally used as a means of controlling microflora. The acid conditions that prevail in most fruit juices will not support pathogens and tend to inhibit organisms in general, although acid-tolerant types may germinate and cause spoilage. The low pH conditions assist pasteurization so that 80°C for 30 sec is adequate for virtually all juices except for the less acid fruits such as apricot and tomato.

Pasteurization is also important for stabilizing the cloud of certain juices (typically orange, grapefruit and tomato) where consumers regard clarified juices as inferior and unacceptable. Such a reaction is based largely on tradition because lemon, lime and apple juices are typically preferred in the clarified state but can be (and sometimes are) made to have a stable cloud. The major enzyme responsible for destabilizing the cloud is pectinmethylesterase and it must be inactivated as soon as possible after extraction of the juice. This is generally done by pasteurizing the juice at 90 to 95°C for 15 to 30 sec, where the precise time depends on the pulp content.

The oil fraction of citrus juices contains many volatiles, which have a major impact on citrus aroma and flavor. These oil-based flavor compounds are relatively easily oxidized resulting in the development of undesirable, terpene-like off-flavors. Removal of O_2 from the juice prior to packaging and avoidance of high pressures during juice extraction so as to limit oil transfer to the juice minimize this form of flavor deterioration.

C. Packaging

From a packaging point of view, there are three categories of juices: single strength (10 to 13°Brix), concentrated juices (42 or 65°Brix) and nectars (20 to 35°Brix).

The traditional packaging procedure for single-strength juices involved heating the deaerated juice to around 90 to 95°C in a tubular or plate heat exchanger, filling the hot juice directly into metal cans, sealing and inverting the cans, holding them for 10 to 20 min and then cooling. This hot-fill/hold/cool process ensured that the juice was commercially sterile, and provided that the seams were of good quality, the cans had an acid-resistant enamel coating, and the juice had been properly deaerated, a shelf life of at least 1 to 2 years was attainable. However, because of the acidic nature of fruit juices, any imperfections or scratches in the enamel coating or tin layer resulted in rapid corrosion, dissolution of metal into the juice, production of hydrogen gas and container failure owing to swelling. The use of glass containers obviated these problems provided that the container closure (typically metal) was resistant to attack by the juice.

The use of glass bottles for the packaging of fruit juices is also widespread, although the hot-fill/hold/cool process has to be applied with care to avoid breakage of the glass containers. Glass is still the preferred packaging medium for high-quality fruit juices. However, over recent years, an increasing proportion of fruit juices and concentrates has been packaged aseptically, generally into laminates of plastic/alufoil/paperboard. These products are then held at room temperature and the shelf life and nutrient composition are greatly influenced by the barrier properties of the carton, the interactions of the juice with the carton and the storage environment.[30] The end of shelf life is typically 6 to 8 months and is related to the extent of nonenzymic browning and the sorption of key aroma and flavor compounds by the plastic in contact with the juice, the latter process being referred to as "scalping."[24] Because of its lipophilic nature, the oil fraction of citrus juices will be absorbed into many nonpolar packaging polymers.[32]

Flexible packaging is also used for juices and especially sports drinks. Two formats are common: the Doy pack is a stand-up pouch constructed (from inside to out) of LDPE–alufoil–PET. The CheerPack was developed in Japan during the 1980s, and is made up of four panels or sections combined to form a stand-up pack with two side gussets. A variety of laminate constructions are available but, for beverages, the most common structure (from inside to out) is LDPE–PET–alufoil–PET. For specific applications, EVOH, OPA or PP can be included in the structure. A HDPE neck and "straw" is sealed into the top portion of the pack, which is filled through the neck and then sealed by a tamper-evident closure. The packs can be cold or hot filled (up to 95°C) and pasteurized after filling if required.[31]

Recent developments in barrier coatings for PET have led to increasing use of PET bottles for fruit juices and drinks, and this trend is likely to accelerate as production ramps up and costs come down. There is still a dearth of independent scientific publications on the performance of the various barrier coatings with respect to gas transfer and shelf life.

In a study[22] comparing the quality of citrus juices aseptically packaged into laminated cartons and glass containers, the *d*-limonene content of the juices in the cartons was reduced by about 25% within 14 days of storage owing to absorption by the polyethylene, and sensory evaluation showed a significant difference after 10 to 12 weeks between juices packaged in glass and cartons stored at ambient temperatures. In contrast, another study[26] reported that an experienced panel did not distinguish between orange juice stored in glass bottles and that stored in laminated aseptic cartons. Absorption of up to 50% limonene and other hydrocarbons, small quantities of ketones and aldehydes had no significant influence on the sensory quality of juice stored at 4°C.

The presence of pulp particles in orange juice decreased absorption of volatile compounds into polymeric packaging materials.[37] The authors suggested that pulp particles hold flavor compounds such as limonene in equilibrium with the aqueous phase, which could be responsible for the decreased absorption of these compounds by the plastics.

Another study[17] determined the amount of *d*-limonene sorbed by three different films as a function of storage time, with the amount sorbed varying with the polymer as shown in Figure 20.2. After 3 days, sorption by LDPE and EVOH plateaued and reached equilibrium, while for Co-PET (a co-polyester developmental film) a slow increase was observed for 24 days. Data on the permeation of three apple aromas in polymer films have also been reported[12]; LDPE was found to be a poor barrier, while PVdC copolymer and EVOH copolymer were found to be excellent barriers, the performance of the latter deteriorating under high humidity conditions.

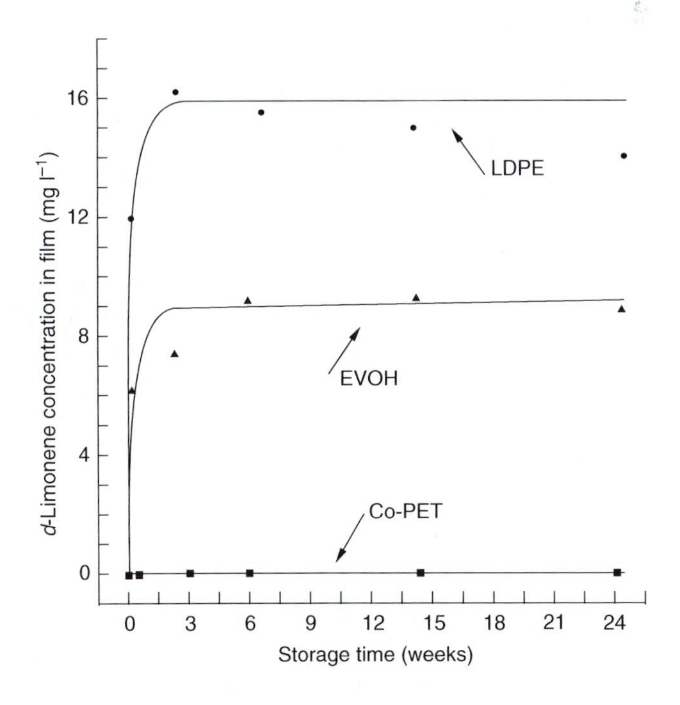

FIGURE 20.2 Sorption of *d*-limonene by LDPE, EVOH and Co-PET. (*Source*: From Imai, T., Harte, B. R., and Giacin, J. R., *J. Food Sci.*, 55, 158–161, 1990. With permission.)

As well as loss of aroma, sorption of organic molecules can also affect the mechanical properties of the film and increase its O_2 permeability. In a study involving the sorption of *d*-limonene by LDPE and ionomer films, rapid absorption was observed with saturation (around 44% of the initial concentration) being reached after 12 days. There was a reduction in seal and tensile strengths, and an increase in O_2 permeability, of two to four times.[15]

Polyesters such as PET, PEN and PC have a more polar character than the polyolefins and therefore show less affinity to the common flavor compounds; that is, they absorb fewer flavor compounds. A recent review[20] concluded that although packaging and flavor interactions exist, they do not influence food quality to the extent that they cause insuperable problems in practical situations. This is evident from the fact that packaging materials in which polyolefins are in contact with juices are widely used commercially.

In the U.S., the production of frozen concentrated orange juice (FCOJ) has become a huge industry. The 42°Brix juice is usually held at -12°C at which temperature it is still liquid. Typical packaging materials for this product consist of spiral wound paperboard tubes with aluminum ends or aluminum cans.

VI. CARBONATED SOFT DRINKS

A. Manufacture

Traditionally, soft drinks were prepared by dissolving granulated sugar in specially treated water or, alternatively, by diluting liquid sugar with this water. A variety of ingredients including flavoring and coloring agents, acidulants (invariably either citric or phosphoric acid) and preservatives were then added. Other constituents such as fruit juice or comminuted fruit, bodying agents, artificial sweeteners, clouding agents, antioxidants and foaming agents were added, depending on the particular product being made. Recently "diet" soft drinks in which the sugar has been replaced with an artificial sweetener (typically aspartame) have become very popular.

Soft drinks are now prepared almost exclusively using the premix system whereby the blended syrup, after flash pasteurization if necessary, is mixed with carbonated, treated water prior to delivery to the filler. Although traditionally the product has been cooled to 1 to 3°C before arrival at the filler in order to minimize loss of carbonation and facilitate filling, fillers and ancillary equipment capable of handling the product at ambient temperatures have recently been introduced.

The degree of carbonation of soft drinks is typically expressed in volumes or $g\ L^{-1}$ of CO_2. One volume equals approximately $2\ g\ L^{-1}$, and at room temperature, each volume produces about one atmosphere (101 kPa) of internal pressure. Temperature has a significant effect on internal pressure, a 4 volume beverage such as a cola rising to 7 atm at 38°C and to 10 atm at maximum storage/pasteurization temperatures.[1] The carbonation level of beverages ranges from 1.5 volumes for citrus and other fruit-based soft drinks to 4 volumes for common cola drinks and 5 volumes for club soda and ginger ale.[1]

B. Deterioration

The two major deteriorative reactions in carbonated beverages are loss of carbonation and oxidation or acid hydrolysis of the essential flavor oils. The first is largely a function of the effectiveness of the package in providing a barrier to gas permeation, while the latter can be largely prevented by the use of high-quality flavorings and antioxidants, and deaerating the mix prior to carbonation.

C. Packaging

1. Glass

From the beginning of the twentieth century, virtually all carbonated soft drinks were packaged in refillable glass bottles, which were sealed with crown cork closures (see Chapter 8 VI.C.1.a). In recent years, nonreturnable glass bottles have replaced refillable glass bottles in many markets. These sometimes have a foam plastic protective label or a paper/polyolefin or all-plastic shrink sleeve, in part as a safety measure to prevent flying glass fragments should the bottle break (see Chapter 9 V.D for more details). On nonreturnable bottles, the crown cork closure is replaced with a roll-on aluminum screw cap on threaded necks with a tamper-evident ring, or a plastic closure (typically PP with or without a liner), which fits and unscrews over the same threads as the roll-on and provides some visible indication of tampering.[1,2]

2. Metal

Three-piece tinplate containers were used for many years for the packaging of carbonated beverages. The highly corrosive nature of carbonated soft drinks (sulfur dioxide should not be used in canned soft drinks packaged in metal cans because it acts as a depolarizer or corrosion promoter) demanded complete protection of the metal container from the product by the use of one or more coatings of an impermeable enamel system. For three-piece cans, this involved spraying an additional coating of enamel (a process known as *sidestriping*) over the inside of the container down the sideseam area after soldering or welding of the sideseam.

Today, most carbonated beverages are packaged in two-piece containers usually made from aluminum. The two-piece container has made it much easier to retain the integrity of the enamel layer inside the can and thus minimize corrosion during storage. The cans must be able to withstand continual internal pressure of up to 5 atm.

Container weight and design have been the subject of much development work, with the weight of a modern 375 mL can (including the end) now less than 17 g. Quadruple necking is quite common on many aluminum beverage cans, and this has resulted in further cost savings in the reduced aluminum required for the end, and smaller overall pack area. Recently, a single-stage process of spin-necking has been introduced, which results in a smooth, conical-shaped top section.

3. Plastics

As early as the 1960s, the Coca-Cola™ and Pepsi-Cola™ companies were considering the use of plastic bottles for soft drinks, and they began to develop their ideas with major polymer manufacturers in the U.S. It soon became apparent that only the polyester and nitrile families of plastics had the necessary physical and chemical characteristics required. PET was the preferred polyester, while acrylonitrile/methylmethacrylate copolymer, methacrylonitrile/styrene copolymer and rubber-modified acrylonitrile/styrene copolymer were also suitable.[33]

Because the nitrile plastics could be made into bottles using existing blow-molding equipment, while PET could not because of its inclination to crystallize and go hazy at higher temperatures, early market development work in the 1970s was carried out with nitrile bottles. Coca-Cola successfully launched a 950 mL nitrile bottle in 1975, but the release of toxicological data in 1977 showing that AN monomer could be carcinogenic at high dosage led to the removal of the nitrile bottle from the market.[33]

Meanwhile, attempts to successfully manufacture PET bottles using a stretch-blow molding process were continuing. In the spring of 1977, the plastic PET bottle for soft drinks was launched by Pepsi-Cola, followed soon after by Coca-Cola and other beverage producers. It has been described as probably the biggest single development in the soft drinks industry since the introduction of the ring-pull can a decade earlier.[33] Today, the greatest volume of soft drinks is packaged in PET bottles, which have achieved their market share mainly at the expense of glass, albeit in an enlarged total market. The early designs had a round base that necessitated a flat base cup, usually injection molded from HDPE and fixed to the bottle with hot-melt adhesives. The increase of PET recycling schemes accelerated the introduction of the petalloid base in which the bottle is formed with four or five extrusions which form feet, thus obviating the need for a base cup, which had disappeared by 1992.[14] Although more material is required for a petalloid base, separation of the base cap is not required prior to recycling of the PET bottle.

The continuing trend for larger and larger containers of soft drink has helped penetration of the PET bottle. The 1 L glass bottle is considered to be near the limit of size and weight, above which it becomes difficult to handle easily, particularly by children. In contrast, PET bottles up to 5 L in size are now available, resulting in considerable savings in container cost per unit volume. In addition, the larger the bottle, the more CO_2 is retained per unit of time because of a smaller surface area to volume ratio (i.e., a reduced area for permeation).

The factors that influence the taste and odor of carbonated beverages packaged in plastic containers are depicted schematically in Figure 20.3. Most of these factors are applicable to many other foods or beverages packaged in plastic containers. Although O_2 pickup is a very critical item with beer (see next section), soft drinks are much less sensitive and a maximum ingress of 20 ppm for citrus-flavored soft drinks and 40 ppm for cola drinks has been suggested, along with a limit for

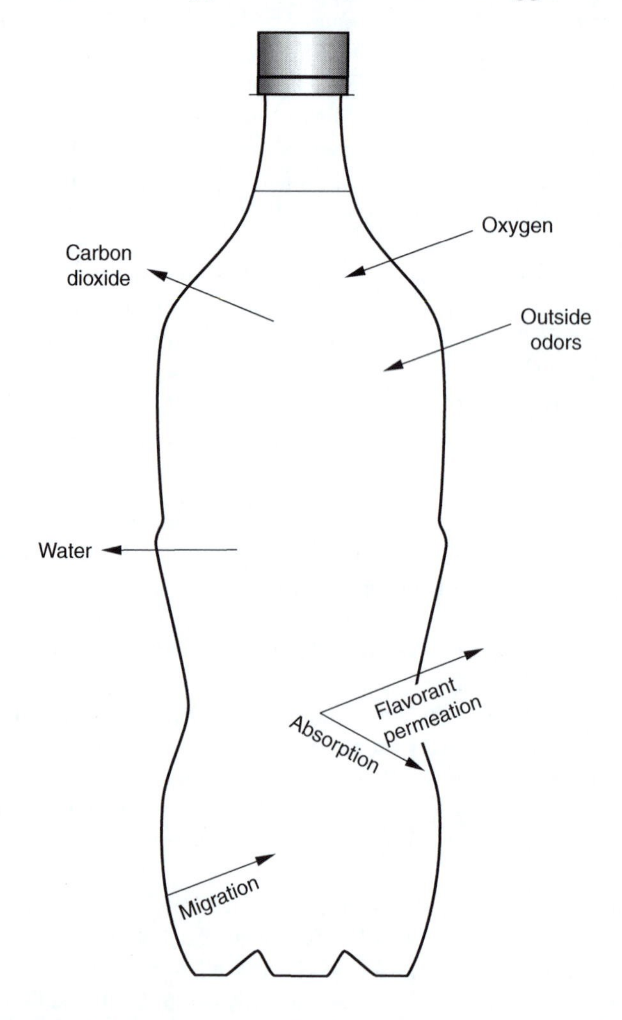

FIGURE 20.3 Factors influencing taste and odor of carbonated beverages packaged in plastic containers.

water loss of 1%, the latter based primarily on maintaining labeled contents. The criteria for flavor absorption and permeation losses are necessarily general because of the great variety of flavorants and wide differences in sensory effects.

Compared with glass, there is a loss of CO_2 through the bottle walls which must be allowed for. While increasing the wall thickness will decrease the rate of CO_2 transmission, this will also increase the cost of the bottle and so a compromise has been sought. In addition, PET (like most thermoplastics) exhibits the phenomenon of creep, which occurs to the greatest extent in the first few days after the bottles have been filled. In addition, there will be an elastic deformation which, unlike the creep effect, will not be permanent but rather will disappear when the bottle is opened and the internal pressure is released. In combination, both these effects will produce an increase of about 2.5% in the volume of the bottle over the first 3 to 4 days after filling, given normal storage conditions.

A typical carbonation loss for 2 L containers shows an initial loss of about 0.3 volumes in 3 to 4 days from the initial 4 volume figure, owing to the volume increase and absorption of CO_2 by the container wall. Thereafter, the rate of CO_2 loss becomes steady at about 0.04 volumes per week, giving a carbonation value around 3.4 after 12 weeks. The maximum shelf life for a 1.5 to 2 L bottle is around 16 to 17 weeks by which time the carbonation will have dropped to a still acceptable level of 3.1 to 3.2 volumes.[33]

The PET bottle is usually fitted with either a standard aluminum roll-on closure or a prethreaded plastic cap, both either in the standard or pilfer-proof form. The cap is ideal for use with PET bottles because its main sealing surface is on the inside bore of the neck finish, and this is very precisely controlled with regard to diameter and smooth surface finish and has almost nil ovality.[33]

VII. BEER

A. Manufacture

Beer is an alcoholic beverage made by brewing and fermentation from cereals (usually malted barley) and flavored with hops to give a bitter flavor. Two major technological steps are required to complete the transformation of the raw material into the finished product: controlled germination (i.e., malting), which allows the ultimate production of a fermentable extract through the activities of enzymes formed during seedling growth, and fermentation.

Malted barley is crushed and "mashed" by mixing with water at temperatures of up to 67°C, resulting in rapid degradation of solubilized starch and less extensive hydrolysis of other high MW substances. This is followed by a leaching process, which completes the separation of the solutes from the spent grains. The resulting "sweet wort" is boiled in a copper kettle together with hops, after which the spent hops and precipitated high MW material are removed. The specific gravity of the wort rises as increasing amounts of sugars are extracted. After adjustment to the desired specific gravity, a selected strain of yeast is added. As the yeast grows and ferments the sugars of the wort, the specific gravity falls.

At the end of fermentation, the yeast is separated from the immature or "green" beer, which is allowed to mature for an appropriate period. Before it is ready for consumption, freshly fermented beer must undergo a number of changes, including the elimination of certain fermentation products, supersaturation with CO_2, separation of yeast cells, and removal of some of the polyphenolic and other material, which will eventually give rise to turbidity in the beer. After centrifugation and filtration to remove residual yeast and assorted detritus from conditioned beer, followed by pasteurization and possibly other treatments, the beer is ready for packaging.[19]

The most common procedure for bulk beer is flash pasteurization where the beer is held at 70°C for approximately 20 sec before cooling and packaging. Tunnel pasteurizers are used with bottled and canned beer where the containers travel through an enclosed area in which they are sprayed with hot water, left for an appropriate time and then treated with a cold water spray.[21]

B. Deterioration

Owing to its low pH (about 4.0), microbial degradation is not usually a problem with beer, and the use of pasteurization and aseptic cold filtration excludes wild (i.e., noncultured) yeasts that could thrive. However, during storage, beer can undergo irreversible changes leading to the appearance of haze, the development of off-flavors and increased color. One of the major oxidative reactions is the oxidation of linoleic acid (introduced into the wort from the malt) to yield 2-trans-nonenal, which gives beer a "cardboard-like" flavor; it has a flavor threshold of as little as 0.1 ppb.[23] Flavor loss is accelerated in the presence of light and certain metal ions.

Because the fermentation process itself consumes O_2, brewing reduces the level of O_2 in beer down to 40 to 50 ppb prior to packaging. However, during the packaging process, atmospheric O_2 enters the package and the level of O_2 contamination reaches 250 to 500 ppb, which corresponds to 0.1 to 0.2 mL of O_2 per 355 mL bottle or can. This results in a shelf life for the beer of 80 to 120 days and even then significant flavor deterioration is likely to occur. Only 30 days after packaging, a discerning beer drinker can tell the difference between a freshly packaged beer and one which has been packaged for 30 days. Thus, brewers would like to be able to reduce the O_2 level in beer to less than 50 ppb immediately after packaging and hold it at that level for as long as possible.

The O_2 consumption of beer varies with the composition of the beer, its age, the presence of reducing agents, temperature and so on, and has been reported as ranging from 0.2 to 0.5 mL L^{-1}.[23] So-called "heavy" beers (i.e., those containing a higher protein and carbohydrate content) can absorb more O_2 before flavor degradation is noticeable.

One method by which flavor compounds can be depleted in beer is from sorption by the packaging material. Some of the more nonpolar flavor compounds in beer are more soluble in soft plastics such as the PVC liners in crown corks than in water. The degree of partitioning into such materials is a function of the polarity of the molecule in question, the volume and chemical composition of the plastic or other lining material, the volume and alcohol content of the beer, and the temperature. For many bottled beers, loss of hop flavor to packaging material occurs and may be a major source of hop flavor loss.

C. Packaging

1. Glass

The traditional packaging media for beer is the glass bottle sealed with a crown cork (see Chapter 8 VI.C.1.a). Pasteurization of the beer in the bottle after sealing is the most common means of securing microbiological stability. The aim is to heat the beer to a high enough temperature and hold it there long enough to destroy any beer spoilage organisms. The brewing industry has developed its own standard measure of the effectiveness of the pasteurization process and uses the term *pasteurization units* (PUs) where one PU is equivalent to holding beer at 60°C for 1 min. About 10 PUs is regarded as a suitable heat treatment for most bottled beers produced under good manufacturing practices.

The crown cork closure is made of tinplate or ECCS and contains a compressible lining material, the composition of which has changed over the years from solid cork to composition cork, plastic and aluminum foil in various combinations. Today, the use of cork-based linings is relatively rare, and most are lined with PVC (often foamed) or sometimes HDPE. Where cork is still used, it is common to laminate it to aluminum foil to improve its barrier properties.

The material properties as well as the shape of the lining have a great effect on the rate of O_2 permeation through crown cork linings. Linings with aluminum foil provide a perfect barrier. The total ingress of O_2 through seals ranges from 0.58 to 1.2 μL day^{-1}. Although PVC linings are normally foamed to different degrees, they appear almost solid after closing and permeation rates are almost identical. Ingress of O_2 does not appear to be measurably influenced by the condition of the glass sealing surface or the amount of pressure used in applying the crown to the bottle.

For a typical brewer using a standard crown closure on a 355 mL bottle, 0.013 mL of O_2 per bottle is dissolved in the beer at the filler head; 0.106 mL per bottle is entrapped in the bottle during filling; and 0.180 mL per bottle leaks in through the crown seal over 3 months at 0.002 mL day^{-1}, making a total O_2 exposure over 3 months of 0.299 mL per bottle.

In an attempt to extend the shelf life of bottled beer, O_2-absorbing materials have been incorporated into the lining material of crown corks to remove O_2 present in the headspace at the time of filling, as well as absorbing O_2 permeating into the bottle through the crown closure. The linings containing the O_2 absorber can be used in metal crowns, as well as roll-on metal and screw-on plastic caps. They also find limited application in the fruit juice and carbonated beverage industries.

2. Metal

The first successful canning of beer took place in 1933 in Newark, New Jersey when 2000 cans were produced for a test market. In January 1935, the first beer cans went on sale — Kruegers Finest Beer and Kruegers Cream Ale. By the end of 1935, no less than 36 American breweries were canning beer.

The greatest problem with using cans for beer packaging was preventing the pickup of metal ions from the tinplated steel and lead solder of the early three-piece cans. The metal ions resulted in undesirable metallic flavors and the rapid onset of haze or "metal turbidity," as it was called. Although several reasonably successful enamel coatings were developed and used, it was not until the development of epoxy-phenolic resin linings around 1960 that a truly effective lining capable of eliminating metal pickup over a long term was found. These linings resulted in less than 0.3 ppm iron pickup over a 6-month period.

Until the late 1950s, tinplate was used almost exclusively. However, three-piece aluminum cans were launched in 1958 by Primo of Hawaii and in 1959 by Coors. This was followed in 1963 by some test markets of two-piece D&I aluminum cans. This latter type of can began to be mass produced in 1966.

Aluminum was used for beer can ends by Schlitz in 1960 when the "soft top" can was introduced to facilitate opening with a special tool known as a "church key." A detachable pull-tab aluminum end was developed in 1962 by Ermal Fraze of Dayton, Ohio (see Chapter 7 V.A). It was test marketed and emerged in 1965 as the ring-pull tab. However, because it was detachable, it resulted in litter and so, in 1975, a patent was granted for a can end with an inseparable tear strip that is the familiar stay-on tab in use today.

The use of foam-producing widgets, originally developed for stout beer packaged in metal cans, has been described in Chapter 14 III.E.

3. Plastics

The use of PVdC copolymer-coated PET bottles for the packaging of beer commenced in the early 1980's in the U.K., but was not adopted very widely by brewers in other countries. The PVdC copolymer coating was necessary to provide an acceptable barrier to oxidation and to prevent flavor degradation in the beer. It also lowered the CO_2 permeability of the bottle. From a consideration of surface area to volume relationships and the O_2 permeability of the bottle material, most brewers tended to use larger sized bottles (typically 2 L) in order to obtain a satisfactory shelf life. A clean filling technique was used to handle the sterile filtered beer. This bottle never achieved widespread commercialization and alternative barrier coatings were developed.

Today, a small but increasing quantity of beer is packaged in PET bottles with an O_2 barrier to provide acceptable shelf life. A variety of barriers are offered commercially including amorphous carbon on the inside, SiO_x on either the inside or outside, an epoxy-amine organic coating on the outside and O_2-absorbing compounds as the middle layer in the bottle wall. Shelf lives of up to 9 months are now being claimed with some of these bottles.

VIII. WINE

A. Classification

Wine is a beverage resulting from the fermentation by yeasts of the juice of the grape with appropriate processing and additions. Wines can be assigned to one or the other of two major groups—table wines and fortified or appetizer wines. In the first group, the fermentable sugar has been largely consumed so that further fermentation by yeasts is prevented; spoilage organisms should not develop if the wine is kept anaerobic. Owing to the natural limit of sugar present in wine grapes, the alcohol content of such wines is normally about 12%. The major subclass of these wines is the still wines, which have no evident CO_2. Further subdivision depends on color and residual sugar. It is possible to produce microbiologically stable wines with a moderate or high amount of residual sugar. The other major subclass of table wines is the sparkling wines. These have undergone two complete fermentations, where the CO_2 of the second fermentation is retained. Sometimes, cheap sparkling wines are produced by carbonation of still wines.

The second major group includes the dessert and appetizer wines which contain 15 to 21% alcohol. The nature and keeping quality of these wines depends heavily on the addition of spirits distilled from wine, the amount of extra alcohol added being sufficient to prevent spoilage by further growth of yeasts or other micro-organisms. There are three major subclasses in this group: sweet wines in which the addition of wine spirits serves to arrest the yeast fermentation with part of the sugar unfermented; sherries, which may be sweet or dry and are characterized by flavors induced by various types and degrees of oxidation; and flavored wines such as vermouths and trade-named proprietary specialty wines.

B. Winemaking

After harvesting at the correct stage of maturity, the grapes are destemed, crushed and sulfur dioxide is added to inhibit the growth of undesirable micro-organisms, inhibit browning enzymes and serve as an antioxidant. If white wine is being made, then the next step is the separation of the fluid must from the pomace (skins, seeds and part of the pulp). This is usually achieved by holding the crushed grapes in a large tank fitted with a screen through which free-run juice can pass. Juice is removed from the remaining material using a press. If red wine is being made, then the separation occurs during fermentation; after the fermented juice has been run off, the remaining material is pressed to increase the yield and concentration of extractives.

Fermentation commonly occurs in large tanks in which the must is inoculated with selected yeasts. After fermentation is complete, the wine is allowed to settle so that yeast cells and other fine suspended materials (known as *sediment* or *lees*) can collect at the bottom of the container. The relatively clear wine is then racked off its lees and stored in tanks or casks. Several additional rackings, along with finings, filtrations and stabilization treatments, may be required to produce wine that is and will stay clear until consumed. It is usual to age red (and some white) wines in wooden barrels for varying periods of time (up to 2 years in some instances) prior to blending and filling into the final package.[7]

Sparkling wines are made by fermenting for a second time wine that has been stabilized and fined beforehand. Yeast is added, together with fermentable sugar (typically sucrose), a rule of thumb being that 4 g of sugar per liter of wine will produce one atmosphere of CO_2 pressure on fermentation; a final pressure of 6 atm is usually desired. For bottle-fermented sparkling wine, the wine is filled into champagne-type bottles, which are sealed with large wired- or clamped-on corks especially made for sparkling wines, or with plastic-lined crown cork closures. The sealed bottles are then stacked horizontally in a cool place to ferment (a process that typically takes several weeks) and left undisturbed for 1 to 4 years to allow the wine to age.

The bottles are then placed on racks and a process (known as *riddling*) begins to remove the yeast sediment. By a process of turning and tilting each bottle, the sediment is moved into the mouth of the bottle. Following this, the neck of the bottle is placed in a bath at − 15°C, which results in the formation of an ice plug containing the sediment. The bottle is then turned about 45° from the vertical and opened, whereupon the ice plug is pushed out by gas pressure. The *dosage* (syrup to adjust the sweetness of the wine) is then added, the level of wine is replenished if necessary and the bottle recorked with its final natural cork or plastic stopper, which is wired in place. Today, automatic equipment is used to perform these tasks.

C. Deterioration

The major deteriorative reaction in white wines is caused primarily by oxidation, the O_2 gradually changing the wine's character, leading to the development of browning and undesirable flavors. With red wines, the situation is more complicated and involves condensation polymerization reactions between tannins and anthocyanins, resulting in loss of pigmentation and color changes.

D. Packaging

Immediately prior to packaging, a small amount of sulfur dioxide may be added to the wine, which is then given a final filtration. The potentially harmful effects of oxidation by air in the headspace of the package and dissolved in the wine are minimized or prevented by various procedures. The wine may be stripped of dissolved O_2 by passing bubbles of N_2 through it, and the package may be purged with N_2, Ar or CO_2 and filled so as to displace the gas without introducing air. Other antioxidants (such as ascorbic acid) or other microbial inhibitors may be used, and the wine may be pasteurized or filtered through a membrane so as to be effectively sterile when it is packaged.

1. Glass

The most common form of packaging used for wines is the glass bottle sealed with natural cork. Cork is the outer bark of the holm oak, an evergreen species *Quercus suber*, which grows mainly in Spain and Portugal. Cork is tough, light and elastic, and is typically sterilized by exposure of damp corks in sealed plastic bags to SO_2, which seems to inactivate any micro-organisms of concern.[7] However, this treatment has no influence over the development of cork taint. A variety of alternatives to the traditional cork closure have been developed and these were discussed in Chapter 8 VI.C.2.

Because many types of wine are damaged by sunlight, the bottles are usually of colored glass, commonly dark green or brown. Bottled wine is normally stored in the horizontal position so that the cork is kept moist, thereby providing a better barrier to the ingress of O_2.

For approximately 120 years, bottles containing high-quality wines had a protective top capsule of tin–lead applied as a closure around the cork. The tin–lead foil comprised a thin layer of lead sandwiched between much thinner layers of tin roll-bonded to both sides; the tin content was approximately 1.5% w/w. The purpose of the tin coating was to prevent contact between the lead and the wine, and to provide a good surface for high quality decoration, including printing and embossing.

In recent years, concern has been expressed in various countries about the use of lead in capsules because of its known toxic effects (see Chapter 21). Contamination of wine by lead is known to occur sometimes as a result of capsule corrosion and the formation of soluble lead salts; this may have various causes, including tin coatings of inadequate thickness that are discontinuous or become disrupted. The disposal of lead-based capsules represents an avoidable burden of a toxic metal into the waste stream, and the use of lead-based capsules was phased out in most countries in 1993.

A replacement capsule made solely from tin has been developed; it weighs 5 g (half the weight of the tin–lead capsule) but costs considerably more. Research is continuing into capsule materials of tin or tin with some alloying additions in order to improve the cost-effectiveness of the product and its efficiency in production and use. Today, capsules made from aluminum or PVC are widely used. As well as providing decoration, the capsule also protects the cork from mold growth, worms and so on, acts as an additional barrier to O_2 ingress and functions as a tamper-evident seal.

2. Plastics

a. *Bag-in-Box*

The most significant change in the packaging of wine resulted from the development of the bag-in-box package, a flexible, collapsible, fully sealed bag made from one or more plies of synthetic films, a closure and tubular spout through which the contents are filled and dispensed, and a rigid outer box or container.[4]

The bag-in-box concept appeared in the U.S. in the late 1950s, being introduced into the dairy industry in the form of a disposable, single-ply bag for bulk milk in 1957. By 1962, it had gained acceptance as a replacement for the returnable 19 L can used in institutional bulk milk dispensers.[4] It also found use in nonfood applications such as the packaging of sulfuric acid to activate dry-charge batteries.

Interest grew during the mid-1960s in Australia and a PVdC copolymer–coated PA–LDPE laminate inside a fiberboard box entered the Australian market in the 1970s as the first bag-in-box package for wine. A similar structure is still used today, although other materials are used such as a triplex (three-ply) construction of EVA copolymer–PET–LDPE with the PET being metallized. More recently, barrier coextruded materials have been used, where these rely on the barrier properties of EVOH copolymer, the strength of nylon and the sealability of LLDPE.[4]

The physical strength of the bag is of prime importance and it must remain intact during distribution and subsequent storage. Under normal circumstances, the pack is subjected to two forms of stress—hydraulic shock (normally caused by sudden acceleration/deceleration of the pack) and flexing. During transportation, the bag is subjected to vibration, which is transmitted to the wine through the materials that flex. Because all flexible materials have a fatigue life, eventually a hole appears and the bag fails. Methods of overcoming this problem include the use of polymers which have high flexure resistance, improving the adhesive between the films and hence increasing the bond strength between the laminated webs, ensuring that the bag and box volumes are close, and finally using an inner bag which provides a cushioning effect.[11] Table 20.2 indicates the stress flex resistance of three wine bag materials.

O_2 permeates into a bag-in-box package in three ways:

1. Through the film of the bag
2. Through the material of the tap
3. Through the gaps and crevices between the tap and its gland

One problem associated with the packaging of wine into bag-in-box packs was the decrease in shelf life compared with that obtained using traditional glass bottles. In the 1970s in Australia, significant loss of free SO_2 accompanied by the appearance of oxidized flavors was observed only 3 months after filling into bag-in-box packs. A minimum shelf life of 6 months was required. It was not certain whether the decrease in shelf life was attributable to permeation of O_2 into the wine or permeation of SO_2 out of the wine. In an elegant piece of research,[11] it was shown that the rate of O_2

TABLE 20.2
Gelbo Flex Resistance of Wine Bag Materials

	Average Number of Pinholes		
Number of Flexes	LDPE–Metallized PET–EVA, 91 μm	PVdC Copolymer–Coated PA–LLDPE Laminate, 65 μm	Metallized PVdC Copolymer–Coated PA–LLDPE Laminate, 65 μm
100	0	0	0
250	0	0	0
500	0.5	0	0
1300	13	0	0
1500	>30	0.5	0.25

uptake by wine (4.5 L inside a pack of surface area 0.25 m^2) was as high as 0.17 ppm day^{-1} and that the valve contributed 33 to 52% of this amount. Under "dry" conditions, O_2-passed through the valves by two mechanisms—true permeation through the valve material, and diffusion through imperfections in the sealing surfaces. Theoretical considerations demonstrated that loss of SO_2 from the wine by permeation was negligible.

As a result of this research, considerable improvements were made to the O_2-barrier properties of the bag and especially to the design of the spout and closure. Today, typical designs consist of a simple, one-piece, flexible valve, which opens and closes as a lever is activated.

The O_2 transmission rates through wine bag material and taps are shown in Table 20.3. The lower values for the wet taps suggest that a considerable amount of O_2 is transmitted through the taps in the dry state due to imperfect sealing of the tap and gland. For a standard 3 L pack with unflexed PVdC copolymer–coated PA–LLDPE laminate and press tap (assumed wet), 73% of the O_2 entering the pack would pass through the bag and 27% through the tap.[11]

b. Bottles

When PVC clear plastic bottles were first produced, it was assumed that they would find application as a container for fermented beverages. However, short-term sensory tests showed unmistakable changes in the wine aroma properties as a result of plasticizers being extracted into the beverage by the ethanol. Their use was never approved for wine in the U.S.

Two separate studies (each extending over a year) indicated that wine in either a 3 L PET container with a PVdC copolymer layer or in an uncoated 4 L PET container maintained quality for 10 to 12 months when compared with the same wine in 1-gal glass containers.[25] Changes in SO_2 and color were the most obvious measurable changes resulting from O_2 permeation into the bottle.

TABLE 20.3
Oxygen Transmission Rates through Wine Bag Material mL m^{-2} day^{-1} atm^{-1} at 75% RH and 23°C

Laminates	Unflexed	Flexed
Coated nylon–LLDPE	5–7	8–10
Metallized PVdC copolymer–coated PA–LLDPE	0.1	0.6
Metallized PET–EVA	1–1.5	5–7

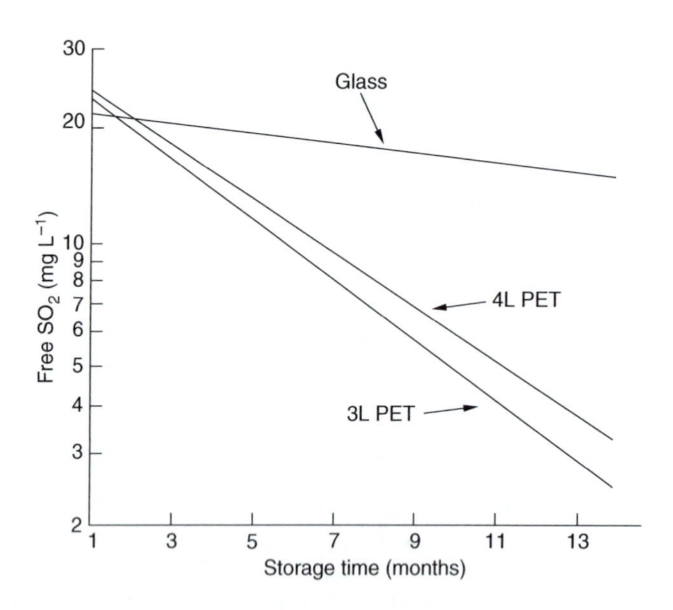

FIGURE 20.4 Decrease in free SO_2 in white wine during storage at 20°C in glass and PET containers over a 13-month period. Initial free SO_2 level at zero time was approximately 30 mg L^{-1}. (*Source*: From Ough, C. S., *Am. J. Enol. Vitic.*, 38, 100–104, 1987. With permission.)

The decrease in free SO_2 with time in the various containers is shown in Figure 20.4. Keeping the level of free SO_2 at about 30 mg L^{-1} was effective for the preservation of the samples of white wine studied. A hypothetical calculation on the shelf life of wine in a PET container was presented in Chapter 11 and indicates that the key feature of a successful plastic bottle for wine is a low O_2 permeation rate through the bottle wall. With the new barrier materials presently available (see Chapter 5), PET bottles are now being used to package wine with smaller sizes (187 mL) predominating.

3. Metal

Wine has been canned in small quantities since the 1960s in Europe, using beverage cans made from aluminum or occasionally tinplate. When packaging still wines, it is necessary to increase the internal pressure in the cans by injection of N_2 in order to prevent collapse of the can body. This is because beverage cans rely on a significant internal pressure to augment the inherently low strength of the can body itself. The two main requirements of the successful packaging of wine in metal containers is first, the nature and integrity of the enamel lining on the inside walls of the can and second, the O_2 content of the wine at the time of filling. The O_2 content should be as close to zero as possible to minimize undesirable degradative reactions; this can be achieved by using N_2 gas flow closure.

4. Miscellaneous

Since the early 1980s, increasing quantities of popular still wines have been packaged aseptically in paperboard–aluminum foil–plastic cartons and free-standing aluminum foil–plastic sachets. It is not expected that the wine in these packages would be held by consumers for aging over long periods of time, and thus a shelf life of 6 to 12 months is all that is required. This can be achieved, although scalping of flavors by the plastic in contact with the wine could be expected to present problems in certain situations similar to those discussed in Section V.C for fruit juices.

REFERENCES

1. Anon. Carbonated beverage packaging, In *The Wiley Encyclopedia of Packaging Technology*, 2nd ed., Brody, A. L. and Marsh, K. S., Eds., Wiley, New York, pp. 158–161, 1997.
2. Anon. Closures, Bottle and Jar, In *The Wiley Encyclopedia of Packaging Technology*, 2nd ed., Brody, A. L. and Marsh, K. S., Eds., Wiley, New York, pp. 206–220, 1997.
3. Anon. Vacuum-bag coffee packaging, In *The Wiley Encyclopedia of Packaging Technology*, 2nd ed., Brody, A. L. and Marsh, K. S., Eds., Wiley, New York, pp. 948–949, 1997.
4. Arch, J., Bag-in-box, liquid product, In *The Wiley Encyclopedia of Packaging Technology*, 2nd ed., Brody, A. L. and Marsh, K. S., Eds., Wiley, New York, pp. 48–51, 1997.
5. Ashurst, P. R., Ed., *Production and Packaging of Noncarbonated Fruit Juices and Fruit Beverages*, Blackie Academic, London, England, 1995.
6. Ashurst, P. R., Ed., *Chemistry and Technology of Soft Drinks and Fruit Juices*, 2nd ed., Blackwell Publishing, Oxford, England, 2004.
7. Boulton, R. B., Singleton, V. L., Bisson, L. F., and Kunkee, R. E., *Principles and Practice of Winemaking*, Aspen Publications, Gaithersburg, Maryland, 1998.
8. Cardelli, C. and Labuza, T. P., Application of Weibull hazard analysis to the determination of the shelf life of roasted and ground coffee, *Lebensm-Wiss u-Technol.*, 34, 273–278, 2001.
9. Clarke, R. J., In *Handbook of Food and Beverage Stability*, Charalambous, G., Ed., Academic Press, Orlando, Florida, 1986, chap. 13.
10. Clarke, R. J., Packing of roast and instant coffee, In *Coffee*, Vol. 2, Clarke, R. J. and Macrae, R., Eds., Elsevier Science Publishing, New York, 1987, chap. 7.
11. Davis, E. G., The performance of liners for retail wine casks, *J. Food Technol.*, 13, 235–241, 1978.
12. DeLassus, P. T., Tou, J. C., Babinec, M. A., Rulf, D. C., Karp, B. K., and Howell, B. A., Transport of apple aromas in polymer films, In *Food and Packaging Interactions, ACS Symposium Series #365*, Hotchkiss, J. H., Ed., American Chemical Society, Washington, DC, 1988, chap. 2.
13. Giles, G. A., Packaging materials, In *Chemistry and Technology of Soft Drinks and Fruit Juices*, 2nd ed., Ashurst, P. R., Ed., Blackwell Publishing, Oxford, England, 2004, chap. 9.
14. Gunning, P., Packaging of beverages in polyethylene terephthalate (PET) bottles, In *Handbook of Beverage Packaging*, Giles, G. A., Ed., CRC Press, Boca Raton, Florida, 1999, chap. 4.
15. Hirose, K., Harte, B. R., Giacin, J. R., Miltz, J., and Stine, C., Sorption of *d*-limonene by sealant films and effect on mechanical properties, In *Food and Packaging Interactions, ACS Symposium Series #365*, Hotchkiss, J. H., Ed., American Chemical Society, Washington, DC, 1988, chap. 3.
16. Hocking, A. D. and Jensen, N., Soft drinks, cordials, juices, bottled waters and related products, In *Spoilage of Processed Foods: Causes and Diagnosis*, Moir, C. J., Andrew-Kabilafkas, C., Arnold, G., Cox, B. M., Hocking, A. D. and Jenson, I., Eds., AIFST (NSW Branch) Food Microbiology Group, Sydney, Australia, pp. 84–100, 2001.
17. Imai, T., Harte, B. R., and Giacin, J. R., Partition distribution of aroma volatiles from orange juice into selected polymeric sealant films, *J. Food Sci.*, 55, 158–161, 1990.
18. Koseki, M., Nakagawa, A., Tanaka, Y., Noguchi, H., and Omochi, T., Sensory evaluation of taste of alkali-ion water and bottled mineral waters, *J. Food Sci.*, 68, 354–359, 2003.
19. Lea, A. G. H. and Piggott, J. R., *Fermented Beverage Production*, Blackie Academic & Professional, Glasgow, 1995.
20. Linssen, J. P. H., van Willige, R. W. G., and Dekker, M., Packaging-flavour interactions, In *Novel Food Packaging Techniques*, Ahvenainen, R., Ed., CRC Press, Boca Raton, Florida, 2003, chap. 8.
21. MacLeod, A. M., Beer, In *Alcoholic Beverages*, Vol. 1, Rose, A. H., Ed., Academic Press, New York, 1977, chap. 2.
22. Mannheim, C. H., Miltz, J., and Passy, N., Interaction between aseptically filled citrus products and laminated structures, In *Food and Packaging Interactions, ACS Symposium Series #365*, Hotchkiss, J. H., Ed., American Chemical Society, Washington, DC, 1988, chap. 6.
23. Moll, N. and Moll, M., Additives and endogenous antioxidants countering the oxidation of beer, In *Handbook of Food and Beverage Stability*, Charalambous, G., Ed., Academic Press, Orlando, Florida, pp. 97–112, 1986.
24. Nielsen, T. J. and Jägerstad, I. M., Flavour scalping by food packaging, *Trends Food Sci. Technol.*, 5, 353–356, 1994.

25. Ough, C. S., Use of PET bottles for wine, *Am. J. Enol. Vitic.*, 38, 100–104, 1987.
26. Pieper, G., Borgudd, L., Ackermann, P., and Fellers, P., Absorption of aroma volatiles of orange juice into laminated carton packages did not affect sensory quality, *J. Food Sci.*, 57, 1408–1411, 1992.
27. Ratke-Granzer, R. and Piringer, O. G., Quantitative analysis of volatile compounds in roasted coffee for quality evaluation, *Deutsche Lebensm-Rundschau*, 77, 203–210, 1981.
28. Senior, D., Bottling water — maintaining safety and integrity through the process, In *Technology of Bottled Water*, 2nd ed., Senior, D. and Dege, N. J., Eds., Blackwell Publishing, Oxford, England, 2004, chap. 6.
29. Sivetz, M. and Desrosier, N. W., *Coffee Technology*, Avi Publishing, Westport, Connecticut, 1979.
30. Sizer, C. E., Waugh, P. L., Edstam, S., and Ackermann, P., Maintaining flavor and nutrient quality of aseptic orange juice, *Food Technol.*, 42(6), 152–157, 1988.
31. Tacchella, A., Packaging of beverages in foil pouches, In *Handbook of Beverage Packaging*, Giles, G. A., Ed., CRC Press, Boca Raton, Florida, 1999, chap. 9.
32. Tawfik, M. S., Devlieghere, F., and Huyghebaert, A., Influence of d-limonene absorption on the physical properties of refillable PET, *Food Chem.*, 61, 157–162, 1998.
33. Turtle, B. I., The polyester bottle, In *Developments in Soft Drinks Technology-2*, Houghton, H. W., Ed., Elsevier Applied Science Publishers, Essex, England, 1984, chap. 3.
34. Warburton, D. W., The microbiological safety of bottled waters, In *Safe Handling of Foods*, Farber, J. M., Ed., Marcel Dekker, New York, 2002, chap.16.
35. Warburton, D. W. and Austin, J. W., Bottled water, In *The Microbiological Safety and Quality of Food*, Vol. 1, Lund, B. M., Baird-Parker, T. C. and Gould, G. W., Eds., Aspen Publishers, Gaithersburg, Maryland, 2000, chap. 32.
36. White, C. H., Gough, R. H., McGregor, J. U., and Vickroy, V. V., Ozonation effect on taste in water packaged in high density polyethylene bottles, *J. Dairy Sci.*, 74, 96–99, 1991.
37. Yamada, K., Mita, K., Yoshida, K., and Ishitani, T., A study of the absorption of fruit juice volatiles by the sealant layer in flexible packaging containers (The effect of package on quality of fruit juice, part IV), *Pack. Technol. Sci.*, 5, 41–47, 1992.
38. Yamanishi, T., Chemical changes during storage of tea, In *Handbook of Food and Beverage Stability*, Charalambous, G., Ed., Academic Press, Orlando, Florida, 1986, chap. 12.

21 Safety and Legislative Aspects of Packaging

CONTENTS

I. INTRODUCTION

A. Package Selection Criteria

A number of criteria must be considered when selecting a packaging system for a food. These include:

1. The stability of the food with respect to the deteriorative chemical, biochemical and microbiological reactions which can occur. The rates of these reactions depend on both intrinsic (compositional) and extrinsic (environmental) factors.
2. The environmental conditions to which the food will be exposed during distribution and storage. The ambient temperature and humidity are the two most important environmental factors and they dictate the barrier properties required of the package.
3. The compatibility of the package with the method of preservation selected. For example, if the food is being thermally processed after packing, then the packaging must obviously be able to withstand the thermal process. Likewise, if the food is to be stored at freezer temperatures after packing, then the packaging must be able to perform at these temperatures.
4. The nature and composition of the specific packaging material and its potential effect on the intrinsic quality and safety of the packaged food as a consequence of the migration of components from the packaging material into the food.

The latter consideration — namely, the migration of potentially toxic moieties from the packaging material to the food — is of major concern in the selection and use of plastic materials for food packaging.[72] However, migration of components from the packaging to the food occurs with other materials as well, and these will also be discussed in this chapter.

B. Migration

In food packaging terminology, *migration* is generally used to describe the transfer of substances from the package to the food. Substances that are transferred to the food as a result of contact or interaction between the food and the packaging material are often referred to as *migrants*. However, it is important to note that migration is a two-way process because constituents of the food can also migrate into the packaging material.[6] An example is the "scalping" of flavor compounds from fruit juices by plastics and this was discussed in Chapter 20. In addition, compounds present in the environment surrounding the packaged food can be sorbed by the packaging and migrate into the food. For example, perfumes from soaps can be picked up by fatty foods under certain circumstances, which depend, among other factors, on the nature of the packaging materials used for the soap and the food, as well as on the proximity of the two products and the time of exposure.

It is important to distinguish between *overall migration* (OM; originally referred to as *global migration*) and *specific migration* (SM). OM is the sum of all (usually unknown) mobile packaging components released per unit area of packaging material under defined test conditions, whereas SM relates to an individual and identifiable compound only. OM is therefore a measure of all compounds transferred into the food whether they are of toxicological interest or not, and will include substances that are physiologically harmless.

The migration of molecules from the packaging material into the food is a complex phenomenon, and most mathematical treatments of transport processes are derived initially from a consideration of gaseous diffusion as discussed in Chapter 4. It is worth recalling that diffusion in liquids is approximately one million times slower than in gases, and diffusion in solids about one million times slower than in liquids.

Although the transfer of substances from packaging materials into foods is undoubtedly a complex process, diffusion resulting from the spontaneous natural molecular movements that occur without the assistance of external forces, such as shaking, mixing or even convection currents in liquids, is thought to be the main controlling mechanism. Theoretical aspects of migration have been discussed by other authors,[18,20,24,29,65] whose work should be consulted for further details.

II. REGULATORY CONSIDERATIONS

A. General Requirements

Concern about the wholesomeness and safety of foods has increased dramatically over the last century, particularly in those countries where food shortage is not a problem. Increasing understanding of and interest in technological matters on the part of consumers and organized consumer groups, coupled with a recognition that neither government nor industry can guarantee the safety of food, have lent support to this concern.

Safety is an emotive issue, and because everyone must consume food to live, the safety of food is especially emotive. Most concern usually focuses on food additives, both those added intentionally to the food and those ending up in the food from, for example, the packaging material or processing equipment. A detailed discussion of food safety is outside the scope of this book but the basic concepts will be presented to put the discussion which follows into perspective.

It is worth repeating the oft-quoted saying of the Swiss alchemist and physician (and the patron saint of toxicology) Paracelsus that "all substances are poisons; there is none that is not a poison; the dose differentiates a poison and a remedy." The last phrase is particularly important; only the dose makes the difference. Attempts to determine what a "safe" dose is lie at the heart of the problem faced by legislators and regulatory authorities.

Because food safety is a subject of intense study by a large group of highly sophisticated scientists, many consumers think that food safety determination can be solely a scientific process. That this is not so has been pointed out by several authors, including one[85] who stated that:

1. There is no known way to demonstrate absence of risk.
2. Controlled experiments, the most reliable means for assessing risk levels, cannot ethically be applied to humans.
3. Retrospective studies are unreliable.
4. Reliance must therefore be placed principally on animal studies.
5. Unfortunately, the mechanisms for extrapolating from risks to animals at high concentrations to risks to humans at low concentrations are unreliable.

It is important to note that it is not the toxicity of the chemical at the concentrations at which it appears in the packaging material that is at issue here, but rather the toxicity of the chemical at the concentration at which it appears in the food from the packaging material.

The various terms used in toxicology need to be defined. The *toxicity* of a substance is its inherent capacity to produce injury when tested by itself. Almost any chemical substance can be shown to be toxic if tested at some sufficiently high level of consumption in experimental animals. Thus, a chemical may be toxic (i.e., inherently capable of producing injury when tested by itself) without being a *hazard* (i.e., likely to produce injury under the circumstances of exposure as in a diet). The concern, therefore, is not directly with the intrinsic toxicity of a particular chemical component of a food, but rather with the potential hazards of those materials when the foods in which they are present are eaten.

Underlying the idea of food safety is the risk/benefit concept. *Benefit* can be defined as anything that contributes to an improvement in condition, while *risk* can be subdivided into two categories: vital and nonvital. A *vital risk* is one necessary or essential to life, while a *nonvital* risk usually does not involve a threat to life but may lead to injury, loss or damage. Although the difference between vital and nonvital risks is not always clear-cut, the categories of risks are different.

Risks can also be subdivided into voluntary and involuntary risks. An example of a *voluntary risk* is cigarette smoking, where the risk of lung cancer is likely but no one is compelled to smoke (passive smoking is an *involuntary risk*). Another example of an involuntary risk would be a food additive in a staple item of the diet, the additive having being shown from animal tests to be carcinogenic. In this situation, consumers would have difficulty avoiding the risk because the food was a staple item. An investigation of consumer attitudes towards technological risk concluded that the public's willingness to accept voluntary risks is approximately a thousand times greater than that for involuntary risks, and the risk of death from disease appears to be a psychological yardstick for establishing the acceptability of other risks. The consumption of a chemical that had migrated from a packaging material into a food would be classified as an involuntary risk.

The toxicity assessment of food additives usually follows a decision-tree approach. Acute, subchronic and chronic toxicity tests are normally required. The final phase of toxicologic evaluation involves assessment of the potential risk to humans and, in particular, the extrapolation of high-dose experiments with animals to low-dose risk assessments of humans. Various mathematical models have been proposed for the relationship between dose and response, and much debate has resulted, particularly over the appropriateness or otherwise of a linear extrapolation. In connection with carcinogens, debate has centered over whether or not it is possible to have a "no-effect" level of a carcinogen in a food.

Returning to the risk/benefit concept, the evaluation of the benefit arising from the presence of a particular chemical in a food is also fraught with difficulties and is ultimately a subjective decision. Manufacturers package foods in a variety of packaging materials to achieve certain benefits such as extending the shelf life of the food, making its storage and preparation for consumption more convenient or reducing the cost of the food compared to its cost if another type of packaging material were used. The fact that a component of the packaging material may migrate into the food and pose a risk to the consumer requires that the benefits arising from the use of the particular material be balanced against the risk arising from consumption of the component. Quantifying the risks and benefits and then arriving at a decision that a certain packaging material should be permitted for use because the benefits outweigh the risks is extraordinarily complex, and because it is ultimately subjective, there will always be some consumers (and manufacturers) who will disagree with the final decision.

The way in which decisions are made about migration of components from packaging materials into foods differs to varying degrees in various countries around the world. Rather than describe the situation in each country, the approach adopted by the U.S. and the EU will be discussed because, between them, they account for the larger proportion of packaging materials used with foods.

B. United States of America

The 1906 Food and Drugs Act was the first time that the U.S. Government exercised the principle that it has a duty to protect the public health by controlling the adulteration of food. Although adulterated food became less common, the provisions of the legislation were clearly not adequate to insure that the public health was protected.

In 1938, the federal Food, Drug and Cosmetic Act was introduced. Among other provisions, it prohibited foods dangerous to health; prohibited unsanitary packages; established the Food and Drug Administration (FDA); established fines for unauthorized practices; and gave the FDA the authority to close plants and to issue injunctions. However, it did not control food additives except where such additives were already known to be poisonous substances. While this legislation provided an improvement over the previous legislation, the laws regarding toxic additives were handled in an *ad hoc* fashion and only those additives with a long toxic history were acted upon.

In 1958, congress passed the Food Additives Amendment to the 1938 Act. This amendment required the manufacturer to establish the safety of any product about to be marketed and the government had the responsibility to check the evidence of safety supplied. The shift in the burden of proof of safety from government to industry meant that manufacturers were required to demonstrate the safety of additives before they would be allowed to be used. The FDA generally defines "safe" as requiring "a reasonable certainty in the minds of competent scientists that [a] substance is not harmful under the intended conditions of use" (§170.3 (i)). It is not altogether clear what a "reasonable certainty" is, which experts are qualified or even expert, and how the intended conditions of use are to be defined.[39]

Section 409(c)(3)(A) of the amendment (often referred to generally as the "Delaney clause" after congressman George Delaney who was chairman of the House Rules Committee) states in part:

> ... no additive shall be deemed to be safe if it is found to induce cancer when ingested by man or animal, or if it is found, after tests which are appropriate for the evaluation of the safety of food additives, to induce cancer in man or animal ...

At the same time, congress provided for the approval of commonly used food ingredients by defining generally recognized as safe (GRAS) substances as well as prior-sanctioned food ingredients (i.e., those approved before 6 September, 1958). The latter category contains certain substances employed in the manufacture of food packaging materials (§181.22). Such substances are excluded from the definition of a food additive, provided that they are of good commercial grade, are suitable for association with food and are used in accordance with good manufacturing practice.

U.S. legal requirements are published in the Code of Federal Regulations (CFR). The code is divided into 50 titles or broad areas subject to federal control. Title 21 is made up of 7 volumes and contains general regulations for enforcement of the Food, Drug and Cosmetic Act and the Fair Packaging and Labeling Act.

The most important provisions relating to packaging materials can be found in Volume II, Parts 100 to 199, including lists of specific antioxidants, plasticizers, release agents, stabilizers, and so on, as well as copolymers and resins. Material composition is controlled by specifying the amount of additive that can be used, as well as the types of polymer to which it can be added. Other regulations contain specifications for the residual monomer content and minimum molecular mass, and many of them limit global migration from the polymer or the final food-contact article. The regulations contain time–temperature–solvent conditions for short-term migration simulations. Selection of extractability conditions depends on the type of food and the conditions of use and, in particular, the thermal treatment applied to the package after filling with food.

Subpart E §110-80(h) states:

> Packaging processes and materials shall not transmit contaminants or objectionable substances to the products, shall conform to any applicable food additive regulation (Parts 170 to 189) and should provide adequate protection from contamination.

The definition of a *food additive* can be found in §201(s):

> Any substance the intended use of which results or may reasonably be expected to result, directly or indirectly, in its becoming a component or otherwise affecting the characteristics of any food (including any substance intended for use in ... packing, ... packaging, ... or holding food ...).

A *food-contact substance* (FCS) is specified in §348(h)(6) as:

> Any substance that is intended for use as a component of materials used in manufacturing, packing, packaging, transporting or holding food if the use is not intended to have any technical effect in the food.

The FDA identifies any FCS that is reasonably expected to migrate to food under conditions of intended use to be a food additive in §170.3, which includes the following statements:

> A material used in the production of containers and packages is subject to the definition [of a food additive] if it may reasonably be expected to become a component, or to affect the characteristics, directly or indirectly, of food packed in the container.
>
> If there is no migration of a packaging component from the package to the food, then it does not become a component of the food and thus is not a food additive.

In addition to the inherent toxicity of the packaging material components, it is the extent of migration that comprises the parameters of risk assessment of packaging materials.[58] While the FDA has not provided definitive criteria for determining the point at which a substance may reasonably be expected to become a component of food, guidance has been provided by a 1979 U.S. Court of Appeals opinion in Monsanto v. Kennedy,[57] which was concerned with migration of acrylonitrile monomer into food. In this case, the court essentially said that the FDA was required to determine with a fair degree of confidence that a substance migrates to food in more than insignificant amounts for the substance to be classified as a food additive. This case has been cited as authority for the FDA's adoption of what has come to be called the *de minimis* policy.[52] The words *de minimis* refer to the legal maxim *de minimis non curat lex*, which is commonly interpreted as meaning that the law does not care for, or take notice of, very small or trifling matters. Under this policy, the FDA has permitted substances that contain low levels of carcinogenic impurities to remain on the market when the amounts expected to become a component of food have been found to be of no toxicological significance or *de minimis*.[39] An example is the FDA's decision to continue to permit the use of methylene chloride for decaffeinating coffee on the basis that the risk from the use of methylene chloride (no greater than 1 in 1 million) is so small as to be effectively no risk.

Because migration is the principal mechanism by which components of packaging materials enter food, the focus of the premarket safety evaluation by the FDA is a prediction of the amount and nature of the migrants from the packaging material under the proposed conditions of use. These predicted levels of migration are then translated into an *acceptable daily intake* (ADI), which is defined as the maximum intake of a substance in mg per kg of body weight per day that the FDA considers safe on a per person basis. The FDA derives its ADIs by taking the no observable effect level (NOEL) demonstrated in either 90-day (subchronic) or 2-year (chronic) animal studies and divides the NOEL by an applicable safety factor, generally 1000 for 90-day studies and 100 for chronic data.[39]

Unlike food additives, human exposure to components of packaging materials that have migrated into foods (known as FCSs) is typically very small. Because complete toxicological data sets are not always available for such migrants, the FDA developed a process in the 1990s to make the evaluation of packaging materials more efficient, instead of relying on the extensive review normally required for food additives. This process is used to determine "when the likelihood or extent of migration to food of a substance used in a food-contact article is so trivial as not to require regulation of the substance as a food additive."[10] This trivial level, also known as the *threshold of regulation* (TOR) was based on a large database of carcinogenic potencies.[30] The basic concept is that, because the toxic hazard arising from a substance varies with dose, then for every substance, there must be a level below which the hazard is so low that there is no need for it to be controlled by specific legislation.

The TOR applies when the overall dietary concentration of a packaging material migrant is <0.5 ppb, which equates to an intake of 1.5 μg day^{-1} if it is assumed that the total daily intake of food and drink is 3 kg per person. Substances that are below the threshold value are considered by the FDA to be exempted from regulation as food additives.[58] The exemption is applicable provided the substance does not contain any carcinogenic constituents or impurities with a TD_{50} (the dose that causes cancer in 50% of the test subjects) of less than 6.25 mg kg^{-1} body weight per day. Complete details concerning the criteria to be followed in seeking TOR exemption can be found in §170.39 and in the Federal Register.[27]

As a result of the FDA Modernization Act of 1997 (specifically §309), the FDA implemented from January 2000 the food-contact notification (FCN) system as the primary means to regulate FCSs.[11] The FCN system is applicable to all FCSs and represents a radically new approach to their regulation. Under the FCN system, manufacturers can file a FCN instead of a food additive petition (FAP) with the FDA, and unless formal substantive objections are made by the FDA, the FCS may be marketed 120 days after filing of the FCN. This compares with an average time of 2 to 4 years for the FDA to publish a formal food additive regulation when a FAP is filed.

The requirements for an FCN are substantially similar to those for a FAP and will not be detailed here[4]; neither will all the calculation methods used by the FDA to estimate probable exposure. Just two aspects will be described.

One is the "consumption factor" (CF) which describes the fraction of the daily diet expected to contact specific packaging materials. The CF represents the ratio of the weight of all food contacting a specific packaging material to the weight of all food packaged. CF values for packaging categories (e.g., metal, glass, polymer and paper) and specific food-contact polymers are summarized in Table 21.1. These values were derived by the FDA using information on the types of food consumed, the types of food contacting each packaging surface, the number of food packaging units in each food packaging category, the distribution of container sizes and the ratio of the weight of food packaged to the weight of the package.

When the FDA computes exposure to an FCS, it assumes that the FCS will capture the entire market for which it is intended for use. This approach reflects both the uncertainties about likely market penetration as well as the limitations in the data surveyed. Thus, if a company proposes the use of an antioxidant in PS, then it is assumed that the antioxidant will be used in all PS manufactured for food contact. In certain cases where an adjuvant is intended for use in only a part of a packaging or resin category, a lower CF representing the coverage that is sought may be used. For example, if a stabilizer is intended for use only in rigid and semirigid PVC, then a CF of 0.05 rather than 0.1 could be used in estimating exposure because only about 50% of all food-contact PVC could contain the stabilizer.

When new products are introduced, they will initially be treated as replacement items for existing technology. The FDA generally makes estimates based on the assumption that the new product will capture the entire market. For example, the retortable pouch was initially treated as a replacement for metal cans and was assigned a CF of 0.17. As additional information on actual use of the retortable pouch became available, the CF was lowered to 0.05. In certain cases, the submission of resin or packaging market data may lead to the use of a lower CF.

Before migration levels can be combined with CF values to derive estimates of probable consumption, the nature of the food that will likely contact the food-contact article containing the FCS must be known. To account for the variable nature of food contacting each food-contact article, the FDA has calculated "food-type distribution factors" (fT) for each packaging material to reflect the fraction of all food contacting each material that is aqueous, acidic, alcoholic and fatty. Appropriate fT values for both packaging categories and polymer types are shown in Table 21.2.

For calculating the concentration of the FCS in the daily diet, the concentration of the FCS in food contacting the food-contact article, $\langle M \rangle$, is derived by multiplying the appropriate fT values by

TABLE 21.1
Fraction of Daily Diet in the U.S. Expected to Contact Specific Packaging Materials Consumption Factors (CF)

Package Category	CF	Package Category	CF
	A. General		
Glass	0.1	Adhesives	0.14
Metal—polymer-coated	0.17	Retort pouch	0.05
Metal—uncoated	0.03	Microwave susceptor	0.001
Paper—polymer-coated	0.2		
Paper—uncoated and clay-coated	0.1		
Polymer	0.4		
	B. Polymer		
Polyolefins	0.35	PVC	0.1
LDPE	0.12	Rigid/semirigid	0.05
LLDPE	0.06	Plasticized	0.05
HDPE	0.13	PET[a]	0.16
PP	0.04	Other polyesters	0.05
Polystyrene	0.1	RCF	0.01
Impact	0.04	Nylon	0.02
Nonimpact	0.06[b]	Acrylics, phenolics, etc.	0.15
EVA	0.02	All other[c]	0.05

[a] A CF of 0.05 is used for recycled PET applications.
[b] General purpose 0.02; foam 0.04.
[c] A minimum CF of 0.05 is used initially for all exposure estimates.

Source: From Anon., *Guidance for Industry. Preparation of Food Contact Notifications and Food Additive Petitions for Food Contact Substances: Chemistry Recommendations Final Guidance*, U.S. Food and Drug Administration, Center for Food Safety & Applied Nutrition, Office of Food Additive Safety, 2002, April. With permission.

the migration values, Mi, for simulants representing the four food types. This effectively scales the migration value from each simulant according to the actual fraction of food of each type that will contact the food-contact article.

More refined exposure estimates may be possible with additional information provided in an FCN or FAP. For instance, subdividing packaging or resin categories could reduce the calculated exposure by lowering the CF for the category. The division of PVC into rigid and plasticized categories and PS into impact and nonimpact categories are two examples. Another example is the division of polymer coatings for paper into subcategories, such as PVA coatings, styrene–butadiene coatings, and so on. If an FCS is to be used solely in styrene–butadiene coatings for paper, then use of the CF for polymer-coated paper (0.2) would be a gross exaggeration.

Migration levels in food are typically estimated based on the results of migration testing under the anticipated conditions of use or, in certain cases, under the assumption of 100% migration of the FCS to food. A third alternative accepted by the FDA involves migration modeling. If this approach is taken, then the source of any material constants used in migration modeling should be appropriately referenced, whether the source is the FDA migration database or the open literature. Semiempirical methods have been developed to determine migration levels with limited or, in certain cases, no migration data.[8,47] These diffusion models rely on estimation of diffusion coefficients based on the nature of the migrant and the physical properties of the polymer. The FDA considers that such models may be useful substitutes for, or additions to, experimental data under limited circumstances, but has several caveats that should be considered in the application of such diffusion models.

TABLE 21.2
Food-Type Distribution Factors (fT) for Each Packaging Material to Reflect the Fraction of All Food Contacting Each Material that Is Aqueous, Acidic, Alcoholic and Fatty

Package Category	Food-Type Distribution Factors (fT)			
	Aqueous[a]	Acidic[a]	Alcoholic	Fatty
		A. General		
Glass	0.08	0.36	0.47	0.09
Metal—polymer-coated	0.16	0.35	0.40	0.09
Metal—uncoated	0.54	0.25	0.01[b]	0.20
Paper—polymer-coated	0.55	0.04	0.01[b]	0.40
Paper—uncoated and clay-coated	0.57	0.01[b]	0.01[b]	0.41
Polymer	0.49	0.16	0.01[b]	0.34
		B. Polymer		
Polyolefins	0.67	0.01[b]	0.01[b]	0.31
Polystyrene	0.67	0.01[b]	0.01[b]	0.31
Impact	0.85	0.01[b]	0.04	0.10
Nonimpact	0.51	0.01	0.01	0.47
Acrylics, phenolics, etc.	0.17	0.40	0.31	0.12
PVC	0.01[b]	0.23	0.27	0.49
Polyacrylonitrile, ionomers, PVdC	0.01[b]	0.01[b]	0.01[b]	0.97
Polycarbonates	0.97	0.01[b]	0.01[b]	0.01[b]
Polyesters	0.01[b]	0.97	0.01[b]	0.01[b]
Polyamides	0.10	0.10	0.05	0.75
EVA	0.30	0.28	0.28	0.14
Wax	0.47	0.01[b]	0.01[b]	0.51
RCF	0.05	0.01[b]	0.01[b]	0.9

[a] For 10% ethanol as the food simulant for aqueous and acidic foods, the food-type distribution factors should summed.
[b] 1% or less.

Source: From Anon., *Guidance for Industry. Preparation of Food Contact Notifications and Food Additive Petitions for Food Contact Substances: Chemistry Recommendations Final Guidance*, U.S. Food and Drug Administration, Center for Food Safety & Applied Nutrition, Office of Food Additive Safety, 2002, April. With permission.

C. European Community

1. Background

When the Treaty of Rome was signed in 1956 and the European Economic Community (now commonly referred to as the European Community or EC, or the European Union or EU) was born, one of its main objectives was to set up a common market so that goods could be moved as freely within the community as within national borders. The Commission of the European Communities (hereinafter referred to as the Commission) is the driving force behind community activities, while the Council of Ministers of the European Communities (hereinafter referred to as the Council) is the decision-making body, operating under the powers conferred on it by the Treaty of Rome.

Because different regulations in member states in the EU could constitute a nontariff barrier to trade, it is valid under the Treaty of Rome and the Single European Act for directives to be promulgated which harmonize such legislation. Harmonization of food legislation is usually classified into *vertical directives* (concerned with a specific group of similar products, e.g., coffee extracts) and *horizontal directives* (concerned with subjects of general application to all foods, e.g., additives, materials intended to come into contact with foods, methods of control, etc.).

To assist in the free movement of goods, the Commission has given priority to harmonization work in the horizontal sectors.

Directives do not automatically become Community law; rather, they are instructions to governments of member states to bring their own legislation into line within a certain period (typically, 18 to 24 months). There is a detailed process of consultation, study and preparation leading up to the promulgation of a directive, and although the resultant legislation is intended to represent a consensus, the views of the Commission carry considerable weight.

There is also a Scientific Committee for Food (SCF), an advisory body set up by the commission in 1974, which consists of individuals nominated by member countries in the fields of toxicology, metabolism, mutagenicity and so on. The SCF constituted an *ad hoc* working committee in the area of packaging materials.

2. Directives

The Commission initially drew up a framework directive in order to establish general principles for all materials and articles, and criteria and procedures to be followed in drafting specific directives; that is, directives concerning individual sectors to be regulated (e.g., plastics, ceramics, etc.) or individual substances (e.g., vinyl chloride). The framework Directive 76/893/EEC of 26 November 1976 (since superseded by 89/109/EEC of 21 December 1988) established two general principles:

1. The principle of the "inertness" of the material and the "purity" of the food, whereby the materials and articles must not transfer to foods any of their constituents in quantities which could "endanger human health and bring about an unacceptable change in the composition of the foodstuffs or a deterioration in the organoleptic characteristics thereof." This regulation applied not only to packaging, but to all articles whose surface could come into contact with food at any stage of production, storage, transport or consumption. For practical reasons, covers and coatings, potable water distribution systems and antiques were excluded.
2. The principle of "positive labeling," whereby materials and articles intended to come into contact with foods must be accompanied by the words "for food" or an appropriate symbol, described in Directive 80/590/EEC. At the retail stage, member states have the option not to insist on marking where articles are "by their nature clearly intended to come into contact with foodstuffs."

In 1989, the framework directive was replaced by Directive 89/109/EEC, which laid down the sectors in which the Commission is asked to establish Community rules and the criteria and procedures to be followed in the drafting of specific directives.

Having defined the general framework, the Commission began to study three of the principal materials to be dealt with at Community level, these being regenerated cellulose film (RCF), ceramics and plastics. The most important of these is plastics and is discussed below.

In 1980, the Commission began to draw up rules for what is undoubtedly the most complex and economically important area of packaging, namely plastic materials and articles intended to come into contact with food. Directive 82/711/EEC (last amended by Directive 2004/19/EC) lays down test methodology, in particular simulants and test conditions for plastics, including laminates, where the plastic is in direct contact with the food. RCF, elastomers, rubbers, adhesives, paper and paperboard impregnated with plastic materials are all excluded.

Directive 82/711/EEC lists conditions of time and temperature that simulate actual product use conditions. The migration of substances under these simulating conditions should not exceed the limits given in the positive list. Directive 85/572/EEC establishes the list of simulants to be used in migration tests and includes a table of correlations between food groups and their food-simulating liquids.

The OM limit is a measure of the inertness of the material and prevents an unacceptable change in the composition of the food. Moreover, it reduces the need for a large number of specific migration limits (SMLs) or other restrictions, thus giving effective control.

Directive 90/128/EEC set the OM limit at a level of 10 mg dm^{-2} of food-contact surface area of material or article. On the assumption that a typical package would have a food-contact surface area:volume ratio of 6 dm^2 L^{-1} [i.e., the same as a 1 dm (10 cm) cube], an upper limit for OM of 10 mg dm^{-2} of contact area becomes 60 mg L^{-1}.

The limit was set at 60 mg kg^{-1} of food for articles that are containers or are comparable to containers or which can be filled with a capacity of 0.5 to 10 L; for sheet, film or other materials, which cannot be filled or for which it is impractical to estimate the surface area in contact with food; and for caps, gaskets, stoppers or similar devices for sealing. The directive also provides for general criteria on the way in which tests are to be carried out, although it does not describe the operating details or the analytical methods.

In addition, the directive contains two lists of substances (1340 in total of which 540 are monomers and 800 additives) used in the preparation of plastics for foods. The first, so-called "community list" contains substances on which the SCF has delivered an opinion and which are therefore authorized and harmonized at Community level. The second, so-called "optional national list" contains substances on which the SCF has not yet been able to deliver a final opinion for lack of data and which therefore do not have Community recognition. Thus, the current list of additives is incomplete because it does not contain all the substances which are currently accepted in one or more member states. Accordingly, these substances continue to be regulated by national laws pending a decision on inclusion in the community list. The European inventory list of chemicals used to make plastics intended for food contact contains more than 1500 listed substances, and inventory lists of a similar length exist for chemicals used to make paper, can enamels, inks and adhesives.[20]

As migration experiments are time-consuming, expensive and often complicated to carry out, the use of mathematical models to predict migration is gaining interest.[70] In the sixth amendment of Directive 90/128/EEC (2001), the use of "generally recognized diffusion models" based on experimental data was approved for estimating the migration level of substances in certain types of plastics as an alternative test method. In a review[40] on predictive migration modeling for regulatory purposes, it was concluded that further research is needed to improve current model approaches, including both fundamental research on diffusion in polymers and applied research on real packaging and food or food simulant combinations.

III. PLASTICS PACKAGING

A. Vinyl Chloride Monomer

PVC is widely used today for food-contact applications, not only for bottle and film applications, but also for other uses including liners and sealing gaskets. Vinyl chloride (boiling point − 13.9°C) is a colorless gas at ordinary temperatures and pressures but, in industry, it is usually handled as a liquid under pressure in steel cylinders. The acute toxic effects of vinyl chloride have been known since the 1930s when it was rejected as an anesthetic because it was found to be a cardiac irritant. An investigation in the U.K. in 1970/1971 into the tainting of spirits packaged in miniature PVC bottles for use in aircraft traced the cause to vinyl chloride monomer (VCM). However, the chronic effects were not made public until January 1974 when the B.F. Goodrich Co. in the U.S. voluntarily revealed to federal and state regulatory officials that, since 1971, three workers at its PVC polymerization plant in Louisville, Kentucky had died of angiosarcoma of the liver. Because this form of cancer is extremely rare, this disclosure caused immediate alarm. Workers who were in contact with PVC only and not VCM were apparently unaffected, so the initial hypothesis was that VCM was the culprit.

A few weeks later, at the request of the U.S. Department of Health, Education and Welfare, ICI in England published the results of toxicological studies by Professor Maltoni from the University of Bologna, Italy, who in August 1972 had discovered a liver angiosarcoma in a test rat exposed to VCM. In January 1973, a team of three scientists representing the U.S. chemical industry had visited his laboratory to study his results. By May 1974, a cause-and-effect relationship between VCM and human angiosarcoma was generally accepted, and confirmed cases of cancer in workers at PVC plants had risen to 19, 17 of whom were already dead.

The immediate response of regulatory authorities was to reduce the maximum permitted concentration of VCM in air for plants manufacturing PVC from 500 (imposed because of VCM's inflammability, not its toxicology) to 50 ppm, and the following year, a 1 ppm 8 h average with a maximum of 5 ppm over any period longer than 15 min was required. The use of VCM as a propellant in hair sprays and other products was also eliminated.

In 1973, it was discovered that VCM in PVC packaging material could migrate into foods. Previously, residual monomer levels had been assumed to be reduced to insignificant levels during the processing of the PVC resin and fabrication of the packaging material. In the U.K., the Steering Group on food surveillance sponsored by the Ministry of Agriculture, Fisheries and Food established a working party on vinyl chloride in 1973. It reported[54] on the levels of VCM found in bottles, rigid film and some foods in the period 1974 to 1977 and a summary of some results is presented in Table 21.3. It is clear that there was a marked reduction in the VCM content of PVC manufactured articles during the period 1974 to 1976, and this change was also reflected in the levels found in foods such as fruit drink, cooking oil, butter and margarine. Similar trends in the level of VCM in foods have also been reported in other countries as a result of modifications to the method of manufacture and improved blending and fabrication techniques. Levels in food are determined primarily by the initial VCM content in the PVC, together with storage time and temperature.

The EU Directive 78/142/EEC laid down the maximum permitted quantity of VCM present in plastic materials and articles prepared with this substance as 1 mg kg^{-1}, and stated that the quantity of VCM released to the food should not be detectable by a method of analysis with a detection limit of 0.01 mg kg^{-1} (10 ppb). Directive 80/766/EEC and Directive 81/432/EEC established the methods of analysis for VCM in the finished article and in foods, respectively.

In the U.S., the prior sanctions and the food additive regulations for PVC were all promulgated prior to the discovery that VCM caused liver tumors in humans, and thus there were no limitations on the residual VCM level in PVC. In response to the findings regarding the toxicology of VCM, the FDA published a series of proposals designed to control the potential exposure to VCM from food-contact materials. The FDA did not propose to ban or otherwise limit use of PVC *per se*, however, because there is no evidence that PVC itself is a carcinogen.

TABLE 21.3
Residual Levels of Vinyl Chloride Monomer (VCM) in PVC Products

	VCM Content (ppm)		
Stage of Production	**1974**	**1975**	**1976**
Bottles: level in polymer at time of use	500–1000	100–250	15–50
Level in powder blend (maximum)	100	5	1
Level in bottle wall	100	5	1
Foil: maximum level	150	15	5
Flexible film (extrusion blown) maximum	1	1	1

Source: From Ministry of Agriculture, Fisheries and Food, *Survey of Vinyl Chloride Content of PVC for Food Contact and of Foods*, Food Surveillance Paper No. 2, Her Majesty's Stationery Office, London, England, 1978. With permission of the Controllers of HMSO.

In general, the FDA proposed to limit residual VCM levels to 10 ppb in rigid PVC food-contact articles and 5 ppb in plasticized, flexible PVC. The FDA also proposed to issue new regulations that would have codified some of the prior sanctions for PVC polymers, as well as to establish a new regulation for rigid and semirigid PVC articles, and establish limits on the residual level of VCM permitted in these articles. This new regulation would have established a 10 ppb limit on residual VCM in these articles. However, the FDA proposals were never finalized because of difficulties encountered in preparing an environmental assessment for the proposed actions. PVC continues to be used in food-contact applications in the U.S., subject to the limitations provided in the 1986 proposal. A practical result of the proposal was the removal of PVC liquor bottles from commercial use.

B. Styrene Monomer

Styrene monomer (boiling point 145°C) is metabolized to styrene oxide, which is a potent mutagen in a number of test systems; further metabolism produces hippuric acid. The most frequently observed changes from the toxic effects of styrene in humans are of a neurological and psychological nature; styrene acts as a depressant on the central nervous system, has a toxic effect on the liver and causes neurological impairment.

Levels of styrene monomer in food packaging were reported to range from 60 to 2250 ppm.[83] The monomer migrated into food-simulating solvents from various styrene-containing beverage containers. In studies simulating filling and storage at room temperature, the average values for styrene migrating into 8% ethanol were 27 ppb for foam cups, 52 ppb for HIPS containers and 151 ppb for crystal PS glasses. Using water, coffee and tea, an average of 6.3 ppb of styrene migrated from foam cups under conditions simulating hot filling or pasteurization above 65.6°C.

A U.K. survey in 1994 measured the levels of styrene monomer in a total of 248 samples of food. The highest levels were found in milk and cream products sold as single-serve (ca. 10 g) coffee/tea whiteners with a mean value of 134 ppb and a range of 23 to 223 ppb. In the majority of samples, styrene was detected at levels between 1 and 60 ppb with an average of less than 3 ppb. A U.K. total diet study in 1999 estimated the average daily intake of styrene as 0.03 to 0.05 $\mu g\ kg^{-1}$ body weight for a 60 kg person, well below the maximum tolerable daily intake (TDI) of 40 $\mu g\ kg^{-1}$ body weight set by the Joint FAO/WHO Committee on Food Additives in 1984.

Styrene contamination of foods is generally apparent as a characteristic, unpleasant, plastic-like chemical odor or taste and is discussed further in Section III.F.2.

C. Acrylonitrile (AN) Monomer

Acrylonitrile (boiling point 77.5°C) is a component of several polymers used as food packaging materials, where the basic terpolymer material contains as much as 70% AN in conjunction with styrene or butadiene.

The FDA began to examine PAN containers in 1974 when a polymer manufacturer submitted test results that indicated that there was the potential for significant migration from PAN bottles which were developed for carbonated soft drinks as a replacement for glass. In 1975, the commissioner published a regulation that limited residual AN monomer levels to 80 ppm in the wall of the container and stipulated that monomer migration into the food could not be greater than 0.3 ppm.

In 1976, the FDA established an interim regulation limiting AN extraction in food-simulating solvents to 0.3 ppm and making the continued use of AN copolymers in food applications conditional on additional toxicological testing. The following year, FDA stayed the interim regulations permitting use of AN copolymers in beverage containers and proposed reduction of the extraction limit in other food applications from 0.3 to 0.05 ppm. The 1977 action was based in part on proliferative brain lesions observed in rats at 300 and 100 ppm of AN in the drinking water after

13 months of a 2-year study. The 1977 ban on the use of AN copolymers for beverage bottles was removed in 1984 on the proviso that the level of residual AN in the container was less than 0.1 ppm. By this time, PET had become firmly established for the packaging of carbonated beverages, and even though AN has a significantly higher melting point and thus can be hot filled, it has never gained significant market share for beverage packaging. AN copolymers were not approved for alcoholic drinks.

Acrylonitrile–butadiene–styrene (ABS) resins are used in many food-contact applications, where the levels of the three monomers in the polymer are varied to obtain the different properties desired. The correlation of residual AN monomer concentration in AN-containing polymers with AN migration into food simulants is of interest because the FDA regulates the use of these polymers on the basis of the amount of AN that may migrate into food simulants.

D. Plasticizers

A *plasticizer* is a substance that is incorporated into a material (usually a plastic or elastomer) to increase its flexibility and processability. The vast majority of plasticizers are esters of phthalic acid (phthalates) with a wide variety of long chain alcohols containing up to 13 carbon atoms; next in importance are those based on adipic acid. About 90% of all plasticizers are used to convert PVC into a soft, elastic material. However, copolymers of PVC and PVdC are plasticized with up to 5% of acetyltributyl citrate (ATBC), such films finding widespread use in microwave cooking.

1. Phthalate and Adipate Esters

Surveys carried out in a number of countries have indicated that, over the last few decades, there has been a fall in the quantity and quality of male sperm, although the effects appear to be variable. There is also evidence that, in some countries, there has been an increase in testicular cancer. Much of the concern focuses on synthetic — and mainly organic — chemicals as highlighted in the book *Our Stolen Future*.[25] *Endocrine disruptors* are chemicals that enter the body from the external environment and mimic or interfere with the human endocrine system. Chemicals that mimic the female hormone estrogen or act as antiestrogens are suggested as being responsible for some observed effects on wildlife such as the feminization of fish. It has been further suggested that these same chemicals may be responsible for the above-mentioned male human health problems.[80]

Phthalates are among the chemicals that have been labeled as xenoestrogens. Of the phthalic acid esters, di-2-ethylhexyl phthalate (DEHP) is the most widely used. DEHP is also known as dioctyl phthalate (DOP), the terms dioctyl and di-2-ethyl being synonymous. DEHP, together with diethyl phthalate (DEP) and di-isooctyl phthalate (DIOP), have been granted prior sanction by the FDA as plasticizers in the manufacture of food packaging materials for food of high water content only, and are listed in §181.27 of the CFR. Other phthalate esters have been cleared as plasticizers under §178.3740 of the CFR for various food-contact uses.[61] The majority of PVC used for food packaging does not contain phthalate plasticizers, and many PVC packaging materials are rigid and unplasticized. Minor uses of phthalates in food packaging include use as plasticizers in cap liners made from PVC plastisols.[80]

An analysis of butter and margarine packaged in an aluminum foil–paper laminate revealed concentrations of dibutyl, butylbenzyl or DEHPs up to 10.6, 47.8 and 11.9 $\mu g\ kg^{-1}$, respectively.[62] The same phthalates were present in the packaging material, and it was shown that they originated from the outer foil surface where they were a component of the protective coating. It was suggested that they transferred from the outer foil surface to the inner paper surface during storage of the rolls of packaging material prior to wrapping of butter and margarine, a process commonly referred to as "back migration."

Plasticizers are commonly used in printing inks where they assist adhesion of the ink to the packaging material and improve the ink's flexibility. A survey[60] of printing inks on a selection of packaging materials for products including confectionery, snacks, chips, potatoes, chocolate bars and biscuits in Spain and England found a number of plasticizers, predominantly phthalates. Although the inks were generally applied to the outer surface of the packaging materials and therefore were not in direct contact with the food, it has been shown that they can migrate through plastic layers to the food.[21]

A survey[37] in the mid-1980s concluded that dietary intakes of plasticizers in the U.K. as a result of migration from food-contact materials are, in general, below 2 mg per person per day. An exception was di-2-ethylhexyl adipate (DEHA), which was used as a plasticizer in PVC cling films and resulted in an average dietary exposure of 16 mg per person per day. A government report recommended that its use be reduced, and the maximum intake of DEHA in the U.K. in 1990 was estimated[55] as 8.2 mg per person per day as a direct consequence of the reformulation of films by manufacturers. However, intake of ATBC (see below) had increased owing to microwave usage of PVdC copolymer films.

2. Acetyltributyl Citrate

ATBC is a plasticizer formed by the esterification of citric acid. It is considered a "prior-sanctioned food ingredient" and is not classified as a food additive by the U.S. FDA.

The results of an assessment of ATBC migration into a variety of foods when plasticized film was used for normal domestic applications including cooking or reheating of meals in a microwave oven are given in Table 21.4. The migration levels represent losses ranging from 1 to 51% of the available plasticizer. The lowest level of migration occurred where there was no direct contact between the film and food. An indirect route for plasticizer transfer in this instance is condensation of water from hot soup onto the inner surface of the film, where some extraction of plasticizer followed by condensate can drip back into the soup. Aqueous foods gave low migration, and the apparently high level of 24.7 mg kg^{-1} of ATBC found in spinach was a reflection of the high film surface area to weight ratio used for cooking this product. The highest levels of migration were found where the film was used as a liner or wrap with direct contact with a fatty food surface.[22]

TABLE 21.4
Migration of Acetyltributyl Citrate into Microwave-Cooked Foods

Food	ATBC Migration mg kg^{-1}	mg dm^{-2}	% loss
Soup	0.4	0.1	1
Chicken breasts	12.7	0.3	3
Pork chop	1.4	0.4	4
Hot bread	22.0	0.6	6
Spinach	24.7	0.1	1
Sweetcorn	3.9	0.2	2
Brussel sprouts	0.9	<0.1	<1
Steam pudding	2.7	0.6	6
Cakes (scones)	22.3	1.7	17
Chocolate cake	22.6	2.4	24
Peanut biscuits	79.8	5.1	51

Source: From Castle, L., Jickells, S. M., Sharman, M., Gramshaw, J. W., and Gilbert, J., *J. Food Prot.*, 51, 916–919, 1988. With permission.

3. Epoxidized Soy Bean Oil

Epoxidized seed and vegetable oils such as epoxidized soy bean oil (ESBO) are widely used in a range of food-contact materials to serve as multifunctional additives exhibiting plasticizer, lubricant and heat stabilizer properties. ESBO is produced by the controlled epoxidation of soy bean oil in which the C=C double bonds are largely converted to epoxy groups. PVC gaskets can contain ESBO levels up to 30%, while other materials such as PVdC copolymer and PS used to contain epoxidized oils, but at lower levels. ESBO was also employed as a component of enamels and therefore was present in the food-contact surface of certain metal cans, but its use for this purpose appears to have ceased.

An average level of ESBO of 2 mg kg^{-1} in baby food packaged in glass jars with PVC gaskets has been reported.[36] In order not to exceed the TDI, it was recommended that the ESBO level in baby food should not be higher than 31.5 mg kg^{-1}. A later U.K. survey[56] found similar results; the TDI was not exceeded for the general population of infants in Great Britain. ESBO was found in 66 out of 137 samples at levels up to 105 mg kg^{-1}. Industry has been taking steps to reduce the migration of ESBO from lid gaskets into baby foods.

E. Antioxidants

Antioxidants are used to prevent degradation of the polymer as a result of its reaction with atmospheric O_2 during molding operations at high temperature or when used in contact with hot foods. They are also used to prevent embrittlement during storage. Derivatives of phenols and organic sulfides are most frequently used.

A study[81] of the migration of 3,5-di(*t*-butyl)-4-hydroxy toluene (BHT) from HDPE to foods (skim and whole milk, margarine and mayonnaise) and food simulants found no accumulation of BHT in the aqueous phase. However, when corn oil was the simulant, all the BHT migrated during the 50-day test period. Migration at 4°C to skim and whole milk was less than for corn oil but greater than for migration to water. It was felt that some ingredients from milk could penetrate the HDPE, even at such a low temperature, and modify the migration propensity of the BHT. A summary of the results is given in Table 21.5. The mechanism for BHT transfer would appear to be evaporation from the film surface with vapor phase diffusion from the HDPE and adsorption onto the powder.

In a study[53] into the loss of the antioxidants BHA and BHT from HDPE film, the volatilization of the antioxidant from the polymer surface was found to be the controlling parameter for mass transfer rather than diffusion of the antioxidant through the bulk polymer to the surface layer. Oatmeal cereal packaged in HDPE impregnated with a high level of BHT had an extended shelf life compared to HDPE impregnated with a low level of BHT, as a result of adsorption by the cereal of antioxidant from the package.

F. Odors and Taints

The ISO definition of *taint* is a taste or odor foreign to the product; an off-flavor is defined as an atypical flavor usually associated with deterioration. An important aspect of these definitions is that food taints are perceived by the human senses. Alternative definitions to distinguish between taints and off-flavors are that taints are unpleasant odors or flavors imparted to food through external sources, while off-flavors are unpleasant odors or flavors imparted to food through internal deteriorative change.[45]

Because of the complex structure and chemical composition of package systems, a variety of chemical reactions can occur during package manufacture and use. These reactions occur between some packaging components, with other components acting as catalysts, resulting in the formation of compounds with low odor thresholds. These compounds then migrate through the material during storage and slowly diffuse into the product or package headspace. Identification and analysis

TABLE 21.5
BHT Migration into Semisolid and Solid Foods

Food	Temperature (°C)	Duration (days)	Food Mass/HDPE Surface Area (g dm^{-2})	Migration	
				μg dm^{-2}	%
Margarine	4	4	70	0.5	0.2
		45	130	1.6	0.5
		99	130	2.4	0.8
Whipped topping	4	7	130	0.9	0.4
		14	130	0.9	0.4
		21	130	0.9	0.4
Mayonnaise	21	2	70	4.7	1.8
		8	130	7.4	2.3
		45	130	26.4	8.5
		89	130	36.2	11.3
Vegetable shortening	21	2	70	6.3	2.5
		91	130	58.6	18.4
		206	130	59.2	18.6
Dry milk	21	100	130	17	5.1
Chicken soup mix	21	99	130	50	16.3

Source: From Till, D. E., Ehntholt, D. J., Reid, R. C., Schwartz, P. S., Sidman, K. R., Schwope, A. D., and Whelan, R. H., *Ind. Eng. Chem. Prod. Res. Dev.*, 21, 106–113, 1982. With permission.

of these compounds is very difficult, and many isolation and concentration steps are required because the odor thresholds are lower than analytically detectable.

Unsaturated carbonyl compounds, particularly C_6–C_{10}, constitute an important class of odor-producing substances because they occur in nature, have low threshold levels and are relatively stable owing to their conjugated double-bond structure. Other oxygen or sulfur derivatives and some hydrocarbons are also important off-odor compounds.[66] A general methodology for testing polymer odor and odor contributors has been presented elsewhere.[46] Some examples of odor and taint problems associated with plastic packaging materials used for foods are discussed below; three book chapters on the subject[50,66,79] should be consulted for further examples.

1. Solvents

During printing of plastic packaging materials, the inks may be applied dissolved or dispersed in solvents which are subsequently removed by evaporation, usually in specially designed ovens. However, a certain amount of residual solvent can remain. The solvents may be low MW organic compounds consisting of hydrocarbons, alcohols, glycol ethers, ketones and esters, which can then migrate into the foods by direct contact or via the free space inside the package. Fortunately, the human senses of smell and taste often show very high sensitivity to the presence of such volatiles and, in most cases, the threshold for sensory detection of the solvents used in ink and adhesive formulations is considerably below the toxicologically significant level. The sensory thresholds for toluene, ethyl acetate, various aldehydes and ketones range from the parts per million to the parts per billion level. The regulatory problem is mitigated by the potential for economic damage from off-taste and odor in packaged foods.[46]

The amount of solvent that enters a food depends on the partitioning of the solvent between the package materials and the contained food. The partition coefficients of six printing ink solvents (ethyl acetate, hexane, isopropanol, 2-methyloxyethanol, methyl ethyl ketone and toluene) in high-fat cookies, soybean oil and chocolate liquor have been determined, where the coefficients are greatest in soybean oil and lowest in cookies.[35]

Off-flavor was detected in a fruity soft drink packaged in a laminated pouch, and residual toluene was found to be a major cause of the odor problem. A comparison of headspace samples withdrawn from sealed pouches showed that faulty pouches contained 26 to 28 times higher toluene levels than that found in good ones. The ultimate cause of excessive residual toluene was found to be inadequate drying during lamination.[63]

2. Monomers

An investigation[63] into an off-flavor problem in chocolate and lemon cream cookies packaged in PS trays and overwrapped with printed RCF showed that styrene monomer from the PS trays was the culprit, with levels of residual monomer in the trays of 0.18 to 0.20% (w/w); the amount of styrene in the cookies was not determined.

Another study[48] investigated styrene taint in cocoa powder for drinks and chocolate flakes. The amount of residual styrene monomer in the PS used was about 320 ppm, while the amounts of styrene ranged from 7 to 132 ppb in cocoa drinks and 414 to 1447 ppb in chocolate flakes. The taste recognition threshold concentration of styrene in water is 22 ppb; values for a variety of foods are presented in Table 21.6.

3. Polyethylene

Polyolefins can develop a wax-like odor if the package is overheated.[74] Fatty acid amide, used as a slip additive, can also cause odor problems when it is stored too long or under conditions where the

TABLE 21.6
Taste Recognition Threshold Concentrations for Styrene in Different Food Types

Food Type	Fat Content (%)	Taste Recognition Threshold Concentration ($\mu g\ kg^{-1}$)
Water	0	22
Emulsions	3	196
	10	654
	15	1181
	20	1396
	25	1559
	30	2078
Yoghurts	0.1	36
	1.5	99
	3	171
Yoghurt drinks:		
Natural	0.1	82
Strawberry	0.1	92
Peach	0.1	94

Source: From Linssen, J. P. H., Janssens, J. L. G. M., Reitsma, J. C. E., and Roozen, J. P., Taste recognition threshold concentrations of styrene in foods and food models, In *Foods and Packaging Materials — Chemical Interactions*, Ackermann, P., Jagerstad, M. and Ohlsson, T., Eds., Royal Society of Chemistry, Cambridge, England, 1995, pp. 74–83.[49] With permission.

compound can oxidize. In an unpublished study quoted by Kim-Kang,[46] pouches made from paper/foil/LDPE laminates were incubated at 60°C for 20 min and then analyzed by GC. Three major components were identified as acetaldehyde, allyl alcohol and acrolein. When odorous pouches were compared with nonodorous pouches, a direct correlation between odor, acetaldehyde and allyl alcohol levels was obtained. Those compounds were considered to be thermal oxidative decomposition products of polyethylene.

4. Poly(Ethylene Terephthalate)

PET polymers are widely used in film, foil and bottle form to package foods. Experimental data showing the effect on overall migration into fat simulants of varying temperature, time, thickness and crystallinity of PET have been presented, together with specific migration data of monomers, catalyst residues, colorants and acetaldehyde from PET into food simulants.[5] The major significant volatile compound in PET is acetaldehyde, and this was of concern in odor quality, particularly in cola-type beverages where trace amounts can adversely affect the flavor; however, present manufacturing techniques have dramatically reduced residual acetaldehyde levels in PET packaging. Acetaldehyde is also the major cause of the color change in PET during aging.[46] It can be formed by thermal degradation of PET during polycondensation and melt processing. It has also been suggested that it could possibly form during exposure to simulants at high temperature. Using water as a food simulant, the peak values for acetaldehyde were <50 ppb after 8 days at 55°C and <15 ppb after 10 days at 40°C.[5]

Migration of PET oligomers (consisting mainly of cyclic compounds ranging from dimer to pentamer) has been reported at very low levels from PET bottles into alcoholic and carbonated beverages. However, total levels of migration of PET oligomers from PET trays and susceptors were found to range from 0.02 to 2.73 mg kg^{-1} depending on the food and the temperature attained during cooking.[23] A large number of PET oligomers that migrated into corn oil (a food simulant) from three types of PET packages under microwave use conditions have been identified.[12] In a study of migration into food from PET susceptor packaging, quantities of oligomers ranged from less that 0.012 $\mu g\ g^{-1}$ to approximately 7 $\mu g\ g^{-1}$.[13]

5. Miscellaneous

Development of a "catty" odor [described elsewhere as *ribes* (blackcurrant leaves from the botanical name *Ribes nigrum* for blackcurrant) or tomcat urine] has been identified in many foods as a result of interactions between food and migrants from packaging. Such odors have been attributed to a sulfur-containing compound 4-methyl-4-mercaptopentan-2-one, which is a hydrogen sulfide adduct of mesityl oxide and has an odor threshold of 0.01 ppb. Investigations into catty odor in two different "cook-in-the-bag" ham products[28] demonstrated that the mesityl oxide originated from diacetone alcohol (DAA), which was present as a residual printing ink solvent. Formation of the mesityl oxide was thought to have arisen from dehydration of the DAA, promoted by some property of the ethylene ionomer sealing layer in the multilayer packaging film. Mesityl oxide can also be formed from acetone, which is sometimes used as a solvent in coatings and adhesives. Avoiding the use of mesityl oxide precursors such as DAA in packaging materials can prevent the occurrence of catty odor in sulfur-rich foods. Mesityl oxide is not a true oxide and was assigned this name at an early date because of an erroneous conception of its chemical nature. It is in fact an α,β-unsaturated ketone with the formula $(CH_3)_2C{=}CHCOCH_3$, and has a peppermint-like odor.[79]

An objectionable catty odor was also detected in fruit drinks packaged in PVC bottles; a metal stabilizer used in the bottles was found to contain mercaptide groups that were split off during molding and resulted in the undesirable flavor that migrated into the drink. The use of a stabilizer without mercaptide groups solved the problem.[31]

IV. METAL PACKAGING

A. Tin

When tinplate was first used to make containers for food over 200 years ago, many cases of food poisoning, apparently resulting from ingestion of excessive amounts of metal, occurred. A congress of physicians held in Heidelberg in Germany even went so far as to recommend that "tinplate should be forbidden for the making of vessels in which articles of food are to be preserved."[68] The quality of tinplate has been greatly improved since those days, and foods that are likely to attack tin are packaged in tinplate containers with an appropriate enamel coating (see Chapter 7). Despite considerable research, it has not yet been demonstrated that tin is an essential element in the diet of humans, though a WHO report suggests that tin deficiency can be produced in rats.[59]

The provisional tolerable weekly intake for tin is 14 mg kg^{-1} body weight and recommended maximum permissible levels of tin in food are typically 250 mg kg^{-1} for solid foods and 150 mg kg^{-1} for beverages. Recently, in Europe, EC Regulation 242/2004 limited lead in canned solid foods to 200 mg kg^{-1} and 100 mg kg^{-1} in canned beverages, with levels in canned foods for infants and young children being limited to 50 mg kg^{-1}. Although no long-term health effects are associated with consuming tin, a recent review of published data[16] concluded that there appears to be a small amount of evidence suggesting that consumption of food or beverages containing tin at concentrations at or below 200 mg kg^{-1} has caused adverse gastrointestinal effects in an unknown but possibly small proportion of those exposed. At its 55th meeting, the FAO/WHO Joint Expert Committee on Food Additives (JECFA) assessed the available evidence on the acute toxicity of tin, and concluded that it was insufficient to establish an acute reference dose or to derive maximum permissible levels in canned foods and beverages. However, they reiterated their previously stated opinion that the limited human data available indicate that concentrations of 150 mg kg^{-1} in canned beverages or 250 mg kg^{-1} in other canned foods may produce acute manifestations of gastric irritation in certain individuals.

Under normal conditions following closure during the canning operation, tin concentrations in food increase by only about 50 mg kg^{-1} after several months.[73] However, if excess residual O_2 is present in the can headspace, or if nitrate is present in the food, then increased rates of dissolution of tin may be as much as an order of magnitude. Although the use of enamels can prevent such adverse changes in the food, their use is not always practicable. Even in enameled cans, however, high concentrations of tin are possible; for example, with red fruits containing anthocyanin pigments, tin concentrations are sometimes in excess of 100 mg kg^{-1}, the tin being dissolved from scratches and pores in the enamel. Mean levels of tin in welded cans of fruit and vegetables have been reported as 3 mg kg^{-1} in enamel cans and 90 mg kg^{-1} in plain cans.[43]

There is some concern that while intake of inorganic tin may be at acceptable levels, this may not always be the case with organotin compounds (i.e., those in which a tin–carbon bond exists), which are known to have biological activity. It is considered most unlikely that tin in canned foods is anything other than inorganic tin.[73] However, organotin compounds have been used as stabilizers in household products such as polyurethane, PVC packaging materials and silicon-coated baking parchment paper and could migrate into foods under certain conditions. The major concern with organotin compounds arises from tributyl tin (TBT), which is used extensively as a marine antifouling agent. As a consequence, many boating harbors are highly contaminated with the compound and marine foods taken from such waters can have high levels of TBT.[68]

B. Lead

The toxicity of lead, especially to the neonate, is a matter of great concern to regulatory authorities. Abundant evidence supports the fact that during early life, human infants are particularly susceptible to lead exposure, with a greater portion of the retained lead being distributed to bone and brain in infants than in adults. Subacute ingestion of lead by children results in encephalopathy, convulsions and mental retardation. The JECFA has recommended that for adults, the weekly dietary intake of lead should not exceed 50 $\mu g\ kg^{-1}$ body weight, and for infants and young children, it should not exceed 25 $\mu g\ kg^{-1}$ body weight.

For many years, the sideseams of three-piece tinplate cans were soldered with a lead/tin (98:2) solder, resulting in some lead being taken up by the food depending on the amount of solder exposed to the food, the acidity of the food and the time the food has been in the can. Some lead contamination may also originate from the tin coating, which contains a small but finite proportion of lead at levels around 500 $mg\ kg^{-1}$.[59] Regulatory limits for lead in most countries are now 2 $mg\ kg^{-1}$ in canned foods generally, but only 0.5 $mg\ kg^{-1}$ for baby foods and 0.2 $mg\ kg^{-1}$ for soft drinks. EC Regulation 466/2001 limits lead in cows' milk and infant formulae to 0.02 $mg\ kg^{-1}$ and 0.05 $mg\ kg^{-1}$ in fruit juices.

U.S. canners voluntarily stopped using lead solder in 1991. Despite this and a 1995 FDA ban on lead-soldered cans requiring their removal from shelves by June 1996, this source of lead in the diet has not been fully eliminated. Some countries still use lead-soldered cans for food, and these food items may still occasionally be imported, albeit illegally, into the U.S. In addition, some small vendors may still stock old inventories of food in lead-soldered cans. A 1997 FDA investigation found more than 100 such cans in ethnic grocery stores in California alone.

To obtain the lower lead levels in baby foods, it was common to use a pure tin solder that was considerably more expensive than the conventional solder. The newer welded cans have eliminated solder altogether, which has done much to reduce the lead intake from canned foods, typically to about one tenth.[43] Lead levels in welded cans that were enameled were almost one sixth of the levels in plain cans.[43] This contrasts to soldered cans where slightly higher lead levels are generally found in enameled cans compared with plain cans.

The tin/lead capsules used on wine bottles are produced by bonding extremely thin tin foil on both faces of a lead strip, where the tin acts as a barrier layer and prevents contact between the lead and the wine. However, analysis of wine poured from bottles fitted with tin/lead capsules revealed the occurrence of toxicologically unacceptable concentrations of lead in a proportion of the samples tested.[76] The wine may become contaminated by deposits of lead salts produced by corrosion of the lead closure if the wine is poured from the bottle without first wiping the top. About 20% of the wines tested gave rise to a poured sample containing lead in excess of 1 $mg\ L^{-1}$, the statutory limit for lead in wine in the U.K. with one wine containing 21 $mg\ L^{-1}$ when poured. The use of tin/lead capsules was phased out from 1993, and the FDA banned these capsules in 1996 after a study by the Bureau of Alcohol, Tobacco and Firearms found that 3 to 4% of wines examined could become contaminated during pouring from lead residues deposited on the mouth of the bottle by the foil capsule. U.S. winemakers stopped using lead foils before the ban, but older bottles with the foils may still be around.

Recent analysis of wines[7] revealed levels of lead ranging from 40 to 453 $\mu g\ L^{-1}$, the wines having the highest lead contents being 40-year-old ports. The maximum threshold set for lead by the L'Organisation Internationale de la Vigne et du Vin (OIV) is 200 $\mu g\ L^{-1}$ regardless of the type of wine; EC Regulation 466/2001 limits lead in wine to 0.2 $mg\ L^{-1}$, which is the same concentration, making the sale of many wines (especially ports) illegal, if recent analyses[7] are typical. Several factors can affect the dissolution of lead including acidity of the wine and the presence of O_2. Lead migration from glass decanters into alcoholic beverages is another source of contamination (see Section VI).

C. Aluminum

Aluminum has a long history of safe usage in connection with food and food packaging, and is deemed to be GRAS by the U.S. FDA. Aluminum is present in small amounts in a large number of plant and animal species, which is hardly surprising because aluminum is the most abundant and widely distributed of all metals, constituting 8.4% of the Earth's crust. However, aluminum is not a part of any known animal metabolic process, in contrast to a great many less common metals involved in enzymes and other metabolic processes.

Interest in the aluminum content of foods and diets is related to concerns about the possible association of excessive intake or elevated tissue levels of this element with various disorders such as dialysis encephalopathy, osteodystrophy and microcytic anemia. Besides the unavoidable daily intake of aluminum directly via food, humans may be exposed to additional aluminum migration from cooking utensils, storage containers and packaging materials.[59]

The aluminum content of the American diet has been reported.[34,64] Daily intakes in the U.S. were estimated to be 18 to 26 mg in pre-1980 reports. They probably now range from 9 to 14 mg day^{-1}, with the aluminum concentrations in water boiled in an aluminum pan at pH 3 increasing in 20 min from 0.05 to 8.08 mg kg^{-1}. While the use of aluminum cooking utensils in the home has declined, the use of aluminum packaging materials (in particular, aluminum cans and foil) has increased. Although aluminum forms an oxide layer very quickly on exposure to air after manufacture, it can still be attacked by certain foods, especially those containing acids or salts.

As was discussed in Chapter 7, pure aluminum is not used as a food-packaging material. Instead, to provide strength, improve formability and increase corrosion resistance, various alloying elements are added, including iron, copper, zinc, manganese and chromium. These metals, as well as aluminum itself, may migrate into the food if corrosion takes place. Published data on aluminum migration from food-contact materials are sparse.

In a study using 3% acetic acid (pH 2.5) as a simulant,[32] migration of aluminum at 40°C was 3 mg dm^{-2} after 1 day, increasing to 66.5 mg dm^{-2} after 10 days. At 5°C, no traces of aluminum were detected until day 10 when the concentration reached 0.5 mg dm^{-2}.

In one study, carbonated, nonalcoholic beverages had aluminum contents of 107 to 2084 ppb with an average of 830 ppb; carbonated, alcoholic beverages had aluminum contents of 67 to 1727 ppb. Beer in aluminum cans averaged 300 ppb, which is not much more aluminum than is found in beverages in the same class; carbonated wines and wine coolers averaged 970 ppb of aluminum.[71]

Aluminum soft drink cans enameled on the inner surface were fairly resistant to acidic cola drinks for storage periods of 30 to 120 days, the aluminum content of various batches ranging from 15 to 250 ppb with a considerable variability between batches. High aluminum concentrations of 400 to 800 ppb were detected in these beverages after storage periods >400 days, while low aluminum levels of 15 to 20 ppb were found in colas stored in bottles made of glass or PET. Inconsistent quality of the protective stove-lacquering in aluminum cans was suggested as being responsible for the observed effects.[59]

D. Chromium

As discussed in Chapter 7, the tin layer in tinplate cans undergoes a chromium treatment known as *passivation* in order to make it more resistant to oxidation and to improve enamel adherence. The chromium deposition on the tinplate after passivation can amount to 0.5 $\mu g\ cm^{-2}$, and if all this chromium were to dissolve in the food, this would result in the contamination of the contents of a 454-g can with about 0.4 mg kg^{-1}.[43] In a study of canned fruit and vegetables,[43] the mean level of chromium in enamel cans was 0.018 mg kg^{-1}, and in plain cans, 0.090 mg kg^{-1}. This compares to levels in fresh foods of the same type of 0.009 mg kg^{-1}. No cases of intoxication have emerged and it appears that the level of chromium present in tinplate is probably not enough to cause either toxic or adverse organoleptic effects.[59]

The whole surface of ECCS cans consists of chromium oxide, but because it is only about 1/30 of the thickness of the tin layer on tinplate, it is always enameled prior to use. Although some Italian studies have indicated that the absence of a tin layer renders ECCS less suitable for acid fruit packs, this is only likely to be a problem if a major loss of enamel occurred, which is unlikely because ECCS displays excellent enamel adhesion properties.[59]

E. Epoxy Resin Coatings

Epoxy resins are thermosetting resins that contain two or more epoxide (oxirane) groups per molecule and are obtained by condensation of epichlorohydrin and bisphenol A, which yields bisphenol A diglycidyl ethers (BADGEs) of varying degrees of condensation (n), depending on the reaction conditions and the mole ratio of the reactants.[15] While high MW ($n = 9$–11) epoxy resins are used as can enamels, many commercial epoxy resins are BADGE ($n = 0$) free.[75] Powder formulations of high MW epoxies are used mainly to coat the internal surfaces of two-piece DRD food cans, while UV-curable coatings based on low viscosity aromatic, aliphatic and cyclo-aliphatic epoxy resins are used to coat the exterior and ends of cans. The success of epoxies as coatings for food cans is a result of their desirable flavor-retaining characteristics, their excellent chemical resistance and their outstanding mechanical properties. A report[44] into the mechanisms involved in the migration of bisphenol A from can enamels into drinks found that it was necessary to heat the can to a temperature above the glass-transition temperature of the epoxy resin (105°C) in order for the compound to be mobilized.

In the mid-1990s, the U.S. FDA and the SCF in Europe began investigating human exposure to bisphenol A and BADGE in order to ascertain whether the use of certain epoxy resins might be exposing consumers to estrogenic xenobiotics. The FDA concluded that there is no public health concern regarding these chemicals.

EU Directive 90/128/EEC established a SML for bisphenol A of 3 mg kg^{-1} food or food stimulant, and an SML for BADGE of 0.02 mg kg^{-1}. The SCF proposed in 1996 that the SML for BADGE in foods or food simulants be increased to 1 mg kg^{-1} and a surface area-related maximum permitted quantity (QM) value of 0.16 mg dm^{-2}. In 2002, the SCF suggested that the SML for BADGE and some of its derivatives be extended for another 3 years, pending the submission of further toxicological data for evaluation (2002/16/EC).

F. Miscellaneous

Off-flavor in two-piece cans has been shown to be caused by the lubricant used in their production. Fatty acids and esters (normal constituents of the lubricants) are easily oxidized and can contaminate canned beverages such as beer, causing stale, rancid, woody or cardboard-like flavors. A potent flavor constituent of mineral oil (used to aid forming of the cans), which appeared to be naphthenic in nature, has also been found. To eliminate this problem, additional washings were included employing cleaning materials specifically effective at removing fatty acids, esters and mineral oil.

Beer is particularly sensitive to picking up off-flavors from metal packaging.[14] If the enamel coating is disrupted in a steel can, then a metallic flavor will develop as iron migrates into the beer. With aluminum cans, the off-flavor that develops is sulfur-like rather than metallic because of galvanic reactions.[2]

Catty odor has also been reported in canned pork products[31] where the mesityl oxide was present as an impurity in the ketone solvent of an enamel used to cover the soldered sideseams in tinplate cans to prevent blackening of the meat. It reacted with sulfhydryl groups in the meat protein or with free H_2S that was present in trace amounts in the meat. Reformulation of the enamel solvent to exclude ketones eliminated the problem.

V. PAPER PACKAGING

A. Dioxins

Dioxin is the generic name for members of the family of polychlorinated dibenzo-*p*-dioxins (PCDDs) and polychlorinated dibenzofurans (PCDFs), the structures of which are shown in Figure 21.1. The different amounts and locations of the chlorine substituents in these molecules

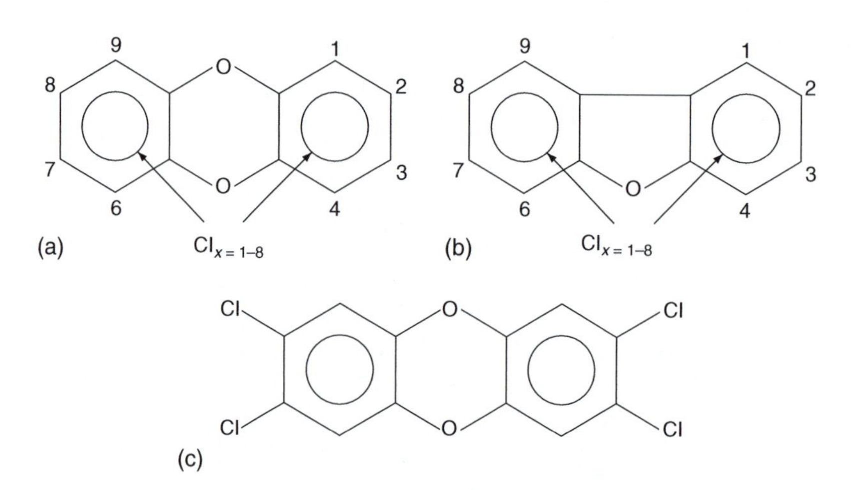

FIGURE 21.1 Structures of: (a) polychlorinated dibenzo-*para*-dioxins (PCDDs); (b) polychlorinated dibenzofurans (PCDFs); and (c) 2,3,7,8-tetrachlorodibenzo-*para*-dioxin (TCDD).

give rise to 75 possible isomers of PCDDs and 135 of PCDFs; these related compounds are known as congeners. There is an enormous body of toxicological information available, although it relates almost entirely to the 2,3,7,8-tetrachlorodibenzo-*p*-dioxin isomer (2,3,7,8-TCDD), the most toxic isomer, which is also depicted in Figure 21.1. In the case of the other isomers, their toxicity is related back to that of 2,3,7,8-TCDD to give the *toxic equivalent* (TEQ) to aid risk assessment. Toxicological and biological data are used to generate a series of weighting factors called *toxic equivalency factors* (TEFs), each of which expresses the toxicity of a "dioxin-like" compound in terms of the equivalent amount of TCCD. Multiplication of the concentration of a compound by its TEF gives a TEQ.[38]

Concern about trace amounts of dioxins in bleached paperboard packaging and the possible migration of the dioxins into milk packaged in paperboard cartons surfaced in North America in 1987. The dioxins arise during the bleaching process to delignify the pulp when chlorine is used. The reported levels of dioxin in paperboard were extremely low and of the order of 4 to 5 ppt (parts per trillion), which is equivalent to 4 to 5 pg kg^{-1}, pg being a picogram that is 1×10^{-12} of a gram. In 1989, concentrations in New Zealand whole milk packaged in chlorine-bleached paperboard cartons were typically 0.1 ppt with the concentration in the fat phase being around 3 ppt. The corresponding figures for milk packaged in glass containers were 0.005 and 0.15 ppt.[3] The figures would be higher in countries where the background levels of dioxins in the environment from sources such as incinerators, car exhausts and so on are greater. Results from a Swedish dioxin survey included data on levels of PCDDs and PCDFs in milk cartons.[67] A carryover rate of about 10% for dioxins from chlorine bleached paperboard cartons into milk and about 28% from coffee filter paper into brewed coffee has been reported.[9]

In 1990, an expert group convened by the WHO Regional Office for Europe recommended a TDI of 10 pg TCDD equivalents per kg bodyweight; in 1999, this was reduced to 1 to 4 pg kg^{-1} bodyweight, the lower end of the range being seen as a target.[38]

Despite the extremely low risk presented by dioxins in milk, suppliers of bleached paperboard have adopted bleaching processes that reduce or avoid the production of dioxins in paper and paperboard.[77] These processes include improving the washing of unbleached pulp to reduce dioxin precursors prior to the bleaching process; avoiding the use of elemental chlorine use in the bleaching process by replacing chlorine with chlorine dioxide or oxygen compounds; and implementing an oxygen or extended delignification process prior to the bleaching process.

B. Benzophenone

Benzophenone (diphenyl ketone, $Ph_2C=O$) is widely used as a photoinitiator for inks and varnishes/lacquers that are cured with UV light. In addition to being a drying catalyst, benzophenone is an excellent wetting agent for pigments and acts as a reactive solvent, increasing the flow of inks. Such inks typically contain 5 to 10% photoinitiator. The use of UV-cured inks for printing cardboard has become widespread because the fast cure permits online cutting and folding, enabling rapid production of finished packaging. Because only a small portion of the initiator is used up during the curing process, benzophenone can remain in the printed material and migrate through the open structure of cartonboard into the packaged food. It may also be present if the cartonboard is made from recycled fibers recovered from printed material.[1]

In an extensive analysis of 350 retail samples for benzophenone residues in cartonboard food packaging and levels of migration to the packaged food itself,[1] 41% had significant (>0.05 mg dm^{-2}) benzophenone with 22% in the range of 0.8 to 3.3 mg dm^{-2}. The highest level was 7.3 mg dm^{-2} found in a high-fat chocolate confectionery product packaged in direct contact with cartonboard. When the mass fraction of benzophenone migration was calculated for different contact and storage regimes, there was a six fold reduction in migration for indirect contact compared with direct contact, a six fold reduction for chilled/frozen storage compared with ambient storage, and a 40-fold reduction for the two contact conditions combined.

Studies of the migration of benzophenone from printed cartonboard have been carried out at freezer temperature and during microwave heating.[42] Benzophenone was found to migrate to the packaged food, even from LDPE-coated board. After 1 week at -20°C, migration was readily apparent with benzophenone being detected in the cartonboard of four out of seven samples at levels of 0.4–3.0 mg dm^{-2}. Benzophenone was also detected in three foods at levels of 0.6–2.9 mg kg^{-1}, which corresponded to a 1 to 2% transfer from the printed board.

The above studies emphasize that the potential for migration to dry and frozen foods cannot be ignored. Although the inks were applied to the outside of the cartonboard, the board itself presented little if any barrier to migration.

In a study[78] on the capability of a PP film barrier to prevent migration of residual contaminants from recycled paperboard into food simulants, benzophenone, anthracene, methyl stearate and pentachlorophenol were chosen as chemical surrogates. Although the concentrations of the surrogates in the food simulants decreased with an increase in PP film thickness, they were still high and generally resulted in dietary concentrations >0.5 μg kg^{-1}, the level that the FDA would equate with negligible risk for a contaminant migrating from food packaging. It was concluded that for an extended time at 100°C, PP would not be an acceptable barrier to migration of contaminants that are expected to be in postconsumer paper/paperboard.

C. Nitrosamines

N-Nitrosamines are a group of environmental carcinogens that have been detected in a wide variety of consumer products including foods and beverages, where they may originate from various food-contact materials such as papers and waxed containers. In these materials, *N*-nitrosomorpholine (NMOR) and morpholine (MOR) are present as contaminants and can migrate into foods that come into direct contact with them for prolonged periods. After ingestion, the migrated amines can form nitrosamines *in vivo* in the acidic environment of the human stomach owing to interaction with salivary or ingested (e.g., through cured meats) nitrite.

Because MOR is widely used as a corrosion inhibitor in boiler feed water, and because large amounts of steam and water are used in the manufacture of paper and paperboard packaging, there is a possibility of finding both MOR and NMOR as contaminants in such products.[41] Of the 34 samples analyzed, nine contained traces (1 to 33 ppb) of NMOR and all 12 samples analyzed contained MOR (98 to 842 ppb). It was recommended that MOR be replaced as a corrosion inhibitor in boiler water with an amine that does not form a stable *N*-nitroso derivative.

D. Chlorophenols and Chloroanisoles

Tainting and off-flavors in foods from contamination by chlorophenols and chloroanisoles originating from packaging material have been well reported and documented.[79] These substances have very low sensory thresholds with the taste threshold in water for 2,4,6-trichlorophenol (described as disinfectant) being 2 $\mu g\ L^{-1}$ and the odor threshold for 2,4,6-trichloroanisole (described as musty) being 3×10^{-8} ppm. Chlorophenols have been used industrially as fungicides, biocides and herbicide intermediates.[79]

The sporadic occurrence of a musty taint in dried fruit packaged in fiberboard cartons and exported from Australia to the Northern Hemisphere results primarily from contamination of the dried fruit with 2,4,6-trichloroanisole (TCA). The corresponding chlorophenol, present in high concentrations in the cartons manufactured from recycled waste materials, was the precursor of the chloroanisole. Fungi present in the fiberboard were implicated in the methylation of the chlorophenol and it has been shown that once the chloroanisole was present in the packaging material, it readily migrated through the LDPE liner into the dried fruit, thereby producing the musty taint. Similar experiences have been recorded with a multiwall paper sack implicated in the tainting of cocoa powder, and jute sacks used to package cereal grains and flour. Seventeen species of fungi with the capacity for methylation of the TCA have been identified, and 10 of these could germinate in packaging materials with a moisture content between 12 and 16%.[82]

A musty cork taint reported in wine was found to be caused by TCA and was capable of producing an off-flavor in good wines at levels as low as 10 ng L^{-1} (ppt). The source of the TCA and other chlorinated compounds was presumed to be from chlorination of lignin-related compounds during chlorine bleaching of the corks.[19,69]

Recently, bromoanisoles have been identified as the cause of musty odor and chemical taste in packaging where, previously, it resulted from the presence of chloroanisoles.[50] The source of these compounds was traced to phenol-based wood preservatives used on wooden pallets, and in particular 2,4,6-tribromophenol, which is converted to 2,4,6-tribromoanisole.[84]

E. Miscellaneous

With the increasing use of microwave-interactive packaging (e.g., susceptors), there is the possibility of degradation products which might adulterate the foods they contact. In an evaluation of the potential dangers of product abuse, the thermal degradation components of a printed popcorn bag, which had been overheated, were identified.[17] The benzene was considered to have arisen from pyrolysis of the PVA used in the adhesive; the source of styrene and α-methylstyrene was the ink, and furfural and acetic acid were degradation products of the paperboard and adhesive.

The possible migration of diisopropylnaphthalenes from recycled paper and paperboard used for food-contact applications has raised health concerns. These constituents are used in the manufacture of carbonless and thermal copy paper, and are not eliminated during the recycling process. As such, they have the potential to migrate into dry foods such as husked rice, wheat semolina pasta, egg pasta and cornflour.[51]

VI. GLASS PACKAGING

Chemically, glass is highly resistant to attack from water, aqueous solutions and organic compounds. Water and acids have very little effect on silica, although they attack some other constituents of the glass. Standard tests have been developed in which glass containers are autoclaved with various test liquids under defined conditions and the liquid analyzed for components present in the glass. Silica and alkali are the main components leached from the glass, and as the initial rate of solution varies approximately with the square root of time, a diffusion mechanism of leaching is suggested.

The main chemicals extracted into aqueous solutions (i.e., silica and sodium oxide) are unlikely to have any significant effect on the organoleptic properties of foods. The danger of contamination by leaching of lead and cadmium from glass into food is remote because these two metals seldom occur in glasses used in food-contact applications. Controlled experiments with lead crystal glasses and decanters[33] found that port wine, originally containing 89 ppb of lead, increased to several thousand parts per billion after 4 months of contact. White wine levels rose steadily from 33 to 68, 81, 92 and 99 ppb after 1, 2, 3 and 4 h, respectively. Lead concentration did not rise in wine poured into nonlead-containing glasses. A wide range of fortified wines and spirits, stored from 6 months to 12 years, showed levels ranging from 11 to 21,500 ppb of lead. By contrast, the U.S. EPA maximum allowable level for lead in drinking water is 15 ppb. However, wine is not considered an important source of lead exposure except by those who drink a lot of it.

Figures presented by Duyvestijn[26] for the total migration of impurities from glass milk bottles were 2.5 mg L^{-1} after 10 days at 40°C in water, and more than 30 mg L^{-1} after 6 months. The migration products consisted mainly of silicon dioxide and metal oxides.

A working document in the EC (DOC 1892/VI/75) proposes that glass and glass ceramic articles meet an overall migration limit of 60 mg L^{-1} or 10 mg dm^{-2}, and lead and cadmium limits of 5 mg L^{-1} or 1 mg dm^{-2} and 0.25 mg L^{-1} or 0.05 mg dm^{-2}, respectively.

REFERENCES

1. Anderson, W. A. C. and Castle, L., Benzophenone in cartonboard packaging materials and the factors that influence its migration into food, *Food Addit. Contam.*, 20, 607–618, 2003.
2. Andrews, D. A., Beer off-flavours — their cause, effect and prevention, *Brew. Guardian*, 116(2), 28–33, 1987.
3. Anon. *A Survey of Some New Zealand Retail Milk Supplies for the Presence of Dioxin*, Department of Health, Wellington, New Zealand, 1989, May.
4. Anon. *Guidance for Industry, Preparation of Food Contact Notifications and Food Additive Petitions for Food Contact Substances: Chemistry Recommendations Final Guidance*, US Food and Drug Administration, Center for Food Safety & Applied Nutrition, Office of Food Additive Safety, 2002, April.
5. Ashby, R., Migration from polyethylene terephthalate under all conditions of use, *Food Addit. Contam.*, 5, 485–492, 1988.
6. Arvanitoyannis, I. S. and Bosnea, L., Migration of substances from food packaging materials to foods, *Crit. Rev. Food Sci. Nut.*, 44, 63–67, 2004.
7. Azenha, M. A. G. O. and Vascobcelos, M. T. S. D., Pb and Cu speciation and bioavailability in port wine, *J. Agric. Food Chem.*, 48, 5740–5749, 2000.
8. Baner, A., Brandsch, J., Franz, R., and Piringer, O., The application of a predictive migration model for evaluating the compliance of plastic materials with European food regulations, *Food Addit. Contam.*, 13, 587–601, 1996.
9. Beck, H., Dross, A., and Mathar, W., PCDD and PCDF levels in paper with food contact. Dioxin-91, *Research Triangle Park*, 152, September 23–27, 1991.
10. Begley, T. H., Methods and approaches used by FDA to evaluate the safety of food packaging materials, *Food Addit. Contam.*, 14, 545–553, 1997.
11. Begley, T. H., Migration from food packaging: regulatory considerations for estimating exposure, In *Plastic Packaging Materials for Food. Barrier Function, Mass Transport, Quality Assurance and Legislation*, Piringer, O.-G. and Baner, A. L., Eds., Wiley-VCH, New York, 2000, chap. 11.
12. Begley, T. H. and Hollifield, H. C., High-performance liquid chromatographic determination of migrating polyethylene terephthalate oligomers in corn oil, *J. Food Prot.*, 53, 1062–1066, 1990.
13. Begley, T. H., Dennison, J. L., and Hollifield, H. C., Migration into food of polyethylene terephthalate (PET) oligomers from PET microwave susceptor packaging, *Food Addit. Contam.*, 7, 797–803, 1990.

14. Bennett, S. J. E., Off-flavours in alcoholic beverages, In *Food Taints and Off-Flavours*, 2nd ed., Saxby, M. J., Ed., Blackie Academic & Professional, Glasgow, 1996, chap. 10.
15. Biles, J. E., White, K. D., McNeal, T. P., and Begley, T. H., Determination of the diglycidyl ether of bisphenol A and its derivatives in canned foods, *J. Agric. Food Chem.*, 47, 1965–1969, 1999.
16. Blunden, S. and Wallace, T., Tin in canned food: a review and understanding of occurrence and effect, *Food Chem. Toxicol.*, 41, 1651–1662, 2003.
17. Booker, J. L. and Friese, M. A., Safety of microwave-interactive paperboard packaging materials, *Food Technol.*, 43(5), 110–118, 1989.
18. Brandsch, J., Mercea, P., and Piringer, O., Modelling of additive diffusion coefficients in polyolefins, In *Food Packaging, ACS Symposium Series #753*, Risch, S. J., Ed., American Chemical Society, Washington, DC, pp. 27–36, 2000.
19. Buser, H. R., Zanier, C., and Tanner, H., Identification of 2,4,6-trichloroanisole as a potent compound causing cork taint in wine, *J. Agric. Food Chem.*, 30, 359–362, 1982.
20. Castle, L., Chemical migration from food packaging, *Food Chemical Safety*, Vol. 1, Watson, D. H., Ed., CRC Press, Boca Raton, FL, 2001, chap. 9.
21. Castle, L., Mercer, A. J., Startin, J. R., and Gilbert, J., Migration from plasticized films into foods. III. Migration of phthalate, sebacate, citrate and phosphate esters from films used for retail food packaging, *Food Addit. Contam.*, 5, 9–20, 1988.
22. Castle, L., Jickells, S. M., Sharman, M., Gramshaw, J. W., and Gilbert, J., Migration of the plasticizer acetyltributyl citrate from plastic film into foods during microwave cooking and other domestic use, *J. Food Prot.*, 51, 916–919, 1988.
23. Castle, L., Mayo, A., Crews, C., and Gilbert, J., Migration of poly(ethylene terephthalate) (PET) oligomers from PET plastics into foods during microwave and conventional cooking and into bottled beverages, *J. Food Prot.*, 52, 337–342, 1989.
24. Chatwin, P. C., Mathematical modeling, In *Migration from Food Contact Materials*, Katan, L. L., Ed., Blackie Academic & Professional, London, 1996, chap. 3.
25. Colborn, T., Dumanoski, D., and Myers, J. P., *Our Stolen Future: Are We Threatening Our Fertility, Intelligence and Survival? — A Scientific Detective Story With a New Epilogue by the Authors*, Penguin, New York, 1997.
26. Duyvestijn, W. J. M., The glass milk bottle is less healthy than cartons, In *Problems in Packaging: The Environmental Issue*, Boustead, I. and Lidgren, K., Eds., Ellis Horwood, Chichester, England, 1984, chap. 13.
27. Federal Register, Food additives: threshold of regulation for substances used in food contact articles (Final Rule), *Federal Register*, 60, 36582–36596, 1999.
28. Franz, R., Kluge, S., Lindner, A., and Piringer, O., Cause of catty odor formation in packaged food, *Package Technol. Sci.*, 3, 89–95, 1990.
29. Gnanasekharan, V. and Floros, J. D., Migration and sorption phenomena in packaged foods, *Crit. Rev. Food Sci. Nut.*, 37, 519–559, 1997.
30. Gold, L. S., Zeiger, E., Eds., *Handbook of Carcinogenic Potency and Genotoxicity Databases*, CRC Press, Boca Raton, FL, 1997.
31. Goldberg, N. and Matheson, H. R., 'Off-flavours' in foods — a summary of experience 1948–74, *Chem. Ind.*, 5, 551–557, 1975.
32. Gramiccioni, L., Cardarelli, E., Milana, M. R., and Denaro, M., An experimental study about aluminum packaged foods, In *Nutritional and Toxicological Aspects of Food Processing*, Walker, R. and Quattrucci, E., Eds., Taylor & Francis, Philadelphia, p. 331, 1988.
33. Graziano, J. H., Slavkovic, V., and Blum, C., Lead crystal: an important potential source of lead exposure, *Chem. Speciat. Bioavailab.*, 3, 81–85, 1991.
34. Greger, J. L., Aluminum content of the American diet, *Food Technol.*, 39(5), 73–80, 1985.
35. Halek, G. W. and Hatzidimitriu, E., Partition coefficients of food package printing ink solvents in soybean oil, chocolate liquor, and a high fat baked product, *J. Food Sci.*, 53, 568–570, 1988; see also p. 596.
36. Hammarling, L., Gustavsson, H., Svensson, L., Karlsson, S., and Oskarsson, A., Migration of epoxidized soya bean oil from plasticized PVC gaskets into baby food, *Food Addit. Contam.*, 15, 203–208, 1998.
37. Harrison, N., Migration of plasticizers from cling film, *Food Addit. Contam.*, 5, 493–499, 1988.

38. Harrison, N., Environmental organic contaminants in food, *Food Chemical Safety*, Vol. 1, Watson, D. H., Ed., CRC Press, Boca Raton, FL, 2001, chap. 8.
39. Heckman, J. H., Safety and regulation, In *Plastics in Food Packaging*, Brown, W. E., Ed., Marcel Dekker, New York, 1992, chap. 10.
40. Helmroth, E., Rijk, R., Dekker, M., and Jongen, W., Predictive modelling of migration from packaging materials into food products for regulatory purposes, *Trends Food Sci. Technol.*, 13, 102–109, 2002.
41. Hotchkiss, J. H. and Vecchio, A. J., Analysis of direct contact paper and paperboard food packaging for N-nitrosomorpholine and morpholine, *J. Food Sci.*, 48, 240–242, 1983.
42. Johns, S. M., Jickells, S. M., Read, W. A., and Castle, L., Studies on functional barriers to migration. 3. Migration of benzophenone and model ink components from cartonboard to food during frozen storage and microwave heating, *Package Technol. Sci.*, 13, 99–104, 2000.
43. Jorhem, L. and Slorach, S., Lead, chromium, tin, iron and cadmium in foods in welded cans, *Food Addit. Contam.*, 4, 309–316, 1987.
44. Kawamura, Y., Inoue, K., Nakazawa, H., Yamada, T., and Maitani, T., Cause of bisphenol A migration from cans for drinks and assessment of improved cans, *J. Food Hygienic Soc. Japan*, 42, 13–17, 2001.
45. Kilcast, D., Organoleptic assessment, In *Migration from Food Contact Materials*, Katan, L. L., Ed., Blackie Academic & Professional, New York, 1996, chap. 4.
46. Kim-Kang, H., Volatiles in packaging materials, *CRC Crit. Rev. Food Sci. Nutr.*, 29, 255–271, 1990.
47. Limm, W. and Hollifield, H. C., Modelling of additive diffusion in polymers, *Food Addit. Contam.*, 13, 949–967, 1996.
48. Linssen, J. P. H., Janssens, J. L. G. M., Reitsma, J. C. E., and Roozen, J. P., Sensory analysis of polystyrene packaging material taint in coca powder for drinks and chocolate flakes, *Food Addit. Contam.*, 8, 1–7, 1991.
49. Linssen, J. P. H., Janssens, J. L. G. M., Reitsma, J. C. E., and Roozen, J. P., Taste recognition threshold concentrations of styrene in foods and food models, In *Foods and Packaging Materials — Chemical Interactions*, Ackermann, P., Jagerstad, M. and Ohlsson, T., Eds., Royal Society of Chemistry, Cambridge, England, pp. 74–83, 1995.
50. Lord, T., Packaging materials as a source of taints, In *Taints and Off-Flavours in Food*, Baigrie, B., Ed., CRC Press, Boca Raton, FL, 2003, chap. 4.
51. Mariani, M. B., Chiacchierini, E., and Gesmundo, C., Potential migration of diisopropylnaphthalenes from recycled paperboard packaging into dry foods, *Food Addit. Contam.*, 16, 207–213, 1999.
52. Middlekauf, R. D., Delaney meets de minimus, *Food Technol.*, 39(11), 62–69, 1985.
53. Miltz, J., Hoojjat, P., Han, J. K., Giacin, J. R., Harte, B. R., and Gray, I. J., Loss of antioxidants from high-density polyethylene, In *Food and Packaging Interactions, ACS Symposium Series #365*, Hotchkiss, J. H., Ed., American Chemical Society, Washington, DC, 1988, chap. 7.
54. Ministry of Agriculture, Fisheries and Food. *Survey of Vinyl Chloride Content of PVC for Food Contact and of Foods, Food Surveillance Paper No. 2,* Her Majesty's Stationery Office, London, England, 1978.
55. Ministry of Agriculture, Fisheries and Food, *Plasticizers: Continuing Surveillance, Food Surveillance Paper No. 30*, Her Majesty's Stationery Office, London, England, 1991.
56. Ministry of Agriculture, Fisheries and Food. *Epoxidised Soya Bean Oil Migration from Plasticized Gaskets, Food Surveillance Paper No. 186*, Her Majesty's Stationery Office, London, England, 1999.
57. Monsanto, v. K., 613 F.2d 947, 955 (DC Circuit 1979).
58. Munro, I. C., Hlywka, J. J., and Kennepohl, E. M., Risk assessment of packaging materials, *Food Addit. Contam.*, 19, 3–12, 2002.
59. Murphy, T. P. and Amberg-Müller, J. P., Metals, In *Migration from Food Contact Materials*, Katan, L. L., Ed., Blackie Academic & Professional, London, 1996, chap. 6.
60. Nerin, C., Cacho, J., and Gancedo, P., Plasticizers from printing inks in a selection of food packagings and their migration to food, *Food Addit. Contam.*, 10, 453–460, 1993.
61. Page, B. D., An overview of analytical methods for phthalate esters in foods, In *Food and Packaging Interactions, ACS Symposium Series #365*, Hotchkiss, J. H., Ed., American Chemical Society, Washington, DC, 1988, chap. 10.

62. Page, B. D. and Lacroix, G. M., Studies into the transfer and migration of phthalate esters from aluminium foil-paper laminates to butter and margarine, *Food Addit. Contam.*, 9, 197–212, 1992.
63. Passey, N., Off-flavors from packaging materials in food products — some case studies, In *Instrumental Analysis of Foods: Recent Progress*, Vol.1, Charalambous, G. and Inglett, G., Eds., Academic Press, New York, p. 413, 1983.
64. Pennington, J. A. T., Aluminum content of foods and diets, *Food Addit. Contam.*, 5, 161–232, 1987.
65. Piringer, O.-G. and Baner, A. L., *Plastic Packaging Materials for Food. Barrier Function, Mass Transport, Quality Assurance and Legislation*, Wiley-VCH, New York, 2000.
66. Piringer, O.-G. and Rüter, M., Sensory problems caused by food and packaging interactions, In *Plastic Packaging Materials for Food. Barrier Function, Mass Transport, Quality Assurance and Legislation*, Piringer, O.-G. and Baner, A. L., Eds., Wiley-VCH, New York, 2000, chap. 13.
67. Rappe, C., Lindstrom, G., Glas, B., Lundstrom, K., and Borgstrom, S., Levels of PCDDs and PCDFs in milk cartons and in commercial milk, *Chemosphere*, 20, 1649–1656, 1990.
68. Reilly, C., *Metal Contamination of Food: Its Significance for Food Quality and Human Health*, 3rd ed., Blackwell Science, Malden, Massachusetts, 2002.
69. Reineccius, G., Off-flavors in foods, *CRC Crit. Rev. Food Sci. Nut.*, 29, 381–402, 1991.
70. Rossi, L., European Community legislation on materials and articles intended to come into contact with food, In *Plastic Packaging Materials for Food. Barrier Function, Mass Transport, Quality Assurance and Legislation*, Piringer, O.-G., Ed., Wiley-VCH, New York, 2000, chap. 12.
71. Schenk, R. U., Bjorksten, J., and Yeager, L., Composition and consequences of aluminum in water, beverages and other ingestibles, In *Environmental Chemistry and Toxicology of Aluminum*, Lewis, T. E., Ed., Lewis Publishers, Chelsea, MI, 1989, chap. 14.
72. Sheftel, V. O., *Indirect Food Additives and Polymers: Migration and Toxicology*, Lewis Publishers, Boca Raton, FL, 2000.
73. Sherlock, J. C. and Smart, G. A., Tin in foods and the diet, *Food Addit. Contam.*, 1, 277–282, 1984.
74. Shorten, D. W., Polyolefins for food packaging, *Food Chem.*, 8, 109–119, 1982.
75. Simal-Gándara, J., Paz-Abuín, S., and Ahrné, L., A critical review of the quality and safety of BADGE-based epoxy coatings for cans: implications for legislation on epoxy coatings for food contact, *Crit. Rev. Food Sci. Nut.*, 38, 675–688, 1998.
76. Smart, G. A., Pickford, C. J., and Sherlock, J. C., Lead in alcoholic beverages: a second survey, *Food Addit. Contam.*, 7, 93–99, 1990.
77. Söderhjelm, L. and Sipiläinen-Malm, T., Paper and board, In *Migration from Food Contact Materials*, Katan, L. L., Ed., Blackie Academic & Professional, London, 1996, chap. 8.
78. Song, Y. S., Begley, T., Paquette, K., and Komolprasert, V., Effectiveness of polypropylene film as a barrier to migration from recycled paperboard packaging to fatty and high-moisture food, *Food Addit. Contam.*, 20, 875–883, 2003.
79. Tice, P., Packaging material as a source of taints, In *Food Taints and Off-Flavours*, 2nd ed., Saxby, M. J., Ed., Blackie Academic & Professional, Glasgow, 1996, chap. 7.
80. Tice, P., A short review of recent migration research, In *Food Packaging Migration and Legislation*, Ashby, R., Cooper, I., Harvey, S. and Tice, P., Eds., Pira International, Leatherhead, England, 1997, chap. 8.
81. Till, D. E., Ehntholt, D. J., Reid, R. C., Schwartz, P. S., Sidman, K. R., Schwope, A. D., and Whelan, R. H., Migration of BHT antioxidant from high density polyethylene to foods and food simulants, *Ind. Eng. Chem. Prod. Res. Dev.*, 21, 106–113, 1982.
82. Tindale, C. R., Whitfield, F. B., Levingston, S. D., and Nguyen, T. H. L., Fungi isolated from packaging materials: their role in the production of 2,4,6-trichloroanisole, *J. Sci. Food Agric.*, 49, 437–447, 1989.
83. Varner, S. L. and Breder, C. V., Liquid chromatographic determination of residual styrene, *J. Assoc. Off. Anal. Chem.*, 64, 647–652, 1981.
84. Whitfield, F. B., Hill, J. L., and Shaw, K. J., 2,4,6-tribromoanisole: a potential cause of mustiness in packaged food, *J. Agric. Food Chem.*, 45, 889–893, 1997.
85. Zeckhauser, R., Social and economic factors in food safety decision-making, *Food Technol.*, 33(11), 47–52, 1979; see also p. 60.

22 Food Packaging and the Environment

CONTENTS

I. INTRODUCTION

It is no longer possible for those involved in the design, development, production or use of packaging and packaging materials to remain oblivious to the environmental demands now placed on them. These demands arise as a consequence of both the materials and processes that are used, and the packaging that is produced, utilized and discarded.

When the public think about packaging, they equate it to waste in their garbage bin, litter in the streets (waste in the wrong place) and excessive or deceptive packaging; these images dominate the public perceptions of packaging, which are created partly by personal experience of its use, partly by personal attitudes to environmental issues and partly by media coverage.[23] The fact that food packaging has almost fulfilled its various functions by the time that the public consumes the food product partly explains why their perceptions are so negative.

In an attempt to understand current attitudes surrounding packaging, Levy[23] has usefully defined myths, facts and realities. *Myths* are fictitious; *facts*, by definition, can be measured and quantified; and *perceptions* are often a mix of myths and facts. A sense of reality is sometimes hard to come by because it can easily be lost in the emotional persuasiveness of perceptions. Realities often relate more to the myths than the facts. Myths persevere because the known facts (quantified data) are sparse or incomplete. Where facts do exist, many incorrect perceptions are perpetrated by those who do not wish to face up to them for whatever reason.

While not all public perceptions of packaging are mythical, most of them tend to be negative and ignore the key benefits and functions of packaging. The most commonly cited myths and perceptions about packaging are[23]:

- It is an unnecessary indulgence on the part of affluent societies.
- It fills the garbage bin and the amount of waste is growing.
- It is the greatest single cause of unrecoverable waste.
- It is disposed of by methods which harm the environment.
- It wastes scarce materials and energy.
- It is not recycled and reused enough.
- It should be returnable for reuse.
- It is a cause of litter.
- It is excessive and products are overpackaged.
- It is deceptive.
- It is only used to promote and help sell the product it contains.
- It should be biodegradable.
- It contributes to pollution.

Rather than rebut each of these myths and perceptions individually, the remainder of this chapter will describe how packaging waste can be, and is being, managed. It discusses how the environmental impacts of packaging can be assessed, the policies on packaging and the environment in the U.S. and Europe and the key issues concerning packaging and sustainability.

A. What is Waste?

Municipal solid waste (MSW) — more commonly known as trash, garbage, refuse or rubbish — is simply what is left of the products that have been used or consumed and are no longer needed. Its consists of everyday items generated by homes, offices, institutions and small businesses such as product packaging, grass clippings, furniture, clothing, food scraps, newspapers, appliances, paint and batteries. MSW does not include construction and demolition debris, nonhazardous industrial wastes, or other nonhousehold and nonbusiness refuse such as municipal wastewater treatment sludge.

Waste is an inevitable product of society and has been described as the "effluence of affluence." Waste generation is directly linked to the economic structure of a country, with MSW being closely linked to demography, urban or rural location and culture. The U.S. and the EU countries contain about one sixth of the world's population, produce and consume more per person than the global average and generate more than one fourth of the world's MSW. However, developing countries produce more waste from a given amount of production and per dollar of GDP. Although they account for less than one half the world's GDP, they produce nearly three quarters of its MSW. Paper is the largest component (by weight) in high-income countries, while food waste predominates in low-income countries.

In 2001, U.S. residents, businesses and institutions produced more than 208 million tonnes of MSW, which is approximately 2.0 kg of waste per person per day, up from 1.22 kg per person per day in 1960 but a decrease of 1.2% from 2000.[15] The packaging fraction of this MSW was approximately 32% w/w if corrugated boxes (secondary packaging) and wooden packaging are included and 16% if they are excluded (i.e., primary packaging made up 16% of MSW). It consisted of 52% paper and paperboard, 15% glass, 15% plastics, 6% metals and 11% wood. Food scraps were 11% of total MSW in 2001 compared with 14% in 1960.

In the EU, packaging waste represents 16% of MSW and 2% of other solid waste streams, making up in total 3% of total solid waste generation.[16] In a detailed analysis of the composition of MSW in 10 cities located in eight EU countries, it was found that packaging varied between 20 and 36% by weight of MSW. The relative quantities of the different packaging materials also varied: paper from 15 to 42%; glass from 5 to 17%; metal from 2 to 11%; and plastics from 5 to 14%, with the organic fraction varying from 19 to 50%. MSW in a specific city varied depending on the season, the housing type (less MSW in high-rise compared to low-rise housing) and the day of the week.

In most countries, a large part of household waste still goes to landfill or incineration without energy recovery. Materials that do not go to these so-called *final disposal options* but instead are recycled, composted or incinerated with energy recovery are often classified as "recovered" or "diverted" and, in the EU, the application of these processes to packaging is referred to as "valorization."

On a per person basis, modern household waste production may not be much higher than early last century, when coal ash and horse manure were significant sources of waste in cities. Coal ash production alone created an estimated 1.5 kg of waste per person per day in Manhattan in the early 1900s. It should always be remembered that nineteenth-century cities were hardly pristine, with trash and human waste routinely dumped into local waterways or primitive sewers that flowed into harbors, creating "foul air" as the tide came in, particularly on warm summer nights. The same is still true today in the large cities of many developing countries.

II. WASTE MANAGEMENT OPTIONS

A. Hierarchy of Waste Management

In its 1989 report entitled *The Solid Waste Dilemma: Agenda for Action*, the EPA outlined what is referred to as a hierarchy of waste management options, with reuse, reduction and recycling at its apex and landfilling and incineration at its base. Several variations of the hierarchy are currently in circulation and one that is widely used is shown in Table 22.1. It is important to note that the hierarchy is not the result of any scientific study of waste management options. It makes no attempt to measure the impacts of individual options or of the overall system. Despite these shortcomings, the hierarchy has become accepted as dogma in some countries and among some policymakers,

TABLE 22.1
A Hierarchy of Solid Waste Management

Waste minimization
Source reduction
Resource conservation
Reuse
Materials recycling (including composting)
Incineration with energy recovery
Incineration without energy recovery
Landfill

politicians and environmentalists who insist, for example, that reuse is always preferable to recycling, despite the realities in a specific geographical location (e.g., the distance that refillable bottles might have to travel to be refilled).

Several MSW management practices, such as source reduction, recycling and composting, prevent or divert materials from the waste stream. Source reduction involves altering the design, manufacture or use of products and materials to reduce the amount and toxicity of what gets thrown away. Recycling diverts items such as paper, glass, plastics and metals from the waste stream. These materials are sorted, collected and processed and then manufactured, sold and bought as new products. Composting involves microbial decomposition of organic waste such as food scraps and yard trimmings, as well as uncoated paper and other biodegradable packaging materials, to produce a humus-like substance.

Other practices address those materials that require disposal. Landfills are engineered areas where waste is placed into or on the land. Landfills usually have liner systems and other safeguards to prevent groundwater contamination. Combustion is another MSW practice that has helped reduce the amount of landfill space needed. Combustion facilities burn MSW at a high temperature, reducing waste volume and, in many cases, generating electricity from the waste heat.

Given the wide variability in MSW composition, it follows that there can be no single, global solution to the issue of packaging recovery and recycling. Specific technical approaches for each waste management program will be required, reflecting geographic differences in both composition and the quantities of waste generated, as well as differences in the availability of some disposal options (e.g., MSW incinerators are rare in many countries). The economic costs of using different waste management options also show large variations between and within countries (e.g., the costs for sorting collected postconsumer packaging).

Since the mid-1990s, the concept of integrated waste management has begun to replace the hierarchy as a more useful, organizing framework for thinking holistically about waste management. It recognizes that all disposal options can have a role to play in integrated waste management and stresses the interrelationships between the options. Today, a mix of waste management options is employed depending on the specific local conditions, the objective being to optimize the whole system rather than its parts, making it economically and environmentally sustainable.[37]

B. Source Reduction

Source reduction (also termed *resource conservation*, *waste prevention* or *lightweighting*) is at the top of the waste management hierarchy and for a very good reason; source reduction translates into less total material use and, consequently, less waste at the end. Lighter packages also require less energy for transportation, thus reducing the environmental impacts from energy production and use.

There have been dramatic reductions in the weight of basic food packages over the last 40 years, driven largely by economic rather than environment reasons (packaging is an extremely cost-competitive industry). Some examples are shown in Table 22.2.

TABLE 22.2
Lightweighting of Packaging

Packaging Type	Year	Unit	Reduction/Comment
Glass milk bottle	1970	340 g	
	1999	220 g	35% weight reduction
330 mL glass bottle	1980	270 g	
	1999	200 g	26% weight reduction
Aluminum beverage can	1970	21 g	
	1998	12 g	43% weight reduction
Steel beverage cans	1985	24,000 cans	Yield of 1 tonne of tinplate
	1990	30,000 cans	25% yield increase per tonne
	1998	32,000 cans	33% yield increase per tonne
1 L paperboard aseptic beverage carton	1980	34 g	
	2003	27 g	20% weight reduction
HDPE milk jug	1965	120 g	
	1990	65 g	46% weight reduction
2 L PET carbonated beverage bottle	1980	68 g	Includes PP base cap
	1999	44 g	No base cap; 35% reduction

More recent data can be obtained at various internet sites, including www.cancentral.com/gacr/ffacts.htm; www.epa.gov/epaoswer/non-hw/muncpl; www.epa.gov/epaoswer/nonhw/muncpl/paper.htm; and the industry websites to which they are linked.

While the potential merits of source reduction are widely acknowledged, less well understood are the potential trade-offs between source reduction and recycling in certain circumstances. A U.S. Office of Technology Assessment report[27] describes the modern snack chip bag to illustrate this point. These bags are now made of thin laminated layers of nine different lightweight materials, each of which serves a different function in assuring overall product integrity and consumer convenience. Although this multilayering makes recycling difficult and economically unattractive, the package is much lighter than an equivalent package made of a single recyclable material and provides longer shelf life, resulting in less food waste.

While lightweighting of primary packaging provides an obvious environmental benefit in reduced material use, it is very important to recognize that the key factor is the amount of material used in the total packaging system (i.e., including secondary and tertiary packaging).

The EPA has estimated that, in 2001, source reduction avoided the production of 50 million tonnes of waste or 20% of total MSW. Source-reduced packaging contributed 14 million tonnes or 28% of this amount.

C. Recycling

Recycling can be defined as the diversion of materials from the solid waste stream for use as raw materials in the manufacture of new products. All common types of food packaging are technically capable of being recycled. However, whether they are actually recycled in practice depends on where they become waste, the local waste management infrastructure and the availability of recycling/reprocessing capacity (Table 22.3).

1. Closed-Loop Recycling

Closed-loop recycling refers to the recycling of a particular material back into a similar product; for example, the recycling of glass bottles back into new glass bottles. Corrugated boxes and glass bottles are typically recycled in a closed-loop system. Among some sections of the community,

TABLE 22.3
Costs of Alternative MSW Programs (2002 dollars per ton)

	Disposal ($)	Baseline Recycling ($)	Extended Recycling ($)
Landfill	34	0	0
Collection and transport	70	155	127
Recyclables processing	0	95	74
Subtotal	104	250	201
Less recovery	0	68	50
Total	104	182	151

To convert tons to tonnes, multiply by 0.9072.

Source: From Franklin Associates, *Characterization of Municipal Solid Waste in the United States: 1998 Update*, Franklin Associates, Prairie Village, KS, 1999.[17] With permission. (Landfill costs have been updated to reflect 2002 actual costs. All other figures are Franklin Associates' estimates, updated to reflect changes in the cost of living between 1998 and 2002.)

closed-loop recycling has taken on a special significance and in their eyes ranks much higher than open-loop recycling. There is no factual basis for this view and the reason is very simple; the physical environment is unconcerned with whether recycling is open or closed loop. What is important are the environmental impacts from the various recycling options, and the processes that result in the lowest impacts should be preferred, regardless of whether or not they are open or closed loop.

2. Collection and Sorting

Before any postconsumer packaging material can be recycled, it first has to be collected and sorted so that a clean stream of material can be delivered to the recycler. In an ideal world, households would source separate their various used packaging materials into individual bins, which would then be collected, consolidated and sent for recycling. In reality, no one wants to have 15 individual bins in their home, let alone remember what day each will be collected. In an effort to reduce the variety and quantity of used packaging collected from homes, drop-off or bring systems have been developed. These are particularly common in many European cities where residents can take their paper and glass to large bins located in their neighborhood for collection and subsequent recycling. However, even where bring systems for paper and glass operate, few cities collect separated used packaging; most run "commingled" systems in which recyclables are collected from the curbside in a single container and sent to a sorting center for separation. Sorting adds significantly to the cost of producing clean streams of materials for recycling.

A sharp increase in the use of one-way, nonreturnable, disposable packaging for beverages began in the 1960s in the U.S. In 1960, refillables accounted for 95% of soft drinks and 53% of beer containers. By 1970, refillable glass bottles accounted for only 49% of soft drinks and 26% of beer containers.[1] The nonreturnable beverage containers became a highly visible type of litter and the public policy response was to introduce so-called bottle bills in several states (Vermont and Oregon lead the way in 1972, being joined by seven more states by the mid-1980s). In these states, consumers paid a deposit for each beverage container at the point of purchase, which could be redeemed at retail outlets or special centers. Refillable containers (primarily glass bottles) were then returned to bottlers and nonrefillable containers (primarily glass bottles and aluminum cans) were recycled. Deposit systems result in relatively pure streams of packaging to recyclers.

3. Materials Recovery Facility

Sorting of commingled packaging materials takes places at a materials recovery facility [MRF (rhymes with "surf")]. The design and operation of MRFs varies widely within and between countries, with developed countries installing more automated sorting machinery in an effort to increase efficiency and reduce costs. Generally, the commingled material is sent down a conveyor belt and "pickers" remove specific items by hand, a typical sorting procedure being shown in Figure 22.1. Steel cans can be removed magnetically and aluminum material by the use of eddy currents. Mechanical systems use air classification or perforated revolving drums to separate lighter fractions such as plastic from heavier fraction such as glass. Of course, not all the material that enters a MRF can be recovered for recycling and the unrecyclable residue which is taken to landfills can exceed 20% in some cases.

4. Benefits

"Why do we recycle?" is a question which many consumers, policymakers and industry have asked. The answers are neither simple nor unanimous. The speed and spread of recycling programs in the U.S. in the early 1990s has been attributed to the perception of an imminent landfill crisis. For many consumers, recycling has been a sensible response to the profligate lifestyle that began in the 1960s and the increasing number of global environmental problems which industrialization has created (e.g., depletion of the ozone layer and global warming).

The EPA gives the following benefits of recycling:

- Conserves resources for our children's future
- Prevents emissions of many greenhouse gases and water pollutants
- Saves energy
- Supplies valuable raw material to industry
- Creates jobs
- Stimulates the development of greener technologies
- Reduces the need for new landfills and incinerators

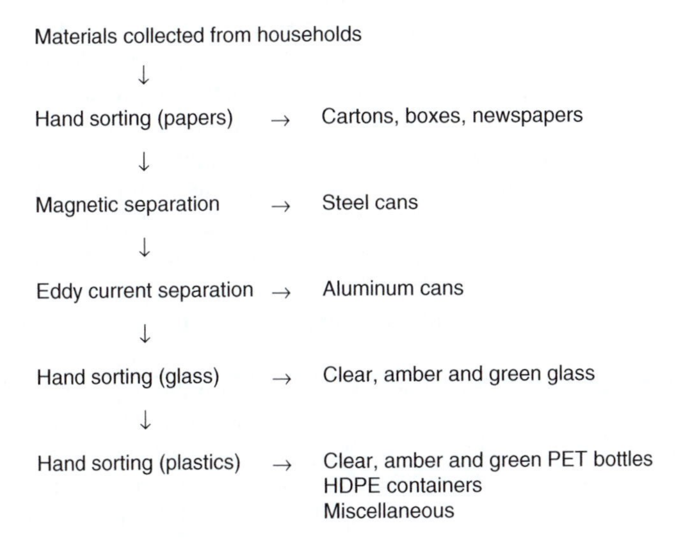

FIGURE 22.1 Typical sorting procedure at a MRF.

Although recycling enjoys broad popular support, there is no consensus about the benefits it is expected to provide. Some people believe that recycling is saving money for their municipality by reducing the costs of garbage collection and disposal. Analysis of U.S. recycling costs, however, suggests that only the best programs consistently save money for their communities. Average programs save money in some years, but add to municipal costs at other times. Another popular reason for recycling is the perceived lack of landfill space, and the fear that there will soon be nowhere left to put garbage.

A scholarly, yet practical, attempt to answer the question "Why do we recycle?" has been provided.[1] It is argued that recycling is a profitable activity only for relatively low-income people (and for a handful of specialized businesses). In the cities of developing countries, countless scavengers pick over the waste found on the streets and in landfills, pulling out anything of value in order to resell it. However, as economies grow and incomes rise, scavenging tends to disappear, displaced by higher-wage occupations. Recycling, as it occurs in developed countries, is a different process, motivated by altruistic concerns for the community, the environment and the future rather than by hope of personal economic gain.

Ackerman[1] suggests that there is still an important meaning and value to recycling, even in the absence of a landfill crisis or an immediate opportunity for profit. Waste disposal is not the whole story; a stronger case for recycling rests on its benefits in resource use and manufacturing. The products that consumers buy and discard are made by industry, and the process of manufacturing most goods has far greater environmental impacts (e.g., much greater toxic emissions) than disposal of the same things in modern landfills. Recycling is good for the environment because making most things out of recycled materials causes lower industrial emissions than making the same things out of virgin material. Using less material in the first place (source reduction) is even better for the environment than recycling. Recycling and waste reduction are also important because they minimize the use of irreplaceable natural resources.

In Ackerman's view, recycling is an attempt to answer at least three questions that the market rarely articulates. First, what new technologies will emerge and become profitable when there is public pressure to reduce, reuse and recycle the material goods that consumers buy? The techniques of production that are profitable today evolved over the last 150 years in a context of cheap, often publicly subsidized, virgin raw materials. Policies that push industry in the opposite direction will, over time, lead to a different set of production techniques and a different calculation of profitability. Industries based on recycling may come to prosper as well, for similar reasons.

Second, what provision should be made for the material welfare of future generations? Economic techniques such as cost-benefit analysis (CBA) are designed to weigh the interests of people who are alive today. Because our obligation to unborn future generations is not expressed in the market, it is not reflected in current prices, costs and benefits. Yet, it is clear that many people care about leaving a livable world for their descendants. Recycling is in part an attempt to conserve materials for the future, even if they appear cheap enough to waste freely at present.

Third, how should the value of nature be described and respected? Air and water pollution, and many other forms of environmental degradation, cause a harm that has no price; they diminish something whose worth was never expressed in monetary terms. Economists have proposed methods for assigning prices to environmental damages, but this has remained largely a theoretical endeavor.

It must be remembered that recycling addresses long-term environmental goals, not an immediate crisis. Thus, it is not necessary — nor is it technically feasible — to recycle everything in the municipal waste stream. Some things remain prohibitively expensive to recycle, and efforts should be concentrated elsewhere. Clarity about the goals of recycling should make it possible to set priorities, selecting areas where the greatest quantities of useful materials can be conserved or recovered, and the greatest environmental gains achieved, at the lowest cost. Clarity about the goals of recycling also leads to the conclusion that waste reduction can be an even better route to the same destination. The objectives that motivate recycling can sometimes be best advanced by minimizing material use, rather than by maximizing recycling.

Ackerman concludes that recycling makes a modest but valuable contribution to the creation of an environmentally sustainable future. It is far from being the most urgent environmental policy initiative under discussion today; it is distinguished by being the most accessible step for millions of people to take in the course of their daily lives. Recycling is the answer to several important questions, including, "What can you, personally, do to help the environment?"

Another academic[36] has argued that in affluent industrial societies, environmental behaviors like recycling are typically classified within the domain of morality in people's minds. Attitudes regarding this type of behavior are not based on a thorough calculation, conscious or unconscious, of the balance of costs and benefits. Rather, they are a function of the person's moral beliefs; that is, beliefs in what is right or wrong. If people are provided with proper information and opportunities, then most are willing to carry some costs if it benefits the environment. However, if an economic incentive is offered to compensate for the private costs from behaving in an environmentally friendly fashion, then this may weaken or destroy the moral obligation.

5. Technologies

a. Paper

Paper packaging can be divided into three main groups: corrugated boxes, cartons and packaging papers. Corrugated boxes are widely recycled, their bulk making them easy and cheap to collect. Cartons and packaging papers used as primary food packaging are also recycled but to a lesser extent. Paper is relatively simple to recycle, the basic process consisting of repulping the fibers by dissolution in water, typically using a hydrapulper (so-called because of the hydraulic forces which are created) as shown in Figure 6.2. Many cartons are coated with a thin layer of LDPE on one or both sides and the paper fibers can also be recovered using a hydrapulper. The hydrapulping process takes from 5 to 20 minutes and is usually continuous. The recycling process shortens the fibers each time they are processed, and after about four cycles, the fibers are too short to stay on the paper-making wire and finish up as a sludge unless they can be used as the middle ply in a multiply board.[27]

A key step in the recycling of paper packaging is removal of the ink, a process known as *de-inking*. The most common technique used for de-inking consists of flotation where soaps are added to the pulp slurry and air is injected into the base of tank. As the air rises to the surface it carries ink particles in the range of 20 to 100 μm suspended on the surface of flotation bubbles. The foam that forms on the surface is then removed. Further washing removes fillers, finely divided ink particles and other colloidal materials from the paper fibers.[19]

Although a large quantity of waste paper is recycled worldwide, it is not generally recommended that packaging material made from it be used in direct contact with food because it cannot be guaranteed to be free from toxic contaminants. There is an increasing lobby that recommends that waste paper packaging should be incinerated with energy recovery[29] and that only pulp made from virgin fiber be used for food contact applications.[28]

b. Steel

Steel cans are either tin- or chromium-plated (ECCS) and there is no simple way of separating them prior to recycling. Because they are magnetic, steel cans are generally automatically sorted at MRFs from other recyclables using a magnetic conveyor belt. These belts usually feed the steel cans directly into automatic baling presses, which produce bales that are ready for shipment. The first step of the recycling process for tin cans is "detinning" the cans. The cans are loaded into a large perforated steel drum and dipped into a sodium hydroxide solution that dissolves the tin from the steel. The detinned steel cans are then drained, rinsed and baled to be sold to steel mills and made into new products. Meanwhile, the sodium stannate solution is filtered to remove scraps of paper and garbage and the tin recovered by electrolysis and formed into ingots.

With a drop in tin prices, a declining percentage of tin in scrap cans and the increasing quantity of ECCS (tin-free steel) cans, detinning has fallen out of favor. In the late 1980s, steel mills began experimenting with direct charging bundled steel cans into their scrap furnaces with positive results.

c. Aluminum

Aluminum cans were the first primary food packaging materials to be recovered and recycled, and recovery rates as high as 85% have been recorded in some countries. After collection and sorting, the cans are crushed and baled for shipment to a recycling plant where they are shredded into small pieces and passed through a magnetic separator to remove any steel present. The ink and enamel coatings are burnt off in a decoating process, where the emissions from this process are recirculated and burned with more fuel to preheat further batches of shredded cans, thus improving energy efficiency and reducing emissions. The hot shreds of aluminum then pass into melting furnaces where, depending on the intended end use, alloying metals are added and the aluminum is then cast into ingots.[28]

Very little aluminum foil is used in the unlaminated state; of the aluminum foil used, very little is recovered for recycling. When aluminum foil is laminated to plastics, it can be recovered either by pyrolysis in the absence of O_2 where the energy in the plastic is captured and used or by dissolving the plastic in a solvent. In both processes, the aluminum foil is recovered in a clean condition but its light weight, coupled with the high cost of separating it at MRFs, has prevented the establishment of any commercial recycling plants.

Aluminum foil used in liquid paperboard cartons (see Chapter 12 for details of the structure) can be recovered as part of the recycling process. The aluminum is sandwiched between layers of LDPE, and after hydrapulping, the combined foil/plastic material is separated from the paper fibers using a simple rotary screen. The combined material can be used "as is" and extruded or injection molded to make a variety of products, or it can be used a fuel. Alternatively, the foil can be recovered using pyrolysis or solvents but this approach is not widespread.

d. Glass

Traditionally, glassworks have used 15 to 20% of cullet (broken or used glass) in their raw material mix, but since the widespread increase in recycling in many countries, the proportion of cullet has increased to 65% or more in many cases, with green glass in Switzerland being manufactured entirely from cullet.[27] Glass for recycling is typically contaminated with a variety of materials including labels, metals, cork and plastics. The glass is crushed in a hammer mill, passed under a magnetic separator and then screened using an air classifier, which removes lightweight material such as paper and plastics.

The key challenge to successful glass recycling is separation of the various colored containers into flint (clear), amber and green. The color distribution of glass in MSW averages 65% clear, 20% amber and 15% green in the U.S.[10] Color separation is generally done by hand at the MRF, although automated processes based on transmittance or reflectance of visible light are used where large quantities of collected glass containers are available. Used glass containers of mixed colors can only be recycled to make amber glass for which there is a finite demand.

A study[20] for the U.S. Department of Energy addressing the question of whether glass container recycling actually saves energy concluded that it does not save much energy or valuable raw material and does not reduce air or water pollution significantly. The most important impacts are the small reduction of waste sent to landfill and increased production rates at glass plants. If used glass containers have to be transported more than 160 km (100 miles) to a MRF then, according to this study, a break-even point is reached and recycling saves no energy.

e. *Plastics*

Plastics packaging is lightweight and functional, and its use in food packaging has increased dramatically over the past 40 years, replacing many of the traditional materials such as glass and metals. Today, plastics account for up to 50% of primary food packaging. Despite the widespread use of plastics films, it is mainly the rigid plastic packaging materials that are recycled, with PET providing the greatest tonnage, followed by HDPE and PP.

Sorting is the critical step in plastic packaging recovery, determining the purity and ultimately the value of the secondary raw materials. The plastics stream, as received from households, consists mainly of films, foamed trays, thermoformed cups and sheets and bottles, which have been used to package food, together with containers for detergents, shampoos and other personal care products. The large majority of the bottles as well as parts of foamed trays and thermoformed materials consist of PET; containers for milk as well as various detergents are composed mainly of HDPE, while LDPE and LLDPE are the main film constituents. Other foamed trays, cups and thermoformed products consist of PP, PS and PVC, the latter also composing a small fraction of beverage bottles.[13]

Most sorting of plastics is still done manually at a MRF as shown in Figure 22.1, although there are large variations between and within countries. The result is a postconsumer recycled (PCR) PET fraction that may or may not consist of bottles of the same color; an HDPE fraction that may be further subdivided into food and nonfood containers (the latter category is subject to heavy odor contamination); and a mixed plastics fraction containing films, trays and other rigid plastic materials. The mixed plastics fraction is typically used for energy recovery, although it can be recycled if a compatibilizer (a rubber-based compound that acts as a binder) is added. One problem with products made from mixed plastics is that they can have variable properties because of the variation in the constituents. The product can be used for fence posts and as a wood substitute.[28] There is now an increasing range of automatic machinery available to perform part, or even all, of the sorting operation. It is based on a variety of technologies which make use of differences in density, optical properties and shape.[4]

After sorting, PCR PET is washed, ground to produce flakes of 4 to 20 mm in size and then washed again. The presence of small quantities of LDPE and PP (mainly from closures) is undesirable and so a separation technique based on flotation in the presence of surfactants is employed. Drying is the final operation, after which the PCR PET flakes are ready to be recycled.

Three types of recycling processes are used for PCR PET flakes: mechanical methods, including melt reprocessing; chemical recycling involving depolymerization, followed by purification and repolymerization; and multilayer extrusion.

Mechanical recycling is defined as reprocessing the PCR PET without changing its chemical structure. It involves washing using a detergent or caustic alkali promoter at temperatures of up to 80°C, depending on the degree of contamination or prior contents of the containers. This is followed by separation involving large tanks or hydrocyclones to remove lower density materials such as PP, LDPE and HDPE, which all have densities of less than 1.0 (PET and PVC have densities >1.33 and cannot be separated from each other using this method). After a final wash in clean water, the flakes are dried and then subjected to a final sort to remove black specks and so on. The flakes may be given a secondary treatment involving further drying (to avoid hydrolytic degradation on melting), followed by extrusion (to remove adsorbed volatile organics) and melt filtration to remove nonmelting, particulate solids and degraded particles. Finally, the PCR PET is extruded and cut into pellets.[26] A major recycling application for PCR PET flakes is the fiber industry, which normally requires lower MW resin than do bottle manufacturers. Typical end-products include fiberfill and carpets for automobiles.

Chemical recycling (also known as *solvolysis*) is much more complicated and energy intensive than mechanical recycling. It requires specialist equipment and, although more expensive than mechanical recycling, can use lower quality feedstock. PET can be recycled by breaking down the polymer chains by treatment with chemicals and two technologies are used commercially: methanolysis and glycolysis. Both cleave the ester linkages in the polymer to form monomers. Methanolysis (using methanol at 250°C) produces stoichiometric amounts of dimethyl terephthalate (DMT) and ethylene glycol. Glycolysis (using an excess of ethylene glycol at 240°C) produces bis-hydroxyethyl terephthalate (BHET) monomer, which is then further treated either via hydrolysis to form terephthalic acid (TPA) or methanolysis to form DMT. These processes effectively depolymerize the PET, with glycolysis having some significant advantages over methanolysis, primarily because BHET may be used as a raw material for either a DMT- or TPA-based PET production process without major modification of the production facility.[26] A third chemical recycling process is hydrolysis using mineral acids or sodium hydroxide to produce TPA and ethylene glycol. However, the subsequent purification steps are very costly and the hydrolysis process is no longer commercial.[10]

An increasingly important use of PCR PET is in the multilayer extrusion of beverage bottles where the PCR PET middle layer is surrounded on both sides by virgin PET. In 1993, the FDA granted a "no objection" for multilayer PET containers, provided there is a 0.025 mm thick inner virgin PET layer to prevent food contact by the PCR PET layer.[14]

The steps involved in recycling HDPE comprise size reduction by grinding, washing, rinsing and drying, with detergents frequently being used to aid label removal. During the rinse cycle, PET and PVC contaminants sink to the bottom of the bath and can be separated from the HDPE.[10] The clean and dry HDPE flakes can be recycled into a number of nonfood applications.

6. Economics

In a landmark national study in 2001, the U.S. Recycling Economic Information (REI)[6] documented the importance of recycling and reuse in the U.S. economy. The report was commissioned by the EPA and several states through a cooperative agreement with the National Recycling Coalition. It involved a comprehensive analysis of both existing economic data and reasonable estimates based on targeted surveys of recycling businesses and sophisticated economic modeling.

According to the study, the U.S. recycling and reuse industry involves more than 56,000 establishments that gross more than $236 billion in annual revenues and employ more than 1.1 million people with an annual payroll of $37 billion.

The study identified 26 different types of recycling organizations, some of which might not commonly be seen as "recycling" businesses, such as steel mills, plastic bottle manufacturers and pavement producers. Four major manufacturing industries account for more than half of the economic activity of the industry: paper mills, steel mills, plastics converters and iron and steel foundries. However, recycling is a diverse industry and includes public sector institutions as well as private businesses. From curbside collection of household recyclables through the brokering or processing of recovered materials to the manufacture of recycled-content products, traditional firms and innovative market sectors support the industry.

Local collection programs play an important role in recycling economics. Although many waste materials are recyclable, their quantities are dispersed across household, business and industry waste streams. Recycling is an integrated system that starts with curbside collection of materials by municipalities, involves processing of recycled materials and leads to manufacturing of new products with recycled content.

While recycling is strong, the report noted that the recycling industry still faces many challenges. Recycling market prices rise and fall in response to many factors that are difficult to manage on a local level. Low-cost manufacturing materials imported into the U.S., global economic downturns and inexpensive disposal rates can impact the cost of recyclable commodities and, therefore, the industry. Many resources are still being thrown away instead of recovered for recycling, and recycled-content products continue to struggle for prominence in marketplaces.

In 1997, the Institute for Prospective Technological Studies[3] was requested by the European Parliament to report on the impediments and prospects for the recycling industry in Europe. The report challenged the current approach of maximizing recycling rates, instead endorsing the need to establish recycling at an optimal rate from both an economic and environmental point of view. It concluded that recycling is not always necessarily the preferable waste management solution because it is limited by the second law of thermodynamics and obeys the law of diminishing returns.

The report noted that the desirability of increased recycling depends on its relative merit compared with other waste management options in a given geographic area. Other options such as prevention, reuse and recovery of energy can offer ecological or economic advantages over recycling depending on the context. As such, the report concludes that the growth of the recycling industry is not necessarily a desirable policy target.

The authors observe that "creating rigid hierarchies of alternative waste management options and dictating solutions without generally acceptable scientific evidence can be suboptimal." The report notes that there must be a framework for "profitable private initiatives," without which substantial recycling cannot be sustained. In their view, the best way to remove all the unfair barriers to the development of recycling would be a transparent procedure to generate realistic recycling targets.

D. Composting

With the exception of plastics, rubber and leather components, the organic fraction of most MSW is composed of proteins, amino acids, lipids, carbohydrates, cellulose, lignin and ash. If these organic materials are subjected to aerobic microbial decomposition, then the remaining end product is a humus material commonly known as *compost*.[35] The EU Directive 94/62/EC has specified that composting of packaging waste is a form of recycling, owing to the fact that the original product (the package) is transformed into a new product (the compost).

During the composting process, the organic material is broken down into CO_2, CH_4 and water vapor. In moving from wet biowaste to normal compost, the weight loss is generally around 50%.[37]

The composting of partially processed, commingled MSW has been suggested as a means of reducing the volume of waste placed in landfills. Use of the composted material as intermediate landfill cover material has also been suggested.[35] In theory, about one third of MSW in the U.S. could be composted: yard trimmings (18% of MSW), food scraps (7%) and mixed paper waste that appears uneconomical to recycle (9%). Although paper and paperboard products make up 41% of MSW in total, many are readily recovered through recycling programs that yield a higher-value product than compost.

The compostable fraction of packaging wastes is the paper, paperboard and wood, because steel, aluminum and glass are totally noncompostable, inorganic materials. Although plastics are organic, they are generally resistant to microbial attack and are therefore noncompostable. However, biodegradable plastics (see Chapter 3) are compostable but at present they make up a negligible part of MSW.[11]

E. Thermal Treatment

Thermal treatment of solid waste within an integrated waste management system can include at least two distinct processes. The most common is mass burning or incineration of mixed MSW in large incinerator plants, either with or without energy recovery. The other involves separating combustible fractions from MSW to form refuse-derived fuel (RDF) and burning them as fuel.

Incinerating MSW reduces the volume of solid waste to be disposed of by, on average, 90%, and the weight by around 70%.[37] Incineration also stabilizes the waste by oxidizing the organic component, making the ash relatively inert and significantly reducing landfill gas and leachate production.

Energy can be recovered from the hot flue gases during the incineration process and used to generate steam for use in electricity generation or district heating schemes (particularly common in northern Europe). The thermal efficiency of modern boilers is around 80%, but if the steam is used to generate electricity, then the overall recovery efficiency (from calorific content of the MSW to electricity generated) is around 20%.[37] Virtually all new incinerators constructed since 1990 employ some form of energy recovery to help offset operating costs and reduce the capital costs of air pollution control equipment.

The energy content of MSW varies according to the country and ranges from 6 to 8 MJ kg^{-1} depending on the proportion of food scraps and green waste. Of the common plastics packaging materials, LDPE has an energy content of 43.6, PS 38.3 and PVC 22.7 MJ kg^{-1}. Beverage cartons (laminates of paperboard, aluminum foil and LDPE) have an energy content of 21.3 MJ kg^{-1}. By comparison, wood chips have an energy content of 8.3, coal 26.0 and oil 41.0 MJ kg^{-1}.

The operation of incinerators results in the production of a variety of gaseous and particulate emissions, many of which are thought to have serious health impacts. In some cases, the cost and complexity of the environmental control systems are equal to (or even greater than) the cost of the combustion facilities.[35]

Several solid residuals are produced by incinerators including bottom ash, fly ash and scrubber product. Bottom ash (about 75% of the total ash produced by an incinerator) is the residue of unburned material from the combustion chamber, and is typically disposed of in landfills after magnets remove any ferrous material (including steel cans) for recycling. A primary concern is that bottom ash may, under certain circumstances, leach contaminants into the groundwater.[35] Electrostatic precipitators and cyclones can be used to remove fly ash (particulates from the flue gases), and acid gases (HCl, SO_2 and HF) can be removed using scrubbers and CaO or NaOH solutions.

Of major concern has been the emission of dioxins from incinerators. Provided that the incineration process is run efficiently (residence time after last air injection of at least 2 sec at 850°C and $\geq 6\%$ O_2), most of the organic pollutants in the incoming waste will be broken down. Although dioxins can be produced *de novo* in the flue gas, the levels of dioxins emitted from MSW incinerators are considerably less that the levels in the input.[37] To meet the latest stringent emission controls for dioxin, further treatment of the flue gases is necessary.

F. Landfill

Landfills are the physical facilities used for the disposal of residual solid wastes in the surface soils of the earth. A *sanitary landfill* refers to an engineered facility for the disposal of MSW designed and operated to minimize public health and environmental impacts.[35] The modern sanitary landfill had its origins in Great Britain in the 1920s and was introduced in the U.S. a decade later.

A popular reason often advanced in support of recycling is the perceived lack of landfill space, and the fear that there will soon be nowhere left to put MSW. In the late 1980s, many believed the U.S. faced a landfill crisis, a perception fueled in part by an EPA study, which focused on the number of landfills (many were closing and few were opening) rather than their capacity (which was growing rapidly), and in part by the publicity surrounding the New York garbage barge Mobro 4000, which had nowhere to unload its cargo of 3000 tonnes of trash. The trash had been headed for a landfill in North Carolina but was rejected as a result of rumors that it contained toxic hospital waste. It traveled to several southern states and Caribbean ports trying unsuccessfully to unload its garbage before returning to New York, where it remained for many weeks before the garbage was unloaded and incinerated.

The great science fiction author Issac Asimov added to the false perception in a 1991 book[5] about environmental issues facing the world claiming that "almost all the existing landfills are reaching their maximum capacity, and we are running out of places to put new ones." The future of traditional waste management was generally agreed to be bleak. However, in most of the U.S., a landfill crisis never arrived, apart from some densely populated parts where a shortage of landfill space is a reality. The same is true for many other countries.

Assuming that Americans continue to produce around 100 million tonnes of garbage each year destined for landfills, it has been calculated[24] that a century of such waste would occupy a landfill only 22.5 km^2 and 33 m high. While this suggests that the U.S. will not become inundated with garbage, it does not imply that landfills will be easy to site because nobody wants to live close to a landfill [which illustrates the 'NIMBY' (not in my backyard) phenomena]. Thus, although landfilling of garbage may be a political problem, it is not a problem of physical space in the U.S.

It is important to put packaging waste into context. Packaging may account for up to one third of the volume of what goes into landfills. In analyzing data from landfill digs for the years 1978 through 1988, it was found that in *per capita* terms, the amount of packaging in MSW in the U.S. experienced a gradual but real decline.[30] In comparison, a year's worth of the *New York Times* newspaper weighs 236 kg and occupies 1.15 m^3 in a landfill. This is equivalent in weight to 17,180 aluminum cans, which is nearly a century's worth of beer and soft drink consumption by an individual.[7]

Landfill regulations in many countries generally include requirements that all landfills have liners, leachate collection and treatment systems, groundwater monitoring, CH_4 control and prefunding of postclosure activities. In addition, these regulations place restrictions on landfill siting. Such regulations have influenced landfill costs in two ways: (1) they have resulted in the closure of a number of existing, noncompliant landfills, thereby reducing available landfill capacity, with the expected upward impact on "price" in the form of tipping fees; (2) they have increased actual construction and operating costs of landfills.

Landfills are constructed according to detailed engineering specifications. A foundation of at least a meter of dense clay is laid down on the site and covered with thick plastic liners known as *geomembranes*. HDPE, very flexible polyethylene (VFPE) and PVC are materials typically used in the manufacture of geomembranes for solid waste landfills. The membrane is then covered by a meter of gravel or sand. As garbage is placed in the landfill, layers of dirt or other inert materials are used to cover it, generally each day. This serves several purposes including prevention of pollution of the surrounding landscape by lightweight materials such as paper and plastic bags, and discourages birds and vermin from living among the garbage.

All landfills produce leachate that must be dealt with. *Leachate* is liquid that has percolated through the layers of waste material. The chemical quality of leachate varies as a function of a number of factors, including the quantity produced, the original nature of the buried waste materials and the various chemical and biochemical reactions that may occur as the waste materials decompose. Thus, leachate may be composed of liquids that originate from a number of sources, including precipitation, groundwater, consolidation, initial moisture content and reactions associated with decomposition of waste materials. Landfills should be designed to prevent any waste or leachate from ever moving into adjacent areas.

The anaerobic decomposition of organic materials in a MSW landfill will generate a combination of gases (collectively called *landfill gas*) at a rate of approximately 0.002 $m^3\ kg^{-1}$ of waste per year. The underground migration of landfill gas can pose serious safety risks in nearby structures when the gas accumulates at concentrated levels. The gas (ca. 50% CH_4 and 50% CO_2) is an excellent energy source. Methane is generally recovered with conventional gas wells and directed to either a gas turbine or internal combustion engine that is used to power an electrical generator. Thus, the energy content of the landfill waste is converted to electricity. As a greenhouse gas, methane has 22 times more effect than CO_2. Consequently, burning CH_4 and converting it to CO_2 reduces the potency of landfill gas as a greenhouse gas.

The most critical factor in determining a landfill's decomposition rate is the moisture content of the waste. Conventional landfills with a tightly sealed cover impede water from entering the waste. Consequently, decomposition is very slow and, in some cases, expected to take up to 1000 years before waste stabilization is completed. Complete stabilization is when the waste material no longer breaks down into by-products that are released into the environment.

Decomposition time can be reduced to a matter of months using solid waste digesters similar to sewage sludge digesters; they can stabilize waste in weeks. These systems, however, have high construction costs, and a more practical approach is to modify an existing landfill operation to accelerate decomposition while maintaining the critical, physical elements of the landfill. Therefore, the current approach to bioreactors is to devise a system in which water is introduced into the waste to wet the material as uniformly as possible. The added moisture then accelerates decomposition, which generates large quantities of landfill gas. With a bioreactor landfill, gas generation begins much faster, and generation rates are much higher.

One myth concerning landfills is that they are giant composters. While some biodegradation does take place (if it did not, then no CH_4 would be produced), biologically and chemically, a landfill is a much more static structure than is commonly supposed. They have been described[30] as vast mummifiers with excavations finding that even after two decades of burial about one third to one half of food and yard waste remains in a recognizable condition, and newspapers can be easily read.

III. LIFE CYCLE ASSESSMENT

Life cycle assessment (LCA) is an environmental management tool, which attempts to consider the resource and energy use and the resultant environmental burdens over the entire life cycle of a package, product or service from extraction of the raw materials through manufacture/conversion, distribution and use to recovery or disposal. It is sometimes referred to as "cradle to grave" analysis and typically compares two or more products that provide the same function or equivalent use. LCAs make it possible to isolate the stages in the life cycle of a process or product that make the most significant contribution to its environmental impacts.[9] The first LCAs were performed on beverage containers in 1972 in the U.S.,[21] and in 1977 in Japan by the Toray Research Institute.

LCA can be subdivided into four phases:

- Goal definition and scoping
- Life cycle inventory (LCI)
- Life cycle impact assessment (LCIA)
- Life cycle interpretation

The ISO 14040 series of international standards provide a framework for conducting LCAs and are intended as nonprescriptive guides because there is no single method or end use for conducting LCAs.

A. Goal Definition and Scoping (ISO 14041)

Here, the purpose and basis of the study, the system boundaries and the procedures for handling data are identified. An important task is to define the "functional unit," which is defined by ISO 14040 as a quantified performance of a product system for use as a reference unit in an LCA study, with all data in the study being related to the functional unit. The chosen functional unit for a system is dependent on the goal and scope definition. It is often expressed in terms of the amount of product (e.g., per kg or L), but should really be related to the amount of product needed to perform a given function (i.e., per equivalent use). For packaging, the functional unit is commonly the packaging required to deliver a given volume of beverage (e.g., 1000 L) or quantity of food (e.g., 1 tonne).

As all of the impacts are calculated per functional unit, any alteration in the size of the unit will have a major effect on the outcome of the assessment. Although many studies have focused on packaging systems in isolation, in most cases, packaging should be considered as an integral part of the product system.[34]

B. Life Cycle Inventory (ISO 14041)

LCI involves materials and energy balances for the entire system; that is, identification and quantification of all the inputs into, and all the outputs from, the system under study. The system must be broken down into individual unit operations and a complete breakdown of all inputs and outputs (both material and energy) must be determined. Material outputs include air, water and solid waste emissions. Difficulties are sometimes experienced in obtaining quality data for each individual unit operation, particularly when these occur in other countries.

Despite the great deal of information provided about energy use and emissions associated with the life cycle of a product or package, the LCI information cannot be generally applied to the product's effect on human health, ecological quality and natural resource depletion. This is because one kilogram of a given emission may provide substantially different effects on human health or the environment than a kilogram of a different emission.[18] The next stage (LCIA) helps to address this issue.

C. Life Cycle Impact Assessment (ISO 14042)

In this stage, the LCI results are converted into a range of environmental indicators in a two-step process. First, the LCI results are organized according to environmental issues or impact groups such as greenhouse gas loading, fossil fuel inputs, ozone depletion potential and nutrient loading. Then, the LCI results for each environmental issue or category are converted using a characterization factor and aggregated into the impact category indicator. The categories chosen for impact assessment should be internationally accepted, scientifically and technically valid, and environmentally relevant.

D. Life Cycle Interpretation (ISO 14043)

This stage involves considerable value judgments and thus needs to be transparent. It involves the evaluation of the different emissions across all impact groups. The lack of a widely accepted and nonsubjective model for impact assessment limits its applicability to industry and in policy-making.

E. Limitations of LCA

Despite the increasing popularity of LCAs by both industry and governments, the technique does have significant limitations which are all too often overlooked. The first is that LCAs are not able to assess the actual environmental effects of emissions and wastes from the product or package. This is because the actual effects will depend on when, where and how they are released into the environment.[34] Releasing an emission from a point source will have a very different environmental effect from releasing it continuously over years from many different sources. Other tools such as risk assessment are able to predict the actual effects that are likely to occur, although they do not link the effects to the functional unit.

A second limitation is that LCAs do not consider the various functions that a package has to perform. Thus, when comparing several different packages for the same product, an LCA takes no account of the convenience function such as whether or not the package is reclosable, easy to pour from and so on. Nor do LCAs consider the protection and preservation function of the packages, so that if, for example, one of the packages provides 50% greater shelf life than the other, this is not taken into account. Finally, the communication function is also ignored; this limitation could be significant from a marketing viewpoint, with some packages providing a much greater shelf-facing surface area for branding and so forth.

A third limitation is that LCAs take no account of economic factors such as the costs of raw materials, manufacturing, transport and recovery or disposal. In the extremely competitive packaging industry, cost is of prime importance.

A fourth (frequently overlooked) limitation is that the conclusions from LCAs are specific to the precise system under study. They cannot be extrapolated to provide universal generalizations that one particular package is always better than another in every situation.

A fifth limitation is the desire to use LCAs in public policy to determine whether a particular package is good or bad for the environment, or whether a particular waste management option is appropriate. The EU Directive 94/62/EC states that LCAs should be completed as soon as possible to justify a clear hierarchy between reusable, recyclable and recoverable packaging. The problem with, and indeed impossibility of, using LCAs in this way is discussed below.

In 1997, the European Commission funded a study[31] to determine the effectiveness of LCA when applied to policy making, and whether a clear hierarchy could be identified between reusable and one-way packaging, and between recycling and incineration. The report made a number of recommendations of which only a few will be mentioned here. It concluded that the recycling targets contained in the 1994 directive cannot be supported by the results of their study. It also found that a significant improvement (reduction) in environmental impacts is unlikely to be achieved by switching from one packaging system to another. It concluded that a clear preference or absolute lowest impact option is never found, and a case-by-case approach is necessary owing to variations in local conditions.

Comparing the environmental impacts of material recycling and incineration with energy recovery, the study concluded that glass recycling should be maximized; aluminum and steel recycling should be high although postincineration recycling could also be acceptable; paper and paperboard recycling should be promoted if, and only if, market conditions are favorable (i.e., a high demand for recycled fibers); and plastics recycling should be analyzed further because the preference for this waste management option depends on several technical factors.

In 2001, the European Commission funded a CBA that investigated the costs and benefits of reusable primary packaging, comparing 330 mL, refillable and one-way glass bottles, and 1.5 L, refillable and one-way PET bottles. The study concluded that from a total social cost perspective, there is no case for universally encouraging the use of returnable packaging.[32]

In 2004, EUROPEN commissioned consultants URS to review LCAs in the public domain conducted between 1993 and 2003 that concerned refillable and one-way packaging. Specifically, they sought to determine whether reusable packaging or recycling of one-way (single use) packaging is the best system. The review[22] found that the environmental benefits of reuse and recycling are indistinguishable. Although some studies showed that reuse is preferable to recycling, others showed the opposite. It was not possible to draw a blanket conclusion owing to the need to take into account many factors that affect the outcomes of the comparisons made.

F. Uses of LCAs

Despite the limitations described above, LCA can be a very useful tool in two major areas. First, package design, development and improvement all benefit from having LCA results available. An LCA can help identify where significant resource use, wastes and emissions occur, and thus suggest where significant changes or improvements can be made. Sometimes, only an LCI is performed, which is used as an initial screening tool when a number of options are being evaluated. In other cases, only life cycle energy is calculated, which is used as a surrogate for environmental impacts. In some situations where a "quick and dirty" comparison is sufficient, only the weight of the competing packages is used because the lightest-weight package almost always has the lowest environmental impacts (PVC would be one exception).

Secondly, LCA is a useful tool for assessing waste management options. The waste management hierarchy (see Section II.A) has no scientific or technical basis and cannot deal with combinations of treatment options. Instead of relying on a rigid hierarchy, waste planners and managers are turning to LCA tools to help them plan integrated solid waste management systems on a case-by-case and regional basis.[34]

IV. PACKAGING AND ENVIRONMENTAL POLICIES

A. North America

Polls of North Americans in the early 1990s showed them ranking MSW high among their list of environmental concerns, though environmental scientists put solid waste near the bottom of important environmental problems. The perception of the solid waste problem was predicated on two beliefs: (1) that disposal capacity could not keep pace with current waste generation rates; (2) that disposal was unsafe, whether in landfills or incinerators. Perceptions of a landfill crisis resulted in legislation in 41 U.S. states to divert waste from landfills.

Examination of the facts confirms that perceptions of a landfill crisis were misplaced. The amount of landfilled MSW in the U.S. peaked in 1987 at about 82%; 9% was incinerated and the balance (9%) recycled. A decade later, 27% of MSW was recycled or composted and less than 60% was landfilled, while the amount incinerated remained flat. Significantly, both the percentage and absolute tonnage of waste landfilled in the U.S. declined over this time. In the U.S. in 2001, 29.7% of MSW was recovered with 22.4% being recycled and 7.3% composted, 15% was burned at combustion facilities and the remaining 56% disposed of in landfills.[15] It has been calculated that more adults participate in recycling on a regular basis than vote for President, which led U.S. journalist Jerry Powell to remark that recycling is more popular than democracy. Landfill capacity does not appear to be a problem today although regional dislocations sometimes occur. The number of landfills in the U.S. has declined from 8000 in 1988 to 1858 in 2001.

In 1976, the Resource Conservation and Recovery Act (RCRA) gave the EPA broad authority over landfills and incinerators and established national environmental standards for landfills which were made stricter by amendments passed in 1984. A major contribution of the EPA has been issuing statistical reports on the management of solid waste, each report appearing about 22 months after the end of the calendar year.[15]

In the U.S., the federal government has not introduced national packaging goals, nor has it embraced the notion of extended producer responsibility (EPR). This contrasts to Europe, which has applied the "polluter pays" principle to packaging and insisted that manufacturers take back the packaging they put on the market. The federal government's most visible role in recycling matters has come through federal procurement policy rather than through any comprehensive national legislation. Paper has been the primary target of these efforts to date, with packaging receiving only occasional consideration.

Dr. Winston Porter, the assistant administrator for solid waste at the EPA, established a national goal in 1989 for recycling 25% of the country's MSW, and this nonbinding goal was reached in 1995. By 1995, nearly all states had implemented laws requiring or encouraging household and commercial recycling. The number of curbside recycling programs in the U.S. jumped from a few hundred to over 9300 in the period from 1989 to 2001.

Legislation at the state level remains more significant than federal legislation. Although some 10 states have deposit systems for beverage cans and bottles (so-called *bottle bills*), there is very little legislation dictating how postconsumer packaging should be handled. Local municipalities have lead the way by introducing curbside collection of recyclables and consumers have generally participated and actively supported these initiatives.

1. Deposits

In response to the increase in one-way (nonrefillable) beverage packages from the 1960s, some cities, states and countries have introduced mandatory deposit/refund systems, typically for soft drink and beer containers. However, these systems were not introduced primarily to increase recycling rates, but rather in response to a rising tide of litter in which one-way beverage containers used to feature quite prominently.[1] Although containers collected under deposit/refund systems appear to achieve higher recycling rates than are achieved in the absence of such programs, they are much more costly than curbside programs, this additional cost ultimately being paid by the consumer. Today, with the widespread access to curbside recycling programs in many cities, deposit/refund systems are no longer seen as necessary or efficient.[2]

2. Extended Product Responsibility

In 1996, the U.S. President's Council on Sustainable Development introduced the term *extended product responsibility* to mean that all participants in the product life cycle (including government, industry with an economic interest in the product, consumers and waste handlers) share responsibility for the environmental effects of products and waste streams. Responsibilities were viewed to shift, depending on the entity that most directly generated an environmental impact and had an ability to mitigate that impact. The concept has not found its way into any legislation and has created some confusion with the quite different concept of EPR, which is widely applied in the EU (see below).

3. Packaging as Part of MSW

Total MSW generation grew from around 80 million tonnes in 1960 to 210 million tons in 1999 and then declined slightly to 208 million tonnes in 2001. The total number of containers and packaging has remained relatively constant over this period at around 32%, but there have been significant shifts in the composition of the containers and packaging during this time. Glass packaging has declined from a peak of 9.8% in 1970 to 4.8% in 2001, while plastics packaging has increased from just 0.1% in 1960 to 4.9% in 2001. Corrugated boxes have also increased from 8.3% in 1960 to 12.6% in 2001. Steel cans have declined from 5.0% in 1960 to 1.0% in 2001. These numbers reflect many interesting changes in packaging design and development, materials substitution and light weighting or source reduction.

B. Europe

Europeans have tended to regard packaging as a form of pollution and applied the "polluter pays" principle to packaging manufacturers, requiring them to take back the used packaging at their cost. However, packaging *per se* is not pollution although it may cause environmental impacts if it is littered or otherwise disposed of improperly. In fact, packaging prevents waste by enabling products to be distributed and consumed long distances from their point of manufacture with a minimum of environmental impacts.

1. Producer Responsibility

The idea that producers should be responsible for the total life cycle of the products they manufacture and that responsibility should not be limited to production but extended to the actual products was first introduced into legislation in Sweden in 1975. Since then, two variations have been introduced: shared producer responsibility, where the producer of the goods may be considered to share responsibility across the life cycle with all the others involved; and EPR, where the producer of the goods alone has responsibility across the life cycle.

A critique of EPR[33] pointed out that packaging is owned throughout its life cycle from production through sale, consumption, recovery and disposal, although who owns the package or the residue changes as it changes hands. With ownership comes clearly defined rights and responsibilities. Relieving consumers of responsibility in the disposal of used products and making producers responsible decouples consumers from the environmental aspects associated with selection. Despite these criticisms, EPR is an established concept in the EU although, today, the term *shared responsibility* is also used.

2. German Packaging Ordinance

The first real application of the EPR concept in Europe was the 1991 German Packaging Ordinance where all the costs for managing packaging waste had to be borne by the economic operators who placed goods into the market. This law led to the formation of the company Duales System Deutschland GmbH (DSD), which operated a separate but parallel municipal packaging waste management infrastructure financed by the producers and importers of packaged goods. To indicate their participation in the scheme, the DSD licenses a mark known as the *Grüner Punkt* (or *Green Dot*), which is placed on packaging for which the appropriate fee has been paid. The DSD contracts with disposal companies for the collection and separation of packaging using a combination of bring and pick-up systems; recycling is coordinated via guarantors for each material waste stream. The law mandates extremely high recycling targets and much of the collected and sorted material is sent to other countries in Europe and Asia for recycling. Germany's packaging ordinance has led to the collection and recycling of more than two thirds of the country's used packaging, albeit at great expense. Other European countries have modified the German approach in the hope of achieving similar results at lower costs.

3. Packaging and Packaging Waste Directive

The EU Directive 94/62/EC on Packaging and Packaging Waste was the first product-specific regulation in the field of EU waste policy. It applies to all packaging placed on the market in the EU and all packaging waste, whether it is used or released at industrial, commercial, office, shop, service, household or any other level, regardless of the material used. The directive laid down quantitative targets for recovery (i.e., material recycling, incineration with energy recovery and composting) and recycling of packaging waste to be achieved by July 2001; 50% at a minimum and 65% at a maximum by weight of the packaging waste must be recovered and between 25 and 45% of total packaging waste must be recycled, with a minimum of 15% of each packaging material.

A substantial increase in the targets for recovery and recycling was contained in the amendment to the directive (2004/12/EC) released in February 2004. By 31 December 2008, 60% as a minimum by weight of packaging waste must be recovered and 55% as a minimum and 80% as a maximum by weight must be recycled, with the following minimum recycling targets for each material (by weight): 60% for glass, paper and board; 50% for metals; 22.5% for plastics (counting exclusively material that is recycled back into plastics); and 15% for wood. Packaging waste that is exported out of the community shall only count for the achievement of the obligations and targets if there is sound evidence that the recovery or recycling operation took place under conditions that are broadly equivalent to those prescribed by the community legislation on the matter. Member states shall, where appropriate, encourage energy recovery where it is preferable to material recycling for environmental and cost-benefit reasons. Furthermore, the directive states that recycling targets for each specific waste material should take account of LCAs and CBAs, which have indicated clear differences both in the costs and in the benefits of recycling.

V. PACKAGING AND SUSTAINABILITY

A. Sustainable Development

The first and most widely-accepted definition of *sustainable development* is the one that appeared in the report of the World Commission on Environment and Development (also known as the "Brundtland Commission") in 1987 entitled *Our Common Future*[38]:

> ... Sustainable development is development that meets the needs of the present without compromising the ability of future generations to meet their own needs....

Contained within this definition are two key concepts: that of needs, particularly the essential needs of the world's poor to which overriding priority should be given, and the idea of limitations imposed by the state of technology and social organization on the environment's ability to meet present and future needs.[12]

An alternative definition is:

> ... Sustainable development is an act that can be maintained indefinitely because it is socially desirable, economically viable and ecologically sustainable....

Imagine being in London 150 years ago and discussing sustainable development. Concerns would have centered on whether there would be sufficient straw to feed the increasing number of horses in the city, or enough whale oil to keep the lamps burning. The subsequent invention of the motor car put an end to concerns about straw production, and electricity replaced oil lamps. These two examples aptly illustrate the difficulty, and indeed the impossibility, of predicting with any certainty what resources future generations may or may not require to enjoy a standard of living at least equal to that which we enjoy today.

Despite these limitations, the term sustainable development has been the underlying principle informing the debate first at the Rio Earth Summit of 1992 and the more recent Johannesburg Summit of 2002. The derivative words sustainable and sustainability have also achieved widespread usage but their meaning has been stretched so far as to become nearly meaningless. Today, other terms have entered the lexicon, with *corporate social responsibility* (CSR) becoming the catch-all term for what was first called sustainable development in 1987.

B. Corporate Social Responsibility

A variety of terms are used — sometimes interchangeably — to talk about CSR: business ethics, corporate citizenship, corporate responsibility, corporate accountability, sustainability and the triple bottom-line. An editorial in *The Times* of London on 8 July 2003 commented that "Corporate Social Responsibility is one of those wonderfully vague terms that are capable of an infinite variety of definitions. One of them is Completely Stupid Rhetoric."

In its simplest sense, CSR is about how companies behave in social, environmental and ethical contexts. Companies are now expected to behave responsibly, not only towards shareholders, but also in all their relationships with people and the planet. Put another way, CSR is about achieving commercial success in ways that honor ethical values and respect people, communities and the natural environment. It is about how companies manage their business processes to produce an overall positive impact on society. Ultimately, CSR is about delivering improved shareholder value, providing enhanced goods and services for customers, building trust and credibility in society and becoming more sustainable over the longer term. It is about companies taking into account their complete impact on society and the environment, not just their impact on the economy. Today, the term *corporate citizenship* appears to be replacing CSR.[25]

As is evident from the above definitions, environmental responsibility is now included within CSR. This should come as no surprise given that environment includes the relationship between people and the natural environment, the use of environmental resources and the issues of equity, social justice and futurity.

C. CSR/Sustainability Reporting

A CSR/sustainability report covers the economic, environmental and social performance of a company: the so-called "triple bottom-line." It provides evidence of whether a company is a good corporate citizen and not just a financial success. Without a CSR/sustainability report, there is no evidence in the public arena to indicate that the company is a good corporate citizen. A wide range of parameters can be found in CSR reports including energy efficiency, community relations, ecodesign, product recyclability, employee relations and materials efficiency.

Today, there are several international guidelines that can assist organizations interested in producing a CSR report. The Global Reporting Initiative (GRI) 2002 Sustainability Reporting Guidelines provide a very detailed guide, and the AA1000 Series of Assurance Standards for Social and Sustainability Reporting have been released by AccountAbility. However, preparing a CSR report is no simple task, and the groundwork has to be laid at least 18 months before the report will see the light of day. Below are some basic activities that need to be in place before starting work on a CSR report.

No company that claims to be serious about reducing its environmental impacts will be taken seriously unless it has in place an environmental management system (EMS) certified to ISO 14001. However, having a certified EMS will not by itself drive continuous improvement. It is necessary to have ecoefficiency indicators across the organization and be able to show that they moving in the right direction over time. Reducing material use and waste can save money, as well as reducing environmental impacts, as can sending zero waste to landfill by finding reuse and recycling outlets for the waste. Focusing on energy efficiency by such measures as upgrading air conditioners and chillers, raising the ambient working environment to a more comfortable temperature and encouraging staff to turn off lights all have very short payback periods and are truly "win–win" situations for both business and the environment. Leading companies are also moving towards a "carbon neutral" position where they offset all the greenhouse gases they generate by, for example, planting an equivalent number of trees in sustainably managed forests somewhere on Earth.

A prerequisite to implementing ecoefficiency programs is a company-wide environmental awareness program so that all staff understand the reasons behind, and benefits of, reducing their impacts on the environment. It is also imperative that time series data indexed to some unit of production are included so that the reader can gauge the magnitude of the improvements that have been made.

Socially responsible companies are now also installing occupational health and safety management systems and having them certified to OHSAS 18001. Apart from the improvement in employee morale and the signal that the organization does care about its workers, OHSAS also delivers real bottom-line benefits including reduced workplace accidents and stoppages.

A key feature of CSR involves the way that a company engages, involves and collaborates with all its stakeholders including shareholders, employees, suppliers, customers, communities, nongovernmental organizations and governments. To the extent that stakeholder engagement and collaboration involve maintaining an open dialogue, being prepared to form effective partnerships and demonstrating transparency (through measuring, accounting and reporting practices), the relationship between the business and the community in which it operates is likely to be more credible and trustworthy.

This is a potentially important benefit for companies because it increases their "license to operate," which enhances companies' prospects of longer-term community support, and improves their capacity to be more sustainable. Companies can use stakeholder engagement to internalize society's needs, hopes and circumstances into their corporate views and decision making. Although there are many questions as to how far a company's responsibilities extend into communities relative to the roles of governments and individual citizens, there is a strong argument that CSR can effectively improve a company's relations with communities and thereby produce some key features that will improve business prospects for its future. The perceptions that stakeholders have of a company's corporate citizenship performance can significantly affect the business's license to operate. Companies with a poor reputation in this area can find themselves continually responding to criticism of their approach to a whole range of environmental and social issues.[8]

D. Supply Chain Management

Managing the environmental impact of suppliers and subcontractors, and taking responsibility for the environmental impact of products and services, are clearly important components of any organization's environmental strategy. For many companies, including those in the packaging industry, the number and diversity of their suppliers and products means they have an environmental impact that extends well beyond their own staff and workplaces. This makes effective supply chain management and product stewardship a complex and challenging process.

The integration of environmental criteria into this process is an important aspect of protecting an organization from environmental risk, liability and poor PR. It also provides the opportunity to work in partnership with key stakeholders to improve environmental performance throughout the entire supply chain and all stages of the product lifecycle, including raw material sourcing, manufacturing, packaging, transport, customer use and disposal. As a result, many suppliers have had to implement ISO 14001 to retain their business.

As a key part of the FMCG industry, the packaging industry could gain real advantages by adopting a CSR approach. Apart from the largely intangible improvements in corporate reputation, the benefits to the bottom-line from increased employee motivation and workplace health and safety, as well as more efficient use of materials and energy are potentially huge. To assure its very future, the packaging industry must take steps to ensure that it is sustainable.

REFERENCES

1. Ackerman, F., *Why Do We Recycle? Markets, Values and Public Policy*, Island Press, Washington, 1997.
2. Alexander, J. H., *In Defense of Garbage*, Praeger Publishers, Westport, CT, 1993.
3. Anonymous, The Recycling Industry in the European Union: Impediments and Prospects, Report for the Committee on Environment, Public Health and Consumer Protection of the European Parliament, Institute for Prospective Technological Studies, Seville, Spain, December 1997.
4. Arvanitoyannis, I. S. and Bosnea, L. A., Recycling of polymeric materials used for food packaging: current status and perspectives, *Food Rev. Int.*, 17, 291–346, 2001.
5. Asimov, I. and Pohl, F., *Our Angry Earth*, Tom Doherty Associates, New York, 1991.
6. Beck RW Inc., U.S. Recycling Economic Information Study, National Recycling Coalition, July 2001.
7. Benjamin, D. K., Eight great myths of recycling, PERC (Property & Environment Research Center), Policy Series 28, September 2003 (see www.perc.org/publications/policyseries/recycling).
8. Bennett, M. and James, P., Eds., *Sustainable Measures: Evaluation and Reporting of Environmental and Social Performance*, Greenleaf Publishing, Sheffield, England, 1999.
9. Berlin, J., Life cycle assessment (LCA): an introduction, In *Environmentally Friendly Food Processing*, Mattson, B. and Sonneson, U., Eds., CRC Press, Boca Raton, FL, 2003, chap. 2.
10. Borchardt, J. K., Recycling, In *The Wiley Encyclopedia of Packaging Technology*, Brody, A. L. and Marsh, K. S., Eds., Wiley, New York, pp. 799–805, 1997.
11. Bosnea, L. A., Arvanitoyannis, I. S., and Nakayama, A., Potential of recycling and biodegradability for food packaging waste, *Curr. Trends Polym. Sci.*, 4, 89–115, 1999.
12. Carley, M. and Christie, I., *Managing Sustainable Development*, 2nd ed., Earthscan Publications, Sterling, VA, 2000.

13. Dainelli, D., Recycling of packaging materials, In *Environmentally Friendly Food Processing*, Mattson, B. and Sonesson, U., Eds., CRC Press, Boca Raton, FL, 2003, chap. 10.
14. De Leo, F., The environmental management of packaging: an overview, In *Environmentally Friendly Food Processing*, Mattson, B. and Sonesson, U., Eds., CRC Press, Boca Raton, FL, 2003, chap. 9.
15. EPA, *Municipal Solid Waste in the United States: 2001 Facts and Figures*, Office of Solid Waste and Emergency Response, Environmental Protection Agency, Washington, 2003, October.
16. Fonteyne, J., Packaging recovery and recycling policy in practice, In *Packaging, Policy and the Environment*, Levy, G. M., Ed., Aspen Publishers, Gaithersburg, MD, 2000, chap. 9.
17. Franklin Associates, *Characterization of Municipal Solid Waste in the United States: 1998 Update*, Franklin Associates, Prairie Village, KS, 1999.
18. Franklin, W. E., Boguski, T. K., and Fry, P., Life-cycle assessment, In *The Wiley Encyclopedia of Packaging Technology*, Brody, A. L. and Marsh, K. S., Eds., Wiley, New York, pp. 563–569, 1997.
19. Franz, R. and Welle, F., Recycling packaging materials, In *Novel Food Packaging Techniques*, Ahvenainen, R., Ed., CRC Press, Boca Raton, FL, 2003, chap. 23.
20. Gaines, L. L. and Mintz, M. M., *Energy Implications of Glass-Container Recycling*, Argonne National Laboratory, Argonne, IL, 1994, March.
21. Hannon, B.M., System energy and recycling: a study of the beverage container industry, ASME Paper No. 72-WA/Ener-3, 1972.
22. Kirkpatrick, N., A review of LCA studies commissioned by EUROPEN, URS Project No. 53990-001, May 2004 (see www.europen.be/reuse.pdf).
23. Levy, G. M., Packaging in the environment — perceptions and realities, In *Packaging, Policy and the Environment*, Levy, G. M., Ed., Aspen Publishers, Gaithersburg, MD, 2000, chap. 3.
24. Lomborg, B., *The Skeptical Environmentalist: Measuring the Real State of the World*, Cambridge University Press, Cambridge, 2001.
25. McIntosh, M., Thomas, R., Leipziger, D., and Coleman, G., *Living Corporate Citizenship*, Pearson Education, Harlow, England, 2003.
26. Matthews, V., Environmental and recycling considerations, In *PET Packaging Technology*, Brooks, D. W. and Giles, G. A., Eds., CRC Press, Boca Raton, FL, 2002, chap. 11.
27. Office of Technology Assessment, *Green Products by Design*, US Government Printing Office, Washington, 1992.
28. Page, P. and White, R., Disposal of used packaging, In *Food Industry and the Environment in the European Union*, 2nd ed., Dalzell, J. M., Ed., Aspen Publications, Gaitherburg, MD, 2000, chap. 8.
29. Pearce, F., Burn me, *New Scientist*, 156(2109), 31–34, 1997.
30. Rathje, W. and Murphy, C., *Rubbish! The Archaeology of Garbage*, HarperCollins Publishers, New York, 1992.
31. RDC and Coopers & Lybrand, Eco-balances for policy-making in the domain of packaging and packaging waste, Ref. No. B4-3040/95001058/MAR/E3, 1997.
32. RDC-Environment and Pira International, Evaluation of costs and benefits for the achievement of reuse and recycling targets for the different packaging in the frame of the Packaging and Packaging Waste Directive 94/62/EC, Brussels, 2001.
33. Scarlett, L., *Extended Producer Responsibility: Theory & Practice*, Reason Foundation, Los Angeles, 1998.
34. Smith, C. and White, P., Life cycle assessment of packaging, In *Packaging, Policy and the Environment*, Levy, G. M., Ed., Aspen Publishers, Gaithersburg, MD, 2000, chap. 8.
35. Tchobanoglous, G., Theisen, H., and Vigil, S. A., *Integrated Solid Waste Management: Engineering Principles and Management Issues*, McGraw-Hill, New York, 1993.
36. Thøgersen, J., Recycling and morality. A critical review of the literature, *Environ. Behav.*, 28, 536–558, 1996.
37. White, P. R., Franke, M., and Hindle, P., *Integrated Solid Waste Management: A Lifecycle Inventory*, Chapman and Hall, London, 1995.
38. World Commission on Environment and Development, *Our Common Future*, Oxford University Press, Oxford, 1987.

Abbreviations, Acronyms and Symbols

ΔH	isosteric net heat of sorption (J mol^{-1})
ΔH_s	heat of solution of permeant in polymer (J mol^{-1})
η	overpotential
θ	time
θ_{doub}	generation or doubling time
θ_s	shelf life
°C	degrees Celcius
A	area (m^2)
Å	Angstrom (1×10^{-10} m)
AA	acrylic acid
ABS	acrylonitrile–butadiene–styrene
ADI	acceptable daily intake
AmPA	amorphous polyamide
AN	acrylonitrile
ANMA	acrylonitrile–methyl acrylate
ANS	acrylonitrile/styrene
AOIR	ambient O_2 ingress method
AOX	adsorbable organic halides
APET	amorphous poly(ethylene terephthalate)
Ar	argon
ASLT	accelerated shelf life testing
ASTM	American Society for Testing and Materials
atm	atmosphere
ATBC	acetyltributyl citrate
ATP	adenosine triphosphate
a_w	water activity
BADGE	bisphenol A diglycidyl ether
B&B	blow and blow
BHA	butylated hydroxyanisole
BHET	bis(hydroxyethyl) terephthalate
BHT	butylated hydroxytoluene
BO	biaxially oriented
BON	biaxially oriented nylon
BOPP	biaxially oriented polypropylene
c	concentration of permeant in polymer
CAP	controlled atmosphere packaging
CAS	controlled atmosphere storage
CBA	cost-benefit analysis
CF	consumption factor
CFR	Code of Federal Regulations

cfu	colony-forming unit
C_2H_4	ethylene
CHDM	cyclohexane dimethanol
CMC	carboxymethylcellulose
cm Hg	centimeters of mercury
CO_2	carbon dioxide
CPET	crystalline poly(ethylene terephthalate)
CPP	crystalline polypropylene
CSR	corporate social responsibility
CTMP	chemithermomechanical pulp
d	diameter of pore in polymer
D	diffusion coefficient ($cm^{-2}\ sec^{-1}$); also decimal reduction time
DEHA	diethylhexyl adipate
DEHP	diethylhexyl phthalate
DFD	dark, firm and dry (meat)
DI	dispersity index
D&I	drawn and ironed
DLC	diamond-like coating
DMA	dimethyl anthracene
DMN	dimethyl naphthalene
DMT	dimethyl terephthalate
DOA	dioctyl adipate
DOP	dioctyl phthalate
DOS	dioctyl sebacate
DP	degree of polymerization
DRD	drawn and redrawn
E	Young's modulus
E_A	activation energy ($J\ mol^{-1}$)
EA	ethyl acetate
EAA	ethylene acrylic acid
EAS	electronic article surveillance
EBM	extrusion blow mold
EC	European Community
ECCS	electrolytically chromium-coated steel
ECF	elemental chlorine free
E_d	activation energy for diffusion ($kJ\ mol^{-1}$)
EG	ethylene glycol
E_h	oxidation-reduction potential
EMS	environmental management system
E_p	activation energy for permeation ($kJ\ mol^{-1}$)
EPA	Environmental Protection Agency
EPS	expanded polystyrene
ESBO	epoxidized soy bean oil
ESC	environmental stress cracking
ESCR	environmental stress crack resistance
ESL	extended shelf life
EU	European Union
EVA	ethylene–vinyl acetate copolymer
EVOH	ethylene–vinyl alcohol copolymer
FAP	food additive petition
FCN	Food Contact Notification

FCS	food contact substance
FCOJ	frozen concentrated orange juice
FDA	Food and Drug Administration
FMCG	fast-moving consumer goods
FND	first noticeable difference
FRH	flameless ration heater
fT	food-type distribution factors
FTC	film transistor circuit
GAB	Guggenheim–Anderson–de Boer
GIP	glazed imitation parchment
GLC	glass-like coating
GPPS	general purpose polystyrene
GRAS	Generally Recognized As Safe
GRI	global reporting initiative
gsm	grams per square meter
HACCP	hazard analysis critical control point
HC	hydrocarbon
HCF	heat–cool–fill
HDPE	high-density polyethylene
HIPS	high-impact polystyrene
HMDSO	hexamethyldisiloxane
HQL	high-quality life
HTST	high temperature–short time
HTW	hole through the wall
I_a	intensity of absorbed light
I_i	intensity of incident light
IBM	injection blow mold
IML	in-mold label
IS	individual section
ISO	International Standards Organisation
J	flux of permeant in polymer
JND	just noticeable difference
L	liter
LCA	life cycle assessment or analysis
LCI	life cycle inventory
LCIA	life cycle impact assessment
L/D	length to diameter ratio
LDPE	low-density polyethylene
LLDPE	linear low density polyethylene
LTLT	low temperature–long time
k	rate constant
kGy	kiloGray
K	Kelvin
M_n	number-average molecular weight
M_w	weight-average molecular weight
MA	methyl acrylate; also modified atmosphere
MAP	modified atmosphere packaging
Mb	myoglobin
MDI	microwave doneness indicator
MetMb	metmyoglobin
mg	milligram (1×10^{-3} g)

mL	milliliter (1×10^{-3} L) = cc = cm^3
MOR	morpholine
MPFVs	minimally processed fruits and vegetables
MRA	metmyoglobin reducing activity
MRE	Meals, Ready-To-Eat
MRF	materials recovery facility
MSW	municipal solid waste
MW	molecular weight
MWD	molecular weight distribution
MXD6	meta-xylylene diamine/adipic acid
NDC	dimethylnaphthalene dicarboxylate
NEB	nonenzymic browning
NMOR	*N*-nitrosomorpholine
NNPB	narrow neck press and blow
NO	nitric oxide
NO_x	oxides of nitrogen
N_2O	nitrous oxide
NOEL	no observable effect level
NOMb	nitrosylmyoglobin
NOMetMb	nitrosylmetmyoglobin
NSSC	neutral-sulfite semichemical
OD	optical density
OLED	organic light emitting diode
OM	overall migration
O_2Mb	oxymyoglobin
OPA	oriented polyamide
OPET	oriented polyester
OPP	oriented polypropylene
OPS	oriented polystyrene
OTR	oxygen transmission rate
p	partial pressure (cm Hg)
P	permeability coefficient (mL cm cm^{-2} s^{-1} (cm Hg^{-1}))
P&B	press and blow
P/X	permeance
PA	polyamide
PAA	peracetic acid
PAN	polyacrylonitrile
PBT	poly(butylene terephthalate)
PC	polycarbonate
PCDDs	polychlorinated dibenzodioxins
PCDFs	polychlorinated dibenzofurans
PCR	post consumer recycled
PE	polyethylene
PECVD	Plasma-Enhanced Chemical Vapor Deposition
PEN	poly(ethylene naphthalate)
PET	poly(ethylene terephthalate) = polyester
PETG	poly(ethylene terephthalate) glycol
PHA	polyhydroxyalkanoate
PHB	polyhydroxybutyrate
PHTC	post-heat treatment contamination
PHV	polyhydroxyvalerate

PLA	polylactic acid
PP	polypropylene
ppb	parts per billion (1×10^{-9})
ppm	parts per million (1×10^{-6})
PPP	product, processing method and packaging
ppt	parts per trillon (1×10^{-12})
PS	polystyrene
PSE	pale, soft and exudative (meat)
PSL	practical storage life
PTFE	poly(tetrafluoroethylene)
PU	Pasteurization Unit
PVA	poly(vinyl acetate)
PVC	poly(vinyl chloride)
PVD	Physical Vapor Deposition
PVdC	poly(vinylidene chloride)
PVOH	poly(vinyl alcohol)
Q	total amount of permeant passing through polymer
Q_{10}	temperature quotient = ratio of reaction rates for 10°C temperature difference
QLF	quartz-like film
R	ideal gas constant (= $8.314\ J\ K^{-1}\ mol^{-1} = 1.987\ cal\ K^{-1}\ mol^{-1}$)
RCF	regenerated cellulose film
RCRA	Resource Conservation and Recovery Act
RDF	refuse-derived fuel
RFID	radio frequency identification
RPET	recycled poly(ethylene terephthalate)
REPFEDs	refrigerated, processed foods with extended durability
RH	relative humidity
RMP	refiner mechanical pulp
RQ	respiratory quotient
RR	respiration rate
RWB	roast whole beans
S	solubility coefficient of permeant in polymer [$mL\ cm^{-3}\ (cm\ Hg^{-1})$]; also circumferential stress
SAN	styrene–acrylonitrile
SAT	standard area of tinplate
SATP	standard ambient temperature and pressure
SB	styrene–butadiene copolymer
SBM	stretch blow mold
SBS	solid bleached sulfite
SCC	side chain crystallizable
SCF	Scientific Committee on Food
sec	seconds
SI	Système International d'Unites
SiO_2	silicon dioxide
SiO_x	oxides of silicon
SITA	Système International Tinplate Area
SM	specific migration
SML	specific migration limit
SMP	skim milk powder
SO_x	oxides of sulfur
SPPF	solid phase pressure forming

STP	standard temperature and pressure
SUS	solid unbleached sulfite
TAPPI	Technical Association of the Pulp and Paper Industries
TBHQ	tertiarybutylhydroquinone
TBT	tributyl tin
TCA	trichloroanisole
TCDD	tetrachlorodibenzodioxin
TCF	total chlorine free
TDI	tolerable daily intake
TEF	toxic equivalent factor
TEQ	toxic equivalent
TFE	tetrafluoroethylene
TFS	tin-free steel
TFTC	thin film transistor circuit
T_g	glass-transition temperature
TiO_2	titanium dioxide
T_m	crystalline melting temperature
TMA	trimethylamine
TMAO	trimethylamine oxide
TMP	thermomechanical pulp
TOR	threshold of regulation
TPA	terephthalic acid
TR	transmission rate
TTI	time–temperature indicator
TTM	time–temperature monitor
TTT	temperature–time–tolerance
UHT	ultra heat treated or ultra high temperature
UPC	universal product code
USP	United States Pharmacopoeia
UV	ultraviolet
VA	vinyl acetate
VC	vinyl chloride
VCM	vinyl chloride monomer
VdC	vinylidene chloride
VFPE	very flexible polyethylene
VLDPE	very low density polyethylene
VOC	volatile organic compound
VOH	vinyl alcohol
VPD	vapor pressure deficit
VSP	vacuum skin packaging
WHC	water-holding capacity
WMP	whole milk powder
WVTR	water vapor transmission rate
X	thickness of polymeric material
μg	microgram (1×10^{-6} g)
μL	microliter (1×10^{-6} L)
μm	micrometer (1×10^{-6} m)
z	temperature change required for 10-fold change in D value

Index

G

H

I

J

K

L

M

N

O

P

Q

R

S

T

X

Y

Z